T0342215

Gas Insulated Substations

Gas Insulated Substations

Second Edition

Edited by

Hermann J. Koch
Electric Power Transmission, drkochconsulting, Germany

This edition first published 2022

Edition History
John Wiley & Sons, Inc. (1e, 2014)

Published by John Wiley & Sons, Inc., Hoboken, New Jersey.
Published simultaneously in Canada.

Library of Congress Cataloging-in-Publication Data

Names: Koch, Hermann, 1954– author.
Title: Gas insulated substations / Hermann J Koch, Energy Transmission, Siemens AG, Germany.
Description: Second edition. | Hoboken, NJ : Wiley-IEEE Press, 2021. | Includes bibliographical references and index.
Identifiers: LCCN 2021027849 (print) | LCCN 2021027850 (ebook) | ISBN 9781119623588 (hardback) | ISBN 9781119623595 (adobe pdf) | ISBN 9781119623618 (epub)
Subjects: LCSH: Electric cables–Gas insulation. | Electric substations.
Classification: LCC TK3441.G3 G37 2021 (print) | LCC TK3441.G3 (ebook) | DDC 621.31/042–dc23
LC record available at https://lccn.loc.gov/2021027849
LC ebook record available at https://lccn.loc.gov/2021027850

Cover design: Wiley
Cover image: © Reproduced by permission of Siemens Energy

Set in 9.5/12.5pt STIXTwoText by Straive, Pondicherry, India
Printed and bound by CPI Group (UK) Ltd, Croydon, CR0 4YY

C9781119623588_110122

Contents

Editor Biography

- **Dr.-Ing. Hermann Koch, IEEE Fellow** received his PhD in 1990 at Technical University Darmstadt in Germany on his research in high-voltage partial discharge measurements. Before he graduated in 1979, in Industrial Control Engineering at Fachhochschule Rüsselsheim, Germany, studied 1980 to 1981 in Automation and Control Engineering at New Jersey Institute of Technology, Newark, NJ, USA, and received his Diploma Engineer degree in 1986 on Electrical Control Engineering from Technical University of Darmstadt, Germany.
- He started his business activities with Hoechst, Frankfurt in 1990 for planning and design of electro-technical processes of chemical and pharmaceutical plants, mainly on gen-manipulated human insulin. In 1991, he started in Siemens Energy, Erlangen, Germany, a 30-year career with sales of high-voltage circuit breakers, in the center of competence for high-voltage gas-insulated switchgear (GIS), he consulted GIS development and project application world-wide. In 1994, he took responsibility for the development, sales, and project execution of gas-insulated transmission line (GIL) with world-wide first installations of this technology in several projects. In 2004, he took responsibility as director of high-voltage gas-insulated technologies (GIS and GIL) in network planning, international standardization coordination as representative of Siemens Energy in several working groups and advisory committees in German DKE/VDE, European CENELEC and international IEC, IEEE, and CIGRE organizations. Since March 2020, he is retired from Siemens and operates his own consulting business "drkochconsulting."
- Since 1996 until 2020, he was active in IEC as secretary of IEC SC 17C High-Voltage Switchgear Assemblies, 2008 until 2013 member of Strategic Group SG2 UHV, and since 2013 until today, he is member of three working groups of IEC TC 122 Ultra High Voltage Systems. In CENELEC, he was secretary of TC 17AC High Voltage Switchgear and Control gear from 2014 until 2021. Starting in 1997, he is active IEEE Power & Energy Society (PES) member and served as board member of IEEE-Standards Association (SA) from 2003 to 2007 and at the PES board from 2012 to 2020. From 2004 to 2010, he served as secretary and chairman of the Substations Committee. Since 2005 he is distinguished lecturer on gas-insulated switchgear (GIS) and transmission lines (GIL). In 2010, he has received the status as IEEE Fellow on high-voltage gas-insulated technology. In CIGRE, he is active member of the Study Committee B3 Substations since 1996. He was convenor of several working groups on gas-insulated technologies from 2006 until 2018 and was

elevated to the status of CIGRE Distinguished Member in 2020. In German DKE/VDE, he is active member since 1996 in national committees on high-voltage switchgear assemblies (K432) and Ultra High Voltage (UHV) systems. From 2012 to 2020, he was chairman of K 432.

- He received more than 50 national and international awards from DKE/VDE, IEC, IEEE PES, IEEE SA, and CIGRE. The most outstanding awards are 2007 CIGRE Technical Committee Award on outstanding contributions to SC B3 Substations. 2008 IEEE-SA International Award on extraordinary contributions to international goals of IEEE-SA. 2010 IEEE PES Distinguished Service Award for outstanding service Secretary, Vice-Chairman, and Chairman of the Substations Committee. 2016 IEEE Charles Proteus Steinmetz Award for leadership in and contributions to the development, standardization, and global impact of gas-insulating technology for substations (GIS) and high-voltage lines (GIL). Certificate of DKE VDE/DIN for his distinguished engagement for more than two decades of standardization in High Voltage Switchgear and Controlgear Assemblies. 2020 CIGRE Distinguished Member in acknowledgement of his long-standing collaboration to the work of the association. 2020 IEC 1906 Award for valuable and sustainable contribution as expert of IEC Technical Committee 122, Ultra High Voltage (UHV) AC transmission systems.
- He has published six books in 2012 on Gas-Insulated Transmission Lines (GIL), Wiley & Sons Publisher and Electric Power Substations Engineering, Chapter 18 Gas Insulated Transmission Line (GIL), CRC Press. In 2014, Gas-Insulated Substations (GIS), IEEE Press and John Wiley & Sons. In 2016, Practical Guide to International Standardization for Electrical Engineers, John Wiley & Sons, Ltd, GIS, GIL. In 2021 the second and extended edition of Gas-Insulated Substations (GIS) will be published by IEEE Press and John Wiley & Sons.
- At this day, he has published 166 national and international papers and conference contributions world-wide in the technical field of high-voltage gas-insulated technology covering AC and DC voltages.
- Since 1992, he has proposed 120 inventions in the technical field of high-voltage power systems. Based on these recorded inventions, 33 patents have been granted to Hermann Koch. Most are related to high-voltage gas-insulated technology.

Contributors

Arnaud Ficheux
GE Grid Solutions
France

Aron Heck
Siemens Energy
Germany

Arun Arora
Consultant
USA

Charles Hand
retired from Southern California Edison
USA

Coboyo Bodjona
PEPCO
USA

Dave Lin
former Siemens
USA

Dave Giegel
MEPPI
USA

Dave Mitchell
Dominion
retired USA

Dave Solhtalab
Pacific Gas & Electric
USA

Devki Sharma
Consultant
USA

Dirk Helbig
Siemens Energy
Germany

George Becker
POWER Engineers Inc
USA

Hermann Koch
drkochconsulting
Germany

James Massura
GE Energy
USA

John Boggess
former ABB, departed
USA

John Brunke
Power Engineers
USA

Jorge Márquez-Sánchez
Burns & McDonnell
USA

Maria Kosse
Siemens Energy
Germany

Mark Kuschel
Siemens Energy
Germany

Noboru Fujimoto
Kinectrics Incretired
Canada

Pathik Patel
Duke Energy
USA

Peter Grossmann
Siemens Energy
Germany

Pravakar Samanta
ABB
USA

Ravi Dhara
ABB
USA

Richard Jones
Consultant
USA

Uwe Riechert
Hitachi ABB Power Grids
Switzerland

Venkatesh Minisandram
National Grid
retired
USA

Vipul Bhagat
National Grid
USA

William Labos
Public Service Electric and Gas Company
retired
USA

Xi Zhu[†]
GE Renewable Energy
USA

Foreword of Editor

Since the first publication of the Gas-Insulated Substations (GIS) book in 2014, the technical development of new solutions and types of applications has increased. The second edition of the GIS book updates the complete text to cover recent technical developments and applications. New chapters have been added on new topics for the use of GIS.

IEEE PES Substations Committee Working Group K10 was charged with the revision work. Many experts from the industry, users, manufacturers and consultants contributed their knowledge and information on the technical changes in gas-insulated high-voltage technology of 52 kV and above.

There are significant changes to GIS technology related to insulating gases, substation resilience, high voltage vacuum switching technology, low power instrument transformers with electronic sensors, digital twin concepts, application of GIS for offshore windfarms and energy collecting platforms, and the operating requirements of digital substations using GIS. These topics are covered by new chapters added to the GIS book.

GIS also plays an increasing role in the development of higher resilience of the electric power network and to provide lower risk of power delivery interruption during extreme weather and climate conditions. GIS can also contribute to faster re-energization of a substation after disasters have caused power delivery interruption. Indoor and even underground substations using GIS technology are one method to increase the resilience of the energy networks. The other method of mobile GIS provides ready to connect high voltage substations assembled on trucks and trailers, which can be energized within a short period of time to replace a damaged substation. These mobile GIS concepts are available for voltages up to 500 kV.

There are many practical examples of GIS installations world-wide that provide examples of successful project in operation. These examples have been extended by case studies for city and network developments to show new possibilities with using GIS in the substation. Underground substations in metropolitan areas and city centers offer technical options to bring electric power to city or industry developments, when only little space and right of way is available. Underground projects are presented from all over the world to show new GIS-related options. With case studies on city development projects, the complex situation to apply power infrastructure is shown and options are explained when using GIS.

Special buildings is a chapter which describes the combination of high-voltage electric power supply installations within the same buildings as offices, apartments, restaurants, public places like cinemas, theaters, or spaces for sports events. The combination of multiple uses of buildings and power supply infrastructure makes GIS a perfect technology to provide the functions of high-power switching capabilities in a small space, such as in the basement of buildings. Examples from all over the world are presented in this chapter including the economic comparison with other technical solutions like classical air-insulated substations. Those cases are explained and the result provides the most efficient and economical solution.

Advanced Technology is dedicated to GIS developments in new technologies and applications. This covers applications of high voltage vacuum circuit breakers of 145 kV up to 500 kV to allow GIS using only technical air and vacuum with no contributing gas to global warming potential.

DC GIS is seen as a new implementation to high voltage direct current (HV DC) transmission systems and provide substantial space and foot print reductions for the converter and transition stations.

Digital substations are key element of the future intelligent or smart electric power network for optimized management and safe control.

Soft technical topics like environmental acceptance, life cycle cost analysis or risk based assessments of high voltage substations are discussed to show advantages which can be reached by GIS.

Advantages of low power instrument transformers for new methods of collecting data in the transmission network, handling of very fast transient overvoltages and electromagnetic fields when using GIS are explained.

Possibilities on replacement of SF_6 to reduce the global warming potential are explained and the advantages of different solutions are shown.

The latest development of GIS technology and applications is covered by this second edition and provides practical information to the reader to understand the advantages of and possibilities with Gas-Insulated Substations (GIS).

Hermann Koch
Editor
Gerhardshofen, Germany, November 2021

Foreword PES Substations Committee

Many years in the making, it is a pleasure to see the second edition of the *Gas-Insulated Substations (GIS) Handbook* being published. The countless hours of many to review, update, and revise the first edition has brought you a more complete and up-to-date glimpse of what is happening in the GIS industry.

Having been involved in many of the committee meetings, where experts of all different backgrounds and companies have met, it is always a great pleasure that these technical leaders can sit down, under the leadership of Dr. Hermann J. Koch, and work through an expansive and broad topic. Even trying to forecast by covering Advanced Technologies – and where this equipment may be used in the future and for what reasons. There are also plenty of sample projects that hopefully the reader can relate to and can use that information to help in his or her own project.

From basic information about GIS, to the different GIS instrument transformers, to the connection types to other pieces of electrical equipment, this book runs the gambit, making sure to inform the reader of all the critical aspects that should be paid attention to during the design of the GIS substation. It has over 100 years of the authors' in-depth expert-level knowledge in this book. Being part of this community has been an honor and look forward to many more years.

It does sadden me that during the publication of this book, we had lost one of the authors to a tragic accident. Dr. Xi Zhu passed away on 29 May 2021. He personally has over 30 years of High Voltage research and development and was an IEEE Senior member and CIGRE member. His impact in the committee meetings will be sincerely missed.

Patrick J. Fitzgerald
Chair, IEEE PES Substations Committee 2020–2021
Boston, November 2021

Foreword GE Grid Solutions

I started my career in GIS activity 24 years ago and at that time I wish I had such a book with me to speed-up my comprehension and expertise on GIS. I must admit that after all these years, this book has found its place on my desk and is frequently opened to enrich my knowledge or to find a specific answer to questions that were never raised before.

GIS is gathering a wide field of technology and expertise within high-voltage world, like switching components, measuring instruments, arresters, bushings, LV equipment, control and monitoring, structures, etc. Many documents, guides, and standards are needed to cover the full scope of GIS. CIGRE, IEC, and IEEE have been working hard for the last 50 years to gather knowledge and best practices for designing, manufacturing, installing, and operating GIS. For newcomers in GIS and even for experienced users, it is sometime difficult to find where is the information they are looking for. This book is just there to simplify their life. It is also the trusted door to access all necessary information they need as it has been written and reviewed by a wide panel of experts in GIS.

Since beginning of GIS in early 1960, technology has continuously changed and GIS got benefit of these changes. This revised edition is considering all recent breakthrough and coming trends in GIS like alternative gases, digital technology, and DC GIS.

I hope you will enjoy the reading and find the information you are looking for.

Arnaud Ficheux
CIGRE, IEC, and IEEE Member
GIS Technical Support and Innovation Manager, GE Grid Solutions
Aix-les Bains, France, November 2021

Foreword Hitachi Energy

Gas-insulated switchgear (GIS) operates invisibly, safely, and reliably and is often hidden in buildings or other structures. No movement can be seen, just a faint hum betrays the flow of bulk AC power. Perhaps that's why there has only been this one comprehensive reference work on GIS technology, further remarkable since it was only first published in 2014. The history of GIS dates back several decades, and its implementation has been very successful. GIS technology originated in 1936, when a Freon-filled GIS assembly, rated at 33 kV, was demonstrated in the United States. Later, in the mid-1950s, sulfur hexafluoride gas (SF6) was discovered, a gas with excellent insulating and arc-extinguishing properties. By the mid-1960s, GIS was sufficiently well developed to be commercially viable and appealing to a broader market. Today, GIS is an established technology that allows for safe, highly reliable operation in confined spaces, owing to its being enclosed and to its significantly reduced footprint compared to air-insulated switchgear (AIS). This revolutionary technology was a key enabler for urbanization. Occupying only 10% of the volume required by the equivalent AIS, it allows flexible placement of substations within cities to supply safe and reliable power. Substations are no longer taking up acres of open spaces but can hide within buildings, underground, or on rooftops.

Although at first glance looking seemly uncomplicated, a closer examination reveals the engineering investment, the complexity in the variety of configurations, and the installation and testing effort found in a typical GIS installation. This handbook helps to decipher GIS technology, to present it in a clear and explanatory way, and gives a comprehensive overview of the subject, making it both interesting and indispensable for anyone with this technology. This includes those who want to start (or have already started) their career with manufacturers or end-users of GIS. Descriptions of practical application examples, as well as operation and maintenance requirements, make the book equally interesting for switchgear operators and decision-makers in energy policy. I also particularly recommend this book to students, as this book provides the most comprehensive overview of the technology currently available.

Today, our power transmission systems are subject to new requirements and far-reaching changes, ushering in a new era of electrical grids and switchgear. On the one hand, an enormous flow of electricity to large megacities often has to cross several thousand kilometers, from the energy sources to the end-user with transmission voltages of 1100–1200 kV having been established. Secondly, a large expansion of the transmission grid is required due to global plans for the extensive integration of new renewable energies into the electrical energy supply. Especially in the case of large-scale (multi-GW) offshore wind farms, the connection to the onshore grid will be made by means of HVDC. The increasing demand for HVDC technology requires the adaptation of gas-insulated switchgear. Finally, the latest developments in the development and use of eco-efficient GIS solutions and the integration of digital techniques in design and diagnostics are worthy of special mention. It combines a low-carbon footprint, superior reliability, and low lifecycle costs in a flexible product layout. These topics are described in the new chapter of this second edition of the *GIS Handbook*, further increasing its interest for all those who want to get to know and follow current and new development trends.

Finally, I would like to add that I am proud to have been able to make a modest contribution to the work of this group of authors, who are well known and esteemed by myself and others in the field.

Uwe Riechert
Switchgear expert, Hitachi Energy Switzerland Ltd
Zurich, Switzerland, November 2021

Foreword Siemens Energy

Today, metal-enclosed gas-insulated switchgear (GIS) is indispensable for safe and reliable power grids. The long experience of more than 50 years of worldwide operation in all climatic zones, onshore and offshore installations, speak for an excellent. GIS is mainly used where space is at a premium or environmental conditions significantly impact the reliability of insulators.

This means above all in urban and coastal areas, where load and the importance of reliability is increasing worldwide. The compact size in combination with modular functions offers a high degree of design diversity, enabling not only outdoor but also more easily indoor and underground installation with the least environmental impact. GIS also stands for more safety; the accessible points of live system parts can be reduced to a minimum. If the connections are realized with gas-insulated lines (GIL) or power cables and without overhead lines, this can even become possible altogether. Today, GIS are available for AC and increasingly also for DC applications.

The GIS success is therefore well-founded; GIS are made to ensure the best possible balance between design, materials used, maintenance effort, and maintenance intervals.

The development continues, on the way to a carbon-free economy, alternative gases to SF_6 without greenhouse effect have become more and more important in recent years and will be one of the main innovation drivers for the future. Climate neutrality is only achievable with greenhouse and F-gas-free technologies enabling electrical power grids with zero emission.

In addition, digital technologies entering including artificial intelligence enabling even higher reliability and more effective operations management.

I hope you enjoy finding all the latest information on GIS in one book, and I am sure that the book will help you to find answers to specific technical aspects and details.

Dr. Mark Kuschel
CTO High Voltage Gas-insulated Switchgear
Siemens Energy, Switchgear Factory Berlin, Germany, 2021

Acknowledgements

This second edition of Gas-Insulated Substations (GIS) is created by power engineering experts from the IEEE Power & Energy Society Substations Committee. In Working Group K10 GIS Book of the K0 GIS Subcommittee, more than 50 experts have contributed to this book with their knowledge and own practical experiences. For more than two years, since 2019, and at four working group meetings the content has been presented and discussed by the contributors. Many ideas have been added to this second edition GIS book to create a practical guide. This book is designed to help with the daily work of engineers in the planning, design, and erection of high-voltage substations above 52 kV using GIS. Focus is given on practical information to find the best, most reliable, and economic solution for substations in the power network.

Thanks to authors and reviewers of the second edition GIS book. Thank you to the 20 authors of the 1st edition Arnaud Ficheux, Arun Arora, Charles Hand, Dave Lin, Dave Solhtalab, Devki Sharma, George Becker, Hermann Koch, James Massura, John Boggess, John Brunke, Jorge Márquez-Sánchez, Noboru Fujimoto, Peter Grossmann, Pravakar Samanta, Ravi Dhara, Richard Jones, Venkatesh Minisandram, William Labos, and Xi Zhu. Thank you for 10 authors Coboyo Bodjona, Dave Giegel, Dave Mitchell, Dave Solthalab, George Becker, Hermann Koch, Maria Kosse, Pathik Patel, Richard Jones, and Xi Zhu, and 7 coauthors Aron Heck, Dirk Helbig, George Becker, Mark Kuschel, Peter Grossmann, Uwe Riechert, and Vipul Bhagat for their contribution of new text to the second edition.

Thank you to the many reviewers of text with their recommendations to improve the writing of the text, making it easier to read and understand. If you find something wrong, please contact the editor for improvement. Thank you to the 29 reviewers of the 1st edition text Arnaud Ficheux, Chuck Hand, Dave Giegel, Dave Solhtalab, David Lin, Devki Sharma, Richard Jones, Eduard Crockett, Ewald Warzecha, George Becker, Hermann Koch, James Massura, John Brunke, Linda Zhao, Markus Etter, Noboru Fujimoto, Patrick Fitzgerald, Peter Grossmann, Phil Bolin, Pravakar Samanta, Ravi Dhara, Ricardo Arredondo, Richard Jones, Scott Scharf, Shawn Lav, Toni Lin, Venkatesh Minisandram, and Xi Zhu.

After the 1st edition had been published some correction and editorial changes were needed to improve the text which was done by Arnaud Ficheux, Bala Kotharu, Coboyo Bodyona, Devki Sharma, Richard Jones, George Becker, Hermann Koch, James Massura, John Brunke, Michael Novev, Pablo Gonzales Toza, Pathik Patel, Patrick Fitzgerald, Pravakar Samanta, Ryan Stone, and Scott Scharf, thank you for your contribution.

The new text of the second edition was review by Arnaud Ficheux, Aron Heck, Bala Kotharu, Denis Steyn, Dirk Helbig, George Becker, Gerd Ottehenning, Hermann Koch, James Massura, John Brunke, Mark Kuschel, Michael Novev, Nick Matone, Pathik Patel, Patrick Fitzgerald, Petr Rudenko, Robert Lüscher, Ryan Stone, Scott Scharf, and Stefan Schedl, thank you for improving the text and eliminate failures.

Thank you to Siemens-Energy, GE, Hitachi-ABB, Mitsubishi, AZZ, Tech S Corp, Kinectrics, Power Engineers, Burns & McDonnell, National Grid, United Illuminating, Pacific Gas & Electric, Dominion, G&E, Southern California Edison the use of photos and graphics.

The authors thank IEEE PES for their permission to reproduce information related to the IEEE Standards C37 series with the main focus on C37.122, C37.122.1, C37.122.2, and C37.122.3 as indicated in the subclause references. Further information on IEC is available from www.ieee.org.

Thanks to International Electrotechnical Commission (IEC) for permission to reproduce information of IEC 62271-203, -203, and -209. All such extracts are copyright of IEC, Geneva, Switzerland. All rights reserved. Further information on IEC is available from www.iec.ch.

The authors also thanks CIGRE (International Council of Large Electric Systems) for their permission to reproduce information from their Technical Brochures as indicated in the subclause references. Further information on www.CIGRE.org.

IEEE, IEC, and CIGRE have no responsibility for the placement and context in which the extracts and contents are reproduced by the authors, nor in any way responsible for the other content or accuracy therein.

Thanks to John Wiley & Sons and the editing team for their professional work and publishing in a nice-looking way.

Thank you to my family with my wife Edith, my children Christian and Katrin, their partners Britta and Peter, and my grandson Lukas for the patience to give me the time to work on the book.

The editor and all authors wish you, the reader of the GIS book to find informative and useable information for your work on high voltage substation design.

1

Introduction

Authors: Hermann Koch, John Brunke
Reviewers: Phil Bolin, Devki Sharma, Jim Massura, George Becker, Scott Scharf, and Michael Novev

1.1 General

This book is based on the tutorial and panel sessions presented by the subject matter experts (SME) of gas-insulated substations in the working group K2 of the Institute of Electronics and Electrical Engineer (IEEE) Substations Committee. Gas-insulated substations (GIS), for alternative current (AC) were invented in the early 1960s with the first projects in the mid-1960s in the United States and Europe. In Japan, research and development for GIS started from 1963 followed by the practical application of 84 kV GIS (1968).

With thousands of installed bays, of GIS today, we can look back to a wide range of experiences gained in very different cases of applications.

The IEEE Substations Committee created the GIS Subcommittee K0 more than twenty-five years ago and since then this subcommittee has continuously worked on standards and guides in the field of GIS technology and application. About twenty-five standards and guides related to the GIS have been published to-date, with continuous revision work in progress on all documents and preparation of new documents.

Around the year 2000, the SME of the GIS Subcommittee started to collect information on GIS and developed a tutorial on Gas-Insulated Substations (GIS) and Transmission Lines (GIL). This working group is numbered as K2 in the GIS Subcommittee and continually works to come up with new tutorials on different subjects associated with GIS.

1.1.1 Organization

The organization of the Substations Committee was developed over the last decades with the focus being on any equipment and systems related to substations. In Table 1.1, the scope of the subcommittees of the substations committee is shown and in Figure 1.1, the organization of the Substations Committee is shown.

IEEE PES Substations Committee has five subcommittees. Three subcommittees are for transmission and distribution (T&D) substations in physical electrical design (D0), physical civil design (E0), and grounding lightning (G0). One subcommittee for power electronic in T&D substations and one for gas-insulated switchgear (GIS) (K0). For the scope of the subcommittees, see Table 1.1. The administrative substation committee (B0) is represented by the chair, vice-chair, secretary, past-chair and holds subcommittees for transactions editor, awards (H0), standards (S0), and meetings (M0).

Gas Insulated Substations, Second Edition. Edited by Hermann J. Koch.

Table 1.1 Scope of subcommittees of the substations committee

D0:	Transmission and distribution substation design for a medium-voltage substation in the range of 1 kV up to and including 52 kV and a high-voltage substation above 52 kV
E0:	Transmission and distribution of substation operations for medium-voltage substations in the range of 1 kV up to and including 52 kV and high-voltage substations for above kV
G0:	Transmission and distribution of substation grounding and lightning
I0:	High-voltage power electronics stations for DC equipment above 1.5 kV to be installed in a substation or converter station like AC/DC converters, coils, filters, grounding, and software for control and protection
K0:	Gas-insulated substations for AC high-voltage equipment above 1 kV of switchgear, disconnectors, and ground switches (GIS) and power transmission (GIL)

The standardization work is split into working groups of the five subcommittees which carry the responsibility for standards and guides. D0 for physical electrical design of substations has seven active working groups: Electrical Clearance D1 (IEEE 1427), Cable System in Substations D2 (IEEE 525), Bus Design D3 (IEEE 605), Seismic requirements to Substation Equipment D4 (IEEE 693), Flexible Bus works in Substations D5 (IEEE 1527), Turnkey Substations D8 (IEEE 1267), and AC/DC Power Supply Systems in Substations D9 (IEEE 1818).

The physical civil design subcommittee E0 is active in six working groups on Community Acceptance of Substations E1 (IEEE 1127), Oil Spill prevention to the substation soil E2 (IEEE 980), Fire Protection E3 (IEEE 979), Animal Deterrents E5 (IEEE 1264), Measuring Earth Resistivity, Ground Impedance, Earth Surface Potentials of a Grounding System E6 (IEEE 81), and Electric Power Substation Physical and Electronic Security E7 (IEEE 1402).

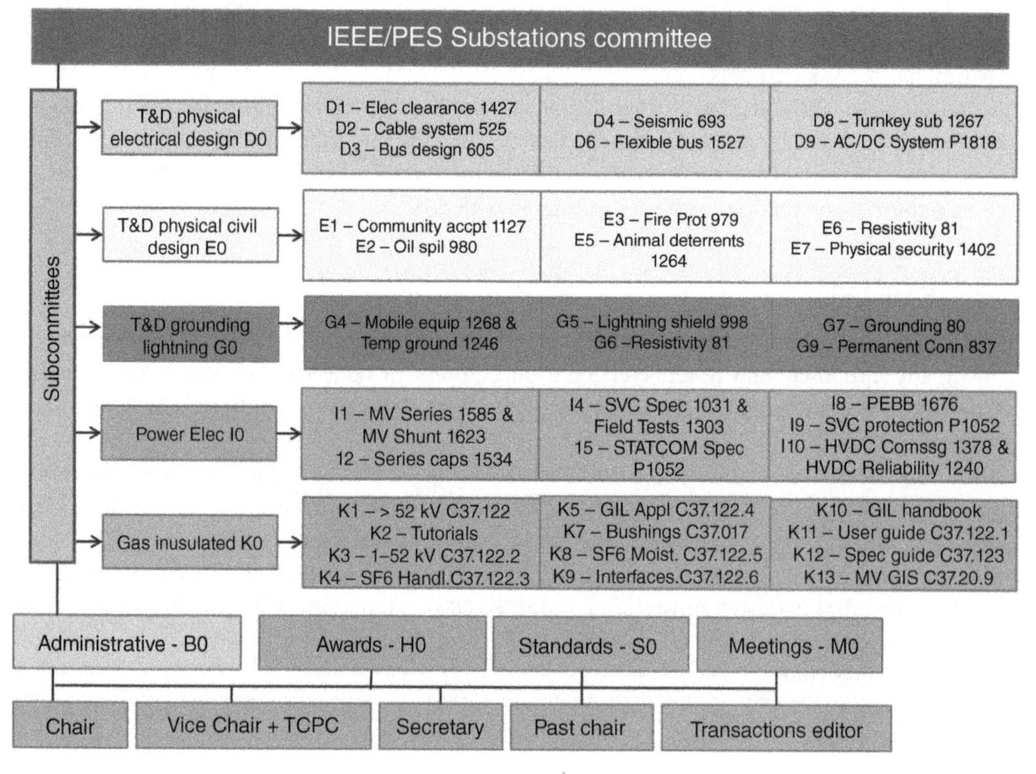

Figure 1.1 Organization of the Substations Committee

The grounding and lightning subcommittee is active in five working groups. G4 is responsible for two standards on Mobile Substation Equipment (IEEE 1268) and Temporary Grounding (IEEE 1246). Lightning Stroke Shielding of Substations of working group G5 (IEEE 998), Grounding Resistivity G6 (IEEE 81), Substation Grounding G7 (IEEE 80), and Qualifying Permanent Connections Used in Substation Grounding G9 (IEEE 837).

The power electronic equipment subcommittee is active in seven working groups. I1 is responsible for two guides, the Functional Specification of Medium Voltage (1–35 kV) Electronic Series Devices for Compensation of Voltage Fluctuations (IEEE 1585) and Electronic Shunt Devices for Dynamic Voltage Compensation (IEEE 1623). For Specifying Thyristor-Controlled Series Capacitors, working group I2 is responsible. The Functional Specification of Transmission Static Var Compensators (IEEE 1031) and the Field Tests of Static Var Compensators (IEEE 1303) are in the scope of working group I4. The Specification of Transmission Static Synchronous Compensator (STATCOM) Systems (IEEE 1052) is covered by I5; for Control Architecture for High Power Electronics (1 MW and Greater) Used in Electric Power Transmission and Distribution Systems (IEEE 1676), working group I8; for Specification of Transmission Static Synchronous Compensator (STATCOM) Systems (IEEE 1052), working group I9; and for Qualifying Permanent Connections Used in Substation Grounding (IEEE 837), working group I10 are responsible.

The gas insulated switchgear and switchgear assemblies subcommittee covers 13 standards and has 5 more upcoming activities. For High-Voltage Gas-Insulated Substations Rated Above 52 kV (IEEE C37.122), K1 is responsible. In K2, GIS Tutorials and Panel Sessions are prepared and presented at IEEE conferences. These tutorials and panels are the basis for this GIS Handbook. K3 is active on the Application of Gas-Insulated Substations 1–52 kV (IEEE C37.122.2), K4 on SF_6 Gas handling (IEEE C37.122.3), K5 on the Gas Insulated Transmission Line (GIL) Application Guide (IEEE C37.122.4), K7 on SF_6 to air Bushings for GIS (IEEE C37.017), K8 for the Moisture content in SF_6 used in GIS (IEEE C37.122.5), and K9 on Interfaces to GIS (IEEE C37.122.6). Working group K10 is responsible for this GIS Handbook. K11 is active on the GIS User Guide (C37.122.1), K12 on the GIS Specification (IEEE C37.123), and K13 on Medium Voltage GIS for Voltages above 1 kV and up to 52 kV (IEEE C37.20.9).

There are six new standardization activities in the K subcommittee now in work. K14 Condition Assessment of GIS (future IEEE C37.122.9), K15 Recommended Practice of Field Test for GIS (future IEEE C37.122.7), K16 Panel Session on Advanced Sensors for GIS, K17 Panels Session for Physical Security and Resilience, K18 Mobile GIS Applications (future C37.122.8). The main content of this new standards is already covered in the second edition of this GIS Handbook.

The gas-insulated switchgear and switchgear assemblies (GIS) has a total of 19 working group twelve are listed in Table 1.2.

1.1.2 Experts of K2 and Location of Tutorial Presentations

Over the years, a wide range of experts have left their footprint in the tutorial and enriched it with a wide range of information. The members of the tutorial working group K2 have, over the last decade, all contributed with their experiences and knowledge accumulated over many years in many executed projects as users of GIS, manufacturers, or consultants. The working group includes members from the United States of America (US), France (FR), and Germany (DE) to give an international outlook. The active members of K2 with status of 2020 are listed in Table 1.2. The past members of K2 are listed in Table 1.3.

Working group K2 has presented the GIS/GIL tutorial at IEEE meetings several times, sometimes as a half- or full-day tutorial or in other cases as a panel discussion, always with the focus of

Table 1.2 Active members of the tutorial in update the table to be reflect members as of 2020

Name	Affiliation	Country	Time
Arun Arora	Consultant	US	Since 2002
George Becker	POWER Engineers	US	Since 2002
Phil Bolin	Mitsubishi	US	Since 2002
Hermann Koch	Consultant – previous Siemens	DE	Since 2002
John Brunke	POWER Engineers	US	Since 2005
Pat Fitzgerald	AZZ-CGIT	US	Since 2006
Ryan Stone	Mitsubishi	US	Since 2006
Arnaud Ficheux	GE	FR	Since 2008
Michael Novev	Burns & Mc Donalds	US	Since 2008
Peter Grossmann	Siemens Energy	DE	Since 2008
Charles L Hand	SCE	US	Since 2010
Steven Scott Scharf	TVA	US	Since 2012
Dave Mitchel	Consultant – previous Dominion	US	Since 2014
Bobby A. Rich	Dominion	US	Since 2016

Table 1.3 Past active members

Name	Affiliation	Country	Time
Lutz Boettger	ABB	US	2002–2006
Hugues Bosia	AREVA	FR	2002–2007
Wolfgang Degen	Consultant – previously Siemens	DE	2002–2008
Mel Hopkins	CGIT	US	2002–2005
Deborah Ottinger	EPA	US	2002–2005
Joseph Pannunzio	AREVA	FR	2010–2013
Richard Jones	Tech S Corp/Energy Initiatives Group	US	2003–2018
Venkatesh Minisandram	National Grid	US	2002–2018
Mark Etter	ABB	US	2002–2020

the information to the engineers participating from the electric power industry. In the present case, the tutorial content will be revised and finally published as a GIS Handbook, see Table 1.4.

Until today, the GIS/GIL tutorial has been presented 35 times around the globe since 2003. The attendance is spread between 10 and 100 participants. A total of 1270 engineers in the electric power business have attended the GIS/GIL tutorial. Most of the presentations were in USA, but also India, Central and South America, and Europe have been visited for tutorial presentations. Attendees could receive credit point to keep their professional status as electric power engineer.

With the Corona crisis in 2020, the tutorial needs to be stopped from presenting attendance and the next tutorial will be presented at the Transmission and Distribution Conferene in New Orleans in April 2022 (if Corona regulations will allow so).

Table 1.4 Locations of GIS/GIL Tutorial Presented

Conference	Location	Year	Attend
Substations Committee Meeting	Sun Valley, USA	April 2003	20
T&D Conference and Exhibition	Dallas, USA	Sept. 2003	50
Substations Committee Meeting	New Orleans, USA	April 2004	20
PES General Meeting	Denver, USA	July 2004	10
Switchgear Committee Meeting	Tucson, USA	Sept. 2004	40
Substations Committee Meeting	Tampa, USA	April 2005	20
PES General Meeting	San Francisco (Panel), USA	June 2005	30
IEEE Distinguished Lecturer Program	Delhi, Kolkata, Cheney, India	August 2005	50
Substations Committee Meeting	Scottsdale, USA	April 2006	15
PES General Meeting	Montreal, Canada (Panel), USA	June 2006	20
Substations Committee Meeting	Bellevue, USA	April 2007	15
PES General Meeting	Tampa (Panel), USA	June 2007	15
Substations Committee Meeting	San Francisco, USA	April 2008	20
T&D Conference and Exhibition	Chicago (Panel), USA	April 2008	100
PES General Meeting	Pittsburgh (Panel), USA	July 2008	20
IEEE DLP	Lima, Peru and La Paz, Bolivia	August 2008	50
IEEE DLP	Pune, Kolkata and Kanpur, India	Sept. 2008	70
Substations Committee Meeting	Kansas City, USA	May 2009	15
PES General Meeting	Calgary, Canada	July 2009	10
UHV Test Base State Grid	Beijing, China	March 2010	40
T&D Conference and Exhibition	New Orleans, USA	April 2010	30
PES General Meeting	Detroit, USA	July 2011	10
T&D Conference and Exhibition	Chicago, USA	July 2012	50
ISGT Conference	Berlin, Germany	August 2012	15
IEEE PES ICPEN	Arunachal Pradesh, India	December 2012	45
IEEE PES Austrian Chapter	Graz, Austria	March 2013	45
IEEE PES Costa Rica Chapter	San Jose, Costa Rica	June 2013	55
IEEE PES El Salvador Chapter	San Salvador, El Salvador	June 2013	65
IEEE PES GM	Vancouver, Canada	July 2013	10
IEEE PES CATON	Kolkata, India	December 2013	75
T&D Conference and Exhibition	Chicago	April 2014	65
T&D Conference and Exhibition	Dallas, Tx, USA	April 2016	50
Humbolt Kolleg Kolkata	Kolkata, India	February 2018	45
IEEE T&D Conference	Denver, Colorado, USA	April 2018	25
IEEE T&D Latin America	Lima, Peru	September 2018	55
		Total	1270

1.1.3 Content of the Tutorial

The experts of K2 work group are responsible for the content and presentation of the GIS/GIL tutorials. The content is under constant revision of working group K2 of the GIS subcommittee. The following topics have been presented:

- Overview: Gives an overview of the content and organization of the tutorial.
- GIS Basics: Here the basic knowledge of GIS is explained for practical applications.
- GIS Applications: Here a wide area of applications is given to show the large variations.
- GIL Basics: Here the basic knowledge of GIL for practical applications is given.
- GIL Applications: Here many applications in typical surroundings and laying methods are explained.
- Mixed Technology Switchgear: Here the compact or hybrid type of partly gas-insulated and air-insulated technology is given.
- SF_6: Here the knowledge of handling, atmospheric impact, and recycling is given.
- GIS Overloading: Here the specific conditions and rules for overloading of GIS are explained.
- Theory: Here the physical theory and gas-insulated systems with SF_6 is explained.
- Life Cycle Assessment: Here the impact for the lifetime of GIS is explained.
- Future Development: Here the next steps in development are explained for GIS.
- GIS Specification: Here the rules and conditions for correct specification of GIS are explained.
- GIS Monitoring: Here the monitoring systems for control and supervision are explained.
- Gas Handling: Here details of correct gas handling when dealing with SF_6 are given.
- Digital Communication IEC 62271-3: Here the impact of digital communication in substations is explained.
- Revision of C37.122 GIS: Here information of the latest revision of the GIS standard is given.
- Moisture Guide: Here information is given on how to monitor and handle the moisture of SF_6.
- Alternative Gases: Here information is given on alternative gases to SF_6 with lower or no global warming potential.
- Determining When to use GIS: Here information is given to find the right decision when to use GIS instead of AIS.
- Specification Development: Here information is given on how to write a good specification for GIS.
- Examples for GIS interfaces: Here information is given on the different interfaces to GIS and how to specify.
- Guidelines for GIS: Here information is given on aspects for using GIS based on IEEE C37.122.1.
- Guidelines for on-site tests: Here information is given on how to test GIS after erection on-site.
- Mobile GIS: Here information is given on the use of GIS for mobile substation installations.
- Moscow street house project and Toronto Junction Point GIS need case studies for the use of GIS and GIL.
- Underground Substations and Special buildings: Here information is given on special applications only possible with GIS.
- Condition Assessment: Here information is given on the possibilities to evaluate the condition of GIS after years of use.
- Substations Resilience: Here information is given to show the possibilities for a resilience use of GIS in the power network.
- Vacuum HV Switching: Here information is given on the use of vacuum switches and circuit breaker for voltages up to 500 kV.
- Low Power Instrument Transformers (LPIT): Here information is given for the use of LPIT in GIS and its advantages.

- Digital Twin of GIS and GIL: Here information is given of how GIS and GIL can be digitalized.
- Offshore GIS: Here information is given on the offshore applications of GIS.
- DC GIS: Here information is given on the use of direct current (DC) GIS.
- Digital substation: Here information is given on digital substations using GIS.

The last nine topics have not been presented in tutorial until now, they are innovative part of this GIS book.

1.2 Definitions

This GIS Handbook is based on definitions used in IEEE and IEC standards. Some of the most important definitions for better understanding of the book are listed below.

1.2.1 In IEEE and IEC Standards

There are two definitions used in IEC "Gas-Insulated Switchgear" and IEEE "Gas-Insulated Substations." The reason for this, different names for GIS has an historical background, where IEC started in SC 17A to develop circuit breaker standards and later started a new subcommittee SC 17C on high-voltage switchgear assemblies, so the link was made to switchgear. In IEEE, the substation committee developed standards on GIS in the substation subcommittee, so the link of GIS was taken to substations.

In IEEE C37.122, GIS is defined as follows:

Gas-insulated switchgear (GIS): a compact, multicomponent switchgear assembly, enclosed in a grounded metallic housing in which the primary insulating medium is gas, typically SF_6 and which normally includes buses, switches, circuit breakers, and other associated equipment. Also commonly known as gas-insulated switchgear.

Gas-insulated switchgear enclosure: A grounded part of gas-insulated metal-enclosed switchgear assembly retaining the insulating gas under the prescribed conditions necessary to maintain the required insulation level, protecting the equipment against external influences and providing a high degree of protection from approach to live energized parts.

Compartment (GIS): A section of a gas-insulated switchgear assembly that is enclosed, except for openings necessary for interconnection providing insulating gas isolation from other compartments. A compartment may be designated by the main components in it, e.g., circuit breaker compartment, disconnect switch compartment, bus compartment, etc.

Partition: Part of an assembly separating one compartment from other compartments. It provides gas isolation and support for the conductor (gas barrier insulator).

Power kinematic chain: Mechanical connecting system from and including the operating mechanism up to and including the moving contacts.

In IEC 62271-203, GIS is defined as follows:

Metal-enclosed switchgear and controlgear: switchgear and controlgear assemblies with an external metal enclosure intended to be earthed, and complete except for external connections.

Gas-insulated metal-enclosed switchgear: metal-enclosed switchgear in which the insulation is obtained, at least partly, by an insulating gas other than air at atmospheric pressure.

Gas-insulated switchgear enclosure: part of gas-insulated metal-enclosed switchgear retaining the insulating gas under the pre-scribed conditions necessary to maintain safely the highest insulation level, protecting the equipment against external influences and providing a high degree of protection to personnel.

Compartment: part of gas-insulated metal-enclosed switchgear, totally enclosed, except for openings necessary for interconnection and control.

Support insulator: internal insulator supporting one or more conductors.

Partition: support insulator of gas-insulated metal-enclosed switchgear separating one compartment from other compartments.

Note 1 to entry: A compartment may be designated by the main component contained therein, e.g., circuit-breaker compartment, bus bar compartment.

1.2.2 Gas-Insulated Metal-Enclosed Switchgear

Metal-enclosed switchgear in which the insulation is obtained, at least partly, by an insulating gas other than air at atmospheric pressure, as defined in IEC 62271-203. This term generally applies to high-voltage switchgear and controlgear. Three-phase enclosed gas-insulated switchgear applies to switchgear with the three phases enclosed in a common enclosure.

Single-phase enclosed gas-insulated switchgear applies to switchgear with each phase enclosed in a single independent enclosure.

1.2.3 Compartment of GIS

A section of a gas-insulated switchgear assembly that is enclosed except for openings necessary for interconnection provides insulating gas isolation from other compartments. A compartment may be designated by the main components in it, for example, circuit breaker compartment, disconnect switch compartment, bus compartment, and so on, as defined in IEEE C37.122.

A compartment of GIS as defined in IEC 62271-203 as part of a gas-insulated metal-enclosed switchgear is totally enclosed, except for openings necessary for interconnection and control.

A compartment may be designated by the main component contained therein, for example, circuit breaker compartment or bus bar compartment.

1.2.4 Design Pressure of Enclosures

The maximum gas pressure to which a gas-insulated switchgear enclosure will be subjected under normal service conditions, including the heating effects of rated continuous current, as defined in IEEE C37.122.

1.2.5 Gas Monitoring Systems

Any instrumentation for measuring, indicating, or giving remote warning of the condition or change in condition of the gas in the enclosure, such as pressure, density, moisture content, etc., as defined in IEEE C37.122.

1.2.6 Gas Leakage Rate (Absolute)

The amount of gas escaping by a time unit expressed in units of Pa m^3/s, as defined in IEEE C37.122.

1.2.7 Gas Leakage Rate (Relative)

The absolute leakage rate related to the total amount (mass or volume) of gas in each compartment at the rated filling pressure (or density). It is expressed in percentage per year, as defined in IEEE C37.122.

1.2.8 Gas Pass Through Insulator

An internal insulator supporting one or more conductors specifically designed to allow the passage of gas between adjoining compartments, as defined in IEEE C37.122.

1.2.9 Gas Zone

A section of the GIS, which may consist of one or several gas compartments that have a common gas monitoring system. The enclosure can be single-phase or three-phase, as defined in IEEE C37.122.

1.2.10 Local Control Cubicle (or Cabinet) (LCC)

A cubicle or cabinet typically containing secondary equipment including control and interlocking, measuring, indicating, alarm, annunciation, and mimic one-line diagrams associated with the primary equipment. It may also include protective relays if specified by the user, as defined in IEEE C37.122.

1.2.11 Support Insulator

An internal insulator supporting one or more conductors, as defined in IEC 62271-203.

1.2.12 Partition

Part of an assembly separating one compartment from other compartments. It provides gas isolation and support for the conductor (gas barrier insulator), as defined in C37.122.

A partition as defined in IEC 62271-203, which is a support insulator of gas-insulated metal-enclosed switchgear separating one compartment from other compartments.

1.2.13 Power Kinematic Chain

A mechanical connecting system from and including the operating mechanism up to and including the moving contacts, as defined in C37.122.

1.2.14 Design Pressure of Enclosures

Relative pressure used to determine the design of the enclosure. It is at least equal to the maximum pressure in the enclosure at the highest temperature that the gas used for insulation can reach under specified maximum service conditions. The transient pressure occurring during and after a breaking operation (e.g., a circuit breaker) is not to be considered in the determination of the design pressure, as defined in IEC 62271-203.

1.2.15 Relative Pressure Across the Partition

Relative pressure across the partition is at least equal to the maximum relative pressure across the partition during maintenance activities. The transient pressure occurring during and after a breaking operation (e.g., a circuit breaker) is not to be considered in the determination of the design pressure, as defined in IEC 62271-203.

1.2.16 Operating Pressure of Pressure Relief Device

Relative pressure chosen for the opening operation of pressure relief devices, as defined in IEC 62271-203.

1.2.17 Routine Test Pressure of Enclosures and Partitions

Relative pressure to which all enclosures and partitions are subjected after manufacturing, as defined in IEC 62271-203.

1.2.18 Type Test Pressure of Enclosures and Partitions

Relative pressure to which all enclosures and partitions are subjected for type test, as defined in IEC 62271-203.

1.2.19 Rated Filling Pressure p_{re}

Insulation and/or switching pressure (in Pa), to which the assembly is filled before putting into service. It is referred to at the standard atmospheric air conditions of +20 °C and 101.3 kPa (or density) and may be expressed in relative or absolute terms, as defined in C37.122.

1.2.20 Bushing

Device that enables one or several conductors to pass through a partition, such as a wall or a tank, and insulate the conductors from it, as defined in IEC 62271-203.

1.2.21 Main Circuit

All the conductive parts of gas-insulated metal-enclosed switchgear included in a circuit which is intended to transmit electrical energy, as defined in IEC 62271-203.

1.2.22 Auxiliary Circuit

All the conductive parts of gas-insulated metal-enclosed switchgear included in a circuit (other than the main circuit) intended to control, measure, signal, and regulate. The auxiliary circuits of gas-insulated metal-enclosed switchgear include the control and auxiliary circuits of the switching devices, as defined in IEC 62271-203.

1.2.23 Design Temperature of Enclosures

Maximum temperature that the enclosures can reach under specified maximum service conditions, as defined in IEC 62271-203.

1.2.24 Service Period

The time until a maintenance, including opening of the gas compartments, is required, as defined in IEC 62271-203.

1.2.25 Transport Unit

Part of gas-insulated metal-enclosed switchgear suitable for shipment without being dismantled, as defined in IEC 62271-203.

1.2.26 Mixed Technologies Switchgear (MTS)

Mixed technology switchgear concerns the following combinations:

AIS in compact and/or combined design
GIS in combined design
Hybrid GIS in compact and/or combined design

As defined in CIGRE Technical Brochure of Study Committee B3 Working Group 20 from November 2008.

1.2.27 Disruptive Discharge

Phenomena associated with the failure of insulation under electric stress, in which the discharge completely bridges the insulation under test, reducing the voltage between the electrodes to zero or almost zero, as defined in IEC 62271-203.

1.2.28 Fragmentation

Damage to enclosure due to pressure rise with projection of solid material, as defined in IEC 62271-203.

1.2.29 Functional Unit

Part of metal-enclosed switchgear and controlgear comprising all the components, as defined in IEC 62271-203.

1.3 Standards and References

1.3.1 Standards

Standards are technical documents that allow the manufacturer to develop equipment to meet the majority of user applications, and users to specify equipment that meets their needs in most cases. There are always circumstances that fall outside typical cases covered by standards, but they are few. Although there are many national and regional standards, the primary standards that apply to GIS are the International Electro-technical Commission (IEC) and the Institute of Electrical and Electronic Engineers (IEEE) standards. In recent years, great effort has been made to harmonize these standards. This effort continues, but differences between them remain. These reflect the differences in the nature of systems, applications, and practices between different parts of the world.

Gas-insulated switchgear, components, and related equipment fall under a large number of standards. Both IEC and IEEE have standards for GIS, circuit breakers, switches, bushings, testing, instrument transformers, controls, cabinets, pressure vessels, and so on. The difference in the equipment built under the two sets of harmonized standards is small and an understanding of how the equipment is designed and tested can usually allow the user to specify equipment under either set of standards. Most manufacturers design the equipment to meet either set of standards, but often limits on testing capability or cost can leave some areas covered by one set of standards only. This requires review of the application requirements and the tested performance of the equipment to determine if it meets the requirements. There has been, and continues to be, efforts made in the standards community to harmonize the requirements between IEEE and IEC. For high-voltage

GIS, efforts on 62271-203 (2020) and recently C37.122 (2021) have resulted in a high level of harmonization. Progress has also been made on high-voltage circuit breaker standards, but here many differences remain. Still, by understanding the differences, a user can use either standard.

An example of differences between IEEE and IEC GIS standards is in North America, where safety requirements for maintenance personnel mandate a visible check to verify that the circuit is not energized before it can be approached for maintenance. This requires a view port or camera to verify the disconnect switch blade position. In other countries, safety requirements allow verification of the position of the disconnect switch linkage as confirmation that the switch is open, kinematic chain.

There are standards other than IEEE and IEC that cover requirements related to GIS, for example, the American Society of Mechanical Engineers (ASME), the American Society for Testing and Materials (ASTM), the European Committee for Electrotechnical Standardization (CENELEC), European Standards (EN), the National Electrical Manufacturers Association (NEMA), to name a few.

1.3.2 Current Standards Most Relevant to GIS

The following is a list of the most relevant standards that may be used for specification of a GIS. This list was developed in 2012 and revised in 2020. Historically, standards can be withdrawn or their numbering changed, but usually only every decade or so.

1.3.2.1 General

IEEE C37.122: IEEE Standard for Gas-Insulated Substations

IEEE C37.123: IEEE Guide to Specifications for Gas-Insulated, Electric Power Substation Equipment

IEEE C37.122.1-2014: IEEE Guide for Gas-Insulated Substations Rated Above 52 kV

IEEE C37.24: Evaluating the Effect of Solar Radiation on Outdoor Metal-Enclosed Switchgear

IEC 62271-203: Gas-Insulated Metal-Enclosed Switchgear for Rated Voltages above 52 kV

IEC 62271-1: High-Voltage switchgear and controlgear Part 1: Common specification for alternating current switchgear and controlgear

IEC TS 62271-5 ED1: High-voltage switchgear and controlgear – Part 5: Common specifications for direct current switchgear

CIGRE Brochure 125: User Guide for the Application of Gas-Insulated Switchgear (GIS) for Rated Voltages of 72.5 kV and Above

1.3.2.2 GIS Enclosures

In some jurisdictions in the United States, local building codes require the use of the following standards for GIS enclosures:

ANSI/ASME B31.1: Power Piping

The ASME standards are not specifically intended for use for electrical enclosures, but are required in local building codes and are therefore relevant.

In Europe and in Canada, the standards developed by the European Committee for Electrotechnical Standardization (CENELEC) for GIS enclosures are commonly used:

CENELEC EN 50052 (2017): Specification for Cast Aluminum Alloy Enclosures for Gas-Filled High-Voltage Switchgear and Controlgear

CENELEC EN 50064-1989 (2017): Specification for Wrought Aluminum and Aluminum-Alloy Enclosures for Gas-Filled High-Voltage Switchgear and Controlgear

CENELEC EN 50069 (2019): Specification for Welded Composite Enclosures of Cast and Wrought Aluminum Alloys for Gas-Filled High-Voltage Switchgear and Controlgear

CENELEC EN 50089 (1994) revised version in 2022: Specification for Cast Resin Partitions for Metal-Enclosed Gas-Filled High-Voltage Switchgear and Controlgear

In other regions, often other standards are in place for GIS enclosures.

1.3.2.3 GIS Systems Above 52 kV

IEEE C37.122: Standard for High-Voltage Gas-Insulated Substations Rated above 52 kV

IEC 62271-203: High-Voltage Switchgear and Controlgear – Part 203: Gas-Insulated Metal-Enclosed Switchgear for Rated Voltages above 52 kV.

1.3.2.4 Gas-Filled Bushings

IEEE Std. C37.017: IEEE Standard for Bushings for High-Voltage Circuit Breakers and Gas-Insulated Switchgear

IEC 61462: Composite Hollow Insulators – Pressurized and Unpressurized Insulators for Use in Electrical Equipment with Rated Voltage Greater than 1000 V – Definitions, Test Methods, Acceptance Criteria, and Design Recommendations

IEC 62155: Hollow Pressurized and Unpressurized Ceramic and Glass Insulators for Use in Electrical Equipment with Rated Voltages Greater than 1000 V

IEC 60507: Artificial Pollution Tests on High-Voltage Insulators to be Used on a.c. Systems

1.3.2.5 Common Clauses for Switchgear

IEEE Std. C37.100: IEEE Standard Definitions for Power Switchgear

IEEE Std. C37.100.1: IEEE Standard of Common Requirements for High-Voltage Power Switchgear Rated above 1000 V

IEC 62271-1: High-Voltage Switchgear and Controlgear. Part 1: Common Specification

1.3.2.6 Sulfur Hexafluoride (SF6) Gas

ASTM D2472-00: Standard Specification for Sulfur Hexafluoride

IEC 62271-4 ED2: High-voltage switchgear and controlgear – Part 4: Handling procedures for gases and gas mixtures for interruption and insulation

IEC 60376: Specification of Technical Grade Sulfur Hexafluoride (SF_6) for Use in Electrical Equipment

IEC 60480: Guidelines for the Checking and Treatment of Sulfur Hexafluoride (SF_6) Taken from Electrical Equipment and Specification for its Re-use

1.3.2.7 High Voltage Testing on Control Systems

IEEE Std. C37.90.1: Surge Withstand Capability (SWC) Tests for Relays and Relay Systems Associated with Electric Power Apparatus

IEC 61180-1: High-Voltage Test Techniques for Low-Voltage Equipment – Part 1: Definitions, Test, and Procedure Requirements

IEC 61180-2: High-Voltage Test Techniques for Low-Voltage Equipment – Part 2: Test Equipment

1.3.2.8 High-Voltage Circuit Breakers

IEEE Std. C37.04: IEEE Standard Rating Structure for AC High-Voltage Circuit Breakers Rated on Symmetrical Current Basis

IEEE Std. C37.06: High-Voltage Circuit Breakers Rated on a Symmetrical Current Basis: Preferred Ratings and Related Required Capabilities

IEEE Std. C37.09: IEEE Standard Test Procedure for AC High-Voltage Circuit Breakers Rated on a Symmetrical Current Basis

IEC 62271-100:2008 + AMD1 and 2:2017: High-Voltage Switchgear and Controlgear – Part 100: High-Voltage Alternating-Current Circuit-Breakers

IEC 62271-101+AMD1:2017: High-Voltage Switchgear and Controlgear – Part 101: Synthetic Testing

1.3.2.9 Disconnect and Grounding (Earthing) Switches

IEEE GIS switches are included in IEEE Std. C37.122 (previous IEEE GIS switch standard C37.38 has been withdrawn)

IEC 62271-102:2018: High-Voltage Switchgear and Controlgear – Part 102: Alternating Current Disconnectors and Earthing Switches

1.3.2.10 Safety and Grounding

ANSI/IEEE C2: National Electrical Safety Code

IEEE Std. 80: IEEE Guide for Safety in AC Substation Grounding

IEEE Std. 367: IEEE Recommended Practice for Determining the Electric Power Station Ground Potential Rise and Induced Voltage from a Power Fault (ANSI)

IEC 61936-1 (2021): Power installations exceeding 1 kV AC and 1,5 kV DC - Part 1: AC

1.3.2.11 Application Guides for Circuit Breakers

IEEE Std. C37.010: IEEE Application Guide for AC High-Voltage Circuit Breakers Rated on a Symmetrical Current Basis

IEEE Std. C37.011: IEEE Application Guide for Transient Recovery Voltage for AC High-Voltage Circuit Breakers Rated on a Symmetrical Current Basis

IEEE Std. C37.012: Application Guide for Capacitance Current Switching for AC High-Voltage Circuit Breakers Rated on a Symmetrical Current Basis

IEEE Std. C37.015: IEEE Application Guide for Shunt Reactor Current Switching

CIGRE 304: Guide for the Application of IEC 62271-100 and IEC 62271-1. Part 1: General Subjects

CIGRE 305: Guide for the Application of IEC 62271-100 and IEC 62271-1. Part 2: Making and Breaking Tests

1.3.2.12 Application Guides for GIS

CIGRE 125: User Guide for the Application of Gas-Insulated Switchgear (GIS) for Rated Voltages of 72.5 kV and Above

IEEE Std. 1300: IEEE Guide for Cable Connections for Gas-Insulated Equipment

1.3.2.13 Application Guides for SF_6

CIGRE 234: SF_6 Recycling Guide

CIGRE 276: Guide for the Preparation of Customized Practical SF_6 Handling Instructions

IEEE C37.122.3-2011: IEEE Guide for Sulphur Hexafluoride (SF6) Gas Handling for High-Voltage (over 1000 Vac) Equipment

IEEE C37.122.5 Guide for moisture measurement and control in SF6 Gas-insulated equipment

IEEE Std. 1416: IEEE Recommended Practice for the Interface of New Gas-Insulated Equipment in Existing Gas-Insulated Substations

1.3.2.14 Reliability Evaluations

CIGRE 83: Final Report of the Second International Inquiry on High-Voltage Circuit-Breaker Failures and Defects in Service

CIGRE 150: Report on the Second International Survey of High-Voltage Gas-Insulated Substations (GIS) Service

CIGRE 319: Circuit-Breaker Controls, Failure Survey on Circuit Breaker Control Systems

1.3.2.15 Cable and Transformer Interfaces

IEEE C37.122.6-2013: IEEE Recommended Practice for the Interface of New Gas-Insulated Equipment in Existing Gas-Insulated Substations Rated above 52 kV

ANSI/NEMA CC 1: Electric Power Connection for Substations

IEEE Std. 48: IEEE Standard Test Procedures and Requirements for Alternating-Current Cable Terminations 2.5 kV through 765 kV

ANSI Std. C63.2: American National Standard Specifications for Electromagnetic Noise and Field-Strength Instrumentation, 10 kHz to 40 GHz.

IEC 62271-211:2014, Edition 1.0 (2014-04-24), High-voltage switchgear and controlgear – Part 211: Direct connection between power transformers and gas-insulated metal-enclosed switchgear for rated voltages above 52 kV

IEC 62271-209:2019, Edition 2.0 (2019-02-08) & Amendment 1, High-voltage switchgear and controlgear – Part 209: Cable connections for gas-insulated metal-enclosed switchgear for rated voltages above 52 kV – Fluid-filled and extruded insulation cables – Fluid-filled and dry-type cable-terminations

1.3.2.16 Seismic Design

IEEE Std. 693: IEEE Recommended Practices for Seismic Design of Substations

IEC 60068-3-3: Environmental Testing – Part 3: Guidance. Seismic Test Methods for Equipment

IEC 62271-207: High-voltage switchgear and controlgear – Part 207: Seismic qualification for gas-insulated switchgear assemblies for rated voltages above 52 kV

1.3.2.17 Control Cabinets

IEC 62262: Degrees of Protection Provided by Enclosures for Electrical Equipment Against External Mechanical Impacts (IK code)

ANSI/IEC 60529: Degrees of Protection Provided by Enclosures (IP code)

NEMA 250: Enclosures for Electrical Equipment

IEEE Std. C37.21: IEEE Standard for Control Switchboards

IEEE Std. C57.13: IEEE Standard Requirements for Instrument Transformers

IEEE Std. C37.24: IEEE Guide for Evaluating the Effect of Solar Radiation on Outdoor Metal-Enclosed Switchgear

IEEE C37.301: Standard for High-Voltage (above 1000 V) Test Techniques – Partial Discharge Measurements

IEEE Std. C62.11: IEEE Standard for Metal-Oxide Surge Arresters for AC Power Circuits (>1 kV)

IEEE Std. 4: IEEE Standard Techniques for High-Voltage Testing

IEEE Std. 315: IEEE Standard, American National Standard, and Canadian Standard Graphic Symbols for Electrical and Electronics Diagrams (Including Reference Designation Letters)

For Canada CAN/CSA C61869 series standards for instrument transformers apply. ASTM publications are available from the Customer Service Department, American Society for Testing and Materials, 1916 Race Street, Philadelphia, PA 19103, USA, CSA Webstore: https://www.csagroup.org/ or ANSI Webstore: https://webstore.ansi.org/.

CENELEC publications are available from the Sales Department, American National Standards Institute, 11 West 42nd Street, 13th Floor, New York, NY 10036, USA, European Standards Store: https://www.en-standard.eu/ or IEC Webstore: https://webstore.iec.ch/.

IEC publications are available from IEC Sales Department, Case Postale 131, 3 rue de Varembe, CH-1211, Geneva 20, Switzerland/Suisse. IEC publications are also available in the United States from the Sales Department, American National Standards Institute, 11 West 42nd Street, 13th Floor, New York, NY 10036, USA, IEC Webstore: https://webstore.iec.ch/.

IEEE publications are available from the Institute of Electrical and Electronics Engineers, 445 Hoes Lane, P.O. Box 1331, Piscataway, NJ 08855-1331, USA, IEEE-SA Standards Store: https://www.techstreet.com/ieee.

CIGRE publications are available from CIGRE, 21 rue d'Artois, 75 008 Paris, France.

ANSI Standards are available from the American National Standards Institute, 11 West 42nd Street, 13th Floor, New York, NY 10036, USA, ANSI Webstore: https://webstore.ansi.org/.

1.3.2.18 Gas-Insulated Transmission Line (GIL)

IEC 62271-204 ED2: High-voltage switchgear and controlgear – Part 204: Rigid gas-insulated transmission lines for rated voltage above 52 kV.

IEEE C37.122.4-2016 – IEEE Guide for Application and User Guide for Gas-Insulated Transmission Lines, Rated 72.5 kV and Above

1.4 Ratings

1.4.1 General

The purpose of ratings is to correctly specify GIS equipment based on electric system topology and characteristics while reducing the variety of technical possibilities and guiding manufacturers. These ratings provide standardized solutions recognized across the industry and reduce cost. The main rating parameters are voltage, insulation level, frequency, current, short time and peak withstand current, duration of short circuit and auxiliary voltages, and frequencies.

In high-voltage switchgear, the rating structures are defined for devices such as circuit breakers, and disconnect switches, grounding (earthing) switches, and connecting conductors/bus. In general, they are covered in IEEE C37.100.1 or in IEC 62271-1 for switchgear products.

For assemblies of high-voltage switchgear like GIS, the standards are used to satisfy design criteria and applications in the field. The design of the GIS must consider the very high cost for developments and manufacturing of different types of pressurized metal enclosures. For this reason, some designs of GIS are grouped to cover multiple voltage ratings. One technical criterion of the equipment, for example, rated voltages of 110, 123, 138, and 145 kV, is covered by the same class of GIS with the same enclosure. Once you reach higher voltages of 230, 345, 500 kV and up, you have GIS in separate enclosures.

Within this range of voltage classification, only different gas densities of SF_6 differentiate between the different voltage levels. In terms of current ratings, the difference between 2000, 2500, and 3000 A might only be a different number of contact fingers or different wall thicknesses of conductors.

1.4.2 Rated Maximum Voltage

The high voltage (HV) levels in standards start at ratings above 52 kV in both IEC and IEEE. Below these voltage levels, the equipment is classified as medium voltage (MV). The typical GIS high-voltage ratings can be grouped into four design classes, even if the split may vary somewhat. The lower high-voltage ratings are in the range from 52 to 72.5 kV as the first level range. The second level range includes 100, 123, 145 kV, and in some cases the 170 kV rating. The third level range covers 245 and 300 kV. The fourth level range includes the voltages 362 and 420 kV. The 345 kV voltage level is considered as 362 kV and is no longer recommended by standards today. The third and fourth level have been historically developed in North America and Europe, typically as 245 and 420 kV in Europe and 300 and 362 kV in North America. The reason behind this is the availability of technical solutions like insulators at the time when the new voltage levels were established. An overview of the voltage ratings is given in Table 1.5.

Only two voltage ratings left in the IEEE and IEC standards, which have different options for power frequency switching and lightning impulse values, are 245 and 362 kV. 550 kV rated voltage offers two insulation levels for the rated power frequency withstand voltage.

1.4.3 Rated Insulation Level

The ratings for insulation levels are derived from the network to which the GIS is connected. Network conditions, like lightning strokes into overhead lines, their local probability, and their expected strength, are indicators for the overvoltage that may occur. In the case of cable networks, the length of cables and their related overvoltage during switching operations will influence this rating.

Rated insulation levels are key parameters for the design of GIS and do have a direct impact on the enclosure diameter and, with this, a high-cost impact in development and manufacturing cost. Each rated voltage in IEC and IEEE has the choice of two or even more insulation levels. In GIS, the choice is usually made in favor of the highest requirement for the GIS.

As shown in Table 1.5, in most cases the listed rated power frequency withstand voltage, the rated switching impulse withstand voltage, and the rated lightning impulse withstand voltage for the related rated maximum voltage is the highest value from IEC and IEEE standards. Only the rated maximum voltage classifications of 245 and 362 kV have the choice of two voltage levels. The reason behind these choices is that, in North America, many such GIS are in operation from the past, while the rated insulation levels of today's GIS offer higher values.

Table 1.5 Rated voltages of IEEE and IEC

		Rated max voltage U_m	**Rated power frequency withstand voltage**	**Rated switching impulse withstand voltage**	**Rated lightning impulse withstand voltage (BIL)**
IEC	**IEEE**	**kV rms**	**kV rms**	**kV peak**	**kV peak**
×	×	72.5	140	—	325
×	×	100	185	—	450
×	×	123	230	—	550
×	×	145	275	—	650
×	×	170	325	—	730
	×	245	425	—	900
×	×	245	460	—	1050
×	×	300	460	850	1050
	×	362	500	850	1050
×	×	362	520	950	1175
×	×	420	650	1050	1425
×	×	550	710	1175	1550
×	×	550	740	1175	1550
×	×	800	960	1425	2100
×	×	1100	1100	1800	2400

1.4.4 Rated Power Frequency

Most power frequencies for GIS are 50 Hz originated in Europe and 60 Hz originated in North America. Apart from 16 ⅓ Hz and 25 Hz for railroad applications, the majority of GIS applications are with 50 and 60 Hz. These two frequencies are distributed world-wide and form regions and countries with one or the other frequency. Some countries, for example, Japan and Saudi Arabia, have both frequencies.

The dielectric impact to the GIS design of these frequencies is negligibly low. The thermal impact needs to be considered when the current rating is approaching the limits, because at 60 Hz the power density is higher leading to higher thermal rise. Temperature limits should not be exceeded because of possible damage to insulators or contact systems.

1.4.5 Rated Continuous Current

The continuous current rating is a basic design criterion of GIS for bus and contact sizing. The complex structure of GIS allows close influence of the different devices such as circuit breakers, ground switches, disconnect switches, current transformers, voltage transformers, and bus bars in terms of heat dissipation and temperature rise. For this reason, the IEEE and IEC standards require temperature rise tests to confirm the correct function of all devices included in GIS. A so-called typical bay configuration will be used for this test.

One of the factors specific to a GIS installation may show that the rated continuous current may be different for the bus bar or the feeders depending on the substations' scheme. Typical rated continuous currents are shown in Table 1.6.

1.4.6 Rated Short Time Withstand Current

The rated short time withstand current (I_K), the peak-withstand current (I_p), and the duration of the short circuit (t_K) are basic dimensioning parameters for GIS design (see Table 1.7).

These values have a great impact on the electromechanical forces to the insulators and conductors, and on the thermal rise, mainly of the contact system. These values are also tested by specific

Table 1.6 Typical current ratings of GIS related to voltage classes

	52–72.5 kV	**100–170 kV**	**245–300 kV**	**362–550 kV**	**800 kV**	**1100 kV**
5000–8000 A				×	×	×
4000–5000 A				×	×	×
3150–4000 A			×	×	×	
2500–3150 A	×	×	×	×		
1250–2500 A	×	×				

Table 1.7 Typical short-circuit current ratings of GIS related to voltage classes

	52–72.5 kV	**100–170 kV**	**245–300 kV**	**362–550 kV**	**800 kV**	**1100 kV**
63–100 kA				×	×	×
50–63 kA		×	×	×	×	×
31.5–50 kA		×	×	×		
25–31.5 kA		×	×			
16–25 kA	×					

Table 1.8 Typical factors to calculate rated peak withstand currents (I_p)

Networks	Factor to calculate I_p	DC constant (ms)
50 Hz up to 500 kV	2,5	45
60 Hz up to 500 kV	2,6	45
50/60 Hz 800 kV and above	2,7	60, 75, 120

Table 1.9 Rated duration of short circuit (t_K)

	Rated duration of short circuit (t_K) (s)
Short	0.5
Standard	1.0
Long	2.0
Very long	3.0

Table 1.10 Rated supply voltages

	Rated supply voltages
DC	48, 110, and 125 V
AC	208/120 V three-phase, 400/230 V three-phase and 230/115 V single-phase

type tests to confirm the satisfactory function of the different devices of a GIS, such as the circuit breaker, disconnect, ground switch, and bus bars.

1.4.7 Rated Peak Withstand Current

The rated peak withstand current (I_p) is defined by the DC time constant of the network. The rated peak withstand current is defined as a factor of the rated short time withstand current. Typical values in the network are 45 ms in most voltage classes, and up to 120 ms in ultra-high voltage (UHV) networks. The related factors are shown in Table 1.8.

GIS equipment is designed to fulfill these requirements.

1.4.8 Rated Duration of Short Circuit

The rated duration of a short circuit (t_K) depends on the network protection and is symmetrical. Over the decades of network development, this value has typical decrease, because of faster protection relay equipment. A typical value today is 1 s, but also 0.5 s can be used. In some cases, 2 s or 3 s may be required. The duration of a short circuit has a significant impact on the GIS design, and it is recommended to keep this time as short as possible (see Table 1.9).

1.4.9 Rated Supply Voltages

There are many different supply voltages used and covered by the standards. This high variation is costly for substation design and should be reduced. Therefore, the standards give some preferred values. For existing substations, this might not be economical but new substation design should follow these recommendations (see Table 1.10).

2

Basic Information

Authors: 1st edition Hermann Koch, John H. Brunke, and John Boggess, 2nd edition Dave Giegel, Hermann Koch, George Becker, Peter Grossmann, and Pathik Patel
Reviewers: 1st edition Phil Bolin, Hermann Koch, Devki Sharma, Markus Etter, Scott Scharf, George Becker, Noboru Fujimoto, Ed Crocket, Shawn Lav, Jim Massura, Tony Lim, Ricardo Arredondo, Chuck Hand, and Dave Giegel 2nd edition Arnaud Ficheux, George Becker, Pathik Patel, John Brunke, Michael Novev, Scott Scharf, and Nick Matone

2.1 History

2.1.1 General

The development of switchgear is guided by the need for safe and reliable devices to operate voltage switching and current breaking. The rating of voltages and currents increased in parallel with new technologies available. The earliest technology for interruption used air under greater than atmospheric pressure. The interruption method design used the atmospheric pressure by blasting air into the arc in order to cool and quench it. This design used air to blast into the arc to cool the arc and quench it. This allowed for an arc quenching at higher current ratings than previously obtained. The next design used electric contacts, immersed in oil for cooling and arc quenching. The initial design big oil tanks with the complete switchgear inside. Later, the oil compartment was much smaller and only held the interruption unit and the contact system of the circuit breaker. The rest of the equipment was air insulated and connected by bushings. Subsequently, a significant change occurred when SF_6 was used for the switchgear insulating medium. Air and different kinds of insulating gases, for example, N_2, had been used for insulating purposes inside metallic enclosures. In the 1960s, sulfur hexafluoride (SF_6) was the preferred choice for switchgear because of its high insulation and arc-quenching capability. Since about 2010 the search for an alternative insulating and switching gas with a much reduced global warming potential brought solutions based on Flournitrile, Flourketone and technical air combined with vacuum circuit breakers. For details see sections 2.9 and 9.13.

Reliable and economical power transmission and distribution are key functions for the future electric power supply. High-voltage switchgear and equipment for voltages above 1 kV up to 1100 kV are critical elements within the electrical energy supply and are therefore subjected to a very high standard of availability and reliability. Industrial areas have high energy demands and requirements for space saving and economical design. These requirements are fulfilled by the use of SF_6-insulated switchgear.

For more information about gas-insulated switchgear refer to References [23–32], while information on the technology can be found in References [33–35]. Operational experiences and

Gas Insulated Substations, Second Edition. Edited by Hermann J. Koch.

recommendations for maintenance have been published in a series of CIGRE technical brochures and publications in Electra [36–41].

2.1.2 Steps of Development

The first insulated switchgear assemblies for high-voltage levels were designed in the 1920s using oil as the insulating medium. These oil-insulated switchgear assemblies had higher switching capabilities than the air-insulated switchgear with an air blast circuit breaker previously used. The current rating went increase, also increased the danger for fire and thus posed significant risks for the substations and personnel. To minimize the risk of flammable oil gaseous insulations were developed.

The first gas-insulated metal enclosed switchgear used freon as the insulating gas; this technology had already been known since the 1930s. GIS installation containing SF_6 with voltage level between 60 kV and 100 kV are first available since early 1960s.

With the introduction of SF_6 in modern switchgear technology the world's first SF_6 high-voltage gas-insulated switchgear was introduced on the market in 1968, using SF_6 as the insulating and arc-quenching medium for the first time. One of the first GIS installed in Berlin, Germany, in 1968, at Wittenau Substation, for 110 kV rated voltage is shown in Figure 2.1. Later, a SF_6 switchgear rated 550 kV with 100 kA was installed in Canada, which has the highest breaking capacity ever achieved. This was the highest breaking capacity ever achieved. Continuous research and development and technological innovations has led to the sixth generation of compact and overall optimized switchgear today.

The GIS design of the first generation of SF_6 technology used steel as an enclosure. At this time gastight aluminum molding processes had not been developed. In addition, aluminum cast mold

Figure 2.1 One of the first GIS installed in Berlin, Germany, in 1968 at Wittenau Substation for 110 kV (Reproduced by permission of Siemens AG)

Figure 2.2 Steel enclosure – inductive flange production (Reproduced by permission of Siemens AG)

Figure 2.3 Circuit breaker enclosure – three-phase design of 110 kV (Reproduced by permission of Siemens AG)

was also limited in size of the enclosure. Figure 2.2 shows how a steel enclosure had been treated to connect a flange. Using inductive heating, the steel was heated enough mechanically form it for the flange connection. In Figure 2.3, the complete three-phase enclosure of a 110 kV GIS is shown in the manufacturing process.

The steel design offered high stability, but high weight of GIS. The design and functionality of this types of GIS was highly reliable from the beginning and some of these installations are still in operation world-wide to this day. The gas tightness of the steel enclosures using O-ring sealings manufactured by leading technology manufacturers has been proven to be highly reliable.

The 110 kV three-phase insulated GIS with the double bus bar shown in Figure 2.4 and the 380 kV three-phase insulated bus bar and single-phase insulated circuit breaker, disconnector and ground switches shown in Figure 2.5 are from this steel type of design and have been in operation in Germany since the 1970s.

Figure 2.4 110 kV GIS – double bus bar, three-phase encapsulated, horizontal circuit breaker (Reproduced by permission of Siemens AG)

Figure 2.5 380 kV horizontal GIS – three-phase double bus bar and single-phase vertical circuit breaker (Reproduced by permission of Siemens AG)

The manufacturing process of steel enclosures consisted of multiple factory assembly steps and production of single parts and elements. The enclosure, the fixing structures inside, and any single parts required a customized handmade manufacturing process. There were very few automation processes available, as can be seen in Figure 2.6, which shows the amount of manual labor required in a manufacturing hall in the 1970s.

In the 1980s, aluminum enclosures were increasingly as of the manufacturing process. The higher functionality of the enclosures made in casting molds reduced the manual labor required and allowed for automation of the production process. Figure 2.7 shows an aluminum enclosures manufacturing process as it can be seen today. Multiflange enclosures are produced in large machinery stations and the process is fully automated. The machined enclosures are ready for the assembly process.

The GIS design of aluminum enclosures had a large impact on the size of the equipment and its visual appearance. In Figure 2.8, a 145 kV GIS with single encapsulated enclosures and a horizontal circuit breaker enclosure as the basis of each bay is shown. The relatively small module enclosures are made for separate functions such as the disconnector and ground switch, voltage, and

Figure 2.6 Steel enclosure manufacturing process (Reproduced by permission of Siemens AG)

Figure 2.7 Aluminum enclosure manufacturing (Reproduced by permission of Siemens AG)

current transformers, and the circuit breaker. The bus bar enclosure in this design is three-phase insulated. Additionally, this compact design of the GIS allowed for the integrated control cubicle for the bay control system, the SF_6 gas metering, and the circuit breaker operation drive.

The three-phase encapsulated design of a 110 kV GIS with a vertical circuit breaker is shown in Figure 2.9. The improvements of aluminum-casting technology led to reduced sizes of enclosures and this allows for the three-phase design. Advantages of this design are reduction of material and a smaller overall switchgear size. In Figure 2.9, the vertical circuit breaker enclosure is connected to the three-phase encapsulated bus bar (left side) and a three-phase encapsulated solid-insulated cable connection (right side). The control cubicle is placed on top of the bus bar and holds the bay control system operation drives for the circuit breakers and SF_6 gas monitoring.

At higher-voltage levels aluminum-casted enclosures are still single-phase encapsulated due to the large size of the equipment. Figure 2.10 shows an example of a 500 kV GIS. The horizontal basis of the circuit breaker enclosure with two interruption units is shown with the operation drive mechanism in the middle below the enclosure. The bus bar on the left side is three-phase encapsulated as well as the out- and ingoing connections on the right side of Figure 2.10.

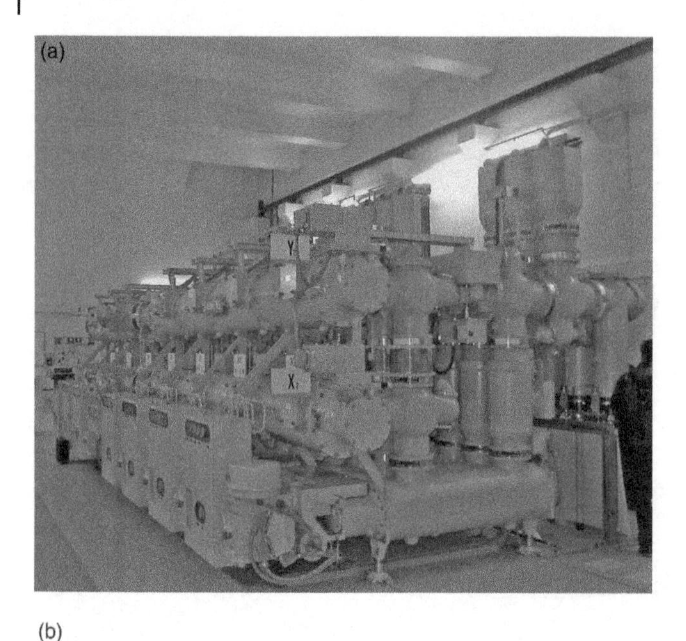

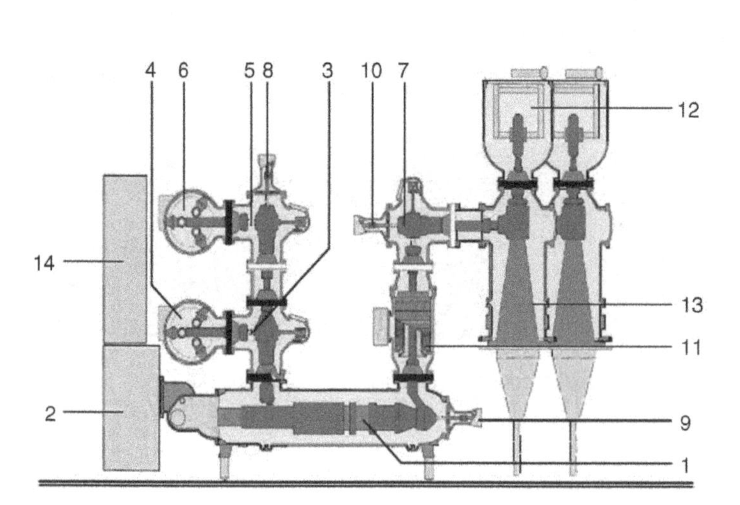

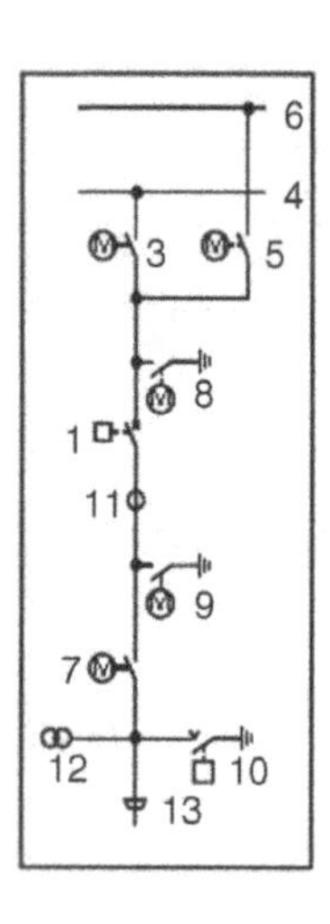

Figure 2.8 (a) 145 kV GIS – single-phase encapsulated and horizontal circuit breaker. (b) Gas-insulated switchgear 245 kV switchgear bay. (1) Circuit breaker interrupter unit, (2) spring-stored energy operating mechanism with circuit breaker control unit, (3) bus bar disconnector I, (4) bus bar I, (5) bus bar disconnector II, (6) bus bar II, (7) outgoing disconnector, (8) earthing switch (for work in progress), (9) earthing switch (for work in progress), (10) make-proof earthing switch (high speed), (11) current transformer, (12) voltage transformer, (13) cable sealing end, and (14) integrated local control cubicle (Reproduced by permission of Siemens AG)

The control cubicle at these voltage levels is not integrated in the bay and the cubicles are located underneath the in- and outgoing line on the right side of Figure 2.10. In some case, control cubicles are located in a separate room.

The technology inside the GIS enclosures has also been developed over the years. In the beginning, a self-blasting interruption unit was developed, as shown in Figure 2.11, in the 1970s a blast cylinder with compressed SF_6 gas inside the cylinder and guided the additional SF_6 through an opening directly into the interruption gap to cool the arc. This increased the interruption capability

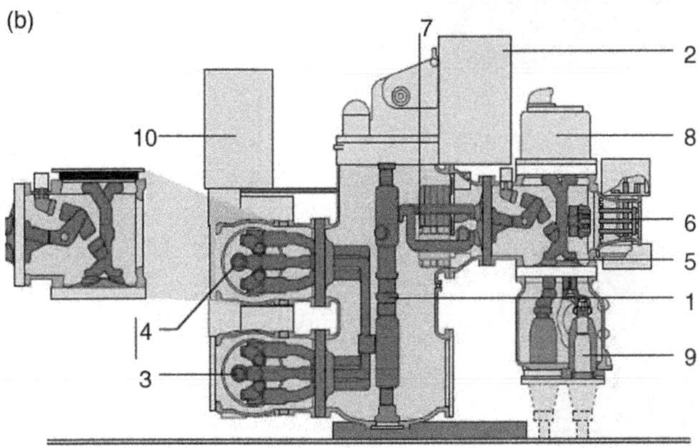

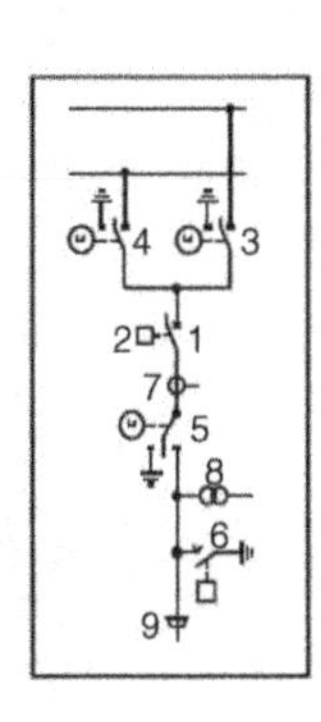

Figure 2.9 (a) 110 kV GIS – vertical circuit breaker and three-phase enclosure. (b) (1) Circuit breaker interrupter unit, (2) spring-stored energy operating mechanism with a circuit breaker control unit, (3) bus bar I with disconnector and earthing switch, (4) bus bar II with disconnector and earthing switch, (5) outgoing module with disconnector and earthing switch, (6) make-proof earthing switch (high speed), (7) current transformer, (8) voltage transformer, (9) cable sealing end, and (10) integrated local control cubicle (Reproduced by permission of Siemens AG)

of the circuit breakers up to 8000 A rated current and 100 kA short circuit currents for 0.5 second. The first patents of this technology appeared in 1973 (see Figure 2.11).

The next improvement in circuit breaker interruption units came in the late 1970s with anticompression cylinders to further increase the cooling of the arc, as shown in Figure 2.12. Here, the compression cylinder had a separate driving mechanism, which was moving against the direction of the interruption unit and therefore doubled the speed of compression. This mechanism improved the switching capability, whoever did not result in higher current rations; further improvements focused on the reduction in size since the current ratings were sufficient. The first patents appeared in 1977.

The main advantage of this interruption technology was the fast interruption time so that a true two cycle circuit breaker was available. In the 1980s, the next development step brought the double

(a)

(b)

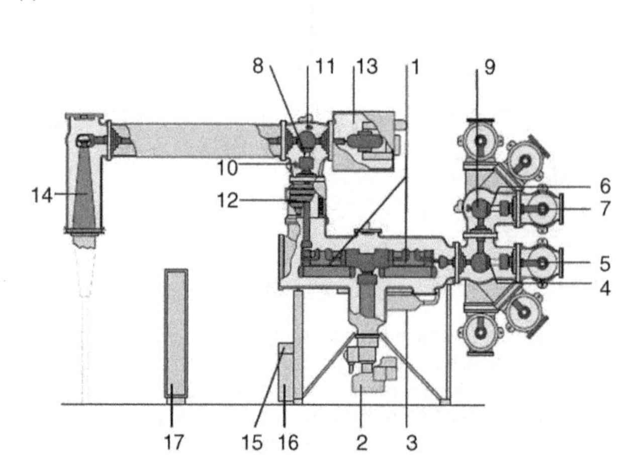

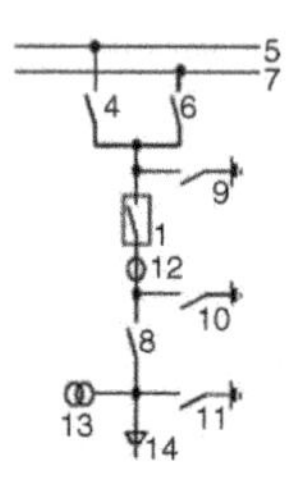

Figure 2.10 (a) 500 kV GIS – horizontal single-phase circuit breaker and two interruption units of the circuit breaker. (b) (1) Circuit breaker interrupter unit, (2) electrohydraulic operating mechanism, (3) hydraulic storage cylinder, (4) bus bar disconnector I, (5) bus bar I, (6) bus bar disconnector II, (7) bus bar II, (8) outgoing feeder disconnector, (9) earthing switch (work-in-progress), (10) earthing switch (work-in-progress), (11) make-proof earthing switch (high speed), (12) current transformer, (13) voltage transformer, (14) cable sealing end, (15) gas monitoring unit, (16) circuit breaker control unit, and (17) local control cubicle (Reproduced by permission of Siemens AG)

valve interruption unit (see Figure 2.13). This technology improved the number of operations of the circuit breaker before maintenance was required. Rated current interruptions and short circuit-rated interruptions were improved by the use of carbon nozzles in the arc burning region. Additionally, hot gas flow was used additionally to support the operation drive and to reduce the power of the drive mechanism. The first patents appeared in 1985.

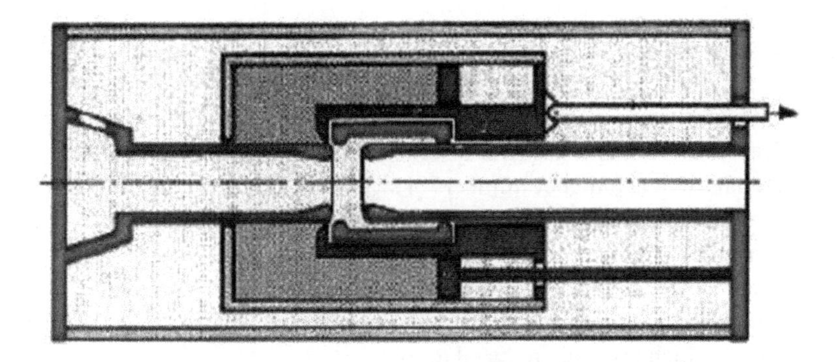

Figure 2.11 Self-blast interruption unit – 1973 first patent (Reproduced by permission of Siemens AG)

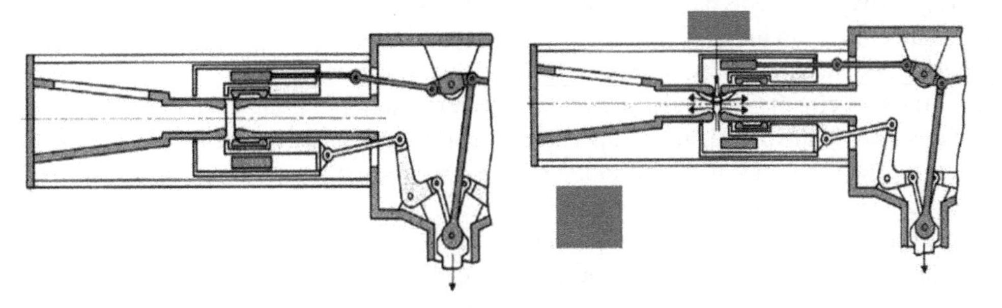

Figure 2.12 Two-cycle circuit breaker – anticompression cylinder – 1977 first patent (Reproduced by permission of Siemens AG)

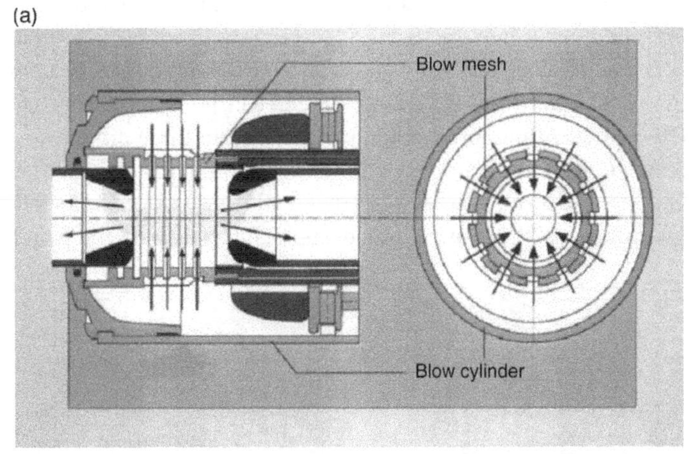

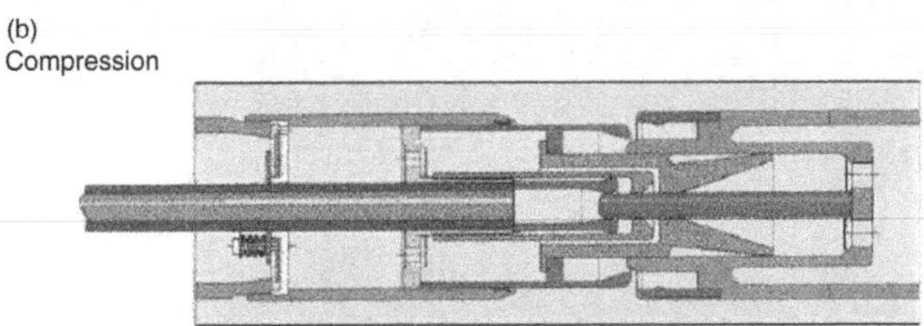

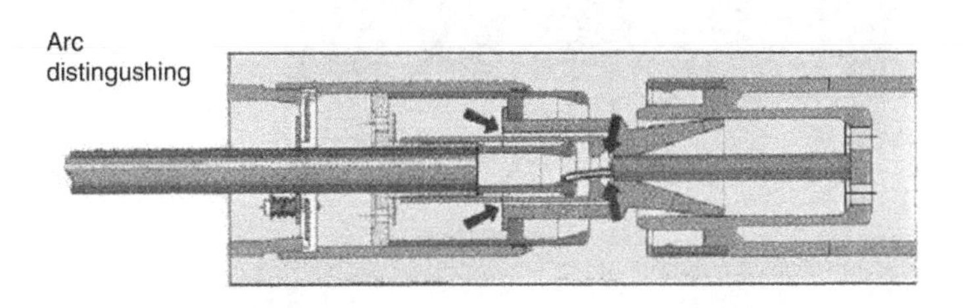

Figure 2.13 Double valve interruption unit – 1985 first patents – low power drive mechanism – high switching numbers (Reproduced by permission of Siemens AG)

The latest steps of development are the self-blast circuit breakers with an insulation material valve. This technology uses the advantage of the insulating material valve, such as carbon or Teflon materials, to reach higher numbers of switching, high current ratings, and a self-blast mechanism to use the hot gases to support the interruption unit movements. This technology allows for smaller interruption chambers, lower numbers of interruption units at higher voltage levels, and reduction in sizes of the circuit breaker operation drives. Today single interruption units are used for up to 500 kV rated voltage; at the early stages of GIS development this voltage level required four interruption, which was then reduced to two.

Technological improvements have reduced the mechanical forces required to move the irruption and driving unit of the circuit breaker. Further, the use of hot gases of the arc and the reduction in weight due to new materials, has allowed for a reduction in driver sizes. The heavy and robust hydraulic drives for circuit breakers have gradually been replaced by spring-operated drives. This started at the lower high-voltage ratings and since the early 2000s spring drives are available at all voltage levels. The main advantage of spring-operated drives is the much simpler technology which results in significant cost savings.

Figure 2.14 shows a hydraulic drive. The main technical features are the hydraulic high-pressure storage with a pressure above 30 MPa to operate the cylindrical drive and to release the high-pressure hydraulic oil into the low-pressure compartment. The cylinder drive is moved in two directions for the on and off positions of the interruption unit.

The spring-operated drive is shown in Figure 2.15. Since about the year 2000, the spring-operated drives have been used at all voltage levels of GIS. The mechanism is relatively simple with one close spring and one open spring connected to the drive rod. An electric motor is used to bring the operation forces into the springs.

The technology of improved aluminum-casting processes allows the production of more optimized design forms fulfilling the dielectric, mechanical, and thermal requirements of a GIS. Contours and forms could be produced in three dimensions, designed with computer-aided design (CAD) tools and three-dimensional (3D) machinery to produce the molding forms for an aluminum-casting process. Together with new materials for all the internal functions, such as interruption of the main current or short circuit current, the switching of disconnectors, or ground/

Figure 2.14 Hydraulic drive for circuit breakers – until 2000 (Reproduced by permission of Siemens AG)

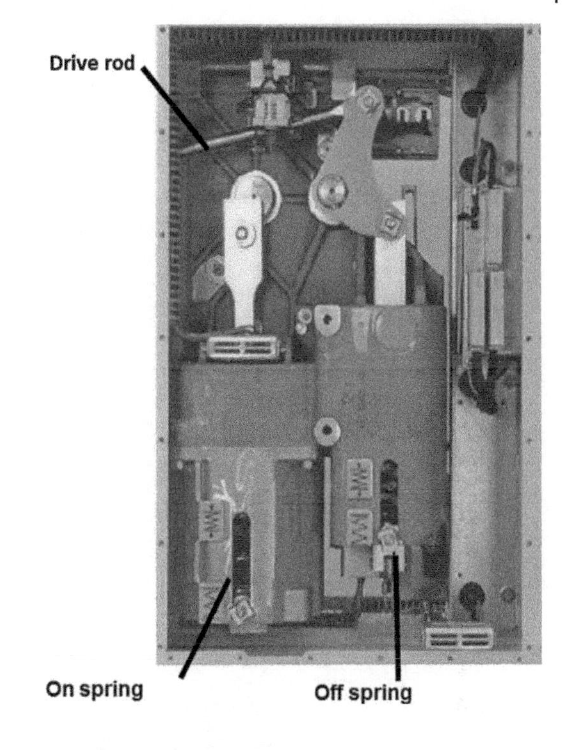

Figure 2.15 Spring drive for circuit breakers – since 2000 (Reproduced by permission of Siemens AG)

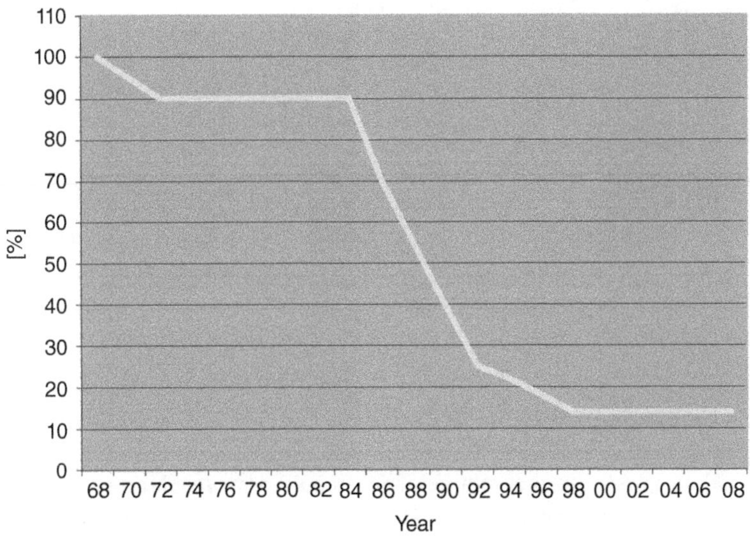

Figure 2.16 Reduction in size of GIS – 145 kV rated voltage (Reproduced by permission of Siemens AG)

earth switches, the size, and volume of GIS has been reduced by each step of development. Figure 2.16 shows the reduction in size of GIS since the start of this technology in 1968. If the GIS design of 1968 is set to 100%, a reduction to 90% is observed in the 1970s, and a further significant reduction down to 30% in the 1980s and 1990s; the size of GIS in the 2000s has dropped to 20% of its original size. The graphic in Figure 2.16 shows the size reduction of a 145 kV GIS.

To gain a better comprehension of this significant size reduction, Figure 2.17 shows a first-generation GIS from the 1960s extended by a fourth-generation GIS of the 1990s.

Figure 2.17 Extension of 110 kV GIS – 110kV rated voltage. Extension of first generation from 1960s with fourth generation of 1990s (Reproduced by permission of Siemens AG)

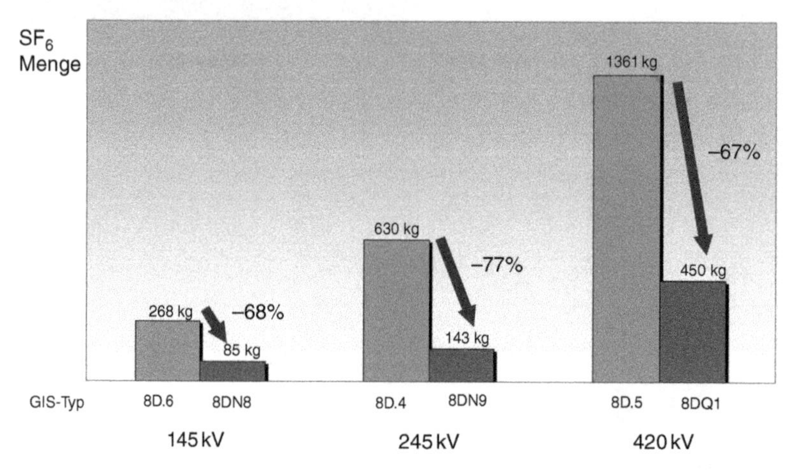

Figure 2.18 Reduction of SF_6 – gas weight reduction (Reproduced by permission of Siemens AG)

This size reduction also had an impact on the volume of SF_6 required for insulation. An SF6 reduction in the range of 67–77% was achieved, as shown in Figure 2.18; the aim was to achieve a reduction of 80%, however, it was unattainable due to a small increase of operating gas pressure. The typical gas pressure of the first-generation GIS had about 0.2–0.3 MPa for the bus bars and 0.5–0.6 MPa for the circuit breaker interruption unit. Today the pressure in the five-generation GIS in the bus bar is typically 0.4–0.5 MPa and in the circuit breaker interruption unit 0.7–0.8 MPa which accounts for the difference.

From an operator's point of view, these technical developments are of secondary importance. The priority is the reliability, maintainability, and cost efficiency. The operator needs equipment that can fulfill the tasks at the lowest cost and with high reliability.

The investment cost and operation cost are both under the focus of the user of GIS when project evaluations are made. The following are some of the main features that need to be explained.

2.1.2.1 Factory Preassembled and Tested Units

The reduction in size allowed the manufacturers to preassemble complete bays or even double bays in the factory and test the functionality, depending on the rated voltages, many manufacturers are able to use double bay transportation units up to 145 kV, in the voltage range up to 245 kV single bay transportation units, and in the upper voltage levels of 400, 500 kV, and above the transportation units are sections of bays. This factory prefabrication reduces the on-site works, saves cost, and increases the reliability of GIS.

2.1.2.2 Operating Life ≥ 50 Years

The experiences made with GIS since 1968 show high reliability and a long lifetime. After more than 40 years of operating GIS, the expected lifetime has been set to greater than 50 years. Other than some minor setbacks in the early years with gas losses or insulator failures, the GIS proved to be more reliable than air-insulated substations. This is because the high-voltage parts of the GIS are not directly exposed to the environment. And the interior of the GIS is under constant and dry conditions.

It has been proved by CIGRE reports on reliability and failure statistic, that to date, no aging effects are observed on GIS, even from the first-generation installations. This means that in a few years, the lifetime expectation may increase to 60 years. So far, the main reason for replacing older GIS is the increased need for power supply and with this high voltage and current ratings.

2.1.2.3 Major Inspection Not Before 25 Years

The technical developments for the GIS over the last 40 years also improved the maintenance requirements for the user. The first generation had typical maintenance cycles of 5 years, this increased to 10 years in the second and third generations, and reached 25 years with the fourth and fifth generations of GIS. In addition to the fixed maintenance cycles, a condition-based maintenance has been developed for opening the gas compartment on-site. It has been found that opening the gas compartments on-site after a fixed time is not necessary and should be avoided. Manufacturer shall be consulted on an appropriate course of action, should a situation, such as the number of rated current switching or the number of short-circuit current switching being high; opening of the gas compartment could be suggested in this situation if required.

2.1.2.4 Motor Operated Self-Lubricated Mechanisms

The compact design of GIS requires small motor-operated drives for disconnecting and ground/earthing switches. These motor drives are self-lubricated and do not need maintenance. This feature simplifies the operation of a substation for the user. Typically, the switching device operations are made remotely and in case of local operation, the operation initiation is made in front of the GIS bay at the control cubicle. No operator has to go directly to the switch in the bay, this increases personnel safety, because the operator is usually in another room.

The drives are maintenance free for the 25 years cycle and do not need any lubrication. Furthermore, the indoor GIS, which is becoming the preferred option, also has no ambient impact to the drives, such as ice, humidity, or dust.

2.1.2.5 Minimal Cleaning Requirements

Indoor and outdoor GIS are assembled in a factory under clean room conditions and in closed gas compartments with low overpressure (e.g., 0.05 MPa) and then transported to the site before final assembly. The clean room conditions in the factory allow for higher levels of cleanliness of the interior of GIS high-voltage sections during manufacturing. Once the gas compartment is closed, the clean and absolutely dry gas conditions will remain stable for a lifetime of GIS installation.

The required periodic cleaning of insulators and bushings in air-insulated substations are not needed with GIS.

2.1.2.6 Corrosion Resistance

First-generation GIS steel enclosures were affected by the environment and required a color coating for corrosion protection. The aluminum alloys for the GIS enclosures of today are not suspectable to corrosion. This is contrary to the steel enclosures, which needed a paint coating for corrosion protection. The aluminum enclosure protects itself by very thin (only some μm thick) oxide surface layers. A color coating for corrosion protection is not necessary. However, the oxide layer of the aluminum enclosure is not visually appealing, as it is gray and could appear as dirt. Therefore, for aesthetic reasons, a paint coating is used. Most of the first- and second-generation GIS received, the beautiful, industrial color: light gray. Today the GIS world is much more colorful with shining yellow, to impressive turquoise, or a deep blue, depending on the wishes of the substation owner.

2.1.2.7 Low Fault Probability – High Reliability

High reliability is one of the most important requirements. During the 40 years of GIS operation and technical design, the causes of possible failures have been studied and eliminated using new improved design and technology. Reliability figures enhanced year by year due to a better understanding of the GIS dielectrics and improving quality during in the manufacturing process.

World leading GIS manufacturers of today have established a 100% partial discharge measuring system on each single insulator of the GIS. The routine test in the factory includes a functional test and a high-voltage test, including a partial discharge measurement and a 100% gas tightness of the O-ring sealings.

The international standards IEC 62271-203 [27] and IEEE C37.122 [23] are covering these quality requirements to provide the high reliability for which GIS equipment is noted.

2.1.2.8 Protection against Aggressive Environmental Conditions

Air-insulated substations are exposed to the elements. A corrosive or coastal environment can harm the high-voltage parts of the substation as well as all the structures and steel works to fix and hold the high-voltage equipment. Because of the nature of GIS, its gas tightness and no contaminants are entering the high-voltage part to cause damage. Such an aggressive environment is found in coastal regions with salty and wet air, in deserts with dusty and wet (in the night) air, and in industrial regions with all kinds of conductive and nonconductive dust in the atmosphere. The structures and steel works are usually painted or galvanized for protection.

2.1.2.9 Seismic Resistance

A GIS has a compact design with a low point of gravity. The compact metallic structure of the GIS has a high withstand ability for earthquakes. The frequency spectra of seismic waves are at much lower values than the resonance frequency of a GIS bay. In cases of close resonance points, the reinforcement bars in the GIS connected to the flange shift frequencies to higher values and, therefore, out of the resonance range of seismic frequencies.

The seismic design of a GIS today is part of the substation design. The principal knowledge of GIS seismic behavior is type tested and later calculated for the requirement of the specific project.

Seismic standards give information about the seismic requirements and can be found in IEC 62271-207 [42] and IEEE 693 [43].

2.1.2.10 Space Requirements Less Than 20% of AIS

The much smaller footprint of GIS compared to AIS offers opportunities to integrate GIS in densely populated areas and city centers. Examples of GIS substation integration can be found in specific buildings or even in the basements of buildings in city centers; they are connected by underground cables thus being completely obscured from public view. used. There is a large implementation of GIS used at voltage levels of 100 kV and up to 200 kV and in large cities even substations of 400 kV and 500 kV can be found. Table 2.1 shows the main features of GIS in an overview.

There are estimated more than 100 000 bays in over 10 000 high-voltage substations of 52 kV and above installed world-wide. Today with more than 60 years of experience with SF_6-insulated GIS, the knowledge of this technology is extensive. The world-wide leading manufacturers have developed GIS gradually from the first generation in the 1960s to the latest sixth generation. The next development will be the integration of measuring and sensor technology using nonconventional sensors to further reduce the size. The basic functionality will remain the same: breaking, switching, and measuring, but the integration into a smart grid of tomorrow with renewable energy in the network will change the way GIS is used.

In Table 2.2, an overview of the main development steps is shown.

Table 2.1 Main features of GIS

Factory preassembled and tested units
Operating life > 50 years
Major inspection not before 25 years
Motor-operated self-lubricated mechanisms
Minimal cleaning requirement
Corrosion-resistant
Low fault probability/high availability
Protected against aggressive environmental conditions
Seismic-resistant
Space requirement less than 20% of comparable AIS

Table 2.2 History steps of development of GIS

1960	Start of fundamental studies in research and development of SF_6 technology
1964	Delivery of the first SF_6 single-pressure circuit breaker
1968	Delivery of first GIS by major manufacturers
1974	Delivery of first GIL (420 kV)
1976	Delivery of first 550 kV GIS
1983	Delivery of the world's largest GIS for Itaipu, Brazil
1984	Delivery of 550 kV GIS for severe network conditions (rated current 8000 A, rated short-circuit breaking current 100 kA, 17 circuit breakers)
1986	Delivery of first 800 kV GIS
1996	Introduction of the smallest 123 kV GIS
1997	Introduction of intelligent bay control, monitoring, and diagnostics
1999	Introduction of the smallest 145 and 245 kV GIS
2000	Introduction of new compact and hybrid solutions
	More than 50 000 bays in over 5000 substations installed worldwide
	More than 2 500 000 bay-years of operation

2.2 Physics of Gas-Insulated Switchgear

It is important for an engineer to have an understanding of the basic physics of the system and equipment being dealt with. To this end, a very brief and simplified explanation of GIS and gaseous insulation physics is the subject of this chapter. A complete explanation of all the phenomena associated with the breakdown in gases would be too lengthy for inclusion here and for a complete understanding of all the related phenomena, it is recommended that readers refer to one of the reference books available on this subject [33, 34].

2.2.1 Electric Fields

Gas-insulated substation equipment and gas-insulated transmission lines are constructed using a coaxial design with a circular center conductor supported by insulators with a circular enclosure that provides electrical shielding and contains the insulating gas (see Figure 2.19).

For the purposes of this discussion, we will focus on single-phase design, although some designs have all three phases enclosed in a single enclosure. For explanation only, the simpler single-phase design will be discussed.

The electric field within a coaxial system can be determined from Laplace's equation and assuming that the electric field varies only with the radius can be simplified to

$$\frac{1}{r}\frac{\partial}{\partial r}\left(r\frac{\partial V}{\partial r}\right)=0 \tag{2.1}$$

Integrating twice, setting boundary conditions, and then converting to E (volts/meter) we can determine the maximum electric field (E) at the surface of the conductor as

$$E=\frac{1}{r_c}\frac{V}{\ln r_e / r_c} \tag{2.2}$$

where V is the peak applied voltage, r_e is the internal radius of the enclosure, and r_c is the radius of the conductor.

The optimal electric field configuration is when the ratio of the radii of the conductor to enclosure is $1/e$. However, as shown in Figure 2.20, the maximum electric field remains fairly constant over a wide range of conductor to enclosure radii ratios. This may be intuitively obvious as the electric field decreases with an increase in the surface area of the conductor and increases with the

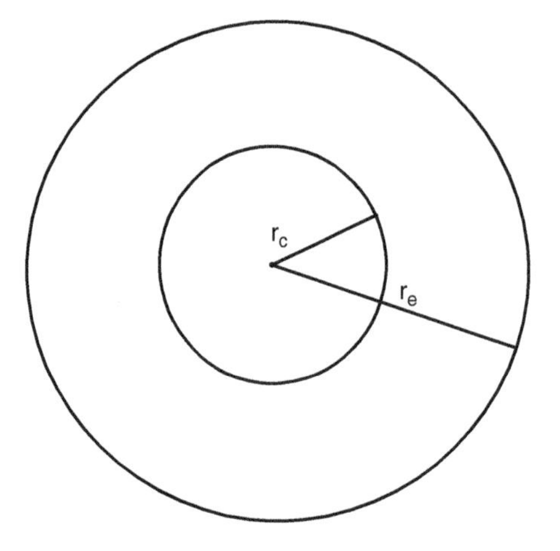

Figure 2.19 A coaxial system such as in GIS (Reproduced by permission of John Brunke)

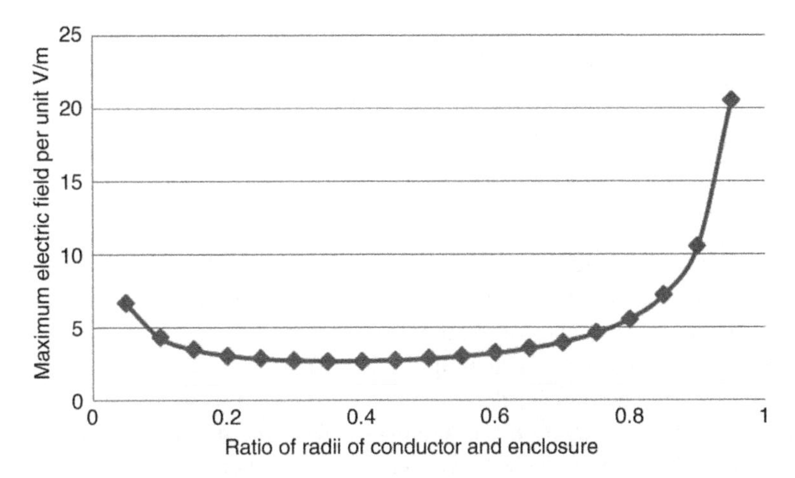

Figure 2.20 The maximum electric field inside a GIS remains near constant over a wide range of ratios of enclosure to conductor radii. When considering other effects this characteristic will be modified, but will still yield the highest electric fields in the center of this region (Reproduced by permission of John Brunke)

decreasing distance from the conductor to the enclosure. It is actually more complex than stated as one must include the mean electric field strength as well as the maximum field strength in the analysis (field efficiency factor) and the electric field is associated with many different configurations of components. This also does not include the effects of the solid support insulator on the electric field.

2.2.2 Breakdown in Gases

Understanding the performance of gas-insulated systems requires a basic understanding of electrical breakdown phenomena in gases. This also answers the question of why SF_6 is a superior insulating gas when compared to air.

2.2.3 Excitation and Ionization

Excitation is an elevation in the energy level of an atom or molecule above its normal energy state. This is accomplished by adding energy and raising the orbit of an electron to a higher, and normally unstable, energy level. When the electron returns to its normal state, the energy is released in the form of a photon.

Ionization is the process of converting an atom or molecule into an ion by either adding or removing an electron. For this discussion, the focus will be on the process of removing an electron and creating a positive ion as it is of the most importance. Negative ions will also be important toward the end of the discussion.

The energy required for excitation or ionization can come from a number of sources. The most common are:

1) Collisions of fast-moving electrons with atoms or molecules
2) Collisions of high energy photons with atoms or molecules
3) Collisions of cosmic rays (ionizing radiation) with atoms or molecules
4) Thermal collisions between atoms or molecules
5) Mechanical/friction between materials
6) Chemical reactions

Whether the result is excitation or ionization depends upon the energy required by different atom or molecule to reach these higher energy states and the energy provided at the source. The excitation potential and ionization potential are both measured in electron volts (eV). These energy requirements are available for most materials from various sources.

2.2.4 Free Electrons

A free electron is any electron that is not associated with an ion, an atom, or a molecule and is free to move, which it will when an electric field is applied to it. Free electrons are created by the methods of ionization listed above. They are everywhere and cannot be avoided.

These free electrons are also present inside the insulating gas in a GIS. There is a time varying electric field inside the GIS and therefore these negatively charged free electrons will experience an electromagnetic force, which will cause them to accelerate away from the negative electrode and toward the positive electrode. The mass of an electron is very small and therefore high velocities can be rapidly reached.

The ions created in the insulating gas (normally SF_6) by the formation of the free electrons have a mass about five orders of magnitude greater than the electrons. As the charge on a positive ion, which has the opposite polarity to that of the electron, is the same magnitude as that of the electron, therefore the force generated by the electric field is the same and the resulting acceleration and motion are relatively very slow.

2.2.5 Mean Free Path

The free electrons are accelerated by the electric field and will continue until they collide with a gas molecule. The velocity of the electron at impact is determined by the magnitude of the electric field (accelerating force) and the time that the accelerating force is applied. The mean distance that an electron can travel before it encounters a gas molecule is called the mean free path. The mean free path is determined by the collision diameters of the electron and the impacted molecule, and the density of the insulating gas.

2.2.6 Electron/Gas Molecule Impact

The kinetic energy at impact depends upon the velocity of the electron. If the energy is sufficient either excitation or ionization will occur. Ionization requires more energy than excitation. The result is that either a photon or an additional free electron is created if the energy is sufficient.

Some of the free electrons find an ion and recombine, releasing the energy that was initially required to ionize the molecule. This energy is released in the form of a high energy photon. These photons travel at the speed of light and will travel farther than an electron before a collision occurs due to their small collision diameter. These photons are a primary source of secondary electrons, which act to propagate the breakdown more rapidly.

2.2.7 Breakdown

A free electron is created by one of the many processes mentioned. It is accelerated in the electric field. If it has sufficient energy (velocity) at the time of collision with a gas molecule, if it has sufficient energy (velocity), it will ionize the molecule. Now there are two free electrons, again accelerated in the electric field, and these collide with gas molecules, forming more free

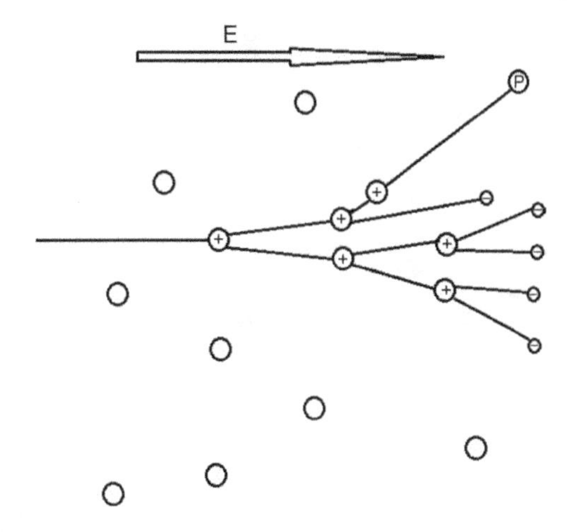

Figure 2.21 An electron in an electric field is accelerated, creating an avalanche breakdown. A negative cloud of electrons accelerates quickly, leaving behind a cloud of positive ions. A recombination results in a high energy photon, which can cause ionization and create secondary electrons (Reproduced by permission of John Brunke)

electrons (see Figure 2.21). This proceeds in an avalanche, with the electrons moving quickly and the ions they have left behind being relatively static due to their relative masses. Some recombination will occur and yield high energy photons, which leap ahead and create additional free electrons (referred to as secondary electrons). This process will continue as long as the accelerating force (electric field) is sufficient to accelerate the electrons to enough energies to cause ionizations on impact.

The cloud of electrons, or the cloud of ions left behind, can modify the electric field if a sufficient number of electron or ions is present. This can assist in the propagation of the avalanche breakdown or can suppress it. In GIS, the electric field is ideally pseudohomogenous and a significant discharge can create a local higher field point, leading to a breakdown. However, other mechanisms in the SF_6 gas can assist in suppressing the propagation of a discharge, as discussed next. The process is complex and many factors influence the outcome.

Electrons have a negative polarity and repel each other, as do the positive ions. A discussion of the effects of polarity and the rise-time of the applied voltage and electric field are beyond the scope of this discussion, but they contribute to the many phenomena that are associated with high-voltage discharges.

2.2.8 Sulfur Hexafluoride

Sulfur hexafluoride (SF_6), or mixtures of SF_6 gas with other gases, is used as the insulating medium in GIS. The first ionization potential of SF_6 is about the same as nitrogen (N_2). The size of the SF_6 molecule (collision diameter) is much larger than a nitrogen molecule. This results in the mean free path for an electron/molecule collision in SF_6 being about a third of the distance that it would be in N_2 (see Figure 2.22). Kinetic energy is a function of velocity squared, and assuming linear acceleration, the average electron to molecule collision in SF_6 would have 16% of the average energy of a collision in N_2.

As gas density is increased, the mean free path is decreased and the electric field strength required to accelerate electrons to sufficient velocity to cause ionization also increases (see Figures 2.23 and 2.24).

SF_6 also has another property that makes it a superior insulating gas. It is electronegative. This means that SF_6 molecules will attract electrons to form negative ions. This suppresses the propagation of a discharge.

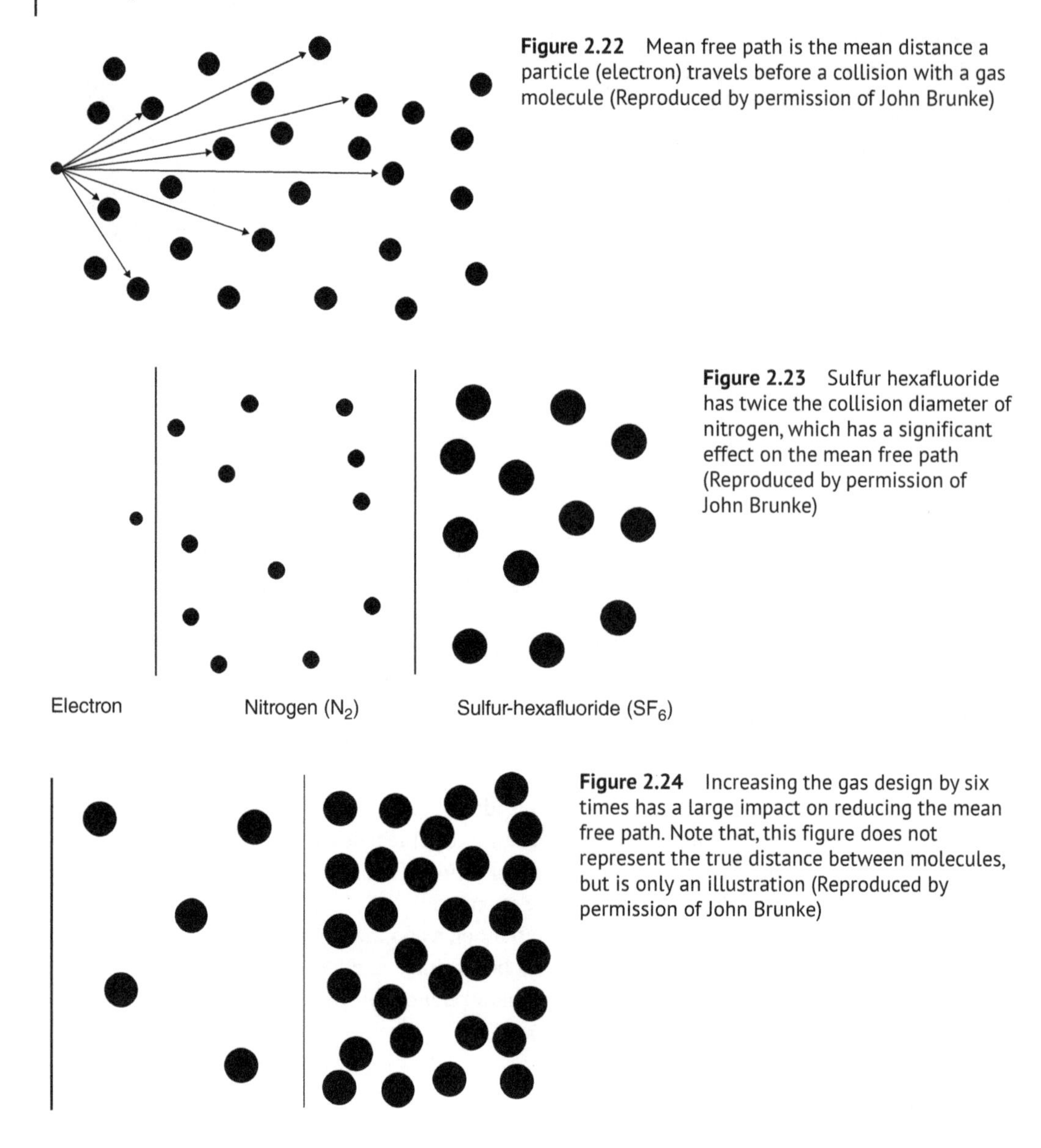

Figure 2.22 Mean free path is the mean distance a particle (electron) travels before a collision with a gas molecule (Reproduced by permission of John Brunke)

Figure 2.23 Sulfur hexafluoride has twice the collision diameter of nitrogen, which has a significant effect on the mean free path (Reproduced by permission of John Brunke)

Figure 2.24 Increasing the gas design by six times has a large impact on reducing the mean free path. Note that, this figure does not represent the true distance between molecules, but is only an illustration (Reproduced by permission of John Brunke)

2.2.9 Electric Field Control in GIS

In order to prevent breakdown in the GIS, the electric field is controlled to a level where the electromotive force on an electron cannot accelerate it to a velocity in the mean free path distance established by the collision diameters and gas density to a kinetic energy level sufficient to cause further ionization. The electric field must be controlled for all applied voltage conditions, including the lighting impulse. It must also be controlled at all points in the GIS and under all reasonable service conditions. GIS designs incorporate shields to maintain large radius surfaces to control fields. Additionally, the surfaces in a GIS should not have any sharp points or defects that can cause a field enhancement. The surface itself has an impact on the breakdown voltage and some manufacturers use surface coatings inside the GIS enclosure.

Conductive particles created during manufacturing or by abrasion and wearing of contacts can create discharges and result in flashover. These are controlled in manufacturing by careful quality control, which also extends to commissioning and maintenance. GIS designs typically incorporate

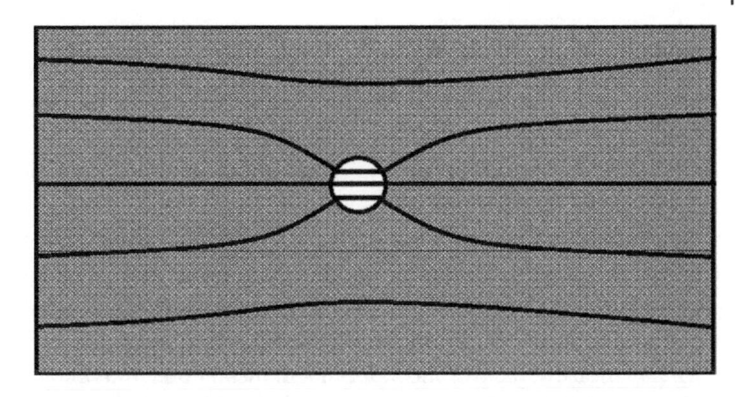

Figure 2.25 A void in a solid insulator will cause a local field enhancement (Reproduced by permission of John Brunke)

zero field areas (particle traps) to which the particles migrate, driven by electric field vibrations. Some manufacturers use internal coatings to reduce the effects of particles or to make them more visible to assist in cleaning.

All aspects of the internal design must be concerned with the electric fields. This includes internal solid insulators. Voids within the solid material cause field enhancements and can lead to failures (see Figure 2.25). The points at which the solid insulator, enclosure or conductor, and SF_6 gas meet (referred to as a triple point) can create field enhancements if not properly designed. The differences in dielectric constants distort the electric field similar to the way lenses distort light. A sort of focusing effect can occur unless care is given in design and manufacturing to avoid it.

2.2.10 GIS Circuit Breakers

The GIS comprises switchgear, circuit breakers, and switches, connected with bus bars and bushings. The circuit breaker is the centerpiece of any GIS. Interruption phenomenon in GIS circuit breakers is the topic of many books [35] and is beyond the scope of a book concentrating on other aspects of GIS. It is enough to say that the properties that make SF_6 a good insulating gas, along with the relationship between its thermal and electrical conductivity properties as it changes from plasma to an insulating gas, contribute to SF_6 being an excellent interrupting medium.

As many SF_6 interrupter designs are commonly used in live tank, dead tank, and GIS circuit breakers, it is appropriate to discuss the circuit breaker as applied in a GIS.

The conditions of an interrupter in a dead tank and those in a GIS application are very similar. The volume of the enclosure, is nearly the same in both, and as such the pressure and gas flow do not interfere with the interrupter's performance. This is not the case for a live tank breaker as the smaller gas volume influences with the interrupter performance. This means that an interrupter tested in a live tank breaker needs to be retested in a dead tank/GIS enclosure to verify performance.

The transient recovery voltages (voltages seen across the circuit breaker contacts as it interrupts) see the same recovery voltages in either a dead tank/live tank or GIS application. Of course, the transient recovery voltage must be considered in any application.

2.3 Reliability and Availability

2.3.1 General

The cost of electricity is dependent on the availability and reliability of the power supply. Power supply interruptions are very costly to the suppliers due to high penalty payments, as such, suppliers need highly reliable and readily available equipment.

Reliability and availability are two important figures for the operator of the electric power supplier for public, private, or business. All devices in the power system show high levels of availability due to the high quality of the products and its high cost. The requirements of reliable power delivery are part of the electricity pricing, and the cost of damages caused by power supply interruptions are getting more and more into the focus of penalty payments to compensate for financial losses [44–46].

Besides the quality of the device, the impact of ambient conditions such as humidity, temperature, dust, salty air, ice, and many others are the key parameters of reliability figures. In the case GIS, the impact of ambient conditions is not affecting the high-voltage part directly, which improves the reliability figures for GIS. For a 110 kV GIS, the mean time between failure (MTBF) is more than 10 000 years based on GIS equipment in service.

The small size of the GIS allows for indoor installation, thus allowing therefore even the nonhigh-voltage parts of the GIS are under constant ambient indoor conditions.

2.3.2 Historical View

The perception of reliability of GIS varies greatly between North America, particularly the United States, and the rest of the world. The United States adapted the emerging GIS technology early on in the 1970s. The first GISs were put in operation in 1967 and 1968 in France, Switzerland, and Germany. The GIS in Germany is still in operation. The GIS in Switzerland was in operation until recently without a major fault or gas leak. The utility made an assessment of the gas leak over the lifetime of this first GIS and concluded that the overall leakage rate was about 0.5% per year which is in compliance with IEEE and IEC standards.

Reliability, economic advantages of the life cycle cost, and physical compactness have resulted in the widespread use of GIS over the last 60 years. Even though GIS technology is mature now, many users approach to its application is still rather unique. GIS is used in specific types of applications only.

There are tens of thousands of GIS bays installed world-wide today, and in operation for decades, at all voltage levels from 52 kV up to 1100 kV. The graphic in Figure 2.26 shows the numbers of GIS installations in Germany, Switzerland, and Austria, which can be seen. The illustration compares 1998 to today and shows that many installations have reached their estimated lifetime.

The GIS is assembled in factories under clean conditions to avoid dust and particles in the high-voltage sections. This clean condition cannot be found on-site. Therefore, any opening of the GIS

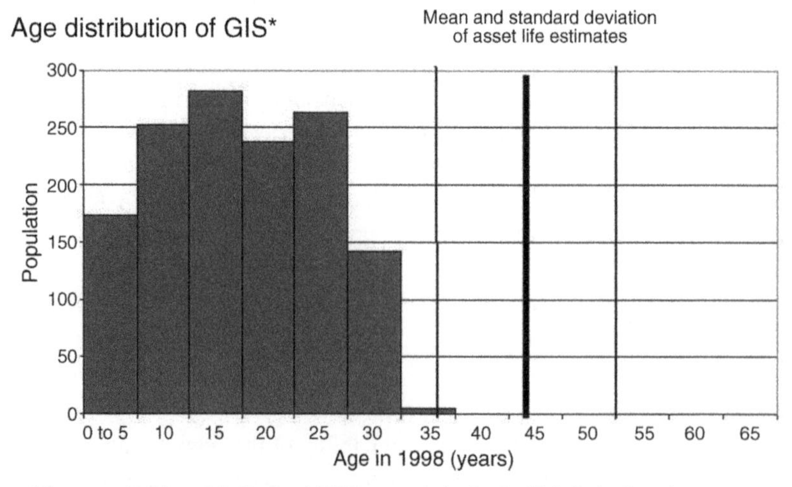

Figure 2.26 World-wide installed GIS (Reproduced by permission of CIGRE)

should be minimized. The high reliability of GIS does not give any indication of necessary repair or replacement under normal operation, which the expected lifetime of 50 years. Operators of GIS, along with the manufacturer have changed the maintenance cycle from time-based maintenance to a condition-based maintenance as a consequence of the high reliability of the GIS. Replacements of GIS are usually done when the substation is changed to higher current or voltage ratings because of an increase in power requirements. Some early GIS installations, mainly in the United States, have had gas leakage caused by sealing problems. In most cases, these installations from the 1970s have been replaced. The most reason for replacement of GIS was increasing power transmission and short circuit interruption requirements, in some cases was leaking sealings the reason. For most of the installed GIS there no real end of life time seen so that the expected life time based on experiences can be shifted to 60 years.

2.3.3 CIGRE Survey

The CIGRE Study Committee substations SC B3 (in former times SC 23) has studied the reliability of the GIS survey based on data from 1974 to 1977, and published in 1981 and 2012. The 1985 survey was focused on the switching devices used in a substation for air-insulated switchgear (AIS) and gas-insulated switchgear (GIS) technology on the basis of 7000 circuit breaker years. The second survey is based on data collected from circuit breakers installed in the period 1988–1991 [37]. The 2012 survey has had a separate GIS part. GIS focused surveys have been carried out by CIGRE in 1992–1994 and were published in 2000 [38, 94]. All surveys are on a global base of information. Most of the GIS installations are in Japan, Middle East, and Europe. In the later years, many GIS installations have been energized in China, Korea, South Africa, and increasingly also in the United States.

2.3.3.1 CIGRE GIS Survey 1994

The reliability of GIS has markedly improved since its introduction 45 years ago. CIGRE distinguishes between GIS commissioned before 1985 and after 1985. The overall trend has a reduced failure rate for GIS commissioned after 1985. Table 2.3 of the CIGRE report gives an overview of the data. Table 2.4 gives the voltage class and corresponding voltage levels. The outdoor GIS population is about 43% of the total CB-Bay-Years. Tables 2.5 and 2.6 of the CIGRE report provide the major failure characteristics of the GIS.

One increasing trend is the circuit breaker failure involvement for the newer GIS observed at all voltage levels.

2.3.3.2 CIGRE GIS Survey 2012

The final report of the 2004–2007 international inquiry on reliability of high-voltage equipment has been published in 2012 by CIGRE Study Committee A3 "High Voltage Devices" in cooperation with B3 "High Voltage Substations" in seven Technical Brochures. TB 509 "Summary and General Matters," TB 510–513 "Details of Switching Devices," and TB 514 "GIS Practice." The goal of the survey is to present reliability data and trends on a world-wide basis. It gives information on the types of failures occurring within a four-year time span of 2004–2007. This survey span is chosen to compare similar technologies without the failures of the first year of technical design. The number of GIS bays investigated in the survey are shown in Table 2.7 for each survey year.

The large majority of GIS bays are installed at the voltage classes of 60–100 kV and from 100 kV to 200 kV. This has not changed when compared to previous studies. The distribution of GIS according to countries has a high level of GIS applications in Japan, Korea, Europe, and Arabic countries. North America shows an increasing use of GIS. The total use of GIS increases from 89 000 circuit breaker bay years to 119 000 from the survey in 1994 to the survey in 2012. The newer breaker technology used in the 2012 survey improves the reliability as the oldest breakers in GIS had extreme high failure numbers. The comparison of the second survey from 1994 to the third survey in 2012 is shown in Table 2.8. In Figure 2.27, the comparison is shown in a graphic.

Table 2.3 Major failure frequency (FF) – second GIS survey total population and comparison between the first and the second surveys results [37, 38].

		GIS in total			
		Second GIS survey		First GIS survey	
Voltage class	No. of failures	CB-bay-years	FF	CB-bay-years	FF
1	27	56 884	0.05	38 471	0.13
2	465	32 048	1.45	23 845	1.1
3	138	16 040	0.86	12 955	1.1
4	179	6371	2.81	4735	4.3
5	49	4525	1.08	3453	4.2
6	12	200	6.00	80	14.0
1 to 5	855	115 868	0.74	83 459	0.96
TOTAL	867	116 068	0.75	83 539	0.97
GIS commissioned before 1.1.1985					
		Second GIS survey		First GIS survey	
Voltage Class	No. of failures	CB-bay-years	FF	CB-bay-years	FF
1	16	28 669	0.06	21 304	0.17
2	351	19 504	1.80	16 035	1.3
3	100	10 362	0.97	8596	1.5
4	110	3694	2.98	3287	4.4
5	32	3252	0.98	2532	3.7
1 to 5	609	65 481	0.93	51 754	1.18
GIS commissioned after 1.1.1985					
		Second GIS survey		First GIS survey	
Voltage Class	No. of failures	CB-bay-years	FF	CB-bay-years	FF
1	11	28 215	0.04	9792	0.06
2	114	12 544	0.91	4605	0.6
3	38	5678	0.67	2636	0.4
4	69	2677	2.58	970	4.0
5	17	1273	1.34	654	1.8
1 to 5	246	50 387	0.49	18 657	0.51

Notes: Failure frequency (FF) = No. of Failures per 100 CB-bay-years.

Table 2.4 Voltage classes used in CIGRE survey [37, 38]

Voltage class kV	1: $60 \leq U_n < 100$	3: $200 \leq U_n < 300$	5: $500 \leq U_n < 700$
	2: $100 \leq U_n < 200$	4: $300 \leq U_n < 500$	6: $U_n > 700$

Table 2.5 Identification of main component involved in the failure from a GIS voltage class point of view

Main component involved in the failure	GIS in total	Class 2	Classes 3+4+5
(Whole period)	%	%	%
Total number of answers (reported failures)	801 = 100%	435 = 100%	335 = 100%
Circuit breaker or switch	43.4 (30.1)	54.7	29.9
Disconnector	17.9 (19.2)	17.2	18.2
Grounding switch	4.4	5.3	3.6
Current transformer	0.9	0.7	1.2
Voltage transformer	5.6 (7.7)	6.2	4.8
Bus bars	5.5 (7.3)	3.7	6.9
Bus ducts and interconnecting parts	11.9 (17.2)	4.1	22.4
SF_6 gas-to-air bushing	3.6	0.9	6.9
Cable box/cable sealing	3.5	4.4	1.8
Power transformer interface chamber/bushing	0.9	0.2	1.8
Surge arrester	0.7	0.5	1.2
Other	1.7	2.1	1.5

Table 2.6 Identification of main component involved in the failure from a GIS age point of view (five most involved components) [27, 28, 47]

Main component involved in the failure	GIS in total	GIS before 1.1.1985	GIS after 1.1.1985
(Whole period)	%	%	%
Total number of answers (reported failures)	801 = 100%	562 = 100%	239 = 100%
Circuit breaker or switch	43.4	42.2	46.2
Disconnector	17.9	18.5	16.3
Voltage transformer	5.6	4.4	8.4
Bus bars	5.5	5.7	5.0
Bus ducts, interconnecting parts	11.9	14.4	5.9

Table 2.7 Number of GIS population (CB bays) per survey year [27, 28, 47]

	Number of GIS CB/bays population collected in reference year			
Voltage class	**2004**	**2005**	**2006**	**2007**
$60 \leq U < 100$ kV	10047	10071	10116	10170
$100 \leq U < 200$ kV	6144	6263	6993	7002
$200 \leq U < 300$ kV	2005	2049	2093	2160
$300 \leq U < 500$ kV	2560	2434	2665	2672
$500 \leq U < 700$ kV	776	797	807	807
≤700 kV	85	85	85	85
Total	**21617**	**21699**	**22759**	**22896**

Table 2.8 Comparison of collected GIS service experiences of 1994 and 2012 surveys [27, 28, 47]

	Collected GIS service experience – second survey (CB-bay-years)			Collected GIS service experience –third survey (CB-bay-years)	
Voltage class	**All data**	**All data without the worst utility**	**Data without country 14 and the worst utility**	**All data**	**Data without countries 14 and 23**
$60 \le U <$ 100 kV	56 884	56 884	5114	40 404	113
$100 \le U$ < 200 kV	34 060	29 415	20 999	26 402	3677
$200 \le U$ < 300 kV	16 040	16 040	9576	8307	1349
$300 \le U$ < 500 kV	6774	6371	6371	10 331	1680
$500 \le U$ < 700 kV	4525	4525	1101	3187	170
≤700 kV	200	200	200	340	192
Total	**118 483**	**113 435**	**43 361**	**88 971**	**7181**

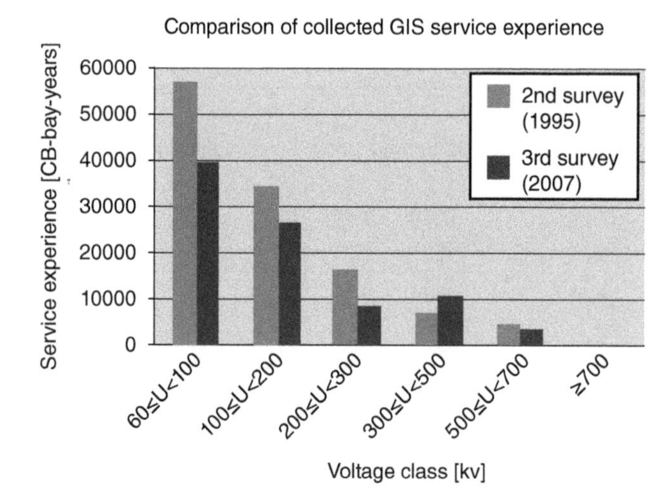

Figure 2.27 Comparison of the GIS service (Reproduced by permission of CIGRE)

The base data of the survey is very much dominated by information collected in Europe and Japan, two traditional users of modern design technologies of high, market leading quality. This should be kept in mind when reading and interpreting the results of the survey.

Two technologies have been investigated: fully GIS, which means all high-voltage sections are gas insulated, and hybrid GIS, which means sections are gas and air insulated (typically the bus bar). The dominating design used is the fully GIS type.

There are single-phase and three-phase design types used for GIS. The lower voltage ranges up to 300 kV are mainly in three phase design and for the higher voltages above 300 kV, the majority is single phase. The highest voltage ranges above 500 kV are only single-phase insulated (see Figure 2.28).

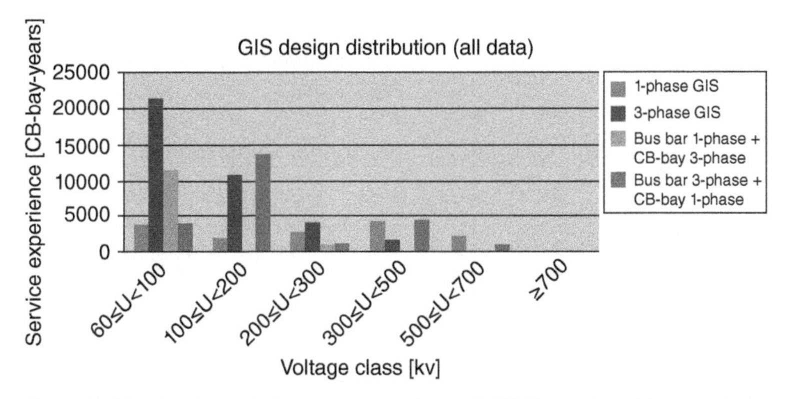

Figure 2.28 Single- and three-phase designs of GIS (Reproduced by permission of CIGRE)

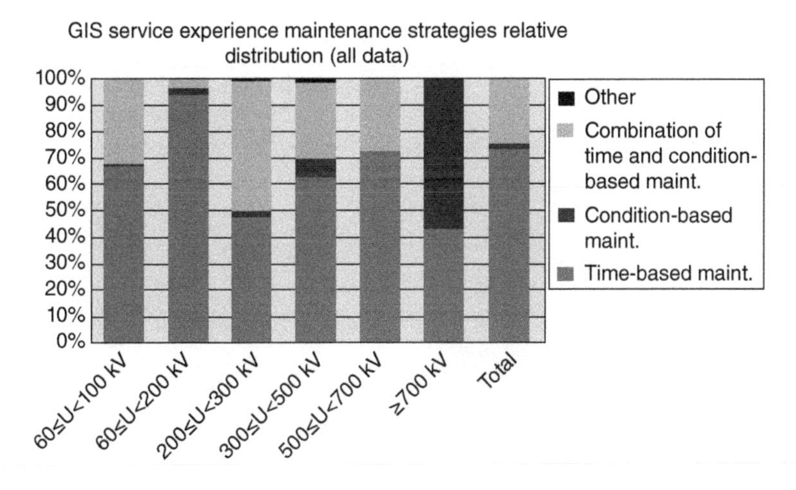

Figure 2.29 GIS maintenance strategies (Reproduced by permission of CIGRE)

The indoor and outdoor types of GIS are used in different regions of the world differently. It is like a philosophy. In Asia, the outdoor application dominates, in Europe, the indoor application. Because of the large amount of data from Japan in the survey, the outdoor dominates the overall data. The trend today for new installations in North and South America goes for indoor GIS.

The age of the GIS covered in this survey shows a majority installed in the years 1984 to 2003 (one year before the survey starts). This means that the GIS is increasingly used as a solution.

The maintenance strategy shows that time-based maintenance was used in most cases with the condition-based maintenance being on the rise. In many cases, both strategies are combined, as shown in Figure 2.29.

GIS Failures

The report covers major (MaF) and minor (MiF) failures. The expected result of GIS failure is that most failures are minor, which do not cause an interruption of power supply, and that the smaller number are major failures, as shown in Figure 2.30. The second finding is that the failure occurrences are increasing in number or percentage with the increase in voltage levels.

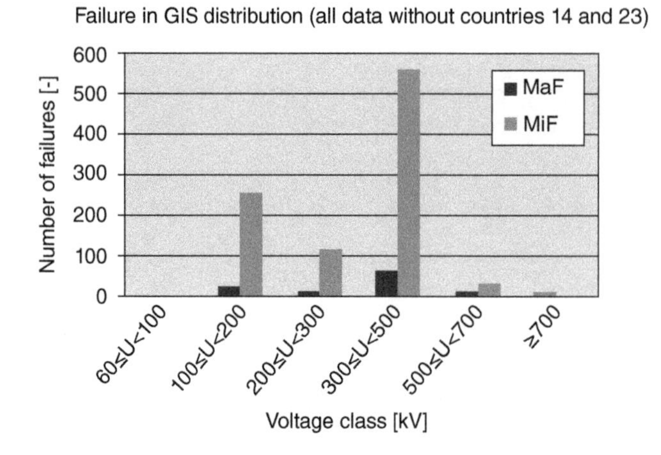

Figure 2.30 Failure distribution on voltage levels (Reproduced by permission of CIGRE)

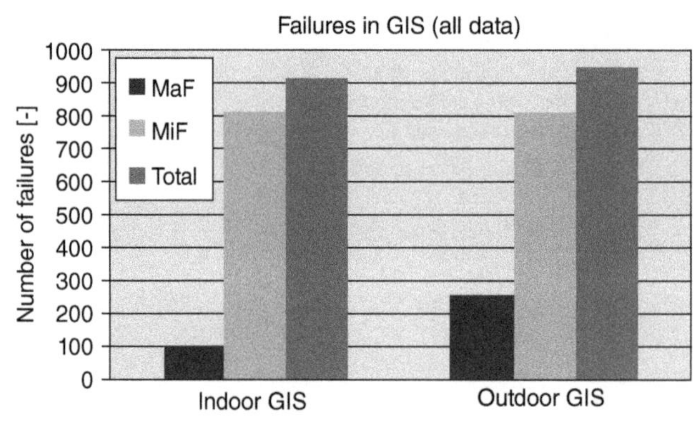

Figure 2.31 Indoor and outdoor GIS failure distribution (Reproduced by permission of CIGRE)

The failure distribution of full GIS installations is the same as for hybrid GIS installations. No difference has been found for MaF and MiF besides the different absolute number of installations. The single-phase and three-phase insulated GIS also do not show any significant difference between the distribution of MaF and MiF. The indoor and outdoor GIS do not show a significant difference in failure distribution for MaF and MiF, as shown in Figure 2.31.

The failure distribution of the single components of GIS have been investigated for circuit breakers (CB), disconnect and earthing/grounding switches (DS/GS), and instrument transformers (IT). For all others, like bus bars, surge arresters, joints, bushings and cable boxes, the label GI is used. Figure 2.32 shows the failure distribution in an overview for major failures. The main portion of failures in GIS overall voltage classes are linked to circuit breakers and disconnection and earthing/grounding switches.

The number of failures related to the age of GIS shows a significant reduction in failures with newer equipment. The technical development process has led to more reliable GIS (see Figure 2.33).

The CIGRE survey in TB 513 [38] discussed many details on failure modes, characteristics of failures, impact on outer conditions like weather, field assembly, environmental stress, service conditions, and the primary cause of the failure.

In a conclusion, it can be stated that hybrid GIS covers only 8% of the applications of GIS. Three phase design is only found below 300 kV. Indoor GIS is preferred in all countries except Japan and Korea. The oldest GIS is from the 1960s for voltage class 1.

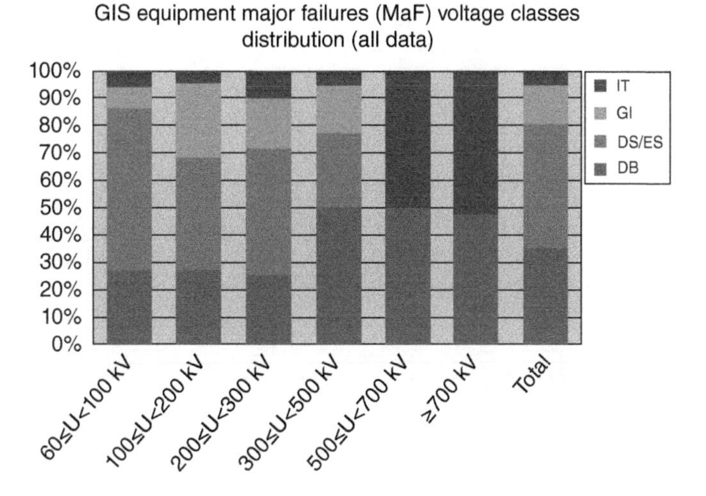

Figure 2.32 GIS component failure distribution (Reproduced by permission of CIGRE)

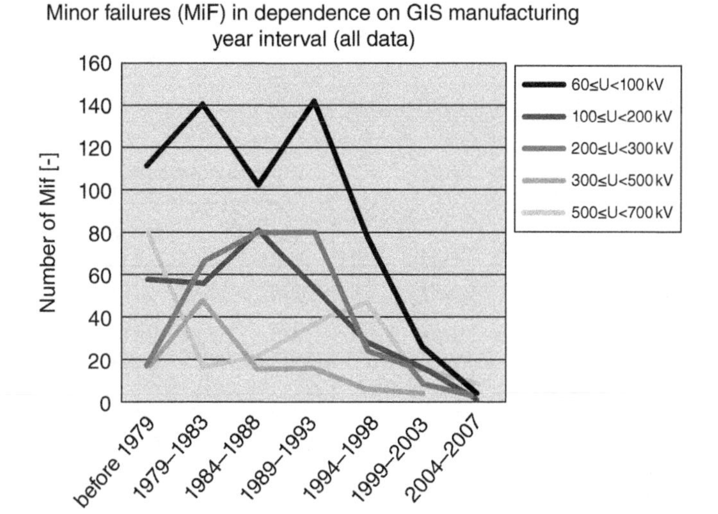

Figure 2.33 Failure numbers of GIS related to age (Reproduced by permission of CIGRE)

2.3.4 Reliability GIS versus AIS

The world-wide experience with GIS as documented by several organizations including CIGRE, IEEE, indicates that GIS has a lower failure rate than a comparable AIS substation. Some users in North America might dispute this fact. Because many first-generation GIS in operation, did not deliver to expectation and as such have hindered GIS adaptation. However, even in North America, the GIS commissioned after 1985 have similar failure rates as observed by CIGRE during the last world-wide GIS survey.

There is no question that a GIS compared with the same configuration in an AIS will always be more expensive if looked at the cost of installation only; however, this does not take into account special conditions such as land availability and/or cost, soil conditions, environment, and so on. However, as a substation builder, one should not compare a GIS and an AIS in the same configuration. One should define the requirements of the substation first and then look for the substation configuration that meets these requirements.

Based on published failure rates for AIS and GIS by international organizations like CIGRE, it can be shown that the failure rate of a six-breaker ring bus in GIS has a lower failure rate than a nine breaker AIS, breaker, and half scheme. There are several commercially software programs

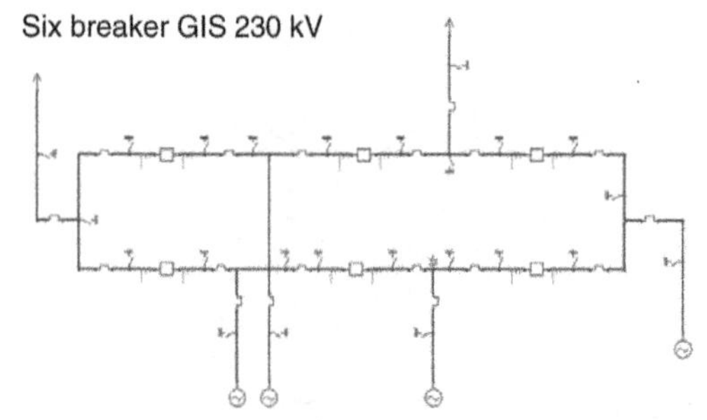

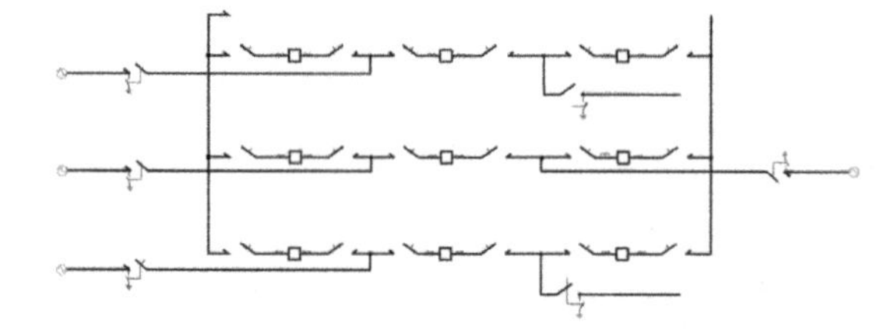

Figure 2.34 Comparison of the failure rates of the GIS and AIS (Reproduced by permission of CIGRE)

Table 2.9 Results of the reliability study [37, 38]

		Failure outages	
		OF (1/yr.)	**OD [h/yr.]**
AIS 1 1/2 cb	Line 1	0.0214	0.117
	Line 2	0.0214	0.117
	T1	0.0214	0.117
	T2	0.0214	0.117
	T3	0.0214	0.117
	T4	0.0214	0.117
			0.702
GIS ring	Line 1	0.0081	0.072
	Line 2	0.0081	0.072
	T1	0.0081	0.072
	T2	0.0081	0.072
	T3	0.0081	0.072
	T4	0.0081	0.072
			0.432

available to make these calculations. Figure 2.34 shows a comparison of the failure rates of a six feeder 230 kV substation and a nine breaker AIS.

The expected outage frequency of the GIS feeders is 2.5 times less than in the AIS, with the AIS feeder at only 47 years compared with the GIS feeder at 123 years. The expected outage duration of the GIS feeders is 1.6 times less than in the AIS.

The result of the reliability study is shown in Table 2.9. Based on the failure rate, the six breaker GIS ring bus substation is superior to the nine AIS breaker and one-half scheme and should be the preferred choice due to all its advantages.

2.4 Design

2.4.1 Overview

Design of the GIS is explained at an overview level by describing the typical choices made by manufacturers and users during the last 60 years. SF_6 gas was first used for high-voltage circuit breakers in 1955. As SF_6 became the preferred interrupting medium for high-voltage circuit breakers, it was natural to extend the use of SF_6 to make compact disconnect switches, ground switches, surge arresters, interconnecting bus, voltage transformers, current transformers, and specialized terminations for interfaces to air-insulated lines, cables, and direct transformer connections. Each of the basic design elements will first be described (SF_6 gas, conductors, enclosures, solid support insulators, conductor contacts, and enclosure joints). Next each functional component will be described. The arrangement of the components to match typical electrical one-lines will then be explained.

2.4.2 SF_6 Gas

SF_6 characteristics, handling procedures, and strong greenhouse gas aspects are covered in Section 2.8. The SF_6 gas pressures and dimensions used in circuit breakers have proven suitable for GIS. The dead tank type of circuit breaker, where the interrupter is enclosed in a grounded metal tank and connections are made to air-insulated lines or a bus bar using SF_6-to-air bushings via nozzles in the tank, can easily be converted for use in GIS by using interfaces consisting of cast epoxy gas barrier/support insulators at the nozzles of the tank. Interruption technology is covered in detail in several books and will not be dealt with here, except to note that a trend toward high interrupting ratings has led to a higher SF_6 pressure in the circuit breaker interrupter (typically about 0,8 MPa or 90 PSIG) than is needed for the other functional components. The higher SF_6 pressure in the circuit breaker results in a low temperature limit of about −20 °C – below this would require the tank to be heated or gas mixtures used to prevent SF_6 condensation. For GIS, these measures are not needed, as GIS is small enough for installation indoors. Only those parts of the GIS extending outside the enclosure need to be of a lower pressure design for ambient temperatures down to −50 °C, where a pressure of 0,4 MPS or 50 PSIG is suitable. For indoor parts, or outdoor parts without extremely low ambient temperatures, pressures from 0,4 MPa or 50 PSIG to 0,6 MPa or 70 PSIG are commonly used. Experience and testing have shown that a purity of SF_6 98% is enough, but the SF_6 gas must be very dry to avoid condensation of water vapor, as liquid severely reduces the dielectric strength. The SF_6, and the entire interior of the GIS, must be "clean" in relation to particles that severely reduces the dielectric strength. The GIS designer can be confident that commercially available SF_6 is reasonably priced, very pure, extremely dry, and clean. The design thus starts with the use of SF_6 at pressures from 0,4 MPa or 50 PSIG to 0,8 MPa or 90 PSIG.

Through experience and tests, the electrical stress levels that are adequate for reliable service have been established for SF_6 as about 5 kV/mm rms for power frequency and about 15 kV/mm peak for lightning impulse (BIL, or basic insulation level). Theoretical limits are much higher, but from a practical viewpoint it has not been worthwhile to pursue higher stress designs. A reasonably large physical size is needed to carry the usual continuous currents of high-voltage substations ranging from 1000 A to 8000 A. Dielectric required sizes given these performance levels for SF_6 are shown in Section 1.4 (Ratings) for the simplest configuration of a cylindrical conductor and enclosure. For more complex shapes, computer field plots can be used for design optimization. Full scale testing is required to confirm that the design is adequate for the type of test specified by the standards. Experience has shown that for the SF_6 gas the lightning impulse test is critical; if that passes, the others are not a concern.

The heat transfer capability of SF_6 is an important design parameter. For theoretical, and as a matter of practice, temperature limits are set by standards for conductor contacts (bolted and plug-in) at 105 °C total temperature. The cast epoxy support insulator materials typically used start to lose mechanical strength at around 120 °C. The exterior of the GIS is limited to a safe to touch temperature of 70 °C. The result is a temperature difference between the conductor and enclosure of about 35 °C. SF_6 gas transfers the heat generated by current in the conductor by conduction and convection and is transparent to most radiation. Theoretically and empirical formulas can be used to calculate the temperature rise, including the enclosure heating due to circulating currents and solar radiation (if outdoors). The calculations must be confirmed by continuous current tests under realistic conditions and thorough instrumentation so that no hot spots are missed. The SF_6 gas is stable at these temperatures and not subject to aging – the SF_6 gas will be good for the life expectance of the GIS (50 years). When the dielectric required sized of the GIS is too small for very high currents, the conductor and enclosure size, materials, and surface treatments can be changed as needed without any need to change the SF_6 gas.

Another aspect of SF_6 that is important to the designer is the pressure rise due to internal faults and decomposition due to interruption arcs, switching sparks, and internal faults. The pressure rises from interruption and switching sparks is negligible. The pressure rise from an internal fault is predictably slow and without shock waves. SF_6 will be decomposed by the heat of the arc, spark, and/or fault. Most of the decomposed SF_6 will very quickly recombine into SF_6, but some of the reactive molecules (S, F, etc.) will react with impurities in the SF_6 (such as water vapor, H_2O) to form decomposition by-products such as HF, SOF, and so on. The design of the GIS takes the behavior of SF_6 and arcs into account by using absorbents to control accumulation of decomposition by-products, pressure vessel standards to provide guidance to achieve adequate enclosure strength in relation to internal fault pressure rise (with or without rupture disks), and use of an epoxy resin formulation for support/barrier insulators that will be stable even if significant levels of HF are expected in the gas compartment (such as a circuit breaker gas compartment). It is not necessary to plan for replacement or replenishment of the SF_6.

2.4.3 Enclosure

The most common enclosure material is an aluminum alloy chosen for a desirable combination of mechanical strength, good electrical conductivity, resistance to atmospheric corrosion, and reasonable price. Cast, extruded, and wrought production methods are used depending on the application. For example, a complex switch enclosure may be cast aluminum and a bus enclosure may be extruded. Enclosure parts may be welded together.

The circuit breaker is also the largest and heaviest physical part of the GIS and is often the main physical attachment and fixed point in relation to environmental and GIS thermal expansion forces. In this role, the circuit breaker is designed to have strong tanks, nozzles, and support structures.

2.4.4 Principles

Gas-insulated switchgear is completely encapsulated, that is impervious to and separate from the external atmosphere. This is a great advantage in environments such as ocean-based oil rigs, particle or mist pollution sources, and coastal saltwater sites. However, because the gas-insulated switchgear is completely encapsulated, a visible disconnecting means, usually required, particularly in the United States, cannot be done directly. World wide the kinematic chain of disconnect and ground/earth switches for correct switch position indication is used instead of view ports. The disconnect and grounding switches, required in both air- and gas-insulated designs, will have view ports in gas-insulated equipment. These inspection windows will be discussed in detail in a subsequent section.

One principal advantage of GIS and may be the most important one is the compact design. A gas-insulated switchgear will have a smaller "footprint" than a comparable air-insulated substation, typically less than half the area. Although a gas-insulated substation will typically cost more initially than a comparable air-insulated substation, the economics may justify its use where real estate is expensive, such as centers of major cities. GIS may also be justified when a low-profile substation is needed to satisfy neighbors' wishes to "hide" a substation. In Chapter 8, Applications many examples of projects are given.

2.4.5 Operation

As a gas-insulated switchgear section is isolated for maintenance, it will be necessary to confirm the positions of the disconnect and grounding switch positions. Since these switches are completely encased within the aluminum housing, it is necessary for manufacturers to provide view ports. The view ports help ensure, by visual inspection, the position of the various disconnect and grounding switches. In some cases, this can be done with the naked eye and a flashlight. In other cases, especially at tall or awkward access points, a camera with a light source provided by the manufacturer is convenient.

Gas-insulated equipment is usually supplied with a local control cabinet (LCC). Typically, this cabinet includes control switches for the operation of one circuit breaker and its associated disconnect and grounding switches, and breaker alarm points. The protective relays associated with the GIS equipment may or may not be in this location. Because the SF_6 gas acts as a crucial insulator, it is necessary to maintain sufficient density within the GIS equipment. Therefore, there will be alarm and trip contacts from sensors for each gas zone to warn personnel or isolate equipment when the insulation integrity is inadequate.

One of the merits of gas-insulated equipment over its air-insulated counterparts is the minimal maintenance that is required of the GIS. This is primarily due to the separation of the conductors and insulators from the outside atmosphere. Newer versions of GIS equipment have very low leakage rates of SF_6 gas. The operation counter may aid in determining whether any maintenance will be required on the mechanisms, but this is typically many years between servicing.

2.5 Safety

2.5.1 General

Personnel safety holds a very high priority status when operating a GIS. The metallic encapsulation of the high-voltage components grounded or earthed where no direct contact is possible, except at the external connections. This safety aspect is inherent in the GIS design.

Moving parts such as operation rods or motor drives are usually covered with protective plates or indicated by coloring for enhanced safety.

In case of an internal failure, pressure relief devices open the enclosure to release the hot gas to the surrounding, to avoid a breaking of the metal enclosure. These pressure relief devices are designed to guide the gas stream away from operation personnel to protect them and help ensure their safety. So even in the very seldom case of an internal arc, the safety level maintained is high. This has also been tested in specific type tests of the GIS design.

To install GIS inside, a substation outdoor or indoor safety rules are further defined in IEC 61936-1 [48].

Here installation rules are given to integrate factory assembled and type-tested GIS. Requirements of grounding, earthing, fire protection, accessibility, safety of walkways and other areas are defined.

GIS is designed and tested according to IEEE C37.122 [23] or IEC 62271-203 [27]. All tests must be satisfied and complete for the GIS to be qualified. Prior to testing, GIS must be manufactured and assembled in a factory under clean particle-free room conditions. The design must pass all type test and routine test requirements. Once installed, the GIS is subject to the required on-site tests as specified in the aforementioned standards.

Additional requirements for the GIS to be installed are related to external connections, the erection process, and operational requirements in service. External connections are usually made with overhead lines, cables, transformers, reactors, or capacitor banks. The installation and erection should be organized in such a way as to avoid danger to all personnel or damage to other equipment.

2.5.2 Design and Erection Requirements

The GIS should be clearly arranged to allow a good overview for operators about the bay structure. Essential parts for the erection, operation, and maintenance should be accessible easily and without danger for the operator. If necessary, ladders and walkways may be provided. For the assembly process, the arrangement and access for handling the delivered components such as a crane, ropes, and hocks should be available.

Appropriate arrangements to connect the GIS to external connections are required to work safely on-site. Sufficient working space is needed and all metallic structures should be earthed or grounded.

2.5.2.1 Platforms and Ladders

The large size of high-voltage GIS mainly at the 420 kV and 550 kV voltage levels may need platforms and ladders installed for operation and maintenance purposes. For example, platforms or ladders may be needed to confirm the position of a disconnect switch or ground switch through the view port. Therefore, platforms or ladders should be attached or built into the GIS, as shown in Figure 2.35.

These ladders and platforms need to be designed in such a way that safety is ensured from an operational standpoint. Platforms are usually fixed to the GIS while ladders may be permanently attached or removable.

Figure 2.35 Ladders and platforms in GIS (Reproduced by permission of Siemens AG)

2.5.2.2 Monitoring Devices

Monitoring devices in a GIS shall be designed and installed in such a way that they can clearly be identified (color and/or numbering). Monitoring devices are used for gas density indication and are directly mounted to the gas compartment. The older method of using gas pipes to connect the gas compartment to a central gas density control cubicle is obsolete because of the increased risk of gas leakages from these pipes and their fittings.

In current GIS designs, gas density indicators typically provide only a red or green indication, and in some cases, an interim yellow indication. Green means all ok, no gas loss, yellow means ok, gas loss, and red means not ok, gas loss. Switchgear will be automatically switched off by clearing the section from high voltage (see Figure 2.36).

It is extremely important that there is clear labeling of each gas compartment for the operation and maintenance personnel. This ensures that the gas compartment can clearly be identified between the two gastight insulators (portions/form) of the gas compartment. Usually, the portions are indicated by outside coloring, for example, yellow.

When required, other monitoring devices are the operation counter or UHF antennas for partial discharge measurement. Usually when the GIS is in normal service, it is not monitored by partial discharge (PD) measurement because it is not needed. However, during commissioning, mainly at the higher voltage levels of 420 kV, 550 kV, or above, an on-site temporary PD measurement is used. The PD measuring equipment is then connected to the UHF antennas of the GIS where safe access is necessary. After the commissioning is successfully concluded, the PD measuring equipment will be dismantled (see Figure 2.37).

2.5.2.3 Thermal Expansion

When current is flowing through the GIS the temperature of the conductor and thus of the enclosure will increase. This can easily reach temperature differences of 40–50 K and will result in a thermal expansion separation of the enclosure and conductor metallic materials. Thermal expansion forces are very strong and can reach 160 t of mechanical force.

Figure 2.36 Gas density monitoring indicator (Reproduced by permission of Siemens AG)

Figure 2.37 UHF antenna connector for PD monitoring (Reproduced by permission of Siemens AG)

To prevent the GIS from mechanical stresses caused by thermal expansion, bellows should be provided to decouple the GIS bays mechanically. In the case of direct transformer connections, bellows are also required to decouple the GIS and the transformer mechanically. The technical requirements are stated in IEC 62271-211: "Direct connection between power transformers and gas-insulated metal-enclosed switchgear for rated voltages above 52 kV" [49] (see Figure 2.38).

2.5.2.4 Cable Connection to GIS

Direct connection of an oil- or solid-insulated cable will need a special enclosure joint to transfer the cable insulation to the gas insulation of the GIS. For this reason, internal insulator cones are used, which are insulated with SF_6 gas on the GIS side and with oil or solid insulation fittings on the cable side. The internal insulator cones are a pressure device built to withstand the GIS gas pressure, typically 0.6–0.8 MPa. The copper conductor of the cable and the aluminum conductor in the GIS are connected by an integrated conductor in the internal insulator cone. The outer enclosure is connected to the cable shield or in some cases high-voltage cable shielding is not connected to the GIS enclosure to avoid induced currents in the cable shield, which may heat up the cable. Another reason for insulating the cable shield from the GIS grounding is in the case of cathodic corrosion protection of the cable. In these cases, an insulating ring is used at the GIS [26].

If such insulation rings are used, in the case of disconnector switching in the GIS, the high-voltage transient overvoltage released may cause sparking across the insulation ring. This causes noises and light flashing and may result in a personnel accident, for example, falling from a ladder because of a shock. The sparking is technically not dangerous for the GIS or the cable, but they are

Figure 2.38 Compensation of thermal expansion. (a) Thermal expansion between GIS bays. (b) Single-phase transformer connection with thermal expansion bellows (Reproduced by permission of Siemens AG). (c) Single-phase transformer connection with thermal expansion bellows in an overview (Reproduced by permission of Meppi)

for people around the GIS. Therefore, in such cases it is recommended to use surge arresters across the insulation ring to bypass the high-voltage transients. In Figure 2.39, the GIS cable connection is using such surge arrestors between cable shield and GIS enclosure. The cable connection housing on the right side in yellow is connected to the cable that comes from the basement through the ceiling wall.

Because of the high frequency of the transient voltage of up to some 100 MHz, it is necessary to locate the surge arresters around the insulating ring. A minimum of four arresters are recommended.

Figure 2.39 Direct cable connection to GIS at 110 kV and single-phase insulated (Reproduced by permission of Alstom)

2.5.3 Building Requirements

In general, building requirements and fire regulations for buildings are regulated on national or regional levels. The following requirements and recommendations should be followed for areas and locations around high-voltage switchgear assemblies in accordance with IEEE C37.122 [23] and IC 62271-203 [27].

2.5.3.1 Load and Ceiling

Modern designs of GIS are delivered in large units. Typical for the voltage levels of up to 145 kV, two complete bays are transported on-site. For voltages up to 420 kV, one complete bay may be shipped as well as anything above that, with sections of a bay being brought on-site for assembly. In any case, weights in excess of multiple hundreds of kilograms or even few tons need to be moved on-site in the building.

Therefore, the ceilings and structure need to be strong enough to carry the load. Crane with the capacity to carry maximum single shipping unit is recommended. This helps both with installation and future maintenance of the GIS. The floor finish must be capable of withstanding the forces coming from fork lifters or air lift devices carrying the GIS bays. The weights need to be given by the manufacturer. A typical floor is shown in Figure 2.40.

2.5.3.2 Air Conditions

The building for an indoor GIS should be waterproof to provide indoor conditions as required per the standard. Depending on the local climatic conditions, it may be necessary to air-condition the building. This is the case when temperatures are very high (e.g., desert climate), humidity is very high (e.g., at the coast), or if high air pollution of dust or industrial dirt is nearby.

Water condensation due to temperature change, mainly when the building is climatized, should be avoided to prevent corrosion. Cold air dryers, usually used with air-conditioning, are recommended. If this cannot be obtained, precautions should be taken to prevent the consequences of water leaking or condensation affecting operating safety. Handrails or slippery-save walkways may be necessary.

2.5.3.3 Arc Fault Overpressure

In the case of an internal arc within the GIS, the pressure inside the enclosure could increase to a point where a disc ruptures, resulting in the burst disc falling into the building room. The wall, ceilings, and floors should be strong enough to adequately withstand the increase in pressure. The

Figure 2.40 View of a typical floor inside a GIS building (Reproduced by permission of Alstom)

pressure load depends on the enclosure gas volume and the short circuit rating of the equipment and can be calculated by the manufacturer.

2.5.3.4 Pipelines

If pipelines for water or other fluids are allowed into the GIS building, they should be installed in such a way that they do not affect the GIS in case of rupture.

2.5.3.5 Walls

The external wall of the building should have sufficient mechanical strength to withstand rain, sun, wind, snow, and ice to prevent the GIS from environmental impact. The walk passages that connect the indoor GIS to outdoor equipment should not affect the mechanical stability of the walls. If metal parts are used for wall passes, they need to be grounded. Any panels or parts accessible from the outside by the public need to be fixed in such a way that they cannot be removed.

2.5.3.6 Windows

Windows on the external walls of the GIS building are not needed and if used should be constructed in such a way that makes any entry is difficult. Therefore, the windows should be located more than 1.8 m above the ground, according to IEC 61936-1 [48]. The glass may be comprised of unbreakable material or the window protected with an iron curtain.

2.5.4 Grounding/Earthing Requirements related to Safety

Grounding/Earthing of GIS has to cover the requirements of technical functionally safety e.g., protection system and the safety of personnel. In this chapter, these aspects are explained. Rules for the design of grounding/earthing specific for GIS is explained in Section 2.6.

Personnel safety

For personnel safety, the grounding/earthing system needs to follow the requirements of IEEE Guide 80 [50]. In this standard, the design requirements for power frequency are given for the different electrical voltage and current ratings to fulfill the limitations given for touch voltages in the substation.

Functional safety

The functional safety of the grounding/earthing system is related to the correct reaction of protection equipment in case of failure. For this correct grounding, impedances are necessary which depend on the correct grounding/earthing system as defined in IEEE Guide 80 [50].

Specific grounding/earthing for high-frequency currents

In GIS disconnector, ground switch and circuit breaker operations in SF_6 cause high-frequency transient voltage of high-voltage levels which may disturb the protection and control system. Also, these transient voltages may be from danger for personnel when touching the GIS enclosure. To avoid any danger for personnel requirements are defined in IEEE C37.122 and IEC 62271-203 [23, 27].

In the following, some principles of HV GIS grounding/earthing are explained.

The specific grounding or earthing requirement of the GIS is related to the high transient voltages when any switch in the GIS is operated and the very compact design of the GIS. The very fast transients of a high-frequency nature need a high frequency, low impedance to ground/earth. This low impedance is reached by having multiple connections made between the concrete reinforcement steel grid and the earthing system of the building at various points in the GIS floor. Typical solutions for the GIS are to use a steel bar in the floor (Halfen bar) to fix the GIS or a bolt fixing with copper bar grounding connections (see Figure 2.41).

At the building wall, a multiple connection for the GIS to air bushing is needed between the GIS enclosure and the building wall. To prevent good conductivity in the building wall steel panels are usually integrated with multiple connections to ground/earthing of the building.

Secondary equipment used with the GIS should be adequately designed and tested regarding about their immunity against to transient overvoltage's of the secondary circuits.

Version (a) in Figure 2.41 shows a bolt fixing in the concrete floor with a separate grounding connection using a copper bar. Version (b) in Figure 2.41 shows a floor anchoring with leveling channels. The steel bar is grounded and the GIS equipment is fixed directly to the steel bar.

2.5.5 Burn Through of Enclosure

Safety at burn through of GIS includes ventilation, gas detection, and personal access. Precaution has been taken by type tests of internal arcs according to IEEE C37.122 [23] and IEC 62271-203 [27] to protect persons, operators, and other equipment.

2.5.6 Work Behind Pressurized Insulators

Safety work instructions to work behind pressurized insulators are defined in maintenance rules between user and manufacturer.

2.5.7 SF_6 Release to Buildings

SF6 is a nontoxic gas, however if large quantity is released Safety of release of SF6 in low areas of building and this may present a danger for personnel, due to the low level of oxygen content in ambient air. When GIS is installed inside buildings or in basements care shall be take when entering such rooms to avoid danger for personnel. The international standards IEC 62271-4 [51] and IEEE C37.122.3 [52] give detailed information on how to build, design, operate, and monitor such GIS installation locations. In these standards, requirements for ventilation, sensors the content of SF_6 and oxygen in the room, monitoring, and indication devices for warning are defined.

In case of internal arc failures and the opening of the enclosure, special care needs to be taken to avoid contamination of personnel. Safety rules and detailed process definitions for room

Figure 2.41 Ground/earth connection of GIS using bolts or a steel bar (Halfen bar) in the floor. (a) Bolted GIS fixing with grounding connection (Reproduced by permission of Alstom). (b) Steel bar (Halfen bar) in the floor including grounding connection (Reproduced by permission of Siemens AG)

contaminations with SF_6 by-products are given in IEC 62271-4 [51] and IEEE C37.122.3 [52] including the handling of the decomposition products. The rules about safety to release SF_6 into buildings are part of agreement between user and manufacturer. These rules have to take into account local regulations to avoid low oxygen content in ambient air, and health hazards of SF6 by-products [53, 54].

2.6 Grounding and Bonding

2.6.1 General

GIS is inherently safe due to the fact that all live parts, with the exception of SF_6-to-air bushings terminals, are enclosed in grounded enclosures and are not subject to accidental contact. Furthermore, to ensure personnel safety and protect equipment, GIS installations employ various grounding practices to accomplish system and protective grounding [50].

2.6.2 GIS Versus AIS Grounding

GIS are installed under the same system parameters as AIS. However, with respect to grounding, one significant difference is that GIS are usually installed on much smaller sites than AIS. Consequently, GIS do not have the same advantage as large AIS switchyards where the station ground grid helps to dissipate fault currents. Therefore, in order to provide low impedance paths to ground for fault currents, reduce magnetic field intensities, and minimize transient overvoltage's, GIS installations use multipoint grounding systems.

Multipoint grounding systems consist of short grounding conductors that interconnect the GIS at numerous points along the enclosures. These multiple grounding conductors provide parallel paths to the GIS main ground bus or GIS grounding mesh.

2.6.3 GIS Enclosure Currents

In most GIS installations, each module is electrically bonded either via flange connections or external shunts. This results in a continuous enclosure throughout the GIS, allowing enclosure currents to flow during normal operation and under fault conditions. Enclosure currents are a result of voltages induced in the metallic enclosure by effects of currents flowing in the enclosed conductors, and can be categorized as induced, return, circulating, or fault currents.

During normal operation, the return current on a GIS enclosure can reach up to 90% of the operating current. In the case of three-phase faults, the return current can achieve up to 90% of fault currents. Therefore, GIS return current conductors, as well as flanges and shunts, are designed for a full return current and fault conditions without exceeding the conductor's thermal and mechanical limits.

Enclosure currents in three-phase GIS applications are not susceptible to circulating enclosure currents because all phase conductors are located inside one enclosure, and the phase conductor's electromagnetic fields essentially cancel each other out. Three-phase enclosure currents are illustrated in Figure 2.42.

In single-phase GIS, each conductor is contained within its own grounded enclosure. As a result, enclosure currents during normal operation are made up of circulating currents. GIS manufacturers typically provide conductors to interconnect each single-phase enclosure at multiple locations. These grounding connections are interconnected across each phase enclosure at intervals along the GIS, as well as the ends of the enclosures to promote circulating currents and reduce magnetic fields. The phase enclosure interconnections also keep heavy circulating currents from passing through grounding conductors and into the substation ground grid. Single-phase enclosure currents are illustrated in Figure 2.43.

2.6.4 General Rules for GIS Grounding

As described in IEEE Standard 80 [50], touch voltages can be more dangerous than step voltages in AIS. With even more equipment within reach in GIS, and longitudinal induced voltages present on GIS enclosures, management of touch voltages is even more critical in GIS. To evaluate maximum touch and step voltages occurring on GIS enclosures during a fault, it is necessary to perform an analysis on the substation grounding system. Commercially available grounding software should be used to perform simulations to evaluate maximum touch and step potentials, as well as ground potential rises.

In most GIS applications, there are two grounding grids that make up the grounding system: (1) the station grounding grid, which is similar to a typical AIS installation; (2) the GIS grounding mesh, which is a narrowly spaced grounding grid (typically 3–5 m) embedded into the concrete slab

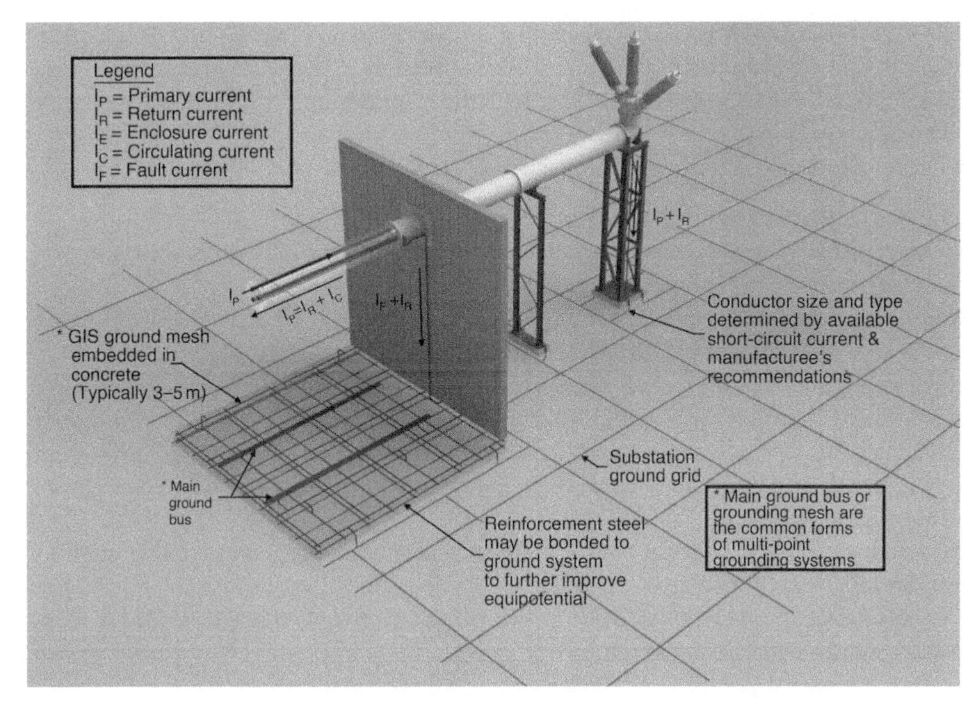

Figure 2.42 Three-phase GIS enclosure currents (Reproduced by permission of ABB)

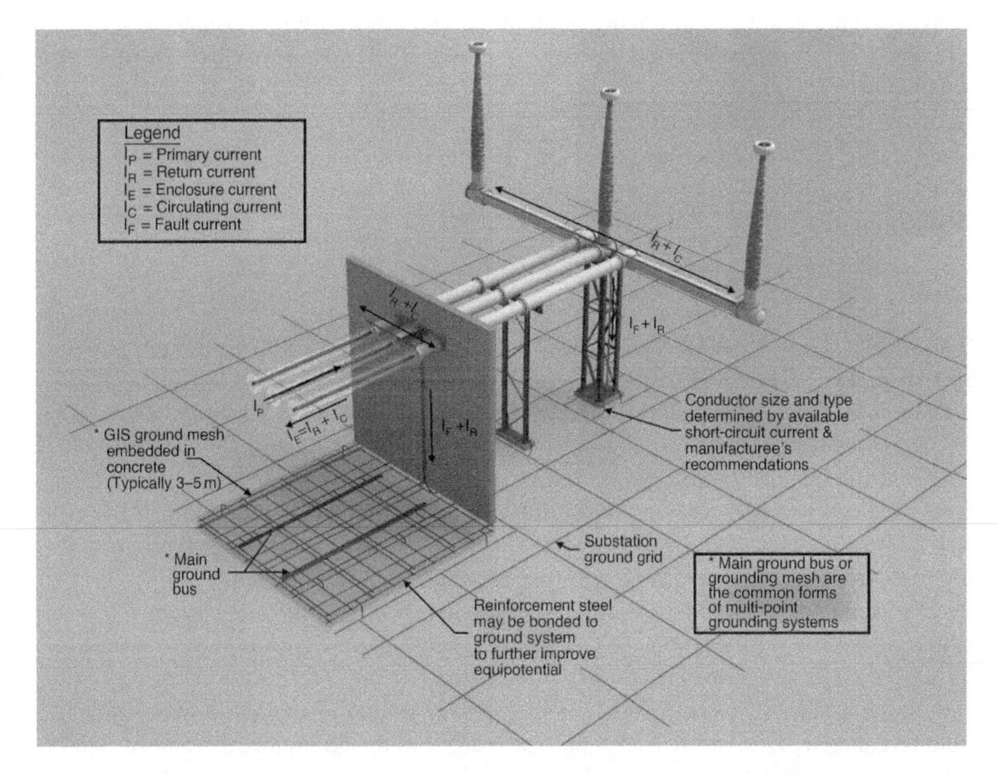

Figure 2.43 Single-phase GIS enclosure currents (Reproduced by permission of ABB)

in which the GIS is installed. A perimeter grounding conductor around the building or enclosure connects the two ground grids. Short interconnections between the GIS enclosures and grounding mesh at intervals of around 10 m, or in accordance with the manufacturer's requirements, make up the multipoint grounding system.

Best practices for GIS grounding and bonding include the following:

1) All grounding conductors should be as short as possible.
2) The grounding mesh and interconnections should be capable of carrying the system's fault currents without exceeding the thermal and mechanical limits.
3) All exposed grounding conductors should be protected against mechanical damage and located so as not to present a "trip hazard" to operation personnel.
4) Proper grounding and bonding techniques, such as multiple conductors or voltage limiters, are required at all discontinuities within the GIS. This includes SF_6 gas-to-air connections, SF_6 gas-to-cable connections, SF_6 gas-to-oil connections, and where GIB exits the building.
5) Ensure all metallic building components, GIS support structures, and GIS maintenance platforms are properly grounded.
6) Reinforcement steel in the building floor should be connected to the GIS grounding mesh to further equalize ground potentials.
7) All secondary cables should be shielded with both ends of each cable shield grounded to mitigate possible electromagnetic interference.

2.6.5 Very Fast Transients

Very fast transients (or VFTs) are generated as a result of switching operations inside the GIS or a dielectric breakdown that causes a voltage collapse within the GIS. The voltage collapse produces traveling waves that propagate away from the disturbance. The traveling waves propagate throughout the GIS with various reflections and combine to produce VFTs or overvoltage's with a very steep rate of rise.

VFTs can cause electromagnetic interference in the local environment and in secondary circuits. As VFTs approach discontinuities, they can cause transient enclosure voltages (i.e., TEVs). TEVs do not present a direct hazard to operations personnel, but may cause electrostatic sparks if the GIS multipoint grounding system is not installed properly.

2.6.6 GIS Grounding Connection Details

Circuit breakers and GIS bay. Each circuit breaker should have two connection points for grounding at each bay of the GIS (see Figure 2.44).

SF_6 gas-to-air bushings. Special attention should be given to SF_6 gas-to-air bushings where high-frequency effects are most prevalent. A minimum of two grounding conductors should be installed around the connecting flange to GIS.

SF_6 gas-to-cable connections. It is important to evaluate the method of grounding at the GIS cable end unit. Multiple conductors may be required if solidly bonded and voltage limiters might need to be considered if single point bonding is used (see Figures 2.45 and 2.46).

Steel structures. All steel structures should be grounded. Normally, a steel structure can be grounded to the nearest grounding point or GIS flange. Several ground connections between steel structure and GIS are recommended for each bay (see Figure 2.47).

Buildings. Building slabs should include an embedded GIS grounding mesh and steel reinforcement should be concrete. Reinforced steel should be bonded to the grounding mesh every 3 m in both directions (see Figure 2.45).

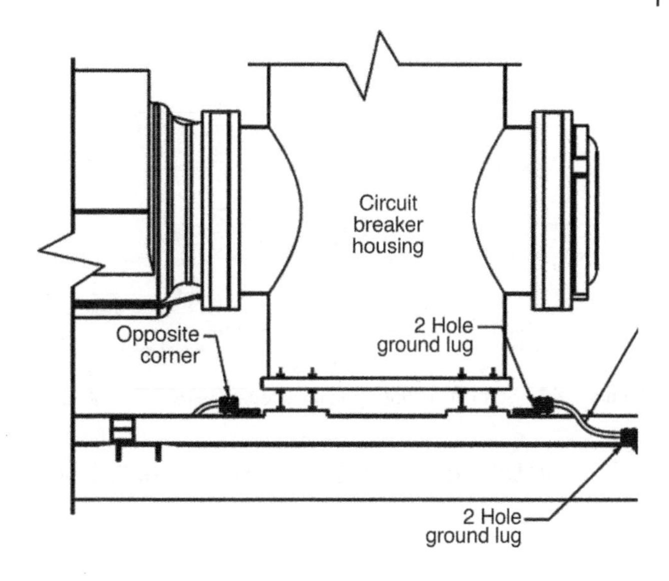

Figure 2.44 Circuit breaker and GIS bay grounding (Reproduced by permission of ABB)

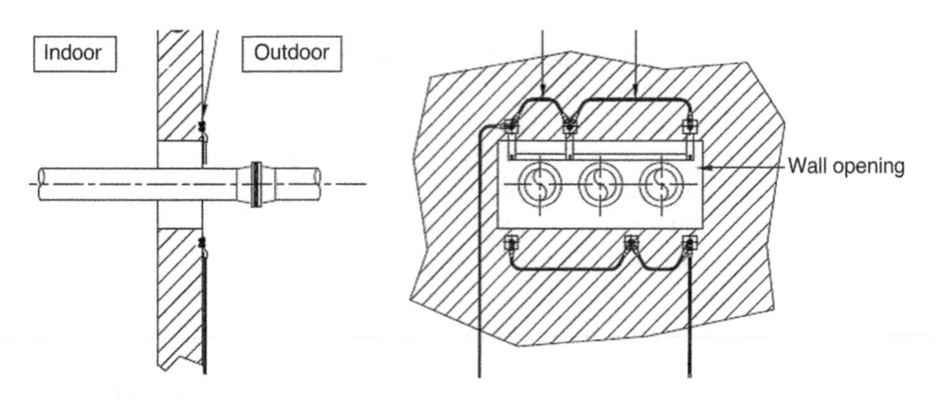

Figure 2.45 Buildings (Reproduced by permission of ABB)

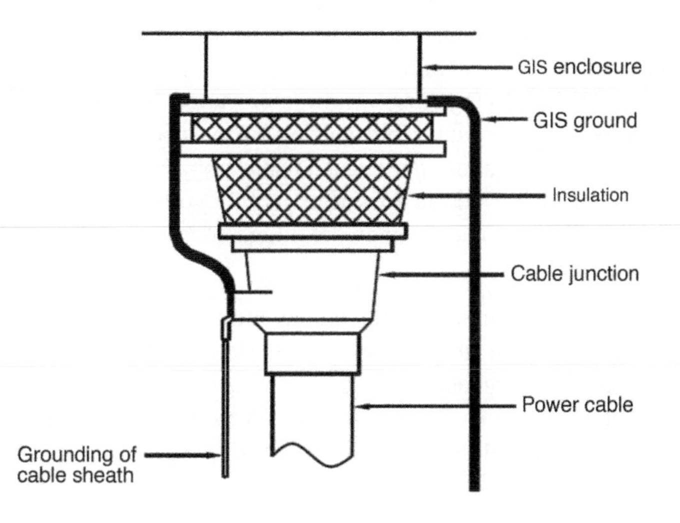

Figure 2.46 SF6 gas-to-cable connections (Reproduced by permission of ABB)

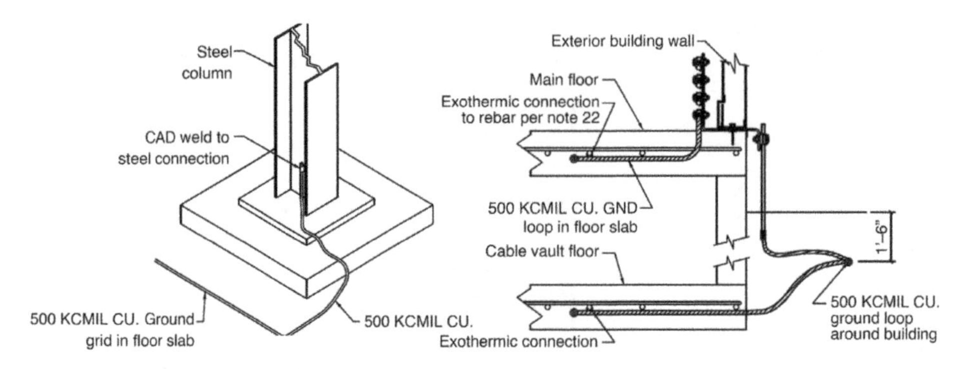

Figure 2.47 Steel structures (Reproduced by permission of ABB)

2.7 Factors for Choosing Gas-Insulated Substations

2.7.1 General

The deployment of GIS in substation applications has been around world-wide since the 1960s and is becoming an increasingly prevalent asset for substation users. Utilities, municipalities, IPPs, and other nonutility power producers are discovering the many advantages of GIS. However, when deciding whether to choose conventional AIS, GIS, or a mixed-technology solution there are many factors that should be evaluated. There is such an abundance of data, which can be interpreted from different respects, that it can make a user's decision quite difficult. In fact, the amount of selection criteria for AIS and GIS available can hinder a user's decision-making ability. Therefore, once a substation's functional requirements are defined, it is important to determine, and even quantify, which factors are important to the user's specific application.

2.7.2 Factors

In order to help in the decision-making process, it is important to define and organize the advantages and disadvantages of AIS and GIS in a hierarchical manner, and subsequently analyze the factors according to the user's needs.

GIS solutions are most noted for space-saving advantages, with substation footprints 15–25% that of an equivalent AIS substation. However, users are capitalizing on additional advantages of GIS, such as improving critical infrastructure reliability, avoiding permitting obstacles, and reducing planned outage durations.

For a user replacing aging infrastructure, upgrading obsolete configurations, increasing capacity or developing new capital, recognizing the most pertinent factors empowers the user to determine the optimal substation solution. All factors may not directly impact investment or life cycle costs, but may still be critical in the selection of the substation location and construction planning.

Following are some common definitions of factors with respect to evaluating AIS and GIS substation solutions:

Aesthetics. Appearance considerations and community acceptance may have a major influence on the area, height, and visibility of a substation. AIS can be more difficult to disguise than GIS.

Altitude. Elevation above sea-level, where equipment that depends on air for its insulating and cooling medium, will have a higher temperature rise and a lower dielectric strength when operated at higher altitudes. Both AIS and GIS designs may need adjustment based on actual substation altitude.

Atmospheric contamination. Airborne contaminants, such as salt, dust, debris, and industrial pollution, can compromise exposed electrical insulation. GIS are typically installed indoors. Furthermore, most GIS components are hermetically sealed inside an enclosure, which makes GIS a superior solution in poor atmospheric conditions.

Availability. The fraction of time that the service is available and the (steady-state) probability that power will be available.

Audible noise. Sound levels produced by electrical equipment may be of concern to the public. Both AIS and GIS noise will have to be studied for the user's requirements.

Automation. Provisions for controlling and monitoring substation equipment can be local (within the substation) or remote (at another location, typically an operation center). GIS may have more opportunities for control because most switches tend to be motor operated. However, in GIS, these functions are collected inside an LCC (local control cabinet) and have to be integrated with the customer's existing automation.

Capacity. A substation's load-carrying ability, usually with reference to a power transformer's MVA and defined by a system load study. The AIS and GIS switching equipment must be coordinated with the substation's capacity, and many times with increased capacity projects there may be added benefits using GIS.

Commissioning. Procedure including inspection, testing, and documenting all primary and secondary components (systematically, as much as possible) required prior to placing a substation into service. Preassembled and pretested GIS shipping units can reduce inspection and testing efforts.

Construction. Preassembled GIS shipping units typically reduce field installation costs and time. However, preassembly is usually less with higher voltages, and most GIS are installed indoors, which requires building construction.

Community impact. Public interest of substation installations generally revolves around safety, aesthetics, land use, environmental impacts, and sometimes electromagnetic field (EMF) concerns. GIS can be appealing when public approval is required due to its compact size and ability to blend with the existing environment.

Cost. GIS equipment is more expensive than AIS equipment. However, consideration of life cycle costs shows many times that GIS is less expensive and provides higher performance. Cost comparisons should be based on the total life cycle costs, including equipment, land, site development, normal operating and maintenance costs, and forced outage costs based on reliability.

Cutovers (planned outages). During substation commissioning, one of the last steps is to switch lines and loads into the new equipment. GIS has an advantage over AIS because of its reduced footprint and interface flexibility (i.e., air, cable, or oil). Therefore, many times GIS construction can occur closer to connection points and thus reduce cutover durations.

Environment. GIS inherently reduces land and space use impacts. However, GIS has more SF_6-enclosed components and thus receives more attention due to potential climate change impacts. Despite this concern, potential overall contributions of SF_6 to global climate is miniscule relative to carbon emissions.

Emissions (SF_6). Additional handling of SF_6 gas is required due to larger quantities used in GIS. Since SF_6 is a greenhouse gas, it needs to be managed properly.

Expandability. AIS can be more easily expanded. On the other hand, GIS is a great solution for expanding existing AIS when space is limited and planned outages are difficult to obtain. Expanding existing GIS requires the original design to include provisions for future plans and may at times require future infrastructure to be installed early.

EMF (electromagnetic field). Magnetic fields due to conductor currents are reduced by GIS enclosure currents and are typically less than those of AIS. Even near GIS exits, where EMF levels can be highest, exposure levels are generally well within tolerable limits. With respect to public safety, studies have shown EMF levels are typically orders of magnitude below safe levels.

Failure rate. The average number of failures of a component or unit of the system in a given time (usually a year). MTBFs are typically less with GIS.

Flexibility. The ease of operation and time needed to perform switching operations within a substation. This varies with AIS and GIS configurations.

Interruption. A cessation of service to one or more customers, whether power was being used or not. Interruptions can be classified as instantaneous, momentary, temporary, or sustained.

Initial capital. All initial costs associated with land acquisition, construction, permits, engineering design, site work, civil construction, purchase of equipment, training, and installation of a substation. This varies between AIS and GIS configurations, but AIS is typically less expensive.

Land size. Property sizes are almost always smaller when using GIS. However, if land is cheap at the proposed site, this may not be a significant consideration.

Location. Small GIS footprints allow locating substations closer to loads (e.g., downtown), reducing permitting requirements, offsetting high land costs, and hiding substation equipment from the public (e.g., inside a building or underground).

Life cycle cost. A cost analysis that integrates capital investment, land acquisition, site preparation, reliability impacts, operation, and maintenance expenditures for the life cycle of the project with a certain interest rate. This methodology allows for an analysis of the pertinent factors, weighted by the user, to calculate the total life cycle cost.

Maintenance. Resources required to upkeep substation equipment. Frequency of GIS maintenance is much lower than AIS (typically every 8 years versus every year) because most components are protected from the environment. However, GIS maintenance procedures require additional training and GIS replacement parts may not be as readily available.

Operation and maintenance (O&M). This cost includes all the fixed costs associated with a substation, which are costs for property taxes, insurance, planned operation and maintenance, and planned service interruptions.

Permitting issues. Indoor GIS may require enclosure or building permits, but overall permitting is usually accelerated and/or reduced when compared to AIS. In addition, GIS can minimize impacts on environmentally sensitive areas, such as wetlands, agricultural lands, cultural resource sites, and so on. In the case of wetlands, costs associated with replicating these areas may be eliminated.

Reliability. The fraction of time that a component or system is capable of performing the required function. The (steady-state) probability that it will be in service where it can function. The four main indices for measuring reliability are System Average Interruption Frequency Index (SAIFI), Customer Average Interruption Duration Index (CAIDI), Average Duration of Interruptions per Customers duration per year (SAIDI) and Customer Average Interruption Duration Index divided by total duration of interruption (CTAIDI).

Restoration. The return of electric service after an interruption, because of repair of the outage that caused the interruption, or because of reswitching of the supply, or the starting of an alternate source.

Safety. Safety is paramount to substation design and involves protecting the public, as well as operation and maintenance personnel, by means of design, construction, security, training, and work procedures. It is also important to note that safety begins at the engineering and equipment selection stage. "Safety by Design" is becoming more integrated in substation engineering.

Security. A threat such as vandalism, terrorism, or unauthorized persons entering the substation. Typically, threats are reduced with GIS because the substation is located indoors and energized parts are enclosed.

Seismic. The ability of substation equipment to withstand forces generated by earthquakes. GIS typically has better seismic withstand capability than comparable AIS.

Site preparation. Site development including all earth work such as cut, fill, grading, and drainage. GIS substations typically reduce the extent of earth work and civil work.

Soil conditions. Characteristics of the surface and subsurface where a substation will be constructed. These characteristics will help define the foundation requirements for the substation equipment. Detailed analysis of resources required to prepare the specific site is recommended to determine whether AIS or GIS is more beneficial.

Stability. The ability of a power system to return to a normal state after a disturbance.

Unique devices. Capacity coupled voltage transformers (CCVTs), wave traps, and load break switches are usually not installed inside a GIS. Such devices are required to be air-insulated once the circuit leaves the GIS.

Weather. Factors such as temperature, wind, ice, rain, snow, storms, and humidity may affect a substation's operation. GIS tends to withstand extreme environmental conditions since it can be installed indoors and GIS components are hermetically sealed inside enclosures.

Workforce training. Resources required teaching the workforce the correct procedures of operating and maintaining substation equipment. GIS requires additional training.

The aforementioned factors for selecting AIS or GIS can be organized into three major categories: power system requirements, such as reliability and availability; environment, such as location and climate; and economics, with respect to installation, maintenance, and outage expenditures. In addition, these factors can be further categorized as quantitative or qualitative (i.e., hard or soft data) because some are more easily quantified than others.

2.7.3 Power System

Factors for choosing AIS or GIS begin to appear in the asset planning stage. Once system or transmission planning has determined the need for a retrofit or green-field substation, various factors have already been studied, such as system strength, stability, reliability, and load flow requirements. The choice to select AIS or GIS at this point may be premature, but information developed during the planning stage may already start guiding the owner to a proposed solution.

2.7.4 Environment

Areas with harsh conditions, such as poor soil conditions, high air contamination, high seismic, or high storm surge, are possible factors created by the natural environment of a possible substation site. However, other qualitative factors related to the environment, such as aesthetics to the local community, permitting issues, and/or urban area impacts can play into the optimal substation solution.

2.7.5 Economics

Economic evaluation of an asset depends on the weight a substation user assigns to each determining factor. In addition, prevailing factors can vary with each substation application. Therefore, in order to understand the overall investment of a substation project, many users are turning to life cycle cost analysis. This analysis provides the user with information on how much a substation will cost over the lifetime of the equipment. When AIS is compared to GIS with respect to the upfront investment cost, most of the time AIS will result in the least-expensive solution. However, factoring in land acquisition, permitting, site preparation, and O&M costs, just to name a few, provides a more holistic approach and may change the optimal substation solution.

Innovation or nontraditional solutions, such as GIS, can buy more reliability for the same budget under the right circumstances. That is why it is important for traditional engineering tools, such as N-1 criterion, to be augmented with a reliability-based life cycle cost planning approach.

2.7.6 Conclusion

When deciding between AIS or GIS, some factors alone, for example aesthetics, may be an overwhelming influence on a user's decision. However, most of the time, the best decision requires evaluating many factors and soliciting input from many departments within a power producer's organization. It is also important to note that the optimal substation solution may not be purely AIS or GIS. Sometimes a combination of the two (i.e., hybrid or mixed-technology solution) might be the best configuration.

In general, a majority of the factors will favor GIS, but it is the value or weight assigned to the factors that determines whether the return on investment is justifiable to select GIS over AIS. Factors that can be quantified should be given weights based on the user's requirements. By applying weights to the various factors, a user can evaluate different substation configurations to determine whether AIS or GIS is more cost-effective.

This analytical process is actually implemented with various commercially available software applications, by using dynamic state enumeration to compute rankings of AIS and GIS configurations. These software applications allow users to quantify life cycle costs based on owner-specific data [i.e., mean time to repair (MTTR), interruption cost, etc.] and historical equipment failure rates (published by the IEEE, CIGRE, and others). Next, life cycle costs, along with other user-specified factors, sometimes referred to as "intangible or soft factors" (i.e., safety or aesthetics), are weighted. This methodology can be implemented using the common principle of weighted averages or by using algorithms based on user-specified preferences. The results of these analyses can also provide a recommended list of substation alternatives, ranked in order, based on clear technical and economic information. Additional information can be found in Section 9.3, Life Cycle Cost Analysis.

2.8 Sulfur Hexafluoride (SF_6)

2.8.1 What is Sulfur Hexafluoride?

Sulfur hexafluoride (SF_6) is a colorless, odorless, nontoxic, and nonflammable gas. It is five times heavier than air and has an extremely stable molecular construction (see Figure 2.48). The gas provides high dielectric strength and excellent arc-quenching properties. However, the high heat

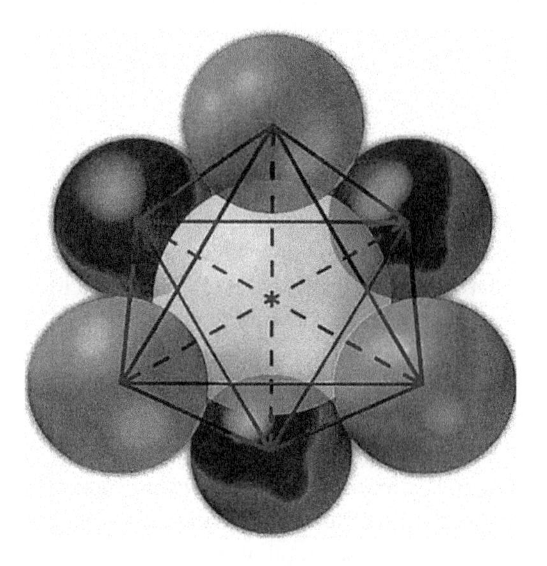

Figure 2.48 Molecular structure of sulfur hexafluoride (SF_6) (Reproduced by permission of Solvay)

absorbing ability makes sulfur hexafluoride a strong greenhouse gas with a CO_2 equivalent contribution to the global warming potential by a factor of 23 900 [51–56]. That means that 1 kg of SF_6 released into the atmosphere contributes to the global warming as much as 23 900 kg of CO_2. For this reason (and because of the high cost of SF_6), the use of SF_6 in a gas-insulated substation (GIS) is organized in a closed cycle of use, from production to filling into the GIS, during maintenance, and, finally, to closed storage when the GIS is dismantled and decommissioned. This process is regulated by international standards IEC 62271-203 and IEEE C37.122 [23, 27]. Further information about SF_6 and its use can be found in References [51–54].

2.8.1.1 Greenhouse Effect

The greenhouse effect, as stated in the Kyoto Protocol, explains the effects of the sun sending light and heat to earth through the atmosphere. Some of the sunlight and heat is absorbed by the earth and some is reflected back toward space. The reflected solar (infrared) radiation is then partially reflected back toward the earth by the atmosphere and contributes to the heating of the atmosphere (see Figure 2.49). This delicate balance of radiation trapped and heating the earth and the radiation lost through reradiation into space creates temperatures that makes life on the planet earth possible (the "greenhouse effect"). However, with more "greenhouse gases" in the atmosphere, a greater proportion of solar radiation is trapped, leading to an overall increase in temperatures on earth, leading to global warming. This can have dramatic impacts on weather patterns and have been blamed for melting ice in polar regions and extreme weather conditions like heat, storms, and droughts. The main contributor to the greenhouse effect is CO_2, which is produced, not only by natural causes, but also by all burning processes of carbon-based materials, such as natural gas, coal, oil, and others. Relative to CO_2, SF_6 has a very small overall contribution to the greenhouse effect (due to the relatively small volume used world-wide), however, SF_6 has a high global warming potential of 23 900, and can potentially have a strong impact even when small amounts are released into the atmosphere. In addition, the half-life of SF_6 in the upper atmosphere is greater than 3000 years [51, 52, 55, 56] with no natural means of reduction. This is much longer than that for CO_2 – the influence of SF_6 on the greenhouse effect will persist much longer than an equal quantity of CO_2. The consequence of this for the high-voltage energy industry is to control and track SF_6 usage by keeping the use of SF_6 in a closed cycle and keeping the release to the atmosphere as low as possible. Unfortunately, for dielectric and arc-quenching applications no other gas has yet been identified, despite intensive research efforts, that allows for the production and operation of such highly efficient high-voltage equipment.

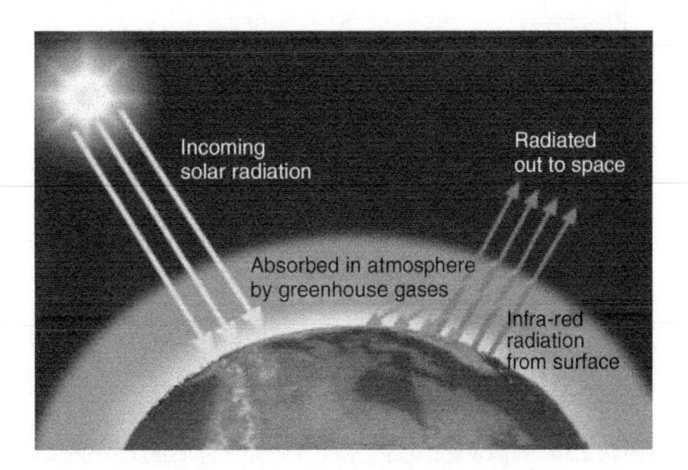

Figure 2.49 Greenhouse effect (Reproduced by permission of Siemens AG)

2.8.1.2 Features of SF_6

The main characteristic of SF_6 useful for the design of high-voltage equipment is the high dielectric withstand capability, which is about 3 times the dielectric withstands of air. Used in high-voltage equipment with gas pressures of up to 8 bar, the size of equipment using SF_6 can be reduced by up to ten times as compared to equivalent air-insulated installations.

SF_6 gas also effectively quenches arcs in circuit breakers, disconnectors, and ground switches. Pure SF_6 increases strongly the arc-quenching capability with increasing pressure, as shown in Figure 2.50. This is the reason why the gas pressure in breaker compartments of a GIS has the highest gas pressure compared to the bus bar gas compartment or to gas compartments of disconnectors and ground/earth switches. If the SF_6 gas is mixed with air, the resulting arc-quenching capability is strongly reduced (see Figure 2.50). The SF_6 related arc currents of air are much lower, as shown in Figure 2.50.

The metal encapsulation of GIS makes the equipment very safe to operate because all high-voltage parts are contained and properly insulated and the metallic enclosure is grounded and can be normally touched without injury.

The SF_6 insulation gas inside the GIS does not show any aging effects and is protected by the metal enclosure from ambient influences such as humidity, dust, salt air, and others. Therefore, the maintenance required is very low. Today's state-of-the-art GIS have recommended maintenance cycles of 25 years. The main features of SF_6 to be used in high-voltage equipment are shown in Table 2.10.

2.8.2 Background Information

The general knowledge required for handling SF_6 and to understand the related procedures and instructions defined is provided here as background information [51–59]. This knowledge has been collected and formulated by CIGRE and was published as technical report CIGRE TB 276 [54].

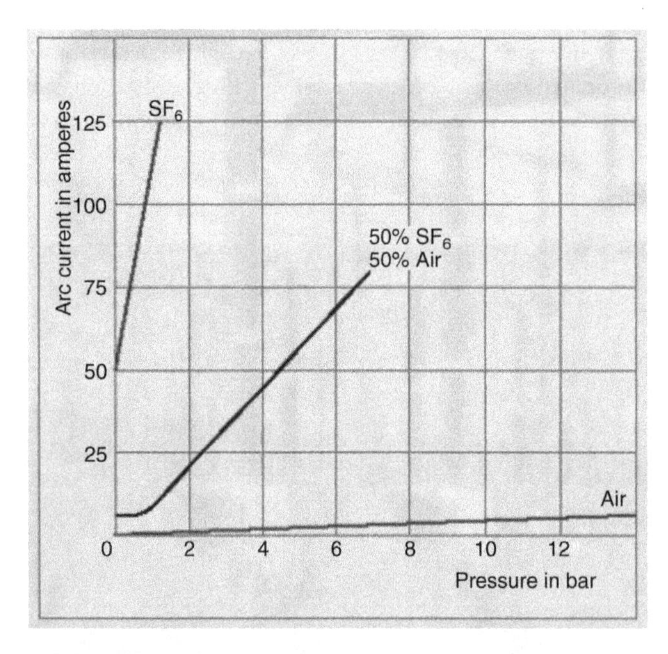

Figure 2.50 Arc current of SF_6, with SF_6 gas-to-air gas mixture and air (Reproduced by permission of Siemens AG)

Table 2.10 Features of SF_6 for high-voltage equipment

High dielectric withstand capability
Effective arc-quenching in circuit breakers, disconnectors, disconnectors, and ground/earth switches
Allows high safety for operational personnel because of grounded/earthed metallic enclosure
Allows compact equipment design and almost maintenance-free for 25 years

2.8.2.1 Sulfur Hexafluoride

Sulfur hexafluoride is a synthetic gas formed by 6 atoms of fluorine gathered around a centrally situated atom of sulfur. The chemical formula is SF_6, the molecular weight is 146.05 g/mol and the gas is identified by CAS Number 2551-62-4.

2.8.2.2 The Chemical Bond

The bonds between fluorine and sulfur in SF6 are known to be some of the most stable existing atomic bonds. Six of these give the molecule very high chemical and thermal stability. In addition, the compatibility of SF_6 with material used in electric constructions is similar to that of nitrogen, up to temperatures of about 180 °C.

2.8.2.3 Use of SF_6

Since the early 1960s, SF_6 has been successfully used by the power electric industry for HV transmission and MV distribution equipment. Typical equipment used are gas-insulated substations, ring main units, circuit breakers, transformers and cables.

2.8.2.4 Use in Electrical Equipment

For electrical equipment SF_6 offers, like no other gas, excellent electric insulation and arc-quenching properties. No other gas can allow today's switchgear to reach high-voltage levels with current switching capabilities.

2.8.2.5 Other Gases

All other gases identified for application in high-voltage switchgear (e.g., He, N_2, COs, CFs) may have a better insulating or switching performance, but not both. In addition, most of these other gases are not stable in the long term, are toxic or extremely expensive. Because of the high global warming potential alternative insulating and switching gases have been developed and are in practical use today. These gases are Flourketone, Flournitrile and technical air (N_2/O_2, or N_2/CO_2) with vacuum switches and circuit breaker. For more details see section 2.9 and 9.13.

2.8.2.6 Nonelectrical

Other industrial applications of SF_6 not related to the electric power industry include metallurgy, electronics, scientific equipment, ocular surgery, and military applications.

2.8.2.7 Physical

The SF_6 molecule is very stable and will last under atmospheric conditions statistically several 1000 years before it will break by thermal or radiation impact. This makes SF_6 to long time reliable insulating and arc quenching gas and will not lose its dielectric and molecule recombination ability.

2.8.2.8 Thermodynamic

At normal room temperatures and pressures (20 °C and 100 kPa), SF_6 is about 5 times heavier than air (density: 6.07 kg/m^3). As the gas is heavier than air, areas below ground level, poorly ventilated, or unventilated areas (i.e., cable ducts, trenches, inspection pits, drainage system, etc.), may collect and remain full of SF_6. Personnel must be aware of the danger of asphyxiation in such places.

2.8.2.9 Liquefaction

As the critical temperature and pressure of SF_6 are 45.54 °C and 3.759 MPa, respectively, it can be liquefied by compression and is usually transported as a liquid in cylinders or containers.

As the gas is delivered in the form of compressed liquid, the temperature of both the gas and the container can fall quickly if large quantities of the gas are released rapidly. Frost and ice may form on metal parts. If this occurs, gas filling has to be immediately stopped until ice and frost are gone. Filling of SF_6 must always be performed slowly. Personnel must be aware of the danger of freeze burns when touching iced and/or frozen metal parts.

2.8.2.10 Electric

SF_6 is strongly electronegative (i.e., it tends to attract free electrons) and has a unique combination of physical properties: high dielectric strength (about 3 times that of air), high thermal interruption capabilities (about 10 times that of air), and high heat transfer performance (about twice that of air).

2.8.2.11 Eco-toxicity

SF_6 does not harm the ecosystem: biological accumulation in the food chain does not occur. It is an inert gas with very low solubility in water so it presents no danger to surface and/or ground water and/or the soil.

2.8.2.12 Greenhouse Gas

SF_6 is a potent greenhouse gas with a global warming potential (GWP) of about 23 900 times that of CO_2. It will stay persistent in the atmosphere with an atmospheric lifetime (ALT) of statistical 650–3200 years, depending on the calculation model [54–56].

SF_6 has no impact on the stratospheric ozone layer (ozone depletion potential (ODP) = 0).

2.8.2.13 Environmental Impact

The environmental impact of SF_6 is depending on its release to the atmosphere. The less the gas losses from GIS equipment the less the environmental impact. Specific applications shall be evaluated and/or compared using the life cycle assessment (LCA) approach, as regulated by ISO 14040 [60]. See also chapter 2.8.6, Life Cycle Assessment Case Study: Würzburg, were the overall advantages over the live time of GIS comparing with AIS is documented.

It is also necessary for a low environmental impact to follow the instructions given for SF_6 gas handling as explained in section 2.8.4 SF_6 Gas Handling.

The GWP of SF_6 by itself is not adequate to measure the environmental impact of electric power equipment based on SF_6 technology.

2.8.2.14 Emissions

SF_6 must be used in a closed cycle. When gas removal from containment is needed, a proper handling procedure should be implemented to avoid any deliberate release into the atmosphere.

The yearly SF_6 emission rate from the overall electric industry represents 0.1% of the yearly emission rate of man-made global warming gases. As just one example, emissions from European manufacturers and users contribute only 0.008% [51, 56, 58].

2.8.2.15 Standards and Guides for Related Equipment

The main applications in electric power equipment utilizing SF_6 are defined by the current IEEE Standards C37.122 [23] and C37.122.1 [25] for HV GIS, C37.122.2 [61] for MV GIS, and C37.100.1

[62] Common Clauses; in IEC Standards 62271-200 [30] for MV GIS and 62271-203 [27] for HV GIS and 62271-1 [28] for Common Clauses; for switching equipment in IEEE several standards of the C37 series and in IEC 62271-100 [9] for circuit breakers and 62271-102 [32] for disconnectors.

There have been efforts for harmonization of IEEE and IEC standards so that the content is very close with minor differences.

2.8.2.16 Tightness

The tightness of certain old installed gas-insulated power equipment, especially for HV systems, could be a significant issue for the environment due to a higher leak rate. Nevertheless, it has to be kept in mind that handling SF_6 during installation, on-site testing, and maintenance activities may contribute significantly to the overall emissions and steps should always be taken to minimize these.

In order to achieve very low leak rates, the quality of the encapsulation, including the materials used, the machining process, the design of gaskets, the sealing material itself, and the factory testing procedures are of major importance. In order to achieve very low handling losses during gas handling, it is important to consider smaller gas compartments, reduced maintenance frequency, more sophisticated tools and instruments to handle and to check the gas quality, and specific training of designated personnel.

2.8.2.17 Closed Pressure Systems

In practice, on-site SF_6 handling is already minimized, as it is normally only required for installation, extension, and/or end-of-life disposal/dismantling of equipment. It is recommended that:

- The leakage rate is kept lower than 0.5% per year, per gas compartment.
- The typical time between two consecutive maintenance works is up to 25 years.
- The SF_6 conditions are checked only after a filling operation.
- Appropriate record-keeping procedures are used.

Today, closed pressure systems are used for high-voltage GIS.

2.8.2.18 Sealed Pressure System

A sealed system uses a volume for which no further gas or vacuum processing is required during its expected operating life. Sealed pressure systems are completely assembled and tested in the factory.

SF_6 is handled only twice – for gas filling at the beginning and for gas recovery at the end after 40 years ("sealed for life"). Today, a typical leakage rate is lower than 0.1% p.a. per gas compartment.

Today, sealed pressure systems are used for medium-voltage GIS.

2.8.2.19 Controlled Pressure Systems

A volume can be automatically replenished from an external or internal gas source. The volume may consist of several permanently connected gas-filled compartments.

Controlled pressure systems are no longer used in new equipment, because of their high leakage rate. It is recommended that controlled pressure systems in old equipment are replaced by closed pressure systems, because of the unacceptable leakage rate (to limit the emissions of SF_6 and the contribution to the greenhouse effect).

2.8.2.20 Monitoring System

It is required that the gas pressure/density of each compartment is monitored whenever technically reasonable, to enable early detection of small leaks. State-of-the-art monitoring systems continuously monitor gas pressure/density, allowing for early detection of small leaks. In addition, appropriate corrective measures to locate and eliminate the leak should be immediately arranged.

2.8.2.21 Toxicity

Pure SF_6 is not toxic. However, toxic gaseous and/or solid decomposition products may be produced during high temperature arcing or electrical discharges in gas-insulated electric equipment. These are fully described in a previous CIGRE document [51] and also in IEC Technical Report 62271-303 [62], which will be transferred to International Standard 62271-4 [51].

Design rules and operational procedures are implemented to handle both the gas and the equipment according to safety regulations to eliminate any potential harmful effects.

2.8.2.22 Gas Categories

Sulfur hexafluoride gas might contain contaminants. These originate from the industrial manufacturing process as well as from use of the gas in electric power equipment. Depending on the nature and the amount of the contaminants, the following gas categories have been defined:

- New gas
- Technical grade SF_6
- Nonarced gas
- Normally arced gas
- Heavily arced gas

2.8.2.23 New SF_6 Gas

New gas is made new in a factory that has not been used before (see Table 2.11).

2.8.2.24 Technical Grade SF_6 Gas

Technical grade gas has been used and was cleaned on-site or in a factory to reach the required values (see Table 2.12).

2.8.2.25 Nonarced Gas

This is gas that has been used or handled in any way and has not experienced arcing. In practice, if the volume concentration of the indicator gases $SO_2 + SOF_2$ is lower than 100 ppmv, then the gas is nonarced. Nonarced gas is to be expected from:

- Insulation testing in the factory
- Insulation testing on-site during erection/commissioning
- Routine maintenance of insulation compartments

Table 2.11 Maximum acceptable impurity levels for new gas are given in (IEC 60376 ed. 2 [64])

Impurity	Specification
Air	0.05% wt.
CF_4	0.05% wt.
H_2O	15 ppmw
Mineral oil	See note
Total acidity expressed in HF	0.3 ppmw
Hydrolysable fluorides, expressed as HF	1.0 ppmw

Note: SF_6 should be substantially free from oil. The maximum permitted concentration of oil and the method of measurement are under consideration.

Table 2.12 Maximum acceptable impurity levels for technical grade SF_6 (IEC 60376 ed. 2 [64])

Impurity	Specification
Air	0.2% wt. (note 1)
CF_4	2400 ppmw (note 2)
H_2O	25 ppmw (note 3)
Mineral oil	10 ppmw
Total acidity expressed in HF	1 ppmw (note 4)

Note 1: 0.2% wt. is equivalent to 1% volume under ambient conditions (100 kPa and 20 °C).
Note 2: 2400 ppmw is equivalent to 4000 ppmv under ambient conditions (100 kPa and 20 °C).
Note 3: 25 mg/kg (25 ppmw) is equivalent to 200 ppmv (200 μl/l) and to a dew point of 36 °C, measured under ambient conditions (100 kPa and 20 °C).
Note 4: 1 ppmw is equivalent to 6 ppmv measured under ambient conditions (100 kPa and 20 °C).

- Repair of insulation compartments after malfunction without arcing
- Retrofitting of insulation compartments
- Decommissioning of insulation compartments in which arcing has not occurred
- Any kind of compartment after filling prior to energizing

2.8.2.26 Normally Arced Gas

This is gas recovered from switchgear compartments after normal switching operations. In practice, if the volume concentration of the indicator gases $SO_2 + SOF_2$ is between 100 ppmv and 1%, then the gas is normally arced. Normally arced gas is to be expected from:

- Maintenance and repair of switching devices after normal (load or fault) operation
- Interruption testing during switchgear development
- Decommissioning of switchgear

2.8.2.27 Heavily Arced Gas

This is gas recovered from equipment in which failure arcing has occurred. In practice, if the volume concentration of the indicator gases $SO_2 + SOF_2$ is greater than 1%, then the gas is heavily arced. Heavily arced gas is to be expected from:

- Circuit breakers after interruption failure
- Insulation compartments after internal arcing failure
- Any kind of arcing failure

2.8.2.28 Reuse of SF_6

SF_6 gas, arced or not, can be reconditioned and reused in equipment provide that the specifications listed in Table 2.13 are met.

2.8.2.29 Not Suited for Reuse

This is used SF_6 gas not complying with a standard for used gas such as IEC 60480 [53]. This gas requires further treatment, usually off-site and/or eventually final disposal.

2.8.2.30 General Safety Rules and Recommendations (see Table 2.14)

Table 2.13 Maximum acceptable impurity levels for reuse of SF_6 with a low range of use pressures (IEC 60480) [42]

Impurity	Specification
Air and/or CF_4	3% volume (note 1)
H_2O	95 ppmw (notes 2 and 3)
Mineral oil	10 ppmw (note 4)
Total reactive gaseous decomposition products	50 μl/l total or 12 μl/l for ($SO_2 + SOF_2$) or 25 μl/l HF

Note 1: In the case of SF_6 mixtures, the equipment manufacturer should specify the levels for these gases.
Note 2: Converted to ppmv, these levels should also apply to mixtures until a suitable standard becomes available.
Note 3: 95 mg/kg (95 ppmw) is equivalent to 750 ppmv (750 μl/l) and to a dew point of 23 °C, measured at 100 kPa and 20 °C.
Note 4: If gas-handling equipment (pump, compressor) containing oil is used, it may be necessary to measure the oil content of the SF_6. If all equipment in contact with the SF_6 is oil-free, then it is not necessary to measure the oil content.

Table 2.14 General measures when working with SF_6 switchgear [51, 52]

Item	Work in the vicinity of switchgear (operation of SF_6 switchgear, visual check, room cleaning)	Filling, recovering, evacuation of SF_6 gas compartments	Opening of SF_6 gas compartments, work on open compartments
SF_6 material safety data sheet/ operational manuals		Mandatory	Mandatory
Training	Mandatory (note)	Mandatory	Mandatory
Gas-handling equipment cleaning/ neutralizing equipment		Mandatory	Mandatory
Personal protection equipment Flames		Not allowed	Mandatory, not allowed
Welding/smoking drinking/eating		Not allowed	Not allowed, not allowed

Note: General information must be specified according to the type of work and installation.

2.8.2.31 Protection of Personnel (see Table 2.15)

2.8.2.32 Training of Personnel Handling SF_6 is Required

- Physical/chemical/environmental characteristics of SF_6
- Application of SF_6, used in electric power equipment (insulation, arc quenching)
- Standards
- Personnel safety: asphyxiation, contamination, and gaseous and solid decomposition products
- Environmental impact
- Disposal of SF_6 and its gaseous and/or solid decomposition products
- Knowledge about gas-handling procedures (filling, recovery)
- Regulations

2.8.2.33 Storage and Transportation

With respect to storage and transportation, five gas categories need to be distinguished:

- New gas or technical grade SF_6, complying with IEC 60376 [64]
- SF_6 suited for reuse in electric power equipment, complying with IEC 60480 [53]
- SF_6 not suited for reuse containing no toxic or corrosive products, not complying with IEC 60480 [53]
- SF_6 not suited for reuse containing toxic gaseous decomposition products
- SF_6 not suited for reuse containing both toxic and corrosive gaseous decomposition products, containing CF_4 (carbon tetrafluoride) and/or air and/or nitrogen

Table 2.15 Safety at work when accessing/entering gas compartments in electric power equipment utilizing SF_6 [51, 52]

Item	Open compartment before first SF_6 filling	Open compartment which contained non-arced SF_6	Open compartment which contained either normally arced or heavily arced SF_6
Potential risk	• Fumes of cleaning material • O_2 starvation • Remaining SF_6 or other gas from production process	• Fumes of cleaning material • O_2 starvation • Remaining gas	• Fumes of cleaning material • O_2 starvation • Remaining gas • Residual reactive gaseous decomposition products • Switching dust and adsorbers
Safety precaution	• Ventilation • Measurement of O_2 concentration when entering	• Ventilation • Measurement of O_2 concentration when entering	• Removal of switching dust and adsorbers • Ventilation • Measurement of O_2 concentration when entering • Wear personal protective equipment
Safety equipment and tools	• Suction ventilator or vacuum cleaner • O_2 concentration measuring device	• Suction ventilator or vacuum cleaner • O_2 concentration measuring device	• Suction ventilator or vacuum cleaner • O_2 concentration measuring device • Single-use protective clothes, shoe covers, hair cap • Acid proof safety gloves • Full face mask (preferred) or, at least, breathing protective mask • Protective goggles

2.8.2.34 Methods for Storage of SF_6 (see Table 2.16)

2.8.2.35 Containers for Transportation of SF_6 (see Table 2.17)

Table 2.16 Methods for storage of SF_6 [51, 52]

Method	Requirements	Features
Gaseous	Typical pressure lower than 2 MPa. Gas remains in the gaseous state	Requires a relatively small recovery pressure differential (typically 100:1) but needs larger storage volumes. Gas cannot be liquefied in cylinders for transportation. Therefore, it is limited to small quantities (200 kg) and stationary use
Liquid: cooling assisted	Typical pressure equal to 3 MPa. Employs additional cooling system to cool SF_6 after compression, which allows SF_6 to be stored in liquid form	Requires a relatively small recovery pressure differential (700:1) but needs a cooling aggregate. Performance of the cooling aggregate can influence processing speed. Additional maintenance requirements. Limited storage volume required and generally not suitable for transportation
Liquid: pressure only	Typical pressure equal to 5 MPa. Gas compressed to 5 MPa liquefies by pressure only	Requires a recovery differential of 1000:1 but eliminates the need for additional aggregates. Can be used with any storage vessel rated 5 MPa or higher

Table 2.17 Containers for transportation [51, 52]

(a)

Gas category	Container type	Container labeling
New gas or technical grade SF_6	**Suitable for liquefied gas up to a pressure of 7 MPa** **Note:** The filling factor for new gas is up to 1.04 kg/liter. **Recommendation:** Containers should be marked with a green label or the container should be painted green according to DIN EN 1089-3	**Stenciled on container:** UN 1080, sulfur hexafluoride **Danger label 2.2**
SF_6 suited for reuse	**Same type of container as for new or technical grade SF_6** **Note:** Due to the inert gas content (N_2, O_2, etc.), the filling factor is smaller than 0.8 kg/liter (note 1). **Recommendation:** Containers should be specially colored to avoid confusion between used and new gas (an orange band on the upper third of the cylinder is suggested)	**Stenciled on container:** UN 3163, sulfur hexafluoride, carbon tetrafluoride or air or nitrogen (note 2) **Danger label 2.2**

(b)

Gas category	Container type	Container labeling
SF_6 not suited for reuse and containing neither toxic nor corrosive gaseous decomposition products	Same as for SF_6 suited for reuse	**Stenciled on container:** UN 3162, sulfur hexafluoride, carbon tetrafluoride or air or nitrogen (note 2) **Danger label 2.2**

Table 2.17 (Continued)

SF_6 not suited for reuse and containing toxic gaseous decomposition products	Same as for SF_6 suited for reuse	**Stenciled on container:** UN 3162, sulfur hexafluoride, hydrogen fluoride, thionyl fluoride (note 2) **Danger label 2.3**
SF_6 not suited for reuse and containing both toxic and corrosive gaseous decomposition products	**Special containers approved for storing and transportation of corrosive gases** (such as hydrofluoric acid and HCl) with a corrosion-proof valve and adapter	**Stenciled on container:** UN 3308, sulfur hexafluoride, hydrogen fluoride, thionyl fluoride **Danger labels 2.3 + 8**

Note 1: The filling factor is the weight of SF_6 contained in the container divided by the container volume and is usually specified in kg/liter.
Note 2: Only the two most abundant contaminants need to be specified.

2.8.3 Producer and User of SF_6

SF_6 is an artificial gas originally developed for the chemical industry as a nonreactive gas to be used as an oxidation stopper. SF_6 today is produced in industrial countries of North America, Europe, Russia, and China and Japan in Asia.

There are SF_6 users in many industries such as electric utilities, original equipment manufacturers, magnesium industry, electronic industry, and others (a small percentage) such as in car tires, noise-reduced windows, and sports shoes (see Figure 2.51).

With the listing of SF_6 as a global warming gas in the Kyoto Protocol most of the industry has replaced SF_6 with other gases. However, for the high-voltage electric energy industry, no other gas is presently available to replace SF_6 without increasing the physical size of equipment by a factor of ten. In some cases, other power transmission equipment would need to be reduced in a way not possible with modern infrastructure requirements. Modern requirements often force the reduction in the size and volume of the equipment and with this the need of SF_6 to be used in the equipment. In Figure 2.52, the reduction of the use of SF_6 in GIS of different voltage levels is shown. Comparing the GIS equipment of the year 1980 with the GIS equipment of the year 2005, in the lower voltage ranges of 145 kV, the reduction is 68%. In the middle voltage ranges of 245 kV the reduction is 77% and in the higher-voltage ranges of 420–550 kV the reduction is 67%.

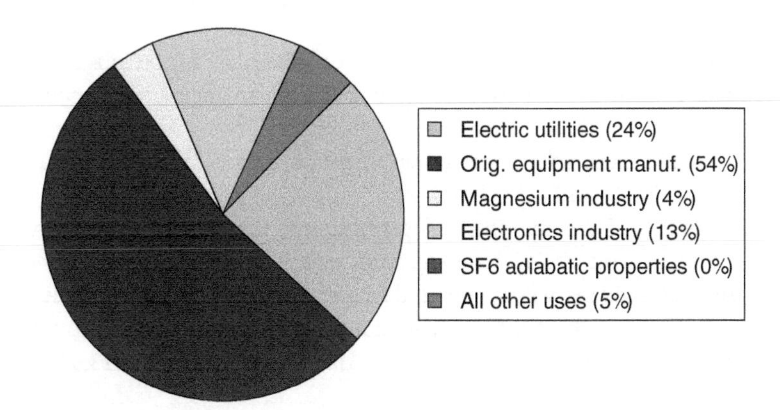

Figure 2.51 SF_6 sales by end use of 2003 (Reproduced by permission of Siemens AG)

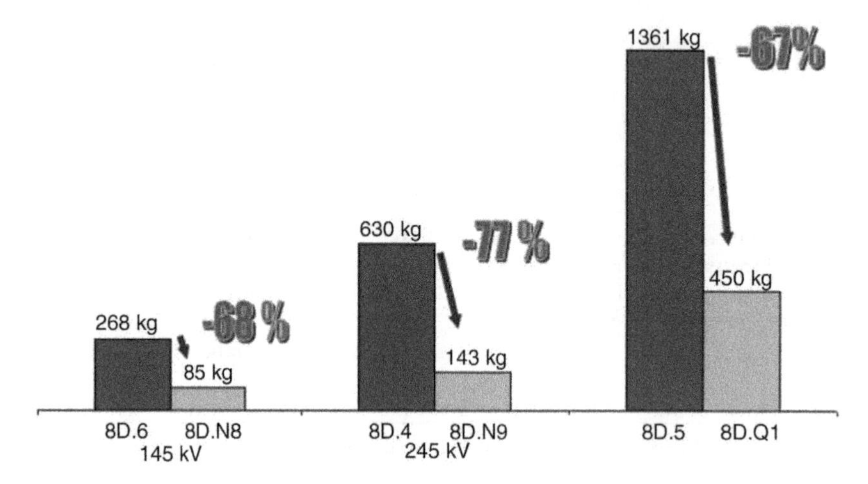

Figure 2.52 Reduction of the use of SF_6 (Reproduced by permission of Siemens AG)

The largest user of SF_6 is the original equipment manufacturer with more than half of the used SF_6 in 2003. Original equipment manufacturers use SF_6 during the development and testing in the factory. SF_6 is filled into the equipment and then high-voltage tests are made. At the end of the test, the SF_6 gas is then recovered and stored in gas compartments for reuse in the next tests to avoid a release to the atmosphere. The largest volume of SF_6 is filled into the GIS when the equipment is installed on-site and commissioned for operation. This volume of SF_6 is filled in only once into the GIS equipment and stays inside until decommissioning at the end of the lifetime of the equipment. On decommissioning the SF_6 is recovered and stored in gas compartments, cleaned, and then reused in other equipment. No release to the atmosphere occurs.

The use of SF_6 in GIS equipment shows some gas leakages. The international standards IEC 62271-203 and IEEE C37.122 [23, 27] allow gas leakages of a maximum of 0.5% per year for each gas compartment. This means that for the expected lifetime of a GIS of 40 years, no refill of the GIS is necessary (assuming that 80% of the initial fill is acceptable for operation). The long-time experience of major GIS producers shows real gas leakages in the range of only 0.1–0.2%.

There are rules for maintenance work when the opening of a gas compartment is needed. International standards give guidance to minimize the release of SF_6 to the atmosphere [23, 27].

The total amount of stored SF_6 in high-voltage equipment is increasing with the use of this technology, as shown in Table 2.18.

The stored SF_6 in GIS world-wide is estimated to be 27 000 tons in 1995, 30 000 tons in 1999, and 45 000 tons in 2010. This development shows an increase of about 1000 tons/year from 1995 to 2010.

To relate these volumes to other greenhouse gases based on CO_2 equivalents some values are shown in Table 2.19.

Table 2.19 shows that in the contribution to the greenhouse effect on the basis of CO_2 equivalents is CO_2 with 60% of the share, followed by CH and CF gases with 40%. The share of all fluoride gases including SF_6 with a potential factor of 22 500 is 0.16% and the use in the high-voltage energy sector is below 0.1%. With the reduction of building size and improved SF_6 handling the contribution of SF_6 used in electrical equipment in 2010 is only 0.05% [23, 27].

Table 2.18 Stored SF_6 in GIS [51, 52]

Stored SF_6 in GIS	
Until 1995	27 000 tons
Until 1999	30 000 tons
Until 2010	45 000 tons
Increase of SF_6 use	
Rate of increase in 1995	2000 tons/year
Rate of increase in 2005	1200 tons/year

Table 2.19 Man-made greenhouse emissions in 1999 [51, 52]

Greenhouse gas	(tons/year)	($Gtons_{eq}$/year)	(%)
CO_2		26	60
CH_4, N_2O, CFCs, etc.		16	40
CF and SF_6		0.07	0.16
Total emissions		**~40**	**100**
"Electrical" SF_6	2200	0.05	0.1
"Electrical" SF_6 in 2010	1000	0.02	0.05

2.8.4 SF_6 Gas Handling

2.8.4.1 General

The unique ability of SF_6 for excellent insulation and arc extinguishing properties makes it indispensable in high-voltage electric power equipment. On the other hand, the high global warming potential makes SF_6 a gas that should not be released to the atmosphere and gas losses must be minimized during the total lifetime of the products using SF_6.

Gas losses can occur from leaks dependent on the level of gas tightness of equipment and from all kinds of gas-handling tasks. The high gas tightness requirements of SF_6 products are defined in standards with a maximum allowed gas loss of less than 0.5% per year and per gas compartment (see Table 2.20). However, practical experience indicates that the actual measured gas losses from modern, state-of-the-art equipment exceeds the standard and is in the range of 0.1–0.3% per gas compartment per year.

The gas-handling requirements are now regulated in international standards such as IEEE C37.122.3 [23] and IEC 62271-4 [51], which will be explained in the following subsections.

2.8.4.2 Introduction

The goal of gas-handling procedures is the minimization of gas losses. Practical recommendations and instructions for customized SF_6 handling are formulated in standards and guides [23, 27]. Standardizing information and procedures for all steps of the closed loop SF_6 handling processes helps to avoid any unnecessary gas release to the atmosphere over the lifetime of the product. The processes of SF_6 handling, explained and defined in standards, are shown in Figure 2.53.

The standards and guides [23, 27] cover all the required background information to understand the handling procedures defined. It is recommended that the standards and guides [23, 27] should

Table 2.20 Maximum acceptable impurity levels for technical grade SF_6 (IEC 60376 ed. 2) [64]

Commissioning or recommissioning
Topping-up
Refilling
Checking gas quality on-site
Sampling and shipment for off-site gas analysis
Recovery and reclaiming
Recovery and reclaiming at the end-of-life when the electric power equipment is dismantled

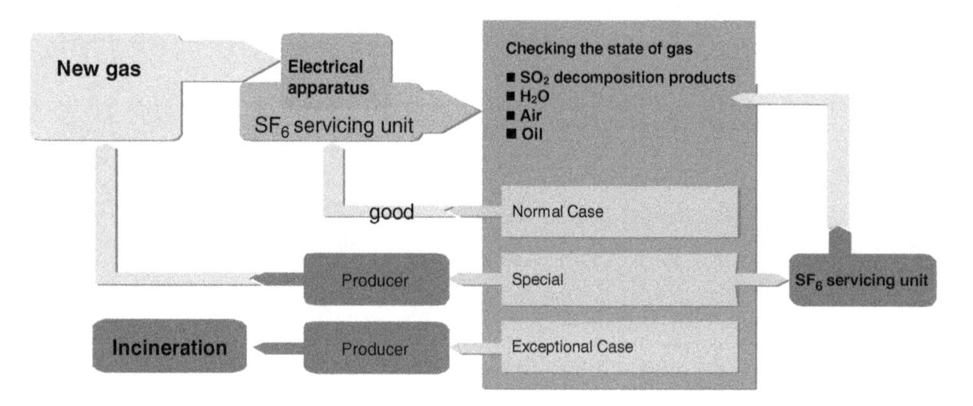

Figure 2.53 SF_6 handling and recovering processes (Reproduced by permission of CIGRE)

Table 2.21 Goals for the SF_6-handling process

Safe operation of the equipment
Optimization of resources and tools required
Minimization of out-of-service time for equipment
Standard training of personnel handling SF_6
Reduction of the amount of gas released during handling operations down to the functional physical limit
Avoidance of any deliberate release, for example, flushing to the atmosphere
Minimizing SF_6 losses and emissions during commissioning, service, and operation.

be followed in order to achieve operational safety at work and to address environmental issues. The objectives of this approach of using the standards and guides [23, 27] are shown in Table 2.21.

2.8.4.3 Recovery/Reuse of SF_6

The industry has developed an SF_6 full cycle reuse program that covers the normal case of recycling as well as some special cases, that is, after an internal arc fault. The process is shown in an overview in Figure 2.53.

Usually, new SF_6 gas is filled into the gas compartments of new GIS. In the case of SF_6 recovery (i.e., SF_6 removal from the GIS) the SF_6 gas will be transferred from the GIS gas compartment to the SF_6 servicing unit (gas recovery cart) or another storage tank. The condition of the SF_6 stored in the cart or storage container gas should be checked for decomposition products, such as SO_2, humidity (H_2O), air, and oil.

In the special case when the contaminants in the SF_6 gas exceed the limit values for used gas as defined in standard IEC 60480 [53], the contaminated SF_6 should be sent to the SF_6 producer for reconditioning if the SF_6 servicing unit does not have this capability. The clean SF_6 will then be brought back into the cycle as new SF_6. In cases where an appropriate SF_6 servicing unit for SF_6 handling is available on-site, the gas does not need to be sent to the SF_6 producer but can be reconditioned locally by the servicing unit. In the exceptional case that SF_6 gas is strongly contaminated and cannot be cleaned adequately, the SF_6 gas can be destroyed by burning in a very high temperature incineration process. The incineration process is carried out by the gas manufacturer.

The three cases of SF_6 handling can be defined as the normal case, special case, and exceptional case, as shown in Table 2.22.

A detailed explanation of the gas-handling process for normal and special case SF_6 is shown in Figure 2.54. The graphic shows three horizontal dotted lines. The lower is for 1 mbar or 10 kPa vacuum, the second lowest is for atmospheric pressure, and the upper horizontal-dotted line is for the filling pressure of the SF_6 compartment. The graphic is to be read from left to right. The first step (at the top left) is to prepare the gas-handling equipment. This step is intended to get the gas-handling device into operation and connect it to the GIS gas compartment. The next step is to connect filters for gas cleaning for gas released from the GIS. The gas recovering starts by releasing the gas pressure in the GIS compartment, through the filters of the gas-handling device into the

Table 2.22 Three cases of SF_6 handling [51, 52]

Normal case	New gas After normal operation
Special case	Normally aged gases in circuit breaker compartments
Exceptional case	Heavily arced gas Moisture content Acidity

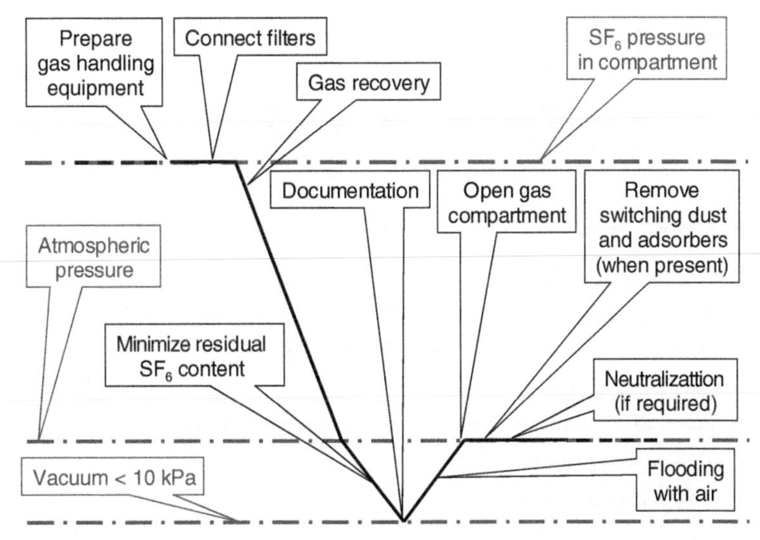

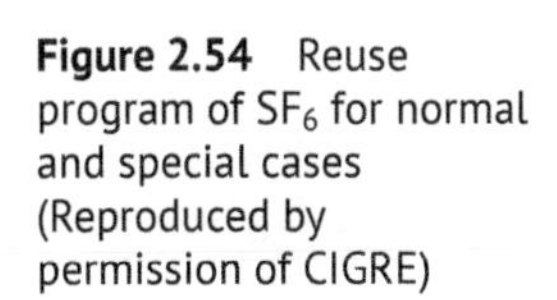

Figure 2.54 Reuse program of SF_6 for normal and special cases (Reproduced by permission of CIGRE)

storage containers. When the pressure equilibrium is reached a vacuum pump is used to further reduce the gas in the GIS compartment to a minimum residual SF_6 content. When 1 mbar (vacuum pressure) has been reached, this event should be noted and documented.

The GIS gas compartment is then opened to the ambient atmosphere and filled with air until atmospheric pressure is reached in the GIS compartment. The GIS compartment is opened and dust and humidity absorbers, if they exist in the compartment, are removed. If necessary, the gas compartment is cleaned and neutralized to deal with residual SF_6 decomposition by-products. In this case it is necessary to use special tools and to provide protective apparatus for personnel [51, 54]. Maintenance work can be carried out on the GIS gas compartment. The gas-handling personnel should be properly trained and certified in order that all these detailed working steps are followed to avoid SF_6 release to the atmosphere as much as possible.

The reuse program may involve an SF_6 producer. In these cases, gas quality analysis could be included by the SF_6 producer (e.g., Solvay). This process is shown in the closed loop handling symbol of Figure 2.55. The analysis covers the steps listed in Table 2.23.

The SF_6 producer will send a sample list when contacted to check the SF_6 gas quality on-site. Based on this analysis, the SF_6 gas needs to be determined as being recyclable and suitable for reclaiming. The reclaimed SF_6 is sent to the producer and will be stored until a sufficient SF_6 quantity is available to process the contaminated SF_6 in the gas production process. In this process used, gas is converted into virgin SF_6 and brought back into the SF_6 use cycle.

As an alternative to sending the SF_6 gas to the gas producer it can be processed and purified on-site, if the gas-handling devices are available and if the gas is not too heavily contaminated. The devices used and the principle of this process are shown in Figure 2.56 for purifying the SF_6 on-site.

Figure 2.55 Closed loop handling symbol (Reproduced by permission of CIGRE)

Table 2.23 Steps of SF_6 reuse program involving an SF_6 producer [51, 52]

Steps
Contact SF_6 producer to receive sampling list
Gas is analyzed and determined suitable for reclaiming
Send used SF_6 to producer
Sufficient SF_6 quantity is collected to bring to production facility
Used SF_6 to be processed to virgin SF_6
Virgin gas to be used in GIS

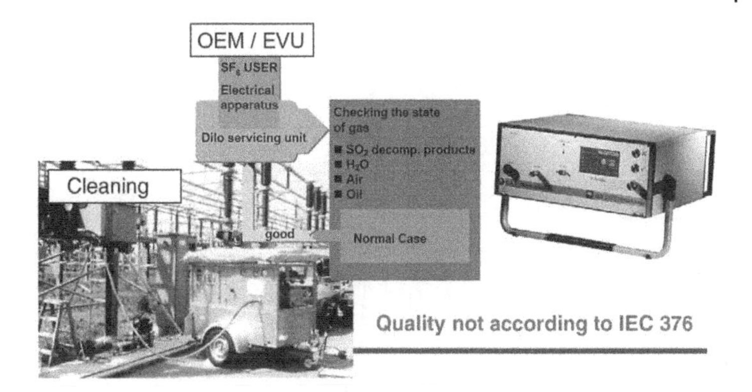

Figure 2.56 SF_6 purifying on-site (Reproduced by permission of Siemens)

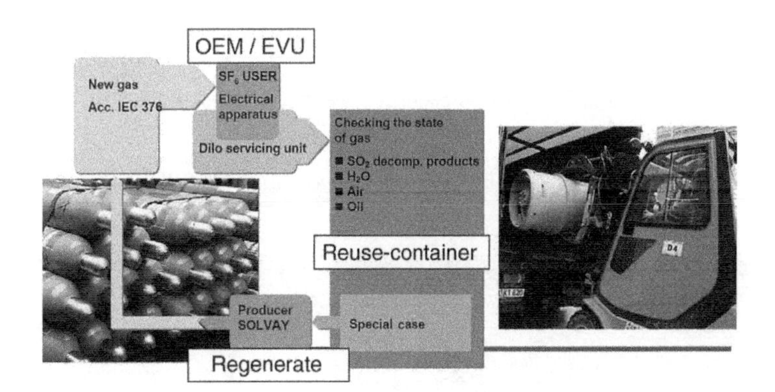

Figure 2.57 SF_6 recycling to ASTM D2472 [65] new gas (Reproduced by permission of Solvay)

The service unit for gas handling is connected to the SF_6 gas compartment on-site, such as a high-voltage circuit breaker. The quality check is then made and the gas quality can only reach the limiting values of used SF_6 gas. To achieve the quality of new SF_6 gas, treatment in the facility of the SF_6 producer is needed, as shown in Figure 2.57. To reach the quality of virgin SF_6 gas, the used gas is stored on-site in transport containers and is brought to the producer's facility for cleaning and recovery to restore the specifications required in the standards.

In the case of exceptionally contaminated gases, for example, after heavy arc switching or internal arc faults, the reuse and cleaning of SF_6 might not be possible and the gas needs to be destroyed in a thermal process. The steps for recovery and reclaiming are shown in Figure 2.58. The first step is to prepare the gas-handling equipment and transfer the SF_6 gas into a storage container for contaminated gases. After atmospheric pressure is reached in the GIS gas compartment, a vacuum pump is used to a vacuum of 1 mbar to reclaim most of the SF_6. The vacuum value is documented. The GIS gas compartment is then flooded with air. To give time for the dust inside the GIS gas compartment to settle, a minimum waiting time of 1 hour is needed before the compartment is physically opened. Remaining dust and absorbers are removed and the compartment is neutralized for cleaning. The various steps should be documented with photos and explanatory text for a report.

To transport nonrecyclable SF_6, special containers are used to avoid contamination with new SF_6 gas. The SF_6 is then incinerated at very high temperatures in special facilities to destroy the SF_6 molecule (see Figure 2.59).

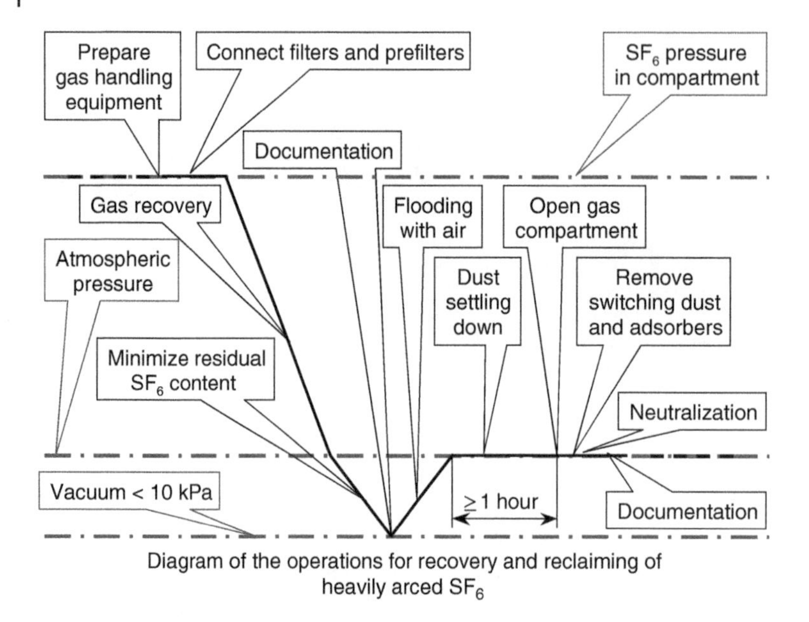

Figure 2.58 Reuse program – exceptional case (Reproduced by permission of CIGRE)

Figure 2.59 Incineration of SF_6 (Reproduced by permission of Solvay)

Gas-Handling Procedures

The following gas-handling procedures are defined for commissioning or recommissioning:

- Topping-up of prefilled compartments
- Refilling
- Measurement of the moisture content/dew point
- Measurement of the SF_6 content/quantity
- Measurement of the of gaseous decomposition products
- Gas sampling and shipment
- Recovery and reclaiming of non-arced and/or normally arced gases
- Recovery and reclaiming of heavily arced gases
- Recovery and reclaiming of SF_6 at the end-of-life disposal

Procedure Description Modules

The following figures explain diagrammatically the various gas-handling procedures for a vacuum of 300 Pa:

- Diagram of the operations for commissioning or recommissioning of SF_6 (see Figure 2.60)
- Diagram of the operations for topping-up of SF_6 prefilled compartments to the nominal pressure/density (see Figure 2.61)
- Diagram of the operations for SF_6 refilling to SF_6 to the nominal pressure/density of leaking compartments (see Figure 2.62)
- Diagram of the operations for the measurement of the moisture content/dew point of SF_6 on-site (see Figure 2.63)
- Diagram of the operations for the measurement of the SF_6 content/quantity of inert gases on-site (see Figure 2.64)
- Diagram of the operations for the measurement of the residual quantity of reactive gaseous decomposition products/residual acidity content on-site (see Figure 2.65)
- Diagram of the operations for gas sampling and shipment (see Figure 2.66)
- Diagram of the operations for recovery and reclaiming nonarced and/or normally arced SF_6 from compartments of controlled and/or closed pressure systems (see Figure 2.67)
- Diagram of the operations for recovering and reclaiming heavily arced SF_6 from compartments of controlled and/or closed pressure systems (see Figure 2.68)
- Diagram of the operations for recovery and reclaiming of SF_6 at the end-of-life disposal when the sealed pressure system is dismantled (see Figure 2.69)

2.8.4.4 Reporting and User Agreements

There can be several user agreements established for the closed loop usage of SF_6 with requirements for an active reporting system to track the SF_6 used. Governmental bodies, regulatory agencies, and international standardization organizations like IEC and IEEE offer rules and processes suitable for this process.

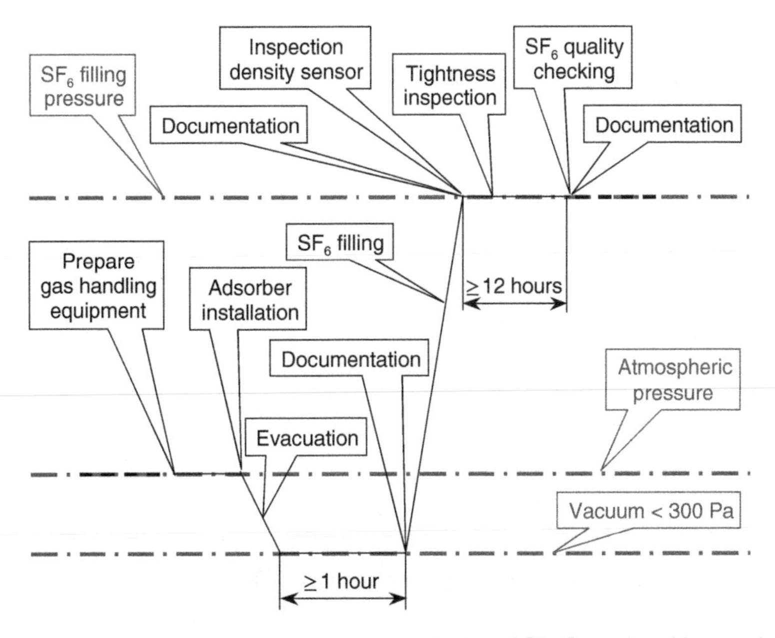

Figure 2.60 Commissioning or recommissioning of SF_6 (Reproduced by permission of CIGRE)

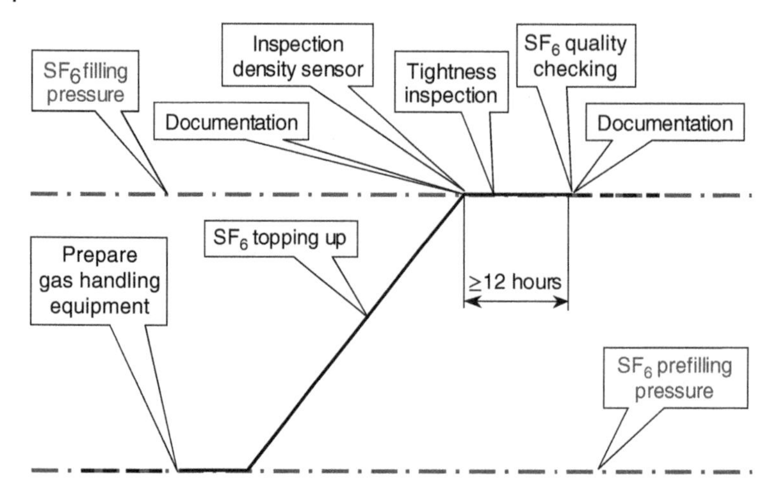

Figure 2.61 For topping-up of SF$_6$ (Reproduced by permission of CIGRE)

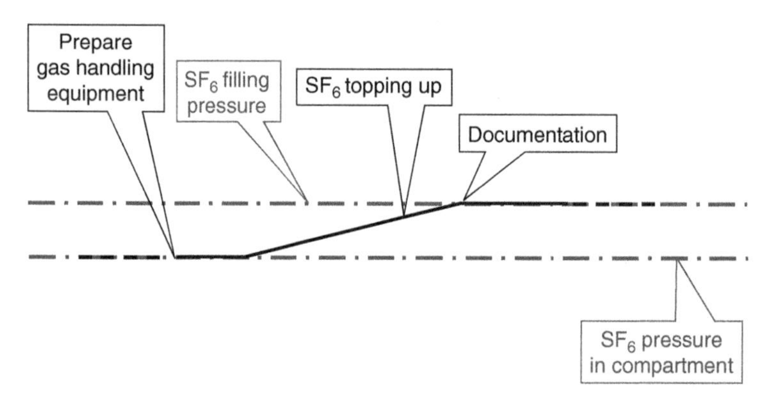

Figure 2.62 SF$_6$ refilling of SF$_6$ (Reproduced by permission of CIGRE)

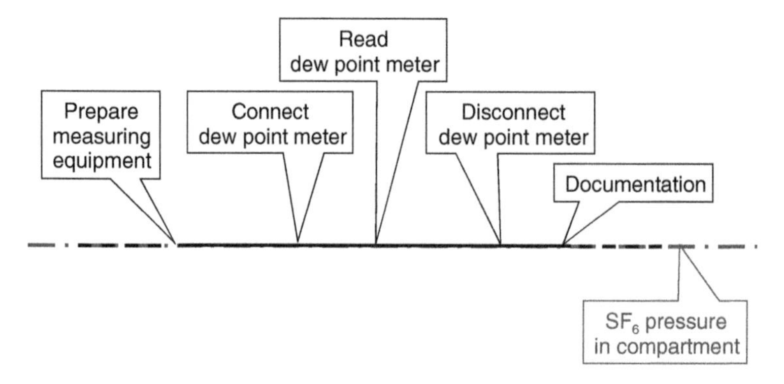

Figure 2.63 Moisture content/dew point of SF$_6$ (Reproduced by permission of CIGRE)

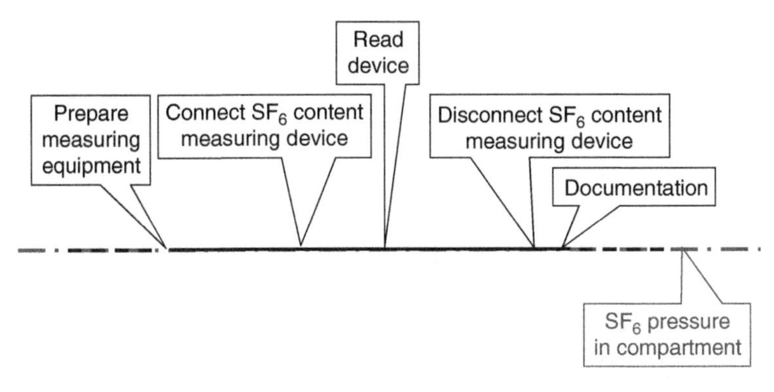

Figure 2.64 Measurement of SF_6 (Reproduced by permission of CIGRE)

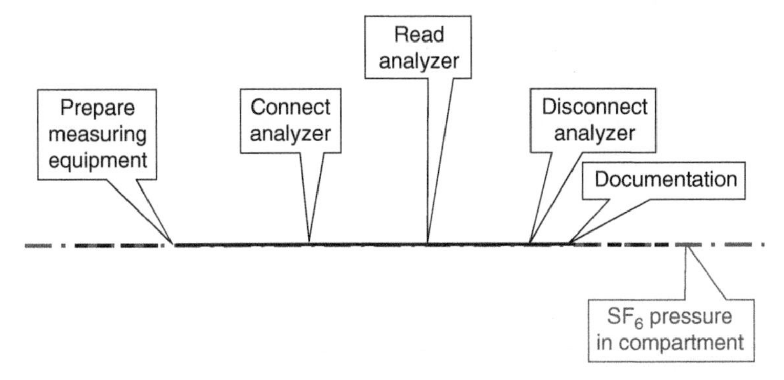

Figure 2.65 Quantity of reactive gaseous of SF_6 (Reproduced by permission of CIGRE)

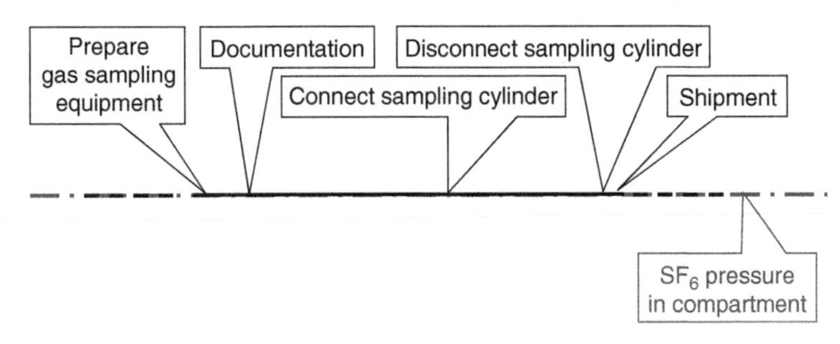

Figure 2.66 Sampling and shipment of SF_6 (Reproduced by permission of CIGRE)

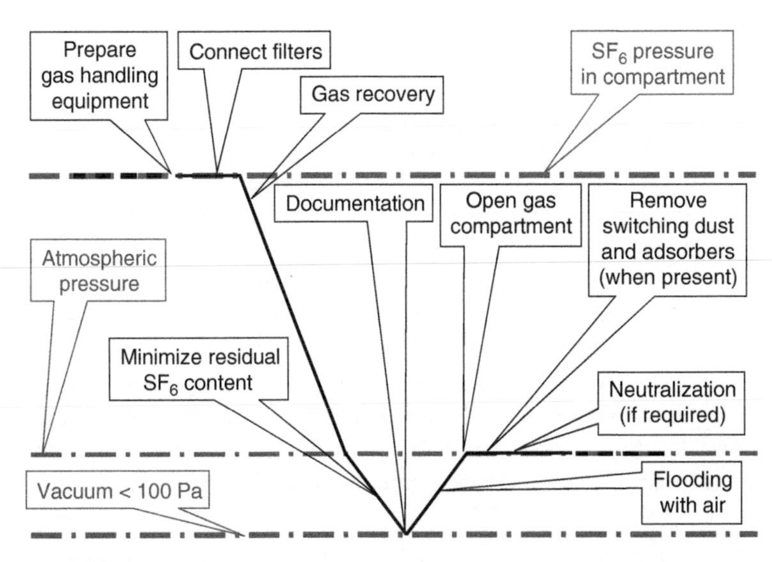

Figure 2.67 Recovery and reclaiming of nonarced and/or normally arced of SF_6 (Reproduced by permission of CIGRE)

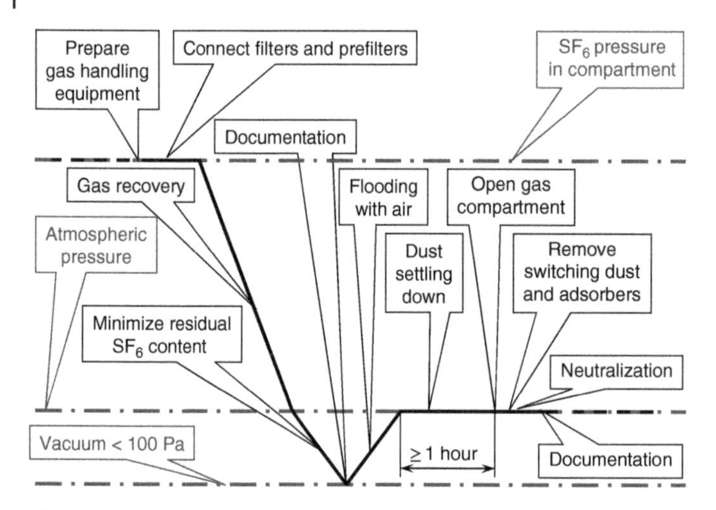

Figure 2.68 Heavily arced of SF_6 (Reproduced by permission of CIGRE)

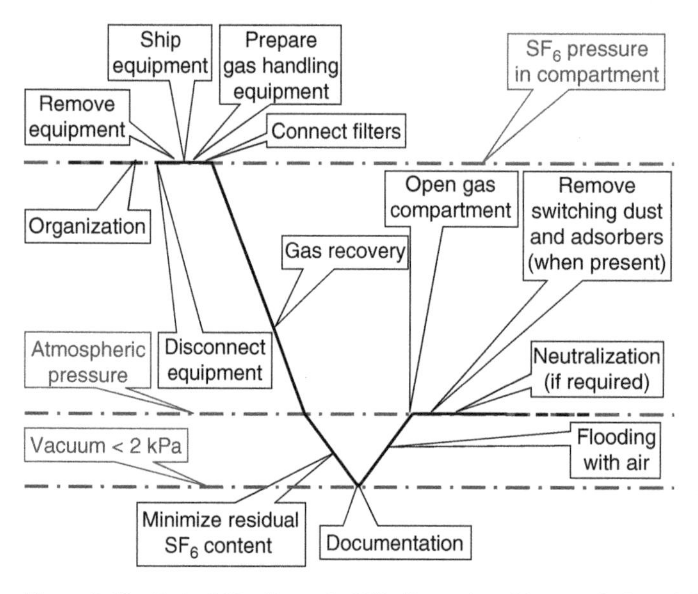

Figure 2.69 End-of-life disposal of SF_6 (Reproduced by permission of CIGRE)

On a global basis CIGRE publishes the "SF_6 Tightness Guide," which then was the basis for the IEC 62271-4 [51] and the IEEE C37.122.4 [66] for SF_6 handling. In Europe, the manufacturer organization T&D Europe published "SF_6 Handling Guidelines." In the United States, the Environmental Protection Association (EPA) published rules for SF_6 handling and reporting.

Participation in these programs has been voluntary with the goal of minimizing the emissions of SF_6 to the atmosphere. All aspects of the industry are involved, including operators of power transmission and distribution networks, manufacturers of electrical equipment with SF_6 for the power transmission, and distribution of above 1 kV and the supplier of SF_6. The "SF_6 Tightness Guide" gives a wide spectrum of information about SF_6, related standards, methods of gas tightness, and recommendations for handling. T&D Europe covers all national manufacturer associations and sets up a reporting system, the recovery process, and offers training to certify handling personnel for SF_6.

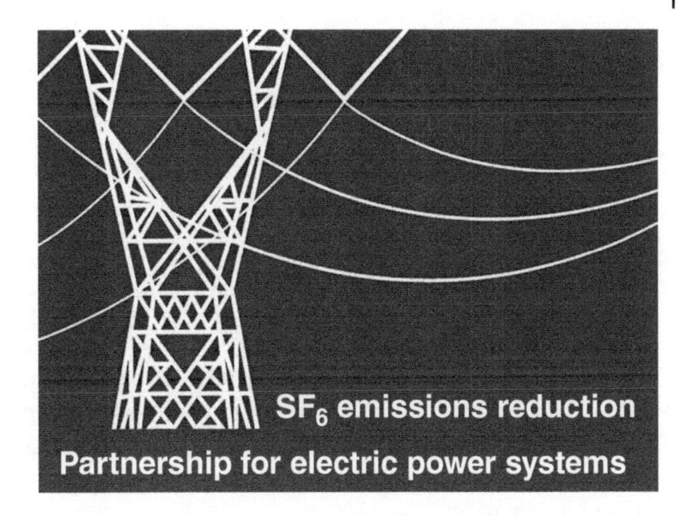

Figure 2.70 EPA Memorandum of Understanding (Reproduced by permission of IEEE)

The EPA has organized more than 80 SF_6 (status 2010) users in the United States committed in emission abatement strategies, based on both technical and economic feasibility to identify the best strategy to minimize the SF_6 emissions to the atmosphere for each participant. The commitment comes in the form of a memorandum of understanding (see Figure 2.70).

2.8.4.5 Labeling

The correct labeling of each gas compartment is an important part of good monitoring and control of SF_6 in use. In Figure 2.71, a label of a gas compartment is shown, which gives the information of how many kg are filled into a GIS gas compartment.

2.8.4.6 Training and Certification

The handling of SF_6 requires special knowledge by personnel. This knowledge needs to include theoretical aspects in addition to the practical use of the equipment. Manufacturers of GIS offer such training to personnel and many authorities require certificates before gas handling can be carried out on the equipment. In Figure 2.72, the essentials of responsible use of SF_6 and the certification of personnel are shown.

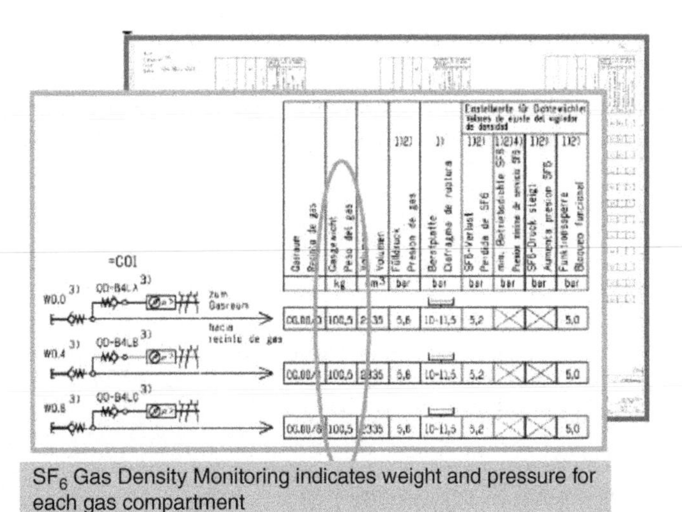

Figure 2.71 Gas-control scheme (e.g., with information about volume and pressure per compartment) (Reproduced by permission of Siemens AG)

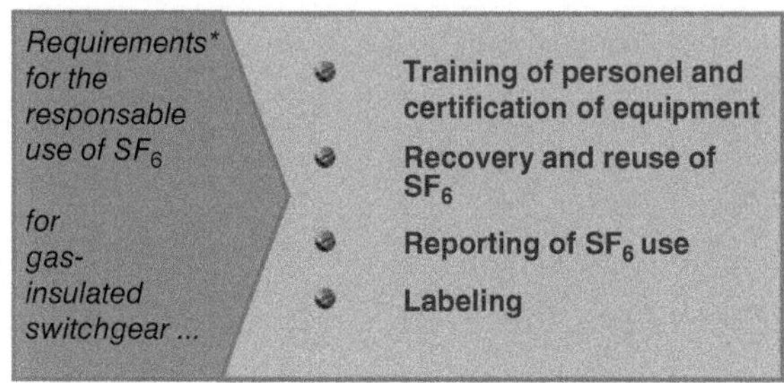

Figure 2.72 Training of SF_6 handling (Reproduced by permission of Siemens AG)

The training of personnel also includes the SF_6 recovery process and the preparation of SF_6 for reuse – cleaning and drying of the gas to conform to the purity requirements of reused SF_6. Documented protocols and procedures are required for the gas-handling process and a reporting system needs to be implemented. All gas compartments need to have clear labeling about the volume and weight of SF_6.

The use of SF_6 in GIS will require, for each user and operator, an action plan to reduce future SF_6 emissions and to identify the reduction potential, as shown in Figure 2.72.

The first step of such an action plan is to improve the SF_6 handling practice by following, for example, the IEC 62271-4 [51] gas-handling requirement by educated and trained personnel. If leakages are detected on the equipment, the leakage needs to be quantified by leakage detectors and precise measuring equipment. Leaks need to be repaired on-site when identified and when a repair is possible. If no repair can be carried out, the replacement of the equipment is required. Any SF_6 gas taken from the equipment needs to be reused and the gas quality needs to be checked by a gas analysis made by qualified methods. Finally, a correct disposal of SF_6 is needed when no reuse is possible on-site. For such a case, the gas can be sent to the SF_6 producer.

The normal case of SF_6 gas handling can be carried out by a gas-handling device, as shown in Figure 2.56. This gas-handling device offers measuring equipment to determine the gas quality and to remove SF_6 from the electrical enclosure down to a pressure of at least 20 mbar. The technically achievable gas pressure is 1 mbar and requires some hours of vacuum pump operation, depending on the size of the gas compartments. The gas-handling device also offers gas filters to clean the processed gas from particles and humidity. After reanalyzing the SF_6 gas stored in a connected gas compartment, the gas-handling device can be used to fill the SF_6 into the gas compartment of the GIS (see Table 2.24).

Example of SF_6 Handling

The gas-handling process is explained by means of an example of a three-phase circuit breaker gas compartment of 7000 l volume. The normal pressure of this gas compartment is 0.66 MPa. The total weight of SF_6 is about 200 kg, which represents 43 g/l, as shown in Table 2.25.

Table 2.24 Normal case of SF_6 handling [51, 52]

Determine gas quality
Remove SF_6 from GIS gas compartment down to 20 mbar (1 mbar possible)
Filter contaminants
Reanalyze
Reintroduce SF_6 to GI gas compartment

Table 2.25 Gas handling at a GIS circuit breaker gas compartment [51, 52]

Three-phase circuit breaker	
Volume	7000 l
Normal pressure	0.66 MPa
SF_6 weight	300 kg
SF_6 density	43 g/l

In Figure 2.64, the remaining SF_6 in the gas compartment is shown for 50, 20, and 1 mbar of evacuation pressure, respectively. The related SF_6 weight released to the atmosphere is 2 kg at 50 mbar, 1 kg at 20 mbar, and almost 0 at 1 mbar. The evacuation time for 50 or 20 mbar is in the range of some hours while the time for 1 mbar goes up to a day. This time needs to be planned for the handling process.

2.8.5 Gas-Handling Equipment

2.8.5.1 Gas Reclaimer

The appropriate type and size of the reclaimer should be chosen according to the gas quantity to be handled. Typical functions of a standard SF_6 reclaimer are as follows:

- Evacuation of air from the gas compartment
- Filling of SF_6 in the gas compartment
- Recovery of SF_6 from the gas compartment
- Storage and filtering of SF_6
- Flooding of the gas compartment with ambient air

For a functional scheme of a general purpose SF_6 reclaimer, see Figure 2.73. The vacuum pump shown in the graphic is only for evacuation of air prior to the gas filling.

2.8.5.2 Gas Filters

For typical filter types used during SF_6 reclamation, see Figure 2.74.

2.8.5.3 Compressors

- The main compression stage usually employs a piston-type compressor, which operates between a gas inlet pressure of about 100 kPa (typically higher than 50 kPa) and the pressure in the gas storage container.
- To reach the evacuation level of 2 kPa (20 mbar) inside the gas compartment a compressor of 0.1 kPa (1 mbar) is needed.
- Compressors need to be dry-running and hermetically sealed to avoid the possibility of SF_6 leaks and oil contamination.

2.8.5.4 Measuring Devices

For technical data, see Table 2.26.

For equipment for SF_6 measurement, see Figure 2.75. The on-site used SF_6 sniffing device is shown in Figure 2.76. The density meter is shown in Figure 2.77.

2.8.5.5 Evacuation and Filling

For equipment for evacuation, filling, and refilling, see Figures 2.78 and 2.79.

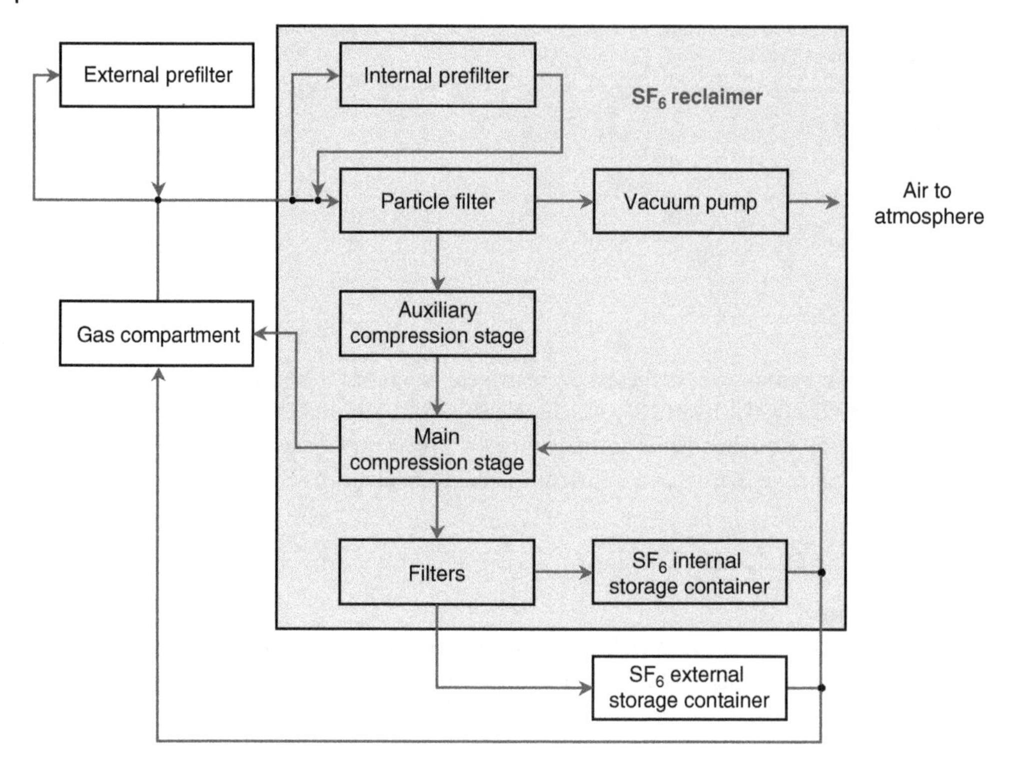

Figure 2.73 General purpose SF_6 reclaimer of SF_6 (Reproduced by permission of CIGRE)

Filter Type	Tasks	Major characteristics
Particle filter	Removes solid decomposition products and other particles at the reclaimer inlet.	Pore size 1 μm.
Gas/moisture filter	Removes reactive gaseous decomposition products and moisture.	Residual moisture lower than 100 ppmv. Residual SO_2+SOF_2 lower than 12 ppmv. Particle retention ability.
Oil filter	Removes oil when required.	Special filter utilising active charcoal.

Figure 2.74 Filter types of SF_6 (Reproduced by permission of CIGRE)

Table 2.26 Technical data for on-site SF_6 measuring devices [51, 52]

Device	Quantity	Range	Minimum accuracy
SF_6 pressure gage	Pressure	0–1 MPa	±10 kPa
Thermometer	Temperature	25–50 °C	±1 °C
Dew point meter	Moisture	Dew point: 50–0 °C	±2 °C
SF_6 content measuring device	SF_6/N_2, SF_6/air	0–100% by vol.	±2% vol.
Reaction tubes	Oil mist	1–25, 0.16–1.6 ppmv	±15%

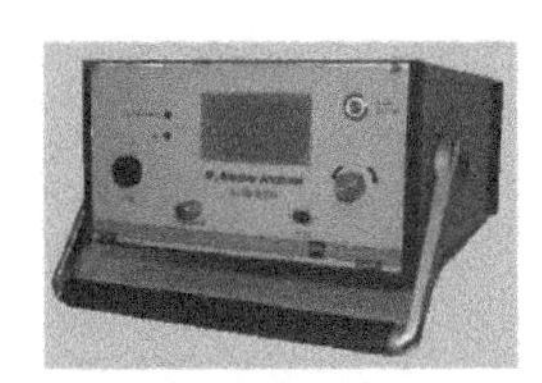

Figure 2.75 Equipment for SF_6 measurement of SF_6 (Reproduced by permission of Dilo)

SF_6-tightness confirmation for routine testing and after installating on site
Various sniffing devices available....

Infrared thermography cameras assisting leak location detection on site

... for the routine testing and on site

Figure 2.76 Sniffing device of SF_6 (Reproduced by permission of Siemens AG)

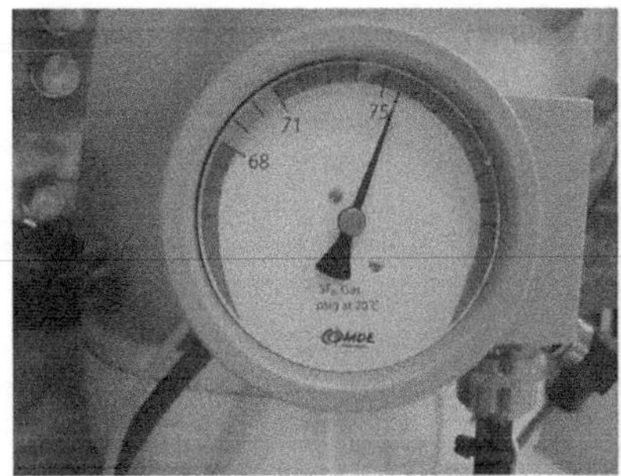

Figure 2.77 Density meter of SF_6 (Reproduced by permission of Siemens AG)

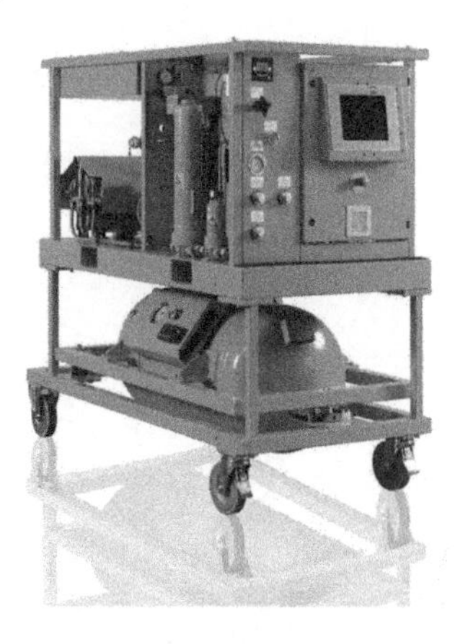

Figure 2.78 Small equipment for Evacuation, Filling and Refilling of SF_6 (Reproduced by permission of Dilo)

Figure 2.79 Large equipment for evacuation, filling and refilling of SF_6 (Reproduced by permission of Siemens AG)

2.8.5.6 Storage

Equipment for storage is shown in Figure 2.80.

2.8.5.7 Gas Recovery Process

The gas recovery process for a small and a large enclosure is shown in Figure 2.81 and a measuring diagram in Figure 2.82, where the conversion monogram between the dew point (°C) and the moisture volume concentration (ppmv) is a function of the SF_6 rated filling pressure (bar). The diagram in Figure 2.82 shows lines drawn for given temperatures and gas pressures that are used to find the gas molecular content.

2.8.6 Life Cycle Assessment Case Study: Würzburg

This Life Cycle Assessment (LCA) looks beyond the environmental view limited to the greenhouse potential of using air-insulated versus SF_6-insulated substations. The LCA provides a comprehensive consideration of the entire power supply system from the high-voltage transmission level down to the medium-voltage distribution voltage level. As an example, the mid-size German city of Würzburg has been chosen to be as realistic as possible. The study was carried out by the Germany industrial organization ZVEI to provide a wider view.

The idea behind this LCA is to bring away the one-sided view on pure SF_6 impact to global warming and provide a wider view of the power supply system in total and its global warming impact. The system-related view includes all relevant environmental criteria in the context of the application of SF_6.

Figure 2.80 Storage of SF_6 (Reproduced by permission of Solvay)

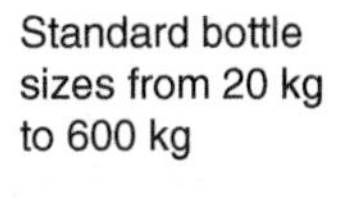

For larger volumes several bottle can be connected

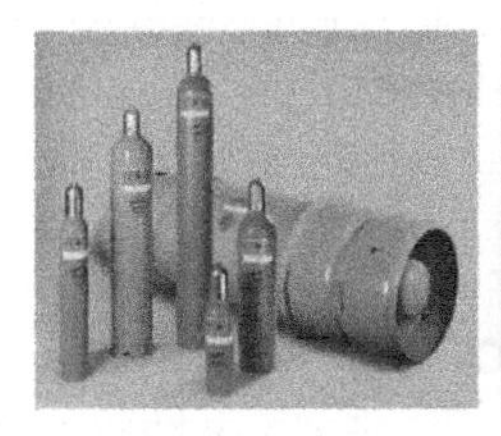

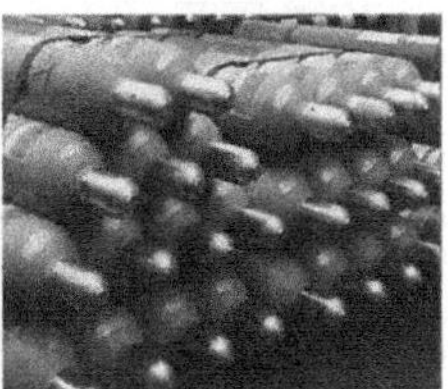

With State-of-the-art-handling equipment SF_6 recovery of each gas compartment till very low pressure (1–20 mbar) is possible, thus securing losses of at least less than 2% during maintenance and end of life.

SF_6-residual quantity (emission) dependence on the SF_6 rated filling pressure / compartment size / SF_6 residual pressure

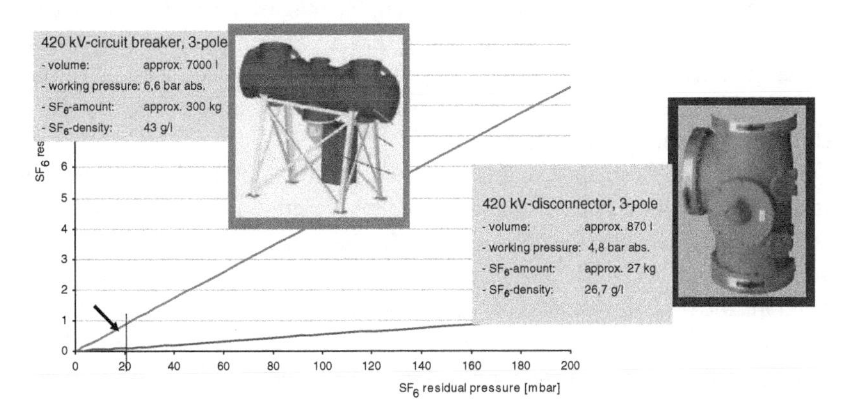

Figure 2.81 Gas recovery of SF_6 for a small and a large enclosure (Reproduced by permission of CIGRE)

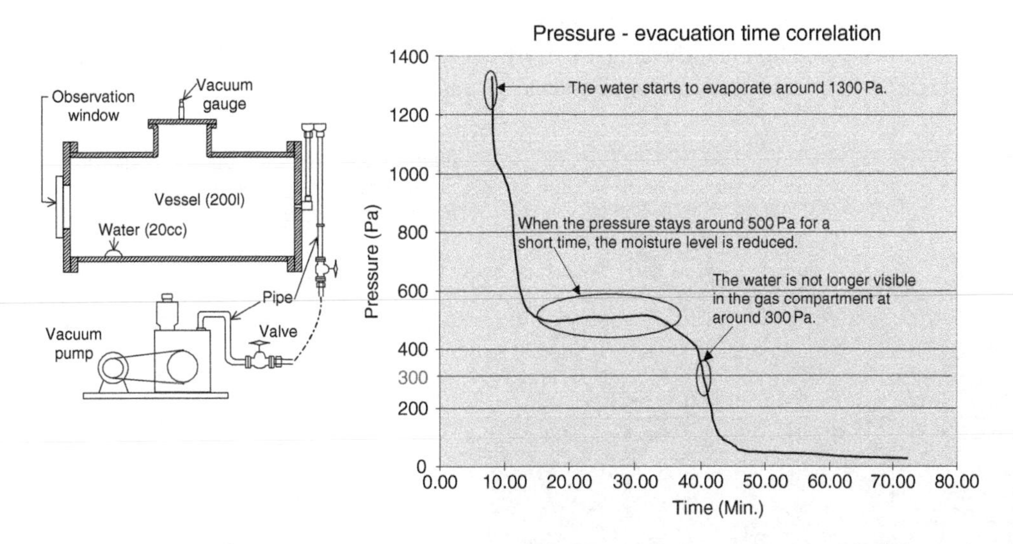

Figure 2.82 Measuring protocol of gas recovery of SF_6 (Reproduced by permission of CIGRE)

All steps of use and handling of SF_6 in GIS and switchgear from production, installation, service, maintenance, leakage repair until decommissioning, and disposal must be considered. Comparative life cycle assessment studies for electric power supply systems using various switchgear technologies arise from the possible conflict between the known advantages of sulfur hexafluoride (SF_6) technology, such as a reliable and economical power supply, and the environmental profile of the use of SF_6 in power supply systems, which has not been quantified in detail until now. This LCA is directed to all relevant environmental criteria.

2.8.6.1 Object of the LCA

The object of the LCA is the supply of electric power using different switchgear technologies. One is based on air-insulated substations with SF_6 circuit breaker, the other on gas-insulated substations (GIS). The LCA code to the 380 kV transmission voltage level and the local 110 kV distribution at the city limits and in the city depends on the switchgear technology (AIS outside the city, limits because of the size, and GIS inside the city) [66].

The power distribution voltage levels in the city of 10 and 20 kV bring the electric energy to the consumers and therefore are distributed by substations in the city limits. The praxis-oriented power supply was studied in the city of Würzburg (see Figure 2.83). This mid-size city of about 130 000 inhabitants comprising an area of 40 km^2 has a load density from 0.1 MW/km^2 in the outskirts of the city and up to 20 MW/km^2 in the city center. The power demand of the city is about 400 GW h in the first year, with an estimated load increase of 1.5% p.a. The lifetime of the electric power equipment was taken as 30 years.

The study is based on Germany's requirements on environmental rules and laws, which might be different in other countries. The technologies used for the study are in both cases AIS and GIS from state-of-the-art equipment and the network planning has been taken as a greenfield plan to have the optimum structure for AIS and GIS solutions. The power supply in both cases is taken to be the same; therefore, the environmental impact of the power generation has not been taken into account.

The principle of the AIS solution is that the substations are located at the city limits in order to find enough space for an air-insulated substation and from this to bring the power supply to consumers using medium voltage cables at 10 and 20 kV depending on the power density (see Figure 2.83).

At the substation UW1 in upper left, the 380 kV transmission line brings in the electric power and feeds the 110 kV ring line with the substations UW2, UW3, and UW4. The connecting lines between the 110 kV substation and the ring are overhead lines. Distribution to the consumers in the city is made by 10 kV and 20 kV underground cables, mainly laid in streets.

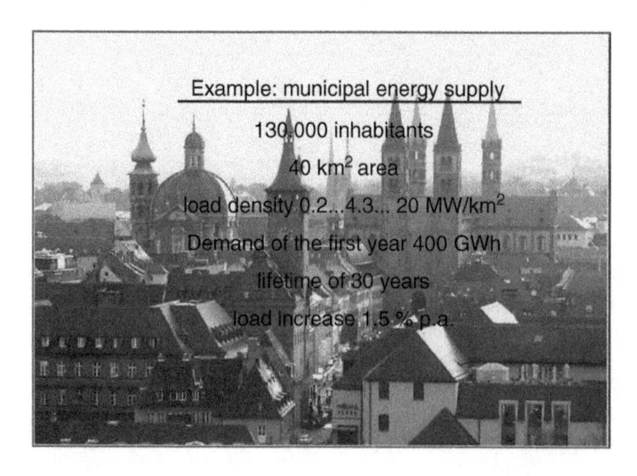

Figure 2.83 Energy supply without/with SF_6 technology – city of Würzburg (Reproduced by permission of Siemens AG)

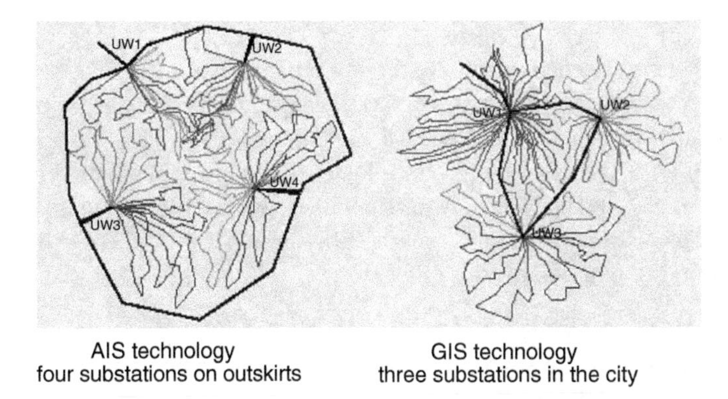

Figure 2.84 AIS and GIS network structures (Reproduced by permission of Siemens AG)

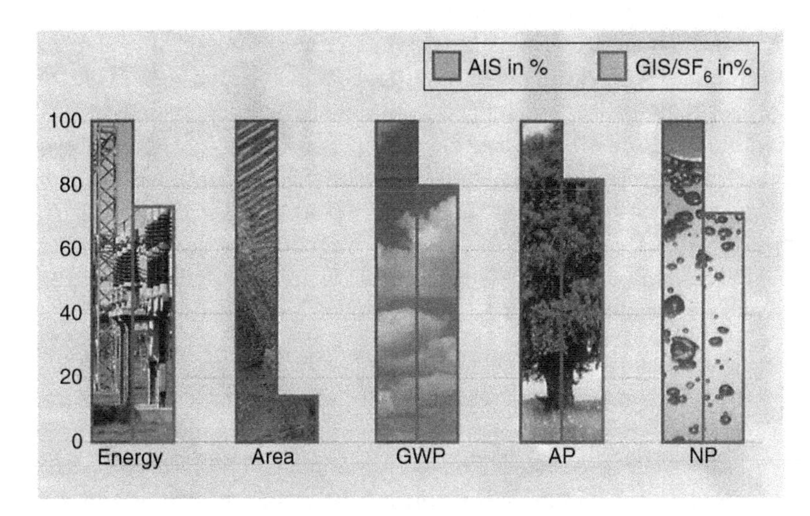

Figure 2.85 Result of environmental impact of AIS and GIS solutions (Reproduced by permission of Siemens AG)

The principle of the GIS solution is that because of the small size of GIS substations the location of the 110 kV substation can be in the city limits (see Figure 2.84). The in feeding 380 kV overhead line is connected to the substation U1 and transformed to 110 kV. The 110 kV ring line now is made by underground cables and connects the other 110 kV substations UW2 and UW3. Because of the shorter distance of the 110 kV voltage level to the power consumers only three substations are required and the total length of 10 kV and 20 kV underground cables is shorter. This reflects in results of much lower transmission losses.

The result of the study is concluded in Figure 2.85. Five aspects have been evaluated in the study:

- Primary energy
- Space consumption
- Global warming potential
- Acid rain potential
- Nitrification potential

The principle of the graphic sets the value of the AIS solution to 100% and shows the deviation of the GIS solution.

2.8.6.2 Primary Energy

In this valuation, the total amount of energy needed to produce the equipment, to install it, to operate it, and to dismantle it after 30 years of usage has been recognized. For the manufacturing, the GIS value is higher than for AIS because more material is needed, but compared to the additional energy needed to operate the system for 30 years and to produce losses this value is much larger and results in an advantage for GIS of 27%. The power transmission losses dominate the primary energy. The big difference in total transmission line length from the AIS solution toward the much shorter line length of the GIS solution is the reason for this difference.

2.8.6.3 Space Consumption

The space reduction of GIS versus AIS is typically 60–70% less space for GIS. In addition, the GIS solution needs only three substations versus four for the AIS solution. This makes a big difference.

2.8.6.4 Global Warming Potential

In this comparison, the SF_6 losses as defined in the international standards of 0.5% gas loss per gas compartment per year have been taken. In a 30-year lifetime of the equipment, a total of 15% of SF_6 has been released to the atmosphere. This makes a higher value for GIS because of a higher volume of SF_6 gas used. In AIS, only the circuit breaker has a small volume of SF_6. Besides the gas losses of the equipment, the global warming potential also takes the transmission losses into account and the fact that additional electric energy needs to be generated and cannot be used by the consumer. The Germany power generation mix is 40% coal, 30% gas, 20% nuclear, 10% hydro, and others, which gives an equivalent value for CO_2 produced for each kW h produced. Again, the higher transmission losses of the AIS solution outweigh the higher global warming potential related to the high SF_6 volume in GIS and give an advantage of 21% toward the GIS solution.

2.8.6.5 Acid Rain and Nitrification Potential

These environmental factors are related to materials released to the atmosphere (acid rain) or the ground water (nitrification) during the manufacturing processes and the operational lifetime. In both cases the need of materials is somehow balanced between AIS and GIS because GIS needs more material for the enclosures and AIS needs more equipment for the large installation and one more substation. The main difference, which gives an advantage to the GIS solution of 19% for the acid rain potential and 29% for the nitrification potential, is related to much lower total transmission losses during the 30 years of operation. This seems to be the advantage of GIS from an environmental view: lower losses with GIS solutions.

2.8.6.6 Conclusion

The use of SF_6 leads to considerable environmental advantages when the total electric power supply system is evaluated. To achieve any benefits for the global warming potential from using GIS, it is required to have a very strict SF_6 management according to the rules defined in international standards. Very low emissions during service and maintenance of the equipment, recovery, and reuse of SF_6 after decommissioning are mandatory requirements.

The results of the study also show that the result depends on the local conditions, rules, and requirements, but the study gives a good orientation on how to generate an environmental advantage for electric power supply system when GIS is used.

2.8.7 Kyoto Protocol

The Kyoto Protocol from December 1997 was the result of the Environmental Protection Conference held in Kyoto, Japan. Today this protocol is still the only contractual intercountry agreement in limiting global warming. The countries agreed on reductions of greenhouse gases for the year 2012 based on the year 1990 (see Figure 2.86).

The main contribution to global warming is CO_2, which is released in any burning process like power plants with coal or gas. Including CO_2 as a contribution to global warming there are six major gases listed in the Kyoto Protocol. Three are related to energy and human activities (CO_2, CH_4, and N_2O) and three are related to industry (HFCi, PFCs, and SF_6). The Kyoto Protocol was empowered in February 2005 after 55 states signed the document.

The Paris Agreement from 2018 approves the requirements of the Kyoto Protokol, also concerning SF_6 and formulates goals to limit the climatice changes to reduce temeperature inclease to 1,5 °C. This also gives pressure on replacement of SF_6. For details see also section 2.9.

2.8.7.1 Requirement for Each Party

The signing nations agreed that a national system for estimation of emissions should be installed by 2007. They must identify the sources of emissions in their countries (see Table 2.27).

They also agreed to reduce the average emissions of all gases in total by 5.2% below the 1990 levels in the period of 2008–2012. An annual report of inventory by source and the way to reduce the emissions was to be produced. The countries also agreed to establish stocks for CO_2 equivalents for the base of 1990 and 1995 for SF_6. Changes of increases or reductions for the following years needed to be estimated.

2.8.7.2 Environmental Aspects of SF_6

The Kyoto Protocol gives information about the impact of SF_6 on the global warming potential (GWP). The most important fact for SF_6 is its very high potential compared to CO_2. The Kyoto

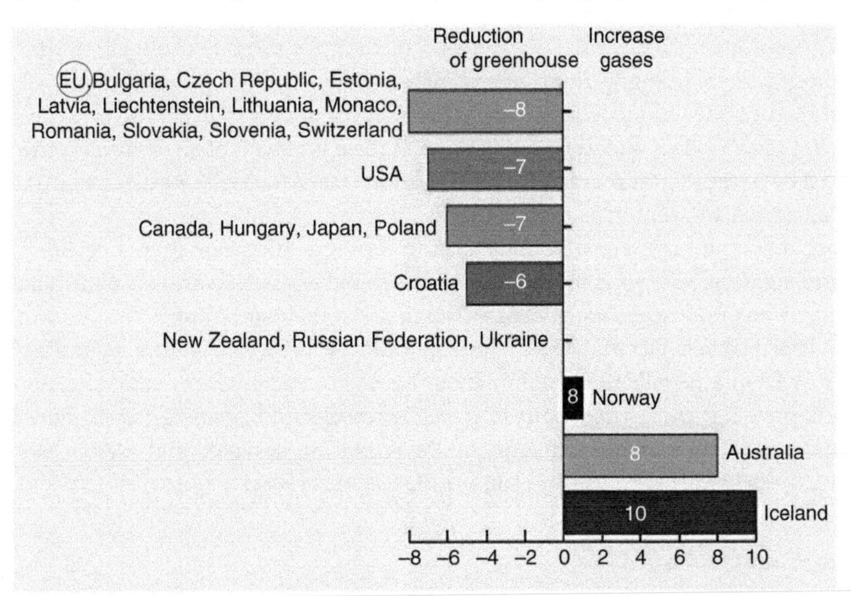

Figure 2.86 Kyoto Protocol – change from 2008 to 2012 in comparison to 1990 (Reproduced by permission of Siemens AG)

Table 2.27 Kyoto Protocol requirements [51, 52]

National system for emission control by 2007 to be installed
Reduce average emissions by 5.2% below 1990 value until 2012
Report annually on emissions
Establish stocks for CO_2 trade

Protocol fixed this value to 23 900. In later published documents, the number 22 500 is also listed, but whichever is correct, these values are high. That means one ton of SF_6 released into the atmosphere is equivalent to a global warming potential of 23 900 tons of CO_2.

The Kyoto Protocol found that approximately 8500 tons are produced each year and in consequence will be released to the atmosphere. This then would be an equivalent emission of 203 million tons of CO_2 per year. The world-wide emission of CO_2 at that time was 22 billion tons. A fossil power station of 500 MW produces CO_2 equivalent to 15 tons of SF_6 released to the atmosphere per year.

The absolute contribution of SF_6 to the global warming potential was not very high, but what made the nations alarming is the high potential of SF_6. As a consequence, governments started actions to reduce SF_6 from being released into the atmosphere and have been successful in many applications, like sport shoes, car tires, and insulated windows. The solution for the electric power industry was the closed loop use of SF_6, because no replacement was found.

Due to the high potential of SF_6 and the very long molecular stability of 800–2300 years, the consequences to limit SF_6 release to the atmosphere are the right way. The present contribution of SF_6 to global warming is low, far below of 0.1%, and it should stay there. The increase in the atmosphere had its high peak with 7–9% per year in the year 1983 to 1994. Since the measures were established based on the Kyoto Protocol the values are going down.

2.8.8 Requirements for the Use of SF_6

The main requirements for use of SF_6 in service are to minimize the SF_6 losses due to leakages and handling of electrical equipment filled with SF_6 (see also Table 2.28). In no case should SF_6 be released into the atmosphere, for example, from storage containers on-site during construction. All quantities of SF_6 not used on-site need to be sent back to the manufacturer for further use in equipment. The SF_6 does not age and even after the used lifetime of the equipment of 30 or 40 years the SF_6 should be recycled by filling SF_6 into containers for transportation. After cleaning using the SF_6 handling and filtering device, SF_6 should be reused.

Any personnel, operators, and maintenance staff need to have qualifications from a certified training person before handling SF_6. With appropriate handling and leakage control the contribution of SF_6 to man-made greenhouse gas and global warming will remain negligible.

SF_6 will remain indispensable in electric power equipment for the foreseeable future. Recycling is essential in order to avoid atmospheric contamination. Recycling equipment is commercially available to allow economic recycling. The quality of correctly recycled SF_6 is satisfactory to allow its reuse in electrical power equipment. Standards for recycling are available and established (CIGRE User Guide TB 125 [29], IEC 62271-4 [51] and IEEE C37.122.4 [66]).

Table 2.28 Main requirements for use and handling SF_6 [51, 52]

Minimize SF_6 leakages
Minimize SF_6 handling losses
Never release SF_6 deliberately
Recycle and reuse SF_6
Educate operators by certified training

2.8.9 Moisture in SF_6 gas

Moisture in the form of H_2O can have detrimental effects on equipment filled with SF6 gas in two ways:

1) Moisture can combine with dissociated SF6 during normal arcing (such as during switch operation) and create hydrofluoric acid and other toxic compounds. These compounds can deteriorate internal polymeric materials, cause accelerated aging, and reduce surface dielectric strength
2) Moisture can also accumulate on internal dielectric surfaces as condensation and reduce the dielectric creep strength of insulators. This can occur when the dew point of the gas is high (0 °C).

Moisture can exist in the vapor phase in the SF6 gas, in the adsorbed phase on the surface of the enclosure, and in the absorbed phase inside the polymeric materials such as epoxy insulators. There are many techniques to accurately measure the moisture content of the gas, in both the liquid state and after it is entered into the gas compartment.

Moisture can be easily managed by proper preparation of the GIS before the SF6 is entered into the gas compartments. This can be accomplished by thoroughly drying the compartment interior by evacuating air to low vacuum levels (such as 0.027 kPa), or purging with a dry gas (usually nitrogen).

Moisture (and arc byproducts) can also be managed by installing desiccants in or attached to the gas compartments. The quantity of desiccant (typically zeolites), will be determined by considering the mass of moisture in the SF6 gas in the initial fill, plus mass of moisture estimated in the epoxy components and on interior surfaces, and the estimated moisture which could enter through seals and enclosures during the expected lifetime of the equipment, or to the first required maintenance. The moisture absorption capability of desiccant is usually 10–15% of their total weight.

Further information can be found in IEEE C37.122.5 "IEEE Guide for Moisture measurement and Control in SF6 Gas Insulated Equipment" [57].

2.9 Alternative Gasses to SF_6

Authors: Peter Grossmann, Hermann Koch

Review: James J. Massura

2.9.1 Introduction

The replacement of SF_6 with a more environmentally friendly gas of lower global warming potential (GWP) has been discussed for several years. Manufacturer investigations and governmental authority requirements are driving this process.

There are different characteristics to be considered which an alternative gas shall provide like: low GWP, high insulation capability, high arc quenching capability, no toxicity, long lifetime of the molecules, no decompensation by-products, and low cost. So far, this spectrum of requirements has been best fulfilled by SF_6 except the low GWP requirement.

Many investigations have been carried out in laboratories of insulating gas producers and in test fields of GIS manufacturers over the last years. First investigations used synthetic air and CO_2 gases beginning in the 1970s, followed by N_2 gases in 2010 until the latest investigations after 2015 using new gas molecules like flouronitrile, flouroketone, and synthetic air, a mixture of O_2/N_2 or O_2/CO_2.

In this chapter, the status of the development of alternative gases to SF_6 is explained and the characteristics of new alternative gases are discussed. The motivation to use alternative gases will be described with the information on the availability of alternative gases. Examples of field experiences are given and the possible transition phase from SF_6 to an alternative gas is described. Further activities in standards and associations conclude this chapter.

There is a wide discussion on alternative gases to SF_6 within many international and national organizations which influence the application alternative gases and define requirements for the future use of SF_6. An overview is given on organizations being active in this topic.

2.9.2 Motivation to Use Alternative Gases

2.9.2.1 Why Alternative Gases to SF_6

The primary reason to look for an alternative to SF_6 as high-voltage insulating and high current interruption gas is the high global warming potential (GWP) of SF_6. One kilogram of the potent greenhouse gas SF6 contributes to the global warming by a GWP equivalent of 23 500 kg CO_2.

Even with the low absolute number of emitted SF_6 to the atmosphere, this high GWP value is an accelerator on global warming effects. The stable SF_6 gas molecule will remain in the atmosphere more than 1000 years and will reach the outer atmosphere where the global warming effect is even higher.

In 2014, these two aspects, high GWP and long remaining time in the atmosphere, were the main reasons to list SF_6 in the Kyoto Protocol of fluorinated substances. The Kyoto Protocol was taken as basis for further regulations, for example, EU-F-Gas-regulation 517/2014 [68] or the EPA regulations 69 in the USA.

The latest development was initiated by the California Air Resources Board (CARB) [7] where the discussion has brought the decision on SF_6 phase-out dates for new GIS applications. The latest version from August 2019 defines phase-out dates for high-voltage equipment containing SF_6 to be used in California depending on the rated voltage levels. Equipment for voltages up to 145 kV by the year 2025, above 145 kV up to 245 kV by the year 2029 and for rated voltage above 245 kV by the year 2031 (Table 2.29).

The principle of the phase-out process according to CARB will give cost benefits to high-voltage equipment with reduced GWP contribution. The more GWP reduction, the higher the percentage of cost benefit for project offers. These requirements are in continuous development and getting adapted to follow more strict environmental goals for reducing the global warming.

Table 2.29 Phase out dates for high-voltage equipment containing SF_6 in California according to CARB requirements [71]

Voltage level (kV)	Short-circuit rating	Year
Up to 145	<63 kA	2025
	≥63 kA	2028
145–245	<63 kA	2029
	≥63 kA	2031
Above 245	All	2033

2.9.2.2 Alternative Gases to SF_6 to Cut CO_2 Emissions

To show the impact on the GWP contribution of using SF6 equipment in relation to compensate the SF6 GWP contribution by green plants, the following example is given.

Technical standards of IEC and IEEE require high gas tightness for GIS and AIS equipment using SF_6. For gas sealings, the leading manufacturers are using an O-ring system to ensure the gas tightness and a gas density monitoring for each gas compartment.

The type and routine test requirements for gas compartments of each gas compartment limit the gas leakage rate to 0.5% per year per gas compartment. This value is related to the limitation of technical measurement of gas tightness in a relatively short time period of type and routine tests. The practical experiences of GIS in operation show a much lower gas leakage rate for the high-quality O-ring sealings of 0.1% per gas compartment and year.

Even this high gas tightness rate for a modern, typical substation scheme of a 145 kV SF_6 GIS with 7 bays with an annual leakage rate of 0.1% would correspondent to 0.6 kg SF_6 emissions, see Figure 2.87. Multiplied by the GWP equivalent for CO_2 of 23 500 kg the loss of 0.6 kg SF_6 will contribute to the global warming in the same way as 14 100 kg CO_2.

To balance this, GWP contribution 1200 beech trees are needed to grow to offset these 14 100 kg CO_2 [72].

Figure 2.88 gives an impression that it needs a small forest to offset the GWP impact described in the example above. It is understandable that the pressure on users and manufacturers will increase to look for technical solutions providing a lower or no GWP.

The consequences of stated governmental authority regulations will give limits for GWP contributions of high-voltage substations and switchgear. The technical development and the experiences collected with new solutions combined with acceptable equipment costs will determine the schedule of the use of alternative insulating gas solutions in high-voltage equipment.

Figure 2.87 145 kV SF_6 GIS with 7 bays (courtesy of Siemens Energy)

Figure 2.88 Impression on the influence of high GWP values: 1200 beech trees to compensate 14100 kg CO_2 (*source* public internet [72])

2.9.2.3 SF_6 Emission Rate and SF_6 Capacity

As shown in Figure 2.89, the SF_6 emission rate was reduced between 1999 and 2016, while at the same time, the total SF_6 capacity has increased.

The total SF_6 emissions in pounds show a continuous decline from less than 800000 lbs., in 1999, to a value of some 1000 lbs., in 2016. During the same period, the total amount of SF_6 stored in the high-voltage equipment increased from 6.000.000 lbs. to more than 10.000.000 lbs. The calculated emission rate is reduced from about 12% down to less than 2%.

The monitoring data of the SF_6 emissions, the stored capacity as indicated on equipment name plates, and the calculated emission rates were published by the US EPA in August 2018 [73].

These emission losses include the gas losses related to leakages at gas compartments and any other gas handling losses during maintenance works or any other gas works on the equipment. This graphic shows that today's SF_6 gas tightness of high-voltage equipment has reached a higher quality level. One source of emission reduction is related to the replacement of old, more leaking equipment with new high gas tight equipment. A second source of emission reduction is related to the improvement of gas handling processes where SF_6 gas is handled in a closed loop without release to the atmosphere.

2.9.2.4 Possible SF_6 Emissions in the Lifecycle Process of Gas-Insulated Equipment

There are two main sources of SF_6 emissions. One is during the development and manufacturing process of the manufacturer; the other is during the operation of the equipment at the user substation. The SF_6 emissions in the lifecycle process of switchgear have been investigated by the EPA for the Electric Power Systems in the USA and was published in August 2018 [73].

During the development of GIS equipment, the type tests require several set ups to fulfill the requirements given in IEC and IEEE standards [4, 27, 23, 52]. See Figure 2.90. With each change

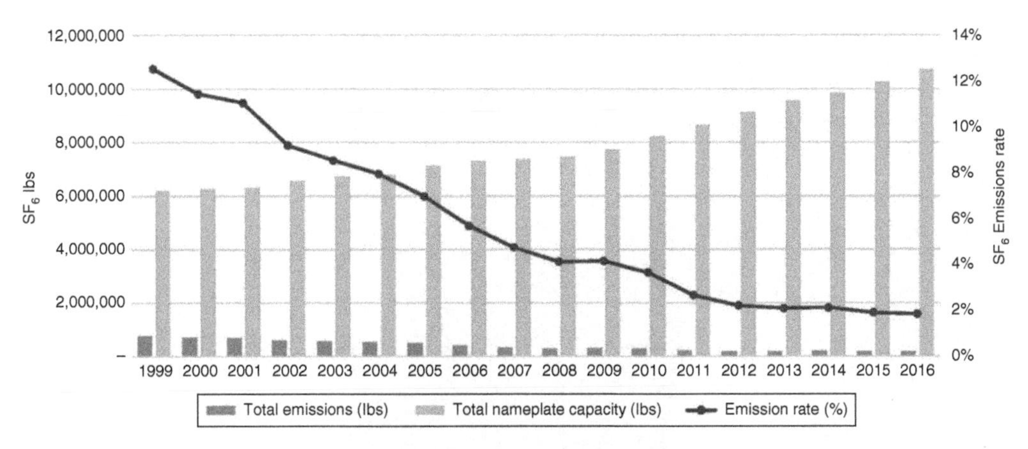

Figure 2.89 SF_6 emission rate reduced but total SF_6 capacity increases (*source* Figure 1 of [73])

Figure 2.90 Typical type test set up for high-voltage test using SF_6 (courtesy of Siemens Energy)

of the test set up, SF_6 gas works are required. The standard requirements for gas handling of GIS are given in IEEE C37.122.3 and IEC 62271-4 [28, 51, 52]. These requirements also apply to the routine tests at the factory and the on-site installation tests.

Following these definitions, the handling of SF_6 is required to be in a closed cycle without releasing SF_6 to the atmosphere. Evacuation of compartments is down to a vacuum of only 20 mbar and the used SF_6 is cleaned in moisture and particle filters before stored in gas tanks. The SF_6 gas can be reused for the next type test.

This closed loop cycle of SF_6 use in laboratories has reduced the gas losses to the atmosphere. The certification of workers handling SF_6 has improved the awareness and is the basis for the application of the correct gas handling process.

Figure 2.91 GIS installation with SF_6 filling on-site (courtesy of Siemens Energy)

Manufacturing GIS requires a routine or factory test for each GIS bay. This includes also high-voltage test where SF_6 is required in the equipment at the rated functional gas pressure. After the routine tests in the factory, the SF_6 pressure is reduced to a low-pressure value for transportation, typically 0.5 bar. The used gas will be stored in a SF_6 storage container. For this gas handling, the rules defined in IEEE C37.122.3 and IEC 62271-4 apply [51, 52].

The installation on-site requires again SF_6 gas works. The required gas pressure needs to be refilled for prefilled bays. For new on-site assemblies as bus bars, the complete evacuation and gas fill process is required following the IEEE C37.122.3 and IEC 62271-4 standard requirements [51, 52], see Figure 2.91.

The principle of the gas handling standards IEEE C37.122.3 and IEC 62271-4 is to keep the SF_6 in a closed loop from production to use, and even final disposal (recapture). In Figure 2.92, the graphic shows the steps during the life cycle use of SF_6 in which possible leakages and emissions are identified.

The cycle of the graphic is divided in two areas. The upper half in green displays the manufacturing cycle and the lower, grey half the user cycle.

The manufacturer of the electrical equipment is using SF_6 for the mechanical and high-voltage routine tests of each GIS bay in the factory. The gas handling of the GIS includes the evacuation process before filling with SF_6 up to the rated pressure and, after the test are performed, to refill the SF_6 gas from the GIS bay into the storage container in the factory. The gas will be cleaned by filters from moisture and dust for reuse with the next routine high-voltage test. Gas losses may be possible in conjunction with handling the SF_6 gas into the GIS bay and back into the factory gas storage container. This includes also the gas handling devices, the flexible tubes, and valves. IEEE C37.122.3 and IEC 62271-4 give detailed information on correct handling processes to avoid SF_6 gas emissions to the atmosphere.

The shipping of the equipment and/or SF_6 gas cylinders from the factory to the installation site requires additional gas handling work and is subject to possible gas losses which need to be avoided. For lower high-voltage levels of GIS, the transportation is done with complete bays or bay modules.

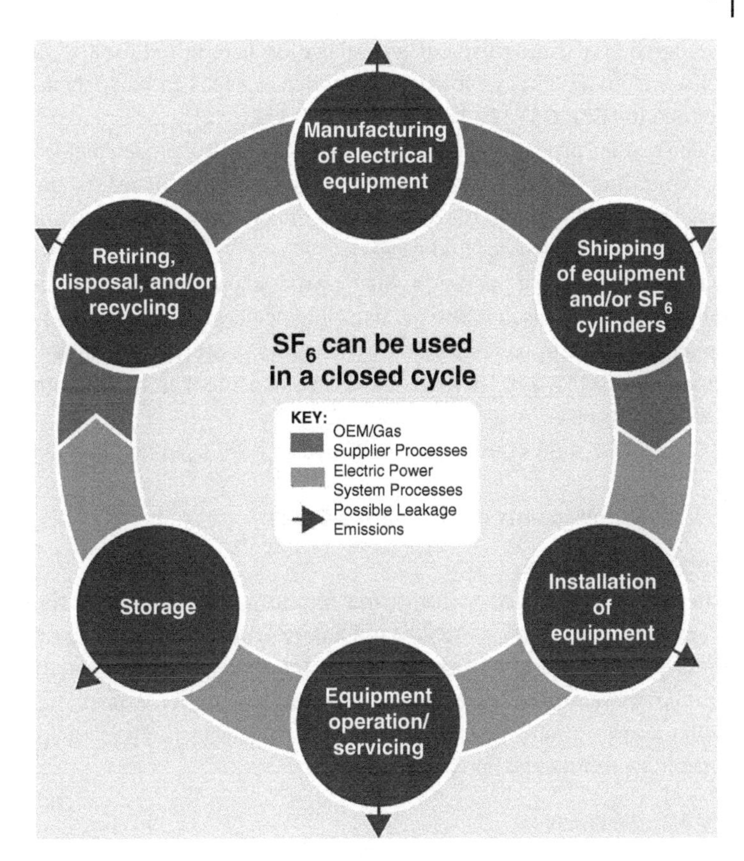

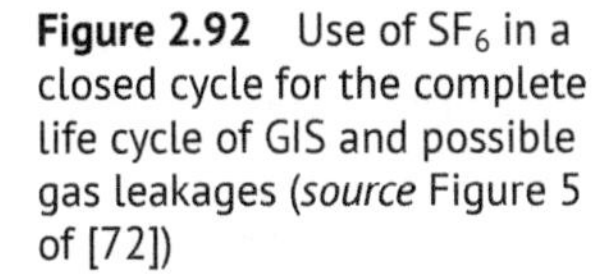

Figure 2.92 Use of SF_6 in a closed cycle for the complete life cycle of GIS and possible gas leakages (*source* Figure 5 of [72])

For GIS up to 145 kV, it is typical today to ship double bay configurations; up to 300 kV single bay configurations; and for voltages of 420 kV and above, modules of a bay are shipped. In most cases, the GIS for transportation is filled with N_2 of a low gauge pressure of about 0.5%.

The lower part of the cycle is linked to possible gas emissions during the assembly on-site and during service works of the GIS. At the substation on-site, the GIS needs to be filled with SF_6 before the GIS can go into operation. After the bays are assembled by connecting the bus bar flanges, the gas compartments need to be evacuated and filled with SF_6. This process is defined in the IEEE C37.122.3 and IEC 62271-4 standards [51, 52]. Following the processes described in the standards, the gas losses will be reduced to a minimum, including the gas handling devices, pipes, and valves.

This concludes the on-site assembly of the GIS. Before the GIS will go into operation as part of the commission tests, the GIS flanges are checked with a gas detecting device (e.g., an SF_6-gas detector) to be sure of no gas leakage.

GIS equipment in operation and service needs to have gas tight sealing to keep the operational gas pressure of typical 0.4–0.8 MPa inside the gas compartment. Gas leakages are monitored by gas density meters and alarm signals will be generated to indicate gas losses.

Mechanical damages which could create holes to the enclosure with fast gas loss are extremely rare and may have other reasons (shooting, explosions, fire, earthquakes). In most cases, the gas losses are very small and develop over a long period of time. The reasons for such gas losses can be found in incorrect sealing surface treatments, flange corrosion, degraded O-ring materials, and severe climatic condition at the substation location.

To find such failures, it might be necessary to provide a long-time measurement over a period of several months with plastic bags around the flanges where the leakage is suspected.

Storage of SF_6 gas on-site at the substation is required for any gas works to capture the SF_6 gas release from a GIS gas compartment when it needs to be opened. How to store SF_6 gas onsite is defined in IEEE C37.122.3 and IEC 62271-4 [51, 52].

The storage process includes a cleaning and drying process before the SF6 is filled into a gas storage container. The SF_6 gas to be stored needs to be analyzed for the content of by-products which may be toxic. The standards [51, 52] give information on the percentage of toxic molecules allowed in the SF_6 gas for storage and reuse.

In normal cases of operation, the by-products are captured by the aluminum oxide (AlO_3) filter material which is used in the gas handling devices. The storage of new gas, used gas, and contaminated gas needs to be identified by the colors of storage container according to the definitions in the standards [51, 52]. These markings are also important for requirements if the storage container need to be transported.

Figure 2.92 shows the complete life cycle of SF_6 used in GIS and the possibilities of gas losses.

2.9.3 Requirements of Alternative Gases

2.9.3.1 Introduction

Alternative gases need to fulfill the requirements of the environment, the insulation and arc quenching performance, health and safety aspects, and GIS equipment applications. SF_6 covers these parameters sufficiently and only the global warming potential brings it to a risk of being prohibited by authorities for environmental protection reasons. The criteria of the alternative gases with lower GWP show disadvantages in certain aspects. Here, the parameters are explained which need to be fulfilled by the alternative gases.

2.9.3.2 Environment

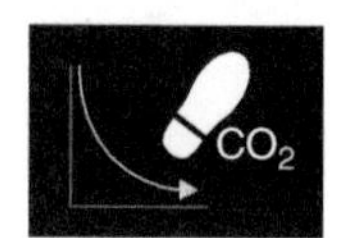

Source public internet

The environmental aspect asks for a CO_2 foot print as small as possible.

A low global warming potential below SF_6 can be reached with many alternative gases. GWP values between of 0 and about 400 are possible. Other parameters like high-voltage insulation capability, arc interruption capability, gaseous temperature range, operational gas pressure, and toxicity need to be considered too.

SF_6 molecules are very stable and show a lifetime of about 1000 years. This makes SF_6 gas used in GIS a very reliable gas for a long time. To reach similar or even better reliability values with alternative gases, it is an important requirement to avoid gas exchanges during the 50+ years operational lifetime of the GIS.

The inert gas molecules of SF_6 are easy to recycle and SF6 can be reused in a new installation [53]. In very rare cases, the contamination of SF_6 requires a final disposal of the molecules by high burning temperatures. A similar recyclability of the alternative gas is required to provide a close life cycle using the gas for insulating and current interruption purposes.

2.9.3.3 Performance

Source public Internet

The performance of the alternative gas needs to provide three basic requirements: dielectric strength, current switching capabilities, and stable behavior over the GIS lifetime.

High dielectric strength is required to keep the size of the GIS equipment small. Over the last four decades the GIS size and volume was reduced by 70–80%. The alternative gas needs to allow that the design can be made with similar physical dimensions used today.

The switching capabilities for short-circuit currents of the compact design of today has developed from about 31.5 kA at the first generation to 63–80 kA as of today, depending of the voltage level. This allows very powerful GIS substations to operate as a small footprint. Alternative gases also need to provide such high switching capabilities to use similarly small substations as seen today.

The stable SF_6 molecule provides a long-time electric insulation capability so that for the GIS lifetime no replacement of the insulating gas is required. Alternative gases need to provide this stable behavior for the 50+ years lifetime of the GIS as well.

2.9.3.4 Health and safety

Source public Internet

The health and safety considerations are important for handling insulating gases during manufacturing, operation and service activities. The inert SF_6 gas is nontoxic in new condition. For contaminated gases, there are established procedures defined [52, 74].

Alternative gases also need to have a low toxicity for new and arced gas conditions for safe gas handling during installation, operation, and maintenance. This is also of importance for service activities, e.g., taking gas probes.

The toxicity is given by the chemical formula of the molecule and there is a range of gases with good dielectric performances but certain toxic parameters too. These gases are not suitable to be used in GIS applications.

2.9.3.5 Applications

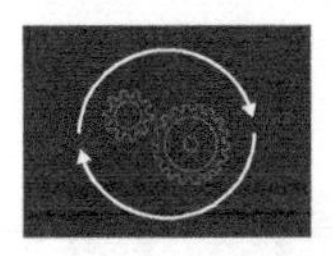

Source public Internet

To use SF_6 in GIS installations, the only restriction is usually coming from the ambient low temperatures. In most cases, SF_6 can be applied without any limitations of insulating and switching capabilities down to −30 °C. GIS installations below this temperature are using a reduced gas pressure or SF6 gas mixtures with N_2.

Alternative gases need to cover suitable low temperature applications as well. To operate substations in a network, it is important for users to have standardized application requirements and typically ambient temperatures between −25 °C and +40 °C is one of them.

Another important requirement comes for the adaptability to the existing design. Most users operate a fleet of substations and GIS installations. Operation and handling of the existing and the new alternative GIS equipment needs to be coordinated. It very likely will be operated by the same crew.

Handling of alternative gases onsite needs to follow practical and easy to handle procedures. Gas-handling works are important for the operation and maintenance of the GIS equipment and can cause critical failures if gas handling processes are too complex.

2.9.4 Characteristics of Different Alternative Gases for GIS

Alternative gases as replacement of SF_6 have been investigated over the last several years. The alternative gases with significant considerations as of today and their gas characteristics are shown in Figure 2.93 [53, 71, 78].

The alternative gases to SF_6 in focus today are synthetic air, CO_2/O_2 mixture, flouronitrile mixtures, and flouroketone mixtures.

The mixing ratios vary from the type of gas and the additive gas for electric insulation purposes. The main gas content in all cases of alternative gases is N_2 or CO_2. Additives are O_2, or types of CF_3 to improve the dielectric insulation capability.

The CO_2 equivalent of the alternative gases is in the range from 0 for synthetic air up to about 400 for flouronitrile, depending on the percentage of the additive gases.

The operating temperatures vary for natural gases as synthetic air or CO_2/O_2 between −50 °C and +55°C. For synthetic gases, the range is smaller for low temperatures, with −30 °C for flouronitrile and −5 °C for flouroketone.

The alternative gases or gas mixtures provide a lower dielectric strength. For synthetic air and CO_2/O_2 gas mixtures, the reduction is about 55% compared to SF_6. Flouronitrile and flouroketone provide about 30% less dielectric insulation capability. This reduced dielectric capability can be compensated by higher gas pressure and/or slightly larger dimensions.

For the arc quenching capability, SF_6 and synthetic air or CO_2/O_2 gas mixtures provide reversible dissociations while flouronitrile and flouroketone mixtures are irreversible at temperatures above 920 K or 970 K. This means for flouronitrile and flouroketone that the molecules will not recombine above these temperature limits.

The decomposition products of SF_6 are HF, SO_2, or any other sulfur compounds. Synthetic air may produce nitrogen oxides and ozone and CO_2/O_2 produces HF and CO. Flouronitrile and flouroketone produce HF, CO, and flouro-organic compounds.

	SF_6	Synthetic Air	CO_2+O_2	Fluoronitrille mixtures	Fluoroketone mixtures
Chem. Formula	SF_6	N_2+O_2	CO_2+O_2	$(CF_3)_2CFCN$	$(CF_3)_2CFC(O)CF_3$
Mixing ratio	20%...100% SF_6 0%...80% N_2 or CF_4	79.5% N_2 20.5% O_2	CO_2: 70% to 90% O_2: 10% to 30%	89%...96% CO_2 4%...6% $(CF_3)_2CFCN$ 0%...5% O_2	83% CO_2 6% $(CF_3)_2CFC(O)CF_3$ 11% O_2
CO_2–Equivalent	≤ 23,500	0	< 1	327...462	< 1
Operating temperature	–30°C...+55°C	–50°C...+55°C	–50°C...+55°C	–30°C...+55°C	–5°C...+55°C
Dielectric strength at same pressure	100%	~ 43%	40% to 45%	> 70%	> 70%
Arcing impact					
Dissociation / decomposition	~ 2000K (reversible)	N_2 ~7000K (reversible) O_2 ~4000 K (reversible)	~2500K (reversible)	> 920K (irreversible)	~ 970K (irreversible)
Gaseous decomposition products	HF, SO_2, sulfur compounds	Only in case of failure: nitrogen oxides, ozone	HF, CO	HF, CO fluoro-organic compounds	

Figure 2.93 Characteristics of different alternative gases for GIS (*source* [53, 71, 78])

2.9.5 Field Experience with SF_6 Alternatives

Since 1970, high-voltage switchgear installations with no SF_6 went into operation at voltage levels between 72.5 kV and 420 kV.

In 1970, an air-insulated 170 kV GIS with an air-insulated circuit breaker was used. See Figure 2.94.

In 2004, two installation went into service using synthetic air and CO_2.

The rated voltages of one installation were 72.5 kV and the other 84 kV. The equipment was of dead tank breaker design type with a circuit breaker using vacuum interrupting (VI) technology. The gas insulation was using synthetic air. See left photo of Figure 2.95.

The other installation was a dead tank breaker design for 72.5 kV using a CO_2 circuit breaker for interruption of rate and short-circuit currents. The dead tank insulation is also using CO_2. See right photo of Figure 2.95.

In 2010, an air-insulated 72.5 kV live tank VI circuit breaker using N_2 for insulation went in operation as part of an air-insulated substation to fulfill the requirement of low ambient temperatures. See Figure 2.96.

In 2015, the search and development activities for SF_6 alternatives focused on replacing SF6 for environmental reasons, not only on such special applications related to very low temperatures.

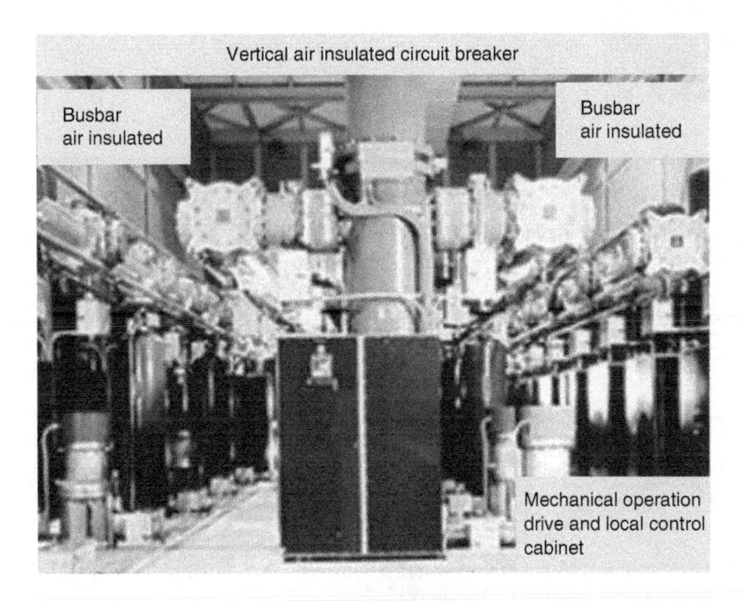

Figure 2.94 170 kV GIS air-insulated circuit breaker in an air insulation design, 1970 (courtesy of Siemens Energy)

Figure 2.95 Left: 72/84 kV dead tank VI circuit breaker with synthetic air insulation (*source* [75], right: 72 kV dead tank CO_2 circuit breaker with CO_2 insulation, 2004 (courtesy of Toshiba [76, 77])

Figure 2.96 145 kV live tank VI circuit breaker with N_2 insulation, 2010 (courtesy of Siemens Energy [77, 94])

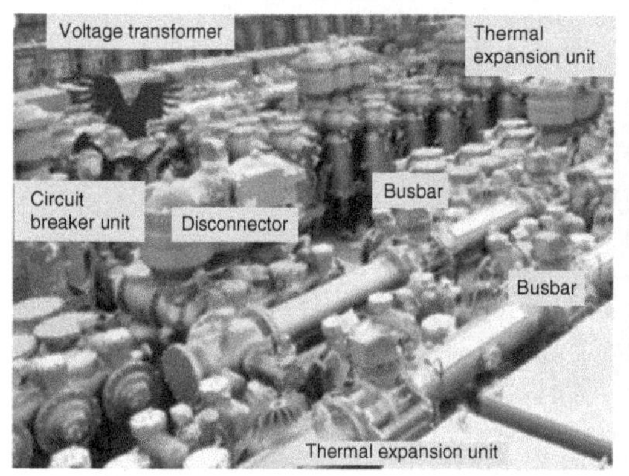

Figure 2.97 Left: 170 kV GIS with flouroketone for the circuit breaker and for Insulation, right: 420 kV GIL components with flouronitrile, 2015 (courtesy of left photo Hitachi ABB, right photo GE)

The first GIS installation using flouroketone for both the insulation and the circuit breaker, was energized at 170 kV rated voltage. See left photo of Figure 2.97.

The first installation using gas-insulated transmission line (GIL) components with flouronitrile was installed in a 420 kV GIS substation to connect the overhead line at outdoor climatic conditions. See right photo of Figure 2.97.

In 2019, the first 145 kV GIS using flouronitrile including the circuit breaker went into operation. See left photo of Figure 2.98.

Also, in 2019, the first 72.5 kV GIS using vacuum interrupter circuit breaker with synthetic air insulation was energized. See middle photo of Figure 2.98.

Later that year, the 145 kV GIS vacuum interrupter circuit breaker with synthetic air insulation followed. See right photo of Figure 2.98.

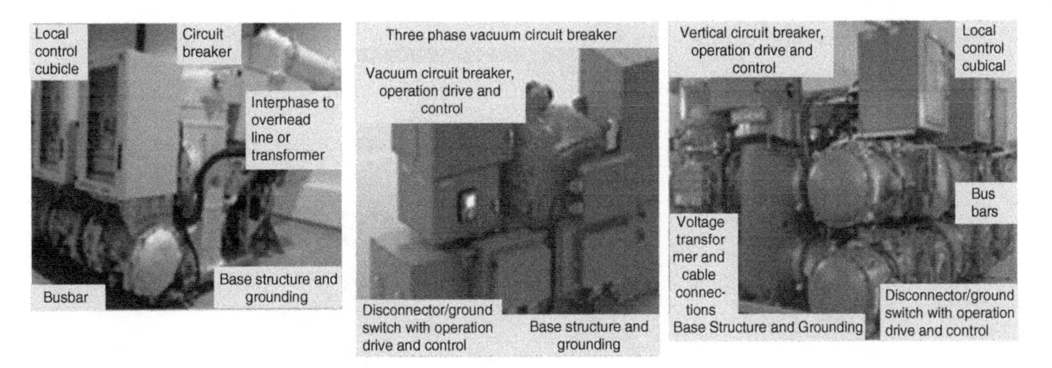

Figure 2.98 Left: 145 kV GIS using flouronitrile for insulation including the circuit breaker, middle: 72.5 kV GIS vacuum interrupter circuit breaker with synthetic air insulation; right: 145 kV GIS Vacuum interrupter circuit breaker with synthetic air insulation, 2019 [105] (courtesy left photo GE, middle, and right photo Siemens Energy)

2.9.6 Transition Period SF_6 to Alternative Gases

The transition period from SF_6 to alternative gases requires different handling procedures and different equipment. The SF_6 handling procedures for GIS are well established. IEEE and IEC standards [51, 52] give detailed instruction on how to handle SF_6 gas. Also, the SF_6 handling equipment such as vacuum pumps, evacuation and filling devices and SF_6 monitoring is available at users of GIS substations.

The alternative gases have different requirements on gas handling and gas mixture processes. This requires separate devices and skilled workers to handle the new procedures. If more than one alternative gas is used by one user, several gas handling procedures need to be developed, workers need to be trained, and suitable devices need to be available. Some devices, such as the vacuum pump, could remain the same.

Figure 2.99 shows the transition phase from an SF6 to several alternative gases related to the gas carts and other gas work equipment.

Today in 2020, there are several gases in use at the transmission and distribution level as shown in Figure 2.100. On the transmission side, left of the line, SF_6, flouroketone gas mixtures, flouronitrile gas mixtures and synthetic air are in use.

At the distribution side, right of the line, SF_6, flouroketone gas mixtures and synthetic air are in use.

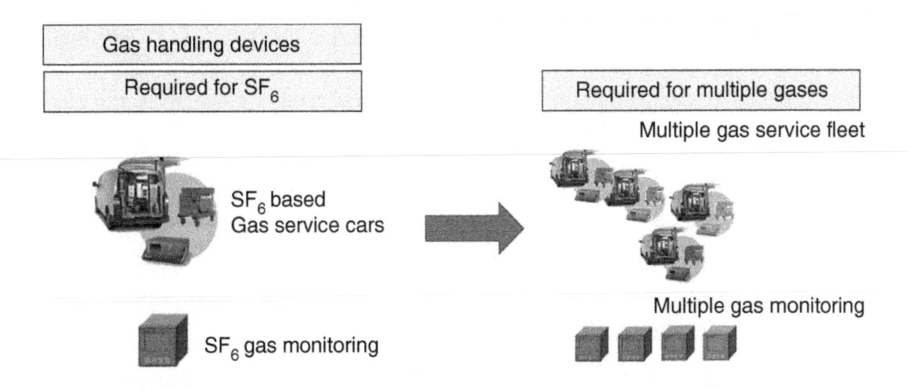

Figure 2.99 Transition from only SF_6 to multiple gases

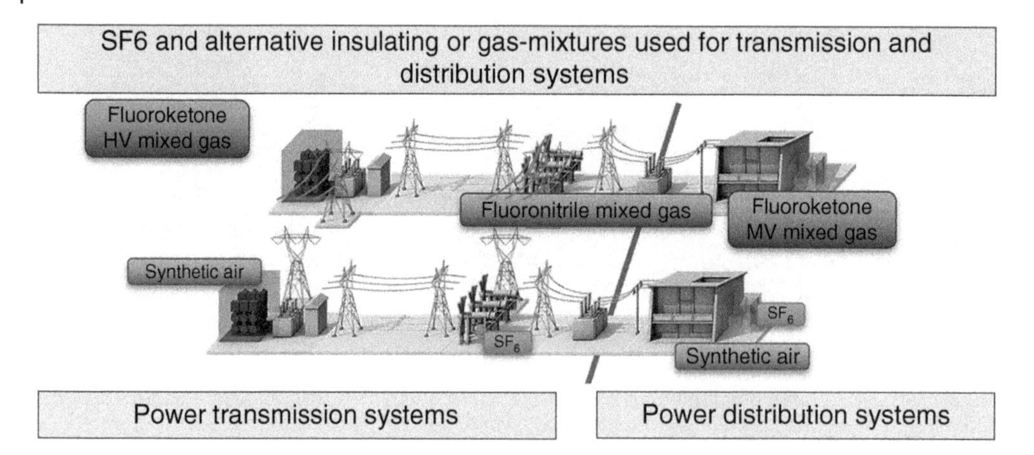

Figure 2.100 Insulating gases used in transmission (left of line) and distribution (right of line) substation 2020

2.9.7 Organizations Dealing with SF_6 and Alternative Gases for GIS

There are several organizations dealing with SF_6 and alternative gases for GIS. The IEEE and IEC are active in several working groups to develop international standardization on handling and application of alternative high voltage insulating gases.

IEEE has started in the Power & Energy Society several activities within the Switchgear and Substations Committees to evaluate the impact of alternative gases [82].

- IEEE Alternative Gases TR 64 (ADSCOM Subcommittee) Impact of Alternative Gases on Existing IEEE Standards [79]
- IEEE Substation Committee" WG K19 "Taskforce of handling non-SF6 mixtures for HV equipment" [80]
- IEEE Switchgear Committee Working Group" Performance of SF6 Alternatives" [81]

In IEC TC 17, SC 17A and SC 17C all standards are reviewed and adapted to cover SF_6 and any alternative insulating or switching gas. The alternative gas properties and handling procedure are treated in IEC 62271-4 in the main text for general requirements and specific features are handled in separate annexes [82].

- IEC 62271-4 "High-voltage switchgear and controlgear: Handling procedures for gases and gas mixtures for interruption and insulation" [51]

CIGRE investigates alternative gases to SF_6 in several working groups with investigations of their physical and technical aspects under the conditions of high voltage [84].

- Cigré WG B3.45 "Application of non-SF_6 gases or gas mixtures in medium voltage and high voltage GIS"

- Cigré WG D1.67 "Dielectric performance of new non-SF_6 gases and gas mixtures for GIS"
- Cigré WG A3.41 "Interruption phenomena with SF_6 alternatives"

T&D Europe is a manufacturer organization which has worked out recommendations and guidelines on alternative high voltage insulating gases [85]

- "T&D Europe Guide to validate alternative gas for SF_6 in electrical equipment" (2016) [86]
- Technical report on alternative to SF_6 gas in medium voltage & high voltage electrical equipment (2018) [87]

ZVEI a manufacturer organization and FNN a transmission and distribution system operator organization have published guidelines and recommendations on alternative gases to SF_6 [90].

- ZVEI Publication (2021) 'SF_6 emission reductio by 60% since 2005' [88]
- ZVEI Publication (2020) 'Scenario for reducing SF_6 operating emissions from electrical equipment through the use of alternative insulating gases [89]

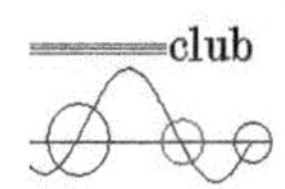

Current Zero Club is an international research group on interruption phenomena of high voltage circuit breaker. They work on SF_6 and alternative insulating and switching/breaking gases and also on vacuum technology [92].

NEMA is a North American based manufacturer organization active in the SF_6 & Alternatives Coalition. The SF_6 & Alternatives Coalition is an industry organization for discussion of SF_6 and alternative insulation technologies as used in electric transmission and distribution equipment, as well as a forum for industry interaction with public officials surrounding emissions reporting and reduction regulations [93].

2.9.8 Summary

The development of alternative gases to SF_6 will significantly lower the GWP for insulating gases used in GIS. This will be a transition phase over several years because in most cases, older equipment cannot be replaced by new alternative gases without changes on the design. For example, different O-ring sealings are necessary and some silver contact material might need to be changed.

The first positive experiences of HV GIS with alternative gases have been made with several GIS in operation since 2015. Operational experience with GIS using alternative gases will increase and will allow to give advice on future applications.

Alternative gas GIS developments are in process to enter higher voltage ratings up to 550 kV. With the vacuum switching technology or alternative gas mixtures for circuit breakers, SF_6 free solutions for voltage levels up to 550 kV are feasible.

2.10 When to Use Gas-Insulated Substations

A gas-insulated high-voltage substation (GIS) is a compact metal encapsulated switchgear and substations equipment arrangement consisting of high-voltage components such as circuit-breakers, disconnect switches, grounding switches, current transformers, voltage transformers, surge arresters, overhead line terminations, underground line terminations, transformer connections, or reactive device connections, which can be safely operated in confined spaces. GIS is primarily used where space is limited, for example, substation expansions, in urban areas, on roofs, on offshore platforms, industrial plants, and hydro power plants.

But there are many other considerations than space for when to apply gas-insulated substation equipment. Since its introduction back in 1968 indoor and outdoor gas-insulated substations have been applied and installed in a variety of applications. Continuous research and development have improved today's generation of gas insulated substations. Some of the general benefits of using gas insulated substations are as follows:

- Economic efficiency
- High reliability
- Safety due to encapsulation
- High degree of gas tightness
- Long service life
- Low maintenance costs
- Easy access and ergonomic design
- High availability
- Reliable operation even under extreme environmental conditions

However, it is important to note that the application of gas-insulated substations as a tool for substation design requires analysis of the specific project requirements. Gas insulated substations offer many advantages over air insulated substations (AIS) especially when specific project requirements and lifecycle considerations are considered [1, 2].

2.10.1 Factors to Consider for GIS as a Substation Design Tool

2.10.1.1 Personnel Safety

GIS substations provide a higher degree of personnel safety than conventional AIS. This is since there are no exposed "live parts" in the substation that can come in contact with operations and maintenance personnel. GIS substations are effectively "arc proof" since there are no exposed live parts and all interruptions and load switching occur in a metal enclosed gas insulated environment.

2.10.1.2 Physical Space

GIS substations require less area as compared to the area required to construct an equivalent AIS. One of the biggest challenges utilities and electrical equipment users face in urban and highly developed environments is finding suitable property for new substations or substation expansions. GIS provides a solution to the space problem, as it requires about 25–30% of the footprint and 30–50% of the height needed for an equivalent AIS. This is especially beneficial in locations where real estate costs are very high or real estate does not exist for construction of a substation. GIS can be installed within a building, on a roof-top, underground or on small plots of land very efficiently, especially when interconnecting lines enter and exit the site underground.

2.10.1.3 Public Safety

GIS substations provide a higher degree of public safety. Since the GIS can be contained within a securely locked and monitored enclosure or building, the potential for accidental or intentional public access is greatly reduced. Second, if an unqualified person or persons were to gain access to the GIS enclosure or building, then since there are no exposed "live parts" in the substation; the potential for injury due to high-voltage (1000 volts or above) contact is significantly reduced.

2.10.1.4 Physical Security

GIS substations provide a higher degree of physical security. Since the substation can be located within an enclosure or building and since the enclosure or building can be positioned within a significantly smaller fenced or walled area than an AIS, then the ability to monitor the substation with surveillance equipment is more easily done and less expensive. Likewise, elaborate perimeter security barriers may not even be needed based on the enclosure or building design. GIS located in an enclosure or building is inherently projectile resistant.

2.10.1.5 Climatic Conditions and Wildlife

GIS substations are less affected by climate conditions and weather events. Since the GIS substation can be located within an enclosure or building, then the effects of routine wind, snow, ice, tropical storm, rain, etc. are significantly reduced. Additionally, the probability of outages due to wildlife contacting "live parts" is extremely low.

2.10.1.6 Extreme and Severe Weather Events

GIS substations are less affected by extreme and severe weather events because of compact equipment configurations and completely enclosed "live parts," except for AIS interfaces (i.e., gas-to-air bushings). This is especially true where extreme or severe weather events generate tornadic activity, high straight-line winds, hail, flying projectiles, or ice formation and heavy rains. Since GIS is generally located within a fortified enclosure or building, the GIS and its associated local control cabinets (LCC's) have additional protection from extreme and severe weather events.

2.10.1.7 Earthquakes

GIS is inherently resilient with respect to the forces of an earthquake. The GIS has a low center of gravity, minimized structure heights and extensions which reduce the effects of external forces and moments. GIS is a highly standardized modular building block design that is seismically qualified as a complete system capable of withstanding at least 0.2 times the equipment weight applied in one horizontal direction, combined with 0.16 times the weight applied in the vertical direction at the center of gravity of the equipment and supporting structure.

2.10.1.8 Pollution

GIS substations are inherently resistant to the effects of air pollutants and naturally occurring contaminants such as salt spray, acids, scales, etc. Since all live parts and operating mechanism live parts are completely sealed and insulated by SF_6 gas, then there are no possibilities of these parts coming in contact with pollutants.

2.10.1.9 Corrosion

GIS substations are inherently corrosion resistant. Since the GIS is located within a climate-controlled enclosure or building, the probability of contamination and ultimately corrosion of equipment is significantly reduced. Additionally, GIS equipment is designed to be installed indoor or outdoor, so the design of the operating mechanisms, linkages, electrical connector, cables, and seals are already corrosion resistant.

2.10.1.10 Availability

Based on average availability factors calculated across the electric utility industry by GIS manufacturers and by IEEE, it is estimated that GIS has a 99.8% availability factor as compared to AIS having a 91.2% availability factor, over the predetermined life cycle of the equipment being 30 years.

2.10.1.11 Maintainability

GIS substations are more easily maintained than AIS substations. This is because the preventive maintenance of GIS as compared to AIS does not require maintenance of "live parts" unless a significant short-circuit event occurs. AIS equipment requires more frequent preventive maintenance of "live parts" as compared to GIS. This is because the AIS equipment is exposed to climate conditions, wildlife, pollution, and corrosion.

2.10.1.12 Meantime of Maintenance and Maintenance Cost

GIS substations require major maintenance of switching equipment only after a significant short circuit event occurs or after 25 years. AIS substations require major maintenance of switching equipment other than circuit breaker internal components every 6–8 years based on electric utility practices.

2.10.1.13 Meantime of Maintenance or Repair

GIS substations will require a longer time to maintain a specific component than AIS substations, but the extent of the outage will be less impactful and the occurrence of maintenance or repair event significantly less frequent. This is due to the fact that when AIS is maintained and/or repaired the electrical clearance areas in the substations are greater due to the required observation of "minimum approach distances" by personnel. It is entirely possible that an entire substation or a portion of a substation would have to be de-energized to repair a single switch in an AIS substation depending on the design.

2.10.1.14 Life Cycle Cost

Based on the fact that the average life of transmission and substation assets is approximately 50 years, the Total Life Cycle Cost of GIS is less than AIS of the same electrical design even though the initial cost of the GIS may be greater. GIS equipment carries an initial premium over the equivalent AIS equipment. While that premium varies greatly based on the voltage class of the GIS, commercial factors, global markets, and manufacturers, a rough approximation (average over all voltage classes) is 125% of the cost of the AIS equivalent. While this is a significant increase when considering just equipment cost, it can turn out to be a minor factor on a project when all other project considerations are evaluated. In fact, the application of GIS may lower the overall cost of the project. This is primarily due to the following factors:

1) GIS offers a significantly reduced footprint for the equivalent amount of AIS equipment. This can translate to savings on real-estate property acquisition. It also allows for additional sites to be considered as options for "Greenfield" projects. GIS has the potential to reduce transmission line costs for a substation installation if a more ideal site location from a transmission access perspective can only be implemented using GIS. Land

acquisition and development costs are greater for AIS substations due to the greater overall area required, including required buffers, site preparation, aesthetics and possible remediation concerns.

2) In the case of "Brownfield" projects, GIS offers flexibility in arrangement, allowing for unconventional layouts and the ability to accept additional transmission lines or transmission lines with an abnormal dead-end locations or cable terminations that an AIS station may not be able to accommodate.
3) Site preparation cost are greater for AIS substations due to the greater overall area required and may require extensive grading, soil structure enhancement, drainage, and retaining walls.
4) Physical security of AIS is costlier as compared to GIS due to the requirement of much more extensive barriers, barrier heights, physical buffers, and surveillance area.
5) Scheduled maintenance of AIS occurs earlier and more frequently in the total Life Cycle of the substation than it does for GIS.
6) Reinvestment for the replacement of individual components of AIS due to wear, corrosion and exposure to environment occurs earlier and more frequently in the total Life Cycle of the AIS substation than it does for GIS.

2.10.2 When to Consider GIS

2.10.2.1 Greenfield Projects

2.10.2.1.1 EHV Transmission Substations

In general, GIS should be considered for any extra high voltage (EHV) substation because of the large land area required to construct EHV switchyards. GIS is a viable option for switchyards that are composed of voltage classes of 345 kV and above. The space reductions of the switchyard area are significant, which affects costs associated with land purchase, permitting, area of grading, erosion, and sedimentation controls, area of stoning, ground grid area, etc.

2.10.2.1.2 When Significant Grading and/or Soil Structure Enhancement Required

Sites that require significant grading can benefit from the reduced footprint of GIS. "Significant" can vary from project to project, but typically if there is a cut and fill of more than 10–15 feet over a large area of the switchyard then GIS should be considered. If any significant soil structure enhancement is required for the switchyard then GIS should also be considered to lower site development costs.

2.10.2.1.3 When Nonstandard Transmission Rights-of-Way Exist

GIS offers flexibility in the layout of dead-ends, which can facilitate nonstandard transmission line entrance locations at the site. This is helpful if the transmission ROW cannot easily accommodate the standard dead-end locations of a greenfield AIS substation. With GIS, the dead-ends can be placed to meet the transmission lines in more favorable locations on the site, or eliminate the need for line crossings, which offers a reliability benefit. For greenfield substations, this could translate to less space required outside the fence for transmission line exits and low transmission access costs.

2.10.2.1.4 When Aesthetics and Visual impact are Key Factors

In areas where aesthetics is a strong consideration, such as in cities, high-cost real-estate areas, resort locations or next to high-traffic areas, GIS located in a building can blend into the aesthetic characteristics of the location and provide an unassuming facade to deter attention from onlookers. GIS buildings can also be designed to match surrounding buildings, this can provide a substation that is more aesthetically pleasing by eliminating the "wire tangle" associated with AIS.

2.10.2.1.5 When Transmission Access Locations are Complicated

If multiple substation sites are being considered, then the total project costs should be investigated when determining the preferred site. If a smaller or more expensive site requiring GIS can reduce transmission routing costs, the difference in Site cost could may be worthwhile.

2.10.2.1.6 When Reduction of Schedule Risks is Necessary

When reducing schedule risk is an important consideration on a project, GIS may provide a viable solution. Since GIS provides almost the entire substation from one source, it avoids the schedule risks associated with the construction of a conventional substation such as an AIS, which requires sourcing of equipment and services from multiple sources. GIS avoids the user having to coordinate between many suppliers, equipment and services. It is essentially one stop shopping for a substation. Also, since GIS requires significantly less site preparation, foundation, and civil works, project schedule risk is also reduced in these areas, this can save significant costs. GIS provides for a reduced time of construction of the substation since the design is modular and essentially "plug and play."

2.10.2.1.7 When Resiliency and Physical Security are Key Factors

The use of gas-insulated substation designs is an effective method to protect a substation and its associated electrical equipment from severe damage from a naturally occurring catastrophic event or a nefarious human-caused physical intentional attack. The use of gas-insulated substation designs is a cost-effective strategy that results in enhanced substation resiliency and higher reliability due to a reduction in substation area, perimeter, the ability to shield equipment and minimized exposure to external threats.

Mobile GIS equipment can also be used to provide complete temporary substation switching arrangements as a proactive solution for the following scenarios:

- Critical loads that develop in short time periods
- By-passes for transmission switching equipment during repairs or maintenance
- Major storm response/recovery/restoration options
- Temporary switching arrangements to aid in expediting construction of substations and lines
- A "ready backup" for critical infrastructure facilities (i.e., government and public services)

The benefits employing GIS designs for use as a hardening tool and a response /recovery/restoration tool for substations, contributes to the overall resiliency of the transmission system.

2.10.2.1.8 When All Price Competitive Solutions are Considered

The general rule has been that a rough approximation (average over all voltage classes) of costs comparing GIS to AIS is that GIS is 125% of the cost of the AIS equivalent. This is a generalization. Each project needs to be analyzed on its own merit. GIS equipment costs for all voltage classes are becoming more in line with AIS of the same configurations, especially for breaker and ½, double bus double breaker and ring bus arrangements. This being the case, it is important to note that just considering the cost of equipment can result in a skewed cost evaluation for the project when all other project considerations are evaluated.

2.10.2.1.9 When Reliability and Availability are Key Factors

There is direct correlation between equipment reliability and maintenance strategies. Typically, the more reliable the equipment the less maintenance is required. GIS is more reliable and requires less maintenance than AIS, since all live parts and operating mechanism live parts are completely sealed and insulated by SF_6 gas. GIS substations are more easily maintained than AIS substations, because the preventive maintenance of GIS as compared to AIS does not require maintenance of "live parts" unless a significant short circuit event occurs.

2.10.2.2 Brownfield Projects

The same considerations for greenfield substation sites apply to brownfield substation sites. There are several additional considerations related to brownfield sites.

2.10.2.2.1 When Expanding an Existing Substation

For any voltage 69 kV and above, when there is a need to expand an existing substation beyond its original ultimate design, or bring more lines in than can normally be accommodated, GIS can often be utilized to allow the expansion to fit in the existing yard footprint. GIS can provide cost savings with respect to building a new substation elsewhere or expanding the existing fence line which will require additional permitting, grading, erosion and sedimentation controls, area of stoning, ground grid area, potential soil structure stabilization, and enhancement etc.

2.10.2.2.2 When New Transmission Access is Needed

GIS offers flexibility in the layout of dead-ends, which can facilitate nonstandard transmission line entrance locations at the site. This is helpful if the transmission ROW cannot easily accommodate the standard dead-end locations. With GIS, the dead-ends can be placed to meet the transmission lines in more favorable locations on the site, or eliminate the need for line crossings, which offers a reliability benefit. For brownfield substations, this could translate to less expansion space required outside the existing fence line for transmission line exits and lower transmission access costs.

2.10.2.2.3 When Constructability is Difficult

In situations where an AIS option will physically fit, sometimes the constraints of equipment outages will make the construction sequence extremely complicated or not executable. In these situations, a GIS option may require fewer equipment outages due to a reduced footprint or different location for the equipment and provide an executable construction sequence.

2.11 Comparison High Voltage and Medium Voltage AIS, MTS and GIS

Author: Michael Novev
Reviewer: Hermann Koch

From the user's perspective, the technology of the equipment used is one of the most important decisions during the planning stage of each project. This decision is based on multiple factors. The purpose of this chapter is to provide comparative information, in order to aid in the selection of the most effective equipment solutions for a new or an extension and refurbishment of an existing substation. The selection of the site is impacted by the type of technology chosen, thus a comparison between air insulated switchgear (AIS), mixed technologies switchgear (MTS), and gas insulated switchgear (GIS) is key. This shall provide the necessary tools to make an informed assessment and help determine the most effective means of developing the substation project with respect to the user, the public, applicable system requirements, and state and local regulations, leading to the best equipment option for the specific application. The goal is to provide a balanced approach in determining whether AIS, MTS, or GIS technology provides the best overall value to the users and their customers.

The decision to either select traditional AIS, GIS, or in some cases MTS technology solution is based on the assessment of many elements. Due to the multitude of requirements, such as public acceptance or cost, to name a few, as well as the abundance of information available, the user's decision may be very difficult. The sheers quantity of selection criteria for AIS and GIS can block a user's decision-making process. To avoid this, once a substation's functional demands are determined, it is essential to find out, and even try to quantify, which factors are crucial to the application.

Some factors, which may seemingly not have a direct impact on the investment or life cycle costs, may still be critical factors in the selection of the equipment technology, for example due to security concerns or public acceptance.

GIS and AIS use proven technologies to provide safe and reliable power. These technologies differ in their approach, with GIS being optimal for limited space requirements, AIS is driven by lowering front-end investment cost, while MTS combines both and is useful in some application. To prove a comparison of these different technologies, a brief characteristic of each one will be provided.

2.11.1 AIS Technology Characteristics

The main characteristic of AIS technology is the use of insulation properties of ambient air at atmospheric pressure as an insulating medium. High-voltage AIS installations are available as outdoor or indoor arrangements. Indoor installations are used significantly less frequently due to the large clearance requirement. Medium-voltage AIS installations are also offered as indoor or outdoor arrangement, with the outdoor arrangements being very similar to those of the high-voltage AIS. Medium-voltage AIS mostly utilizes metal clad. Due to its relatively smaller size, it is typically installed indoors.

Both high-voltage and medium-voltage outdoor AIS installations are constructed in an open, non-enclosed, and noncontrolled environment. The individual components of the installation, such as interconnecting bus, circuit breakers, disconnect switches, grounding switches, instrument transformers various terminations for connections to air-insulated cables, and overhead lines, are individually specified, designed, delivered, and installed at the site. Typical installations use concrete foundations or screw piles with galvanized steel supports for each individually installed component. All equipment is exposed to the elements and mostly visible from nearby surroundings. Due to the nature of these installations, modular design solutions are difficult to implement and although some elements, for example a circuit breaker with disconnect switches, do exist, they are not adaptable, and rarely used. See Figure 2.101 for typical AIS outdoor substation.

Indoor AIS installation are available for both high and medium voltage. Due to the large clearance requirements, especially at voltage levels 230 kV and above, such installations are extremely rare. Medium-voltage AIS installation are usually comprised of metal clad switchgear. They are relatively smaller in size and provide some protection for the equipment and operation personnel.

2.11.2 GIS Technology Characteristics

The main characteristic of GIS technology is the use of sulfur hexafluoride (SF_6), or other gasses and even dry air as an insulating medium. Major characteristics, advantages, and disadvantages of GIS technology are valid irrespective of the gas used, with slight variation due to the specific gas properties. Although there are different gasses used in GIS installations, so far, most installations use SF_6 as an insulating media. The comparison will be based on SF_6 installations, unless specifically mentioned otherwise.

Major components such as buses, circuit breakers, disconnect switches, grounding switches, instruments transformers, etc. and other associated equipment are built as a multicomponent

Figure 2.101 Typical Outdoor AIS substation (*Source* [96])

assembly, enclosed in a grounded metallic housing. High-voltage GIS installation use modular type design, which makes them more adaptable and significantly decreases site installation time. Medium-voltage GIS installations are arc-resistant metal clad type design. For both high- and medium-voltage GIS major components are contained in a sealed environment. They are prefabricated and in the case of high-voltage GIS partly preassembled, while the medium-voltage GIS switchgear is completely preassembled.

High-voltage GIS installations are available as outdoor or indoor arrangement. As a rule, GIS up to 170 kV voltage level is installed indoors. Due to the features of this technology, the required space is significantly reduced, which allows installations up to 500 kV to be mounted indoors. Outdoor high-voltage GIS installations are also utilized, but in lesser extent. Medium-voltage GIS installations are typically installed indoors.

In addition to AIS and GIS technologies described above, there is third type of equipment available for high-voltage installations, namely hybrid technology switchgear. This type of switchgear is also known as dead tank compact switchgear (DTC) or mixed technology switchgear (MTS). This type of switchgear represents a mixture of both technologies. Hybrid Substations are the combination of both conventional AIS and GIS. The main components such as circuit breakers, disconnectors, grounding switches, and instrument transformers are based on GIS technology, while busbars and bushings are the conventional AIS.

Hybrid technology switchgear provides adaptive modular solutions, which are mostly used for refurbishment and/or extension of existing air-insulated indoor and outdoor installations. It is best fitted for applications where additional feeders must be added to existing installations with limited space availability. It is also applied in industrial plant systems. Hybrid technology switchgear is typical used for systems of voltage levels of up to 230 kV. An example of a hybrid substation would be the use of AIS equipment in substations with gas-insulated busbars; standalone GIS modules with air-insulated busbars; or when some bays in a substation are the gas-insulated type and some are the air-insulated type. Figure 2.102 shows typical DTC installation.

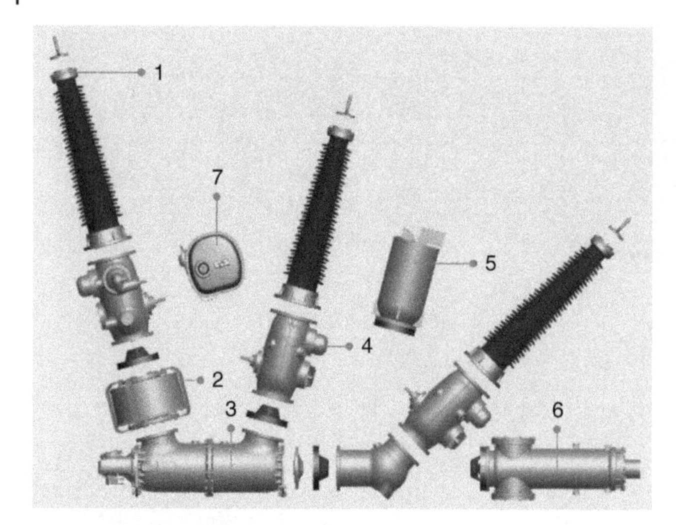

Figure 2.102 Typical Dead Tank Compact Switchgear (DTC) components. 1. Bushing; 2. Current transformer; 3. Circuit breaker with self-compression principle; 4. Three-position disconnector/earthing switch; 5. Voltage transformer; 6. Cable connection assembly; 7. High-speed earthing switch (*Source* [108])

This approach combines the advantage of the high reliability of GIS technology and the high maintainability of AIS-based equipment. Easy accessibility and quick replacement of the defective devices of AIS results in modular solutions for overall substation design.

The design and construction of each technology differs significantly due to the different characteristics as described above.

2.11.3 Design and Construction of AIS

A common factor between outdoor and indoor AIS installations is the use of air at atmospheric pressure as an insulating medium. In AIS power substations busbars and connectors are visible. In these power substations circuit breakers and disconnect switches, instrument transformers, etc. are installed either outdoors or indoors. Conventional AIS design is well established, general guidelines are also provided by CIGRE WG B3-03 [3].

A significant factor for the proper operation and maintenance of equipment and for auto-reclosing systems is the ability of air to restore its insulating properties after disconnection of the voltage, which requires large clearances. This is especially important for high-voltage installations. The design of AIS stations is contingent on having sufficient clearances and area for the installation.

In addition, air insulation is dependent on environmental conditions such as ambient temperature and altitude. In the case of medium-voltage AIS, with installation mainly indoors, space requirements difference is not that great; however, these installations are also dependent on environmental conditions especially high altitudes.

The individual electrical and mechanical components of AIS installation, for both indoor and outdoor types, are assembled predominantly on site. In open design, all parts are insulated by the air and thus not covered.

Air-insulated outdoor high-voltage electrical equipment is generally covered by standards based on assumed ambient temperatures and altitudes. Ambient temperatures are generally rated over a range from −40 °C to +40 °C for equipment that is air-insulated and dependent on ambient cooling. Equipment rated to −50 °C is also available as required, although circuit breakers typically required tank heaters at lower temperatures.

Reliance on ambient air dielectric properties requires de-rating of the equipment at altitudes above 1000 m, due to the air density. Operating clearances must be increased to compensate for the reduction of the ambient air dielectric strength. Ambient air is also a cooling medium for dissipation of the heat generated by the load losses associated with load current levels. Therefore, the current ratings generally decrease at higher elevations.

There are different types of substation bus/switching arrangements used in air-insulated substations. These arrangements affect the reliability of a substation. AIS components and bus arrangements are well known in the industry and will not be subject to detailed review in this chapter.

Outdoor high-voltage or medium-voltage AIS installation use foundations and galvanized steel structures to support the equipment. The profile of the substation structures and equipment consist of either large lattice and box-type structures supporting overhead strain buses or low-profile structures and rigid busbar. Lately, substations use predominantly low-profile designs. Busbars are supported on posts or strain insulators with all busbars and connections visible. To satisfy minimum safety clearances, the height limitations of busbars may cause the use of low-profile construction, which sometimes result in arrangements of increased area, particularly for the lower voltage levels.

Indoor installations for high-voltage AIS are relatively rare, medium-voltage AIS metal clad is installed mostly indoor in either prefabricated or site-installed buildings.

2.11.4 Design and Construction of GIS

Both high-voltage and medium-voltage GIS installations use the same components as the conventional AIS substations, such as circuit breakers, disconnect switches, instrument transformers, and surge arresters. The difference is that GIS uses either SF_6 or other pressurized gasses as an insulation media, so the active parts are not exposed to the elements. The atmospheric air insulation used in conventional AIS requires meters of air insulation. Pressurized SF_6 gas allows for clearances much smaller than those of AIS substations.

In any GIS installation, both high and medium voltage, SF_6 gas transfers the heat produced by the current in the conductor by conduction and convection. The cast epoxy support isolator materials mechanical strength does not decrease until to around 120 °C. As such, temperature boundaries are defined by the conductor contacts at 105 °C. As a standard requirement, the GIS exterior temperature is limited to a safe to touch temperature of 60 °C or of 70 °C with touch protection. The temperature difference between the conductor and enclosure is about 25 or 35 °C. If required, theoretical and empirically derived calculations are used to determine the temperature, including the enclosure heating due to circulating currents or solar radiation in the case of outdoor installations.

The first-generation high-voltage GIS installations used multiple elements in a single large gas compartment. However, in case of failure of any equipment, substantial time and effort was necessary. GIS repair required highly specialized skills level, and access to primary elements was difficult. Thus, the removal of adjacent elements was sometimes necessary to reach the element to be worked on.

Newer GIS installations use modular components comprised of different elements, such as circuit breakers, disconnect, and grounding switches, installed into separate gas-tight compartments. Parts are supported by cast resin barriers or partitions, which serve multiple purposes and are enclosed in metal housing. They support the primary conductor, serve as insulation of the conductor from the ground, and can withstand the differential pressures that occur between compartments during maintenance. This allows the entire installation to be subdivided into compartments, which are gas tight with respect to each other. Modular technology gives modern GIS installation much greater flexibility and substantially decreases installation and repair time. Some of the features of high-voltage GIS are presented in IEEE white paper [95]. Figure 2.103 represents a cross section and physical assembly of 363 kV GIS.

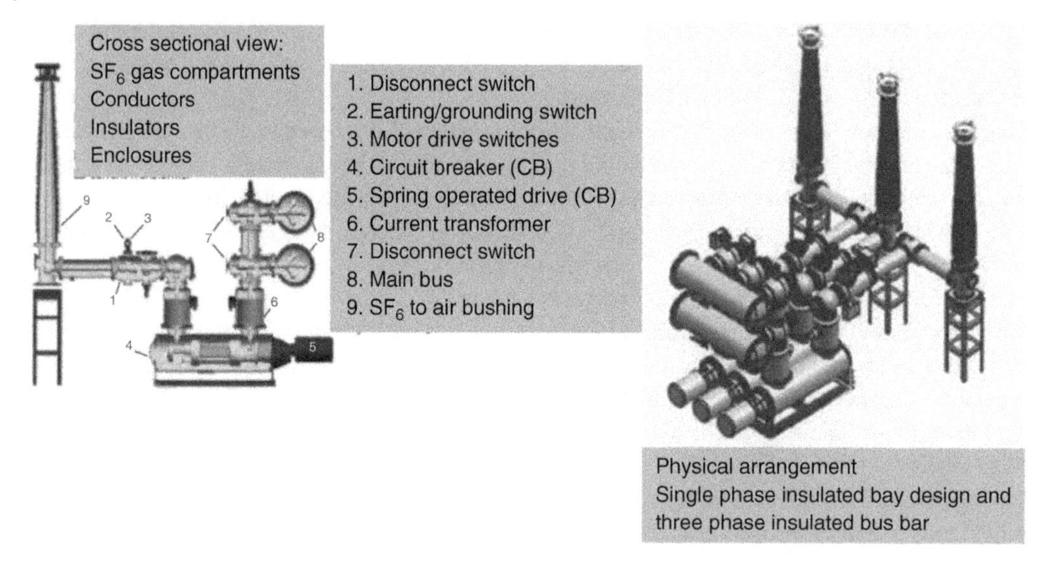

Figure 2.103 Cross section and physical assembly of 363kV GIS (*Source* [96, 109])

The active parts are not exposed to atmospheric air, moisture, contamination, etc., which protects them from deterioration. Modular components can be assembled in a variety of ways, which allows for flexibility in the switching schematic arrangement and substation needs. Medium-voltage GIS switchgear also uses separate gas compartments for the different elements.

Although there are different gasses used in GIS installations, up to date most installations use SF_6 as insulating media. Properties of other gasses used are covered in Section 2.9.

Pure SF_6 has no known toxic properties and contains no dangerous components. Its density is about five times higher than air and one of the heaviest known gases. SF_6 is about a hundred times better than air for interrupting arcs. SF_6 has two to three times the insulating ability of air at the same pressure. At high temperatures of electrical arc, SF_6 disintegrates, however, after arc dissipates the disintegrated gas turns back into SF_6. Due to this physical property, no replacement of SF_6 is necessary in GIS. Because of moisture, air, and other contaminants, there are some reactive decomposition by-products present in very small quantities. Such by-products are eliminated using molecular filter absorbents inside the GIS enclosure. For details see Sections 2.2 and 2.8.

Another important characteristic of SF_6 to consider when designing GIS is the pressure increase due to internal faults and degradation due to interruption arcs and switching sparks. The pressure growth from breaking and switching sparks is small and can be neglected. In case of an internal fault, the SF_6 gas will be degraded by the arc heat. Most of the degraded SF_6 will fast recombine into SF_6, but some of the free molecules (S, F, etc.) may react with impureness in the SF_6 gas (mostly water vapor) to create degradation by-products such as HF, and SOF. The GIS design considers the use of absorbents to limit the quantity of degradation by-products. The pressure increase inside the enclosure during an interal arc fault can reach high levels. To avoid rupture to the enclosure, both the high- and medium-voltage GIS, is equipped with pressure relief devices. During the design of the enclosure release of pressurized gas shall be considered and pressure relief devices directed either in areas not accessible to personnel or in some cases outside the building, see also Section 2.2.

The enclosures used in modern GIS installations are of nonmagnetic materials such as aluminum or stainless steel and are grounded. Aluminum is predominately used due to its light weight and superior corrosion properties. Modern GIS switchgear has a high degree of versatility due to the use of modular design. Components are housed in separate modules either individually or combined, depending on the respective function. Modules are connected by flanges. The gas tightness is provided using static "O" ring seals, so each module is in a separate pressure resistant gas tight enclosure.

Each GIS has a gas monitoring system. Gas pressure and density inside each compartment is monitored continuously. The pressure of the SF_6 gas varies with temperature, so a temperature-compensated pressure switch is used to monitor the equivalent of gas density.

Aluminum is almost exclusively used as a conductor in GIS installations. Copper is used in some special applications. Support insulators are epoxy resin cast. Post, disc, and cone type support insulators are used. Each GIS manufacturer develops their own material formulation and insulator shape to optimize the insulator in terms of mechanical strength, electric field distribution, and surface electric discharges resistance.

Different components in each bay are joined by connection modules. They are also used to connect equipment, which may be at a distance from the GIS, such as power transformers and overhead lines. Attaching different components to the bus, such as a voltage transformer, surge arrester, or outgoing feeder module, are provided using T modules. To turn a bus at an angle, angular modules are utilized. Each manufacturer's design provides standardized connection modules. Termination modules provide a separation between pressurized gas-tight enclosure of GIS and other insulation media. They are used to connect GIS to other station equipment such as cables, overhead lines, transformers, and reactors.

High-voltage GIS uses similar dead tank SF_6 puffer circuit breakers as the ones used in AIS. Modern high-voltage GIS circuit breakers use the self-compression quenching system principles exclusively. Medium-and high-voltage GIS utilizes vacuum-type circuit breakers. Current transformers, generally, are conventional induction current transformers. Primary winding is formed by the GIS high-voltage bus. Voltage transformers are conventional induction types with iron core. Low power instrument transformers using Rogowski coil, capacitive dividers or optical sensor offer high performance sensor and measuring and are used more often, specially in the case of smart grid requirements with renewable energy supply. For details see Section 9.14.

Disconnect switches in both high- and medium-voltage GIS are used for electrical isolation of parts of the circuit. The functions of a disconnecting switch and a grounding switch are often combined in a three-position switch. The moving contact closes the gap or connects the high-voltage conductor to the contact of the grounding switch. Both functions are mechanically interlocked. The moving contact is relatively low acting and the disconnect switch is used only for off load. In high-voltage GIS, the disconnect switch can interrupt low capacitive current (switching of unloaded busbar sections) or inductive currents (transformer magnetizing current). Contacts are usually silver plated and lubricated to avoid release of microparticles during switching. Ground switches use a moving contact to open or close a gap between the high-voltage conductor and the enclosure. When fully closed, it can carry the rated short-circuit current without damage, for the specified time period (one or three seconds). For service in North America, in both high- and medium-voltage GIS installations, the enclosure will be provided with view ports (or viewing cameras), through which the switching position of all three phases is visible.

At high-voltage GIS, the fast-acting ground switch has a high-speed spring drive and can be closed twice during a full-rated short-circuit current without significant damage. In this type of switch, the grounding pin at earth potential is pushed into the stationary contact. The fast-acting grounding switch is equipped with a spring-operated mechanism. Grounding switches, which have a short-circuit-making capability, are also used for grounding line or cable entrances and busbars. Fast-acting ground switches are usually installed at the connection point of the GIS to the rest of the station.

To protect against any overvoltage that may occur, ZnO_2 resistor discs surge arresters are used. For high-voltage GIS, the arrester is usually flange-jointed to the switchgear via a gas-tight bushing. The housing incorporates an inspection hole to inspect the internal conductor. Medium-voltage GIS uses plug in surge arresters typically installed outside the gas compartment.

High-voltage GIS installations are predominantly installed indoor. CIGRE B3-02 [15] provides guidance about design and applications of HV GIS. Even in the case of an outdoor installation, the size is much smaller compared with a similar AIS installation. Since enclosure is grounded, there is no need for safety clearances, which allows for lower profile. Medium-voltage GIS is installed indoors, and the smaller size allows for use of prefabricated buildings.

2.11.5 Considerations for Selection AIS or GIS Technology

In order to simplify the process, selection factors can be classified into three major groups: power system demands, such as availability, reliability, and load flow; environment, such as location and climate; and economic, with respect to installation, operation, and maintenance costs.

Whether it is substituting older infrastructure, upgrading existing out of date arrangements, expanding capacity, or building new facilities, recognizing the most relevant elements helps the end user to find the best technological option and thus optimizing substation design. Each element may not have a direct impact on the investment or life cycle cost, however, may still be a determining factor in the choice of substation location and construction design. In some cases, even a single condition can be a determining factor, such as space limitation for example.

2.11.6 Power System Considerations

Considerations for selecting AIS or GIS technology are evaluated in the asset planning stage. At this stage, different elements such as availability, reliability, and load flow needs are assessed. Each one of these factors can be evaluated by comparing the advantages and disadvantages of AIS and GIS technologies, respectively [16].

The selection of the right technology predominantly depends on location. For rural areas, space is not a concern, which make AIS technology a more economical solution. Being able to select substation sites, which satisfy above conditions in urban areas, especially major cities, is increasingly difficult [11]. Sites that are large enough for AIS substations are seldomly available, and the costs associated with the land are extremely high. GIS provide a solution to the space problem, as it requires about 15–25% of the footprint and 30–50% of the height needed for AIS, depending on the voltage with higher voltages being lower percentage. GIS technology also offers further benefits, which are sometimes harder to quantify, but which can be a decisive factor in selecting the type of technology. For example, if required the size of GIS enables a HV substation to be built in an existing building with minor modifications when no extra site area is available. AIS is also available for indoor applications, although due to the large clearances required, it is difficult to find the space to construct a large enough building. Even if the space is available, there are other disadvantages for high-voltage AIS indoor installations, namely, the need for periodic maintenance, as the insulators require cleaning, which is typically done by the rain in outdoor conditions. Figure 2.104 shows space savings using GIS.

Medium-voltage AIS metal clad is better suited for indoor installation, although it also requires a larger space in comparison with GIS. Additionally, a medium-voltage GIS helps save even more space. Although not on the same scale as high voltage, medium-voltage GIS provides about 30% size reduction compared with AIS.

Most challenges are faced due to the load increase and consequent need of new substation in major cities. The goal is to locate new facilities as close as possible to the load, allowing a much

Figure 2.104 Space savings using GIS instead of AIS (*Source* [108])

more efficient configuration for both the high-voltage system and the medium-voltage distribution network. Minimizing the need for large and costly upgrades to both systems leads to overall higher reliability and system stability. Failure to carefully consider these factors can result in excessive investment in the number of substations and associated transmission and distribution facilities. If AIS technology is used, the substation would be in a less-densely populated suburbs of the city due to the larger space required. This will necessitate costly expansion and upgrade of medium-voltage distribution systems and will additionally lead to increase of power supply losses. Figure 2.105 shows a 132-kV transformer substation in a city center in Spain. It is constructed underground and a park built over it which harmonizes with the surroundings. Figure 2.106 shows installation of GIS in an existing historic building in the old part of a city in Switzerland.

Another problem faced by users is that power supply from existing AIS installations at inner cities can rarely be interrupted for a long period of time to allow for major equipment overhaul. The solution is to build high-voltage GIS adjacent to the existing AIS equipment in stages without major interruption. In such cases, the use of GIS may not only be the most cost effective but also the only available solution. CIGRE publications [11, 12] provide specific examples of city projects in different countries. One of the final steps during upgraded substation commissioning is load transfer into the new facility. GIS technology has a benefit over AIS technology due to its smaller footprint and interface flexibility. This means that the GIS building can be constructed closer to the connection points and therefore decrease cutover durations.

Availability and reliability are two crucial factors for the electric power system and its elements. Depending on the location and importance, some facilities are more sensitive to interruptions and require higher reliability. Reliable power supply is an integral element of the electricity cost, power outages due to failure of equipment are not tolerated and subject to financial penalties.

With respect to reliability, there is a difference between both technologies. AIS switchgears are more vulnerable to faults because they are exposed to the elements. They are susceptible to direct lightning strikes, strong winds, heavy rains, etc. All these lead to accelerated aging. GIS, on the other hand, has substantial advantage over AIS, due to the equipment being enclosed and completely independent from the environment equipment. Inside the enclosed gas compartments, the primary conductors have complete protection against all external effects. The influence of ambient considerations does not affect the high-voltage part, which improves the GIS reliability. Because of the smaller space requirements, high-voltage GIS installations are mostly indoors, which provides even more protection from the elements. A number of CIGRE publications compare AIS and GIS technology's reliability [100]. The time between failures in GIS depends on the voltage class, and according to CIGRE publications the trend is an increase in the time between failure and hence an

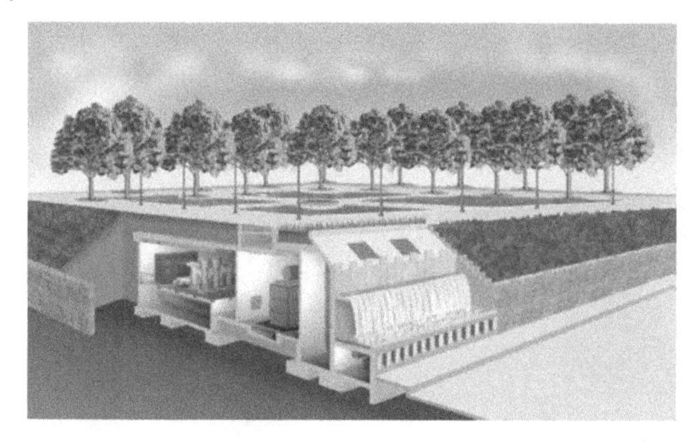

Figure 2.105 GIS substation under a park in a city center (*Source* [98, 111])

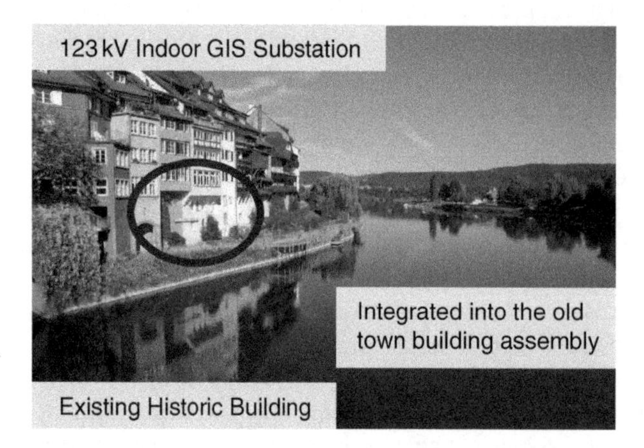

Figure 2.106 GIS substation within a historical building in the old part of city (*Source* 10, 99)

increase in reliability of the new generation of GIS [13–16]. IEEE also has a number of publications showing GIS superior reliability [101].

2.11.7 Environmental Considerations

Environmental conditions are consideration for any electrical equipment. However, some unusual harsh conditions can have significant weight in choosing the technology. Such conditions may be excessive or aggressive air pollution, high altitude, or storm surge.

At normal urban and even city suburb locations, air pollution is not a concern. Based on this condition solely, the use of AIS is more justifiable. However, in some locations such as major industrial areas, airborne pollution can be high. If AIS installations are used, they require periodic maintenance to wash the insulators. GIS equipment is usually installed indoors. Moreover, most GIS elements are hermetically closed inside an enclosure that makes GIS technology a superior solution in harsh atmospheric conditions.

Another location where the use of GIS is preferable due to harsh environmental conditions is desert areas with fine dust contaminants or regions with high humidity and hot climate. Other examples of locations where GIS technology has definite advantage is coastal areas with high saline contamination or aggressive chemical environment in some refineries or chemical plants.

Locations with higher seismic level present unique challenges for the design. The power of substation elements to resist forces produced by earthquakes is related to the equipment design and center of gravity of the installation. Due to the low profile of GIS installations, they have a better seismic capability than similar AIS.

Smaller space requirements combined with high resistance against the elements make GIS the only possible option for offshore windfarm or ocean-based oil rig applications. This was discussed in detail in a different chapter of this book.

Extreme temperatures, both high and low, can be another challenge faced by users in some cases. AIS faces a challenge in dissipating heat, due to its outdoor nature, where there is high solar radiation exposure and almost no wind. On the other hand, GIS, since it is typically installed indoors, provides a much better environment. In such cases, the combination of high-voltage and medium-voltage GIS in the same building provides the best solution. On the other extreme, the use of AIS in extreme low temperatures of up to −50 °C in some locations may lead to an issue with SF_6 circuit breakers condensation. If the temperature is lower than −20 °C, the pressure of SF_6 in the circuit breaker increases, thus requiring the tank to be heated. Using GIS equipment, which is sufficiently small to fit indoors will provide a solution to this problem. Only those GIS elements projected outside the building enclosure have to be designed for extreme outdoor temperature. Figure 2.107 shows a 330 kV indoor GIS substation at −50 °C (−58°) F outdoor temperature.

Storm surges in flood-prone areas along the coast are another potential harsh environment zone. Major hurricanes in the last few years caused storm surges, which flooded large areas along the cost including electrical substation, which lead to design considerations to eliminate or reduce such damage. There are not many available options. Moving the high-voltage stations inland is not an acceptable solution since it moves the whole infrastructure away from the load. Doing so will require extending medium-voltage distribution back to the coast. Rising the whole high-voltage stations above the storm surge is possible, however costly and will create an issue with public acceptance of high structures. The most optical solution is to use high voltage, combined with medium-voltage GIS installations in buildings, which will be mounted on pilons above the water surge level.

Figure 2.107 330 kV indoor GIS substation at −50 °C (−58°) F outdoor temperature (Used with permission of GE Power)

High altitude above sea level is another unique challenge. AIS installations depend on ambient air as its isolating and cooling medium. Due to the greater temperature increase and a lower dielectric strength at higher elevations, AIS installation equipment must be derated accordingly. On the other hand, GIS is totally encapsuled and its properties do not change with increase in altitude.

2.11.8 Economic Considerations

When AIS is compared to compatible GIS technology with respect to the equipment price only, AIS technology will be a less-expensive solution. However, all factors must be taken into consideration including life cycle cost such as equipment price, site property purchase, site preparation, construction, quantified reliability, operational costs including planned maintenance and repair maintenance, operation cost including power losses, costs related to equipment failure-driven outages and life expectancy with a defined interest rate on project specific basis. This method provides a more holistic approach for the assessment of the relevant items, weighted by the end user, to calculate the overall life cycle price. This assessment will provide different results based on specific project applications and location.

Detailed analyses on life cycle cost with examples is provided in Section 9.3 of this book, therefore only major points will be mentioned here.

Front end investments cost includes equipment, land acquisition, permitting and licensing, engineering design, site preparation and civil construction, electrical installation, commissioning, and personnel training.

Property size requirements using GIS technology are only 15–25% of the footprint needed by AIS stations, which is one of the most important advantages of GIS. Nevertheless, for rural areas with enough available, relatively low-cost land, this may not be an advantage.

Under the right circumstances, GIS technology can provide better reliability for the same budget. It is paramount that a reliability-based life cycle cost planning approach is used when providing reliability assessment such as N–1 assessment.

Important consideration when deciding on technology is not only the cost of the specific station, but the overall cost to upgrade and increase capacity of the network, especially in the case of major inner-city upgrades. In most cases, the use of both high-voltage and medium-voltage GIS technology provides the best and most economical solution [102].

Planned maintenance cost includes the labor and materials required to perform the work and outage costs. For any high- and medium-voltage installation, planned maintenance plays an important role in achieving maximum life span and minimum failure during operation. GIS-planned maintenance requirements and frequency is discussed in detail in Chapter 7 of this book. Here, maintenance will be reviewed only as comparison between AIS and GIS technologies. As already discussed in the characteristics of each technology, there are major differences between GIS and AIS. In both high-voltage and medium-voltage GIS, the main components are enclosed and protected from the environment. SF_6 does not age or deplete and equipment is encapsulated. It can operate reliably even under extreme environmental conditions. As such, site location has very little influence on maintenance with respect to both frequency and how extensive the work is. AIS on other hand must deal with humidity, dust, and contaminants. Therefore, high-voltage AIS site location has a major impact on frequency and duration of maintenance. Medium-voltage AIS metal clad is installed indoors and somewhat protected; however, it is still dependent upon site conditions although in lesser effect.

The maintenance strategy optimization is of the utmost importance for each electrical installation. Both high-voltage and medium-voltage maintenance strategies for GIS are predominantly condition maintenance. For AIS, installation strategy depends exclusively on-site location, with both scheduled and condition maintenance strategies used. Due to reasons mentioned above, GIS requires much less scheduled maintenance. AIS, on the other hand, require regular maintenance compared to indoor substations, especially if located in a contaminated environment. Either way AIS maintenance is much higher, both as frequency and duration, compared with GIS for both high-voltage and medium-voltage installations. It should be mentioned that GIS maintenance processes requires extra-trained personnel, thus appropriate training programs shall be considered.

Frequency of planned maintenance for GIS is much lower than AIS because most elements are preserved from the environment. One of the advantages of GIS over AIS is the minimal servicing that is needed of the GIS. This is mainly due to the breakup of the conductors and isolators from the outside ambience. Modern GIS devices have very low SF_6 gas leakage rates. The operation counter may help in finding out if any servicing will be needed on the mechanisms, but there are generally many years between maintenance.

There is also direct correlation between equipment reliability and maintenance strategies. Typically, the more reliable the equipment, the less maintenance is required. As discussed already, GIS is more reliable and hence requires less maintenance than AIS [103].

An equipment failure results in penalty costs for undelivered energy (including profit losses) and component replacement costs. Unplanned maintenance and/or repair also differ substantially between both technologies. Failure is directly related to the reliability of the electrical installations, which was discussed already. Statistically, GIS installations are much more reliable [6, 7]. The nature of both technologies influence both failure frequency and impact. AIS is prone to significantly greater failure rate. However, failure in AIS installations is easy and inexpensive to repair and usually require less lead time for delivery of replacement parts, as utilities frequently have readily available parts in storage. The repair and maintenance include the labor and replacement parts as a direct cost. GIS shows a significant advantage over AIS in respect to monetary impact of failures. Modular design of modern GIS allows for less down time during repair. However, it is important to develop a replacement part strategy, as GIS replacement elements may not be as easily available as AIS.

2.11.9 Life Expectancy

AIS installations typically have a life expectancy of 35–40 years. Installed equipment is exposed to environmental conditions, which have an impact on aging. In comparison, GIS installations are enclosed and completely isolated from the environment. SF_6 does not deplete, and equipment is not subjected to harsh environmental conditions.

Some of the first generations of GIS dating back to 1967 are still in service. Expectations for the newer generation GIS are even higher and service life should be longer. A lot of the failures observed in the first GIS generations do not exist with modern GIS, because of numerous design improvements. As such, GIS operating life can easily reach 50 years and with proper maintenance can be prolonged further.

2.11.10 Construction

This chapter will not go into details and analysis of the construction process involved, instead the goal is to provide users with major differences between AIS and GIS technologies related to design and construction.

Site preparation depends to a greater extend on location and includes all earth work such as cut, fill, grading, and drainage. The volume of work involved is directly related to facility layout size with GIS substations being much smaller in size, thus minimize the degree of earth and civil works.

One of the significant advantages of GIS when comparing both technologies is design compactness and the use of the modular system design. The standardized modular structure has the flexibility to match the various project requirements and allows for almost all substation configurations, while keeping the layout much smaller compared with compatible AIS installation.

The modular system using GIS components allows for the creation of any circuit configurations schemes in the most optimized way. Based on the nominal voltage, the bay design and the construction of the GIS have a wide range. Irrespective of the bus arrangement, the GIS modules can be customized to provide the most optimized and smaller site layout.

The difference between both technologies is that AIS requires individual elements to be installed on separate foundations and steel structures. This makes the site civil construction process slower. Prefabricated GIS shipping elements usually decrease field installation prices and time. On the other hand, GIS technology often requires construction of a building to accommodate indoor installation of the equipment. The outdoor GIS has similar amount of work with tighter tolerances.

Depending on the size and voltage level, GIS can be installed in either site-constructed building, typically for all installations above 230 kV or facilities with large number of bays. The requirement of a site-constructed building increases the complexity of the design and construction. Another option for smaller installations is the use of a prefabricated building. Depending on the size, in some cases the whole assembly can be delivered to site in components and put together locally. For medium voltage, both AIS and GIS are typically installed indoors. This will be either part of the site-constructed building in case of GIS, or alternatively pre-engineered enclosure assemblies. Although medium-voltage AIS metal clad is about 30% larger than GIS, it is small enough to be installed in a pre-engineered enclosure, especially one shipped split in multiple sections to site.

2.11.11 Building Requirements

Design and construction of a GIS building shall accommodate for site installation of factory assembled equipment. Individual preassembled sections whose weights exceed several hundred kilograms to a few tons must be assembled on site in the building. This requires the GIS building design to accommodate for sufficiently robust ceilings and for the structure to be able to support the load, especially when an indoor crane is utilized, which is the case with most high-voltage installations. Additionally, the floor construction must withstand the forces produced by forklift equipment transporting the GIS bays. The weights are provided by the GIS manufacturer. Medium-voltage GIS only require sufficient floor baring capacity design and construction.

Another consideration during design and installation of both high- and medium-voltage GIS enclosures is the possibility of an internal fault within the equipment. In case of internal arc, the pressure inside the enclosure rises and triggers the pressure relief device, thus expending gas into the building space. The building needs to have pressure release shafts to the outside. GIS building wall, ceilings, and floors should be designed to withstand pressure increase. The pressure load is determined by calculating the maximum release for the largest enclosure, as well as the short-circuit rating provided by the GIS manufacturer.

Prefabricated enclosures can be used in lieu of conventional concrete buildings, especially in cases of smaller size high-voltage GIS. If separate MV switchgear, control, and protection equipment is used, it is installed predominantly in prefabricated enclosures. There are certain cost benefits for using such an enclosure in lieu of concrete buildings; however, it is project- and site-specific and will not be discussed in this chapter.

Any pre-engineered type enclosures due to their smaller size would need to be designed to withstand possible gas release. In some cases of medium-voltage GIS installations, a plenum extending the whole length of the lineup is installed and a release outside the enclosure is designed.

As an overview when comparing both technologies, AIS installations take longer to complete and, however, have a higher tolerance in construction. GIS has tight tolerance requirements, however requires a lot less time to complete.

2.11.12 Equipment Installation

With respect to equipment installation, GIS is assembled and handed over in large units. Generally, for voltage levels of up to 145 kV, two whole bays can be shipped completely assembled. For voltages up to 500 kV, either whole bay or partially assembled bay may be transported to site for assembly. This is in contrast with AIS, where all components are installed on site, and while the installation is relatively easy, it is much longer. Figures 2.108 and 2.109 below show the difference.

Site installation of GIS is much faster and requires less laydown area. On the other hand, GIS installation is more stringent, requires specialized skills, and is always under manufacturer supervision. However, due to the faster installation time, GIS installation is at about 80% of the cost of AIS station installation.

During equipment assembly, cleanliness poses a major difference. AIS installations are assembled in open air environment, thereby sanitation is not a major concern. In contrast, cleanliness is of major importance during GIS site assembly; this is why GIS sections are shipped pre-assembled to the largest extend possible to minimize site assembly. There are two main reasons, which cause a decrease in dielectric strength of GIS, namely moisture and contaminant particles. Both can be introduced much easier during installation on site versus a factory installation.

If significant amount of moisture is introduced inside any GIS compartment, a liquid film forms on the surfaces of the solid epoxy support insulators and it can lead to a dielectric breakdown. High-voltage GIS design provides enclosures usually equipped with absorbents to keep the moisture level in the gas low and to prevent moisture from developing over the insulation of the internal surfaces. Levels of less than 1000 ppmv of moisture for SF_6 gas are typically required and those are not difficult to achieve during the installation process.

Another concern is the small conducting contaminant particles. Any such particles can be moved by electric fields to either the insulator surface or where the electrical field reaches higher level. In both cases, this will lead to significant reduction in the dielectric strength of the SF_6 gas. A sufficient quantity of contaminant particles can cause a dielectric breakdown even at operation voltage level. High-voltage GIS design provides methods to counter the effect of

Figure 2.108 Typical AIS equipment installation (*Source* [110])

Figure 2.109 GIS installation by a complete bay (*Source* [112])

contamination. Most GIS modules are designed to provide some natural low electric field regions or internal traps to attract contaminant particles; however, due to its limitations, strict cleanliness requirements during high-voltage GIS installation are justifiable. To prove the particle free assembly and installation of GIS partial discharge (pd) measurements are carried out with the high voltage on-site tests.

2.11.13 Grounding

Grounding requirements for any electrical installation are based on system parameters such as short-circuit current and duration and do not change with the equipment used; therefore, the conditions for both GIS and AIS are the same. A major difference with respect to grounding is the much smaller size of the GIS equipment installation. Large AIS switchyards provide enough area to install a ground grid, which can easily dissipate even high fault currents. The use of GIS technology allows for a much more compact substation. Reduction in station size, however, also means a reduction in the grounding area.

This constitutes unique challenges in the design of GIS installation ground grids. Some additional measures are needed during ground grid design. The installation of a high-density grid or more frequent and direct connections from the switchgear elements to the station ground grid is required.

Another issue with GIS technology is that phase conductors are much closer to each other compared with AIS. This generates electromagnetically induced currents in the grounding system. In case of failure due to the small dielectric clearances, a dielectric breakdown can develop within nanoseconds. This is especially true for high-voltage GIS installations. Rapid voltage collapse leads to the generation of very fast transient overvoltage traveling waves and can lead to a transient ground potential rise. To cover the fast transient overvoltage's the GIS is conducted to the ground grid at the foot point of each bay.

The design of grounding systems for high-voltage GIS installations must be coordinated with the limitations of the technology and provide low impedance paths to ground for fault currents, decrease magnetic field magnitudes, and reduce transient overvoltages. In the case of a single-phase GIS enclosure, due to the bonding of all single-phase enclosures, circulating currents are generated. Circulating currents lead to some electrical losses. In three-phase enclosure GIS, there are vortex currents instead of circulating currents within the enclosure.

To alleviate some of the above concerns with GIS installations, the use of grounding arrangements with multiple points consisting of short grounding conductors that link the GIS equipment at various points along the enclosures as straight as possible to ground are used. These numerous grounding connections provide for parallel paths from equipment to ground to relieve the concerns related to transient overvoltages, circulating currents in single phase or vortex current in three-phase enclosures.

Since GIS is installed indoors, in some cases ground grid design uses the connection to the rebar inside the concrete. Rebar and grounding design shall be coordinated in order to avoid problems of overheating due to circulating currents.

2.11.14 Project Complexity

The type of technology has a direct impact on the compactness of the installation. The more compact the site requires, the higher the level of coordination between responsible parties and activities. Any changes in the original design are much more complicated and sometimes impossible to do on GIS and to some extend on MTS high-voltage installations. Specific GIS arrangements are difficult to customized and are very costly, especially if equipment is already being manufactured. In contract, since AIS installations are assembled on site, even late changes are easier to implement.

2.11.15 Expandability

Using GIS technology is better suited for projects where future extension is not expected. If future extension is expected, it should be considered as part of the original design of any GIS installation. Expansions in GIS are more complicated and the original design needs to allow for provisions for future upgrades; at times this may mean future infrastructure to be build earlier to accommodate the future expansion, especially in indoor installations. Since most installations are indoors, larger buildings should be designed to include enough space for possible later stage extension. This is easily achievable for medium-voltage GIS due to the smaller size of the installation.

In comparison, AIS installation allows for relatively easier extension, although available site area should be taken into consideration. If additional area is available, AIS equipment can be easily expanded. Expanding at later stages using AIS components is easier due to the availability and interchangeability of different manufacturer elements. However, GIS technology is a perfect solution for expanding AIS when there is limited space and planned outages are hard to get.

2.11.16 Testing and Commissioning

General commissioning activities such as visual checks, equipment tests, documentation checks, energizing tests, and tightness tests do not differ significantly between different technologies. There are some additional tests that need to be performed, such as a field power frequency high-voltage test for GIS installations. The pressurized gas compartments, which were opened during site assembly work, require leakage tests. If the pressurized gas compartments were not opened during assembly work, no additional tests on-site beyond normal commissioning tests are required.

In comparison to AIS installations, GIS requires cleaner site installation in case that any compartments being opened during the assembly. Therefore, site high-voltage test and pd measurement are of vital importance and are typically performed by manufacturer. Any failure in the early stage of operation is typically due to omissions during assembly.

Prefabricated and pretested GIS elements are also prewired prior to shipping. Connections to local control cabinets (LCC) are sometimes plug-in type, which accounts for less inspection and testing time. Although some activities may differ, there is no significant difference in the testing and commissioning between both technologies.

2.11.17 Environmental Impact

Environment impact of SF_6 is well known and discussed in detail in Section 2.8 of this book. For any electrical installation using SF_6 filled equipment, proper gas handling is a concern. This is true for both GIS and AIS stations. The use of SF_6 within the power energy sector has increased mostly due to the increase in GIS equipment installations. The first GIS installations contained fewer and larger gas compartments. Gas loss was relatively high. Further development allowed for both a decrease in the volume of gas used in similar installations and percentage of leaks.

The modular design GIS equipment available on the market today requires much less amount of SF_6 compared with previous generations. This allows for leaks to be contained to a relatively small compartment. Most manufacturers nowadays are successful in achieving around 0.1% gas loss per year.

As an example, Figure 2.110 shows a comparison of Siemens 145 kV GIS switchgear gas volumes between 1970 and 2016 as shown in Figure 2.110 below, for a model calculation of 145 kV GIS equipment.

2.11.18 Model Calculation for 145 kV

Based on modern design and manufacturing improvements, SF6 contributes to only about 0.06% of the overall human contribution greenhouse effect today. As just one example, emissions from European manufacturers and users contribute only 0.008%.

Gas loss is only one aspect of the environment impact of electrical installations. Except the global warming aspect, electrical systems also result in aggregating acid air emissions, expressed in

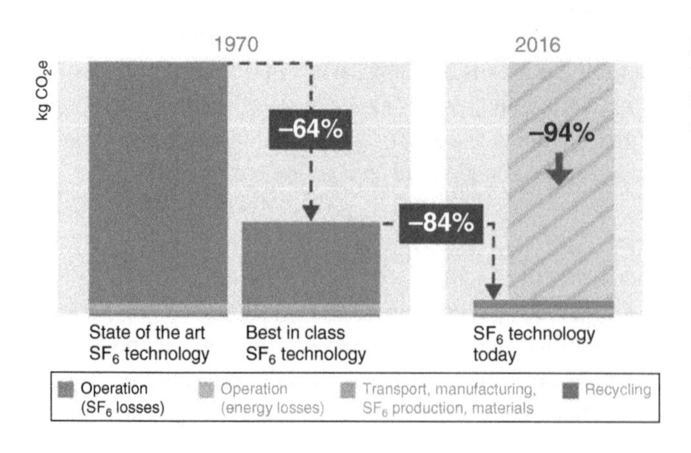

Figure 2.110 Reduction of SF_6 GIS from the year 1970 until 2016 for a model calculation of 145 kV (*Source* [106])

SO_2 equivalents (air pollutant) and nitrification (water pollution) due to increase in either nitrogen or phosphorus. This requires the comparison of total environmental footprint of AIS versus GIS installations throughout their life cycles. All stages from the manufacturing to transportation and installation of all materials on site shall be part of this comparison. Additionally, the different technology equipment has unique power and gas losses. At the end of their service life, the cost of dismantling and recycling needs to be addressed. All above factors are dependent on actual location and size of installation, etc.

IEEE supported research [107] reviews the case study of specific 220 kV station by comparing AIS and GIS technologies. The study is based on 18 environmental indicators. The results shows that the GIS substation has a lower environmental impact than the AIS substation for most environmental indicators with an overall 64% lower environmental impact. The negative effects of SF_6 on the environment are offset by the lower overall impact, due to lesser system loses, less materials required during manufacture, and lower transportation costs. However, GIS installations have a higher impact on global warming due to the higher quantities of SF_6. Analyses are dependent on the source of the generation and its environmental impact, since AIS losses are up to four times higher than GIS.

Additionally, this study assumed that the lifespan of AIS and GIS is the same, which is typically not the case. A reasonable assumption would be that GIS has a longer life span and by taking this into consideration, the results will further support the lower environmental impact of GIS. Therefore, even if we consider some of the GIS advantages in the above study as being overestimated, there is still no doubt that GIS is the more environmentally efficient solution.

2.11.19 Safety

Safety is crucial to substation design, and irrespectively of the technology used, properly constructed electrical installations are safe. Nevertheless, there are differences between AIS in GIS installations with respect to safety. Possible human error is one of the elements to consider during design and installation of high- and medium-voltage equipment. Well-designed AIS substations have interlocks preventing any wrong operation or contact with energized parts. However, even the best designs cannot fully eliminate the possibility of human error, especially since the high-voltage equipment is exposed. Medium-voltage AIS with metal clad switchgear has relatively better protection, as energized parts are enclosed. In comparison, both high- and medium-voltage GIS reduces the accident risk substantially. Integrated disconnects and grounding switches allow for risk-free isolation and grounding of switchgear sections. All live parts are totally enclosed in grounded metal enclosures, so any accidental contact with energized parts is practically impossible, except at the external links. With the GIS installation, compartments are safe to touch even when energized. Even in the case of human error, with the worst-case scenario of equipment failure, GIS will contain the fault internally.

Another safety consideration is equipment failures. Due to the nature of high-voltage and medium-voltage outdoor AIS installations, they are inherently dangerous in case of major equipment faults. Danger is mitigated by remote operations; however, faults can happen even during normal maintenance inspections. Indoor high-voltage AIS installation failure mitigations are difficult to achieve and can raise safety concerns. Indoor medium-voltage AIS can utilize either simple metal clad or arc resistant one, with the latter providing higher degree of safety. In comparison, in the case of major internal fault, both high- and medium-voltage GIS installations are equipped with rupture diaphragms, which relieve the pressure and prevent fragmentation of the enclosure in case of abnormally high-pressure buildup. Rupture diaphragms expel the gas in a defined direction of discharge to guide the gas stream away from personnel to ensure their safety. So even in the very rare event of an internal arc, the safety level is kept high.

Different kinds of wildlife are known to cause damage, especially to outdoor electrical equipment. Aside from the inconvenience and reliability aspects of animal-induced outages, there can be damage to the substation equipment or major outages. Although less common in high-voltage outdoor AIS station, large birds or snakes still present a danger to the equipment. Medium-voltage outdoor AIS installations are especially vulnerable and typically require phase animal protection. Indoor metal clad AIS has a better level of protection, although rodents can still access it in some cases. In comparison, both high- and medium-voltage GIS installations are completely enclosed and out of animal reach.

Health hazards from SF_6 are possible in both AIS and GIS stations. The gas is nontoxic to humans; however, since it has little or no odor, under high concentration it can deplete the air of oxygen and lead to physical asphyxiation. Under normal circumstances, it might be considered rare to be exposed to SF_6 without sufficient oxygen dilution. However, for utility employees, this type of exposure is well within the normal course of duties when installing, servicing, recycling, monitoring, and de-commissioning SF_6-filled switchgear in vaults, basements, buildings, and other enclosed spaces. If a substantial quantity of SF_6 gas leaks in an enclosed area, it can pose a real danger of asphyxiation to personnel, due to oxygen deficiency. This is the reason why oxygen monitors are installed in underground small space GIS installations. In case of failure, any electrical equipment can potentially produce decomposition, these products may be toxic if exposed to. Pure SF_6 is not toxic, nevertheless during high-temperature arcing or electrical discharges, the molecules break down. By-products from the reaction consist of metal fluoride powders and various gases. Cleaning should be done by trained qualified personnel.

2.11.20 Security

Terrorism, vandalism, or unauthorized persons entering the substations have been getting more attention recently, due to some instances of intentional damage to electrical equipment. All high-voltage substations implement some means of protection such as intruder alarms and security cameras. Even then, there is a difference in security mostly because of indoor vs outdoor installations. Usually, threats are minimized with GIS, because the substation is placed indoor and energized elements are sealed. Some installations, especially when cable connections are utilized, are completely hidden from the public without even a resemblance to electrical installation [22].

2.11.21 Community Acceptance

Any new installation or updates to existing electrical installations are subject to public security, especially if located in urban areas. This may also influence and delay the permitting process. Public interest in substation installations typically is related to aesthetics, safety, environmental effects, noise, and sometimes electromagnetic field (EMF) concerns. The choice of technology has a direct impact on all components.

The aesthetics appearance in some cases may have a dominant impact on both public and local permitting authorities' acceptance. Some defining factors in the appearance are substation's area, height, and visibility. Typically, outdoor installations are more difficult to disguise due to its large size and height. Using architectural enhanced fences and landscape to conceal the substation can in some instances be costly and have limited effect. Outdoor GIS technology substations still have an advantage due to the lower profile of this equipment compared to AIS, thus reducing the visual impact. In most cases, the high- and medium-voltage GIS installations are installed in a building, which is easily blended with the surroundings. GIS can even be installed inside an existing building without changing the appearance of the façade. Indoor GIS technology may need enclosure or building licenses, but complete permitting is typically sped up and/or decreased when compared with AIS technology. In addition, GIS technology can reduce effects on environmentally sensitive locations. Figure 2.111 shows GIS installation in architecturally enhanced building.

In some cases, GIS substations can even be installed underground, which would be extremely expensive for AIS installations. Another option for GIS is the installation in multistory existing buildings.

With respect to community acceptance, public safety is paramount. Electrical substation equipment is typically not accessible for the public; however, there are cases where unauthorized entrance has been recorded. In the case of outdoor AIS installations, high-voltage parts are exposed. This is especially a concern with medium-voltage outdoor installations. In comparison, all part of the GIS, which are connected at life-threatening high voltages, are completely enclosed. With these types of installations, indoors access is easily monitored; even in case of unauthorized entrance, the metallic enclosure of the GIS is grounded, and accidental touching of life parts is not possible. Moving elements such as operation rods or motor drives are protected with protective plates for greater safety.

Audible noise levels generated by electrical equipment is another reason for public concern. There are two types of sources of noise in high-voltage electrical installations. The first one is due to electrical noise sources and the second one is mechanical operation of devices. Electrical interferences may be transmitted into high-voltage installations by means of high-frequency interference due to switching in primary circuits or lightning strikes. Properly designed installations employ measures to mitigate both. Indoor installed high- and medium-voltage GIS is immune to lightning strikes, because the installation is fully enclosed and installed in the building.

Low-frequency interferences are generated due to partial discharge in the insulation and discharge in surrounding air or corona effect. Partial discharge levels are low and do not generate audible noise. Corona of the other hand is always a concern for high-voltage open air installations,

Figure 2.111 GIS installation in architecturally enhanced building (Used with permission of GE Power)

especially at voltages 230 kV and above. Even well-designed new facilities are not immune to corona under all weather conditions. In foggy conditions, corona is almost always present in open-air high-voltage stations with audible and sometimes even visual glow effect. On the other hand, GIS installations with its enclosed equipment are unaffected by corona, unless bushing to air connections exist. If installations are supplied by high-voltage cables, corona is totally absent.

Other source of audible noise is transformers and mechanical operation of disconnect switches and circuit breakers. Transformer generated noise and mitigation is not a subject of this chapter. The disconnect switches audible noise level is low and as such not limited in any IEEE standard. For GIS installation, in which disconnect switches are enclosed into gas insulated compartments, audible noise is barely detectable close to the switchgear assembly. This noise will not be measurable outside the building envelope. Standards governing high-voltage circuit breakers provide maximum value for the outdoor circuit breakers only. Maximum values have an impulse noise limit of 105 dB and intermittent noise limit of 90 dB. This may present a concern in some instances. For high- and medium-voltage GIS substations where circuit breakers are installed inside encapsulated metal enclosure, these values are significantly lower. With additional protection of building walls, circuit breaker audible noise levels will be either not at all or barely measurable outside, and as such not a concern to the public.

2.11.22 EMF and a Potential Impact

All electric substations produce electric and magnetic fields (EMF). In a substation, the strongest fields around the perimeter fence come from the transmission and distribution lines entering and leaving the substation if they are installed in open air.

The strength of the fields from equipment inside the fence decrease rapidly with distance, reaching very low levels at relatively short distances beyond energized equipment. Electric fields are present whenever voltage exists on a conductor. Electric fields are not dependent on the current. The magnitude of the electric field is a function of the operating voltage and decreases with the square of the distance from the source. As an example, for 500 kV station the level of the electric field approaches zero at about 7 m (21 ft) distance to the source.

Gas-insulated substation equipment are constructed using a coaxial design with a circular center conductor supported by insulators with a circular enclosure that provides electrical shielding and contains the insulating gas as shown in Figure 2.112 below.

Since the enclosure is grounded, this provides an additional screening effect against electric

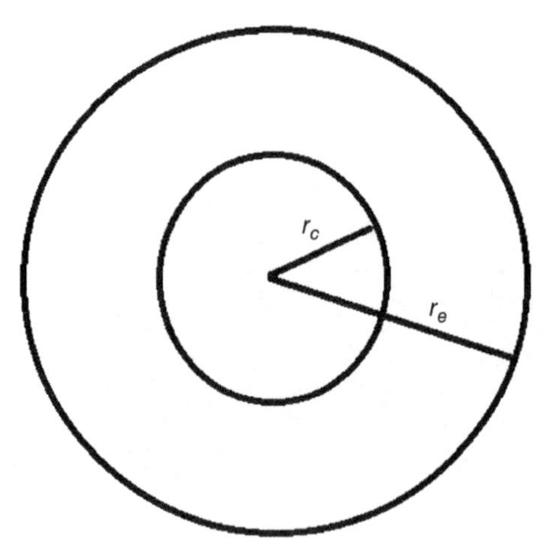

Figure 2.112 Coaxial design principal of gas-insulated switchgear, r_c = radius of conductor, r_e = radius of enclosure

fields. As such, electric fields even in proximity to the GIS installations is negligible. In addition, if installed indoors, the building superstructure contains rebar, which represents another layer of shielding against electrical fields. If the incoming or outgoing lines are underground, the level of the electric fields at the point of crossing the fence is negligible.

Magnetic fields are present whenever current flows in a conductor and are not voltage dependent. Magnetic fields also decrease with increasing distance (r) from the source. The rate is an inverse function, and the function is $1/r^2$. Magnetic fields due to conductor currents are reduced by GIS enclosure currents and are less than those of AIS. Even near GIS where EMF levels can be highest, exposure levels are well within tolerable limits. With respect to public safety, studies have shown that EMF levels in GIS are significantly lower than the typical maximum safety levels.

2.11.23 Conclusion

The decision of what type of technology to use is one of, if not the most important, decision during the planning stage of any project. It is critical to organize specific project requirements and then quantify advantages and disadvantages of both AIS and GIS technologies. In some cases, single elements alone, such as aesthetics or site constrains, may be a deciding factor. However, most often reaching the optimal conclusion requires assessment of many requirements and their impact on the final decision. Sometimes the optimal substation arrangement may not be even solely AIS or GIS, but rather a combination of both, or MTS technology for high voltage or the use of high-voltage GIS with medium-voltage metal clad switchgear.

In this respect, GIS is not always the preferred option; however, it provides some substantial benefits which can provide a solution for specific applications. The best-known advantage of GIS technology is space saving. Though there are also other extra benefits of GIS technology, such as enhanced reliability, a decrease in planned outage durations, due to lower maintenance requirements, indoor installation, enhanced resilience and extended life expectancy, to name a few.

References Chapter 2

1. Public Internet (2020) http://gegridsolutions.com/app/resources.aspx?prod=gis&type=1)
2. Public Internet (2020) http://siemens-energy.com /global/en/offerings/power-transmission / portfolio /circuit-breaker/dead-tank-compact.html
3. CIGRE (2001) Brochure No. 161 "Guidelines for the design of outdoor AC substations" 2nd version, 2001.
4. IEEE (1994) White paper "Gas-Insulated Switchgear GIS – State of the Art" Phil Bolin, Hermann Koch.
5. CIGRE Brochure 125: User Guide for the Application of Gas-Insulated Switchgear (GIS) for Rated Voltages of 72.5 kV and above, 1996.
6. CIGRE Brochure No. 390 'Evaluation of different switchgear technologies (AIS, MTS, GIS) for rated voltages of 52 kV and above.
7. Werner Zimmermann, André Osterhol, Dr. Jürgen Backes (2005) "Comparison of GIS and AIS Systemes for Urban Supply Netwworks."
8. Publication Nuevas Technologias y Tendencias Constructiveas en el Diseno de Substationes de Potentia, Power Transmission and Distribution, Siemens 2007.

9 Public Internet (2020) htttps://www.hitachiabb-powergrids.com/us/en/offering/product-and-system/high-voltage-switchgear-and-breakers/gas-insulated-switchgear
10 Public Internet (2020) https://www.wikipedia/substation/fischergasse, laufenburg
11 CIGRE (2005) Working Group B3-110 Valencia Plan, Model Project to integrate Facilities into the Environment and Reduces Spaces.
12 CIGRE (2012) Working Group B3-203 Receicling, Uprating and Updating of a high voltage substation located at down town of a Big City.
13 CIGRE (2008) Technical Brochure TB 509 Final report of the 2004 – 2007 international enquiry on reliability of high voltage Equipment Part 1 – Summary and General Matters.
14 CIGRE (2000) Technical Brochure TB 150 Report on the second International Survey on high voltage gas insulated substations (GIS) Service Experiences.
15 CIGRE (2008) Technical Brochure TB 513 Final report of the 2004 – 2007 international enquiry on reliability of high voltage Equipment Part 1 – Gas Insulated Switchgear.
16 CIGRE (2008) Technical Brochure TB 514 Final report of the 2004 – 2007 international enquiry on reliability of high voltage Equipment Part 1 – Gas Insulated Switchgear Services.
17 IEEE (197) White paper Three phase Enclosed Gas Insulated Technology up to 245 kV and its impact for improved power system reliability, D. Fuechsle, K. Kutlev Sr. Member IEEE, R.Dhara Memebr IEEE, Th. Schulz.
18 Public Internet search (2020) air-insulated substation/installation
19 Public Internet (2020) http://siemens-energy.com/in/en/offerings/power-transmission/transmission -products/gas-insulated.html
20 Public Internet (2020) http://siemens-energy.com/in/en/news/key-topics/ecotransparency.html
21 IEEE (2000) White Paper "Environmental Impact Comparison between a 220 kV Gas-Insulated Substation and a 220 kV Air-Insulated Substation", Laruelle Elodie, Ficheux Arnaud, Kieffel Yannick, Huet Isabelle.
22 IEEE (1998) White Papare "Improving the Physical Security and Availability of Substations by using new Switchgear Concepts, Hermann Koch, Gerd Ottehenning, Ernst Zöchling.
23 IEEE C37.122 (2017) IEEE Standard for Gas-Insulated Substations.
24 IEEE C37.123 (2021) IEEE Guide to Specifications for Gas-Insulated, Electric Power Substation Equipment.
25 IEEE C37.122.1 (2019) IEEE Guide for Gas-Insulated Substations.
26 IEEE C37.1300 (2011) IEEE Guide for Cable Connections for Gas-Insulated Substations.
27 IEC 62271-203 (2017) Gas-Insulated Metal Enclosed Switchgear for Rated Voltages above 52 kV.
28 IEC 62271-1 Edition 2.0 High-voltage switchgear and controlgear – Part 1: Common specifications for alternating current switchgear and controlgear.
29 CIGRE Brochure 125: User Guide for the Application of Gas-Insulated Switchgear (GIS) for Rated Voltages of 72.5 kV and Above.
30 IEC (2003) 61276-200, 1st edition. A.C. Metal-Enclosed Switchgear and Controlgear for Rated Voltages above 1 kV and up to and Including 52 kV.
31 IEC 62271-100 (2012) High-Voltage Switchgear and Controlgear – Part 100: Alternating-Current Circuit-Breakers.
32 IEC 62271-102 (2001) High-Voltage Switchgear and Controlgear – Part 102: Alternating Current Disconnectors and Earthing Switches.
33 Kuffel, E., Zaengl, W.S., and Kuffel, J. (2000) *High Voltage Engineering Fundamentals*, Newnes, ISBN: 0 7506 3634 3.
34 Khalifa, M. (1990) *High-Voltage Engineering, Theory and Practice*, Dekker, ISBN: 0-8247-8128-7.

35 Garzon, R.D. (2002) *High Voltage Circuit Breakers, Design and Applications*, Marcel Dekker, ISBN: 0-8247-0799-0.

36 CIGRÉ (1994) WG 13.06: Final Report of the Second International Enquiry on High Voltage Circuit-Breaker Failures and Defects in Service, CIGRÉ Technical Brochure 83.

37 CIGRÉ TB 509: Final Report of the 2004–2007: International Enquiry on Reliability of High Voltage Equipment, Part 1 – Summary and General Matters.

38 CIGRÉ TB 513: Final Report of the 2004–2007 International Enquiry on Reliability of High Voltage Equipment, Part 5 – Gas Insulated Switchgear (GIS).

39 Welch, I.M. GIS Experience Survey and Database (ELECTRA No. 157, December 1994, pp. 81–83).

40 Welch, I.M., Jones, C.J., Kopejtkova, D., *et al.* (1994) GIS in Service – Experience and Recommendations, CIGRÉ, Paris, 1994, Paper SC 23 No. 23–104.

41 Molony, T., Kopejtkova, D., Kobayashi, S., and Welch, I.M. (1992) Twenty-five-year Review of Experience with SF_6 Gas Insulated Substations (GIS), CIGRÉ, Paris 1992, Paper SC 23 No. 23–101.

42 IEC 62271-207:2012 High-voltage switchgear and controlgear – Part 207: Seismic qualification for gas-insulated switchgear assemblies for rated voltages above 52 kV.

43 IEEE 693-2018 – IEEE Recommended Practice for Seismic Design of Substations.

44 Willis, H L., Welch, G.V., and Schrieber, R.R. (2001) *Aging Power Delivery Infrastructure*, Marcel Dekker, New York.

45 RUS Bulletin 1724E-300, Rural Utilities Service, Design Guide for Rural Substations, June 2001.

46 Koutlev, K., Pahwa, A., Wang, Z., and Tang, L. (2003) Methodology and Algorithm for Ranking Substation Design Alternatives, IEEE 0-7803-8110-6, February 2003.

47 CIGRE Brochure 125: User Guide for the Application of Gas-Insulated Switchgear (GIS) for Rated Voltages of 72.5 kV and Above.

48 IEC 61936-1:2021 CMV Power installations exceeding 1 kV AC and 1,5 kV DC – Part 1: AC.

49 IEC 62271-211:2014 High-voltage switchgear and controlgear – Part 211: Direct connection between power transformers and gas-insulated metal-enclosed switchgear for rated voltages above 52 kV.

50 IEEE Standard 80 (2013). Guide for Safety in AC Substation Grounding.

51 IEC 62271-4:2013 High-voltage switchgear and controlgear – Part 4: Handling procedures for Sulphur hexafluoride (SF_6) and its mixtures.

52 IEEE C37.122.3 (2011). Guide for Sulfur Hexafluoride (SF_6) Gas Handling for High-Voltage (over 1000 Vac) Equipment.

53 IEC Standard 60480. Guide to the Checking of Sulfur Hexafluoride (SF_6) taken from Electrical Installations.

54 CIGRE Guide 276: Handling of SF_6 and Its Decomposition Products in Gas Insulated Switchgear (GIS).

55 CIGRE (2003) Technical Brochure TB 234: SF_6 Recycling Guide, Task Force B3.02.01.

56 CIGRE (2002) Publication Electra, SF_6 in the Electric Industry, Status 2000 ELECTRA No. 200

57 IEEE C37.122.5 (2013) Guide for Moisture Measurement and Control in SF_6 Gas-Insulated Equipment.

58 CIGRE Technical Brochure TB 276 (2005): Guide for the Preparation of Customized Practical SF_6 Handling Instructions.

59 Maiss, M. and Brenninkmeijer, C.A.M. (1998) Atmospheric SF_6, trends, sources and prospects. *Environmental Science and Technology*, **32** (20), 3077–3086.

60 ISO 14040:2006 Environmental management — Life cycle assessment — Principles and framework.

61 IEEE C37.122.2 (2011) Guide for the Application of Gas-Insulated Substations 1 kV to 52 kV.

62 IEEE C37.100.1 (2018) Standard of Common Requirements for High Voltage Power Switchgear Rated Above 1000 V.
63 IEC 62271-303 (2008) (former IEC TR 62271-303 and replace by IEC 62271-4) High-voltage switchgear and controlgear - Part 303: Use and handling of sulphur hexafluoride (SF_6).
64 IEC 60376 ed. 2.0 (2005). Specification of technical grade sulphur hexafluoride (SF_6) and complementary gases to be used in its mixtures for use in electrical equipment.
65 ASTM D2472 (2020) Standard Specification for Sulfur Hexafluoride.
66 IEEE C37.122.4 (2016) – IEEE Guide for Application and User Guide for Gas-Insulated Transmission Lines, Rated 72.5 kV and Above.
67 ZVEI, Germany (German language) (2021) Study on live cycle assessment including inpact of SF_6 https://www.zvei.org/verband/fachverbaende/energietechnik/sf6-in-der-energietechnik
68 EU Commission, Regulation (EU) No 517/2014 of the European Parliament and of the Council of 16 April 2014 on fluorinated greenhouse gases and repealing Regulation (EC) No 842/2006 Text with EEA relevance, 2021.
69 EPA Fluorinated Greenhouse Gas Emissions and Supplies Reported to the GHGRP, Greenhouse Gas Reporting Program (GHGRP), United States Environmental Protection Agency, 2021.
70 CARB Regulation for Reducing Sulfur Hexafluoride Emissions from Gas Insulated Switchgear, California Air Resource Board, USA, 2020.
71 Proposed Amendments to the Regulation for reducing Sulfur Hexafluoride Emissions from gas insulated switchgear, subchapter 10. Climate change , State of California AIR RESOURCES BOARD, May 05, 2021.
72 Co2online gGmbH, https://www.co2online.de/, Johannes Hengstenberg, Berlin, 2021.
73 SF_6 Partnership accomplishments, 1999 to 2016 (USA), Overview of SF_6 Emissions Sources and Reduction Options in Electric Power Systems, EPA August 2018.
74 Lutz, B., Kuschel, M., Glaubitz, P. “Future Challenges for the Grid – Integration of environment friendly Gas-insulated Substation”, Cigré SC B3 Colloquium 2017, Brazil.
75 R. Smeets et al. “The Impact of the Application of Vacuum Switchgear at Transmission Voltages”, Cigré TB 589, July 2014.
76 T. Uchii, “Development of 72 kV Class Environmentally-Benign CO_2 Gas Circuit Breaker Model”, Electrical Insulation News in Asia, No. 14, November 2007.
77 T. Uchii, Y. Hoshina, T. Mori, H. Kawano, T. Nakamoto, H. Mizoguchi, “Investigations on SF_6-free Gas Circuit Breaker Adopting CO_2 Gas as an Alternative Arc Quenching and Insulating Medium”, Gaseous Dielectrics X, Springer, ISBN 0-387-23298-2, pp. 205–210 (2004).
78 IEEE PES Substation Committee K2 Tutorial Subcommittee, Schiffbauer, D., ‘CO_2+O_2 Data’. February 2020.
79 IEEE PES Switchgear Committee TR 64 ’Impact on Alternative Gases on switchgear standads’, Alternative Gases Task Force, IEEE PES resource center, 2018.
80 IEEE C37.122.10 (expected to be published in 2023) Guide for Handling Non-Sulphur Hexafluoride (SF_6) Gas Mixtures for High Voltage Equipment, PES Substations Committee GIS Subcommittee, Working Group K19, 2020.
81 IEEE C37.100.7 Working Group: Performance Evaluation of Sulfur Hexafluoride Alternatives, IEEE PES Switchgear Committee Working Group, 2019.
82 IEEE Power & Energy website: https://www.ieee-pes.org/, 2021.
83 IEC TC 17 High Voltage Switchgear and Controlgear: http://tc17.iec.ch/about_tc17/basics.htm, 2021.
84 CIGRE Study Committee B3 Substations, D1 Materials and A3 Switchgear: https://www.cigre.org/, 2021.
85 T&D Europe, Transmission and Distribution: https://tdeurope.eu/transmission-distribution/the-electricity-grid.html, 2021.

86 T&D Europe (2016) Technical guide to validate alternative gas for SF_6 in electrical equipment, Brussels.

87 T&D Europe (2018) Technical report on alternative to SF_6 gas in medium voltage & high voltage electrical equipment, Brussels.

88 ZVEI (2021) SF_6 emissions reduced by 60 percent since 2005.

89 ZVEI (2020), Scenario for reducing SF_6 operating emissions from electrical equipment through the use of alternative insulating gases.

90 ZVEI German Manufacturer Oganisation, www.zvei.org/power-engineering-dicvision

91 FNN Forum Netztechnik/Netzbetrieb im VDE, German Power Network Operator Organisation, https://www.vde.com/de/fnn

92 Current Zero Club, International Research Group on Interrupting Phenomena of High Voltage Circuit Breaker: http://currentzeroclub.org/

93 NEMA SF_6 & Alternatives Coalition: https://www.nema.org/directory/products/the-electric-transmission-and-distribution-sf6-coalition

94 J. Brucher, S. Giere, C. Watier, A. Hessenmüller und P. Nielsen, "3AV1FG – 72.5 kV Prototype Vacuum Circuit Breaker (Case Study with Pilot Customers)," in Cigre Session, 2012.

95 IEEE white paper "Gas Insulated Switchgear GIS - State of the Art" Phil Bolin, Hermann Koch, 1994.

96 Public internet https://www.gegridsolutions.com/app/resources.aspx?prod=gis&type=1), 2020

97 Public internet https://www.siemens-energy.com/global/en/offerings/power-transmission/portfolio/ circuit-breakers/dead-tank-compact.html, 2020

98 Publication Nuevas Tecnologias y Tendencias Constructivas en el Diseno de Substaciones de Potencia, Power Transmission and Distriburtion, Siemens, 2007.

99 Public internet https://www.wikipedia/high voltage/gis/substation/laufenburg

100 IEEE white paper "Improving the Physical Security and Availability of Substations by Using New Switchgear Concepts" Hermann Koch, Gerd Ottehenning, Ernst Zöchling, 1998.

101 IEEE white paper "Gas Insulated Switchgear GIS - State of the Art" Phil Bolin, Hermann Koch, 1994.

102 "Comparison of GIS and AIS systems for urban supply networks" Werner Zimmermann, André Osterholt, Dr. Jürgen Backes, 2005.

103 IEEE white paper "Environmental impact comparison between a 220 kV Gas-Insulated Substation and a 220 kV Air-Insulation Substation" Laruelle Elodie, Ficheux Arnaud, Kieffel Yannick, Huet Isabelle, 2000.

104 Public internet https://www.hitachiabb-powergrids.com/us/en/offering/product-and-system/high-voltage-switchgear-and-breakers/gas-insulated-switchgear, 2020.

105 Public internet https://www.siemens-energy.com/in/en/offerings/power-transmission/transmission-products/gas-insulated.html

106 Public internet https://www.siemens-energy.com/in/en/news/key-topics/ecotransparency.html

107 IEEE white paper "Three-phase enclosed Gas Insulated Substation technology up to 245 kV and its impact for improved Power System reliability" D. Fuechsle, K. Kutlev Sr. Member IEEE, R. Dhara Member IEEE, Th. Schulz, 1997.

108 Public internet air insulated Public internet https://www.siemens-energy.com/global/en/offerings/power-transmission/portfolio/circuit-breakers/dead-tank-compact.html, 2020substations/installation, 2020.

109 Public internet https://www.gegridsolutions.com/app/resources.aspx?prod= gis& type=1, 2020.

110 Public internet air insulated substations/installation, 2020.

111 Public internet https://assets.new.siemens.com/siemens/assets/public.1541967200.479e53977d19eee396bdd9535864fedb511dd9e9.gis-72-550-d.pdf

112 Technical brochure 'Gasisolierte Schaltanlagen von 72,5 bis 550 kV', Siemens AG, 2016

3

Technology

Authors: 1st edition: Hermann Koch, George Becker, Xi Zhu, Devki Sharma, Arnaud Ficheux, and Dave Lin; 2nd edition: Dave Solhtalab, George Becker, Xi Zhu, and Vipul Bhagat
Reviewers: 1st edition: Phil Bolin, Hermann Koch, Devki Sharma, Markus Etter, Scott Scharf, Patrick Fitzgerald, George Becker, Toni Lin, Chuck Hand, Xi Zhu, Noboru Fujimoto, Dave Giegel, Eduard Crockett, Pravakar Samanta, John Brunke, 2nd edition: Scott Scharf, Michael Novev, and Nick Matone

3.1 General

The GIS technology is constantly evolving with the use of new materials, technical functionality, manufacturing processes, quality, and reliability improvements. The result of these innovations is GIS with higher performances in voltage ratings (today up to 1100 kV UHV systems), current ratings (today up to 8000 A), and short-circuit ratings (today up to 100 kA). Simultaneously the size and volume of gas in GIS has been reduced leading to reductions in cost. This process is still ongoing but the steps are getting smaller in which the ratings are being increased or the size and volume is being reduced. Improvements are expected to continue although not at same rate.

There are ongoing studies investigating new technical principles, such as vacuum switching for high voltages 52 kV and above, alternatives to SF_6, for insulating purposes in gas insulated transmission lines (GILs), electronic switching, or short-circuit limitations to name a few.

The design of a GIS follows the rules of pressure vessel design with the exception of the repetitive pressure test after installation. The design and operation rules are defined in a series of European Standards depending on the material and type of design (see References 1–6). The general quality assurance and management system of ISO 9000 is the basis for the design and manufacturing process [7].

In this section the different aspects in technical development of GIS are explained for GIS.

3.1.1 Materials

The basic materials to in GIS manufacturing are metals to form the enclosure and the conductors, epoxy resin insulating materials, insulating gas, contact materials for switches and breakers, and several metallic and insulator materials to fix and operate the GIS functionality.

When invented in the 1960s, the GIS materials were dominated by a steel enclosure, aluminum conductor, SF_6 insulating gas, and silver-plated contact materials. In some cases, copper for the conductor was used to reduce the transmission losses of aluminum conductors.

Later, the enclosure material of the steel was replaced by aluminum alloys. The insulating gas SF_6 is still used because of its excellent performance in insulating high voltages with high reliability and its arc-quenching ability, which allows current ratings of up to 8000 A and short-circuit ratings of up to 100 kA. No other gas known today can offer such features. The high global warming potential of SF_6 has been taken into account by developing a closed loop cycle for the lifetime of SF_6, which is fixed in international standards like IEC 62271-4 [16] and IEEE C37.122.3 [17].

Gas Insulated Substations, Second Edition. Edited by Hermann J. Koch.

The insulating materials are based on epoxy resin of various mixtures with additives for improvement of mechanical strength, tracking behavior, and other properties to increase reliability. In the following, the basic materials are explained and their use in the GIS is described.

3.1.2 Steel

The first-generation design of GIS utilized steel sheet materials in most cases due to the lower cost of steel compared with aluminum. The two basic tasks of the enclosure: are to be able to handle the gas pressure inside and to be gastight so that SF_6 is not released into the atmosphere. Sheet steel material is capable of handling both of those tasks. Cast steel has porosities that cannot guarantee such high gas tightness values as required for GIS. Today the standards allow only 0.5% of gas pressure loss per year. Most manufacturers have equipment tested to 0.1% loss per year. Sheeted steel has high mechanical stability, allows pressure vessel to handle gas pressures up to 0.8 MPa in accordance with the pressure vessel standards [1–6].

Steel is relatively easy to weld and to form. Therefore, the steel enclosures have been welded by steel sheets that have been mechanically formed to cylinders and then welded. The required fixing points inside the steel cylinders have also been fixed by welded fixing points and plates. There was a lot of hand work required before the circuit breaker, disconnecting, or ground switches were fixed inside the steel enclosure (see Figure 3.1).

Because of the possibility of steel corrosion, it is necessary to coat the steel enclosure inside and outside. Usually, standard painted color coatings are used.

Steel enclosures were used until the 1980s by all manufacturers world-wide. Later, the casted aluminum enclosures were predominantly used in Europe and North America due to a reduction in sizes and better functionality of cast enclosures. Today steel enclosures are still used by some manufacturers in Asia.

The increasing values for continuous current ratings induce higher sheath currents (eddy currents) in the steel enclosures so that these enclosures can be heated up above the thermal limits. This occurs mainly at bus bar enclosures due to added current values there. To overcome these

Figure 3.1 Steel encapsulated GIS (Reproduced by permission of Siemens AG)

thermal limitations, bus bars enclosures have been manufactured in stainless steel and later in aluminum. Stainless steel has the advantage of a lower magnetic induction but is very difficult to manufacture because of its material strength. Also, the material cost of stainless steel is high. For these reasons, the manufacturers switched over to aluminum enclosures.

3.1.3 Aluminum

Aluminum is used for the conductors as an extruded pipe or in cast technology. For enclosures aluminum is used in cast technology or with welded sheet materials using longitudinal welds or spiral welding processes.

Extruded pipes can be used as conductors in a straight bus bar section (see Figure 3.2). When conductors are used inside disconnecting or ground switches, the required shape and design requires the use of casting technology (see Figure 3.3).

Cast aluminum has the advantage in manufacturing of complex structures and designs including a lot of functionalities, for example, fixing points for circuit breakers. The enclosure shape can be designed in accordance with the high-voltage electric field requirements. This can optimize the overall size and volume of the enclosures without increase in the electric field strength. The disadvantage of cast aluminum in the early years was that the insulation gas was leaking to the atmosphere due to its porosity. This was a very slow process but any gas leaks are not acceptable and the gas density is required to be maintained for electrical insulation. The casting process was eventually improved so that aluminum casting could be manufactured to be absolutely gastight. This

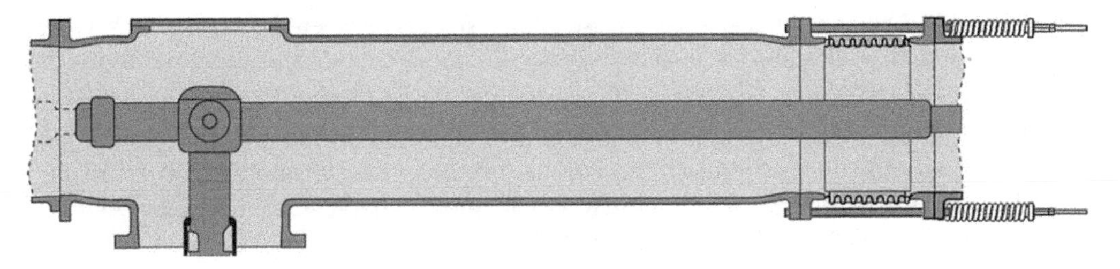

Figure 3.2 Straight conductor graphic (Reproduced by permission of Siemens AG)

Figure 3.3 Three-phase bus bar (Reproduced by permission of Siemens AG)

advancement resulted in significant improvements in the GIS design with ever larger sizes and greater complexity.

Today, most enclosures are manufactured in cast aluminum as the top performer for GIS.

3.2 Modular Components, Design, and Development Process

3.2.1 Modular Design

Metal-enclosed SF_6-insulated switchgear (GIS) has already acquired a long service experience since it was first introduced into the market in 1968, with SF_6 as the arc-quenching and insulating medium, as an interesting alternative to conventional air insulated substations. GIS technology in its infancy started based on extensive fundamental research and since then the service experience together with innovative development work has brought this technique to a safe and environmentally compatible, most reliable alternative. The tremendous progress of development can be seen as an example of the classic three-phase enclosure of a 72.5 kV GIS from 1968 to a modern 170 kV GIS of today in Figure 3.4.

3.2.1.1 Three-Phase Enclosure

The advantages of gas-insulated switchgear are its compact design and the modular system. The standardized modular structure is designed to match the various customers' specifications and allows almost all substation configurations to be realized in compliance with them.

Three-phase design requires relatively large aluminum enclosures because it must house all three conductors. At higher voltage levels, the isolation distances between the phases and between the phase to ground enclosures are larger. The cast aluminum technology economics limits the maximum sizes of enclosures. Over the past few years, the casting technology has improved and with this the voltage levels for three-phase insulated enclosures have increased. At first, only voltages up to 123 kV were of a three-phase encapsulation design; today the levels are at 170 kV and approaching 245 kV. A three-phase encapsulation has fewer parts, less insulating gas, and less enclosure material than its single-phase counterpart.

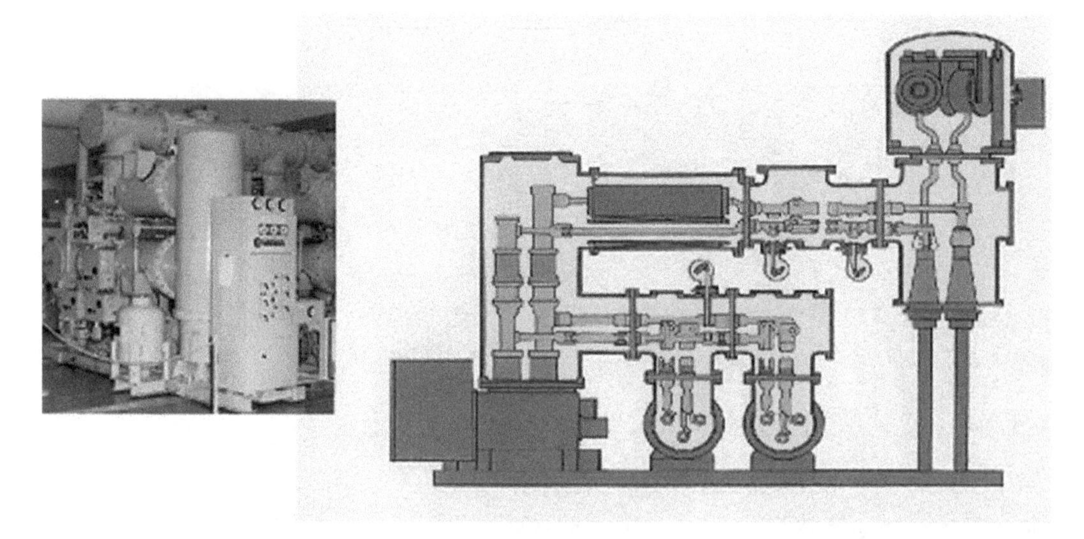

Figure 3.4 Classic three-phase cross section of a 72.5–170 kV GIS and a photograph of a 170 kV GIS shows the progress in development from 1968 up to today (Reproduced by permission of Siemens AG)

Circuit Breaker Module

The three-phase circuit breaker module in most design cases has a vertical (see Figure 3.5) circuit breaker with the operation mechanism at ground level. Some older designs have a horizontal circuit breaker. In a vertical circuit breaker design, the weight of the contact system is reducing drive forces for opening the contact system in a downward movement.

Three-Position Switch, Three-Phase Encapsulated Module

The three-position switch is a mechanical device that can be operated as a disconnect switch and as a grounding switch (see Figure 3.6). In this figure, the moving contact is in the middle position

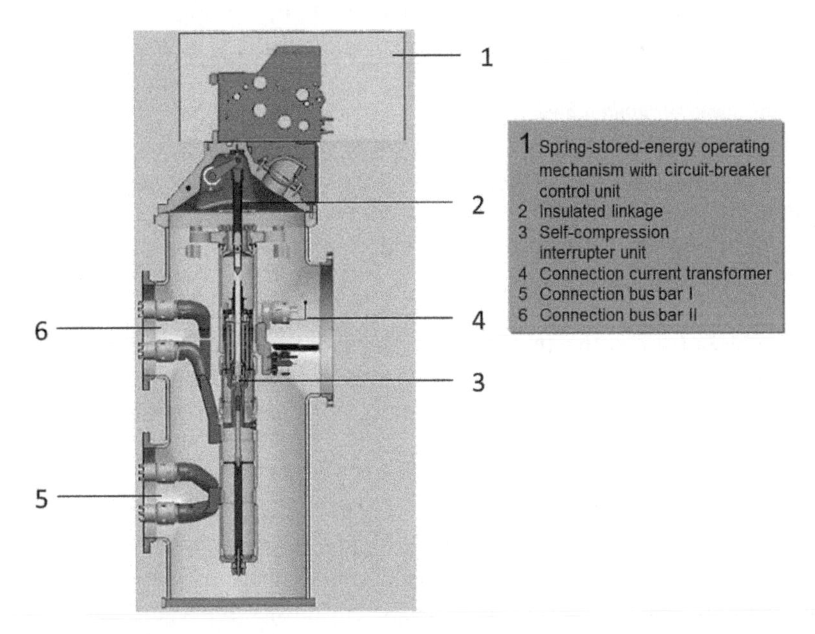

Figure 3.5 Three-phase circuit breaker enclosure up to 72.5 kV (Reproduced by permission of Siemens AG)

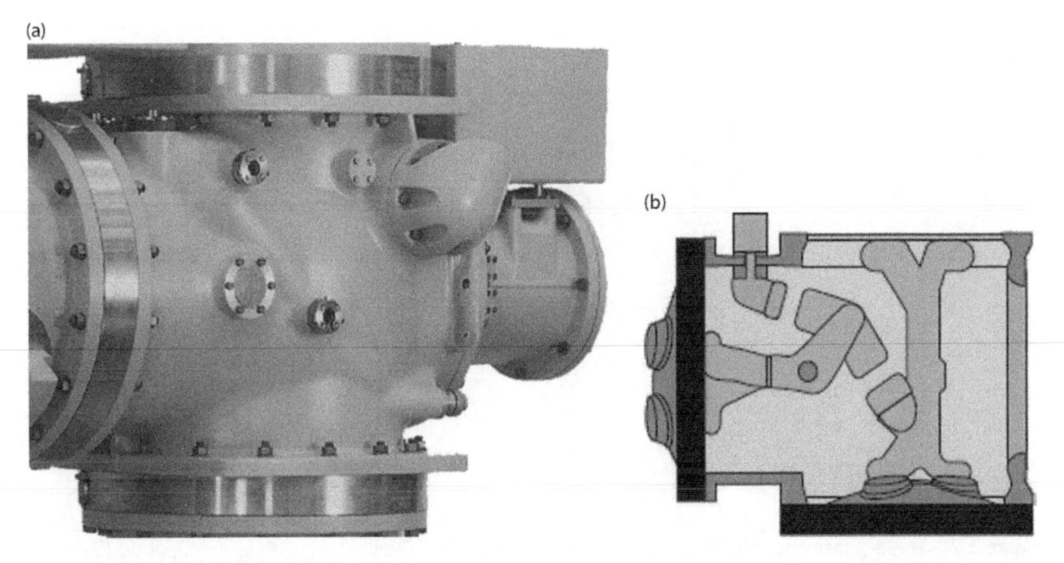

Figure 3.6 Three-position, three-phase encapsulated switch (a) four flange enclosure with view and sensor ports and motor drive, (b) cross section of three position switch (Reproduced by permission of Siemens AG)

and connected to the incoming conductor at the left, the outgoing conductor is on the right, and the ground contact is on the upper left side. In this figure, the device is shown as an open disconnect. If the moving contact is turned to connect to the ground/earthed contact then the incoming conductor on left is grounded/earthed.

If the moving contacts are turned to the right side then the incoming contact on the top is connected to the outgoing conductor on the right side and the switch is in closed the disconnect position.

The operation mechanism has an electric motor drive and operates in a few seconds from disconnect to closed or grounded/earthed. In Figure 3.6a, the three-position switch is connected to the circuit breaker enclosure module and the bus bar module of a GIS bay. In Figure 3.6b, the internal design is shown in principle.

Three-Phase Encapsulated High-Speed Grounding Switch Module

The three-phase encapsulated high-speed ground switch is used in emergency cases when it is necessary to ground a live conductor. Such cases might happen if a connected overhead line or cable is energized at the other end without notice. The high-speed drive operates the ground switch in some hundred milliseconds. The contact system is designed in such a way to withstand at least two making operations with short-circuit ratings (see Figure 3.7). It is not able to interrupt rated or short-circuit currents.

The operation mechanism has a spring drive that is charged by an electric motor. The spring-driven contact closes in typically 0.1–0.2 s. In Figure 3.7a, the photo shows the three-phase insulated ground switch attached to the circuit breaker module of the GIS. In Figure 3.7b, the graphic shows the internal design in principle.

Three-Phase Encapsulated Voltage Transformer Module

Three-phase encapsulated voltage transformers have a compact design in separate gas compartments at a higher gas pressure than the rest of the GIS. In many cases, voltage transformers are manufactured in separate factories from GIS and are delivered directly on-site (see Figure 3.8).

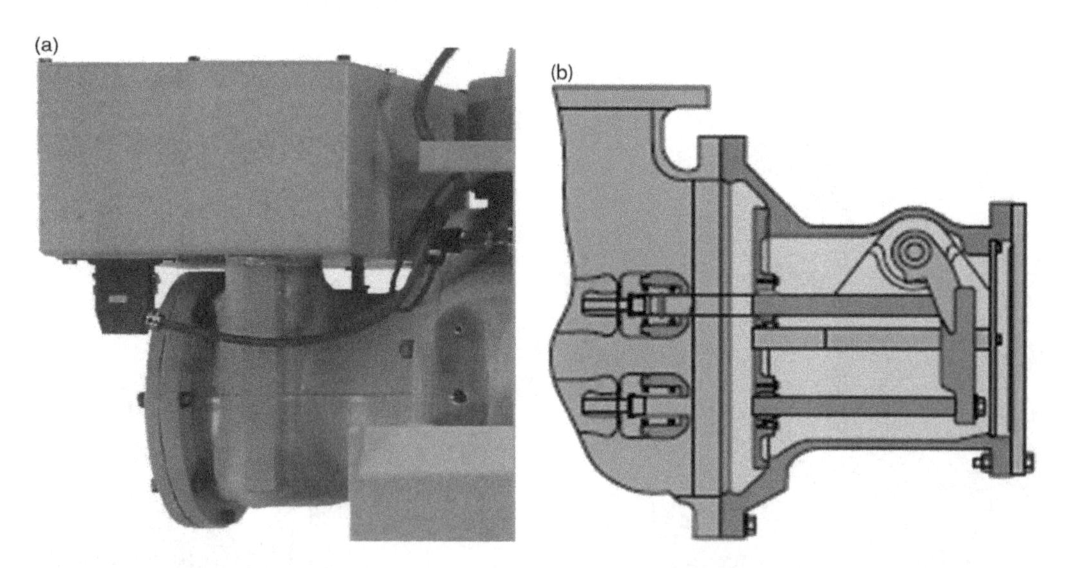

Figure 3.7 Three-phase encapsulated high-speed ground switch (a) enclosure with spring operated drive on top, (b) cross section with ground contacts and spring operating drive (Reproduced by permission of Siemens AG)

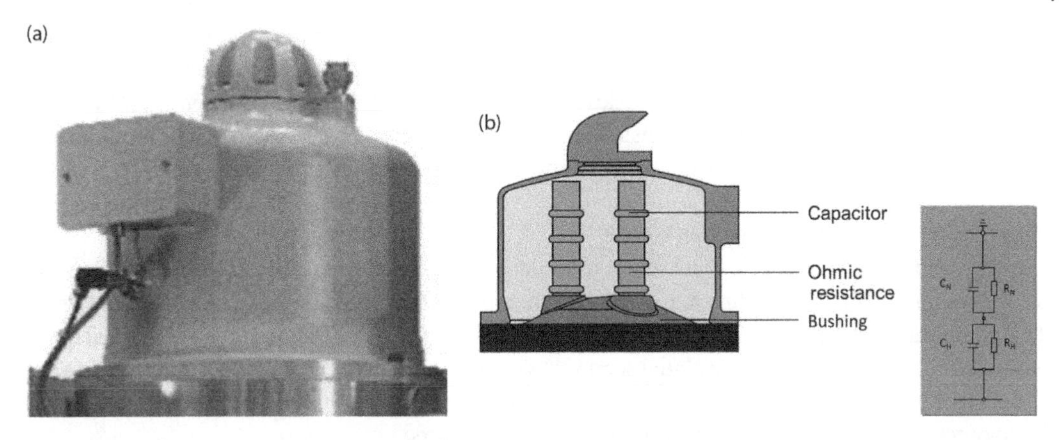

Figure 3.8 Three-phase encapsulated voltage transformer of 145 kV (a) enclosure with secondary terminal box, (b) cross section of voltage divider capacitors and resistances (Reproduced by permission of Siemens AG)

Voltage transformers are usually attached after the on-site tests to avoid damage to the secondary side of the transformer due to high test voltages. Some designs provide for either removable link, or fir some extra high-voltage disconnect switch. In cases when the voltage transformers cannot be disconnected from the GIS the higher frequency test voltage (up to 300 Hz) used to get the iron core into inductive saturation [15]. In Figure 3.8a, the three-phase encapsulated voltage transformer module is shown attached on top of the bus bar disconnector enclosure of the GIS. In Figure. 3.8b, the graphic shows the internal design.

Three-Phase Encapsulated Current Transformer Module

Three-phase encapsulated current transformers in GIS are usually completely encapsulated with the coil inside the gas zone (see Figure 3.9), which is differs from dead tank ring current

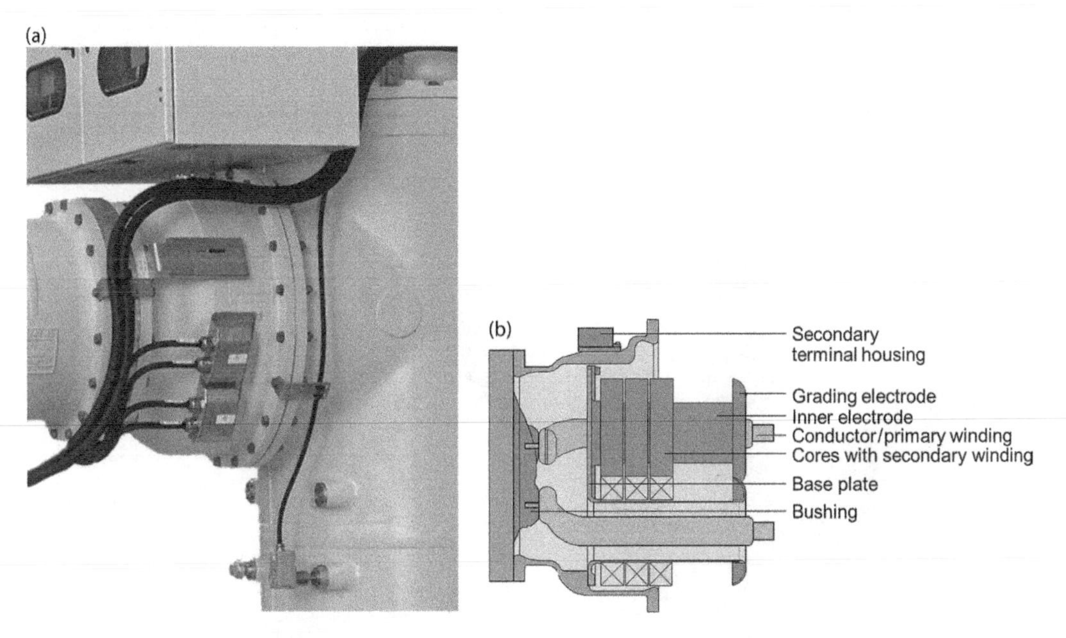

Figure 3.9 Three-phase encapsulated current transformer of 145 kV, (a) enclosure with secondary terminal boxes (b) cross section with conductor and secondary cores (Reproduced by permission of Siemens AG)

transformers, mounted outside the pressurized enclosure. The advantage is that electrically, the GIS is completely grounded/earthed at any location. The modular design also allows the current transformer to have any position at any location in the GIS bay. In Figure 3.9a, the photo shows the three-phase encapsulated current transformer attached to the GIS circuit breaker module. In Figure 3.9b, a graphic shows a cross section of the interior.

Three-Phase Overhead Line Connection Module

The three-phase overhead line connection module is an enclosure with three flanges to connect SF_6 gas-to-air bushings from the gas insulated section of the GIS to the air insulated overhead line. Therefore, the bushings are placed at an angle of about 30° to each other to increase the distance between the phases for the air-insulated section, as shown in Figure 3.10. The bushings used are of a porcelain or composite type. In the last few years, composite bushings performance is very similar to porcelain one this type of bushings are increasingly used.

Figure 3.11 shows the three-phase connection module in cross section. The three bushings are fixed by bolts and nuts to the flanges of an aluminum cast enclosure. The conductors inside the aluminum enclosure are under SF_6 gas pressure including the internal part of the bushings.

Figure 3.10 Three-phase overhead line connection module (Reproduced by permission of Siemens AG)

Three-Phase Encapsulated Cable Connection Module

To connect three-phase cables to the GIS, a module, as shown in Figure 3.12, is used. Dimensioning of the three-phase cable connection is given in IEC 62271-209 [18] for dry-type cables or oil cables. This standardized cable connection reduces the installation time on site and gives freedom of choice to the cable manufacturer.

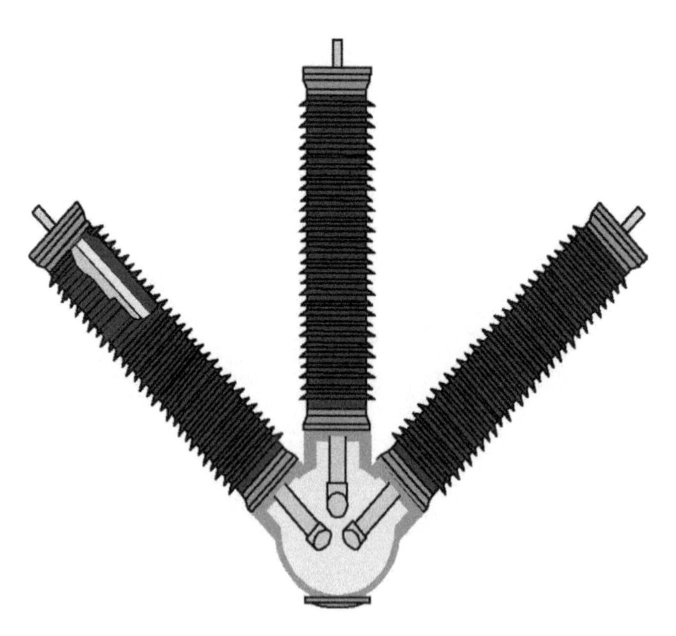

Figure 3.11 Three-phase overhead line connection module (cross section) (Reproduced by permission of Siemens AG)

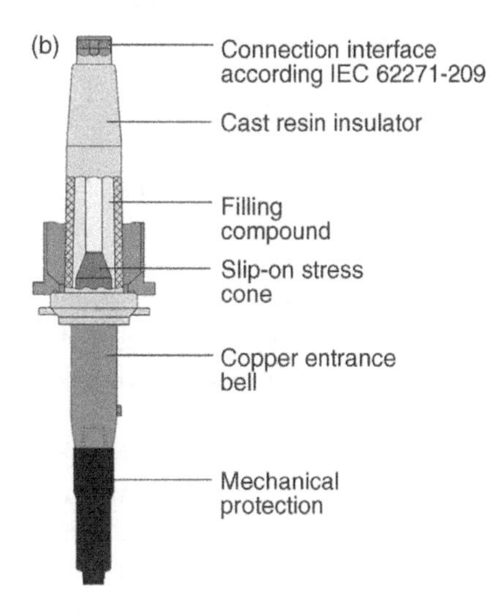

Figure 3.12 Three-phase (3a) and single phase (3b) encapsulated cable connection (Reproduced by permission of Siemens AG)

Three-Phase Encapsulated Surge Arrester Module

Three-phase encapsulated surge arrester modules are used to protect the GIS from overvoltages coming from overhead lines as lighting impulse overvoltage, at cable connections with switching impulse overvoltages, or by protecting direct connected power transformers from very fast transient overvoltage coming from disconnector or ground/earthing switch operations in the GIS (see Figure 3.13).

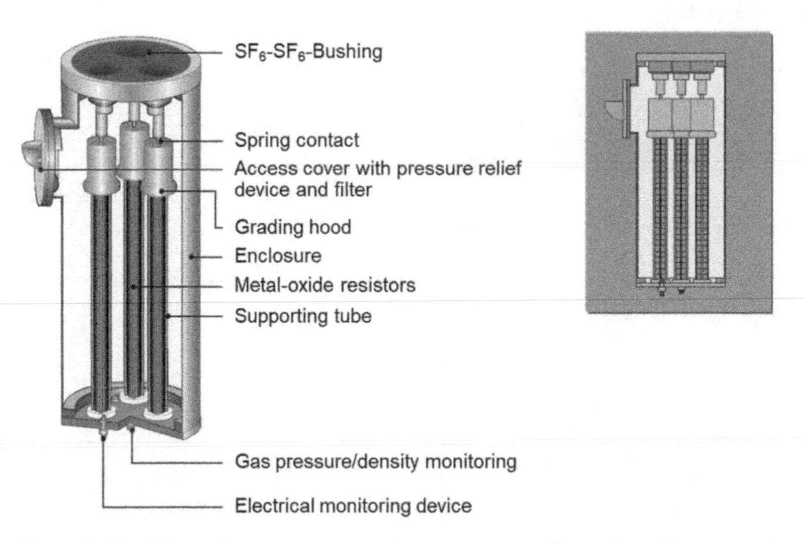

Figure 3.13 Three-phase encapsulated surge arresters (Reproduced by permission of Siemens AG)

Three-Phase Encapsulated Bus Bar

Bus bars in GIS are single phase or three phase. In some cases, they are so-called passive gas compartments if there is no switching equipment in the same gas compartment, for example, a grounding or disconnection switch. The photo in Figure 3.14a shows the three-phase insulated bus bar section with a single-phase insulated conductor connection to the single-phase insulated bay of the GIS. In Figure 3.14b, the internal schematic of the three-phase bus bar is shown.

3.2.1.2 Single-Phase Enclosure

The single-phase encapsulated GIS has a high level of standardized enclosure. Each module is basically kept at only one function, for example, switching, measuring, and connecting. The main modules are circuit breakers, disconnectors, ground/earth switch, current and voltage transformers, bus bar, extension modules with different angles, surge arresters, thermal expansion joints, cable connection, transformer connection, and outdoor connection to overhead lines or transformers (see Figure 3.15).

In the following, major components of the GIS in schematic cross sections are reviewed in detail. A wide product range of bay variations can be built with only 20 different modules. This applies even to unconventional arrangements such as the triplicate bus bypass or the 1½ circuit breaker arrangement.

The modular design of GIS components allows the creation of any single-line arrangements including circuit configurations and bus bar schemes in a most effective way corresponding to the

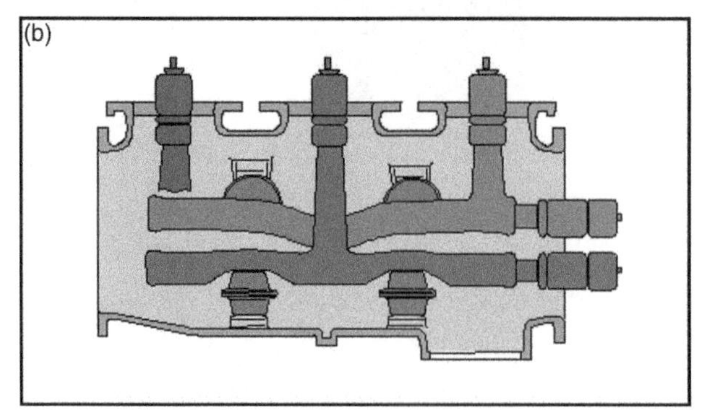

Figure 3.14 Three-phase encapsulated bus bar for upto145 kV, (a) enclosure with three phase conductors, (b) cross section with bus bars (Reproduced by permission of Siemens AG)

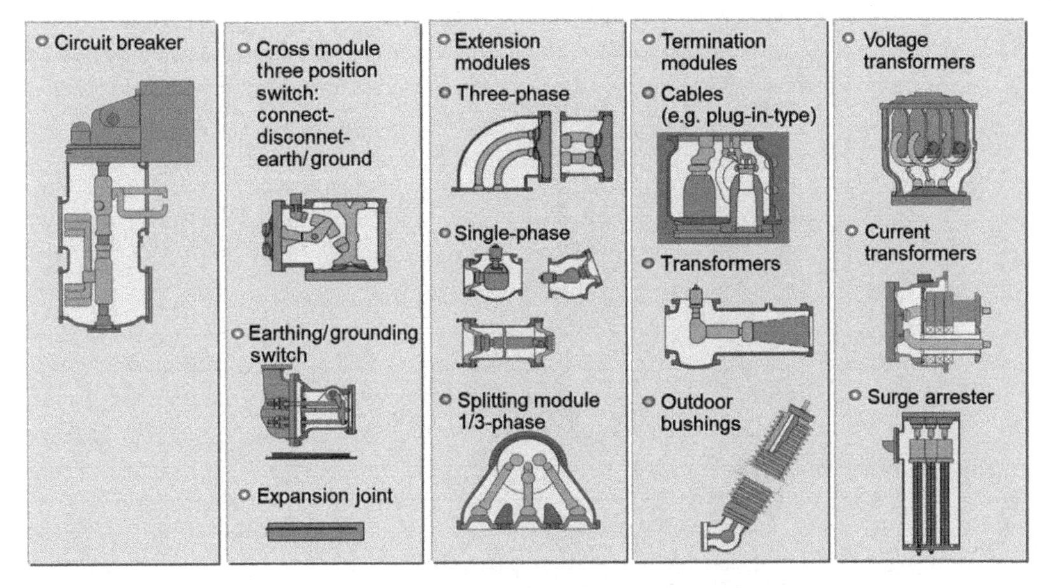

Figure 3.15 Typical GIS modules of single-phase encapsulation of 245 kV and above (Reproduced by permission of Siemens AG)

specific conditions of each individual construction. In a short time, a new three-position type of integrated disconnection and earthing switch with a common moving contact and a common drive was also introduced among the single-phase enclosed constructions.

Circuit Breaker Module

The circuit breaker is usually in a separate enclosure because of the higher gas pressure requirement for its arc distinguishing capability. The circuit breaker module may be horizontal or vertical oriented and is the base module of a bay. Other modules are connected to it.

This compartment usually has a higher operational pressure (0.7–0.8 MPa) than the other modules (0.4–0.6 MPa) because of the need to distinguish the switching arc for interruption of rated currents (typically 2000–5000 A) or in case of short circuits (typical 25–100 kA). In Figure 3.16a, a typical single-phase encapsulated 245 kV GIS bay is shown. The circuit breaker housing is used as the basis of the bay to which disconnectors, ground switches, and bus bars are connected. In Figure 3.16b, a cross section is shown in a graphic with the internal circuit breaker unit.

Disconnector and Ground Switch Module

Ground and disconnecting switches are usually in the same gas compartment. They are typically motor operated. The disconnector contact can only be operated after the circuit breaker is open with only charging or induced current on the line.

The ground/earthing switch is available in two versions. The standard ground switch with electric motor operation can handle a ground charged or induced currents. The high-speed ground switch, which is spring operated, can close the contact within few hundred milliseconds and is designed to make at least two times the short-circuit rating (see Figure 3.17). In Figure 3.17a, a photo shows a view into a cut open enclosure with the contact system in the center and a non-gastight insulator on the right. In Figure 3.17b, the cross-sectional graphic gives a side view of the ground switch on the left side and the disconnector gap on the right side.

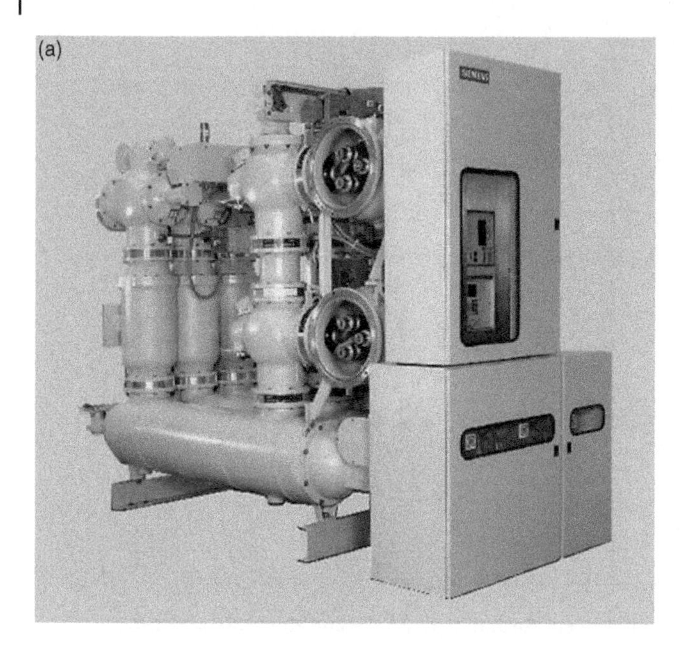

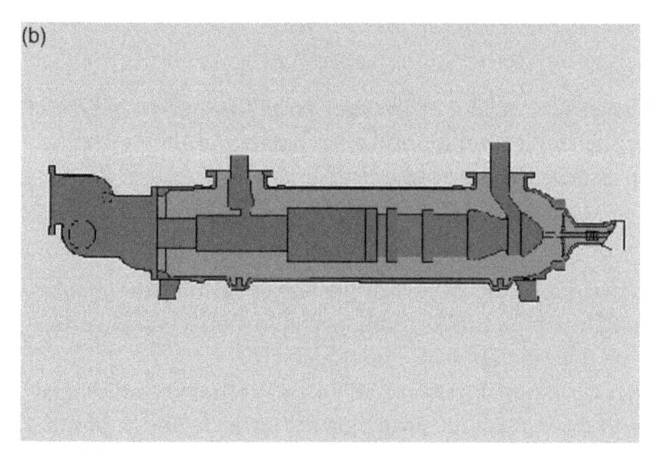

Figure 3.16 Circuit breaker single-phase encapsulated module for 245 kV and above, (a) GIS bay with circuit breaker, disconnector, ground switch and bus bar modules and attached control cubical, (b) cross section of circuit breaker module with circuit breaker and conductor function (Reproduced by permission of Siemens AG)

Load-Break Disconnector Switch Module

Specially developed "load-break disconnectors," which cannot interrupt a short-circuit current, but are able to handle the load current of the switchgear, can fulfill the functions of the disconnector as well (see Figure 3.18). These load-break disconnector switches are used if a circuit breaker is available in the network to clear short-circuit currents. For normal load current interruption, the load-break disconnect switch can be used.

Single-phase Current Transformer Module

The current transformer module of GIS has secondary coils in the side of the gas compartment with a termination box attached. In Figure 3.19a, a cut open view into the enclosure shows the current transformer secondary winding in the center and the conductor on the top and bottom, fixed by gastight conical insulators. The gas compartment of current transformers may have a different gas pressure than other compartments of the GIS bays. In Figure 3.19b, the graphic shows the cross section with five secondary coils indicated.

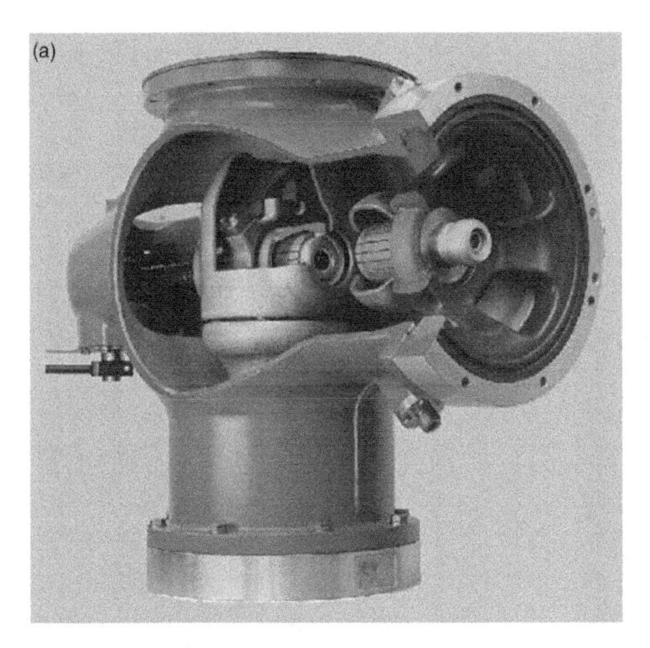

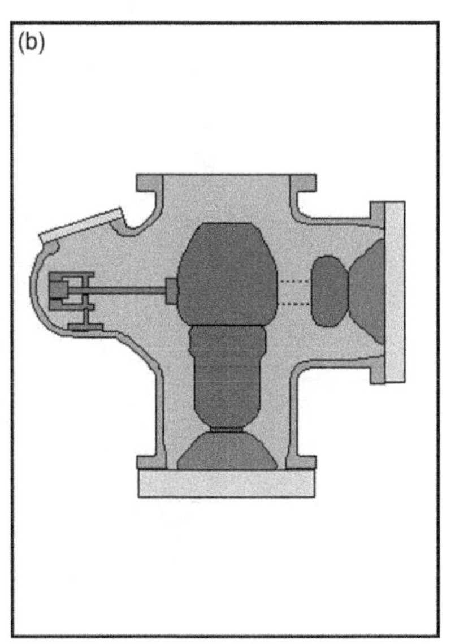

Figure 3.17 Disconnector and ground single-phase encapsulated switch of 245 kV GIS and above (a) cut open module with enclosure disconnector contact system and conical insulator (b) cross section with disconnector and ground switch function (Reproduced by permission of Siemens AG)

Single-phase Voltage Transformer Module

The voltage transformer module, as show in Figure 3.20, has a separate gas compartment, usually of a higher gas pressure than the compartments attached.

Regarding the layout, the GIS earthed modular system with its compactness and minimal dimensions offers, in comparison with AIS layouts, a much wider range of different combinations. These may be characterized, for example, by the following:

Three- or single-phase encapsulation or combinations thereof
Mixed, separated, or coupled phases of bus bars and/or bay arrangements
Single-, two-, or more-line arrangement of circuit breakers
Horizontal or vertical ("U" or "Z") circuit breaker designs
Vertical, horizontal, triangular, or upper or lower flange-connected bus bar arrangements

Figure 3.18 Load-break disconnector single-phase encapsulated switch of 245 kV and above (Reproduced by permission of Siemens AG)

The bay design and the construction of the GIS switchgear has a wide variation depending on its voltage range from 72.5 kV to up to 1100 kV. It has been tailored to the demands of all kind of substations with its different circuit arrangements, as, bypass and ring-bus systems as well as sectionalizers and bus couplers. Figure 3.21 provides a comparison of the dimensions of a 145 kV bay of 1968 versus 1991 which shows the reduction of the required space to 26.5%.

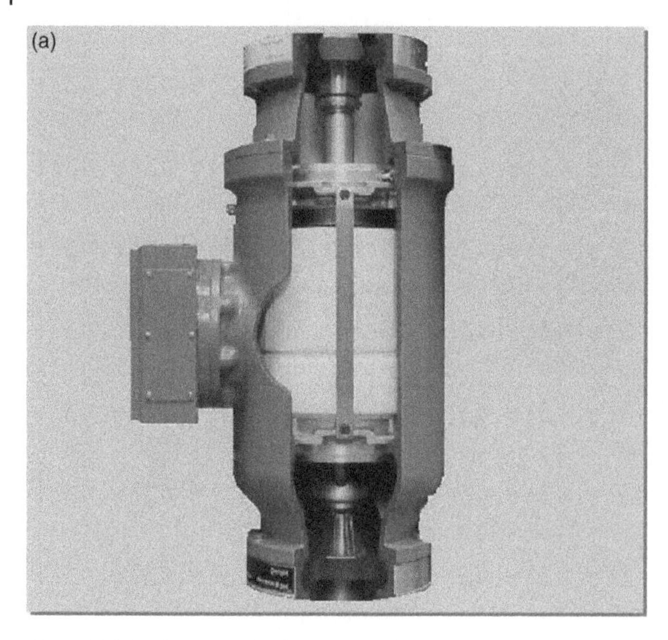

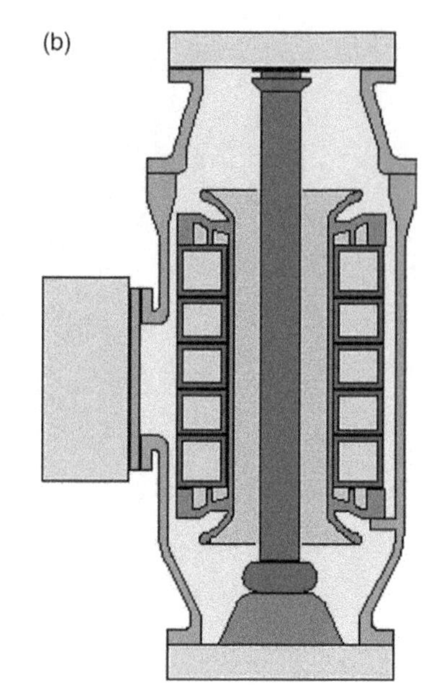

Figure 3.19 Current transformer module (a) cut open module with conductor, secondary cores, conical insulators and secondary terminal box, (b) cross section with conductor, secondary cores, enclosure and secondary terminal box (Reproduced by permission of Siemens AG)

3.2.2 Design Features

The use of SF_6 within the power energy supply is mainly driven by the gas insulated switchgear. This includes single-phase and three-phase encapsulated designs. For the distribution voltage level, mainly three-phase enclosures are used. For higher voltage levels, single-phase encapsulation is the standard.

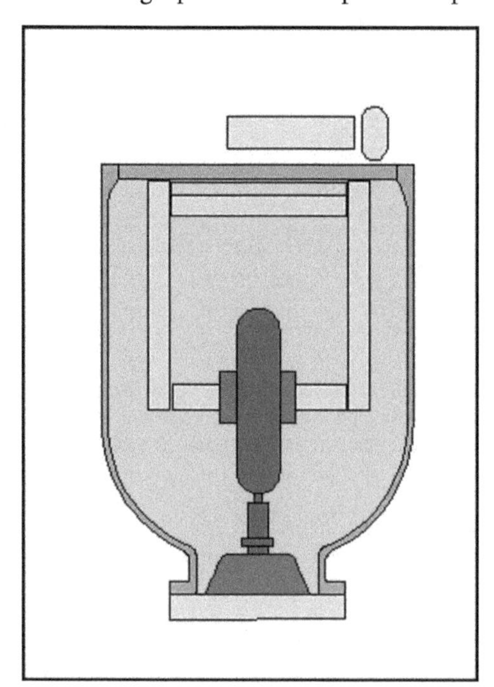

Figure 3.20 Voltage transformer module (Reproduced by permission of Siemens AG)

In the last few years, the development of SF_6-insulated switchgear was mainly driven by the objective to reduce the use of material and costs, while still maintain extremely high reliability (Figure 3.21).

The main steps of the development were as follows:

- Progress of circuit breaker technology, which allows to reduce the number of interrupter units, despite increasing breaking capability
- Progress of casting and machining technology of aluminum-casted parts, which allow the use of minimized shapes and volumes
- Use of computerized production and testing equipment with high-quality standards
- Design of integrated components with several functions like disconnectors and grounding switches within one gas compartment
- Use of intelligent monitoring and diagnostic tools to postpone maintenance activities and to avoid unnecessary tasks

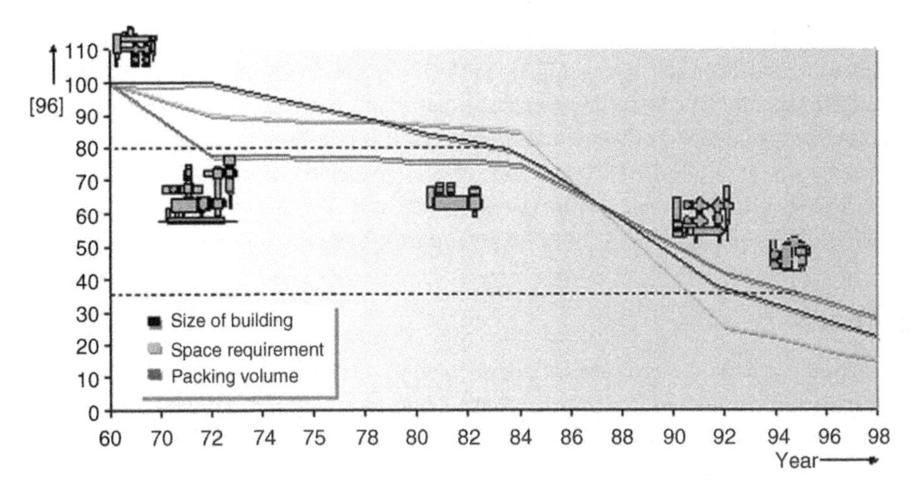

Figure 3.21 Progress of GIS development of a 145 kV GIS for size, space, and shipping volume reduction (Reproduced by permission of Siemens AG)

As a result, new very compact substation designs are on the market with the following changes compared to older design:

- Up to 98% of space reduction in comparison to air insulated switchgear
- Up to 75% reduction of SF_6 volume
- Delivery of completely sealed and tested bay units up to 245 kV
- Leakage rates down to less than 0.5% per compartment per year. Lately most reputable GIS manufacturers have designs tested to 0.1% leaks.

The development process of the last decades and its progress in size, space, and shipping volume reduction is shown in Figure 3.21. A mean time between failures has reached levels of 400–1000 years according to international statistics (IEC and CIGRE) depending on the kind of switchgear and its voltage level.

The importance of quality and reliability of all kinds of switchgear equipment has become an ever-increasing issue in recent years. Quality and reliability are the result of a complex process, which includes manufacturing, delivery and erection, and after-sales service.

Quality and reliability already are an integral part of development activities. Reliability starts with the right design, followed by the choice of suitable material, relevant testing procedures and appropriate manufacturing techniques, all of which reviewed by a stringent quality control. Modern methods of development, such as latest procedures for computer-aided design (CAD), optimization of parts and components, failure mode effective analysis (FMEA), dynamic calculations of arc extinction and drive behavior, accompanied by rigors quality checks, are well-known methods to achieve the customers' expectations.

After the design stage, materials and components are subjected to thorough development tests. In this respect, particular significance is given to long-term strength, even at extremely high numbers of operations, as well as resistance to all kinds of environmental influences. Due to the dominance of mechanical failures, major components like drive mechanisms are tested on a hydraulic test rig independently of the switchgear but with the previously measured loads given as the stress level. This allows thousands of operating cycles to be performed in just a few hours. In this way, the components can be subjected intentionally to a higher load than encountered in the switchgear itself in order to determine their safety factor. This method of testing has been proven for contact systems, operating rods, drive components, and so on, and has enabled considerable improvement in their reliability.

Development tests is followed by prototype testing comprising mechanical, power, dielectric, heat run, and environmental tests, even seismic testing, up to the limits, on several test objects in parallel.

Applicable standards specify type tests however some manufacturers often test their equipment beyond the standard requirements. For some manufacturers, it is common practice to conduct as mechanical endurance testing up to 10 000 operations or even more on one or several test objects. Furthermore, it is also specified by some utilities to extend the number of successful short-circuit interruptions beyond the requirements by the actual standards for type testing.

3.2.3 Design Process

The testing during development and type tests are performed in certified test facilities of manufacturers or, if requested, in independent laboratories. Special attention is given to the lifetime behavior of insulating material. Especially developed of GIS insulators have been a subject to an extensive test program since the first GIS delivery.

As a result of tests and more than 50 years of service experience, it can be concluded that the lifetime of the GIS insulators due to the design and the dielectric working stress will reach more than 50 years without a failure and that they are unaffected by aging. Many years of experience in switchgear design and use allow test and measurement techniques to be optimized to the degree of severity appropriate for the duty of the components with its material and functional characteristics. All of these activities were accompanied by theoretical advances and more precise definitions of requirements.

Manufacturing process has also changed. Any customer order will enter into an electronic data processing system, which details the substation down to the components and finally into the single parts. The steps of the manufacturing processes are defined by a process map and accompanied by quality assurance milestones to assure the same level of reliability is reached with every single piece of equipment. The manufacturing is a highly complex processes in terms of technical specifications, used materials, available machinery, logistics, and personnel. The manufacturing process has reached a point when the module meets all the technological demands while being produced in few and simple manufacturing steps.

The long-time service experience and extensive tests have shown no difference between triple-pole and single-pole bus arrangements. In general, the bus conductors are arranged symmetrically with insulating and supporting elements of cast resin material.

The enclosures nowadays are no longer made of steel but of aluminum alloy with several advantages like lightweight, excellent gas tightness due to outstanding surfaces of seal areas, high corrosion resistance, and negligible resistive and eddy current losses.

The enclosure design is able to withstand the electrical arc. By extensive research and using technologies of 3D CAD and the FEM (finite element method), the enclosure design has been optimized and tested for the worst-case scenarios. The result is a level of safety far greater than that required by the IEC standards.

The design of a modern GIS substation is as described above without significant differences between the major manufacturers. The following paragraphs give a few highlights about the trends of the further development of GIS.

Techniques such as 3D CAD as a computerized design tool, finite element method for the mechanical safety of enclosures and housings, stereolithography to product test models, and field distribution calculation for predict the dielectric stress allow the components of the HV equipment to be optimized very precisely.

The use of 3D CAD systems allows for a simple three-dimensional modeling is and the data can be applied for mechanical and electrical optimization. The same data are also the basis for the computerized machining process and measurement system of quality control.

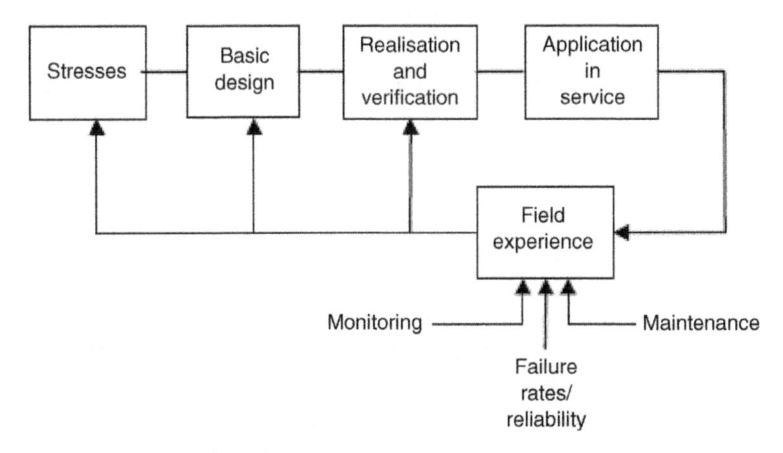

Figure 3.22 Field experiences – feedback for the development

The finite element method is a perfect tool for design of components or shapes and for the calculation of critical stress areas, either on internal loads like pressure or on external loads like, for example, seismic activities.

Stereolithography is a method for the manufacturing of testing models based on 3D CAD design. It is made from photopolymer and semicured with a UV laser, which produces the contours. This experimental model can be used for mechanical and dielectric tests and for the manufacturing of the casting pattern. For the dielectric tests, the models are coated with silver.

The effective dielectric design supported by two field distribution analyzing programs: the equivalent charge method and the finite element method. Both programs provide results of field strength values along given contours, and their minimum–maximum distribution. The field distribution analysis programs are linked to the CAD system and allow optimization of the dielectric design by iteration process.

The change to smaller and more compact substations is contradicted by the needs of the utilities for convenient service and maintenance activities. However, the advantages of such small equipment are shipment of complete, factory assembled, and prefilled double bays (245 kV) and triple bays (145 kV) with the shortest site erection time.

A further possibility of the space saving is the installation within a container for mobile use at different places.

The circuit diagram in Figure 3.22 prescribes the relation between to development and production and is the basis for an efficient cost–benefit optimization between the user and manufacturer of high-voltage switchgear.

3.3 Manufacturing

3.3.1 General

To manufacture GIS, special requirements related to high-voltage conditions need to be fulfilled. Design tolerances are small, gas tightness of the enclosure needs to be very high, sealings of flanges need high precise surface treatment, and not least the manufacturing process needs to be efficient to keep the equipment cost low.

From a manual-oriented manufacturing at the beginning of GIS manufacturing in the 1960s and 1970s, the works have developed into high-level automated factory processes. From computer-aided design (CAD) in the development the electronic drawings are directly transferred to computer-aided manufacturing (CAM) in the workshop.

The progress in aluminum molding technology has contributed strongly to the design changes of GIS and in a reduction in size and volume. The aluminum molding technology of today allows integration of functional parts, for example, to fixing of an insulator or a switch inside the enclosure and in order to simplify the assembly process later on. The aluminum molding technology also allows design of a complex enclosure forms for better use of functional parts like disconnectors, ground switches, or circuit breakers and to optimize the electric field strength inside the enclosure into acceptable limits.

The results are three-dimensional optimized forms between a sphere (best distribution of the electric field) and a cylinder (best form of functional design). The design results of such optimized enclosures do have a lower maximum field strength as has been the case in older enclosures with larger dimensions. To archive this, the manufacturing of GIS enclosures equipment is able to produce three-dimensional forms, possible only with computer-aided manufacturing (CAM).

The assembly process of GIS has high requirements for cleanliness. Particles, mainly metallic, inside an assembled GIS could cause an internal fault during operation, which could have a high impact and cause great damage and high-cost impact. Therefore, the GIS assembly process is set up to avoid any dangerous particle inside the enclosure, including a 100% routine test procedure on each insulator and with each assembled GIS bay.

Following section explains some important aspects of GIS manufacturing are explained.

3.3.2 GIS Factory

GIS manufacturing requires a large facility. A typical manufacturing facility is shown in Figure 3.23. The manufacturing process includes the equipment for casting of aluminum enclosures, the pressure and gas tightness test of the enclosure, the production of insulators and operation rods, the preassembly of disconnectors, ground/earth switches and circuit breakers, the preassembly of drives, assembly of GIS bays in a clean room environment, the assembly of control cubicles at each bay, and, finally, the routine testing before shipment.

The photo in Figure 3.23 shows the tall building in the front housing the management and development offices. The first factory hall is placed behind, where the premanufacturing takes place, including the enclosure machinery and mechanical works on metal parts, as well as fixing and connecting parts for the internal functional devices, such as disconnectors, ground/earthing

Figure 3.23 Typical GIS factory (Reproduced by permission of Siemens AG)

switches, circuit breakers, and instrument current and voltage transformers. The following three sections of the factory hall hold in parallel the preassembly of functional units, testing of enclosures, coloring of enclosures inside and outside, and a final cleaning process of all preassembled parts before they are transported into the next section for final assembly.

The final assembly hall at the upper end of the photo is designed as a clean room assembly hall with special floors and permanent overpressure to avoid dust from entering.

The access of all personnel is controlled and only approved wear is allowed for workers inside. The right side of the photo shows separated factory halls for the manufacturing of insulators and operation rods and the assembly of surge arrestors.

In the upper far left side, the logistic area for shipment of the GIS bays can be seen. GIS is sent around the world as preassembled bays or even double bays at voltage ratings up to 245 kV. For higher-voltage levels, bays are usually shipped in sections. The GIS bays or sections of bays are packed in containers in most cases, but they have wooden boxes are used in case that the containers do not fit.

As previously mentioned, large manufacturing facilities are needed to produce the GIS in an efficient way. High-quality requirements are linked to the manufacturing process to fulfill the needs of high-voltage equipment.

Today, the manufacturiung tendency is to focus some pre-manufacturng works like mechanical machinery, welding or coloring are out-sourced to speciallized sub-suppliers.

3.3.3 Insulating Parts

Insulators are the key elements of GIS. There are three principal types: post type, conical and gastight, and conical nongastight. Insulators of GIS are manufactured in a cast resin process under vacuum. Figure 3.24 shows a view into the manufacturing hall for insulator vacuum casting. The process is completely automated. The front of the photo shows mold preparation of the insulator forms. At the back of the photo shows the vacuum unit with the storage containers of the resin components. The mold form is set under vacuum, the resin is filled in and heated to melt the grain, hardener is added, and the heat is controlled for the hardening time before the finished insulators are released from the old form. Each single insulator is then tested using partial discharge measurements. The high stability of the automated cast molding process provides a very high-quality level so that virtually all insulators pass the quality test.

One important part of the insulator is the internal electrode for electric field control. Such electrodes are fixed inside the mold form before casting. This preparation is shown in Figure 3.25.

Before the insulators are molded, the acceptance test on raw materials is carried out to fulfill the quality requirements of the cast resin process. Chemical and mechanical test samples of the insulation material are performed prior to the next set of insulator production. Quality checks of the automated manufacturing process are integrated into the production and delivery to give early warning signals if some parameters are out of tolerance.

Figure 3.24 Vacuum testing of insulators (Reproduced by permission of Siemens AG)

Figure 3.25 Electrode preparation of insulators (Reproduced by permission of Siemens AG)

The conical gastight insulator, as shown in Figure 3.26, has two principal tasks:

1) To hold the conductor in the center.
2) To provide separation between two gas compartments.

Separation of the gas compartment requires pressure stability for the case where one side of the insulator is charged with the maximum filling pressure and the other side is in vacuum. Today's GIS has maximum filling pressure values of up to 0.8 MPa, which then means a maximum pressure difference of 0.9 MPa across the conical insulator. This pressure requirement is tested as a routine test with each insulator reaching the levels according to the pressure standard EN 50089 [5] for cast resin insulators.

The conical insulators in Figure 3.26 are prepared for the partial discharge (pd) test. Therefore, several insulators are mounted on a frame and then inserted into the gastight enclosure and filled with SF_6 at the minimum operation pressure for the pd test. The test method for the insulator routine test is according to IEC standard 60270 [19] and IEC 62271-1 [20].

3.3.4 Operation Rods and Tubes

Operation rods and tubes are used for mechanical force transmission, for example, for the operation of a circuit breaker, disconnector, or ground/earth switch. They insulate the high-voltage parts and to carry the mechanical forces during the operation process. In the case of the circuit breaker

Figure 3.26 Conical gastight insulators (Reproduced by permission of Siemens AG)

these mechanical forces are very large. The rods and tubes are reinforced with glass fibers to increase the mechanical strength and at the same time to fulfill the dielectric requirements of the high voltage applied.

Figure 3.27 shows the casting equipment for operation rods and tubes. After the casting process the operation rods and tubes need to be machined to prepare the functional connection points as required by the circuit breakers, disconnector, and ground/earthing switch (see Figure 3.28).

Each operation rod and tube are finally partially discharge tested to prove the dielectric quality. The manufacturing process is highly automated to ensure the process stability and a continuous high production quality. The parameters of the raw material mix, vacuum for casting, temperatures for the mold, and final hardening of the molded mixture are supervised by sensors and recorded in the computer software for each step of the production.

Figure 3.27 Casting resin equipment for operation rods and tubes (Reproduced by permission of Siemens AG)

3.3.5 Machining of the Enclosure

The GIS enclosures are made predominantly of cast aluminum alloys. These raw molded enclosures are delivered, to the factory. During the factory machining process, the raw enclosures are prepared for the assembly process. Internal fixing points are machined with the most precise machinery work needs to be done for the flanges. The surface of the flange area where the O-ring sealing fits has a very high accuracy requirement in surface leveling, paralyzing, and surface roughness. This is required to guarantee the high gas tightness required in IEC 62271-203 [21] and IEEE C37.122 [9].

The on-time delivery progression of today's manufacturing process requires the different enclosures to be machined just in time as they are needed.

In Figure 3.29, a machinery center for enclosures is shown with different enclosures on the enclosure carrousel. This figure shows one enclosure inside the cubicle in the machining process. The next enclosure is at the center of the carrousel. The two enclosures in the back of the photo are finished and will leave the machinery center with the next step. The two enclosures in the front of the photo are

Figure 3.28 Machinery of operation rods and tubes (Reproduced by permission of Siemens AG)

Figure 3.29 Machinery center for enclosures (Reproduced by permission of Siemens AG)

next in line for machining. The machining data for each enclosure is sent by the central manufacturing computer system to the machining center and is related to the project of the just-machined enclosure. Each enclosure gets a quality control protocol of each step through the manufacturing process.

3.3.6 Cleaning and Degreasing

The machining process in Section 3.3.5 uses special fluids and oils for machining. After completion the remaining metal parts from the machining process are still attached to the enclosure.

Metallic parts of all sizes shall be removed because they could cause high-voltage problems at a later stage, for example, during the routine high-voltage test after a full GIS bay is assembled. This would cause major rework and will be both expensive and time consuming. That is why an intensive cleaning process is used to move away any particles, oil, grease, and other fluids. Alkaline fluid at a 70°C is typically used to remove grease as well as all metallic particles, as shown in Figure 3.30.

Figure 3.30 Cleaning and degreasing of enclosures (Reproduced by permission of Siemens AG)

3.3.7 Pressure and Gas Tightness Test

The pressure test is required for each single enclosure since it is a pressure vessel. There are two test procedures for conducting a pressure test. The pressure test using a gas is shown in Figure 3.31 or using water is shown in Figure 3.32.

If gas is used for the pressure test it is required to have a robust protective housing around the enclosure under test in case it should burst. Typical routine test values are in the range of 1.0–1.5 MPa. Figure 3.32 shows a steel vessel protective housing and two enclosures being prepared for the test. The flanges are closed with a plate and the enclosure under test is connected to a filling pipe to pressurize the enclosure with the required value. This test uses helium since its molecule is much smaller than SF_6 one.

Helium is also a good tracing gas so that in the case of a leak the helium would come into the surrounding steel enclosure and would be detected. This would be a signal that the flange and O-ring system does not fulfill the gas tightness requirement and corrective work would be required before the enclosure could continue its way through the manufacturing process.

The combined pressure and gas tightness test with gas has the advantage of doing two manufacturing steps at once and the enclosures that pass can continue to the next step without additional cleaning and drying. The disadvantage is a necessary steel enclosure around the enclosure in the test for safety.

The hydro pressure test uses water to reach the required internal pressure but does not need any external safety enclosure. As water is noncompressive, in case, the tested enclosure breaks it will not burst, only water will leak and the pressure will be reduced immediately.

The hydro pressure test is much faster than the gas pressure test but leaves the enclosure in a wet condition, requiring drying before the enclosure can continue its way through the production process. This drying is time consuming as time is also required to remove the moisture in the porous aluminum walls. Figure 3.33 shows the hydro pressure test of enclosures.

Figure 3.31 Test equipment for routine pressure and gas tightness test of enclosures (Reproduced by permission of Siemens AG)

Figure 3.32 Enclosures ready for routine tests of pressure and gas tightness (Reproduced by permission of Siemens AG)

3.3.8 Painting Enclosures

Aluminum does not corrode under normal atmospheric conditions. Only a permanent contact with chlorides in the atmosphere could cause a distortion of the natural oxide layer of aluminum. A GIS placed outside in the rain and storm would withstand the condition without corrosion, but it would not look very attractive because the aluminum surface would get different shades of gray. Therefore, most users like to have it painted in a pleasant color, even when the GIS is installed inside a building. The choice of the color in most cases is arbitrary.

Painting of the enclosure from the inside has another technical reason. Corrosion is not possible inside because of the very dry gas condition. Painting inside is used to make it easier in the later assembly process to identify particles that do not belong inside the enclosure. For this reason, usually a light gray color is used. The painting process is shown in Figure 3.34.

Figure 3.33 Hydro pressure test of enclosures (Reproduced by permission of Siemens AG)

Figure 3.34 Painting of enclosures (Reproduced by permission of Siemens AG)

3.3.9 Preassembly of Functional Units

In parallel to the enclosure and insulator manufacturing, functional units like circuit breakers, disconnector, ground/earth switches, and current and voltage transformers are preassembled before the final assembly creates the GIS bay. In Figure 3.35, the preassembly of a circuit breaker interruption unit including a capacitor and a resistor is shown.

In Figure 3.36, the preassembly of a hydraulic drive of a circuit breaker is shown. These are only two examples of the many parallel preassembly parts of a GIS production line. Preassembly of any GIS part is separated from the final assembly process so that there is a clear separation between workstation.

3.3.10 Final Assembly

The final assembly process of a GIS is a resembles the final assembly of a car. Prefabricated parts are delivered just in time to the final assembly hall. Here, these parts are assembled in a sequence starting with the base frame, in most cases the circuit breaker enclosure. Attached to the base are usually disconnector and ground/earth switch enclosures. Then in most cases current

Figure 3.35 Preassembly of a circuit breaker with capacitor and resistor (Reproduced by permission of Siemens AG)

Figure 3.36 Preassembly of a hydraulic drive of a circuit breaker (Reproduced by permission of Siemens AG)

transformers are installed and finally followed the bus bar system. This assembly flow depends on the principal design of the GIS.

In Figure 3.37, a view in the final assembly hall gives an impression of this process. The assembly process starts at the front side of the photo and moves through the photo to the back where the GIS bays are being routine tested. The clean room conditions guarantee a high-quality level of already assembled GIS bays. The dielectric and mechanical routine tests prove the dielectric integrity of the internal insulation and the mechanical functioning of each device for interruption or switch of each GIS bay. The routine tests are automated and computer controlled so that each test measurement can be verified within the acceptable limits.

The high-voltage test includes partial discharge measurements and each flange and O-ring sealing is checked by an SF_6 sniffer device for any gas leakages. Only GIS bays without any flaw indications will pass the tests. The many years of experiences with GIS have developed quality measurement tools and processes that allow sensitive detection of any design and manufacturer mistakes. Repair and correction in the factory are always much more cost efficient than any similar correction on site.

The mechanical testing includes all switching and interruption operation including traveling time diagrams. For more details on routine testing see Section 5.3.

3.3.11 Quality Assurance

Modern quality assurance of GIS has a systematic approach of providing high-quality production integrated into the manufacturing process. This means that quality check measures are implemented for all parts and products within the manufacture facility. This reduces the requirements to check incoming materials to a minimum. The basis for the quality assurance system is ISO 9000 [7].

Figure 3.37 Final assembly hall (Reproduced by permission of Siemens AG)

The experiences in manufacturing and assembly of GIS over the last three decades have shown that the principle is to avoid any quality problem before the GIS has been assembled and is in operation.

3.3.12 Regional Manufacturing

GIS manufacturing has been decentralized over the last few years with manufacturing and assembly facilities around the globe. This development requires new quality assurance measures. The grade of local manufacturing and assembly varies from region to region and from product to product. In all cases, the new borderlines on what parts and elements are manufactured in a region and what parts and elements are assembled in the region need to be defined on a case-by-case basis.

For example, interruption units are assembled in Germany, aluminum cast housings are produced in Switzerland, and final assembly of the GIS is done in China or India, to be delivered to regional markets or to be exported to other countries. This shows that the quality assurance system must be adapted to the process to avoid any problems at the end with the GIS equipment.

Local resourcing of parts of the GIS will need an to adopt quality assurance system, which may not be the standard process with the local subsupplier. Type tests of the GIS will assure the quality of the regional manufacturing and assembly process. Add in IEC Technical Report 62271-312 explains requirements on the transferability of type tests in cases of different manufacturing and testing locations [22].

3.4 Specification Development

3.4.1 Introduction

Gas insulated substations (GIS) are assembled using standard equipment modules to construct a substation that matches the desired project electrical one-line diagram. These standard modules include circuit breakers, current transformers, voltage transformers, disconnect switches, grounding switches, interconnecting bus, surge arresters, and connections to the surrounding power system, such as SF_6 gas-to-air bushings, cable sealing ends, and transformer interface modules.

The modules are joined using a bolted flange arrangement with an "O"-ring seal system for the enclosure and a sliding plug-in contact for the conductor. Internal parts of the GIS are supported by cast epoxy insulators. These insulators may provide a gas barrier between parts of the GIS or they may be cast with holes in the epoxy to allow sulfur hexaflouride (SF_6) gas to pass from one side to the other. The barrier insulators are used to create separate gas compartments in order to limit the gas volume in any one enclosure.

GIS that operates at voltages up to 170 kV are typically designed in such a manner that all three phases are contained within one enclosure. When operating voltages exceed 170 kV, the three-phase enclosure design becomes too large to be practically produced.

Most enclosures are cast or welded aluminum, although some steel enclosures are also used. Steel enclosures are painted inside and outside, while aluminum enclosures are only painted inside.

Conductors are mainly aluminum, but copper can be used and all current transferring surfaces are silver plated. Bolted and sliding electrical contacts are used to join conductor systems.

Support insulators are made of highly processed filled epoxy resin, cast very carefully in order to prevent the formation of voids and/or cracks during the curing phase of production.

Quality assurance programs for support insulators include a high-voltage power frequency withstand test with sensitive partial discharge monitoring. For further information about the GIS specification, standard design, and application guidance see References 8–11.

3.4.2 Specification Documentation

3.4.2.1 General

The specification of the GIS is based upon the single-line diagram, the gas zone diagram, and the physical layout plan. It is possible to prepare sketches of different GIS designs and layouts that are available and to see how these may be used in an actual project with respect to site conditions, civil requirements, space and clearance requirements, environmental aspects, and interface requirements with existing equipment. The different designs may dictate whether the GIS can be an outdoor, indoor, or hybrid (mixed) technology design.

The following is a list of the critical documentation that should form the parts of a GIS specification:

Detailed One-Line Diagram
Site Drawings and Plot Plan (highlighting restrictions and physical imitations)
Primary Equipment Data
Secondary Equipment Data
Engineering Studies (as appropriate)
Maintenance and Operation Requirements
Standards and Regulations
Regional Reliability Criteria
Project Deliverables

It is usual and useful to contact manufacturers prior to generating specification documents, other than the Detailed One-Line Diagram and the Site Drawings and Plot Plan, for pre-specification discussions and to obtain preliminary technical proposals with budgetary prices. At this stage one should note whether the user's basic layout perhaps excludes a certain design that may be more effective from the manufacturer's standpoint.

It is very advantageous to the user to employ the services of an "Owner's Engineer," who has extensive GIS experience. The "Owner's Engineer" can offer expertise that is not readily available on the user's engineering staff and help to produce optimized designs and provide cost savings.

Close and continued contacts with several manufacturers through an "Owner's Engineer" will also yield good information regarding the experience and the design concepts of each potential design and arrangement.

Special attention has to be given to the connections between the GIS and other components of the network, such as overhead lines, transformers, cables, and so on. The type and location of these connections will have a major impact on the overall layout and cost. Internal discussions facilitated between the user's Engineering, Maintenance, and Operations organizations should be ongoing during the exploratory process of gathering information to write the GIS specification.

3.4.2.2 Detailed One-Line Diagram

The detailed one-line diagram is the first major section of the GIS specification. The detailed one-line diagram (one-line) should show all major equipment for the project and any future expansion requirements for the GIS. Figure 3.38 is a typical detailed one-line diagram of a single bay of a GIS.

The one-line should, at a minimum, show the following equipment:

High-Voltage Circuit Breakers
Current Transformers
Primary Disconnect Switches with Operator Type
High Speed Ground Switches with Operator Type

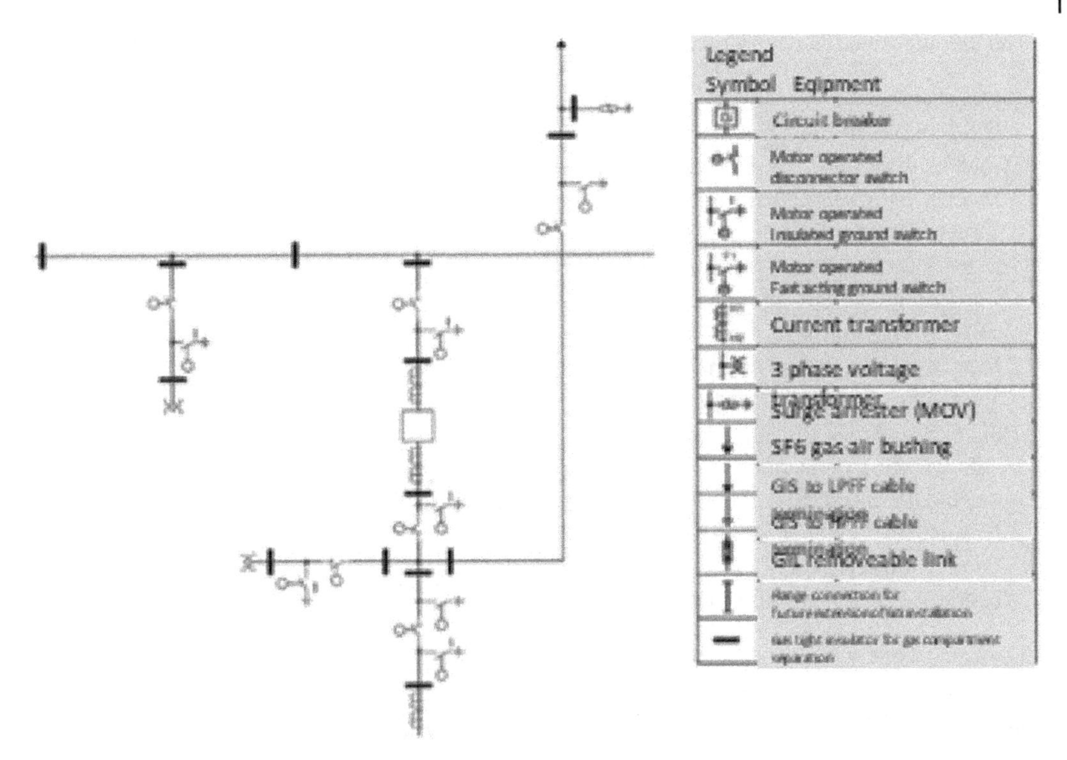

Figure 3.38 Typical detailed one-line diagram of a single bay of GIS (Reproduced by permission of United Illuminating Company)

Grounding Switches with Operator Type
Voltage Transformers
Main Buses and Lines
Terminal Interfaces (i.e., cable terminations, SF_6 gas-to-air bushings, etc.)
Gas Barrier Insulators
Surge Arresters
Removable Bus Links

The most important action taken in the development of the one-line is the consultation between Engineering, Maintenance, and Operations organizations.

It is important to engage the Protection and Controls Engineering organization at this point to ensure that all necessary high-voltage sensing devices for protection schemes are identified and located as well as gas monitoring equipment and supervisory control and data acquisition (SCADA) equipment.

The one-line should be drawn in a semiphysical orientation to give the manufacturer some idea of how the GIS is to be arranged. If the one-line is not in a semiphysical orientation then it should be noted in the specification and any special physical arrangement requirements should be detailed.

3.4.2.3 Primary Equipment Data

The specification of primary equipment is the second major section of the GIS specification. This primary equipment data section outlines in detail the types, ratings, and configuration of the components in the GIS.

The following are the typical sections of the primary equipment specifications for the GIS:

General Criteria
GIS Equipment Ratings and Service Conditions
Enclosure Design – Single Phase or Three Phase
Specific Equipment Requirements
Circuit Breakers
Disconnect Switches, Grounding Switches, and Operators
Gas System and Gas Zone Configurations
Current Transformers
Inductive Voltage Transformers
Metal-Enclosed Surge Arresters
SF_6 Gas-to-Air Bushings
GIS to Cable Connections
GIS to GIB (gas insulated bus)/GIL (gas insulated line) Connections
Power Transformer Bushing Connections
Local Control Cabinets and Marshaling Cabinets
Ladders, Platforms, Stairs, and Walkways

The equipment and materials should be provided in accordance with the latest approved revisions of national and international standards including, but not limited to, those listed below, and made a part of the specification:

- IEEE C37.123 IEEE Guide to Specifications for Gas-Insulated, Electric Power Substation Equipment
- C37.122 IEEE Standard for High Voltage Gas-Insulated Substations Rated above 52 kV
- IEEE Std. C37.122.1, IEEE Guide for Gas-Insulated Substations
- IEEE Std. C37.122.2, IEEE Guide for Application of Gas-Insulated Substations 1 kV to 52 kV
- C37.100.1, IEEE Standard of Common Requirements for High Voltage Power Switchgear Rated above 1000 V
- C37.04, IEEE Standard for Rating Structure for AC High-Voltage Circuit Breakers
- C37.06, IEEE Standard for AC High-Voltage Circuit Breakers Rated on a Symmetrical Current Basis – Preferred Ratings and Related Required Capabilities for Voltages above 1000 V
- C37.09, IEEE Standard Test Procedure for AC High-Voltage Circuit Breakers Rated on a Symmetrical Current Basis
- C37.010, IEEE Application Guide for AC High-Voltage Circuit Breakers Rated on a Symmetrical Current Basis
- C37.011, IEEE Guide for the Application of Transient Recovery Voltage for AC High-Voltage Circuit Breakers
- C37.017, IEEE Standard for Bushings for High Voltage (over 1000 V(ac)) Circuit Breakers and Gas-Insulated Switchgear
- ANSI (US) C2, National Electric Safety Code
- ASTM (US) D2472, Specification for Sulfur Hexafluoride
- IEC 60044 (including amendments), Instrument Transformers
- IEC 60050, International Electrotechnical Vocabulary
- IEC 62271-100, High-Voltage Alternating Current Circuit Breakers
- IEC 60060, High Voltage Test Techniques
- IEC 60071, Insulation Coordination
- IEC 62271-102, Alternating Current Disconnectors and Earthing Switches

- IEC 62271-200, A.C. Metal-Enclosed Switchgear and Controlgear for Rated Voltages above 1 kV and up to and including 52 kV
- IEC 60376, Specification and Acceptance of New Sulfur Hexafluoride
- IEC 60480, Guide to the Checking of Sulfur Hexafluoride Taken from Electrical Equipment
- IEC 62271-203, Gas-Insulated Metal-Enclosed Switchgear for Rated Voltages above 52 kV
- IEC 62271-1, Common Specifications for High Voltage Switchgear and Control Gear Standards.
- IEC 62271-209, Cable Connections for Gas-Insulated Metal-Enclosed Switchgear for Rated Voltages of 72.5 kV and above
- IEC 61129, Alternating Current Earthing Switches-Induced Current Switching
- IEEE 80, Guide for Safety in AC Substation Grounding
- IEEE Std. C57.13, IEEE Standard Requirements for Instrument Transformers
- ANSI/NEMA CC 1, Electric Power Connectors for Substations
- NFPA 70, National Electrical Code
- IEC 60137, Bushings for Alternating Voltages above 1000 V
- IEC 61869-2, Current Transformers
- IEC 61896-3, Voltage Transformers
- IEC 61128, Alternating Current Disconnectors Bus-Transfer Current Switching by Disconnectors
- IEEE 693 - Recommended Practice for Seismic Design of Substations

General Criteria

All equipment and material should be prefabricated, factory assembled, tested, and shipped in the largest practical assemblies. All ratings should be equal to or greater than the GIS standards for this class of equipment except where specifically noted otherwise.

Assembled equipment should be capable of withstanding electrical, mechanical, and thermal ratings of the specified system. All joints and connections should be able to withstand the forces of expansion, vibration, contraction, and specified seismic requirements without deformation, malfunction, or leakage. Equipment should be capable of withstanding the specified environmental conditions.

Optimized arrangements may be employed to reduce installation time, provide ease of operation, minimize maintenance and repair costs, and facilitate future additions.

Sufficient space and access areas should be provided to permit the ready removal and re-installation of each internal and external equipment and component. The access areas around the equipment should accommodate the use of overhead cranes, maintenance fixtures, jigs, and any other required test and maintenance equipment. All gauges, viewports, and gas fill points should be located so as to be readily accessible and viewable by maintenance personnel.

The footprint size of a proposed GIS, hoist (maintenance crane) capacity required, field assembly/erection, recommended maintenance intervals, guaranteed SF_6 gas leakage rate, and cost of operation should be considered in the evaluation of proposals.

GIS Equipment Ratings and Service Conditions

The GIS equipment should be specified with detailed service condition requirements and ratings. Table 3.1 is an example service conditions and ratings table with the specified ratings that may be used to provide to the manufacturer as part of the overall GIS specification.

Enclosure Design – Single Phase or Three Phase

In general, for system voltages up to 170 kV, all three phases are often in one enclosure. Equipment that operates at system voltages above 170 kV are generally designed with individual phases in single-phase enclosures. The GIS design for all three phases in one enclosure at voltages above

Table 3.1 Service conditions and ratings (use the template to specify values)

Service conditions	
Enclosure	Description of the GIS environment (e.g., mostly enclosed in a building; bus to SF_6 gas-to-air bushings and cable sealing ends are outdoors)
Maximum allowable total temperature Main conductor joints	XXX°C
External surfaces	XX°C
Ratings	
GIS	
Rated maximum voltage	XXX kV or higher
Rated insulation level – basic impulse level	XXX kV minimum – peak value
(BIL) (across the isolating distance)	
Rated low-frequency phase-to-ground withstand voltage	XXX kV root mean square (rms)
Rated frequency	50 or 60 Hz
Rated minimum current, main buses	XXXX A (amperes)
Rated minimum current, cross buses, and bus	XXXX A
taps	
Rated short-time withstand current	XX kiloamperes (kA) rms
Rated withstand current	XXX kA peak
Rated duration of short circuit	1 second (minimum)
Circuit breakers	
Rated maximum voltage	XXX kV for circuit breakers defined as "definite purpose for fast transient recovery voltage rise times" per ANSI/IEEE C37.06.1;
	or
	XXX kV for "general purpose" per ANSI/IEEE C37.06
Rated maximum interrupting time	X cycles on a 60 Hz basis
Rated minimum current, all breakers	XXXX A
Rated short-circuit breaking current	XX kA rms
Rated closing and latching current	XXX kA peak
Rated operating sequence	Duty cycle: O-t_1-CO-t_2-CO where $t_1 = 0.3$ second, and $t_2 = 3$ minute
Rated capacitive switching currents	IEC 60056 (Table 5)
Number of mechanical operations	10 000 (minimum)
Number of trip coils	X
Disconnect switches	
Rated minimum current, all disconnect switches	XXXX A
Rated short-time withstand current	XX kA
Rated peak withstand current	XXX kA
Rated duration of short circuit	1 second
Mechanical endurance	1000 cycles (minimum)
Grounding switches	
Rated minimum current, all grounding switches	XXXX A
Rated short-time withstand current	XX kA

Table 3.1 (Continued)

Rated peak withstand current	XXX kA
Rated short-circuit making current (fast-acting ground switch only)[a]	XXX kA[a]
Number of fault closing operations (fast-acting ground switch only)	Two minima, before inspection/contact replacement[a]
Rated duration of short circuit	1 second
Mechanical endurance	1000 cycles (minimum)
Current transformers	
Ratio of rated primary current	Five lead, multi-ratio (IEEE C57.13, Table 8), XXXX A
Rated secondary current	5 A
Rated continuous thermal secondary current	10 A
Rated 1 second thermal equivalent primary current	XX kA
Relay accuracy class	C800 (IEEE C57.13)
Metering accuracy class	0.3 at burdens B-0.1 through B-1.8
Voltage transformers	
Application	Three-phase, phase-to-ground in an effectively grounded system
Rated primary voltage, line-to-ground	XXX XXX Ground Y/XX XXX
Rated secondary voltages	XXX V and XX V (two secondary windings, two taps in each winding)
Rated output	XXX volt-amperes (VA) (minimum), each winding
Accuracy class	IEEE C57.13, 0.3 at M, W, X, Y, Z, ZZ
Metal-enclosed surge arresters[b]	
Type	Metal oxide varistor (MOV), station class
Application	Metal-enclosed, phase-to-ground in an effectively grounded system
Maximum continuous operating voltage (MCOV) rating	XXX kV
Rated frequency	60 Hz
SF_6 gas-to-air bushings	
Application	Outdoor for connection to owner's AIS equipment
Rated maximum system voltage	XXX kV (minimum)
Rated insulation level – BIL	XXX kV (minimum)
Rated short-duration power frequency withstand voltage	XXX kV – rms
Rated frequency	60 Hz
Rated minimum continuous current	XXXX A
Minimum centerline phase-to-phase spacing	XX feet (minimum), unless otherwise noted on the drawings
Minimum external housing insulation leakage distance	XXX inches (minimum)

[a] The manufacturer should provide information with its proposal to ensure the switch will perform in this service as required.
[b] A discharge counter and leakage current monitoring system should be provided for each surge arrester.

170 kV becomes impractical to produce from a design perspective related to dielectric dimension requirements. There are no major established performance differences between the three-phase enclosure and the single-phase enclosure GIS except for the applications of the equipment related to transient recovery voltage (TRV) requirements.

With respect to TRV, in the case of a three-phase common gas insulated switchgear enclosure, an arc between phase and ground will, within a few milliseconds, evolve into a three-phase fault between conductors, owing to the ionization of the gap between the conductors, and at the same time the phase-to-ground arc will extinguish. Consequently, an enclosure burn-through is not likely, which is a positive characteristic of the three-phase common enclosure. However, since within the range of 20–50 ms, the fault has evolved into a three-phase ungrounded fault, and since the breaker contacts typically begin to part at or after 20 ms, then the three-phase ungrounded fault should be considered in the TRV analysis for three-phase common gas insulated switchgear enclosures.

Specific Equipment Requirements

The GIS electrical configuration and equipment rating/features shall be in accordance with all drawings listed in the specification.

All disconnect switches, fast-acting grounding switches, and maintenance grounding switches should be group and motor operated, and capable of interrupting the charging current of the connected GIS bus and associated components.

Fast-acting grounding switches shall be capable of handling the charging currents associated with the GIS bus, high-voltage overhead transmission lines, and high-voltage underground cable termination lines. Fast-acting grounding switches shall be capable of accidentally closing an energized bus without damage to the switch or the enclosure.

Each disconnect switch, maintenance grounding switch, and fast-acting ground switch shall be equipped with mechanism-actuated auxiliary switches for indication of the contact position. These mechanism-actuated auxiliary switches should be accessible from floor level.

Visual inspection means shall be provided to observe the disconnect switch, maintenance grounding switch, and fast-acting ground switch contact positions. A mechanically connected external position indicator should be provided for each switch and each switch mechanism should have provisions for mechanical blocking.

Electrical interlocks must be furnished to prevent incorrect sequential operation or switching equipment malfunction that might result in equipment damage or personnel injury. Electrical interlock schemes must provide the interlock function upon loss of the control power.

The user may require provisions for bypassing the electrical interlocking scheme.

Mechanical blocking devices should be provided to block the operation of all disconnect switches and ground switches, with provisions for padlocking on each blocking device.

Proper grounding for mitigating overvoltages during disconnect switch or circuit breaker operation should be included. Ground pads for connection to the user's ground system should be provided.

All high-voltage conductors should be made of aluminum tubing suitable for the specified current and voltage ratings. The enclosure should be aluminum or steel with all adjoining enclosures to be bolted together and connected to ground.

The GIS equipment should be designed to prevent mechanical failure and withstand pressure buildup if a circuit breaker fails to interrupt a full-rated fault current. The use of rupture discs as a pressure relief device is acceptable. Each SF_6 gas-filled enclosure should be in complete compliance with the requirements of established national pressure vessel code standard requirements. The manufacturer should identify in its proposal the methodology for calculating the thickness and construction of the enclosures and the applicable pressure vessel code standards.

Pressure relief devices should be provided with a shield and should be vented to provide a safe environment for personnel, and for equipment, during operation. The bursting pressure of the relief device should be effectively coordinated with the rated gas pressure and the pressure rise due to arcing as described in IEEE Std. C37.122 [9] and IEC 62271–203 [21].

Expansion and installation alignment should be considered in the design of the bus and the enclosure. If required, expansion joints should be provided with compensators for the enclosure and sliding plug-in contacts used for the conductors.

Support insulators should maintain the conductor and enclosure in the proper relation for the rated maximum voltage, rated insulation level (BIL, or basic impulse level), and rated low-frequency phase-to-ground withstand voltage.

Connections between adjacent conductor sections should be made by means of plug-in type contacts. Shields should be placed to capture metallic particles that may result from a contact rubbing. All welding of high-voltage conductors should be made at the fabrication facilities. Field welding of the conductors is not acceptable.

Shipping sections should be joined in the field using bolted, gasketed, flange connections of the enclosure. Flanged connections should have a gas seal between the flange surfaces. A second seal ring, sealants, or other suitable means are required to protect the gas seal from the external environment. Connections, including bolts, washers, and nuts, should be adequately protected from corrosion and should be easily accessible with the proper tools.

Field welding of the enclosures is not acceptable.

Structures to support the equipment, platforms, and walkways for operation and maintenance access to operating and monitoring devices should be designed to permit access without use of special devices or portable ladders.

GIS grounding pad terminals should be located to permit proper connection to the user's ground grid and to minimize external bus enclosure voltage gradients to a safe limit. All support structures and GIS local control cabinets (LCCs) should be grounded.

Surge arresters/surge suppressors across the insulating joint of enclosures to eliminate very fast transients (VFTs) due to switching or circuit breaker operation should be provided.

All wiring of devices and terminations internal to the switchgear and all shielded control cables and associated raceways, above the foundation, between the equipment and the GIS local control cabinets, and all low voltage raceways and wiring furnished by the manufacturer are to be designed in accordance with NFPA 70 (NEC) [23] for US and IEC 61936 [24] for international applications. Raceways and conduits should be installed so as not to present a safety hazard or require removal by personnel inspecting, maintaining, or servicing the equipment.

Circuit Breakers

The circuit breakers used in GIS are essentially the same type of dead tank SF_6 puffer design or dynamic self-compression design that are used in air insulated substations. The circuit breakers in the GIS, however, do not use SF_6 gas-to-air bushings as the connections to the substation in general. The nozzles and the interrupter assemblies of the circuit breaker are directly connected to the adjacent GIS module via bolted or plug-in connectors and the connections are an integral part of the overall GIS conductor current path.

GIS circuit breakers are designed in the same manner as dead tank SF_6 puffer circuit breakers that are used in air insulated substations. The GIS circuit breakers are subject to the same IEEE and IEC standards as their air insulated counterparts.

The user should incorporate all of the specification requirements, except those related to SF_6 gas-to-air bushings, that are present in a dead tank SF_6 puffer circuit breaker specification, and apply these requirements to the GIS circuit breakers.

Circuit breakers should have ratings described in applicable IEEE and IEC standards. The circuit breakers should be capable of performing the specified duty cycle without derating.

Each circuit breaker should be factory assembled, adjusted, and tested, and should be shipped as a complete three-phase, gang-operated unit where practical.

The circuit breakers should include a suitable operating mechanism to ensure proper opening and closing and should permit checking adjustments and operating characteristics. The mechanism should be capable of reclosing within the time range specified.

Each breaker should have necessary valves and connections to ensure ease in handling the SF_6 gas.

Each circuit breaker should be equipped with an operation counter. The preferred arrangement for this device is to operate only during the opening cycle. An indicator that shows the position of the contacts should be provided. All gauges, counters, and position indicators should be readable by an operator standing near the equipment at floor level.

If pre-insertion resistors are required, they should meet the rated line-closing switching surge factor, as specified in the applicable standards.

A minimum of number of "a" and "b" auxiliary contacts should be specified by the user and be provided such that the contacts are field reversible.

The user should specify the acceptable type of circuit breaker operating system, spring – spring, spring–hydraulic, hydraulic, or pneumatic.

The user should specify the circuit breaker spring charging motor to be a universal AC/DC motor. It is typical to operate the spring charging motor on an AC supply under normal conditions with DC as a backup. The user should specify a manual throwover switch at the circuit breaker to allow maintenance personnel to switch from AC to DC supply.

A circuit breaker trip and close coils should be of the low energy type requiring not more than 6 A direct current (DC) for proper operation.

Each gang-operated circuit breaker should have dual trip coils.

Figure 3.39 depicts a typical GIS circuit breaker and Figure 3.40 shows a typical coupling contact arrangement.

Disconnect Switches, Grounding Switches and Operators

The disconnect switches (also referred to as disconnectors) in GIS have a moving contact that opens or closes a gap between stationary contacts in conjunction with the activation of an insulated operating rod. The insulated operating rod is driven by a sealed shaft that penetrates the enclosure wall.

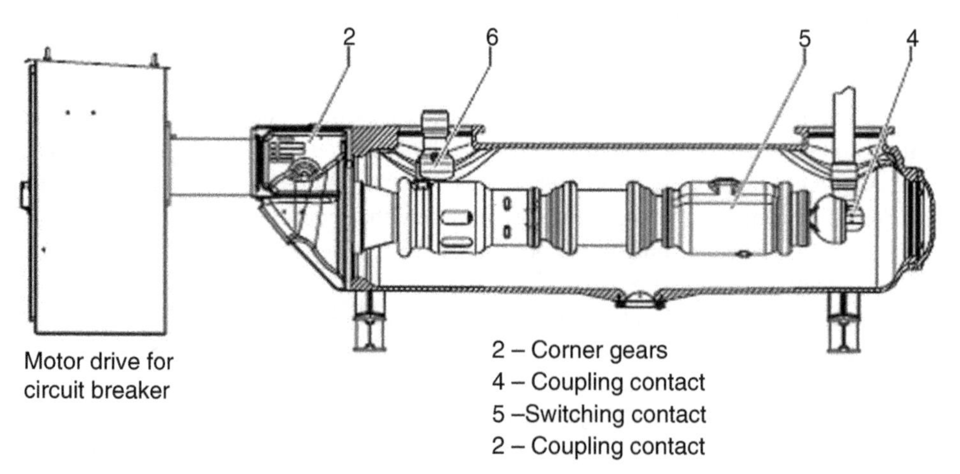

Figure 3.39 GIS circuit breaker (Reproduced by permission of Siemens AG)

The stationary contacts usually have shields that provide required electric field distribution to avoid high surface stresses at discontinuous surfaces.

The velocity of the moving contacts of the disconnect switches is low as compared to the moving contacts of the circuit breaker. The disconnect switches in GIS are only capable of interrupting low levels of capacitive and inductive currents such as disconnecting GIS bus sections and transformer magnetizing currents.

Disconnect switches should be gang- and motor-operated, nonload break, with one operating mechanism per three-pole switch. The operating mechanisms should be provided with a position indication. Figure 3.41 depicts a typical disconnect switch enclosure arrangement.

9 – Insulator
10 – Conductor
11 – Bolt
12 – Circuit Breaker Connection

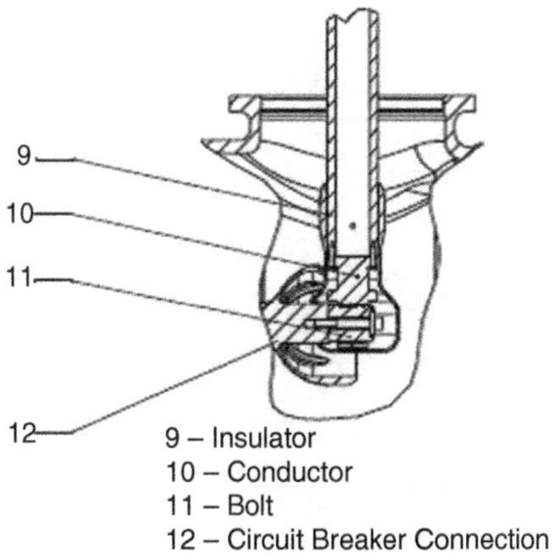

Figure 3.40 Typical coupling contact arrangement (Reproduced by permission of Siemens AG)

The grounding switches (also referred to as earthing switches) have moving contacts that open or close a gap between the high-voltage bus conductors and a ground connection on the enclosure. The grounding switches are equipped with electric field shields at the bus conductor and the ground connection on the enclosure. There are two basic types of grounding switches, maintenance grounding switches and fast-acting grounding switches.

Grounding switches should be gang- and motor-operated, nonload break. Each grounding switch/mechanism enclosure should be painted a different color to clearly distinguish it from a disconnect switch.

The maintenance grounding switch is operated either manually or by a motor mechanism to close or open. When the switch is fully closed it is rated to carry the rated short-circuit current for the specified time period without damage to the GIS. Maintenance grounding switches open and close over a period of several seconds.

The fast-acting grounding switch is operated by a high-speed drive mechanism that uses spring energy to close and open the switch. The switch contact materials are able to withstand arcing and the switch is usually rated to close into the rated short-circuit current of the GIS for two operations without damage to itself or adjacent equipment. The fast-acting grounding switch is usually used at the points where the GIS interfaces with the power system, such as underground cable line terminals, overhead line terminals, and transformer interface connections. The fast-acting grounding switch is rated to handle the discharge of trapped charge and the breaking of capacitive or inductively coupled currents.

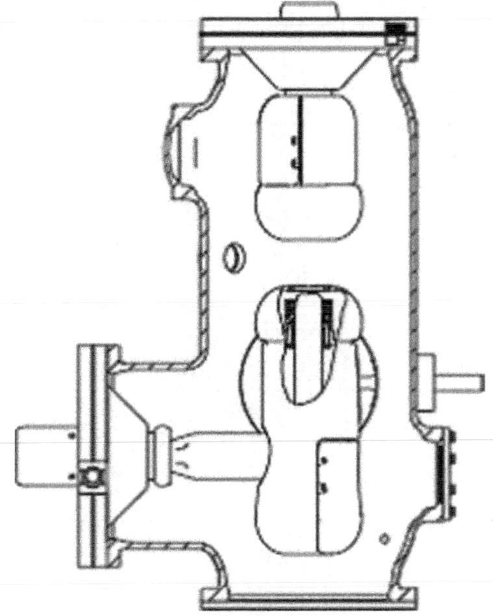

Figure 3.41 Typical disconnect switch enclosure arrangement (Reproduced by permission of Siemens AG)

Grounding switches are designed with an insulated bushing for the connection of the ground conductor. If the user specifies provisions for bypassing the electrical interlocking scheme between the disconnect switch and the grounding switch, the grounding switch may be closed to facilitate voltage

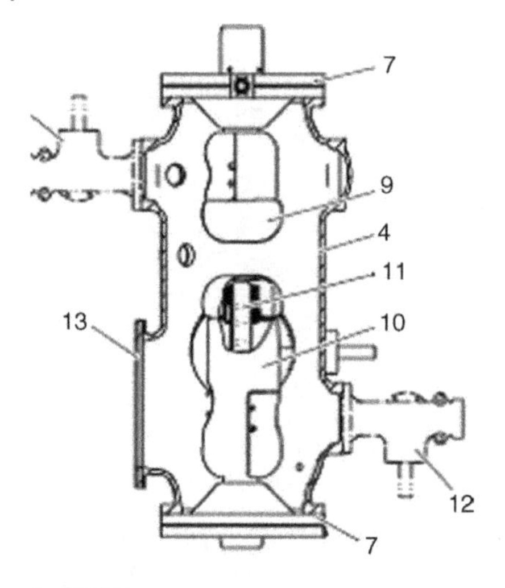

Figure 3.42 Typical maintenance grounding switch enclosure arrangement (Reproduced by permission of Siemens AG)

and current testing of the internal parts of the GIS without removing SF_6 gas or opening the enclosure. A shunt may be removed to allow for the connection of test equipment to the insulated bushing where the ground conductor is connected. The grounding switch may be closed to allow for the performance of contact resistance tests in the circuit breaker, bus resistance tests, primary current injection to test current transformers, and high potential tests, for example.

Figure 3.42 depicts a typical maintenance grounding switch enclosure arrangement and Figure 3.43 depicts a typical fast-acting grounding switch enclosure arrangement.

The motor-operated disconnect and grounding switches should be equipped with electrically and mechanically operated devices to uncouple the motor when the switch is operated manually, to prevent coincident power operation of the switch and the drive mechanism(s).

Each disconnect and grounding switch should open or close only due to manual or motor-driven operation. The switch moving contact should not move due to gravity, or other means, if a part fails. Once initiated, the motor mechanism should complete an open or close operation without requiring the initiating contact to be held closed.

Each disconnect switch and grounding switch should be furnished with electrically independent auxiliary switches. The auxiliary switches should indicate the position of the switch blades and should be provided so that the contacts can be adjusted to be fully engaged and in proper alignment when in the closed position. A minimum of number of “a” and “b” auxiliary contacts should be specified by the user and be provided such that the contacts are field reversible.

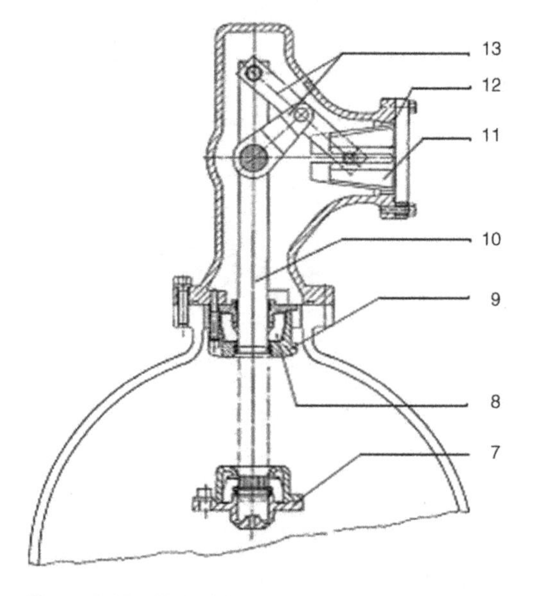

Figure 3.43 Typical fast-acting grounding switch enclosure arrangement (Reproduced by permission of Siemens AG)

Visual verification, via a viewport accessible by personnel, should be provided for each pole of each disconnect switch and grounding switch to permit visual inspection of each switch contact position. External position indicators connected directly to the mechanism shaft should be provided.

In addition to the position indicators following the mechanism shaft discussed above, inspection viewports in the switch GIS enclosure should be provided, and they should have removable covers to prevent damage of the viewport. Operating personnel should not be required to remove or reposition any other devices/mechanisms/hardware in the vicinity or to engage any type of mechanism or tool to access the viewport. All viewports should be accessible from the floor by personnel. If this is not feasible, then dedicated optical cameras at each viewport with user approved viewing monitor and cabling systems should be provided.

As an option to the use of cameras the user should specify an option to furnish maintenance platforms, ladders, and/or stairs to access all of the viewports instead of the dedicated optical cameras. If maintenance platforms are employed, they should comply with all applicable fall prevention regulations.

The viewports should be arranged such that one person has the ability to operate a light source and inspect the main switch contacts or, if a camera system is employed, it should include the necessary light source to allow verification of the main switch contacts.

Low voltage test provisions should be supplied to permit testing of each grounding switch. The provisions should allow the test voltage and current to be applied to the conductor without removing SF_6 gas or other components.

Gas System and Gas Zone Configurations

The user should specify the GIS with sufficient SF_6 gas to pressurize the complete system in a sequential approach, one zone or compartment at a time, to the rated nominal density. The SF_6 gas should conform to ASTM D2472 [25] for the US or IEC 60270 [19] for internal applications and should be at least 99% pure. An amount equal to 1% of the SF_6 gas in the total GIS assembly should be included as spare gas.

The GIS enclosure system should be divided into several sections separated by gastight barrier insulators. Each section should be provided with the necessary piping and valves to allow isolation, evacuation, and refill of gas without evacuation of any other section. The location of gas barrier insulators should be clearly discernable outside the enclosure by a band of a distinct color normally used for safety purposes and acceptable to the user.

For the purpose of gas monitoring and maintenance, each circuit breaker assembly (including circuit breakers, CTs, disconnect switches, and maintenance ground switches), fast-acting ground switch, voltage transformer, metal-enclosed surge arrester, cable sealing end enclosure, bus sections to SF_6 gas-to-air bushings, bus sections to cable sealing end enclosures, and interface connections for future expansion of the GIS should be independent from all other gas compartments, as a minimum requirement.

Gas zones should be arranged such that the boundaries of any electrical outage(s) required to maintain any piece of equipment would be minimized.

Gastight barrier insulators should be designed to withstand the differential pressures to which they may be subjected during preventive or corrective maintenance. This should not require lowering of gas pressure in the adjacent gas zones or disturbing the adjacent compartments.

A gas schematic diagram should be submitted for approval. It should include the necessary valves, connections, density monitors, gas monitor system and controls, indication, orifices, and isolation to prevent current circulation. Figure 3.44 depicts a typical gas zone diagram for one GIS bay.

The gas system should be provided with means of calibrating density monitors without de-energizing the equipment. For three-in-one enclosure type GIS equipment, three-phase gas

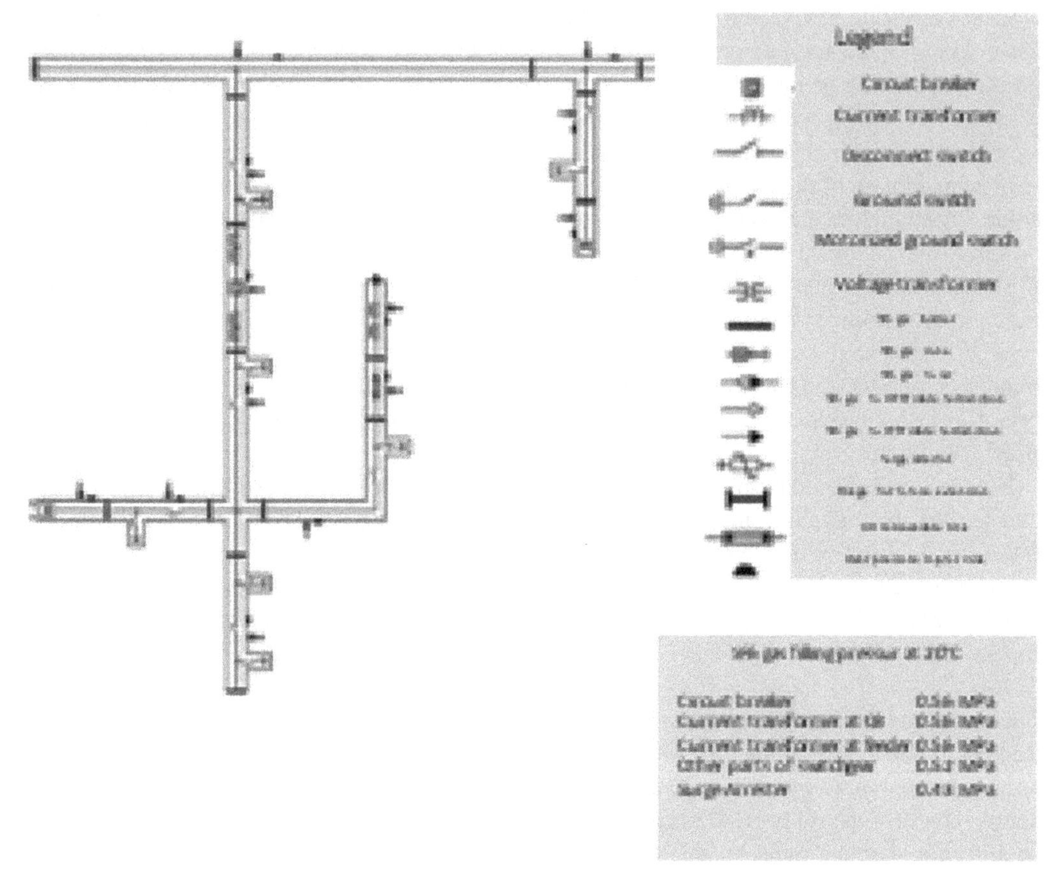

Figure 3.44 Typical gas zone diagram for one GIS bay (Reproduced by permission of United Illuminating Company)

pressure and density monitoring should be provided and for single-phase encapsulated type GIS equipment, single-phase gas pressure and density monitoring should be provided.

Since most manufactures have equipment tested with leakage rate of 0.1%, it is suggested that the leakage rate of SF_6 gas from any individual gas compartment should be designed not to exceed 0.1% per year. The total leakage rate from the total GIS system should be designed not to exceed 0.1% per year. This is beyond current requirements of the standards.

The user should specify connections for a gas density relay/monitoring system, the gas handling equipment, and moisture detection instrumentation to each one of the gas compartments. Facilities utilizing fittings should be included that permit the addition of SF_6 gas while GIS components are in service (energized).

Each gas zone with a switching device should be furnished with a gas density monitoring device capable of signaling two adjustable, independent alarms. Each device should be individually adjustable and should have electrically independent contacts that operate at the following alarm levels:

First alarm – low gas density (nominally 5–10% below the nominal fill density) to the local annunciator and to the user's SCADA RTU.

Second alarm – trip circuit breakers associated with the affected gas zone before the minimum gas density to achieve equipment ratings is reached and block closing of circuit breakers associated with the affected gas zone.

The user may also specify a standalone GDM (gas density monitoring) system using SF_6 gas density transducers. The GDM should have communications capability to transmit gas zone and system alarms to the user's operating system. The user should specify a GDM system that is capable of trending densities and pressures, identifying a leak by gas zone, and maintaining a complete gas inventory. The GDM system should include a gas zone mimic diagram local display that indicates the alarm level by zone on a schematic.

Current Transformers

Current transformers are typically of the inductive ring type. Each current transformer should be provided so that the enclosure current does not affect the accuracy or the ratio of the device or the conductor current being measured. There should be provisions that, for example, prevent arcing across the enclosure insulation.

Current transformer secondaries should be terminated to six-point shorting terminal blocks in the local control cabinets. The user should specify that it is possible to test each current transformer without the removal of gas. The current transformer location, polarity, ratios, and accuracy should be as specified and in accordance with IEEE C57.13 [26]. Figure 3.45 depicts a typical current transformer assembly.

Inductive Voltage Transformers

The inductive voltage transformers are typically iron core-type transformers with the primary winding supported on an insulating plastic film immersed in SF_6. The inductive voltage transformer should have an electric field shield between the primary and secondary windings to prevent capacitive coupling of any transient voltages.

The inductive voltage transformer is a sealed unit with a gas barrier. The voltage transformer should be easily removable to provide a high-voltage test connection point for the GIS and should be provided with a disconnect switch or removable link.

Each voltage transformer should be fabricated to mitigate the possibility of ferro-resonance during operation. If damping equipment connected to a secondary winding is necessary to mitigate ferro-resonant effects, it is preferred that such damping equipment be mounted in the voltage transformer secondary wiring junction box on the voltage transformer.

Transformers should be of either plug-in construction or the disconnect-link type, and attached to the GIS system in such a manner that they can be easily disconnected while the system is being dielectrically tested. The metal housing of the transformer should be connected to the metal enclosure of the GIS with a flanged, bolted, and gasketed joint so that the transformer housing is grounded to the GIS enclosure.

Primary and secondary terminals should have permanent markings, for identification of polarity, in accordance with IEEE C57.13 [26]. The voltage transformer secondary

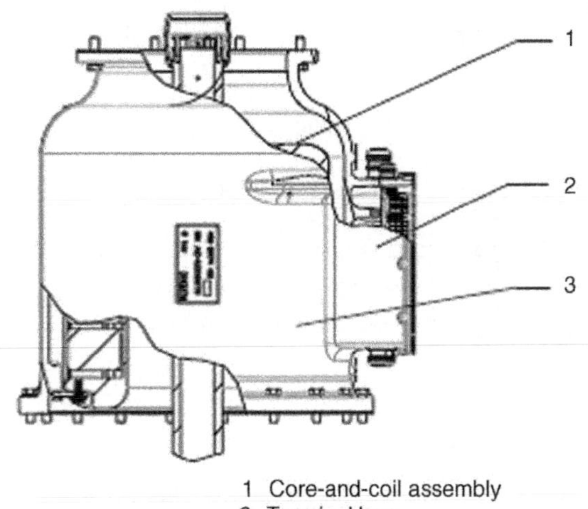

Figure 3.45 Typical current transformer assembly (Reproduced by permission of Siemens AG)

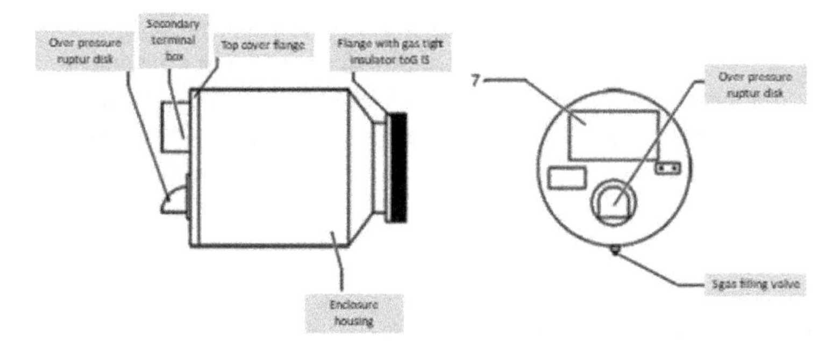

Figure 3.46 Typical inductive voltage transformer assembly (Reproduced by permission of Siemens AG)

should be fused and connected through a visible secondary break (VSB) disconnect device. Figure 3.46 depicts a typical inductive voltage transformer assembly.

Metal-Enclosed Surge Arresters

The type of surge arrester commonly used in GIS is the zinc oxide surge arrester that is suitable for immersion in SF_6. The arrester elements are supported by an insulating cylinder inside a GIS enclosure.

Surge arresters are used for overvoltage control in the GIS. Lightning impulse voltage surges can enter the GIS via SF_6 gas-to-air bushings connected to air insulated equipment. This is generally the only way the lightning surge can enter the GIS since the GIS conductors are housed inside grounded metal enclosures. The user may choose to design the system with surge arresters in parallel with SF_6 gas-to-air bushings to provided adequate protection of the GIS from lightning impulse voltage surges, at a much lower cost than the SF_6 insulated arresters in a GIS enclosure, if space permits.

Surge arresters can be used to protect underground cables connected to the GIS from impulse surges as well. The arresters should be connected to the SF_6 cable sealing end at the junction of the cable connection to the GIS.

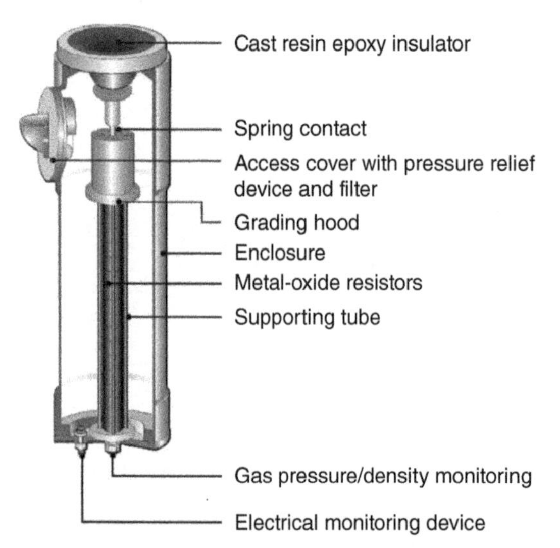

Figure 3.47 Typical single-phase metal-enclosed surge arrester assembly (Reproduced by permission of Siemens AG)

Switching surges are generally not a concern in GIS, because the SF_6 insulation structure provides for withstand voltage levels associated with switching surges that are almost equivalent to the lightning impulse voltage withstand of the GIS.

In air insulated equipment, there is a significant decrease in withstand voltage for switching surges than for lightning impulse because the longer time span of the switching surge allows time for the discharge to completely bridge the long insulation distances in air. In GIS, the short insulation distances can be bridged in the short time span of a lightning impulse, so the longer time span associated with the switching surge does not significantly decrease the breakdown voltage.

Insulation coordination studies should be performed to ensure proper surge arrester application. Figure 3.47 depicts a typical metal-enclosed surge arrester assembly.

SF_6 Gas-to-Air Bushings

SF_6 gas-to-air bushings are designed by connecting a hollow high-voltage insulator cylinder to a flange base and connecting that flange base to a flange at the end of a GIS enclosure. The insulating cylinder is filled with SF_6 gas, which is pressurized to the same level as the enclosures containing the bus conductors. The exterior of the cylinder has insulator sheds to employ the correct insulating distance for the rated voltage and is exposed to the atmosphere (air). The interior of the insulating cylinder is smooth. The conductor extends up through the center of the insulating cylinder and is connected to a metal end plate. The exterior of the metal end plate has provisions for bolting to air insulated conductors or bus bars via standard bolt configurations. Internal metal shields are installed in the cylinder to control electric field distribution, for voltages above a certain level external shields may be necessary. The insulating cylinder may be porcelain, but the more widely used material is a fiberglass epoxy inner cylinder with external weather sheds made of silicone rubber. Figure 3.48 depicts a typical SF_6 gas-to-air bushing assembly.

GIS to Cable Connections

Underground cables connected to GIS are done via a cable sealing end enclosure. The supplied cable is provided with a cable termination kit that is installed on the cable end to provide a physical barrier between the cable dielectric structure and the SF_6 gas in the cable sealing end enclosure. The cable termination assembly is designed to ensure the appropriate electric field distribution at the end of the cable.

The cable conductor is fitted with a compression connector that has standard bolt configurations and is connected to an end plate or cylinder of the cable termination assembly. The GIS cable sealing end has a removable link or plug in the connector that allows current transfer from the cable conductor to the GIS conductor. The cable is disconnected from the GIS when testing is performed on either the GIS or the cable.

Cable sealing end arrangements vary based on the cable system design. There are many different assembly arrangements depending on whether the cables are insulated with solid dielectric insulation, high-pressure fluid-filled pipe enclosures with paper insulation, low-pressure gas-filled pipe enclosures with paper insulation or gas-insulated lines (GILs). Figure 3.49 depicts a typical GIS to cable connection assembly.

GIS to GIB/GIL Connections

The GIS is typically connected to overhead lines and other equipment via the use of gas insulated bus (GIB) sections and/or gas insulated line (GIL) sections.

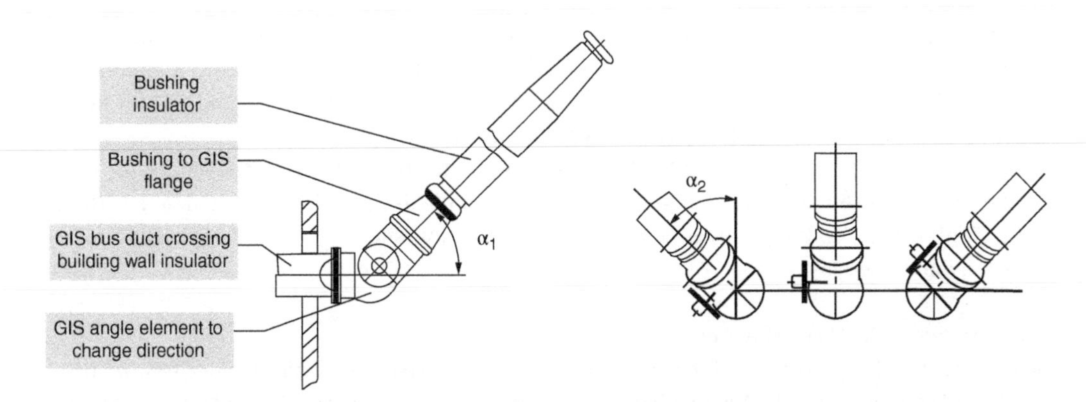

Figure 3.48 Typical SF_6 gas-to-air bushing assembly (Reproduced by permission of Siemens AG)

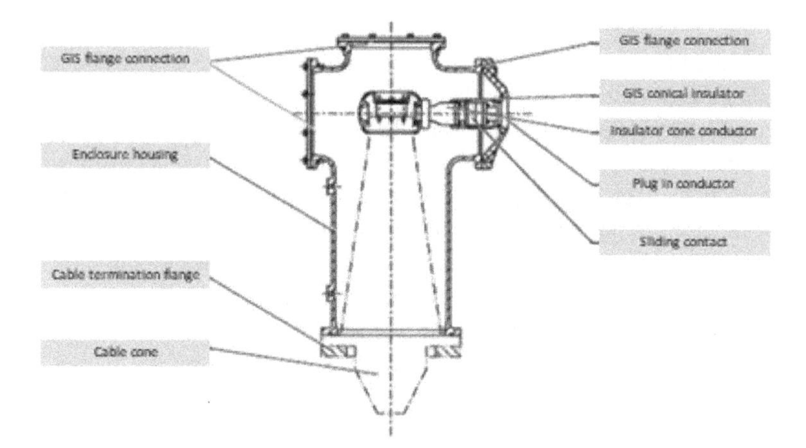

Figure 3.49 Typical GIS to cable connection assembly (Reproduced by permission of Siemens AG)

The bus sections and enclosures that are used to connect to external equipment are the same design as those used to connect between internal modules of the GIS. The GIB can be single-phase encapsulated or three-phase encapsulated. They connect the sections and bays of the switchgear according to the operating requirements and integrate the GIS into the electrical power supply network.

Each bus section module usually consists of the housing, the conductor, and the tie-rods. An expansion joint is usually specified between individual bays. The housings of the GIB modules are usually made of aluminum alloy tubes with welded-on cast headers.

The user should specify that the housings are opposite the flange, for any disconnect switch, an access opening which, if required, is sealed by means of a filter cover. The housings of the end bays are sealed with tie-rod covers.

The conductors are made of aluminum tubes and are supported by the bushings of the adjacent disconnect switch modules. The conductors and the connecting ball type assembly that connects the conductors are mounted on the bushings.

The conductors are connected between the GIB modules via axially routed sliding contacts consisting of the connection journal and the contact pin. The user should specify that there is direct access to the sliding contact via an access opening.

Each GIB module is usually provided with two opposed end tie-rods. The tie-rods are guided in tie-rod supports, which are designed to absorb the bending forces acting on the tie-rods when the gas compartments are evacuated.

The user should specify that GIS to GIB/GIL connections be designed to permit independent testing of the connecting GIB/GIL and should provide means to permit such testing. The connection should be equipped with a removable link to permit separation of GIB/GIL and GIS for high-voltage testing. Figure 3.50 depicts a typical GIB connection assembly.

The user should specify expansion joints to compensate for the changes in the length the GIB housings caused by temperature changes, permissible module design tolerances, and, to a certain extent, axial, lateral, and angular inaccuracies of the erection site. Figure 3.51 depicts a typical expansion joint assembly.

Power Transformer Bushing Connections

The connection of a power transformer to a GIS is accomplished via a transformer termination module. This module provides the link between the conductor path of the GIS bay and the bushing

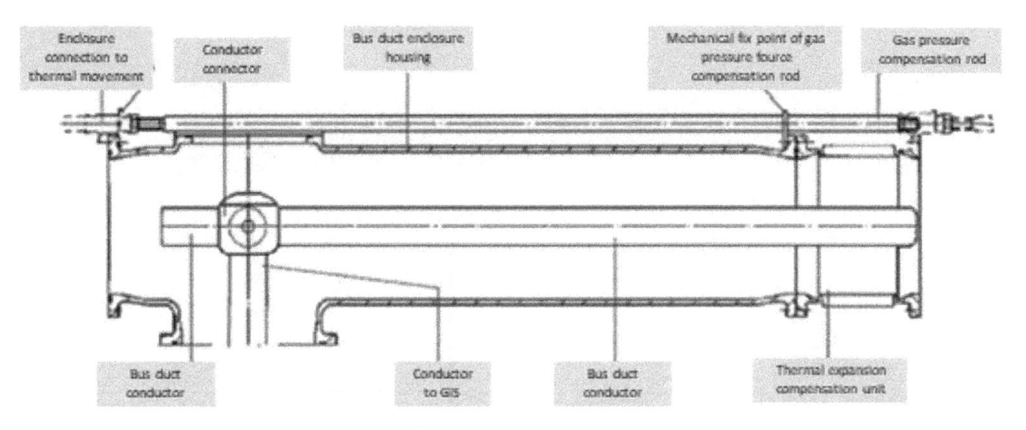

Figure 3.50 Typical GIB connection assembly (Reproduced by permission of Siemens AG)

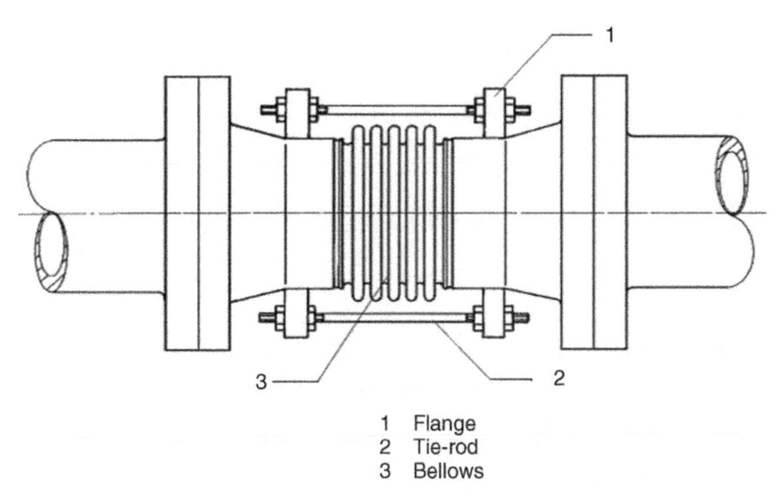

1 Flange
2 Tie-rod
3 Bellows

Figure 3.51 Typical expansion joint assembly (Reproduced by permission of Siemens AG)

of the transformer. The transformer termination module is linked with the GIS bay via matched extension modules, angle modules, rotary flanges, and expansion joints.

The user should specify that the expansion joints compensate for the following:

- Design tolerances of the facility, the building, and the transformer
- Movements caused by differences in the settling of the transformer and facility foundations
- Thermal expansion of component enclosures

The transformer termination module is usually composed of a housing together with an intermediate plate and the conductor components for connecting the transformer bushing to the transformer bay. The user should specify a universal intermediate plate that permits connection of various transformer bushings in the event the transformer is replaced. The housing is placed with the intermediate plate over the transformer bushing and bolted to its flange to provide a gastight joint.

Depending on the configuration of the transformer termination, the remaining openings in the housing are sealed with a flange cover. The current path connection from the GIS bay to the transformer bay is via the transformer bushing, a module conductor, a corner-ball type assembly, and a

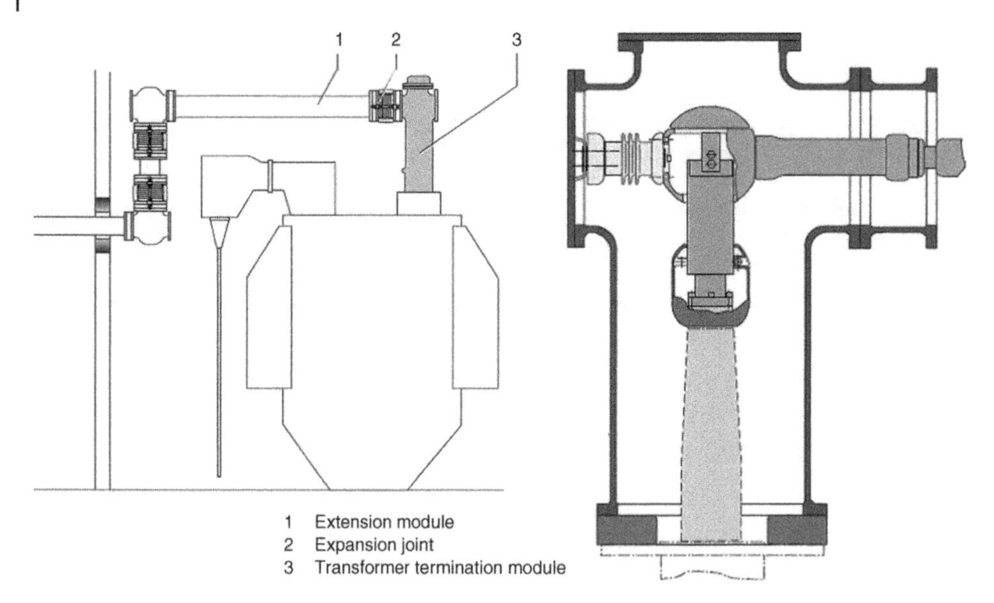

Figure 3.52 Typical transformer termination module arrangement (Reproduced by permission of Siemens AG)

GIS bus conductor with sliding contacts. Figure 3.52 depicts a typical transformer termination module arrangement.

Local Control and Marshaling Cabinets

Typically, the GIS is connected to the substation control room via a local control cabinet (LCC) for each GIS bay position for ease of operation and to reduce the amount of wiring. Marshaling cabinets may also be required as intermediate termination locations between the GIS and the LCC and between the LCC and the substation control room.

The user should specify one local LCC for each circuit breaker bay. Each cabinet should be completely fabricated, wired, assembled, and tested at the factory. Each cabinet furnished should be fully equipped and completely wired to the terminal blocks for termination of the circuit breaker and switch control, instrument transformer, indication, gas alarm, and other miscellaneous cables.

Each cabinet should carry the appropriate NEMA for the US and IEC for international enclosure rating for either indoor or outdoor service. Any outdoor junction boxes or marshaling cabinets should be constructed of stainless steel or corrosion protect steel.

The user should specify stainless steel hinges, door hardware, fastening hardware, and each cabinet should have a rear access door with all external cabling brought into and terminated in the rear of the cabinet.

Heaters and heater circuits should include contact-making ammeters (or other user approved methodology) as required in LCCs and marshaling cabinets to avoid the formation of condensation.

At a minimum each cabinet should contain the following equipment, as a minimum, for control, indication and protection of switches, circuit breakers, and associated components:

- One combined open–close control and remote–local switch for each three-phase circuit breaker.
- One combined open–close control and remote–local switch for each motor-operated disconnect switch.
- One combined open–close control and remote–local switch for each motor-operated or spring-operated grounding switch.

- A mimic diagram showing connections of all furnished equipment and showing location of all gas zones.
- One red light-emitting diode (LED) (for each circuit breaker trip coil) and one green LED for each circuit breaker, each disconnect and grounding (maintenance or fast-acting) switch, or contact position indication on the mimic diagram. LEDs should be visible under all lighting conditions and all indicating lights should be capable of being seen without the necessity of opening any doors on the LCC.
- Control switches, fuse blocks, fuses, and so on, for alternating current (AC) and DC supply to each LCC and for each device being controlled electrically from the LCC.
- Interlock bypass switches with locking provisions approved by the user.
- An annunciator and global positioning system (GPS) time synchronization with the following alarms connected:
  - Gas system alarms for each gas zone
  - Circuit breaker close system discharged alarm
  - Loss of DC alarm for each supply
  - Loss of AC alarm for each supply
  - Charging motor excessive run alarm
  - All other equipment trouble alarms for the GIS bay

Remote I/O (input/output) hardware should be supplied in sufficient quantity for all required inputs and outputs plus space for one additional I/O unit mounted and wired within each LCC and designed to monitor or control the following points, as a minimum:

- Circuit breaker position
- Disconnect switch position
- Ground switch position
- Fast-acting ground switch position
- Interlocking status for each circuit breaker, disconnect, ground switch, and fast-acting ground switch.
- Control switch remote-local status.

Also required are all terminal blocks (including user specified spare termination points per LCC), test switches and terminations, as well as all engraved nameplates for each device in the cabinet.

With the digitalization of substations functions, alarms, measurements or any other information is visualized by monitors integrated to LCC.

Ladders, Platforms, Stairs, and Walkways

Safe, efficient access is required to all viewports, actuator mechanisms, switch operators, gas density monitoring equipment, and so on, for maintenance personnel. In accordance with applicable safety standards, permanent facilities should be provided to meet this requirement. Appropriate railings and toe-plates should be installed on all ladders, elevated platforms, and maintenance walkways. Open grating should be employed for all platforms and stair/ladder treads. Treads should be painted with a bright color. Each structure should have provisions for safety grounding at two points on diametrically opposite sides.

The intention is to eliminate the need for climbing on the apparatus or for portable ladders to perform routine operations or view routine indicating devices, switch positions (viewports), or alarms.

Each ladder, platform, stair, or walkway should be a minimum of 762 mm (30 inches) in width and be clear of overhead or projecting obstructions for a minimum of 1981 mm (6 ft 6 in.) The main operations platform or walkway should have egress means at either end as a minimum. Each stairwell should have a top landing area of at least 1219 mm (48 inches) by 1219 mm (48 inches).

3.4.2.4 Secondary Equipment Data

The specification of secondary equipment is the third major section of the GIS specification. This section lists the area to be specified and used for the control and protection of the GIS, and usually details the system interfaces with other components of the associated power system.

The following are the typical sections of the secondary equipment specifications for the GIS:

Protection, Control and Monitoring Requirements
Relaying and Control Cabinets
Logic Diagrams for Numerical Relays
SCADA Interface Points
Wiring Connections and Interconnections Requirements
Annunciation and Alarms
Mimic Bus Diagram

Protection, Control, and Monitoring Requirements

It is the user's decision either to install protection, control, and metering system separate from the GIS as standalone protection panels. Alternatively, this may be part of GIS supplier scope. The following requirements cover such case.

The user should include a one-line metering and relaying diagram as part of the GIS specification. The one-line metering and relaying diagram should be created in conjunction with the detailed functional one-line diagram that is part of the GIS equipment specification.

The user should include a relay and control device list that details the specific protective relays and control relays necessary for operation of the GIS in accordance with the user's electric system protection and control standards, regional reliability criteria, and system operating requirements.

The user should specify all monitoring requirements including, but not limited to, gas density, device operation, control voltage, communication systems, and so on.

Relaying and Control Cabinets

The user should include as part of the GIS specification panel arrangement sketches for the GIS local control cabinets and for the primary, secondary, and ancillary relaying and control panels. These sketches should at a minimum contain panel elevations and device locations for primary and secondary protection systems.

Logic Diagrams for Numerical Relays

The user should include as part of the GIS specification preliminary logic diagrams for numerical relays, protective relaying, and control schemes associated with the primary and secondary protective relaying schemes. These diagrams should at a minimum show the logic for primary breaker tripping, secondary breaker tripping, reclosing, and gas density level tripping.

SCADA Interface Points

The user should include as part of the GIS specification a SCADA (interface) data table and a list of all associated interposing relays. The SCADA data table should at a minimum show all control and indication points required for the proper operation of the GIS in accordance with the user's system requirements.

Wiring Connection and Interconnection Requirements

The user should include as part of the GIS specification sample wiring diagrams and any CAD drafting standard documents that pertain to the drawing systems of the user's organization. The sample wiring diagrams and CAD drafting standards should at a minimum depict the required symbols, attributes, and layers to be used to create any engineering drawings, as well as the wiring methods to be used (i.e., point-to-point, wiring table, etc.).

Annunciation and Alarms

The user should include as part of the GIS specification data tables and schematics that depict the user's requirements for local and remote annunciator points, as well as any local or remote alarms that are required. The user should also specify the type and details of annunciation and alarms that are to be integrated into the user's SCADA system.

Mimic Bus Diagram

In addition to the mimic diagram showing connections of all furnished equipment and showing the location and details of all gas zones for the local control cabinets, the user should specify in the GIS specification the creation of an overall mimic bus diagram. This diagram should be used to create any operating diagrams and monitor screens that are part of the design of the human machine interface (HMI).

3.4.2.5 Engineering and Logistics Studies

Engineering Studies

Once the user has specified the preliminary configuration of the GIS and the primary equipment data have been determined and specified, further studies related to the engineering and logistics of delivery and erection need to be performed. These studies should cover the following:

Transient Recovery Voltage (TRV) Conditions

The user should specify that the manufacturer perform a TRV study to evaluate the worst-case rate of rise of recovery voltage (RRRV) and the maximum peak voltage across the circuit breakers considering the transient response of the electrical network surrounding the GIS. The calculated TRV values should be compared to the TRV ratings guaranteed by the test report of the circuit breaker and to standard TRV envelopes that can be found in industry standards.

The TRV that a circuit breaker experiences is the voltage across its terminals after current interruption. The TRV wave shape is determined by the characteristics of the electrical network surrounding the circuit breaker. Generally, the TRV stress on a circuit breaker is determined by the fault location, fault current magnitude, and switching configuration of the switchgear.

Since the TRV is a determining parameter for successful current interruption, circuit breakers are normally type tested in a laboratory to withstand a standardized TRV. This standardized TRV is determined by a four-parameter envelope (two-parameter envelope for circuit breakers rated up to 100 kV) with a first period of high rate of rise followed by a later period of lower rate of rise. The slope of the first period of the TRV envelope is defined as the rate of rise of recovery voltage (RRRV). In the case of a very low amplitude of the short-circuit breaking current, two-parameter envelopes have to be considered in order to evaluate the TRV stress on a circuit breaker.

It is the purpose of the study to evaluate the worst case RRRV and maximum crest voltage across the circuit breakers in the GIS in accordance with the transient response of the electrical network surrounding the switchgear.

Transient Over Voltage (TOV) Conditions

The user should specify that the manufacturer perform a review of the GIS design to establish if any transients generated during a switching operation may lead to significantly elevated and damaging voltage levels. The user should provide the manufacturer with system design data to allow for assessment of switching surge conditions.

Very Fast Transient (VFT) Conditions

The user should specify that the manufacturer perform a VFT study. In gas-insulated substations (GIS), very fast transient (VFT) overvoltages with oscillation frequencies in the MHz range may occur during disconnect switch operations because of the fast voltage collapse within a few nanoseconds and due to the length and coaxial design of the GIS.

In the vicinity of the operated disconnect switch, frequencies of up to more than 100 MHz may occur. At more distant locations inside the GIS, frequencies of several MHz can be expected.

The frequencies and the amplitudes of the VFT depend on the length and the design of the GIS. Due to the traveling wave nature of the phenomenon, the voltages and frequencies differ from location to location within the GIS.

High amplitudes occur if long sections of gas insulated buses are switched and if tapped buses exist at the source for the main bus section. If the natural frequencies of the source and the switched end of the bus are similar and if the voltage difference across the disconnect switch is high, then a high-voltage difference occurs during the opening of the disconnect switch. The highest amplitudes of the VFT generally occur on open GIS sections.

The purpose of the study is to simulate the VFT overvoltages within the GIS caused by energizing switchgear segments by means of disconnect switches. VFT overvoltages caused by circuit breaker switching operations should also be calculated.

Insulation Coordination Studies

The user should specify that the manufacturer perform insulation coordination studies. An insulation coordination study should be performed to verify the location and number of GIS metal-enclosed type surge arresters for protection of the GIS equipment and/or any interconnected underground cable circuits and other air-insulated equipment.

The insulation coordination study investigates the overvoltage stresses at the gas insulated switchgear and its bays and cables caused by lightning surges approaching the substation and the lines connected to it. Therefore, the maximum voltage stresses within the GIS and at the bays – caused by typical lightning strokes (remote strokes, direct strokes to conductors, strokes to last towers) to the overhead lines – should be simulated for several specified configurations of the substation, including the normal operation configuration.

The correct insulation coordination level should be verified by comparing the insulation levels of the individual equipment with the maximum overvoltage stresses to be expected when taking into account the maximum correction and safety factors according to industry standards.

Thermal Rating Calculations

The user should specify that the manufacturer provide thermal rating calculations for all equipment and devices in the main current paths. The thermal rating calculations should be determined in accordance with the facility ratings methodology of the user and the Regional System Operating Authority.

Effects of Ferro-Resonance

The user should specify that a study be performed to determine if there is the possibility of ferro-resonance occurring in conjunction with the switching in and out of service of potential transformers in the GIS. The study should indicate the severity of the condition and recommend mitigation, such as tuned inductors.

GIS Resistance and Capacitance

The user should specify that the calculated and measured capacitance and resistance values for each component in the GIS, including, but not limited to, the bushings, bus runs, switches, and circuit breakers, be provided by the manufacturer.

Seismic Calculations

The user should specify that the manufacturer provide all documentation for seismic design testing.

Electromagnetic Compatibility

The user should specify that the manufacturer perform any studies such as shielding and mitigation procedures for interference with control, protection, diagnostics, and monitoring equipment.

Civil Engineering Aspects

The user should request that the manufacturer provide documentation for any special civil designs required to accommodate the GIS due to specific site conditions.

Grounding and Bonding

The user should specify that the manufacturer perform grounding studies in accordance with the current version of IEEE Standard 80 [26] or IEC 61936–1 [24]. The manufacturer should ensure that the GIS equipment grounding is in conformance with National Electric Safety Code C2 and IEEE Standard 80.

All studies should be provided in formal reports forwarded to the user within the specified time frame after the contract is awarded. All documentation including, but not limited to, calculations, curves, assumptions, graphs, computer outputs, and so on, should be provided to support the conclusions reached.

Logistics Studies

- Transport, storage, and erection facilities
- Demands imposed by the service and maintenance of the GIS and possible future extensions
- Quality assurance, testing procedures during manufacture, and especially on-site testing

3.4.2.6 Standards and Regulations

The following are an important list of critical documentation concerning referenced standards and regulatory compliance criteria that should form the parts of a GIS specification:

- Compliance with Utility Operations Criteria
- Good Utility Practices (GUPs)
- Standard Safety Rules and Criteria
- OSHA (Government) Regulations
- Visible Means of Disconnect
- Power Source Isolation Switches
- Decoupling or Blocking of Devices
- IEEE Standards
- IEC Standards

3.4.2.7 Test and Inspections

The following tests and inspections are to be performed on individual shipping units in the factory and then on the complete GIS at the site after installation:

- General assembly inspection
- HV and partial discharge tests
- Gas leakage test
- SF_6 moisture test (at the site only)
- Equipment list check
- Nameplate check
- Component device check
- Point-to-point wiring check
- Control system functional test
- Control wiring HV test
- Overall appearance inspection

A Certified Test and Inspection Report should be provided for each GIS Section and Local Control Cabinet.

3.4.2.8 Project Deliverables

The manufacturer should submit complete drawing packages to the user for review and approval prior to manufacturing. This drawing submittal typically includes the following:

- Physical outline drawings
  - Plan view of the complete GIS
  - Section views of the typical GIS bays
  - Plan and section views of the LCCs
- Foundation plan including grounding connections
- One-line relaying and metering diagram
- AC three-line relaying and metering diagram
- Interlocking logic diagram
- Wiring connection diagram with opposite end designations
- Static and Dynamic Load Plan
- Control Schematics
- Gas Schematic Diagram per ANSI C37.11
- Control Device Designations per ANSI C37.2
- Wiring Connection Diagram with Opposite End Designations
- Instruction Manuals
- Training of User Personnel

3.4.2.9 Project Specific Requirements

- Regional Reliability Criteria
  - Capability for Independent Pole Operation
  - Independent and Isolated Wiring for Dual Trip Schemes
  - Separate and Redundant DC Control Power Sources
- Compliance with State and Local Requirements
  - Permitting
  - Union/Nonunion Labor
  - Visual Impacts
  - Site Construction (Protection of Wetlands).

- Together with the GIS delivery, a detailed packing list, installation instructions and certified copies of all test and inspection reports should be submitted. Upon completion all drawings are to be updated and submitted as as-built drawings for the user.

3.5 Instrument Transformers

3.5.1 Current Transformers

The current transformers (CTs) in a GIS may be three-phase insulated as in Figure 3.53 for voltages up to 145 kV or single-phase insulated for voltages above 145 kV as in Figure 3.54.

In Figure 3.53a, a three-phase insulated current transformer for voltages up to 145 kV is shown. The secondary coils are inside the aluminum enclosure and the coils are connected via pressure-tight contacts to the secondary control cubicle. The cross section of the interior is shown in Figure 3.53b.

In Figure 3.54a, a single-phase insulated current transformer is shown with the terminal boxes for the secondary wiring on the bottom side. The cross section of the interior of the single-phase current transformer with seven secondary coils indicated is shown in Figure 3.54b.

3.5.2 Voltage Transformers

The voltage transformers (VT) in a GIS may be three-phase or single-phase insulated, as shown in Figures 3.55 and 3.56. In Figure 3.55a, a photo shows the enclosure of a three-phase insulated voltage

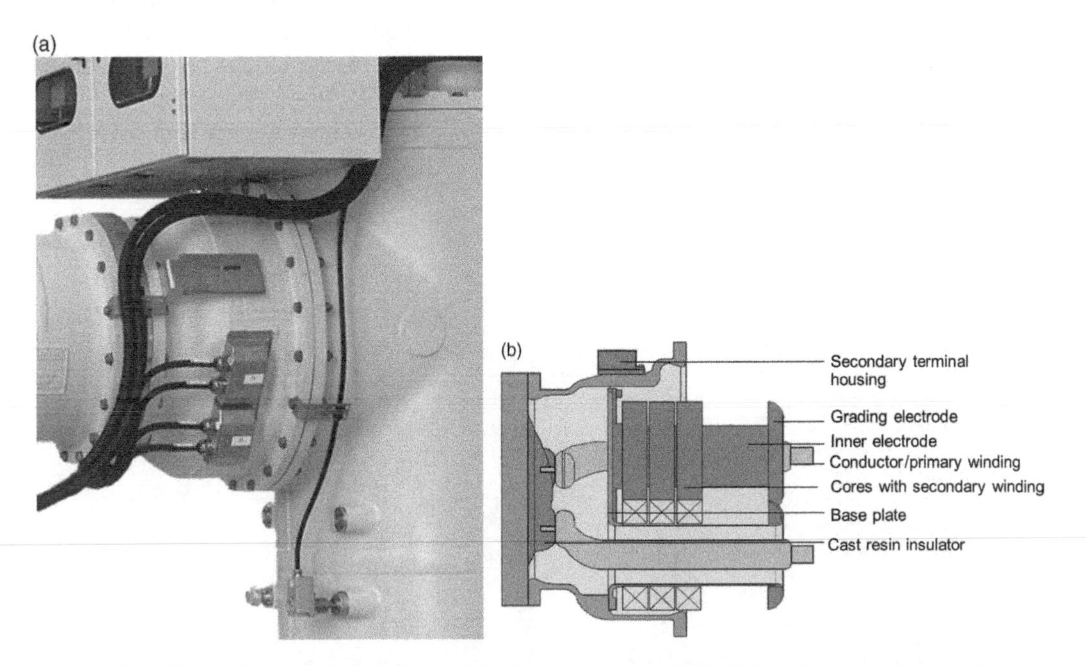

Figure 3.53 Three-phase insulated current transformer for up to 145 kV (a) enclosure with secondary termination (b) cross section with conductor, secondary core, insulator and enclosure (Reproduced by permission of Siemens AG)

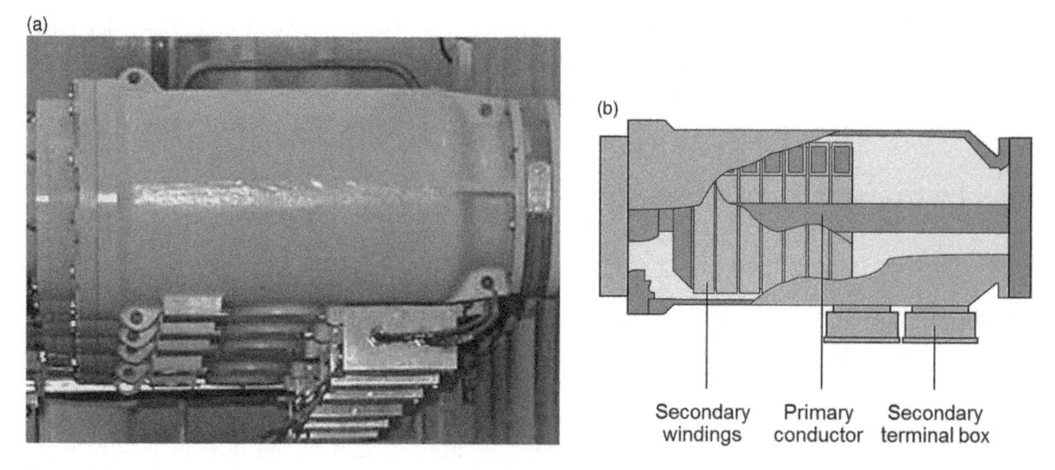

Figure 3.54 Single-phase insulated current transformer of 245 kV (a) enclosure with secondary termination (b) cross section with conductor, secondary core, insulator and enclosure (Reproduced by permission of Siemens AG)

transformer with a pressure-relieve device on the top and a secondary cable termination on the side. The cross section of the inside is shown in Figure 3.55b, with a gastight conical insulator at the bottom to separate the gas compartment of the voltage transformer from other GIS bay gas compartments.

3.5.3 Transient Overvoltages of a CT and VT

Transient overvoltages may occur in GIS during switching operations. In most cases, these transients are generated mainly by disconnector operations. They can be transferred by the installed current and voltage transformers to the auxiliary equipment and may influence negatively the operation of the protection and control devices. According to the relevant standards such transient voltages (U_T) should not be higher than 1600 V at the terminals of the auxiliary equipment.

To fulfill these requirements the installed instrument transformers are equipped with shieldings accordingly.

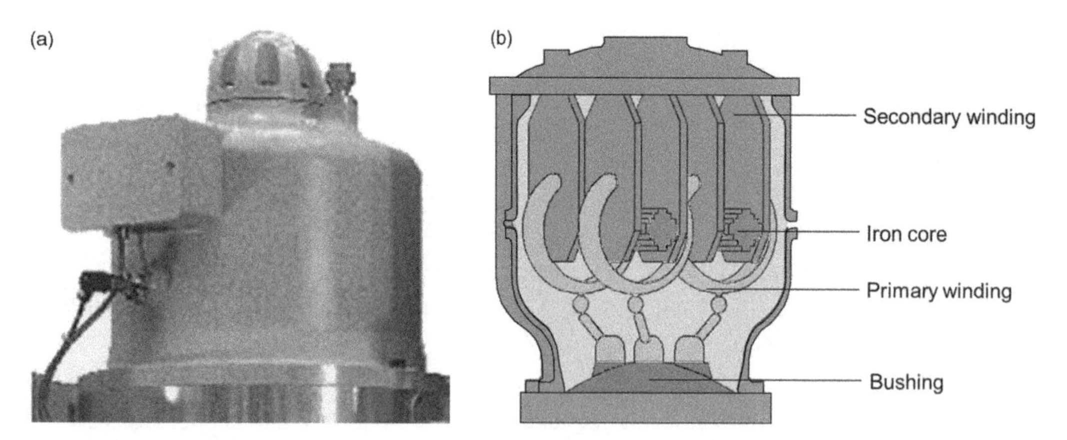

Figure 3.55 Three-phase voltage transformer for voltages up to 145 kV (a) enclosure with overpressure rupture disk on top and secondary termination box, (b) cross section with primary and secondary windings, iron core and bushing insulator (Reproduced by permission of Siemens AG)

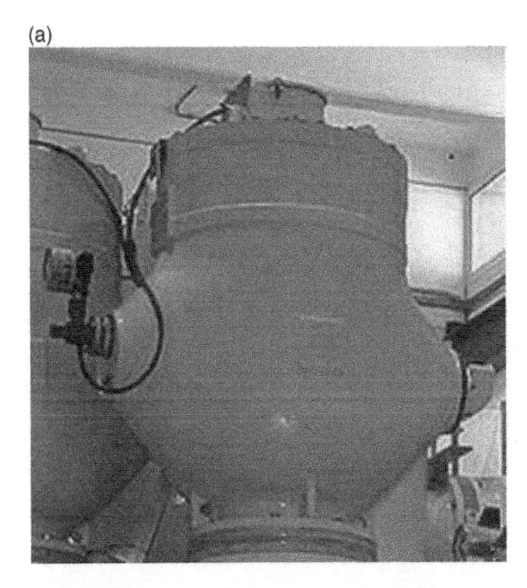

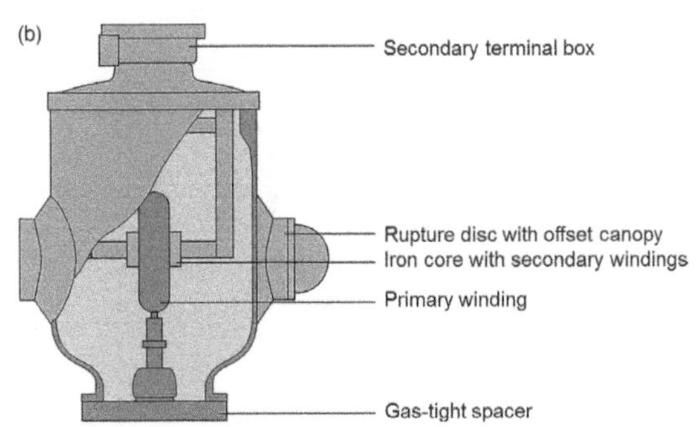

Figure 3.56 Single-phase insulated voltage transformer for voltages of 245 kV (a) enclosure with gas density meter (left) and gas over pressure rupture disk (right) (b) cross section with primary and secondary winding, iron core and rupture disk (Reproduced by permission of Siemens AG)

3.6 Interfaces

3.6.1 Direct Connection Between Power Transformers and GIS

Connections between GIS and power transformers can be done by either connecting GIS SF_6 gas-to-air bushings to transformer bushings utilizing conductor cable or by using the direct connection method utilizing a gas insulated bus. IEC standard IEC 62271-211 [28] illustrates the direct connection method, establishes electrical and mechanical interchangeability, and defines the limits of supply of the transformer connection, which is immersed at one end in the transformer oil or insulating gas and at the other end in the insulating gas of the switchgear. This direct connection method is applicable to single- and three-phase arrangements for GIS with rated voltages above 52 kV.

A typical direct connection is illustrated in Figure 3.57 and the limits of supply for the GIS and power transformer manufacturers are listed in Table 3.2.

In addition to the parts listed in Table 3.2, the GIS manufacturer should supply connections between the GIS enclosures of the different phases, in order to limit circulating currents in the transformer tank. Additionally, the transformer should be designed to carry a continuous circulating current of 1250 A within its tank without exceeding the surface temperatures as specified in the relevant transformer specifications. If circulating currents are expected to exceed 1250 A, an insulated junction should be utilized.

To achieve proper protection schemes for transformer faults, an insulated junction may be added in between the transformer connection enclosure and the bushing flange. The insulation junction should be able to withstand a 5 kV power frequency voltage for 1 min. Nonlinear resistors may be connected in parallel with the insulated junction to limit the very-fast-front transient voltage caused by switching operations.

When dimensioning a switchgear power transformer connection assembly, the following parameters need to be determined: rated voltage, insulation level, normal load current, temperature rise limit, rated short-time withstand current peak, and maximum duration of short-circuit current.

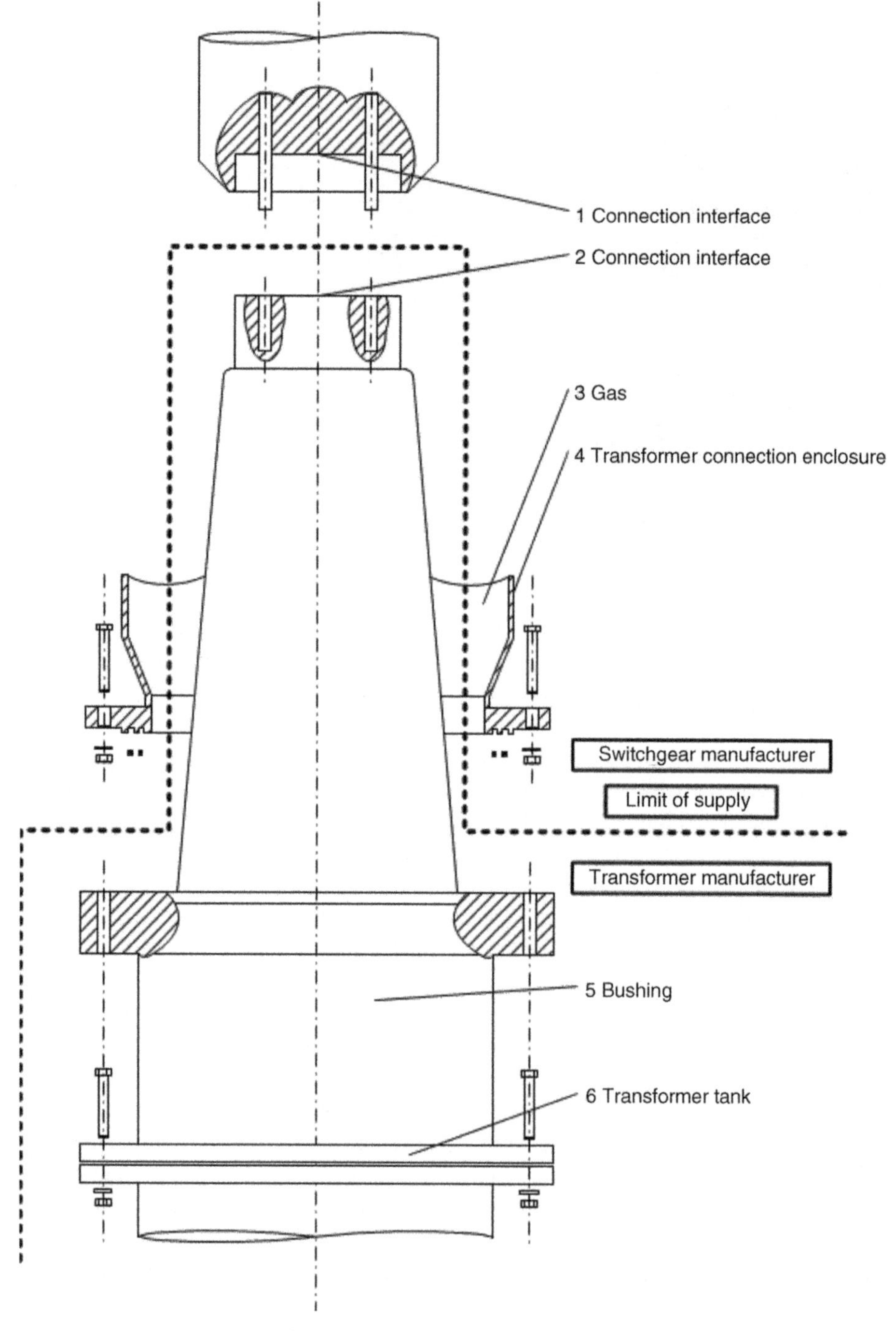

Figure 3.57 Typical direct connection between a power transformer and GIS for limits of supply (Simplified from drawing of IEC 62271-211 [28]) (Reproduced by permission of IEEE)

Table 3.2 Limits of supply referring to Figure 3.57 (Simplified from table of IEC 62271-211)

		Manufacturer	
Description	**Item**	**Switchgear**	**Transformer**
Connection interface	1	x	
Connection interface	2		x
Gas	3	x	
Transformer connection enclosure	4	x	
Bushing	5		x
Transformer tank	6		x

These parameters can be selected from relevant IEC standards. The dimensions of the connection interface defined in Table 3.3 allow a maximum value of 3150 A for the rated normal current. The contact surfaces of the connection interface should be silver-coated, copper-coated, or bare copper. For the rated normal current, the connection between the switchgear and power transformer should be designed such that the temperature of the transformer connection enclosure and the temperature of the connection interface do not exceed the values given in Clause 4.4.2 of IEC 62271-203 [21].

The rated filling pressure of gas for insulating is assigned by the switchgear manufacturer. If SF_6 is used as the insulating gas, the minimum functional pressure for insulation used to determine the design of the transformer-termination insulation should be not more than 0.35 MPa (absolute) at 20°C.

If a gas other than SF_6 is used, the minimum functional pressure should be chosen to give the same dielectric strength while being lower than the maximum operating pressure.

The design of the bushing needs to be able to withstand a maximum operating gas pressure of at least 0.85 MPa (absolute). In addition, the bushing should be capable of withstanding the vacuum conditions when the transformer connection enclosure is evacuated, as part of the gas filling process.

The design of the transformer connection enclosure should satisfy the requirements of the design pressure of the switchgear. The maximum operating gas pressure (absolute) of a direct connection assembly should not exceed the lesser of:

Table 3.3 Typical standard dimensions in millimeters (simplified from table of IEC 62271-211)

Rated voltage (rms value) kV	72.5–100	123–170	245–300	362–550
Ø d_1	about 100	about 100	about 140	about 140
Ø d_2	about 200	about 220	about 450	about 540
Ø d_3	about 196	about 215	about 440	about 500
Ø d_4 minimum	about 315	about 335	about 565	about 690

- The design pressure of the transformer connection enclosure plus 0.1 MPa when the design pressure is lower than 0.75 MPa (gage);
- 0.85 MPa (absolute) when the design pressure equals or exceeds 0.75 MPa (gage).

The design of the bushing connection interface should withstand the composite forces resulting from electrodynamic effects, tolerances of components, thermal expansion or contraction, and the weight of the switchgear main circuit. A minimum withstands of a mechanical force of 2 kN applied to the connection interface either transversely or axially should be assumed. It is the responsibility of the switchgear manufacturer to ensure that this specified force is not exceeded.

The flange of the bushing attached to the transformer connection enclosure is subjected, in service, to the following loads:

- The weight of the switchgear not supported by the switchgear's own supporting structures
- The wind load, if applicable, not supported by the switchgear's own supporting structures
- Expansion or contraction stresses due to the temperature variations of the switchgear enclosures

These loads result in the simultaneous application, at the center of the bushing flange, of:

- A bending moment M_o
- A shearing force F_t
- A tensile or compressive force F_a

The bushing and the transformer shall be capable of withstanding, in service, the values of M_o, F_t, and F_a specified in Table 3.4, and it should be the responsibility of the switchgear manufacturer to ensure that these values are not exceeded.

In addition, the vibrations generated inside the energized transformer are transmitted by the oil and the tank wall of the transformer to the bushing rigidly fixed on this wall and to the switchgear. The transformer manufacturer should identify and present these vibrations to the switchgear manufacturer so that the switchgear can be designed to accommodate them.

Standard dimensions for single-phase transformer connection enclosures, main circuit end terminals, bushing end terminals, and bushing flanges are shown in Figure 3.58. For connecting the gas insulated metal-enclosed switchgear and the transformer in a proper way the transformer bushing should have a positioning accuracy of ±3 mm in lateral directions.

The location of the fixing holes on the flange of the bushing and on the flange of the transformer connection and the fixing holes on the switchgear main circuit end terminal and the main circuit end terminal of the bushing are shown in Figure 3.59.

During manufacture, handling, and storage, provisions have to be made by the bushing manufacturer so that the requirements given in Clause 5.2 of IEC 62271-1 can be satisfied after final assembly of the direct connection between the transformer and the switchgear.

The components and the direction connection assembly should be tested in accordance with IEC 60076 series, IEC 60137, and IEC 62271-203 standards.

The dielectric type tests of the bushing are recommended to be performed in an enclosure filled with insulating gas at the minimum functional pressure. The test setup should represent the most severe in-service stress condition.

The transformer connection enclosure and main circuit end terminal should be type tested to verify that they meet the associated dielectric requirements. The type tests should be performed at the minimum functional levels.

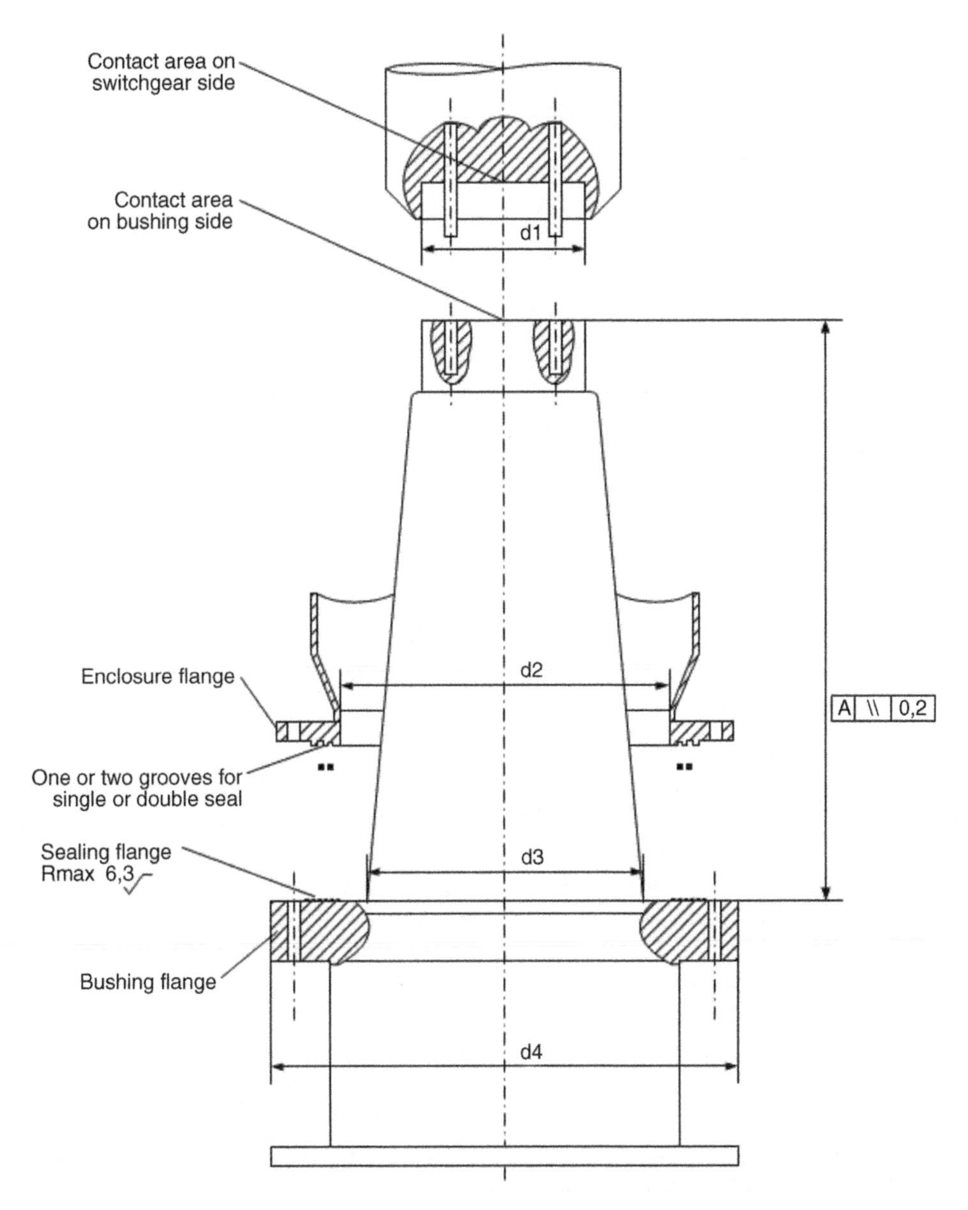

Figure 3.58 Typical standard dimensions for a typical direct connection between a power transformer and GIS for dimensions. (Simplified drawing from IEC 62271-211) (Reproduced by permission of IEEE)

The single-phase test arrangement using a single-phase transformer connection enclosure covers the test requirements of a transformer connection in a three-phase enclosure as it imposes the most severe dielectric stress to the test object. It is therefore the referenced type test.

For cantilever load withstand type tests, the bushing should be tested in accordance with Clause 8.9 of IEC 60137 [29], except that the test load applied at the connection interface should be 4 kN for all ratings.

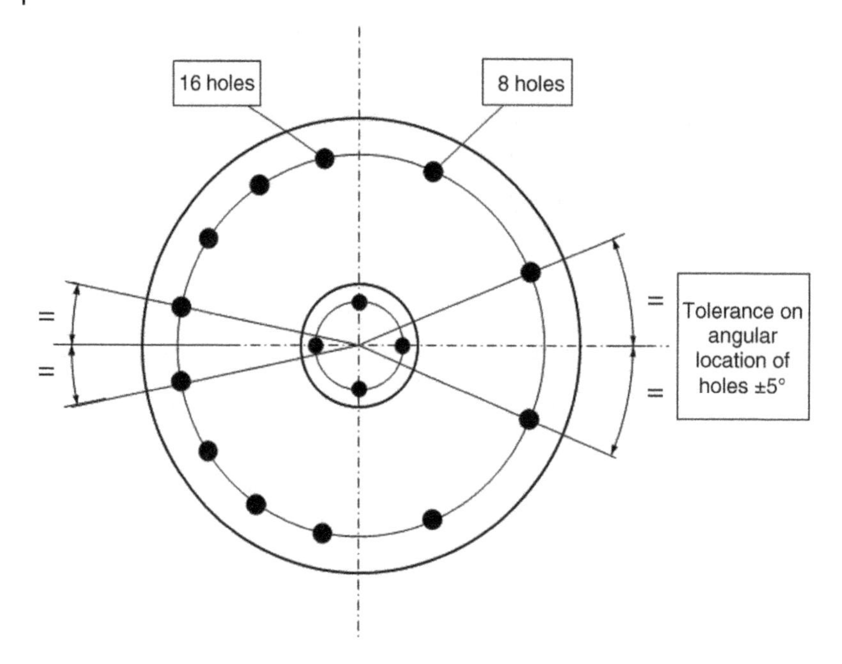

Figure 3.59 Typical standard orientation of fixing holes (Simplified drawing from IEC 62271-211) (Reproduced by permission of IEEE)

To demonstrate how much bend, it can withstand in accordance with Table 3.4, an additional test should be performed as follows. The bushing should be assembled as far as necessary for the test, but there should not be any internal gas pressure. It should be installed vertically with its oil-side flange rigidly fixed to a suitable device. The end for gas immersion should be mounted in a tank as for normal operation, at ambient temperature. The tank should be filled with an appropriate medium at 0.75 MPa (gage) and a test load applied to the tank so that a bending moment equal to two times M_o in accordance with Table 3.4 is produced at the switchgear side flange of the bushing for 1 min. The shearing force applied should be equal to two times F_t as far as possible. The acceptance criteria should be as prescribed in Clause 8.9.3 of IEC 60137 [29].

A routine external pressure test of the bushing should be made before the gas tightness test. The bushing end for gas immersion should be mounted in a tank as for normal operation, at ambient temperature. The tank should be filled with gas or liquid, at the choice of the supplier, at a pressure of 1.15 MPa (gage), for 1 min. The bushing should be considered to have passed the test if there is no evidence of mechanical damage (e.g., deformation, rupture).

Table 3.4 Typical moment and forces applied on the bushing flange and transformer (simplified from table of IEC 62271-211)

Rated voltage (kV)	Bending moment M_o (kN m)	Shearing force F_t (kN)	Tensile or compressive force F_a (kN)
72.5–100	about 5	about 7	about 4
123–170	about 10	about 10	about 5
245–300	about 20	about 14	about 7
362–550	about 40	about 20	about 10

The gas tightness specifications and tests of IEC 62271-1 [20] are applicable to direct connections between power transformers and switchgear. The permissible leakage rate F_p (Clause 3.6.6.6 of IEC 62271-1) is set to 0.5% per year according to IEC 60137 [29] and 62271-203 [21]. The tightness test of each bushing has to be performed according to Clause 9.8 of IEC 60137 [29], with the gas at the maximum operating pressure. Its absolute leakage rate F (Clause 3.6.6.5 of IEC 62271-1 [20]) should not exceed F_p.

The rules for transportation, storage, erection, operation, and maintenance should be followed in accordance with IEC 62271-1 [20].

3.6.2 Cable Connections for Gas-Insulated Switchgear for Rated Voltage above 52 kV

Cable connection is one of the commonly used connection methods for GIS inputs or outputs. IEC standard IEC62271-209 [18] and IEEE Guide 1300 provide the standard dimensions for cable connections to GIS from 72.5 kV to 550 kV, thus establishing the electrical and mechanical interchangeability between cable terminations and GIS. There are two typical cable termination assemblies: fluid-filled and extruded cable assemblies depending on the insulation barrier used to isolate the cable and the GIS. The extruded cable connection is also referred to as a dry-type connection. The connection assembly can accept either a single-phase or three-phase design.

The IEC standard also clearly defines the supply scope between the cable and GIS suppliers to avoid confusion. Figures 3.60 and 3.61 of the IEC Standard 62271 graphically illustrates a typical connection arrangement, division of parts supply scope, and standard dimensions for different voltage ratings for a fluid-filled cable assembly. Figures 3.62 and 3.63 of the IEC Standard 62271 provide the same information for dry-type connections.

When a metallic connection between the GIS enclosure and the cable gland is not feasible, nonlinear resistors are used to limit the magnitude of transient voltages that may appear across the insulated gap.

When determining the design of a cable connection, the following parameters need to be determined: rated voltage, insulation level, normal load current, temperature rise limit, rated short-time withstand current peak, and maximum duration of a short-circuit current. These parameters can be selected from relevant IEC standards. Furthermore, it must be determined whether a single-phase or a three-phase (common) connection will be required. Figures 3.60–3.63 in IEC Standard 62271 illustrate four different voltage classes from 72.5 kV to 550 kV with a rated normal current up to 3150 A. At its maximum current, the maximum temperature is not to exceed 90°C.

If SF_6 gas is used as the insulating gas, the minimum functional pressure should not exceed 0.35 MPa or 0.40 MPa (absolute) at 20°C for voltage ratings up to 300 kV and higher. The rated filling pressure is determined by the GIS manufacturer with the upper limit of 0.85 MPa absolute. This limit is also applicable if another gas is used as the insulating gas.

The mechanical design of the cable connections needs to take into account the dynamic forces generated during short-circuit conditions. The manufacturer of GIS needs to ensure that the stress limits are not exceeded. It is assumed in the standard that 2 kN and 5 kN in the transverse direction are the maximum forces that may appear on the GIS–cable connection interface.

The testing of the cable terminations is done in accordance with IEC60141-1 [30] for fluid-filled cables, IEC60141-2 [31] for gas-filled cables, and IEC 60840 [32] or IEC 62067 [33] for extruded cables. The dielectric type test should be performed at the minimum insulating gas pressure and the test setup should represent the most severe stressful service condition. The single-phase test

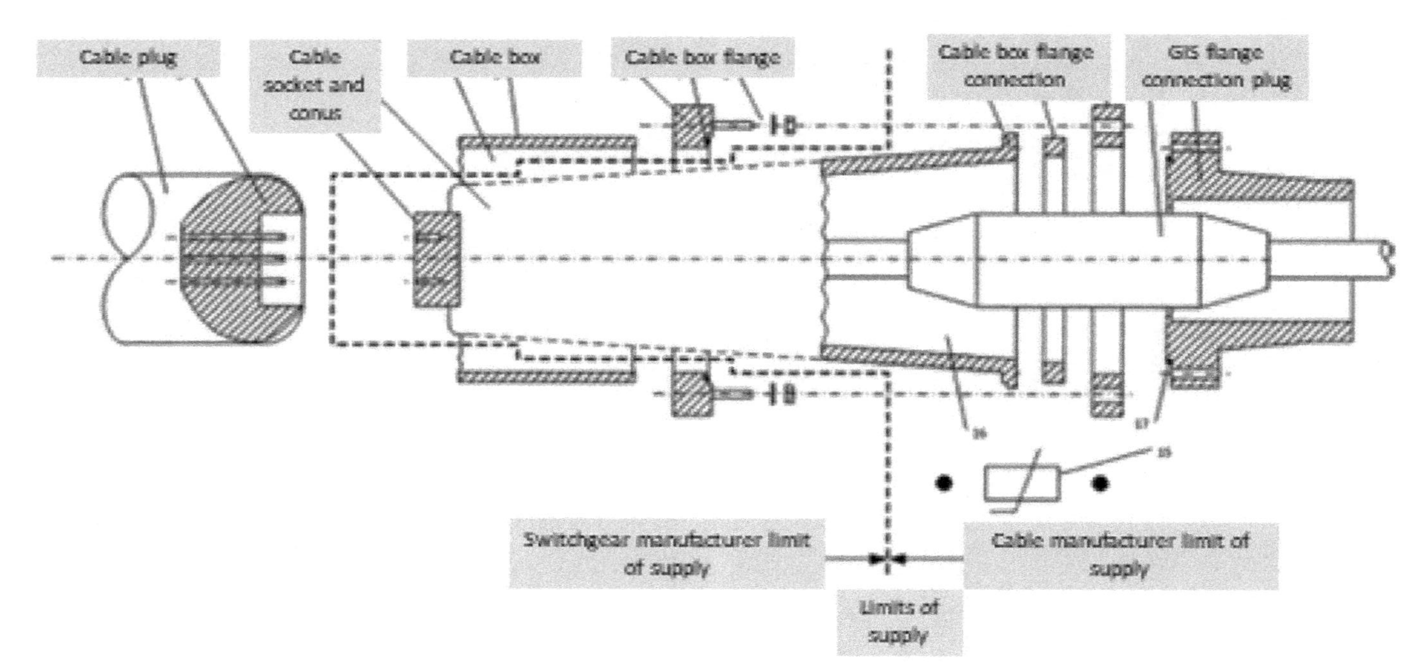

Figure 3.60 Fluid-filled cable connection assembly – typical arrangement. (Simplified drawing from IEC 62271-209) (Reproduced by permission of IEEE)

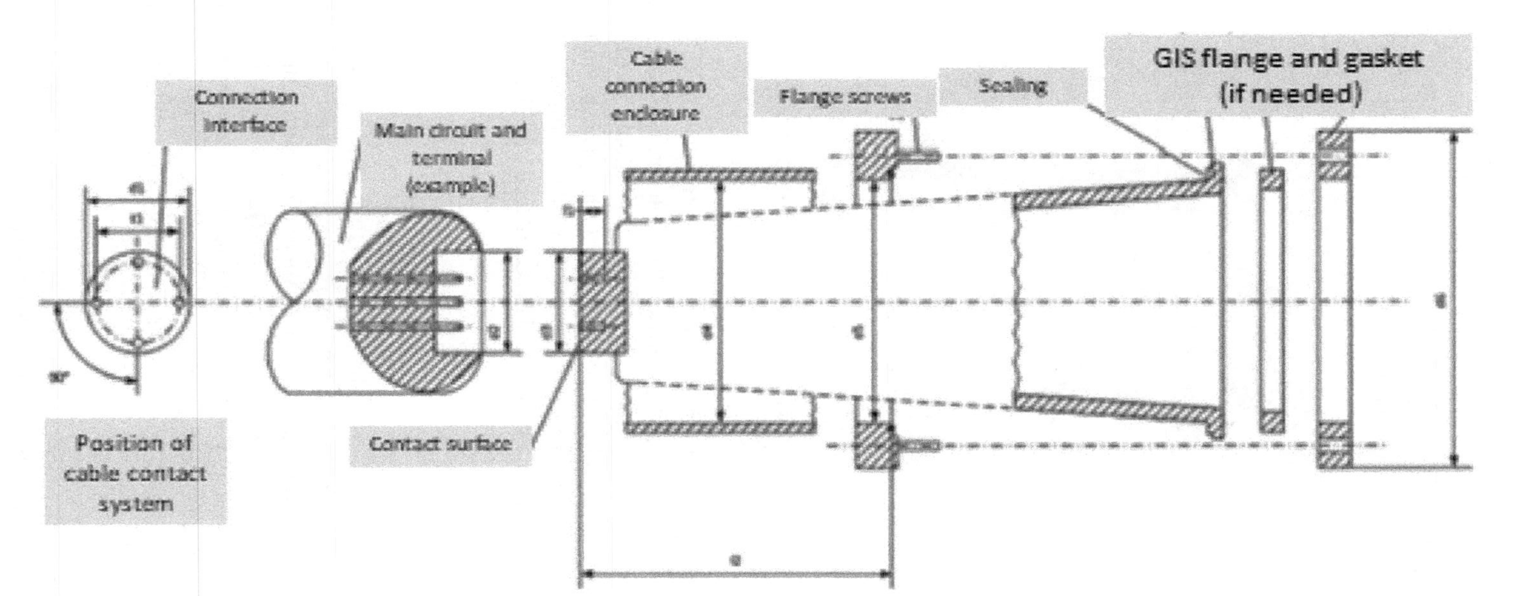

Figure 3.61 Fluid-filled cable connection – typical assembly dimensions. (Simplified drawing from IEC 62271-209) (Reproduced by permission of IEEE)

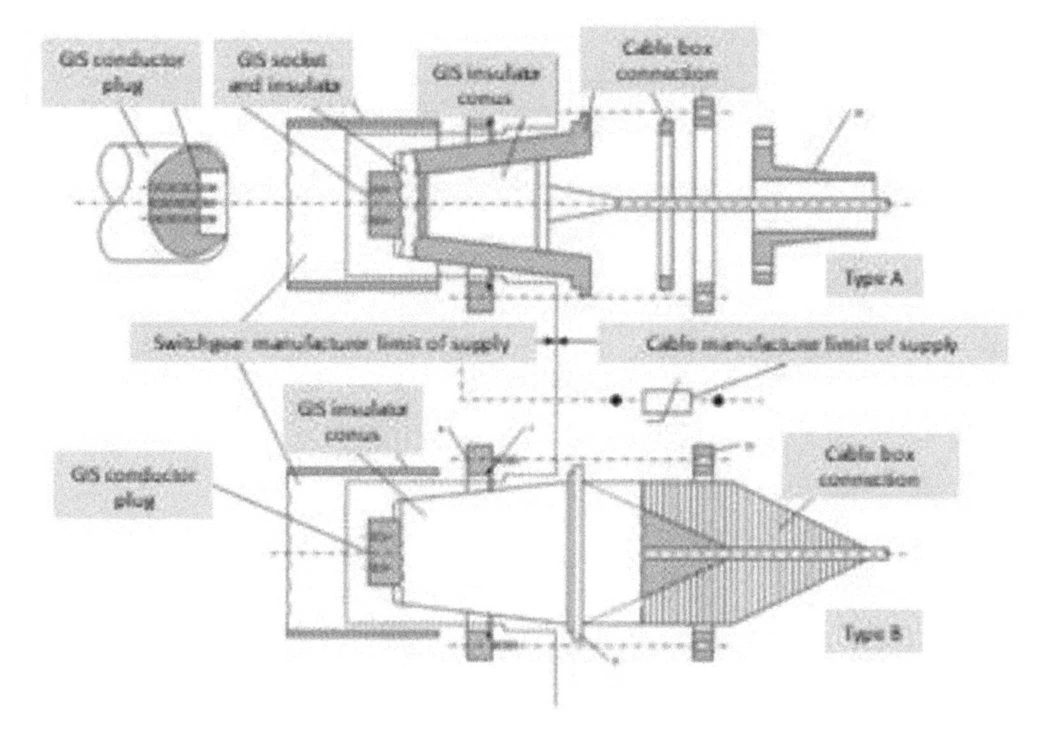

Figure 3.62 Dry type cable connection assembly – typical arrangement. (Simplified drawing from IEC 62271-209) (Reproduced by permission of IEEE)

arrangement using the single-phase termination enclosure represents a more severe stress condition than a three-phase enclosure. Therefore, a single-phase test covers a three-phase test.

Before finalizing the design, the manufacturer and the user should consider the installation requirements. The manufacturer should state the specific civil, electrical, and installation clearances that are required for installation, testing and maintenance. The rules for transportation, storage, erection, operation, and maintenance should be followed in accordance with IEC 62271-1. [20]

3.6.3 Bushings

3.6.3.1 Purpose of Bushings

Bushings are required to provide an interface between the GIS and overhead transmission lines or other substation equipment such as power transformers, disconnect switches, and so on, external to the GIS. It allows connection to the inner bus conductor of the GIS from the external equipment terminals by means of a conductor that is insulated from the GIS bus enclosure. IEEE Std. C37.017 [12] defines a bushing as "an insulating structure, including a through conductor or providing a central passage for such a conductor, with provision for mounting on a barrier, conducting or otherwise, for the purpose of insulating conductor from the barrier and conducting current from one side of the barrier to the other."

3.6.3.2 Bushing Standards

There are two important standards that apply to GIS bushings:

1) ANSI/IEEE C37.017 [12]: IEEE Standard for Bushings for High-Voltage (over 1000 V (ac)) Circuit Breakers and Gas-Insulated Switchgear. This standard is based on the standard practices

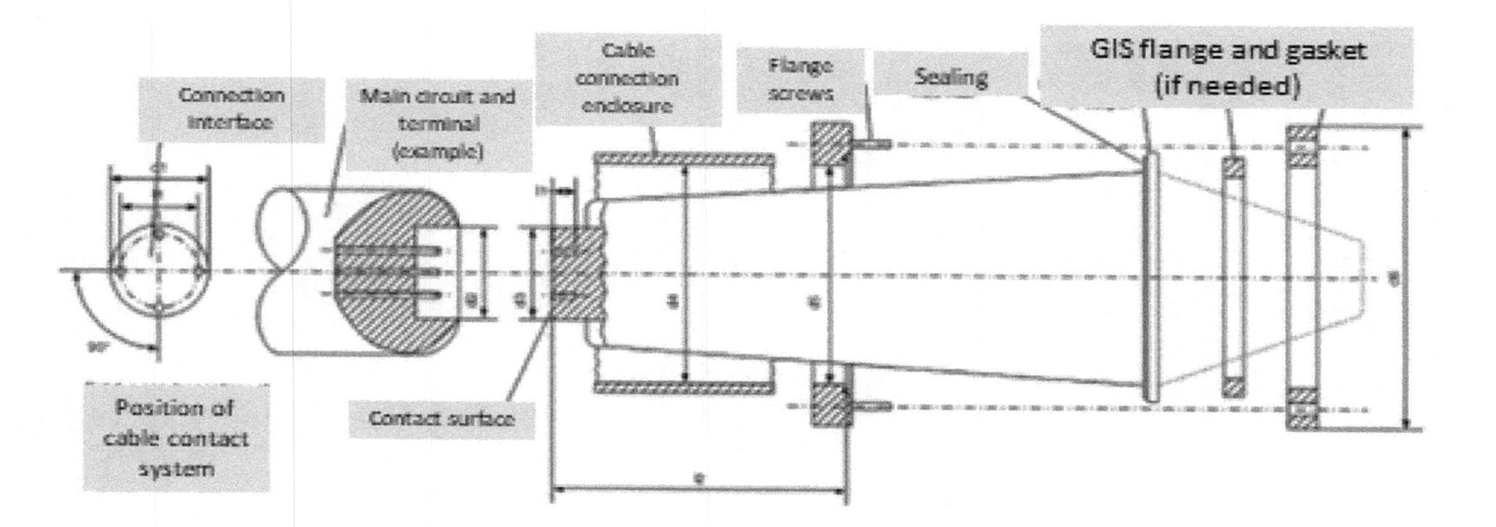

Figure 3.63 Dry type cable connection assembly – typical assembly dimensions. (Simplified drawing from IEC 62271-209) (Reproduced by permission of IEEE)

in the United States for bushings for high-voltage power circuit breakers, gas insulated substation switchgear, and gas insulated transmission lines. This standard defines the special terms, service conditions, ratings, general requirements, test procedure, and acceptance criteria for the gas insulated circuit breakers, gas insulated switchgear, and transmission lines. This standard applies to both ceramic and composite polymeric housing bushings.
2) IEC 60137: Insulated Bushings for Alternating Voltages above 1000V. This standard applies to all types of bushings intended for use in electrical apparatus, machinery, transformers, switchgear, and installations above 1000V AC and at power frequencies of 15Hz up to and including 60Hz. This is an international standard and is widely used in Europe and Asian countries.

3.6.3.3 Bushing Rating

Bushings are rated in terms of withstand voltages and continuous current carrying capability. Withstand voltages include maximum system voltage, lightning and switching impulse voltage withstand, and power frequency voltage withstand. The current carrying capability includes continuous current, short-time withstand current, peak withstand current, and duration of the short-circuit current. The rated maximum voltage rating and rated insulation level of a bushing should meet or exceed the rated maximum voltage and insulation level of the GIS itself. IEEE Std. C37.017 [12] provides recommended voltage and current values.

3.6.3.4 Bushing Construction

All GIS bushings consist of three main components. These include a hollow insulator with metal end fittings, a central conductor with top caps and shields, and SF_6 gas, or a gas mixture to fill the hollow insulator to provide electrical insulation. GIS bushings are commonly "gas insulated" bushings in which the gas is the major insulation. Some bushings are of the "gas-filled" type in which the conductor is provided with a solid major insulation and the space between the inside surface of the hollow insulator and the solid major insulation is filled with gas. Other bushing designs include a capacitance-graded core made of plastic film that is compatible with the SF_6 gas. The insulating gas may be at the same pressure as that of the insulating gas in the GIS, or at some other pressure.

The hollow insulator bushing housing may be either ceramic or composite polymeric. Composite polymeric housings are usually preferred due to their light weight, ease of handling, and nonexplosive failure nature.

The central conductor may be either copper or aluminum and tubular. The conductor diameter depends upon the required current carrying capability. To reduce the losses, the conductor wall thickness is limited to the skin depth.

3.6.3.5 Application in Contaminated Environment

Bushings are the only GIS component with insulation to ground that is exposed to the environment. Consequently, the bushings applied in a contaminated environment should be provided with adequate insulator creepage distance. IEC 60815-1 [14] describes in detail the various types of pollutions, methods to measure contaminants, and a procedure to determine, evaluate, and classify the site pollution severity. The pollution levels are categorized as "very light," "light," "medium," "heavy," and "very heavy." For each of these pollution levels a method to determine recommended creepage distances is provided in IEC 60815-2 [15]. Table 3.5 summarizes the recommended creepage distance values.

3.6.3.6 High Altitude Application

Standard bushings are designed to be adequate for installation at altitudes up to 1000m without the use of an altitude correction factor. For installation at an altitude higher than 1000m, the

Table 3.5 Recommended creepage distance values [12, 13]

Pollution class	Minimum nominal-specific creepage distance	
	Line-to-line voltage (mm/kV)	Line-to-ground voltage (mm/kV)
Very light	12.7	22.0
Light	16	27.8
Medium	20	34.7
Heavy	25	43.3
Very heavy	31	53.7

bushing insulators require a higher insulation withstand level. The required rated insulation withstand level of the bushing should be determined by multiplying the rated insulation level at sea level by the altitude correction factor K_a provided in IEEE Std C37.100.1 [34].

3.6.3.7 Bushing Terminal Loading

IEEE Std. C37.017 [12] provides recommended cantilever operating load values for standard and high strength bushings. In most of the installations these operating load values should be adequate. However, if loads are expected to be higher, these should be specified.

3.6.3.8 Gas Monitoring

Bushings are generally installed in the GIS as a self-contained gas compartment complete with gas-filling facilities and a monitoring system.

3.6.4 Interface Project Applications

Sections 3.6.1, 3.6.2 and 3.6.3 introduce the technologies for three common interface connections in GIS: (1) Direction connection between GIS and a transformer; (2) Cable connections between GIS and incoming or outgoing lines or feeders; (3) Connections of GIS gas-to-air bushings to other devices. This section gives two project examples of two out of the three types of interface connections.

3.6.4.1 Cable Connection to GIS

In this application of a 69 kV project, three cables are connected to the GIS in a three-phase common enclosure.

- This type of connection is used for transition of underground cable circuits into the GIS or GIB.
- The cable can directly interface with the GIS or interface outdoors with the GIB via GIB/Cable interface compartment.
- It may require having cable vault or basement for GIS where the termination point is located at lower elevation. Alternatively, if the termination point is located at proper height, the cable support structure is used and no need for a basement.
- In Figure 3.64, dry-type cables usage has made cable termination easier for direct interface with GIS.

3.6.4.2 Gas to Air Bushing Connection

Gas to air bushing connection is a very commonly used for outdoor GIS. Two project examples are given in this section to demonstration its application.

Figure 3.64 Cable termination three-phase GIS enclosure 69 kV GIS, 1380kcmil HXLMK Cable (Reproduced by permission of National Grid)

- This type of connection is used for the GIS transition (1) from the overhead line circuits and (2) from the equipment with open air bushings.
- It can also be used to interface with underground cable open air terminators.
- Figure 3.65 shows the double conductor 345 kV transmission line is dropped down with insulator support and transitioned into GIB rated for 4000 A, 50 kA.
- Figure 3.66 shows the transition from 345 kV bushing of the transformer to the GIB via two 1590 AAC.
- In both examples, the composite bushings are used as the gas-to-air bushings.

3.7 Gas-Insulated Surge Arresters

Gas-insulated surge arresters may be applied in gas insulated switchgear to limit transient overvoltages due to switching operations within the gas-insulated switchgear. The surge arrester metal oxide valving technology used in air insulated arresters is the same for gas designs. Compact size and a better high-frequency transient response, due to the direct method of connection to the bus, are the primary advantages of using gas insulated arresters when compared with air-insulated arresters. The location of a GIS surge arrester is very important to limit overvolatges due to lightning surge and switching surge.

The gas insulated connection to the GIS does not need any air bushings for connection. This saves space in a substation and might be a reason to choose gas insulated surge arresters. Typically,

Figure 3.65 345kV Transmission Line transition to the 4000A, 50kA gas insulated transmission line (GIL) (Reproduced by permission of National Grid)

Figure 3.66 345kV gas insulated Transmission Line (GIL) transition to the 4000A, 50kA transformer air insulated bushing (Reproduced by permission of National Grid)

Figure 3.67 Gas-insulated surge arresters connected directly to the GIS bus (Reproduced by permission of Siemens AG)

they are used to protect transformers from transient overvoltages generated in the GIS by switching, which takes place mainly with the disconnector. It is of either a plug-in construction or a disconnecting-link type.

The energy rating of the GIS surge arrester is such that it is adequate to dissipate the energy in normal conditions and also that generated in case one of the circuit breaker pole fails to close and the circuit is then opened by the pole discrepancy protection. The gas insulated surge arrester ground connection is insulated from the enclosure in order to permit monitoring of the leakage current.

In Figure 3.67, two systems of 400 kV are shown with directly connected gas insulated surge arrestors.

3.8 Gas-Insulated Bus

3.8.1 General

The gas-insulated bus duct is used within the GIS bay to interconnect the bus duct from bay to bay or to connect external connections from the GIS bay to overhead lines and transformers. There are two principles of bus ducts: a passive and an active types. The passive type of bus duct is without any moving or switching device in the same gas compartment. The active bus duct has a disconnecting and ground switch included in the bus duct gas compartment.

External connections to overhead lines and transformers can reach several kilometers of bus duct length within a substation or a power plant. Bus ducts can be manufactured and assembled in the GIS factory or they may be assembled on site. This depends mainly on the total transmission length of bus ducts to be assembled on site. The longer the length of the bus duct, the more economical is an on-site assembly. A typical bus duct length for an on-site assembly has some kilometers of single-phase pipe length. At low voltage ranges the bus duct design is three-phase encapsulation. At voltages above 200 kV, the design may be single-phase encapsulated. For voltages of 400 kV and greater, any design is single-phase insulated.

Bus ducts may be installed above ground on steel structures, in trenches underground, or in some cases directly buried. As part of the GIS bay, bus ducts are an integral part of the bay. On-site testing of long bus duct sections may be limited because of the maximum capacitive load allowed by the test equipment. In such cases, sectionizers are required to split the total length in the maximum bus duct length of up to 1 km.

3.8.2 Three-Phase Insulated Bus Duct

The three-phase insulated bus duct is used for the lower voltage range up to 170 kV rated voltage. The limitation of the three-phase insulated design is mainly due to the size of the aluminum casted enclosures. Also, the high short-circuit current levels of the higher voltage classes are giving limits because of the mechanical forces between the conductors.

This is mainly the reason in long bus ducts when conductors are running in parallel. These strong forces require strong insulators in relatively short distances. That is why in such cases a single-phase design is preferable.

The principal design of three-phase insulated bus ducts has active and passive versions. In the case of the passive design, the gas compartment of the bus duct does not include any switching devices (see Figure 3.68). The bus duct in the upper part of the photo shows the enclosure of three internal conductors connected to the GIS bay.

The active three-phase encapsulated design of bus ducts includes disconnector switches and ground/earth switches. These disconnect switches connect the bus bar to the GIS bay and the ground/earth switches are used to ground/earth the GIS side when disconnected to the bus duct. In Figure 3.69, such an installation is shown on the top of the GIS bays on the left and right of the vertical positioned enclosure for the circuit breaker.

3.8.3 Single-Phase Insulated Bus Duct

The single-phase insulated bus duct design is the simplest form of GIS. One conical or post type insulator at one end of the bus duct section and a support tool for assembly at the other end is all that is needed to keep the internal conductor pipe in the center of the aluminum pipe. In Figure 3.70, a section of a single-phase bus duct is shown fixed on a crane during the assembly process.

The advantage of a single-phase bus duct is that, in the case of a short-circuit current the mechanical forces of the current in the conductor and reverse, the induced current in the enclosure is superpositioned in a negative way so that the forces to the insulator are reduced. This allows simple insulators and a high current carrying capability.

A single-phase passive bus duct is shown in Figure 3.71. The bus duct is located on top of the GIS bays next to the personnel walkway. Each section of the bus duct has the length of one bay width,

Figure 3.68 Three-phase insulated passive bus duct (Reproduced by permission of Siemens AG)

Figure 3.69 Three-phase insulated active bus duct (Reproduced by permission of Siemens AG)

Figure 3.70 Single-phase insulated bus duct on the crane (Reproduced by permission of Siemens AG)

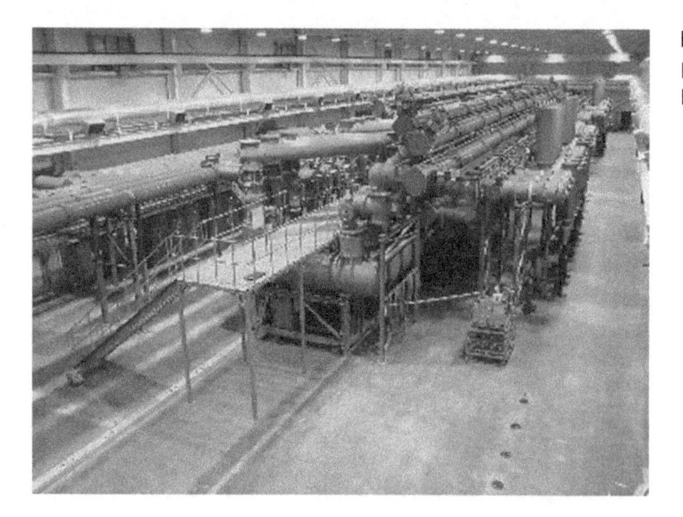

Figure 3.71 Single-phase insulated passive bus duct (Reproduced by permission of Siemens AG)

which is indicated by a flange. The flange has no gastight insulator so that the gas compartments of each GIS bay is connected to one single gas compartment of the single-phase bus duct. This reduces the total number of gas compartments. If needed, to reach high flexibility, the gas compartments of the bus duct can be limited to the bay width of the GIS bay by using gastight insulators.

3.8.4 Bus Duct to Connect Overhead Lines

Overhead lines need a larger insulation distance between phases and towards earth or ground than those of the GIS because of the lower insulating capability at atmospheric air. Therefore, bus ducts of GIS are used to expand the distances of the GIS bays to those of overhead lines. This is shown in Figure 3.72. The single-phase passive bus ducts connect the overhead lines outside the building by using a Z-form. The Z-form has two advantages. First, it allows an expansion to the side of the GIS bay and also a larger distance between phases before leaving the building through a wall passing section. The Z-form also provides the required movement of thermal expansion by simple angle compensators attached to the Z-form. In Figure 3.73, the wall passing section is shown. The grounded/earthed enclosure of the bus duct is connected to a grounded/earthed metal plate at the wall passing section to avoid voltage potential differences between the enclosure and the wall. This

Figure 3.72 Single-phase insulated bus duct connection to overhead lines outside the GIS building (Reproduced by permission of Siemens AG)

Figure 3.73 Single-phase insulated bus duct to connect overhead lines with wall passing (Reproduced by permission of Siemens AG)

Figure 3.74 Single-phase insulated bus duct connection to overhead lines with SF_6 gas-to-air bushings and surge arresters (Reproduced by permission of CGIT)

avoids arc flashing in this section due to transient overvoltage, for example, at disconnector switching of the GIS.

To connect a single-phase bus duct to an overhead line an SF_6 gas-to-air bushing is used. At the extended distance between phases the bushings are then directly connected to the wires of the overhead line. To protect the gas insulated bus duct and the connected GIS from transient overvoltages coming from lightning strokes into the overhead line, surge arresters are used in parallel to the SF_6/air bushings, as shown in Figure 3.74.

3.8.5 Bus Duct to Connect Circuit Breakers

In some cases, bus ducts are used to connect circuit breakers or GIS bays. As this is a connection of SF_6-to-SF_6 insulation only gastight conical insulators are needed to separate the bus duct from the circuit breaker or GIS. This is usually needed if the gas pressure on both sides is different. For mechanical separation, compensators with bellows are used. In Figure 3.75, a single insulated phase bus duct connection to outdoor circuit breakers is shown.

3.8.6 Bus Duct to Connect Transformers

For direct connection of bus ducts with transformers, a separation of the oil insulation of the transformer and the bus duct is needed. For this, a conical insulator is needed to fulfill the oil insulating requirements on one side with the gas insulated requirements on the other side. For technical details, see Section 3.6.1 on direct transformer connection of the interfaces. To standardize the large variations of direct connection between GIS and transformers, IEC 62271-211 [28] gives recommendations in dimensioning.

Figure 3.75 Single-phase insulated bus duct connection to circuit breakers (Reproduced by permission of Siemens CGIT)

Besides the electrical requirements, the mechanical forces and vibrations are also limited between the bus duct and the transformer. For transformer protection of transient overvoltages coming from disconnect operation in the GIS, surge arresters are usually used. In Figure 3.76, a transformer connection to a bus duct is shown.

3.8.7 Bus Duct to Connect Cables

To connect the bus duct with cables, two types of cables are used: oil and solid PE insulated cables. The separation of SF_6 insulated bus duct and the oil or solid PE insulated cables is made by a special conical insulator to meet the different electrical requirements. For technical details, see Clause 3.6.2 on cable connection of the interface section.

To standardize the big variations of connections, IEC 62271-209 [18] on cable connections for GIS with fluid filled and dry-type cable terminations, recommendations of dimensioning are given.

In Figure 3.77, a single-phase bus duct connection to cables is shown. The vertical enclosure attached to the cable holds the conical insulator between the cable side solid insulation and SF_6 gas insulation on the GIS side.

In some cases, the cable insulation and the bus duct and GIS insulation need to be separated to avoid sheet currents in the cable shield. For this reason, an insulating ring (the brown ring in Figure 3.78) is required. Varistors bridge the insulating ring at a minimum of three locations in

Figure 3.76 Single-phase insulated bus duct connection to transformers (Reproduced by permission of Siemens CGIT)

Figure 3.77 Single-phase insulated bus duct connection to cables (Reproduced by permission of Siemens AG)

Figure 3.78 Single-phase insulated bus duct connection to cables with overvoltage protection (Reproduced by permission of Siemens AG)

order to avoid flashovers in the case of disconnector switching. If there were no varistors transient overvoltage coming from the GIS would cause a flashover at the insulated flanges (see Figure 3.78).

3.8.8 Bus Duct to Underpass Overhead Lines

In substations, sometimes the overhead lines do not reach the substation at the location where the GIS bay is located. Then it may be necessary to cross under other lines. A bus duct allows a simple solution to underpass overhead lines, as shown in Figure 3.79. The bus duct in the front of the photo is connected to the GIS bay inside a building and connects three bushings to an overhead line in the background of the photo.

When outside a substation a bus duct needs to be protected with a fence, as shown in Figure 3.80. This access protection is needed to prevent vandalism and general access to unauthorized personnel.

3.8.9 Bus Duct Above Ground

Inside the substations, the above ground installation of bus ducts is often used to connect different sections of the substation, to connect the power transformer of the generator, to connect overhead lines, or to connect transformers. Above ground installations use steel structures to carry the enclosure pipes and to allow thermal expansion.

Figure 3.79 Single-phase insulated bus duct to underpass overhead lines inside a substation (Reproduced by permission of CGIT)

Figure 3.80 Single-phase insulated bus duct to underpass overhead lines outside a substation (Reproduced by permission of Siemens CGIT)

An example of a low steel structure to hold a bus duct is shown in Figure 3.81. The steel portal holds the weight of the bus duct but allows angle movement of the Z-form arrangement with angle compensation bellows.

The disadvantage of low steel structures is that the bus duct may hinder the free movement or ground traffic in the substation. If it is required to allow free movement, the steel structures are elevated to a height of 4–5 m, as shown in Figure 3.82.

Figure 3.81 Single-phase insulated bus duct above ground installed at low steel structures (Reproduced by permission of Alstom)

Figure 3.82 Single-phase insulated bus duct above ground installation at high steel structures (Reproduced by permission of Alstom)

In this case, the bus ducts connect several machine transformers of the gas turbine power plant with the GIS building. The high elevation of the steel structures allows free traffic below.

The disadvantage of high elevated steel structures is that a large amount of steel is needed. This can be optimized to a mixture of high elevated steel structures only at locations with ground traffic; for example, such structures are not needed for substation streets and sections where no free movement is required. Then a jump in elevation can be made at the street with traffic, as shown in Figure 3.83. In addition to straight bus ducts, section angle elements, which are more costly, are needed. This means that a cost optimization is necessary to find the optimum of low and high sections.

In some application cases, it is more economical to use concrete blocks than steel structures. In cases of low elevation, concrete blocks are used to hold the bus duct. In Figure 3.84, the single-phase insulated bus duct is mounted on top of a concrete block using a steel beam and a sliding plate to allow thermal movement of the bus duct.

A special application of a bus duct connection underneath overhead lines is shown in Figure 3.85, where a flooding area is crossed. The concrete blocks are fixed in the dry ring bed, which is temporarily flooded. The water at flooding times is usually not higher than the concrete blocks. However,

Figure 3.83 Single-phase insulated bus duct above ground installation with different steel structure heights (Reproduced by permission of Alstom)

Figure 3.84 Single-phase insulated bus duct above ground installation at concrete blocks (Reproduced by permission of Alstom)

Figure 3.85 Single-phase insulated bus duct above ground installation at concrete blocks in a flooding area (Reproduced by permission of CGIT)

the bus duct enclosure pipes are fixed to the steel structure on the top. The fixing will prevent the bus duct from floating in the case where it is under water during a flooding. The fixing is connected to concrete blocks that are fixed to the ground.

In cases in which the location of the GIS bays and the in- and outgoing lines do not match well many line crossings are required. As shown in Figure 3.86, the bus ducts offer a space saving solution because of the grounded/earthed enclosure potential.

3.8.10 Bus Duct Trench Laid

In cases where free movement on the ground is required, trench-laid bus ducts offer a solution. Trenches can be inside a substation or even outside as they are covered with concrete blocks and, therefore, are not accessible to the public.

The concrete trenches need water treatment to keep the bus duct dry for corrosion protection. The bus duct laid in a trench will have thermal limitations of typical maximum enclosure temperatures of 60°C when touchable and 70°C when not touchable according to GIS standards. Therefore,

Figure 3.86 Single-phase insulated bus duct above ground installation to cross lines (Reproduced by permission of CGIT)

a thermal calculation is needed for the sections in the trenches that differ from the thermal calculations above ground.

In Figure 3.87, a single-phase insulated bus duct is shown laid in a trench inside and in Figure 3.88, outside a substation. A trench can be connected to a low tunnel, as shown in Figure 3.89, where a single-phase insulated bus duct is shown laid in a trench crossing a highway.

3.8.11 Bus Duct Laid in a Tunnel

In power plants, mainly in hydro power plants, energy tunnels with bus ducts are often used to connect the generator transformer in the cavern in the dam or cavern at the high-voltage side with the substation outside the dam or cavern. Such applications have been built all around the world and usually have one thing in common: a high-power transmission capability. High-voltage ratings of 400, 500, and 800 kV are usual. Also, current ratings of 3000, 4000, or 5000 A are often required.

The tunnels are round or squared and usually carry two bus ducts, with three-phase systems of single-phase insulated pipes. In Figure 3.90, a tunnel with a double system of single-phase insulated bus ducts is shown. To follow the vertical changes of the tunnel, angle elements are used. The voltage level is 500 kV.

In Figure 3.91, a tunnel with a double system of single-phase insulated bus ducts is shown. To follow the horizontal bending of the tunnel angle elements are used. The voltage level is 800 kV.

In Figure 3.92, a tunnel with a double system of single-phase insulated bus ducts is shown. To follow the horizontal bending of the tunnel elastic bending of the aluminum pipes of the bus duct is used. The voltage is 300 kV.

In Figure 3.93, a tunnel on a slope with a double system of single-phase insulated bus ducts is shown. To follow the slope and to allow thermal movement the bus duct is laid on rollers. The voltage level is 400 kV.

In Figure 3.94, a vertical shaft is shown, which holds a double system of single-phase insulated bus ducts. The aluminum enclosures are fixed to the shaft walls. The voltage level is 800 kV.

Figure 3.87 Single-phase insulated bus duct laid in a trench inside a substation (Reproduced by permission of CGIT)

Figure 3.88 Single-phase insulated bus duct laid in a trench outside a substation (Reproduced by permission of CGIT)

Figure 3.89 Single-phase insulated bus duct laid in a trench crossing a highway (Reproduced by permission of CGIT)

Figure 3.90 Single-phase insulated bus duct double system laid in a horizontal tunnel with vertical angle elements (Reproduced by permission of CGIT)

Figure 3.91 Single-phase insulated bus duct double system laid in a horizontal tunnel with horizontal angle elements (Reproduced by permission of CGIT)

Figure 3.92 Single-phase insulated bus duct double system laid in a horizontal tunnel using elastic bending (Reproduced by permission of Siemens AG)

Figure 3.93 Single-phase insulated bus duct double system laid in a tunnel at a slope (Reproduced by permission of CGIT)

Figure 3.94 Single-phase insulated bus duct double system laid in a vertical tunnel (Reproduced by permission of CGIT)

3.8.12 Bus Duct Directly Buried

Similar to pipelines or cables, the bus duct can also be laid directly buried into the ground. In such a case, passive and active outer corrosion protection is needed. Soil coverage to a minimum of 1 m is required to fix the bus duct in the soil. No thermal expansion bellows are needed. The directly buried method can be used inside and outside substations. For directional changes elastic bending and directly buried angle elements are used.

In Figure 3.95, a single-phase insulated bus duct single system is directly buried and includes a horizontal angle element for directional change. The installation is inside a substation to cross existing lines. The voltage level is 145 kV.

In Figure 3.96, a single-phase insulated bus duct two-phase loop test circuit shows a vertical angle element. The voltage level is 400 kV.

In Figure 3.97, a single-phase insulated single-phase test section shows a directly buried horizontal and vertical angle element during the laying process. The voltage level is 400 kV.

In Figure 3.98, a single-phase insulated bus duct double system directly buried with elastic bending is shown during the laying process. The voltage level is 400 kV.

3.8.13 Shipment on Site

Bus duct sections are shipped to the site on extended bed trucks. Handling on site can be done by crane or forklift trucks. An 18-m bus section is shown in Figure 3.99 as it is being unloaded from the truck.

3.8.14 Assembly on Site

The support structures are placed in position first and roughly leveled. Perfect alignment is not necessary since the GIL is flexible. The GIL shipping sections are moved into place and connected together, as shown in Figure 3.100. A special clean tent is not normally required except for heavy dust conditions.

Figure 3.95 Single-phase insulated bus duct single system directly buried with a horizontal angle element (Reproduced by permission of CGIT)

Figure 3.96 Single-phase insulated bus duct two-phase test loop directly buried with a vertical angle element (Reproduced by permission of CGIT)

Figure 3.97 Single-phase insulated bus duct single-phase test section directly buried with horizontal and vertical angle elements (Reproduced by permission of Siemens AG)

Figure 3.98 Single-phase insulated bus duct double-phase system directly buried with elastic bending (Reproduced by permission of Siemens AG)

Figure 3.99 An 18-m bus section being unloaded at the site (Reproduced by permission of CGIT)

Figure 3.100 Final assembly of a 550 kV field joint. Locating pins are used to align the flanges. Conductor connection is a plug and socket connection with self-alignment (Reproduced by permission of CGIT)

Flanged or welded joints are done in a similar fashion. The conductor connection is made using a plug and socket connection. It will engage automatically as the two bus duct sections are brought together. Then the enclosure joint is made.

If double O-rings are used on the flanges, the joint can be checked for leaks immediately after assembly by bagging the area around the joint with plastic, pressurizing the area between the two O-rings, and checking for leaks with a hand-held leak detector. Using this technique for long-circuit lengths ensures that each joint is complete and leaktight without first evacuating and filling the system with gas.

3.9 Guidelines for GIS

3.9.1 General

IEEE Guide for Gas-Insulated Substation (GIS) Std C37.122.1 (2014) [8] was developed to assist GIS users with their GIS equipment. It was updated in 2014. This section provides a summary of the guide's content and some of the highlights.

3.9.2 Scope

This guide provides information of special relevance to planning, design, testing, installation, operation and maintenance of gas-insulated substations (GIS) and equipment. This guide is intended to supplement standard IEEE Std C37.122 [9].

This guide is applicable to all gas-insulated substations (GIS) above 52 kV. However, the importance of topics covered varies with application category. For example, the issues related to advanced field test techniques and very fast transients (VFT) are of particular interest for extrahigh-voltage (EHV) gas-insulated substations (GIS).

3.9.3 GIS Guide Content

In addition to the standard scope, normative references, and definition sections, the following are the topics covered in the main body of the guide:

GIS Arrangement; Benefits and Drawbacks of GIS; Primary Components; Design, Installation, and Equipment Handling; Control Wiring; Local Control Cabinet (LCC); Gas Handling; Partitions; Switch View Ports and Viewing Options; GIS Grounding; GIS Seismic Requirements; Partial Discharge (PD) Testing; Field Testing; Maintenance and Repair; GIS and the Environment; Future GIS Extension Considerations, Thermal Overload Capability Requirements of Circuit Breakers and Remaining GIS Equipment; And Special Transient Recovery Voltage (TRV) Issues for Faults in Three-Phase Enclosures.

The next several sections go over some of the topic's highlights.

3.9.4 Different Layouts and One-Line Diagrams

The modular system of GIS allows the implementation of any type of layout and single line arrangement. The following examples in Figure 3.101 are giving typical single line and gas partitioning for various arrangements:

Single bus, Double bus with single breaker, Ring bus, Double bus with double breaker, and Breaker-and-a-half.

In Figure 3.102 a GIS diameter and GIS bay is shown. The GIS diameter represents the connection of bus bar 1 with bus bar 2 of the breaker and one-half scheme. The GIS bay represents the switchgear related to one circuit breaker including disconnectors, ground switches, voltage and current transformers.

3.9.5 Benefits and Drawbacks of GIS

The most common benefits of GIS include the following:

Compact in dimensions, requiring much less footprint allowing a greater number of sites that can accommodate new substation installations, GIS can be located underground, in basement of

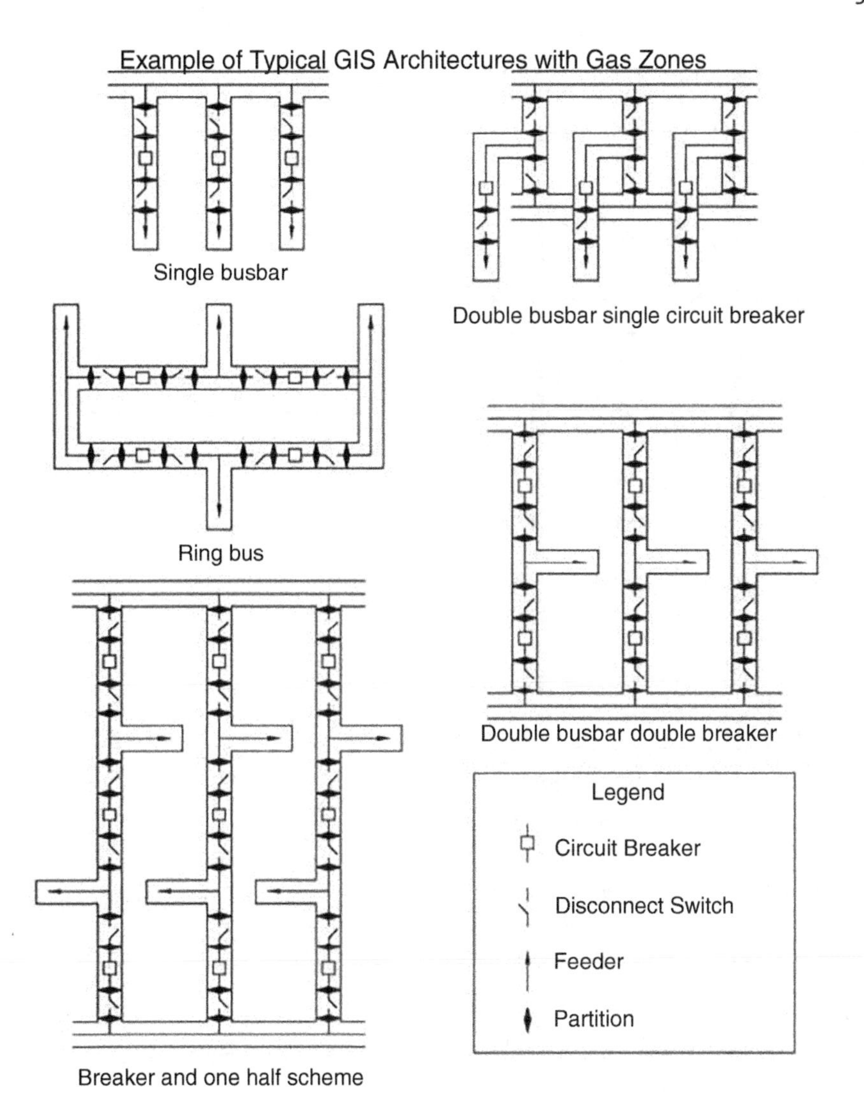

Figure 3.101 Example of typical GIS architectures with gas zones of single bus, double bus single breaker, ring bus, double bus double breaker and breaker and on half scheme (Reproduced by permission of IEEE)

a building, inside a new or existing building, on platform, inside a cavern, on a slope, on or inside a dam, or an unusually shaped site. Extension of AIS for where limited space is available. Better seismic withstand capability. Compact size and aesthetics can mesh with the existing environment to appeal to public approval when required. Ideal for locating near load centers in urban areas. Reduced maintenance requirements over the life of equipment due to the limited number of components exposed to the atmospheric contaminants. A well-defined interlock is provided electrically and/or mechanically, to avoid misoperation and reduce injury risk. Electric fields are shielded by grounded GIS. Magnetic fields due to conductor currents are reduced by GIS enclosure currents. High reliability proven by 50 years of experience with GIS in service world-wide. And GIS can withstand extreme environmental conditions since it has few exposed insulators and can be installed indoors.

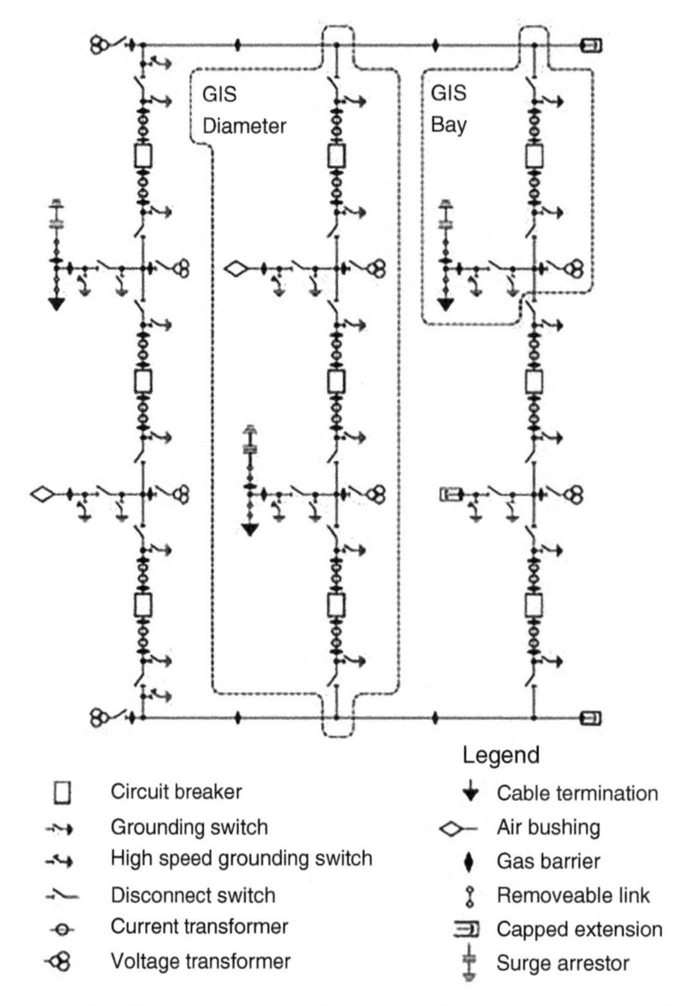

Figure 3.102 Example of GIS diameter and bay in a breaker-and-a-half scheme (Reproduced by permission of IEEE)

The most common drawbacks of GIS include the following:

Requires properly pressurized SF_6 gas for electrical insulation of busses, switches, etc., requiring more extensive gas monitoring. Initial installed cost of GIS equipment may be higher than AIS, depending on the local requirements. Longer downtime in case major repairs. A larger amount of SF_6 gas, which needs to be processed, managed, and monitored. Maintenance and operating crews need special training when the first GIS installation is introduced. And Due to compact design, the access for operating and maintenance can be more difficult.

3.9.6 Local Control Cabinets (LCC)

The SF_6 gas-insulated substation typically has a local control cabinet (LCC) mounted in the immediate vicinity of the GIS bays that may provide the following functions:

A mimic diagram showing the arrangement of all of the electrical equipment in the bay. Control and position indication of all the switching devices, circuit breakers, disconnect switches, and grounding switches in the GIS equipment. Monitoring of the SF_6 gas density in the various gas zones. Monitoring of the drive mechanisms of the circuit breakers. Local-Remote switching for transferring control of the circuit breakers and disconnect switches from the LCC (Local) to the station control system (Remote). Electrical interlocking between the circuit breakers disconnecting switches, grounding switches, line potentials, and apparatus on the opposite side of power transformers. Annunciation of equipment alarms in each bay. And a connection point to the central control equipment such as protective relays, measuring instruments, remote control interfacing, and other devices.

An interlocking scheme is incorporated inside the LCC cabinet to account the following basic requirements: Safeguard of maintenance personnel who may be working on one section of the equipment with other sections energized and prevent incorrect switching sequences which could lead to a hazardous situation to plant, equipment, or personnel.

3.9.7 Partitions

Partitions are used to separate compartments of the GIS and are gas tight such that contamination between adjacent compartments cannot occur. Partitions are made of material having insulating and mechanical properties so as to ensure proper operation over the lifetime of the GIS. Partitions should maintain their dielectric withstand strength at service voltage when contaminated by SF_6 by-products generated from normal load switching or interrupting short-circuit fault current. The design pressure of partitions should be determined in accordance with IEEE Std C37.122-2010 (Figure 3.103).

3.9.8 GIS Grounding

GIS is are subjected to the same magnitude of ground fault current and require the same low-impedance grounding as conventional substations. However, they are typically installed on smaller sites which make it more challenging to dissipate fault current. GIS also require touch and step voltage mitigation; however different techniques are used in GIS applications to control hazardous voltages. Touch and step voltages can develop along the GIS enclosures during internal faults (e.g., flashovers between conductor and enclosure) and by external faults outside the GIS, therefore adequate grounding practices, along with manufacturer's recommendations, should be implemented.

3.9.9 Partial Discharge (PD) Measurements

Partial discharge (PD) is an electrical phenomenon consisting of small magnitude electrical discharges that occur when high voltages are applied. PD could be from localized, corona in the insulating gas or other small discharges in metal- or dielectric-bound contained spaces in the equipment. PD will induce a short-duration, low magnitude current pulses to flow in the system. These current pulses can be detected, measured and/or monitored to provide diagnostic information about the overall apparatus or insulation system. In gas-insulated equipment, PD is generated by the small "defects" in the system.

In order to perform PD measurements, small electrical discharge of PD shall be detected. As PD is an electrical phenomenon, most measurement methods used are electrical techniques. However, the electrical discharge will also generate acoustic energy, therefore acoustic PD methods are also possible. Overtime, electrical discharges can also decompose SF_6 gas, in which case the decomposed byproducts can be detected by chemical means.

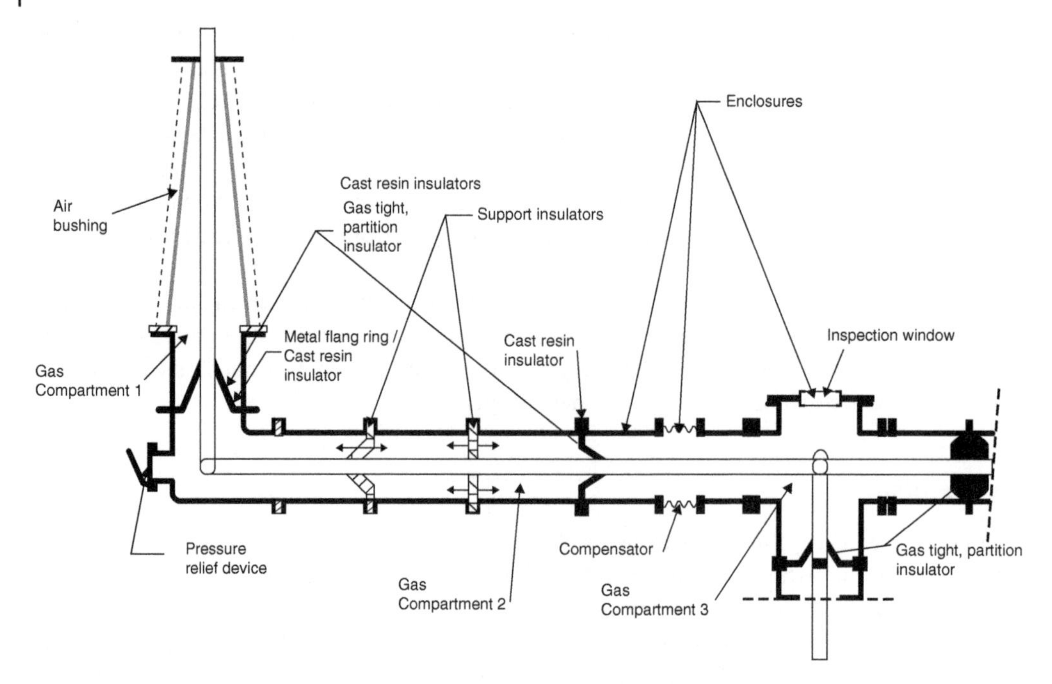

Figure 3.103 Example of arrangement of enclosures and gas compartments (Reproduced by permission of IEEE)

References

1. EN 50052 (2017). High-Voltage Switchgear and Controlgear - Gas-Filled Cast Aluminium Alloy Enclosures.
2. EN 50064 (2018). High-Voltage Switchgear and Controlgear - Gas-Filled Wrought Aluminium and Aluminium Alloy Enclosures.
3. EN 50068 (2018). High-Voltage Switchgear and Controlgear - Gas-filled Wrought Steel Enclosures.
4. EN 50069 (2018). High-Voltage Switchgear and Controlgear - Gas-Filled Welded Composite Enclosures of Cast and Wrought Aluminium Alloys.
5. EN 50089 (1994). Cast Resin Partitions for Metal Enclosed Gas-Filled High-Voltage Switchgear and Controlgear.
6. EN 50187 (1996). Gas-Filled Compartments for a.c. Switchgear and Controlgear for Rated Voltages Above 1kV and up to and Including 52kV.
7. ISO 9000 (2005). Quality Management System.
8. IEEE Std. C37.122.1 (2014). IEEE Guide for Gas-Insulated Substations.
9. IEEE Std. C37.122 (2010). IEEE Standard for Gas-Insulated Substations.
10. IEEE Std. C37.123 (1996). IEEE Guide for Specifications for Gas-Insulated, Electric Power Substation Equipment.
11. McDonald, J. and Bolin, P. (2008). Electric power substations engineering. *Gas Insulated Substations*, 2-1–2-18, CRC Press, Boca Raton, FL, USA.
12. IEEE Std. C37.017 (2010), IEEE Standard for Bushings for High-Voltage [over 1000 V (ac)] Circuit Breakers and Gas-Insulated Switchgear.
13. IEC 60137 Insulated Bushing for a.c. Voltages Above 1000 V.
14. IEC TS 60815-1 (2008). Selection and Dimensioning of High-Voltage Insulators Intended for use in Polluted Conditions – Part 1: Definitions, Information and General Principles.

15 IEC 60815-2 (2008). Selection and Dimensioning of High-Voltage Insulators Intended for use in Polluted Conditions – Part 2: Ceramic and Glass Insulators for a.c. Systems.
16 IEC 62271-4 (2013). High-voltage switchgear and controlgear – Part 4: Handling procedures for sulphur hexafluoride (SF_6) and its mixtures.
17 IEEE C37.122.3 (2011). IEEE Guide for Sulphur Hexafluoride (SF_6) Gas Handling for High-Voltage (over 1000 Vac) Equipment.
18 IEC 62271-209 (2019). RLV High-voltage switchgear and controlgear - Part 209: Cable connections for gas-insulated metal-enclosed switchgear for rated voltages above 52 kV - Fluid-filled and extruded insulation cables - Fluid-filled and dry-type cable-terminations.
19 IEC 60270 (2000+AMD1:2015). CSV, High-voltage test techniques - Partial discharge measurements.
20 IEC 62271-1 Common Specifications for High Voltage Switchgear and Control Gear Standards.
21 IEC 62271-203 Gas-Insulated Metal-Enclosed Switchgear for Rated Voltages above 52 kV.
22 IEC TR 62271-312 (2021). High-voltage switchgear and controlgear - Part 312: Guidance for the transferability of type tests of high-voltage/low-voltage prefabricated substations.
23 NFPA 70, National Electrical Code, National Fire Protection Association2020.
24 IEC 61936-1 (2021). CMV, Power installations exceeding 1 kV AC and 1,5 kV DC - Part 1: AC.
25 ASTM D2472 – 15, Standard Specification for Sulfur Hexafluoride, 2020.
26 IEEE C57.13 (2016). IEEE Standard Requirements for Instrument Transformers.
27 IEEE 80 (2013) IEEE Guide for Safety in AC Substation Grounding.
28 IEC 62271-211 (2014). High-voltage switchgear and controlgear - Part 211: Direct connection between power transformers and gas-insulated metal-enclosed switchgear for rated voltages above 52 kV.
29 IEC 60137 (2017). RLV, Insulated bushings for alternating voltages above 1000 V.
30 IEC 60141-1 (1993). Tests on oil-filled and gas-pressure cables and their accessories - Part 1: Oil-filled, paper or polypropylene paper laminate insulated, metal-sheathed cables and accessories for alternating voltages up to and including 500 kV.
31 IEC 60141-2 (1963). Tests on oil-filled and gas-pressure cables and their accessories - Part 2: Internal gas-pressure cables and accessories for alternating voltages up to 275 kV.
32 IEC 60840 (2020). RLV, Power cables with extruded insulation and their accessories for rated voltages above 30 kV (Um= 36 kV) up to 150 kV (Um = 170 kV) - Test methods and requirements.
33 IEC 62067 (2011). RLV, Power cables with extruded insulation and their accessories for rated voltages above 150 kV (Um = 170 kV) up to 500 kV (Um = 550 kV) - Test methods and requirements.
34 C37.100.1, IEEE Standard of Common Requirements for High Voltage Power Switchgear Rated above 1000 V.

4

Control and Monitoring

Authors: 1st edition Hermann Koch, Noboru Fujimoto, and Pravakar Samanta
Reviewers: 1st edition Noboru Fujimoto, Hermann Koch, Pravakar Samanta, Devki Sharma, Xi Zhu, John Brunke, Arnaud Ficheux, and Michael Novev, 2nd edition Michael Novev, and Arnaud Ficheux

4.1 General

This section on Control and Monitoring covers monitoring topics including gas monitoring, partial discharge measurements, and circuit breaker monitoring. The control topics of bay controllers and control schemes are also covered. A special topic, related to digital monitoring in substations, is also discussed.

In Section 4.2, GIS Monitoring, the specifics of gas monitoring including the related alarms are explained and gas monitoring practices are given. The partial discharge monitoring section also covers the types of defects causing partial discharges and gives information on partial discharge measuring methods. Electric, acoustic, and chemical means of partial discharge monitoring within GIS are explained. The subsection on other monitoring collects information on circuit breakers and gas density monitoring.

Section 4.3, Local Control Cabinet, gives information on different types of GIS solutions in service. Current and voltage or potential transformer wiring including a mimic diagram are explained. The function of the bay controller with all its basic functions is clarified. Control schemes of different mode selections and interlockings are given and examples are shown.

Section 4.4, Digital Communication, provides information about the impact on digital communication to GIS, which is based on IEC 61850. Basic digital communication standards are shown and their relations to other standards explained. Switchgear-related communication standards of GIS are defined and the locations of the controls in the GIS and their timing operations explained. Information is given to measure and test digital communication in GIS.

For advanced GIS technologies on monitoring, digitalization and measurements see sections 9.14 Low Power Instrument Transformers, 9.15 Digital Twin of GIS and GIL and 9.18 Digital Substation.

4.2 GIS Monitoring

GIS can be monitored in different ways and for different purposes. Many of these are not necessarily unique to GIS but can also be found on other switchgear and breaker components. The most common monitoring scheme is typically gas monitoring. Gas monitoring will exist on nearly every apparatus that uses pressurized SF_6 gas and alternative gas mixtures, including GIS and breakers. Another common monitoring scheme involves the circuit breakers monitoring. Breaker monitoring is not restricted to GIS and a number of commercial products are available, both from OEMs and third-party vendors. The simplest form of breaker monitoring (e.g., monitoring the number of

Gas Insulated Substations, Second Edition. Edited by Hermann J. Koch.

operations) will exist on almost all breakers. More sophisticated monitors are also available and commonly installed, although all available functions are not necessarily used by some users. In recent years, partial discharge (PD) monitoring more frequently used, due to its ability to warn about developing insulation problems in the GIS. Relatively few commercial solutions exist since data analysis often requires expert interpretation.

4.2.1 Gas Monitoring

In GIS, the SF_6 gas provides electrical insulation and, in the breakers, arc-quenching capability. Both of these properties depend on the density of the SF_6 gas. However, gas pressure is often quoted in lieu of density. Various documents often use terms such as "fill pressure" or "normal operating pressure." Pressure is used since it is both easy to measure and is intuitive, but the important parameter is gas density. Pressure depends strongly on temperature but density does not (provided gas state change is not involved).

Gas monitoring is done primarily to ensure that an adequate quantity of SF_6 gas is present to meet the equipment's requirements. This usually means some form of gas density monitoring. Gas density measurements, however, have limitations. Gas density is measured at one location within the GIS system, usually at the enclosure (Figure 4.1). While the gas pressure can be considered constant throughout the gas chamber, there may be density variations resulting from variations in temperature. For GIS systems under load, the gas temperature is higher closer to the central conductor as opposed to at the enclosure, where the gas density monitor is usually mounted. Consequently, a measure of density at the enclosure may tend to overestimate the density at the live parts of the GIS. In some cases, there may also be issues such as convective flow within the enclosure, solar gain (in outdoor installations), and so on, which may affect the distribution of temperature and density within the equipment. In most cases, these density variations are considered secondary and SF_6 density thresholds are established with sufficient margins to take these secondary effects into consideration.

In a practical sense, gas density monitoring systems can be set up to provide one or two types of outputs:

1) Continuous output signal. These can be used to establish trend for diagnostic purposes. For instance, past records could differentiate between a slow leak occurring over a long period of time as opposed to a recently developed leak of greater severity. Trending information can also be used to track SF_6 emissions hence help to meet environmental requirements and regulations. As the SF_6 gas is a powerful greenhouse gas [1], the value of monitoring emissions cannot be understated.

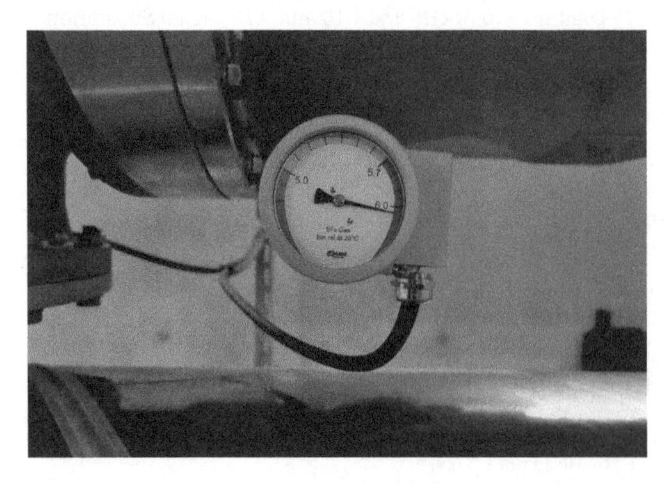

Figure 4.1 Gas density monitor mounted on the GIS enclosure. This type displays temperature-compensated pressure with a simple to read dial and with color codes. Built-in relays are used for signaling when certain thresholds are encountered (Reproduced by permission of Siemens AG)

2) Threshold alarms. An alarm signal is raised when the gas density drops below a certain threshold. Typically, two thresholds are used. The first level is a warning to signal low gas (used to trigger some corrective action). The second level is usually a control signal used to block an operation (of switchgear) or, in some cases, to fully disconnect the affected equipment (depending on the policy of the equipment user). The second level is usually tied to the minimum density required to ensure proper operation of the equipment.

   Different technologies are available to perform these functions. These include:
   - Simple pressure switches. Since pressure is not density, this method is only used in devices with inherently high design margins where the threshold pressures, given the anticipated temperature variation, still ensures sufficient gas density for proper operation. Some medium-voltage devices use such switches, calibrated to one or both of the threshold levels discussed above.
   - Temperature-compensated pressure gauges. A separate temperature indication is used to modify the response of a pressure sensor. These sensors usually have a visible gauge (calibrated to read true pressure or compensated pressure), but the signaling is done via relays or switches set to the two threshold levels of density.
   - Gas monitor with reference gas. Reference gas in a sealed chamber interfaces with the measured gas via a mechanical bellows. Since temperature changes affect both the measured and reference gas equally, changes in pressure also affect both equally and the effect of temperature is eliminated. The bellows will respond to differential pressures related to density and cause microswitches to operate.
   - Direct measure of density. Sensors using tuning forks change their resonant frequency when gas density changes. These sensors provide a continuous (analog) signal, which tracks density. This signal is connected to relays which provide the threshold alarms.

Note that sophisticated monitoring systems that measure pressure and temperature separately could use state equations to compute density. It is also possible to include thermal models to provide better indications of gas density at different parts of the GIS equipment. Such systems have the potential to monitor and quantify small gas leakages more precisely. However, this technique is used with digital substation design, see section 9.18.

4.2.1.1 Gas Monitoring Practices

Most gas density monitors used on GIS are of the "relay" type that use contact closures to signal that certain gas density thresholds have been reached. Visual displays on the devices themselves usually consist of color-coded status indicators, but do not provide a quantified value of density. With these types of devices, low-gas alarms are usually the only form of monitoring available. Manual observation of the indicator status can also be done during inspections. With these systems, SF_6 loss is calculated by tracking the quantity of gas added when low gas is detected. More precise tracking of SF_6 loss is only possible if quantified values of SF_6 density are recorded at regular intervals. Although such records could be maintained manually, the process is only practical with automated systems. With increasing emphasis on tracking SF_6 inventory for environmental reasons, it is expected that automated systems to record numeric values of density will become more common. See section 2.8 and 2.9.

4.2.2 Partial Discharge

Partial discharges (PDs) are small electrical discharge events that occur in some types of electrical insulation systems. In some cases, especially with complex multicomponent apparatus, PD (at a low level) is almost considered a normal occurrence (e.g., in some types of rotating machine insulation). However, in many cases, PD is a symptom of an insulation defect and its presence indicates some form of deterioration. In principle, GIS falls into this category – GIS should be PD free. In practice, limitations in detecting sensitivity factor into this assessment. In addition, some types of low-level PD might not be of consequence within the normal lifetime of the equipment. Consequently, for acceptance testing, upper limits on PD levels are usually adopted. Any PD lower than this level is usually considered to be of little consequence to the long-term health of the equipment [2].

From a monitoring perspective, however, the objectives are slightly different. Monitoring is used to determine whether significant insulation deterioration is developing over time and to help in assessing this situation. Ideally, monitoring is used to help address the following:

- Is there a developing insulation problem?
- Where within the equipment is this problem occurring? What components are suspected?
- How severe is the problem? What are the consequences?
- How much time is available to address the problem?

The first question deals with detection. Since GIS, in principle, should be PD free, any detection of PD above a minimum threshold is construed to be an indication of a developing problem. However, detection needs to be followed up with the second issue of location, especially if the sources of the PD can be identified. PD from different components (e.g., corona shields, insulators, particles) should be interpreted and assessed differently. If the source is identified, an assessment can usually be made (albeit subjectively by an expert) as to its severity. The last question referring to the time availability to solve the problem, would be very valuable to many users (as it allows for planning) but is the most difficult (nearly impossible) to answer, especially as random processes are involved. For instance, some types of insulator defects that generate PD can cause failure almost immediately but might also persist for many months or years without causing problems.

For most monitoring strategies, practical objectives would focus on detection and location, from which some assessment can follow with expert interpretation.

4.2.2.1 Defect Types

In GIS, there are several types of defects that could generate PD [3].

1) Metallic particles. Particles are, by far, the most common type of defect found in GIS. However, particles are mostly an issue during commissioning since they are often introduced during the assembly process. For inservice monitoring, particles are less of a problem. However, on occasion, particles in relatively harmless locations could be physically moved into more active areas through vibration and other mechanical forces resulting from breaker operation. In addition, compartments with moving parts (switchgear) will sometimes generate their own particles over time through wear, especially if some mechanical deficiency (a possible misalignment) exists.

   Metallic particles "free" to move inside the GIS are relatively easy to detect. Movement due to acquired charges in the applied electric field will cause the particles to "bounce" along the enclosure, generating both acoustic and electrical signals on contact. However, particles that become adhered to an insulator surface (either via static charge or excess grease) are particularly dangerous. These particles can, over time, initiate surface tracking on the insulator and lead to failure. Unfortunately, the PD associated with such phenomenon in the early stages are typically low and difficult to detect.
2) Floating components. Internally, many GIS designs utilize shielding to protect certain high-stress areas. These shields must be electrically connected to a conductor of known potential, whether it be the main high-voltage conductor or the grounded enclosure. In some cases, the

contact is established using relatively low force springs, clips, or other means. If the contact is not firm due to damage or contamination, electrical contact resistance will increase (over time), resulting in a shield at floating potential. Partial discharges will often occur between these floating components and one of the other conductors. In the early stages, the PD could be very small and intermittent. However, when fully established, the floating component discharges are usually high and can be detected easily both electrically and acoustically. Floating component discharges, when fully active, are quite hazardous and could cause failure in a relatively short period of time (minutes to days). The constant discharging can cause local decomposition of the SF_6 gas – the resulting corrosive by-products will attack nearby insulators and may cause them to fail. In severe cases, the discharging will also generate conducting and nonconducting particles, as the contact material erodes.

3) Insulator defects. Defects in the insulator could include small manufacturing defects, such as voids, that were missed during manufacturing quality control or were inactive during prior testing. PD from such defects is generally quite small (otherwise they would have been detected previously) but may sometimes cause failure after time in service. Fortunately, manufacturers have taken steps to improve manufacturing quality and defects of this type are rare. Other internal defects, such as internal metallic contamination in solid insulator structures, can also cause premature failure. However, PD resulting from such defects may be extremely small and nearly impossible to detect until some seconds to minutes prior to failure [4]. These defect types are also very rare in modern designs. On occasion, insulators could also become damaged during operation as a result of unusual external forces or thermal stress – depending on the type of damage, detectable PD may also occur.

4.2.2.2 Partial Discharge (PD) Discharge Measurements

PD monitoring requires some form of measurement. In order to perform PD measurements, the small electrical discharge of PD must be detected. As PD is an electrical phenomenon, most measurement methods used are electrical techniques. However, the discharge will also generate acoustic energy and acoustic PD methods are also possible. Over time, the electrical discharges can also decompose the SF_6 gas, in which case the decomposition by-products can be detected by chemical means.

The development of PD measurement systems has been driven by the need to optimize detection sensitivity and the need to interpret the results. The need for greater detection sensitivity and to differentiate a true signal from many sources of noise and interference has led to many of the nonconventional methods discussed below. The most recent developments focus on PD signal classification on a pulse-by-pulse basis in order to separate PD signals belonging to different sources (including interference signals) in order to improve the assessment of the PD measurement. However, some monitoring systems may focus on a simpler detection method intended to detect and indicate a need for more sophisticated measurements done manually.

Electrical Methods

The partial discharge, at the source, occurs very rapidly and can be as little as 1–2 ns in duration. As the coaxial design of GIS can support high frequency signals, it is possible to detect PD pulses with high accuracy. However, as the PD propagates throughout the system, some attenuation and distortion (pulse broadening) may occur.

For high sensitivity, PD detection can be performed using ultrahigh frequency (UHF) methods with detection bandwidths extending to 1000 MHz or more [3]. However, as the highest frequency components suffer the highest propagation loss, UHF methods may be limited to cases where the sensor is in close proximity to the defect source (typically within 10–20 m). Another approach is to use a lower frequency band (up to a few hundred MHz), which offers a good compromise between sensitivity (i.e., signal-to-noise) and sensor placement. Both of these are considered to be advanced methods and may require sensors specially adapted to the GIS.

Conventional PD measurement methods use a reduced bandwidth of about 100 kHz. These methods are fully described in standards (IEC 60480) and are frequently used on GIS components and some subassemblies. However, because of challenges in achieving a good signal-to-noise ratio, these methods are less suitable for large assemblies and field testing.

Acoustic Methods

The micro discharges associated with PD will release acoustic energy in addition to electrical signals. The acoustic waves generated by PD occurring in the SF_6 gas will transfer energy to the GIS enclosure – the signals can be detected on the enclosure using acoustic emission (AE) sensors. Acoustic methods make it difficult to relate PD to commonly used electrical quantities (such as pC). However, the technique has been used successfully to assess the condition of GIS in the field [3].

Acoustic PD measurements will require some expert interpretation as the signal magnitude does not always correlate with the defect severity. For example, discharges occurring within solid insulation (such as in a void) are very difficult to detect, as the insulator will often attenuate acoustic signals. On the other hand, acoustic methods are extremely sensitive to metallic particle contamination. When voltage is applied to the GIS, metallic particles will often elevate and "dance" inside the GIS. Although PD will occur as the particles discharge to other metallic structures, the physical contact of the particle against the enclosure generates an easily detected and distinctive acoustic signal.

A related method involves the use of a portable ultrasonic detector with a contact probe. The detector's metallic probe tip is pressed on the GIS enclosure to pick up acoustic signals. The detector's output is translated electronically to the audible range and fed into headphones for the operator's use. This method still requires expert interpretation for many types of discharge signals but is relatively easy to use for metallic particle detection. It has the added advantage in that the probe is easily moved from location to location, allowing large sections of GIS or GIL to be scanned quickly.

Acoustic PD testing is undertaken during the application of AC high voltage using a test generator. During the test, the probability of test flashover is elevated. Such a breakdown during testing will cause a momentary transient voltage on the grounded enclosure (VFT and TEV, see Section 9.5), which could cause an electrical shock to a person using an ultrasonic detector, whereas the operator of fixed-sensor systems is usually isolated from the GIS enclosure. Previous experience suggests that the danger is primarily one of being startled and not a direct hazard to health and safety. When using a handheld sensor, the risk of a shock can be reduced by minimizing the contact time with the GIS enclosure. If the measurement is applied during conditioning, waiting a few minutes at each voltage level prior to the start of the test is advisable.

Chemical Techniques

Partial discharges that occur in the SF_6 gas will cause the gas to decompose and generate by-products in trace quantities. Consequently, the detection of these by-products can be used to "detect" the presence of PD. As the rate of production of the by-products is small, this form of detection is only suitable for diagnostic purposes in service and not as a short-term testing tool. An extended period of time (weeks, months, years) is usually required to provide measurable results.

The chemistry of SF_6 decomposition can be complex but the most commonly analyzed by-products are [5]:

- Thionyl fluoride (SOF_2)
- Sulfuryl fluoride (SO_2F_2)
- Sulfur dioxide (SO_2)

Measurements can be made either by taking SF_6 gas samples that are sent to a laboratory for analysis or by using some form of portable sensing equipment. The first approach is analogous to the dissolved gas analysis performed on transformer oils. Laboratory analysis will usually provide by-product levels to a few parts per million (ppm).

Portable instruments can be of a kind that makes use of chemically sensitive detector tubes that change color in the presence of certain gases. By controlling the time and flow rates through these tubes, quantitative assessments can be done. Typically, SO_2 detector tubes are used, as most of the other by-products will further decompose into SO_2 in the presence of trace quantities of moisture. More recently, a number of commercial instruments have become available that provide similar functionality. Many of these detect hydrogen fluoride (HF), which is a by-product formed by secondary decomposition of the products, are listed above.

Laboratory analysis, in general, provides better information as individual and specific by-products are analyzed. However, portable instruments offer more rapid assessment at the expense of a simpler measurement. In principle, GIS equipment should be PD free and therefore free of SF_6 decomposition by-products. Detection sensitivity to a few ppm is adequate for this purpose. Chemical methods have numerous issues, which need to be taken into account:

- Gas compartment volume. Large volumes will "dilute" by-product concentrations and the sensitivity will be reduced.
- Switching compartments will generate by-products "normally" making PD detection nearly impossible in these compartments.
- Some compartments are equipped with absorbing materials (desiccants, molecular sieve, etc.). These will absorb the by-products generated by PD and interfere with the analysis.
- Internal failures (faults) will generate large quantities of by-products. PD analysis on faulted chambers cannot be done.

As the by-products generated by PD are similar to those generated by internal failures, the same techniques and equipment can usually be used for both as long as the differences in by-product levels are considered. By the same token, as decomposition by-products are highly toxic, the precautions used for gas analysis for faulted chambers should also be considered for PD detection.

4.2.3 PD Monitoring Strategies

Complete systems used to perform PD monitoring can be complex as multiple sensors are required at close proximity for the best coverage of the GIS. Electrical PD sensors are usually specially designed couplers (antenna) designed as part of the GIS equipment (Figure 4.2). As a compromise between sensitivity and cost, a spacing of no more than 20 m between couplers is usually recommended [3]. If couplers are a part of the initial design, the incremental cost can be low, but retrofitting couplers on existing equipment can be much higher. Some specialized, easy-to-install couplers that take advantage of existing electrical apertures in the GIS system (for instance, at insulated flanges, viewports, etc.) are available but, in general, their response may be less than optimal.

Figure 4.2 Photos showing a variety of couplers used for PD detection in GIS. These couplers are customized for various flange openings and are designed not to compromise the insulation of the GIS. Some designs (not shown) integrate couplers into insulators and other devices (Reproduced by permission of Kinectrics Inc.)

The availability of couplers can be an issue, but a greater issue is how they might be used for monitoring purposes. Each coupler could be:

- Permanently wired into a central measuring device. Owing to the distances involved and signal attenuation, measurement bandwidth might be limited.
- Permanently wired to a nearby local measuring device. Multiple measuring devices would be required for station coverage. In addition, a method of communication from each device to a central hub might also be required.
- Used with portable equipment. Monitoring is done manually at intervals determined by the user.

In all cases, the data collected require expert interpretation to derive value. As with many monitoring systems, gathering data and information has not been an issue, but drawing meaningful conclusions to base operating decisions is challenging.

The above comments generally apply to electrical measurement-based monitoring systems. The same issues apply with acoustic measurement-based systems but, since acoustic sensors are easily mounted almost anywhere on the outside of the enclosure, acoustic systems are primarily used as portable equipment.

A fully implemented monitoring system can be complex and could be costly. Implementation decisions are often based on the perceived value of the expected outputs and the cost/benefit at that particular installation. At the most critical installations, a full monitoring system might be warranted. However, in other cases, a user might choose to only install monitoring on limited portions of the installation. Another approach would be to use a less invasive monitoring approach, such as periodic gas sampling or occasional surveys with portable equipment, and only apply continuous monitoring when problems are suspected. However, with increased reliability of modern GIS equipment, the need for PD monitoring is reduced and many users choose not to implement such systems. With the digital sensors as part of the digital substation PD monitoring is integrated in the substation operation and maintenance concepts, see section 9.15.

4.2.4 Circuit Breaker Monitoring

GIS that includes switchgear components, such as circuit breakers, may implement some form of circuit breaker monitoring. A number of redundant circuit breaker monitors are available commercially, however, are not exclusively for GIS breakers. These devices range from simple add-on devices to sophisticated systems fully integrated into SCADA with web-based user interfaces.

Most breaker monitoring systems focus on parameters related to contact wear or mechanical aspects of breaker operation. Typical monitored parameters may include:

- Operation counter
- Arc interruption time
- Breaker timing (open/close times) values
- Accumulated fault current (cumulative It or I^2t)

Some systems may also integrate other functions, such as gas monitoring, within the same package. In many cases, manufacturers will include a monitoring device with the breaker, which users will integrate (select functions) into their operations. As with many monitoring systems, a tremendous amount of potentially useful data are available. However, the user must analyze the data to extract useful information. An example of this is the I^2t monitoring as a measure of contact wear. Manufacturers might supply guidelines correlating accumulated I^2t values to the need for maintenance. However, for best results, users should develop their own criteria gained through gathering their own data and experience, especially as operational conditions could vary from user to user.

4.2.5 Other Monitoring

In addition to the above, other forms of monitoring could be found in GIS. These could include:

- Monitoring by video camera of disconnector and ground switch positions via viewports. Some users will use video imaging in viewports to confirm open/closed positions of switch contacts. This would be done in addition to other signals normally used with SCADA systems.
- Thermal monitoring (temperature, infrared) of GIS and conductor contacts. Although IR thermography might not be able to resolve contact problems on enclosed parts, general trends in temperature might be capable of detecting some forms of unusual heating patterns. Permanently installed infrared (IR) systems are also useful in locating ground faults in GIS, as high current faults will cause localized heating of the enclosure detectable for several minutes following the fault.
- Monitoring of air ventilation systems for indoor installations. In some cases, SF_6 detectors are used to monitor and automatically trigger forced ventilation for indoor installations. This approach addresses the case of enclosure burn-through or operation of pressure-relief devices (following an internal power arc fault), in which case a large quantity of SF_6 gas and possible toxic decomposition by-products could be released into the ambient.

Many monitoring schemes take the form of a measured quantity "hard-wired" to provide a specific action. However, recent trends involve the acquisition of data and information, used to gather intelligence on a particular aspect of GIS operation. With modern technologies, the latter form of monitoring has become feasible with tremendous potential to provide useful benefit to the user. However, such systems are only useful if time and effort are invested in analyzing the information gathered.

4.3 Local Control Cabinet

4.3.1 General

Each circuit breaker of the gas-insulated substation (GIS) is provided with a control cabinet for local control and monitoring of the respective bay and is generally placed in front or adjacent to their GIS bays depending on the voltage level (see Figure 4.3 for an indoor cabinet and Figure 4.4 for an outdoor cabinet).

Figure 4.3 Indoor local control cabinet (Reproduced by permission of Phoenix Electric Corp.)

Figure 4.4 Outdoor local control cabinet (Reproduced by permission of Phoenix Electric Corp.)

The control cabinet is metal enclosed, free standing, made of sheet steel, and provided with a lockable hinged door and door-operated lights. The local control cabinet has all necessary control switches, local/off/remote lockable selector switches, close and open switches, measuring instruments, all position indicators for circuit breakers, disconnect switches and grounding switches, alarms, mimic diagram, AC and DC supply terminals, control and auxiliary relays, and so on. The cabinet is fully designed as per IEC 60439 [47] or IEEE C37.123 [48].

The control cabinet is designed in such a way as to facilitate full and independent control and monitoring of the GIS locally. All electronic components inside the bay control cabinet are designed to work satisfactorily for the specified project requirement. Arrangement and components in control cabinets can vary depending on user preferences. Below are typical requirements and arrangements. At least 20% of each spare contact (NO (normally open) and NC (normally closed)) are provided with an auxiliary relay for future use.

All CT secondary taps should be wired to the local control cabinet. The CT terminal block is such that it will provide isolation and testing facilities for CT secondaries at the cabinet. For multiratio CTs, the terminal block is provided on the LCC as per IEEE C57.13 [49] to facilitate connection of various taps. Facility is provided in the LCC for shorting and grounding of secondary terminals.

Potential transformer (PT) secondary windings are terminated at the local control cabinet through a terminal box. For PT wiring in the LCC, each phase of each circuit is provided with a miniature knife switch and a high rupturing capacity (HRC) fuse/supervised mini circuit breaker (MCB). Knife switches are located on the PT side of fuses. Separate terminals are provided for PT fuse supervision.

The control cabinet is equipped with a mimic diagram on the front of the cabinet showing (see Figure 4.5):

Figure 4.5 Mimic diagram (Reproduced by permission of Phoenix Electric Corp.)

a) A mimic diagram showing the arrangement of electrical equipment in the bay including bus bar isolating links.
b) Control switches and local/off/remote changeover (lockable) for operation of all circuit breakers, disconnect switches, and grounding switches.
c) Position indicators showing the position of all circuit breakers, disconnect switches, and grounding switches.
d) Overriding interlock switch between disconnects and grounding switches associated with circuit breakers (depending on the user's requirements).
e) SF_6 gas zones.
f) The color of the mimic bus should be according to the user's requirements.

The cabinet is equipped with a thermostatically controlled anticondensation space heater along with a cabinet light, door switch, safety shrouds, and receptacle. The arrangement of equipment within cubicles is such that access for maintenance or removal of any item should be possible with minimum disturbance of an associated apparatus.

All control power circuits are protected by miniature circuit breakers in each cabinet. Other circuits supplying loads, such as heaters, receptacles, or lights, have separate overload protection.

The cabinet is grounded with a suitable copper bus and the hinged door of the cabinet is grounded by a flexible grounding connection.

Alarm/annunciators are of the window type as per IEC 60255 [50] or IEEE C37.1 [27], with a minimum of 20% spare windows for use. The alarm/annunciator system is designed for continuous operation of all alarms independently and simultaneously. A view inside a local control cabinet is shown in Figure 4.6.

Figure 4.6 Local control cabinet, door open view (Reproduced by permission of Phoenix Electric Corp.)

The following minimum alarm is provided as a local alarm in the LCC:

- SF_6 gas pressure Low–Low, Stage 1 alarm for each gas zone/section (in the case of a single phase, an alarm is provided for each phase)
- SF_6 gas pressure Low–Low, Stage 2 alarm for each gas zone/section (in case of a single phase, the alarm is grouped for all phases)
- Excess run time of the motor for the circuit breaker, disconnecting switch, and ground switch
- Spring overcharged for the circuit breaker mechanism
- Loss of DC for the trip and close circuit
- Circuit breaker trip
- VT supply fail (VT MCB trip)
- Loss of AC supply
- Circuit breaker mechanism failure
- Local/remote switch
- Pole discrepancy operated (for single-phase breaker)
- Trip circuit failure
- Loss of DC supply to circuit breaker motor

The cable connections to the local control cabinet are shown in Figure 4.7.

4.3.2 Bay Controller

The bay controller unit is installed in the local control cabinet (LCC). There is hard wiring from the GIS to the LCC/bay control unit including CT and VT wiring.

The high-voltage equipment within the GIS is operated from different places with a predetermined hierarchy:

- Local control panel with mimic
- Bay control unit (control IED)
- Station human–machine interface (HMI) (micro SCADA)
- SCADA master station

Bay level functions include data acquisition and data collection functionality in bay control intelligent electronic devices (IEDs).

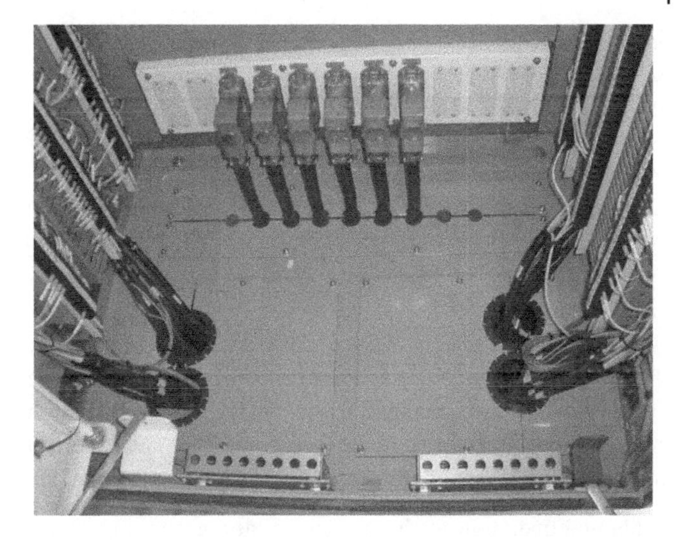

Figure 4.7 Local control cabinet, cable termination view (Reproduced by permission of Phoenix Electric Corp.)

The following basic functions are included in the control unit:

- Control mode selection
- Interlocking and blocking
- Double command blocking
- Auto reclosing
- Synchrocheck and voltage selection
- Motor excessive run
- Monitoring pole discrepancy and trip function
- Measurement display
- Alarm indication
- Display of interlocking and blocking
- Device position indication (circuit breaker, disconnect switch, ground switch)
- Transformer tap change control and indication
- Data storage interface to the station level
- Bay control unit (BCU) interlocking and blocking

Some applications use software/GOOSE (generic object orientated substation event) interlocking control is through bay control IEDs. In the case where an "interlock override" feature is provided as part of the "GOOSE" scheme, privileged users can only access it using a strong password and other security features.

Double operation interlocking is made in a GOOSE design where separately dedicated IP address/subnets are allocated for each voltage level in each substation.

4.3.3 Control Schemes

4.3.3.1 Control Mode Selection

In this mode, the operator receives the operation access at bay level and allows the operation of all switching devices through control IED. Operation is done generally through the local HMI.

Off Mode

In this mode, it is not possible to operate any devices, neither locally nor remotely.

Local (BCU) Mode

Operation is done from the BCU directly and operation from other places (e.g., HMI/REMOTE) is not possible in this operating mode.

Remote Level Mode

Control in this mode is possible from the highest level (SCADA master station) and the installation can be controlled remotely via the station HMI. Operation from lower levels is not possible in this operating mode.

4.3.3.2 Interlocking

The following interlocking scheme is incorporated inside the cabinet for reasons of safety and convenience of operation and also to prevent incorrect switching sequences that could lead to a hazardous situation to plant, equipment, or personnel. The electrical interlocking is effective under both local and remote operations. The following are some typical requirements for interlocking:

- Manual operation of the disconnect and grounding switches is only possible under electrical interlock release conditions. Some applications use a key switch for overriding interlocks between disconnect and grounding switches associated with circuit breakers during maintenance is provided in the control cabinet.
- Mechanical and electrical interlock between disconnect and grounding switch operation is provided.
- The electrical interlock scheme is fail-safe to prevent loss of interlock function upon loss of control voltage.
- Electrical interlock between the line PT secondary voltage and respective high-speed grounding switch operation is provided through undervoltage relay contacts.
- The feeder grounding switch is interlocked with the corresponding circuit breaker and disconnect switch.
- The bus bar grounding switch is interlocked with all disconnect switches on the same bus bar section.
- The high-speed grounding (fast-acting grounding) switch is interlocked with the associated circuit breaker open.

4.3.3.3 Synchronism and Voltage Selection

The synchrocheck function is bay-oriented and depends on voltage, phase angle, frequency, and live line-live bus. Determination of live line/dead line or dead bus/live line is done at the IED level for particular bays with associated circuit breakers and disconnect and grounding switches.

The correct voltage for synchronizing is derived from the auxiliary switches of the associated circuit breaker, disconnect switch, and ground switch and related VTs. Automatic selection is done by the bay control unit IEDs or through VT selection relays in the case of the conventional LCC.

4.3.3.4 Autoreclosing and Related Synchrocheck Functions

Autoreclosing and synchrocheck functions are generally performed through the bay control IED or separate relay or are built into a protection IED.

The autoreclosure is settable for the following modes:

- Three-phase autoreclosure
- Single-/three-phase autoreclosure
- Single-phase autoreclose

The three-phase autoreclosure sequence can perform with or without the synchrocheck.

4.3.3.5 Pole Discrepancy Monitoring (for Single-Pole CB)

All single-pole circuit breakers are equipped with a pole discrepancy protection scheme. The pole discrepancy protection for CB is of the two-stage type. The pole discrepancy monitoring function is provided based on measurement of phase overcurrents and current differences between phases.

4.4 Digital Communication

4.4.1 General

Digital communication in substations covers high-voltage switchgear and controlgear and assemblies. The basic standard for digital communication protocols is IEC 61850 [6]. In this standard, the communication architecture is defined and finds world-wide application. In this GIS-related book, the focus of digital communication is set to the requirement of GIS when specifying on-site commissioning and testing. The information given here shall help the substation planning engineer of GIS to get a better understanding of requirements related to digital communication. Specific information for digital communication in substations IEC 62271–3 [7] has been published.

Digital communication today has spread all over the world and mainly in Asia and Europe almost all GIS is equipped with these technologies. The benefits of digital communication are seen more often in substation projects, where nonconventional instrument transformers will be applied as a consequence to make the best use of digital communication within the near future. Copper wiring will be replaced by optical fibers and field bus systems. Interchangeability between manufacturer designs has been provided for years on the basis of standards and CIGRE coordinating work groups.

GIS will change its face completely and with this the required knowledge of power and communication engineers will need to be adapted. Not only will the hardware and software change but also the language of the engineers will change to adapt this new technology into GIS. The new digital communication world will have to coexist alongside the old conventional way of inductive current transformers and voltage transformers and copper wiring for a long time. The guideline for the new digital communication world is given by the IEC 61850 series as a horizontal standard for the communication equipment, high-voltage switchgear and assemblies, controlgear, and the relevant testing requirements.

Specific rules and requirements mainly for on-site testing of GIS is covered by IEC 62271–3 [7] and will be explained here. This international standard deals with relevant aspects of high-voltage switchgear and controlgear and assemblies thereof, with serial digital communication interfaces according to IEC 61850 [6]. In this standard, those information models used in GIS are explained.

4.4.2 Basic Digital Communication Standard

This standard IEC 61850 [6] covers the communication networks and system requirements as they are used for power utility automation. The series of IEC 61850 has more than 10 parts and more than 1000 pages in total. The field of application is for substation automation systems, and it defines the digital communication between so-called intelligent electronic devices (IEDs) in the substation and the related system requirements. Table 4.1 lists the exemplary requirements related to GIS.

In Parts 1 and 2, a general introduction for digital communication is given. Here the structure of IEC 61850 is explained as well as the principles of data structures and concepts. In Part 3, general requirements for the hardware, the software, and the devices used in the substation requirement are defined. In Parts 4, 5, and 6, the requirements for device models, the configuration language, and system and project management are given. Parts 8, 9, and 10 give information on the basic

Table 4.1 IEC 61850 series on communication networks and systems for power utility automation. Exemplary list of GIS related standards

Part	Title
1	Introduction and overview
2	Glossary
3	General requirements
4	System and project management
5	Communication requirements for functions and device models
6	Configuration description language for communication in electric substations related to IEDs
7-1	Basic communication structure for substation and feeder equipment – principles and models
7-2	Basic communication structure for substation and feeder equipment – abstract communication service interface (ACSI)
7-3	Basic communication structure for substation and feeder equipment – common data classes
7-4	Basic communication structure for substation and feeder equipment – compatible logical node classes and data classes
7-410	Basic communication structure for substation and feeder equipment – hydroelectric power plants – communication for monitoring and control
7-420	Basic communication structure for substation and feeder equipment – basic communication structure – distributed energy resources logical nodes
8-1	Basic communication structure for substation and feeder equipment – specific communication service mapping (SCSM) – mappings to MMS (ISO/IEC 9506, Part 1 and Part 2) and to ISO/IEC 8802-3
9-2	Basic communication structure for substation and feeder equipment – specific communication service mapping (SCSM) – sampled values over ISO/IEC 8802-3
10	Basic communication structure for substation and feeder equipment – conformance testing
90-1	Use of IEC 61850 for communication between substations and technical reports (TRs) with numbers like IEC 61850-80-x and IEC 61850-90-x

information structure of different equipment used in substations. Here the data models are defined with data classification. Special information is given for hydroelectric power plants and distributed energy resources.

Special communication service mapping related to ISO/IEC 9506, Part 1 and Part 2 [8, 9] and to ISO/IEC 8802-3 [10] is explained. The latest Part 90-1 covers digital communication between substations, and there are more parts coming under the series number −80 and −90.

The IEC 61850 series is more than a software protocol; it is a way to model the substations in digital models, an engineering process. On the station bus and process bus levels, the IEC 61850 is used to define models of equipment and models of functions and to document them in a standard format for easy exchange with the client server's interactions.

On the devices level, specific standards are available for high-voltage switchgear (IEC 62271-3) [7], current transformer, and voltage transformer (IEC 61869) [11].

In Figure 4.8, an overview is given on the relationship to other IEC standards for GIS. The graphic in Figure 4.8 shows the principal function of a GIS within the dotted line on the left side.

The connection of the switching devices to the process bus is defined in IEC 62271-3 [7], and the connection of voltage and current transformers is defined in IEC 61869 Part 9 [12] and Part 13 [13]. The switching devices in a substation use Part 8-1 of IEC 61850 for the GOOSE models and the voltage and current transformer use Part 9-2 of IEC 61850 for sampled measuring values. Conventional interfaces are connected by the feeder protection standard IEC 60255 series.

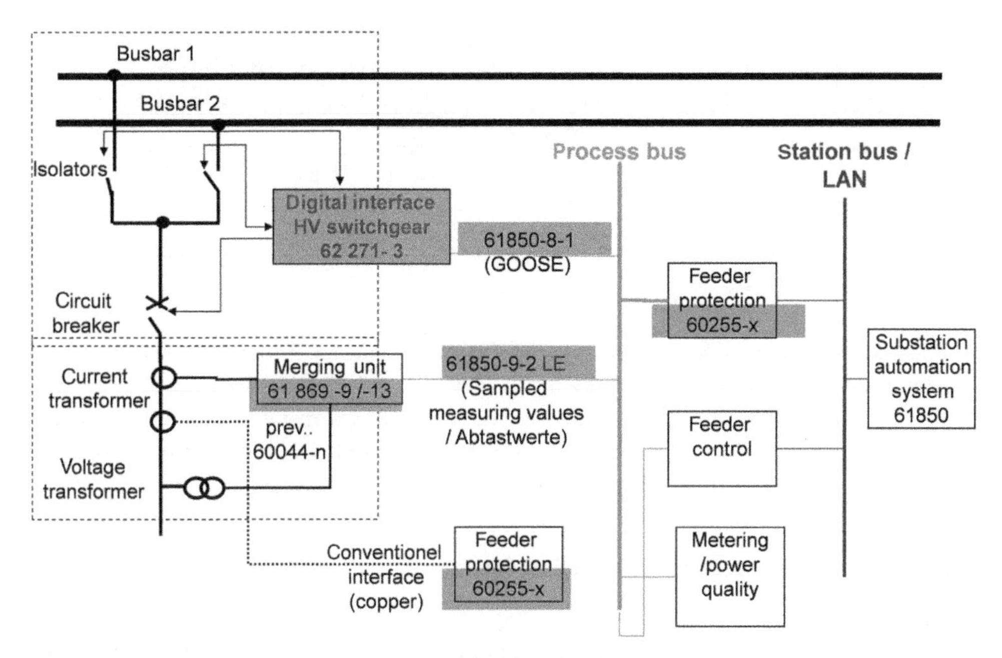

Figure 4.8 IEC 61850 relation to other IEC standards (Reproduced by permission of IEC) (IEC 62271-3 ed.1.0 "Copyright © 2013 IEC Geneva, Switzerland. www.iec.ch")

4.4.2.1 Communication Requirements (Part 5)

In Part 5, the communication requirements of all known functions are defined. A description of functions is used to identify requirements for communication between IEDs within the substations, between substations, and between substations and higher level remote operating places and interfaces to remote technical services. The goal is to have a seamless communication system for the overall power management system and for interoperability between devices of different manufacturers. The categories of functions are shown in Table 4.2.

All functions of a substation need to be identified, and it is required to get the communication requirements within the substation. Complex functions have to be split into pieces with indivisible core functionality. These core pieces are allocated to high-level data objects (logical nodes) containing all data to be exchanged (PICOM) (see Figure 4.9).

Table 4.2 Categories of functions

Category of function	Type
System support functions	Network management, time synchronization
System configuration or maintenance functions	Node identification, software and configuration management, system security management, test mode . . .
Operational or control functions	Access security management, control, synchronous switching (point-on-wave switching), alarm/event management, and recording . . .
Bay local process automation functions	Protection functions (overcurrent, distance), bay interlocking, measuring/metering, and power quality monitoring
Distributed process automation functions	Station-wide interlocking, distributed synchrocheck, breaker failure, load shedding, automatic protection adaption (e.g., reverse interlocking), automatic switching sequences . . .

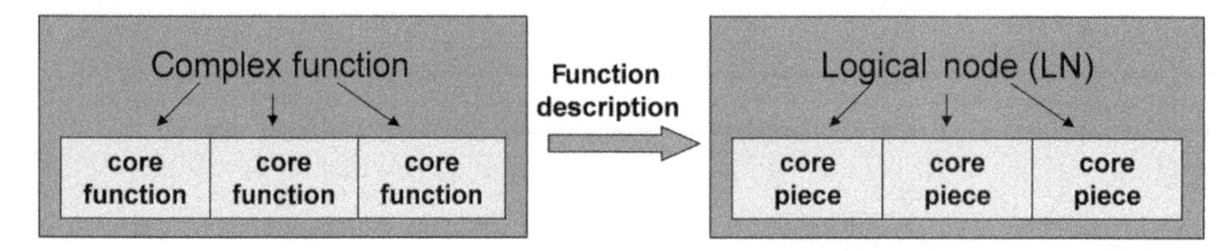

Figure 4.9 Complex functions to be transferred in core pieces of logical nodes (LNs) (Reproduced by permission of Siemens AG)

The function description provides the information shown in Table 4.3. A logical node (LN) function is an abbreviation or acronym as defined in IEC 61850-5 with the systematic syntax in IEC 61850 focused on functional requirements. A logical node (LN) class is an abbreviation or acronym as defined in part IEC 61850-7-4 with the systematic syntax used in IEC 61850 focused on object-oriented modeling. In Figure 4.10, logical nodes are shown in principle for a GIS bay.

4.4.3 Switchgear-Related Communication Standard

4.4.3.1 General

Based on the basic standard for digital substation communication IEC 61850 series [6], a standard IEC 62271-3 [7] for high-voltage switchgear digital communication with other parts of the power utility automation and its impact on testing has been developed. In these, the digital communication

Table 4.3 Information of function description

Function	Description
Task	Names the task
Starting criteria	Gives criteria to start
Result or impact	For example, switching a breaker, trigger another function
Performance	For example, total requested response time
Interaction with other functions	If other function is needed
Function decomposition	How a function can be decomposed in logical nodes (LN)

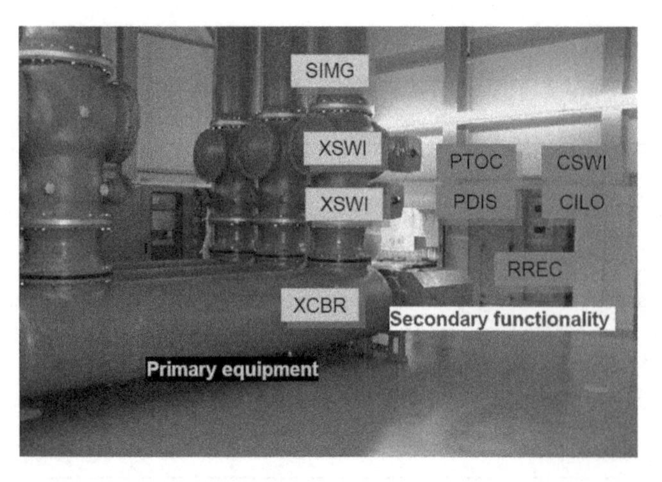

Figure 4.10 Logical nodes of a GIS bay (Reproduced by permission of IEC) (IEC 62271-3 ed.1.0 "Copyright © 2013 IEC Geneva, Switzerland. www.iec.ch")

requirements of circuit breakers, disconnecting switches, and grounding/earthing switches are defined. In Clause 5.3, the timing requirements for the opening and closing function of the circuit breakers are given, and Clauses 7 and 8 cover the type test and routine test procedures. Standard IEC 62271-3 covers all high-voltage switchgear above 1 kV. This includes medium-voltage equipment of 1 kV up to and including 52 kV is included in this standard.

4.4.3.2 Location of Controls

The control and communication device's location may vary with manufacturer design and users' preferences. To simplify the readability of the standard, the design shown in Figure 4.11 has been chosen as an example for the GIS.

A typical configuration of GIS with switchgear controllers and communication devices as shown in Figure 4.11 has a circuit breaker controller (CBC) and a disconnector or ground/earth switch controller (DCC) for the three-phase pole arrangement.

The CBC typically implements the logical node XCBR for the circuit breaker control, and the DCC typically implements the logical node XSWI for the disconnector or ground/earth switch control, as shown in Figure 4.10. Sensors for monitoring and diagnostic for partial discharge are integrated in the GIS.

Bay control functions, such as bay interlocking and local human machine interfaces, may also be located inside the GIS control cubicle. Their communication link to the station or remote level, which is outside the scope of IEC 62271-3, is covered in IEC 61850.

The interconnection between switchgear controllers and other substation equipment is done via serial communication links. A typical example of the communication network inside a GIS is shown in Figure 4.12.

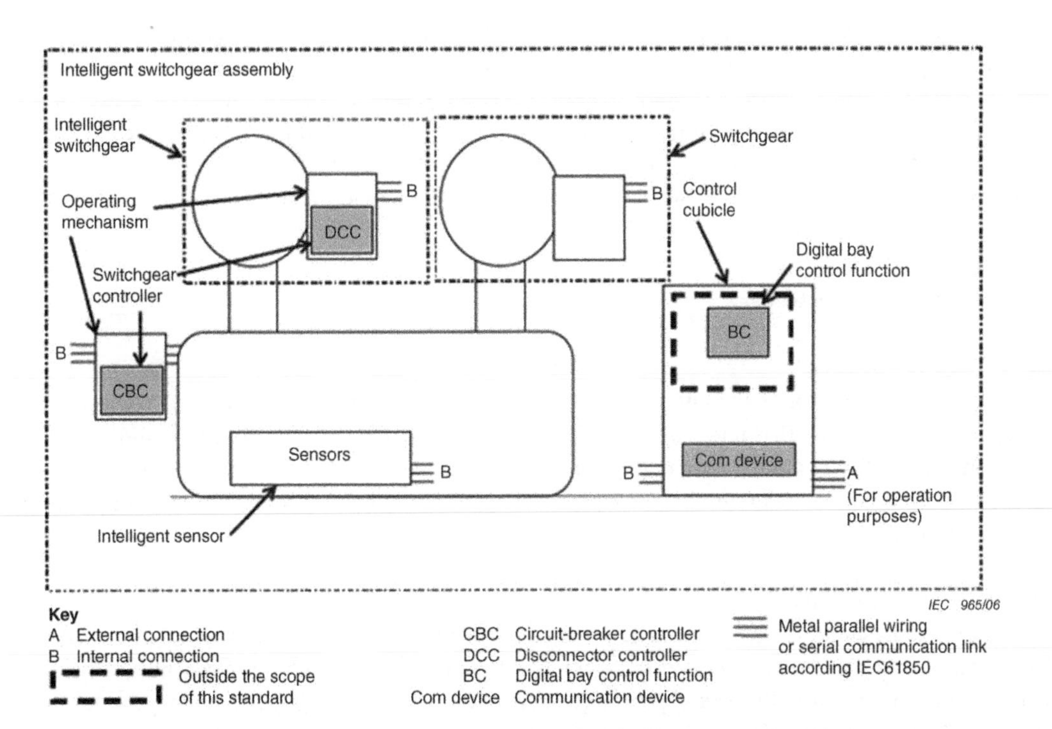

Figure 4.11 Location of control and communication devices of GIS (Example 1) (Reproduced by permission of IEC) (IEC 62271-3 ed.1.0 "Copyright © 2013 IEC Geneva, Switzerland. www.iec.ch")

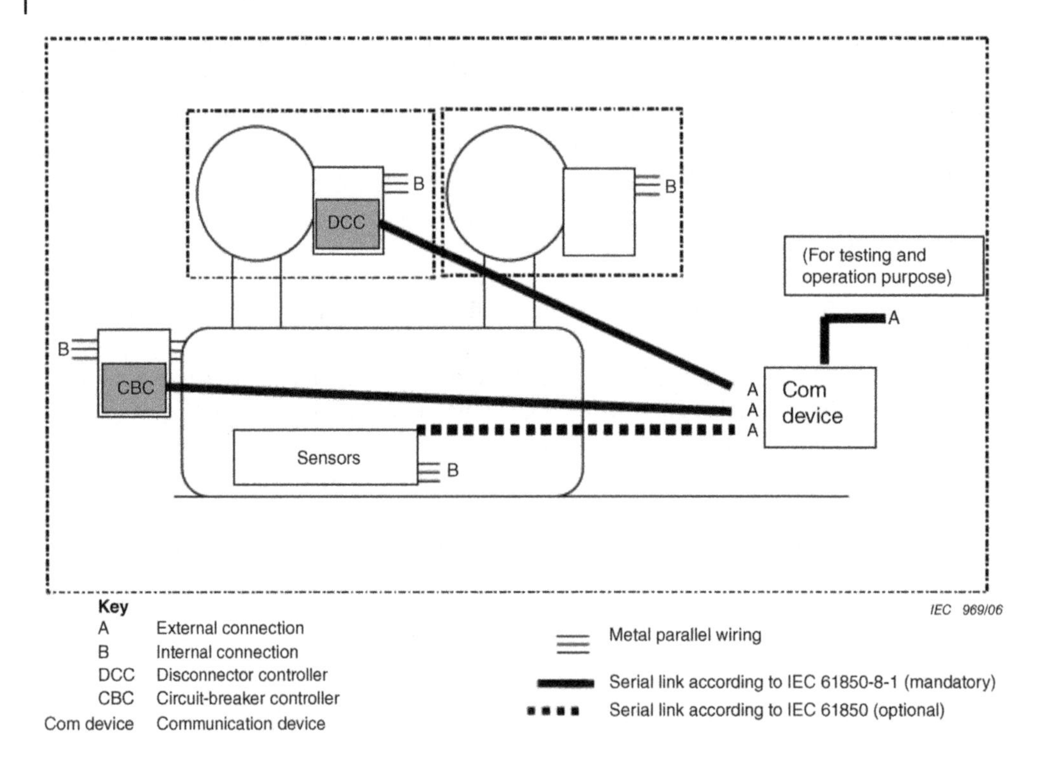

Figure 4.12 Typical example of a communication network inside a GIS (Reproduced by permission of IEC) (IEC 62271-3 ed.1.0 "Copyright © 2013 IEC Geneva, Switzerland. www.iec.ch")

A switchgear controller may have type A or B external or internal connections as defined in IEC 62271-1 [14] (see A and B in Figure 4.12). The interface point "A" can be located on a part of the relevant communication equipment "com device" or directly at the switchgear controller "CBC" or "DCC." Any external connection for testing and operation purposes should be in accordance with IEC 61850-8-1. An internal connection type "B" as defined in IEC 62271-1 for a switchgear controller should be in accordance with IEC 61850-8-1. External connections should be available by means of a communication device.

4.4.3.3 Operation of Switchgear

Opening/Closing Command

The principle of operation of switchgear is shown in Figure 4.13 for a circuit breaker. The serial input generator sends the first telegram of a GOOSE message containing the trip command to the circuit breaker controller at the operating mechanism. The digital telegram is then translated to the electric trip command to operate the circuit breaker of the intelligent switchgear. Opening and closing commands are executed.

Calculation of Operation Time

The principle of calculating the operating times of an intelligent switchgear is shown in Figure 4.14. The total processing time covers the period when the digital telegram reaches the communication device (com device) until the switchgear has been operated. The intelligent switchgear total operating time is a mechanical fixed time, related to the design and the switchgear standards requirements, with IEC 62271-100 [15] for circuit breakers and IEC 62271-102 [16] for

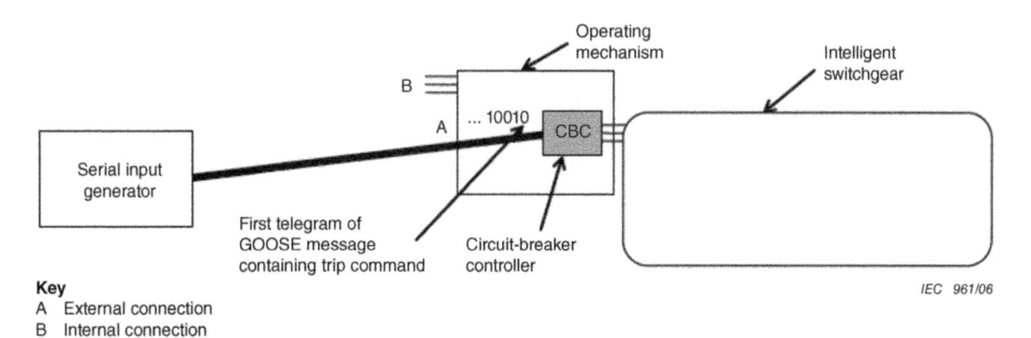

Figure 4.13 Opening/closing command to intelligent switchgear (Reproduced by permission of IEC) (IEC 62271-3 ed.1.0 "Copyright © 2013 IEC Geneva, Switzerland. www.iec.ch")

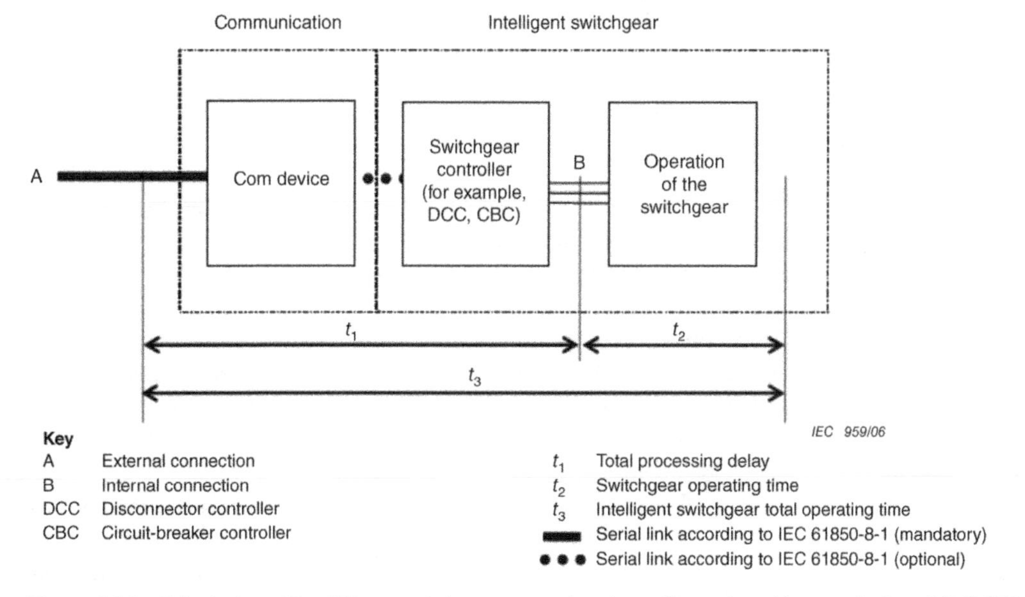

Figure 4.14 Calculation of intelligent switchgear operating times (Reproduced by permission of IEC) (IEC 62271-3 ed.1.0 "Copyright © 2013 IEC Geneva, Switzerland. www.iec.ch")

disconnecting/earthing switches. The total processing delay time covers the time from the arrival of the digital telegram at the com devices until the operation command (trip or closed) arrives at the internal connection point "B."

Measuring Operation Time

The method explained in IEC 62271-3 to measure the operation time of high-voltage switching devices of intelligent switchgear is shown in Figure 4.15.

A switch is connected in parallel to the serial input generator and communication analyzer. The total processing delay time is measured by the communication analyzer (incoming signal) and the time at the internal connecting point "B" (outcoming signal). To determine the operation time of the switchgear Test Message 1 is used, which corresponds to the frame representing the sampled value of one measuring point (four currents and four voltages), as defined in IEC 61850-9-2. A test message 2 may be used as defined in IEC 61850-8-1 where the transfer request is started for a transfer file with a length of 2 MBytes at high background data traffic.

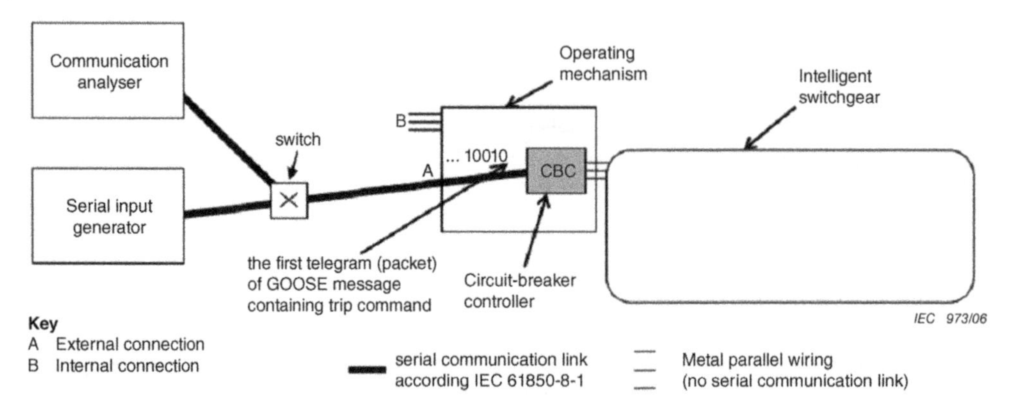

Figure 4.15 Measurement of the operating time (Reproduced by permission of IEC) (IEC 62271-3 ed.1.0 "Copyright © 2013 IEC Geneva, Switzerland. www.iec.ch")

4.4.3.4 Type Tests

The type tests for digital communication of high-voltage switchgear, as defined in IEC 62271-3 [7], explain the test procedures required to prove the full functionality at the required timing for operation of the manufacturer's design GIS.

The tests described in this clause require an involvement of experts familiar with the testing of switchgear, especially the time measurement of circuit breakers, and of experts familiar with serial communication in substations, especially the IEC 61850 series standards. The relevant switchgear product standards of the IEC 62271 series are applicable in general. Digital interfaces should be taken into account where applicable.

This test is done in order to verify the correct behavior of an IED by the use of system tested software under the environmental test conditions corresponding to the technical data of the equipment being tested in IEC 61850-10 [17]. The purpose of the tests described in this clause is to demonstrate that the opening and closing times are within the rated limits [7].

In the configuration shown in Figure 4.14, commands (e.g., opening or closing commands) are directly sent from the serial input generator to the switchgear controller (circuit breaker controller). These commands are sent via GOOSE messages. The relevant return indications are captured by a communication analyzer, via the serial communication link [7].

The input generator for the type test generates the commands and data for background traffic of the communication network. The test message has a rate of 8 kHz, which represents the load of two measuring points sampled at 4 kHz each. Then there is a repeated request of a file upload from the device under test. The background traffic should be applied at least one minute before sending the command [7].

The test consists of two parts. One part compares the opening time rated by the manufacturer against the measured time. The second test compares the closing time rated by the manufacturer against the measured time. Both tests should be repeated five times.

In the configuration shown in Figure 4.10, the test consists of sending commands (e.g., opening or closing commands) from the serial input generator to the switchgear controller of the equipment under test via the communication network. Those commands are sent via GOOSE messages. The relevant return indications are captured by a communication analyzer.

In addition to the background traffic, the test generator initiates command to send a Test Message 1 with 8 kHz, which sends a repeat request to the device for a file upload, while also sending repeat file upload requests to another switchgear controller. The background traffic is applied for at least one minute before sending the command. The command is sent from the serial input generator immediately after the request for a file transfer. This is to implement a worst-case scenario for the device under test [7].

The load generation for the background data traffic is using test message 1 as defined in IEC 61850-9-2, which represents sample values of 1 measuring point (4 currents and 4 voltages). The test message 2 represents the request for a transfer of a file with a length of 2 Mbytes as defined in IEC 61850-8-1 [18].

4.4.3.5 Routine Tests

The routine tests for digital communication of high-voltage switchgear as defined in IEC 62271-3 [7] have the goal to define routine test procedures for the application at the factory; these tests are running after assembly of the GIS for functional and dielectric routine tests to prove the correct manufacture and assembly.

Routine tests are for the purpose of revealing faults in material or construction. They do not impair the properties and reliability of a test object. The routine tests shall be made wherever reasonably practicable at the manufacturer's factory on each apparatus manufactured the goal is to ensure that the product is in accordance with the equipment on which the type tests have been passed. By agreement, any routine test may be made on site in accordance with IEC 62271-1 [14].

The relevant switchgear product standards of the IEC 62271 series are applicable in general. Digital interfaces should be taken into account where applicable. There are two tests that can be applied. The test as described in Clause 7.102.1.2 of IEC 62271-3 should be done once for opening and once for closing. The test as described in Clause 7.102.1.3 of IEC 62271-3 should be done once for opening and once for closing.

4.4.4 Normative References

4.4.4.1 Digital Communication-Related Standards

In this clause, an overview is given on standards released to digital communication and their role for high-voltage switchgear of AIS and GIS. The digital communication in substations covers a wide spectrum of standards. Basic standards for information technology and software architecture (ISO/IEC 7498, ISO/IEC 8802-3, and IEC 61850 series) are forming the basis of information handled in substations.

More specific digital communication standards for substations cover testing, time performances, cyber security, SCADA, and practical information and recommendations for the use in substations (IEC 62271-3, IEEE 1613, IEEE 1613a, IEEE 1615, IEEE 1646, IEEE 1686, IEEE 1711, IEEE C37.1, and IEEE C37.2 [7, 23, 26, 27, 28]). The high-voltage switchgear device-oriented standards are giving information for substation digital communication related to the specific equipment (IEC 62271-1, IEC 60794, ITU-T V.24, and IEC 60870-4 [14, 31, 32, 50]).

In the following, an introduction to the standards is given on basic information technology and software architecture.

IEC 61850

The series of this standard, Communication Networks in Systems for Power Utility Automation, defines the communication in power systems at the process bus level and station bus/LAN level for power systems. It is the basis for intelligent switchgear used in substations [6].

IEC 62271-3:2006

This International Standard, High-Voltage Switchgear and Controlgear – Part 3: Digital Interfaces Based on IEC 61850, is applicable to high-voltage switchgear and controlgear and assemblies thereof above 1 kV. It specifies equipment for digital communication with other parts of the substation and its impact on testing. This equipment for digital communication, replacing metal parallel wiring, can be integrated into the high-voltage switchgear, controlgear, and assemblies thereof, or can be an external equipment in order to provide compliance for existing switchgear and controlgear and assemblies thereof with the standards of the IEC 61850 series.

This International Standard is a product standard based on the IEC 61850 series. It deals with all relevant aspects of switchgear and controlgear, and assemblies thereof, with a serial communication interface according to the IEC 61850 series [7].

In Table 4.4, an overview is given on standards related to digital communication in high-voltage substations for AIS and GIS [6, 7, 10, 20, 21, 26, 27, 28].

4.4.5 Classifications

4.4.5.1 Timing Requirements

Opening and Closing Times for Circuit Breakers

For circuit breakers, the definitions of opening and closing times given in IEC 62 271-100 [15] are applicable, with the following additions.

For intelligent switchgear, the opening time should be the time from the reception of the first bit of the first frame of the trip command via the interface according to the IEC 61850 series, the circuit breaker being in the closed position, to the instant when the arcing contacts have separated in

Table 4.4 Overview of standards related to digital communication

Number	Title (short version)	Remark
ISO/IEC 7498	Information technology • Open systems interconnection • Basic reference model	Defines open model for system interconnection
ISO/IEC 8802–3	Local area network (LAN) methods and physical layer	Defines the LAN to be used for communication in substation
IEC 61850	Communication networks and systems for power utility automation	Basic standard for digital communication in substations
IEC 62271–3	High-voltage switchgear and controlgear – Part 3: Digital interfaces based on IEC 61850	Gives specific rules for high-voltage switchgear (AIS) and assemblies (GIS)
IEEE P1613	Environmental and testing requirements	Defines requirements for communication networking devices in electric power substations
IEEE P1615	Recommended practice for network communication in electric power substations	Gives recommendation for applications
IEEE P1646	Communication delivery time performance requirement for electric power substation automation	Defines delivery times for data exchange
IEEE C37.1	Standard for SCADA and automation systems	Basic standard of SCADA and automation systems in electric substations
IEEE C37.2	Standard for electric power system device function numbers, acronyms, and contact destination	Basic standard on definition and application of function numbers

all poles. The reception of the first bit of the first frame of the trip command can be measured by using a communication analyzer. Opening and closing times are both examples of intelligent switchgear total operating times, explained in Figure 4.16 [7].

Opening Operation

The timing definitions for the opening operation of an intelligent circuit breaker are shown in Figure 4.17.

Closing Operation

The timing definitions for the closing operation of an intelligent circuit breaker are shown in Figure 4.18. For intelligent switchgear, the closing time should be the time from the reception of the first bit of the first frame of the close command via the interface according to the IEC 61850 series, the circuit breaker being in the open position, to the instant when the contacts touch in all poles. The reception of the first bit of the first frame of the close command can be measured by

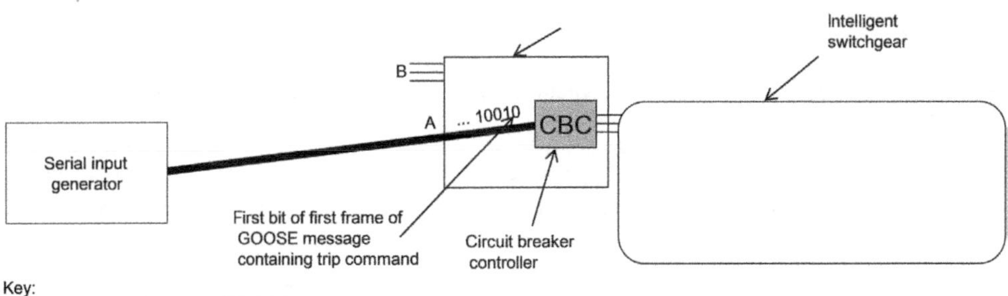

Figure 4.16 Timing of opening/closing command to intelligent switchgear [7] (Reproduced by permission of IEC) (IEC 62271-3 ed.1.0 "Copyright © 2013 IEC Geneva, Switzerland. www.iec.ch")

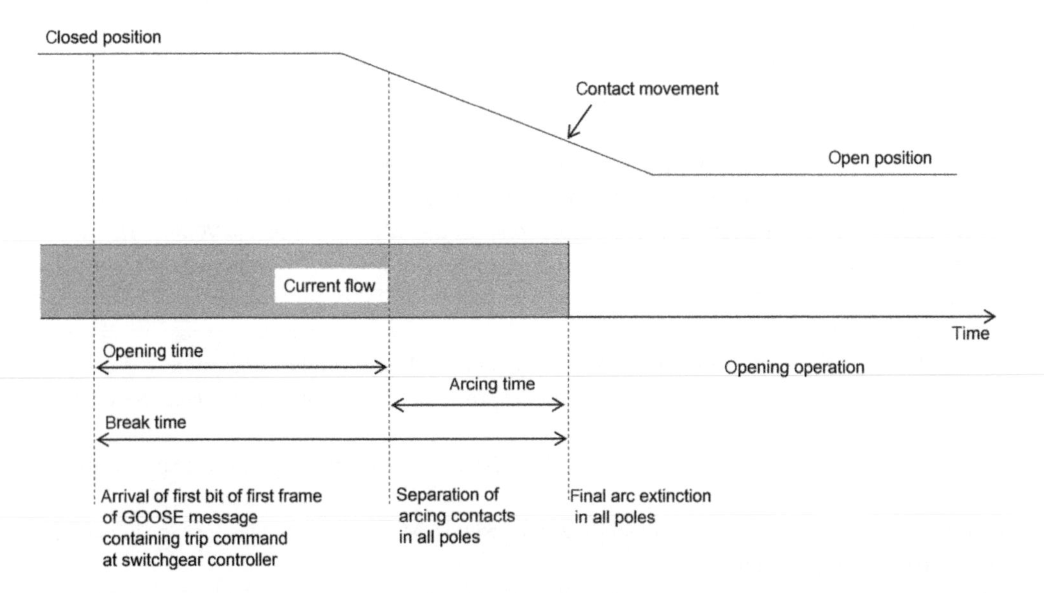

Figure 4.17 Opening operation of an intelligent circuit breaker [7] (Reproduced by permission of IEC) (IEC 62271-3 ed.1.0 "Copyright © 2013 IEC Geneva, Switzerland. www.iec.ch")

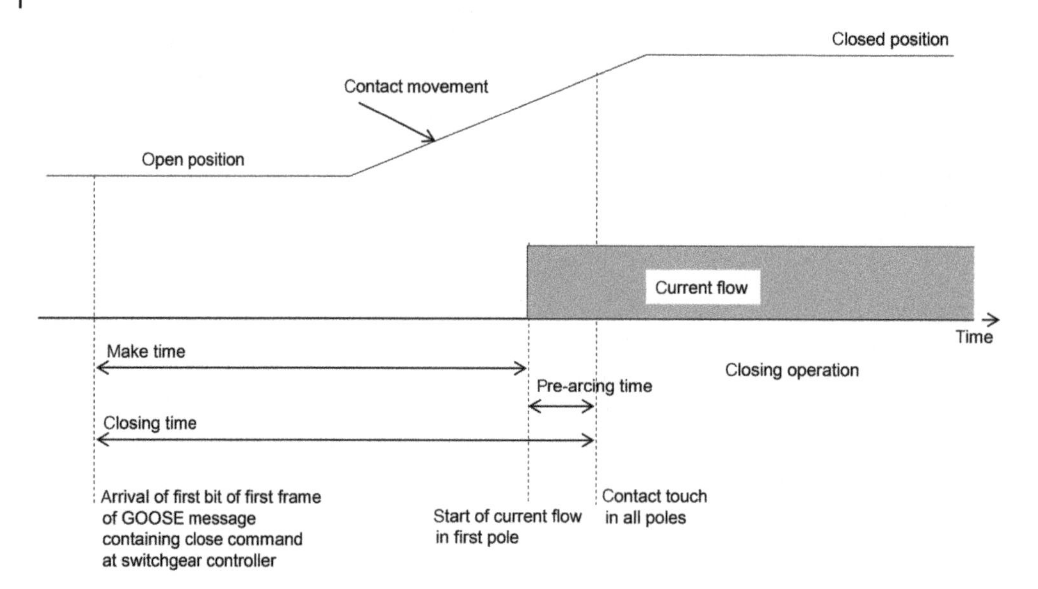

Figure 4.18 Closing operation of an intelligent circuit breaker [7] (Reproduced by permission of IEC) (IEC 62271-3 ed.1.0 "Copyright © 2013 IEC Geneva, Switzerland. www.iec.ch")

using a communication analyzer. In the case of time measurements, coherence should be checked between the position indication via the serial interface in the secondary system and the real position of the intelligent switchgear [7].

References

1 IPCC Fourth Assessment Report – Climate Change (2007) Intergovernmental Panel on Climate Change www.ipcc.ch.

2 Cavallini, A., Montanari, G.C., Puletti, F., and Contin, A. (2005) A new methodology for the identification of PD in electrical apparatus: properties and applications. *IEEE Transactions on Dielectrics and Electrical Insulation*, **12** (2) 203–215.

3 CIGRE WG 15.03 (1992) Diagnostic Methods for Gas Insulating Systems. General Session of CIGRE, Paper 15/23-01.

4 Lopez-Roldan, J., Braun, J.M., Densley, J., and Fujimoto, N. (1996) The development of electrical trees in epoxy insulation – partial discharge pulse characterization by ultra-wideband techniques. Annual Report – Conference on Electrical Insulation and Dielectric Phenomena, Volume 1, October 1996, Paper 8A-5.

5 Chu, F.Y. (1986) Decomposition in gas-insulated equipment. *IEEE Transactions on Electrical Insulation*, EI-21 (5).

6 IEC 61850 (2019) Communication Networks and Systems for Power utility automation.

7 IEC 62271-3 (2015) High-Voltage Switchgear and Controlgear – Part 3: Digital Interfaces Based on IEC 61850.

8 ISO/IEC 9506 (2003) Part 1: Industrial Automation Systems; Manufacturing Message Specification; Part 1: Service Definition.

9 ISO/IEC 9506 (2003) Part 2: Industrial Automation Systems; Manufacturing Message Specification; Part 2: Protocol Specification.

10 ISO/IEC 8802-3 (2003) Information Technology – Telecommunications and Information Exchange between Systems – Local and Metropolitan Area Networks; Specific Requirements – Part 3: Carrier

Sense Multiple Access with Collision Detection (CSMA/CD) Access Method and Physical Layer Specifications.
11 IEC 61869 Series (latest edition) Instrument Transformers.
12 IEC 61869 (2016) Part 9: Instrument Transformers – Part 9: Digital Interface for Instrument Transformers.
13 IEC/TR 61869-103 (2012) Instrument Transformers – The Use of Instrument Transformers for Power Quality Measurement.
14 IEC 62271-1 (2015) High-Voltage Switchgear and Controlgear – Part 1: Common Specifications.
15 IEC 62271-100 (2017) High-Voltage Switchgear and Controlgear – Part 100: Alternating Current Circuit-Breakers.
16 IEC 62271-102 (2013) High-Voltage Switchgear and Controlgear – Part 102: Alternating Current Disconnectors and Earthing Switches.
17 IEC 61850-10 (2012) Communication Networks and Systems for Power Utility Automation – Part 10: Conformance Testing.
18 IEC 61850-8-1 (2011). Communication Networks and Systems for Power Utility Automation – Part 8-1: Specific Communication Service Mapping (SCSM) – Mappings to MMS (ISO 9506-1 and ISO 9506-2) and to ISO/IEC 8802-3.
19 IEEE C37.100 (2010) IEEE Standard Definitions for Power Switchgear.
20 ISO/IEC 7498 (2012) Information Technology – Open Systems Interconnection.
21 IEEE 1613 (2010) IEEE Standard Environmental and Testing Requirements for Communications Networking Devices in Electric Power Substations.
22 IEEE C37.90™ (2005) Relays and Relay Systems Associated with Electric Power Apparatus.
23 IEEE C37.90.1™ (2012) Surge Withstand Capability (SWC) Tests for Relays and Relay Systems Associated with Electric Power Apparatus.
24 IEEE C37.90.2™ (2004) Withstand Capability of Relay Systems to Radiated Electromagnetic Interference from Transceivers.
25 IEEE C37.90.3™ (2001) Electrostatic Discharge Tests for Protective Relays.
26 IEEE 1615 (2007) Recommended Practice for Network Communication in Electric Power Substations.
27 IEEE C37.1 (2007) SCADA and Automation Systems.
28 IEEE C37.2 (2008) Electrical Power System Device Function Numbers, Acronyms, and Contact Designations.
29 IEC 62271-200 (2011) High-Voltage Switchgear and Controlgear – Part 200: AC Metal-Enclosed Switchgear and Controlgear for Rated Voltages above 1 kV and up to and Including 52 kV.
30 IEC 62271-203 (2011) High-Voltage Switchgear and Controlgear – Part 203: Gas-Insulated Metal-Enclosed Switchgear for Rated Voltages above 52 kV.
31 IEC 60794 (2010) Optical Fibre Cables.
32 IEC 60870-4 (2010) Telecontrol Equipment and Systems. Part 4: Performance Requirements.
33 ITU-T V.24 (2010) List of Definitions for Interchange Circuits between Data Terminal Equipment (DTE) and Data Circuit-Terminating Equipment (DCE).
34 IEC 60815 (2007) Selection and Dimensioning of High-Voltage Insulators Intended for Use in Polluted Conditions.
35 IEC 62271-304 (2012) High-Voltage Switchgear and Controlgear – Part 304: Design Classes for Indoor Enclosed Switchgear and Controlgear for Rated Voltages above 1 kV up to and Includuing 52 kV to be Used in Severe Climatic Conditions.
36 IEEE 693 (2005) Recommended Practices for Seismic Design of Substations.
37 IEC 62271-207 (2012) High-Voltage Switchgear and Controlgear – Part 207: Seismic Qualification for Gas-Insulated Switchgear Assemblies for Rated Voltages above 52 kV.

38 IEC 62271-210 (2013) High-Voltage Switchgear and Controlgear – Part 210: Seismic Qualification for Metal Enclosed and Solid-Insulation Enclosed Switchgear and Controlgear Assemblies for Rated Voltages above 1 kV and up to and Including 52kV.

39 IEC 62271-306 (2012) High-Voltage Switchgear and Controlgear – Part 306: Guide to IEC 62271-100, IEC 62271-1 and Other IEC Standards Related to Alternating Current Circuit-Breakers.

40 IEC 60721 Series Classification of Environmental Conditions.

41 IEC 61850-5 (2013) Communication Networks and Systems for Power Utility Automation – Part 5: Communication Requirements for Functions and Device Models.

42 IEC 61850-7-2 (2010) Communication Networks and Systems for Power Utility Automation – Part 7-2: Basic Information and Communication Structure – Abstract Communication Service Interface (ACSI).

43 IEC 61850-7-4 (2012) Communication Networks and Systems for Power Utility Automation – Part 7-4: Basic Communication Structure – Compatible Logical Node Classes and Data Object Classes.

44 IEC 60870 Series Telecontrol Equipment and Systems.

45 IEC 61850-3 (2019) Communication Networks and Systems in Substations – Part 3: General Requirements.

46 IEC 61439 (2010) series, The new standard for low-voltage switchgear and controlgear assemblies.

47 IEEE C37.123-2016 - IEEE Guide for Specifications for High-Voltage Gas-Insulated Substations Rated 52 kV and Above.

48 IEEE C57.13-2016 series - IEEE Standard Requirements for Instrument Transformers.

49 IEC 60255-1:2009 series, Measuring relays and protection equipment.

5

Testing

Authors: 1st edition Peter Grossmann, Charles L Hand, 2nd edition Dave Giegel, Coboyo Bodjona, Reviewers: 1st edition Phil Bolin, Xi Zhu, Noboru Fujimoto, Dave Solhtalab, Jim Massura, Eduard Crockett, Hermann Koch, and 2nd edition Hermann Koch

5.1 General

To ensure the function and safety the GIS has been designed for, the GIS needs to be tested. The testing is determined to confirm the technical data as well as safe operation of the GIS over its life duration.

Related to the GIS use, the main tests are dielectric tests, short-circuit tests, mechanical tests, and temperature rise tests.

Two different test procedures apply: type tests, also called design tests, and routine tests, also called factory or production tests. While type tests verify the performance of one GIS type after product development, routine tests ensure that each unit produced operates according to the technical requirements to which the GIS is supposed to adhere.

5.2 Type Tests

As an important step in the development of a GIS, type tests were carried out to verify the performance of the GIS. The ratings established in the type tests will also be used as default data when the GIS is in the production cycle later on. Type tests at least involve (see IEEE C37.122, Table 6) [1, 6]:

- Dielectric tests
- Measurement of the resistance of the main circuits
- Temperature rise tests
- Short-time withstand current and peak withstand current tests
- Verification of the degree of protection of the enclosure
- Tightness tests
- Electromagnetic compatibility (EMC) test
- Verification of making and breaking capacities
- Low- and high-temperature tests
- Proof tests for enclosures
- Pressure test on partitions
- Tests to prove performance under thermal cycling and gas tightness tests on insulators
- Circuit breaker design tests
- Fault-making capability of high-speed grounding switch

Gas Insulated Substations, Second Edition. Edited by Hermann J. Koch.

- Switch operating mechanical life tests

The following sections describe selected tests extracted from the list above.

5.2.1 Dielectric Tests

Dielectric tests are used to verify the dielectric capability of the GIS under all foreseeable operating conditions, including temporary and transient overvoltages, and therefore involve power frequency tests, lightning impulse tests, switching impulse tests, partial discharge tests, and tests on auxiliary and control circuits. In Figure 5.1, an overview is given of the different dielectric tests. The high-voltage tests require a large size of test equipment to generate the test voltages of some thousands or millions of volts. In Figure 5.2, a test setup for a power frequency and impulse test on a 145 kV GIS is shown [1, 5, 6].

5.2.2 Measurement of the Resistance of the Main Circuits

This test measures the resistance of a set of conduction paths in a GIS. The test will prove the conductivity of conductor material, conductor connections, and contacts. At a current of typically 100 A DC, the resistance or voltage drop of defined layouts will be measured. The test results give

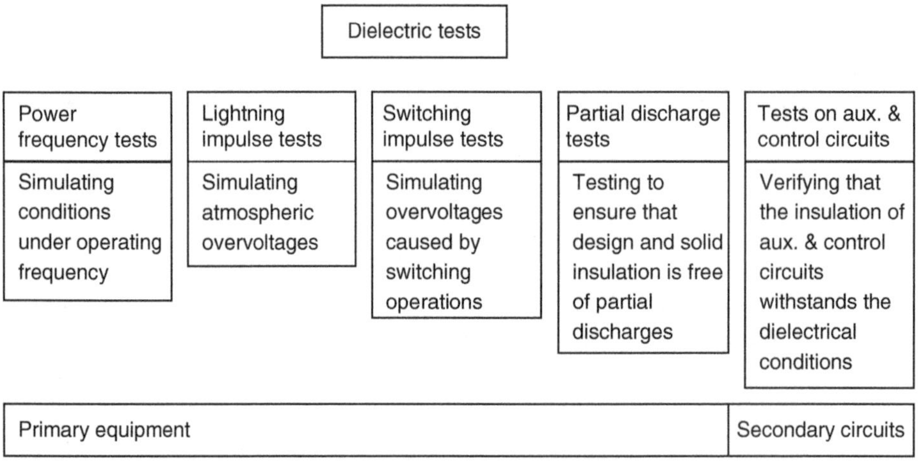

Dielectric tests

Power frequency tests	Lightning impulse tests	Switching impulse tests	Partial discharge tests	Tests on aux. & control circuits
Simulating conditions under operating frequency	Simulating atmospheric overvoltages	Simulating overvoltages caused by switching operations	Testing to ensure that design and solid insulation is free of partial discharges	Verifying that the insulation of aux. & control circuits withstands the dielectrical conditions
Primary equipment				Secondary circuits

Figure 5.1 Overview of dielectric tests for GIS [1, 6]

Figure 5.2 Example of a dielectric type test of a GIS (Reproduced by permission of Siemens AG)

information about the quality of conductor connections and contacts and also provide a basis for a comparison between the three phases. The lower the resistance values, the lower the temperature rise would be when in service. The temperature rise is an important factor for determining the continuous current capability of the product. The test results establish a benchmark for the GIS test later during production [1, 5, 6].

5.2.3 Temperature Rise Tests

To prove at what maximum continuous current the GIS can be operated, a temperature rise test is performed. Thermocouples will be placed at various locations such as conductors, connections, contacts, and insulators to measure the temperature rise at a defined continuous current the GIS is designed for. Other than this discrete measure method, by using thermocouples additional thermographic measures can be used to support the analysis of the arrangement related to the temperature rise, especially during development tests.

The test setup of the GIS including the circuit breaker (right and left side of test setup), disconnect switch (in the back of the test setup), and bus bar (in the front of the test setup) is shown in Figure 5.3. The left part of the figure shows the test setup with thermocouples connected to the GIS enclosure and inside conductors. The right part shows the thermographic measurement of the temperatures of the enclosure [1, 5, 6].

5.2.4 Short-Time Withstand Current and Peak Withstand Current Tests

This test is for verification that main circuits of the GIS will be able to carry the peak withstand current and the rated short-time withstand current. Parts of the main circuit as well as support insulators need to withstand the dynamical stress during the short-time withstand current and the peak withstand current that the GIS needs to carry in the closed position of the circuit breaker and disconnect switches. Typical values of the short-circuit duration are 1 s or 3 s. With a time duration of 45 ms and for a frequency of 60 Hz, the value of the peak withstand current is 2.6 times the rated short-time withstand current. The short-circuit tests require high currents, which are generated in large special generators, as shown in Figure 5.4. In some cases, the short-circuit current is taken from the network, but as this might cause disturbances in the network, it might not be allowed [1, 5, 6].

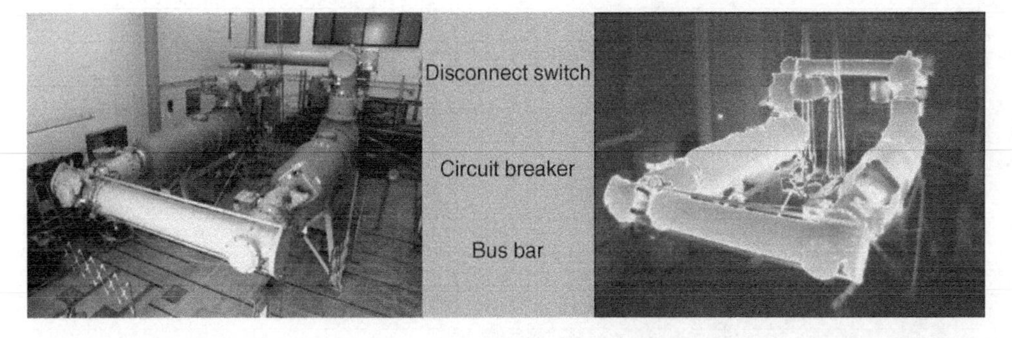

Figure 5.3 Test setup for the temperature rise test: (left) thermocouples are connected to the enclosure outside and conductors inside; (right) thermographic measurement of the enclosure temperature (Reproduced by permission of Siemens AG)

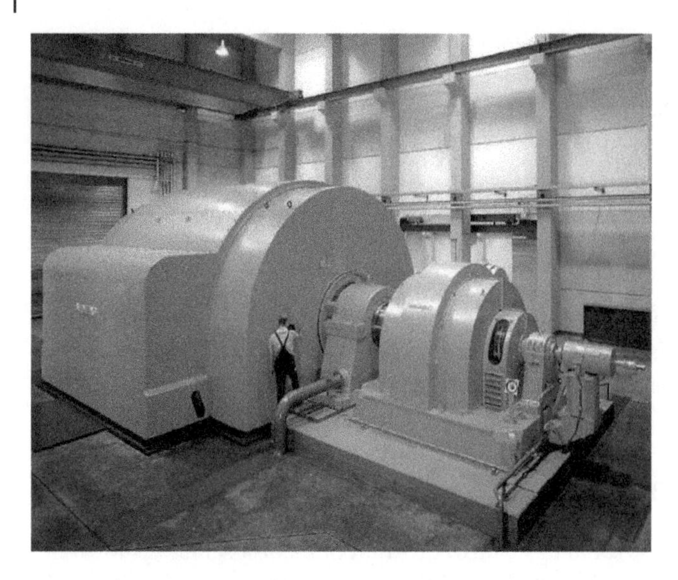

Figure 5.4 Generator for short-circuit testing (Reproduced by permission of Siemens AG)

5.2.5 Tightness Tests

Tightness tests demonstrate that the SF_6 leakage rate of the tested GIS does not exceed a specific value of a permissible leakage rate. According to GIS IEEE and IEC standards, the leakage rate should not exceed 0.5% per year per gas compartment (see IEC 62271-1 [5]). Some GIS manufacturers provide even leakage rates of 0.1% per year per gas compartment. The setup to prove the SF_6 tightness is shown in Figure 5.5.

The plastic barriers are used for long-time measurement inside the plastic enclosure [1, 5, 6].

5.2.6 Low- and High-Temperature Tests

This test is part of the mechanical and environmental tests. All components of the GIS must operate under defined low- and high-temperature conditions. The GIS or components of the GIS will be installed in a climate chamber. At minimum and maximum temperatures, operation tests will be performed. After the test cycles, the SF_6 gas pressure and the SF_6 leakage rate over a period of 24 hours will be observed. The GIS installed in a climate chamber for a low- and high-temperature test is shown in Figure 5.6 [1, 5, 6].

Figure 5.5 SF_6 tightness test (Reproduced by permission of Siemens AG)

Figure 5.6 Low- and high-temperature test in a climate chamber (Reproduced by permission of Siemens AG)

5.2.7 Proof Tests for Enclosures

Proof tests of the enclosures can be destructive or nondestructive. The graph in Figure 5.7 shows the type test pressure as the highest pressure before the burst and rupture pressure of the enclosure. Coordination of the design and test pressure levels for the GIS enclosures is shown in Figure 5.7 [1, 6, 9, 10, 11, 12, 13].

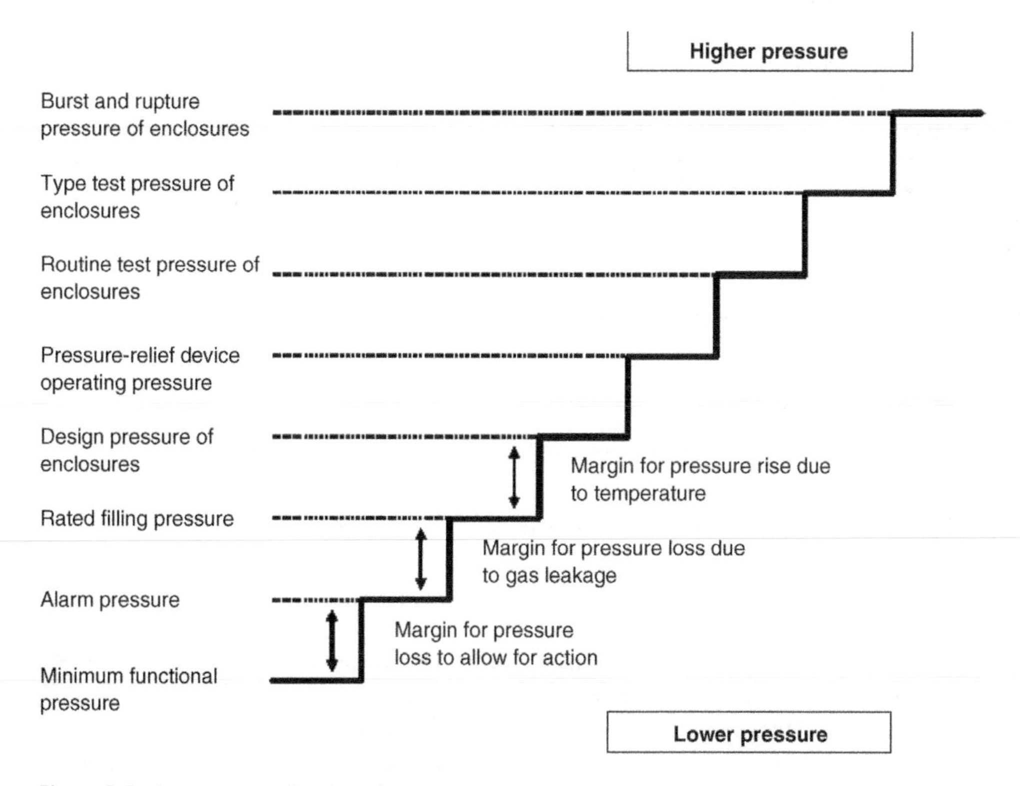

Figure 5.7 Pressure coordination of enclosures and pressure-relief device [1]

5.2.8 Circuit Breaker Design Tests

Apart from the tests already mentioned, such as the dielectrical tests and temperature rise tests, circuit breakers need to be type-tested according to their relevant operation duties. These tests include, but are not limited to, interrupting time tests, transient recovery voltage (TRV) tests, short-circuit current interrupting tests, load current tests, capacitor switching current tests, out-of-phase switching tests, and mechanical endurance tests. These tests are described in the IEEE standard for test procedures for high-voltage circuit breakers (IEEE C37.09 [7] and IEC 62271-100 [8].

5.2.9 Switch Operating Mechanical Life Tests

To test the mechanical durability of the GIS, disconnect and ground switches, these switches are operated with at least 1000close/open operations according to IEEE C37.122 [1]. Fifty close/open operations are included herein to be performed with minimal control voltage and 50close/open operations with maximal control voltage. The test needs to demonstrate that the switch and the operating mechanism do not show excessive wear and that they are in a good mechanical condition. This will be done by an examination of the switch contacts and related parts of the kinematic chain and of the mechanism as well. A measurement of the resistance of the contacts, as described in Section 5.2.2, will confirm the contact capability to carry the continuous current after being stressed by the mechanical operations. To verify that the mechanical operation test does not influence the SF_6 tightness, an SF_6 gas tightness test is performed before and after the mechanical operation test.

5.3 Routine Tests

Routine tests, often referred to as production tests, are performed to ensure that each GIS operates as it has been designed and type-tested for. The routine tests are performed for each GIS after assembly and marks a major quality gate before the GIS leaves the factory. The test parameters are based on the type test results, which means that, within certain tolerances, the routine tests need to reflect the type test data. Tests included in routine tests are (see IEEE C37.122 [1]):

- Dielectric tests
- Tests on auxiliary and control circuits
- Measurement of the resistance of the main circuits
- Tightness tests
- Pressure tests of enclosures
- Mechanical operation tests
- Tests on auxiliary circuits, equipment, and interlocks in the control mechanism
- Pressure tests on partitions

The following sections describe selected tests extracted from the list above.

5.3.1 Dielectric Tests

The dielectric tests are done after the mechanical routine testing and demonstrate the dielectric performance of the GIS, ensuring the correct assembly, correctly manufactured parts from a dielectric point of view, and the absence of particles and other contaminants.

For routine tests, the dielectric test is a power frequency withstand voltage test. Impulse testing, such as lighting and switching impulse, is not typically part of the routine testing. At minimum functional SF_6 pressure, the following conditions are tested: phase-to-ground, phase-to-phase (in the case of three phases in one enclosure design), and across open switching devices. Successfully withstanding the one-minute withstand level without a disruptive discharge is the main criteria to mark that the test passed successfully.

To detect possible material and manufacturing defects, partial discharge testing is also included as part of the dielectric routine tests.

5.3.2 Measurement of the Resistance of the Main Circuits

Typically, the voltage drop or resistance of main circuits is measured using a DC current of 100 A. Correct contact assembly, proper treatment of clean contact areas, and the correct contact materials used will be verified with this test. The test data should be within a 20% tolerance band compared with the type test data.

5.3.3 Tightness Tests

Using devices such as SF_6 leakage detectors, all areas of enclosures assemblies, SF_6 piping, adaptation of SF_6 gauges, and SF_6 density monitoring will be checked for leaks. Correct assembly, including correct use of sealing rings, will also be verified with this test. The measurement of gas tightness using the so-called sniffing device is shown in Figure 5.8.

5.3.4 Pressure Tests of Enclosures

After complete machining of enclosures, pressure tests are made at 1.3 times the design pressure for welded aluminum and welded steel enclosures and at two times the design pressure for cast enclosures. With the state-of-the-art technology today, enclosures are made of cast aluminum. Using 3D CAD systems and FEM calculations, cast aluminum enclosures can be shaped to meet the dielectric and mechanical requirements while providing excellent gas tightness. Automated test stations facilitate the inclusion of a tightness test using helium after the pressure test of the enclosure. The pressure test compartment in which the GIS enclosures are inserted filled with helium under pressure is shown in Figure 5.9 [1, 5, 6, 9, 10, 11, 12, 13].

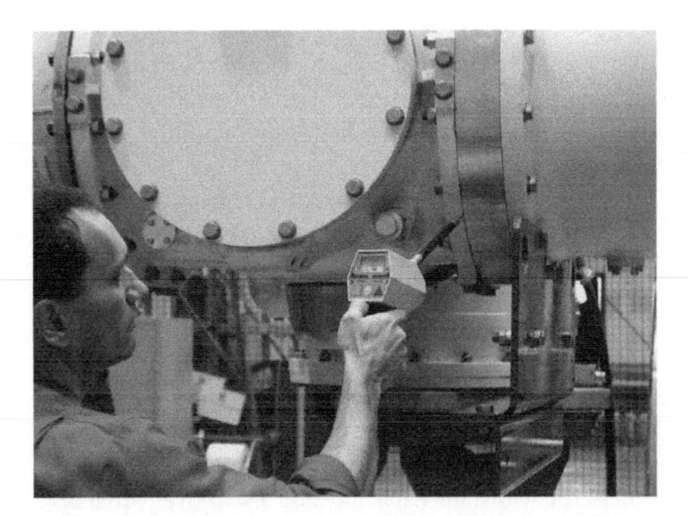

Figure 5.8 Leakage test as part of routine testing (Reproduced by permission of Siemens AG)

Figure 5.9 Pressure and tightness test of cast aluminum enclosures (Reproduced by permission of Siemens AG)

5.3.5 Mechanical Operation Tests

Mechanical operation tests include all devices of the GIS that will be mechanically operated, such as a circuit breaker, disconnect switches, ground switches, and high-speed ground switches. The tests include a certain number of operation cycles at different control voltage levels and the correct function of the related auxiliary equipment; for example, auxiliary switches to indicate the position of the circuit breaker and switches. During these tests, several parameters will be recorded to make sure that the devices operate in their tolerance bands: closing and opening times and pole difference times, travel curves of interrupter units, charging times and currents of motors of a spring-operated mechanism or hydraulic-operated mechanism, as well as running times and currents of motors of disconnect and ground switch mechanisms. These tests verify that the assembly of a circuit breaker and switches has been done correctly and that the proper function of these GIS devices is ensured when the GIS is in service [1, 5, 6].

5.3.6 Tests on Auxiliary and Control Circuits

These tests confirm that, during manufacturing, all wiring has been done correctly according to the related circuit diagrams. Functional tests of all low-voltage circuits and of auxiliary, control, and protection devices verify their correct function and proof of their interconnection with the GIS. Next to functional testing at low and upper voltage levels, dielectric tests are performed to ensure the dielectric withstand capability of the wiring insulation and components [1, 5, 6].

5.3.7 Pressure Tests on Partitions

Partitions are gastight insulators that separate one gas compartment from the other. They allow full pressure on one side and vacuum on the other.

Each partition has to be tested to twice that of the design pressure. It has to be ensured that the weakest mechanical direction of the device is being considered for the test. This test verifies that the partition has been manufactured correctly to withstand the pressure the partition is designed for. Partitions will also be tested to the dielectric withstand capability and a sensitive partial discharge measurement [1, 5, 6, 13].

5.4 Onsite Field Testing

Typically, a gas-insulated substation is only partially assembled in the factory. Major components of a GIS are frequently manufactured in different factories in different countries, sometimes by different manufacturers, and shipped directly to the job site.

The final assembly of the gas-insulated substation is then completed in the field, where all the various components that comprise a GIS meet for the first time. Even if the gas-insulated substation could be completely assembled in one factory, it would still need to be disassembled for shipment, shipped, and then reassembled at the job site.

The purpose of the field tests is to verify that all the GIS components perform satisfactorily, both electrically and mechanically, after assembly at the job site. The tests provide a method of demonstrating that the GIS apparatus has been assembled and wired correctly and will perform satisfactorily [1, 6].

5.4.1 Gas Leakage and Gas Quality (Moisture, Purity, and Density)

After processing and filling each gas compartment to the manufacturer's required nominal rated filling density and verifying the density value, the assembled gas-insulated substation needs to be tested. An initial test is performed to detect any and all gas leaks and ensure compliance with the specified maximum gas leak rate. These gas leak tests need to include all enclosure flanges, welds of enclosures, all gas monitoring devices, gas valves, and interconnecting gas piping that have been assembled at the job site.

The moisture content of the gas in each compartment needs to be measured directly after installation and again at least five days after final filling. These tests are to ensure that the moisture content does not exceed the specified maximum limits. The second test after five days is necessary to take into account the possibility of moisture from components internal to the GIS.

The purity of the gas in each gas compartment needs to be measured directly after installation. These tests are to ensure that any gas impurities (mostly air) do not exceed the specified maximum limits [1, 6].

5.4.2 Electrical Tests: Contact Resistance

Contact resistance measurements of the main current carrying circuits need to be performed on each bus connecting joint, circuit breaker, disconnect switch, grounding switch, bushing, and power cable connection to demonstrate and verify that the resistance values are within specified requirements.

Because the metallic enclosure inhibits accessibility to current carrying parts, it is not usually possible to measure the resistance of individual components. Therefore, the resistance readings are obtained for several components connected in series. These field measurements can then be compared to the expected resistance values supplied by the manufacturer as a basis for verifying acceptable test results in the field.

Contact resistance measurements also need to be made on the GIS enclosure bonding connections, in cases where an isolated (single) phase bus is being used [1, 5, 6].

5.4.3 Electrical Tests: AC Voltage Withstand

The gaseous and solid insulation of the gas-insulated substation needs to be subjected to an AC voltage withstand test. Due to the wide variations of the capacitance of different GIS designs, it is often that a variable frequency hi-pot test unit be used. The variable frequency high potential unit can generate low-frequency (30–300 Hz) voltage applications at magnitudes and durations specified in standards. This one-minute low-frequency voltage withstand test is performed at 80% of the rated low frequency withstand voltage performed in the manufacturer's factory. A conditioning voltage application sequence, with magnitude and durations specified by the manufacturer, should precede the specified one-minute withstand test. The intention of the conditioning test is to drive any small particles, if they exist, to low electric field intensity areas such as particle traps.

The purpose of these high-voltage tests is to verify that the components of the gas-insulated substation have survived shipment, have been assembled correctly, that no extraneous material has been left inside the enclosures, and that the GIS can withstand the test voltage.

The conditioning voltage application sequence and the one-minute low-frequency voltage withstand test need to be performed after the GIS has been completely installed, the gas compartments have been filled to the manufacturer's recommended nominal rated fill density, and the moisture content and purity of the gas have been verified to be within specified limits [1, 5, 6].

5.4.4 Electrical Tests: AC Voltage Withstand Requirements and Conditions

Voltage withstand tests need to be made between each energized phase and the grounded enclosure. For enclosures containing all three phases, each phase needs to be tested, one at a time, with the enclosure and the other two phases grounded. Before voltage withstand tests are initiated, all power transformers, surge arresters, protective gaps, power cables, overhead transmission lines, and voltage transformers need to be disconnected. Voltage transformers may be tested up to the saturation voltage of the transformer at the frequency of the test [1, 5, 6].

5.4.5 Electrical Tests: AC Voltage Withstand Configurations and Applications

When the GIS apparatus being tested is connected to the GIS apparatus that is already in service, the in-service portion needs to be electrically isolated from the tested portion. However, it is highly possible that the test voltage could be 180° out of phase with the in-service voltage, potentially exposing the open gap of a disconnect switch, being used for isolation, to voltages in excess of what can be withstood. Therefore, an isolated section with suitable grounds needs to be applied between the in-service GIS and the GIS to be tested. This ensures that the test voltage cannot cause service disruptions to the electrical system nor can the service voltage cause severe damage to the testing apparatus or danger to the test personnel.

Due to the electrical loading restrictions of the testing apparatus, it may be necessary to isolate sections of the GIS equipment using open disconnects and test each section separately. To do this, it may require that portions of the GIS apparatus be subjected to more than one test voltage application. The sections that are not being tested need to be grounded.

Isolating sections of the GIS apparatus may give an additional benefit of field testing the open gap of some disconnecting switches, although such a field test is not a requirement. In addition, it may be necessary to isolate sections of the GIS to facilitate location of a disruptive discharge or to limit the energy potentially discharged during a disruptive discharge.

The test voltage source may be connected to any convenient point of the phase being tested [1, 5, 6].

5.4.6 Electrical Tests: DC Voltage Withstand Tests

DC voltage withstand testing is not recommended on a completed GIS. However, it may be necessary to perform a DC voltage withstand test on power cables connected to a GIS. These test voltages would, by necessity, be applied from the end of the cable opposite to that of the GIS, therefore subjecting a small portion of the GIS to the DC voltage. It is recommended that the portion of the GIS subjected to this DC voltage be kept as small as possible. The manufacturer should be consulted before performing these tests [1, 6].

5.4.7 Mechanical and Electrical Functional and Operational Tests

The following need to be verified after assembly of the GIS at the job site:

1) The torque value of all bolts and connections assembled in the field need to be verified to be in accordance with the specified requirements.
2) The conformity of the control wiring needs to be verified to be in accordance with the schematic and wiring diagrams.
3) The proper function of each electrical, pneumatic, hydraulic, mechanical, key, or combination of interlock methods needs to be verified for correct operation in both the permissive and blocking condition.
4) The proper function of the controls, gas, pneumatic, and hydraulic monitoring and alarming systems, protective and regulating equipment, operation counters, including heaters and lights, needs to be verified.
5) Each mechanical and electrical position indicator for each circuit breaker, disconnect switch, and grounding switch needs to be verified that it correctly indicates the device's position, both open and closed.
6) The conformity of the gas zones, gas zone identification, gas valves, gas valve positions, and interconnecting piping needs to be verified to be in accordance with the physical drawings.
7) The operating parameters, such as contact alignment, contact travel, velocity, opening time and closing time of each circuit breaker, disconnect switch, and grounding switch, need to be verified in accordance with the specified requirements.
8) The correct operation of compressors, pumps, auxiliary contacts, and anti-pump schemes needs to be verified to be in conformance with the specified requirements.
9) The circuit breakers need to be trip-tested at minimum and maximum control voltages to verify correct operation.
10) The secondary wiring needs to be verified to have correct wire lugs, correct crimping, tightened terminal block screws, correct wire and cable markers, and correct wiring in accordance with the manufacturer's drawings.
11) The polarity, saturation, turns ratio, and secondary resistance of each current transformer, including all connected secondary wiring, need to be verified to be in accordance with the specified requirements.
12) The turns ratio and polarity of each tap of each potential transformer, including all connected secondary wiring, need to be verified to be in accordance with the specified requirements.
13) Dielectric and contact resistance tests need to be performed on all interconnecting control wiring.

5.4.8 Connecting the GIS to the Electrical System

Once the gas-insulated substation has been completely installed, wired, and all field testing has been completed satisfactorily, the new apparatus is ready to be connected to the existing electrical system. This effort involves another series of testing to verify protective relay operation, ability of the circuit breakers to trip on command from remote locations, and proper phase relationships with various transmission lines. This second series of tests is expected to be similar, if not exactly the same, as the tests performed on an AIS substation.

5.5 Guidelines for Onsite Tests

This guideline provides recommendations to help gas-insulated substations (GIS) users to define the on-site field tests or practices for newly installed gas insulated substations. The principles may be applied to isolated phase enclosures or three phases in one enclosure for gas insulated substations rated above 52 kV.

Standards References

- *IEEE C37.122 IEEE Standard for Gas-Insulated Substations* [1].
- *IEC 62271-203 Gas-Insulated Metal Enclosed Switchgear for Rated Voltages above 52 kV* [6].

Tests

Note that the manufacturer and the user should agree on the field tests plan to be used for the gas-insulated substation. These field tests are recommended for new installations with the tests and acceptance criteria documented.

5.5.1 General

_________ Conduct final construction visual inspection
_________ Check control cable insulation resistance
_________ Conduct control cable visual point to point inspection
_________ Conduct SF_6 gas leakage test

- SF_6 leakage rate acceptance criteria: Maximum of 1%/year

_________ Conduct SF_6 gas quality test (moisture, purity, and density)

- New gas acceptance criteria: Moisture:<100 ppmv, Purity: 98–100%

_________ Conduct equipment interlocking test
_________ Conduct electrical continuity tests on GIS grounding and bonding connections
_________ Check primary circuit resistance
_________ Check local alarms
_________ Conduct dielectric, continuity and resistance tests on field installed wiring
_________ Conduct high-voltage withstand test with PD measurements on high-voltage GIS circuits, including voltage transformers up to the saturation voltage of the transformers

- IEC 62271-203 paragraph 10.2.101.2.3 Procedure A recommended for GIS rated 170 kV and below
- IEC 62271-203 paragraph 10.2.101.2.3 Procedure B recommended for GIS rated 245 kV and above

_________Check AC system – heaters, lighting, and receptacles
_________ Conduct final operation counter readings
_________Check final SF_6 gas density

5.5.2 Circuit Breaker Operational Tests from Local Control Cabinet and from Individual Mechanism

_________ Conduct mechanism stroke and wipe measurement test
_________ Conduct electrical/manual operation test – open/close

__________Check anti-pumping operation
__________ Conduct timing and travel test
__________ Conduct GCB low gas trip/close block test

5.5.3 Disconnect Switch (DS), Grounding Switch (GS) Operational Tests from Local Control Cabinet and from Individual Mechanism

__________Check clearance and mechanical adjustment
__________Check electrical/manual device operation
__________ Check device interlocking
__________Check fixed viewport/camera

5.5.4 Current Transformer

__________ Conduct secondary wiring resistance test
__________ Conduct polarity test
__________ Conduct turns ratio test
__________ Conduct saturation test
__________ Conduct excitation test

5.5.5 Voltage Transformer

__________ Conduct secondary wiring resistance test
__________ Conduct double PF/Watt loss test
__________ Conduct turns ratio test
__________ Conduct polarity test

5.6 Best Practice for On-Site Field Testing

Required On-Site Field Tests as described in Section 5.4 above, and in IEEE C37.122 [1] and C37.122.7 [14] are many times difficult to understand and properly perform in the field without some additional background information, proper test preparation, testing techniques, and accommodations for certain real-world situations.

In this section, select field tests will be better described along with recommended practices.

5.6.1 Gas Leakage

This test is performed to demonstrate there are no gas leaks at the field assembled joints of the GIS. Leaks may occur during field assembly as a result of sealing surface damage, improper placement, application, damage or omission of seals, improper application of lubricants and sealants, improper alignment and tightening of the mating surfaces, and contamination. There is no need to check for leakage through the walls of the gas compartments or at other joints that were assembled at the factory because they were already leak tested at the factory. The only exception may be if suspected damage occurred during transportation and assembly or field repairs were made. If factory joints were disassembled for any reason during field assembly, these joints should be tested again.

Upon completion of filling GIS gas compartment(s) with sulfur hexafluoride gas (SF_6) or the required gas mixture to the manufacturer's recommended temperature corrected pressure as indicated on the nameplate, verification that no gas leaks are present must be completed by means of a portable gas leak detector. It is recommended that a portable leak detector that provides the level of leak and leak rate be used. However, a standard hand-held "go/no-go" (i.e., audible) leak detector is acceptable for initial verification that a leak is present.

Gas leaks can be irregular and are known to intermittently leak, and as a consequence, a leak may not be detected by quickly passing a leak detector over the area to be tested. An accumulation-type test should be considered to mitigate this issue, in which the area to be tested can be contained for a period of time, and the leak detector can then be inserted into the trapped volume for testing.

Prior to the gas leak test, a vacuum rise leak test will be effective in identifying large leaks on field assembled flanges/joints prior to SF_6 gas filling, but it may not be effective in correctly identifying leaks when the vessel is pressurized. Users should discuss vacuum processes with the manufacturer and follow the manufacturer's recommendations before filling the equipment.

A vacuum rise test usually consists of measuring how much vacuum (as measured by means of a vacuum gauge) is lost in a gas compartment after the vacuum pump has been disconnected and before gas filling. The manufacturer will provide an allowable value of vacuum loss over a predetermined amount of time. If the vacuum loss is greater, then a leak should be suspected. Certain factors can cause false readings with this test technique, which include leaks caused by the vacuum gauges and vacuum processing equipment, as well as vacuum lost due to interior gas compartment moisture which may be off-gassing as it is pulled from interior epoxy materials.

SF_6 gas leak testing should be performed immediately after filling to the manufacturer's recommended temperature compensated pressure. Testing must be performed on all field-assembled enclosure joints, field welds, field connected monitoring devices, gas valves, and gas piping.

As described earlier, an accumulation-type leak test is recommended on field-assembled items and joints, and consists of wrapping plastic sheets to create "bags" around these areas to capture SF_6 gas molecules that intermittently escape from the leak. See Figure 5.10 below for best bagging practice.

The bags avoid background readings from interfering with test results. Covers or caps should be placed on self-sealing fill valves to avoid trapped gasses from being measured with the test sample. Additional verification of factory assembled joints should also be performed in conjunction with field leak testing if there is a suspected leak.

The leak test will be performed on each bagged joint after 12 hours has elapsed. Without significantly disturbing the bag, a small incision should be made above the pocket (shown in Figure 5.10). The nozzle of the hand-held SF_6 gas detector is inserted into the small incision and fed into the pocket at the bottom of the bag. Depending on the testing equipment used, the operator should consult the manufacturer for what is an acceptable leakage rate passing criteria. The leakage rate in ppmv or go/no-go results for all tested locations on the GIS should be documented. If a leak is detected, it is recommended that the detector be removed from the suspected leak area, recalibrated, and returned to the leak area again to validate the presence of a leak. In the event that a leak is confirmed using the hand-held leak detector, additional investigation to pinpoint the leak is required. The operator has a number of options:

1) Remove the bag and use a liquid leak detection solution or soapy water. Since this method is not as sensitive as using the gas leak detector, it may not be able to show the exact location of the leak.
2) Remove the bag and use the hand-held leak detector around the suspected leak joint. The rate at which the detector is moved around the perimeter of the suspected leak area is to be determined by the manufacturer.

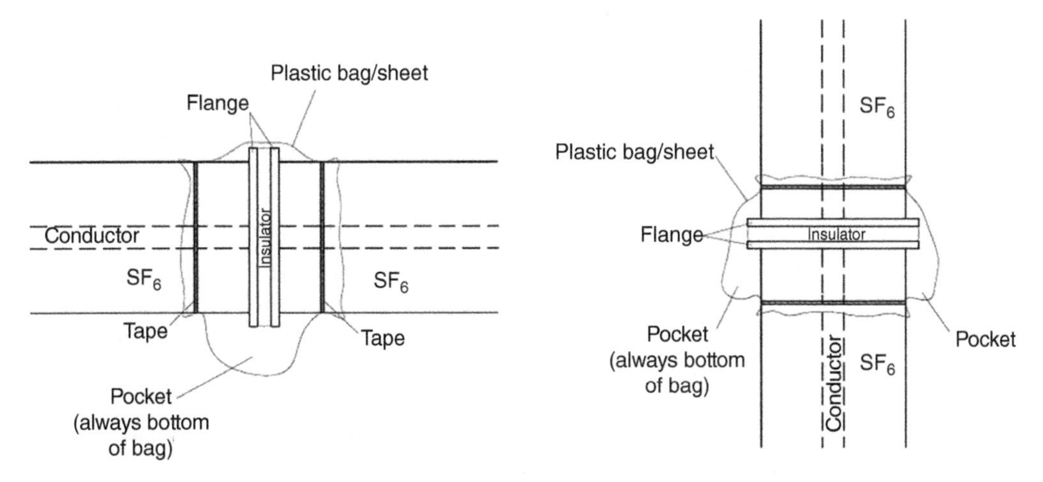

Figure 5.10 Accumulation-type test using plastic sheets or bags (Reproduced by permission of IEEE)

3) Depending on site conditions, infrared cameras may be utilized to pinpoint small leaks following the bag test.
4) Repeat the leakage test as described above, but with compartmentalized bags to isolate the leak area, and reduce the labor necessary for disassembly, correction, and reassembly.

Once a leak is confirmed, then it must be repaired. This is usually accomplished by recovering the SF_6, disassembling the GIS, determining the cause of the leak, and cleaning and replacing components or seals. In certain situations, the customer and manufacturer may agree to a permanent sealing fixture, clamp, or patch. After the repair is made, the gas compartment will be evacuated and filled with SF_6. The leak detection process will then be repeated. It is likely the installation schedule will be affected if a leak is found on the equipment.

Certain chemicals used in sealing/assembling the GIS such as alcohol and silicone sealant may have an effect on the equipment used to detect a leak causing a false reading. Dust, cobwebs, water, and other contaminants are also known to cause a false reading. Prior to leak testing, always ensure the area to be tested is clean and dry.

If a condition-based monitoring/gas trending system is included with the new GIS, it is important to recognize that the sensors take some time to normalize, and therefore may not be effective in providing a true indication of gas leaks immediately after filling the equipment. For gas handling see [3, 15].

5.6.2 Gas Quality

This test is performed to check the moisture content, purity, and density of the gas in the system prior to energization. The moisture content of the gas, gas purity, and gas density must be measured and verified to be in accordance with the manufacturer's rated moisture, purity, and density levels.

Moisture levels within the GIS may adversely affect the dielectric strength of the surface of insulators and significantly increase the possibility of flashover. Corrosive and toxic decomposition by-product may form in a reaction between moisture and dissociated SF_6 found in arcs during normal switching operations. These by-products cause corrosion and may degrade insulators and other components within the GIS, leading to equipment failure.

Gas moisture readings from newly filled equipment should be taken after a minimum of 24 hours from the completion of the filling process. This will allow time for moisture from enclosure walls, internal conductors, and internal insulating materials, if any, to reach equilibrium with the SF_6.

Gas purity is the number of impurities such as water, acidic impurities, nitrogen, oxygen, and oil. There are various industry standards that identify the maximum amount of these impurities. The GIS manufacturer will specify which standards must be followed for the selection and maintenance of purity. The quantities of these impurities must be small so that corrosion (which may lead to faulty equipment operation) or condensation is insignificant.

The correct operation of the equipment depends on the density of the SF_6 gas as specified by the manufacturer. The density must be maintained for the life of the equipment.

Density is usually expressed as a temperature-compensated pressure, typically referring to the pressure of gas at 20°C (the density curve is typically installed on the nameplate of the equipment). It is recommended to fill only to the manufacturer's nameplated density, because overfilling could result in the activation of pressure relief devices. Typical pressures for SF_6 in breakers range between 4 and 6 bar (58–87 psig).

When checking for proper density, use an independent and calibrated digital pressure gauge, not the manufacturer's gas density monitoring gauge that is permanently mounted on the equipment. It is recommended to contact the manufacturer if the field measured density from the digital pressure gauge does not match the gauge mounted on the equipment. In order to be as accurate as possible when checking, thermocouples attached to the SF_6 enclosure should be used to determine the temperature of the gas within the equipment. It is recommended to avoid areas of the enclosure that may be affected by radiant heating from sunlight or onsite heaters. It is not recommended to rely on a gas trending sensor for an accurate density of the compartment during filling and immediately post filling as there is a known delay on the trending equipment.

Measurement of the SF_6 quality on-site is performed using portable equipment. Proper handling of SF_6 gas during testing such as gas recovery, reclamation, and recycling is important in order to keep the gas permanently in a closed cycle and avoid any deliberate release into the environment. These tests should be performed when the equipment is not energized. SF_6 shall be prohibited from releasing into the atmosphere while filling/testing. This can be accomplished by using testing equipment that captures the gas in a closed loop system. A capture bag can also be used. For gas handling see [3, 15].

5.6.3 Power Frequency AC Withstand Voltage

5.6.3.1 Introduction

The power frequency AC withstand voltage test is performed to confirm the dielectric strength of the dielectric gas and solid insulation within the GIS in a safe manner. It also ensures proper installation of the GIS and reduces the probability of an internal flashover once the GIS is commissioned and energized to the rated voltage of the substation. Although this test is performed in the factory prior to shipment, foreign metallic particles can be present or introduced to the system during GIS assembly. Any foreign particle inside the GIS presents a possibility of an internal flashover, and this test is used to verify the existence/nonexistence of a particle. Some examples of common particles include contaminants such as dirt or oil, objects and tools left behind, shavings from bolts during assembly, a worker drops a wrench, and inferior workmanship.

The testing equipment consists of a HV reactor (made up of stacked reactor modules), HV divider (completed by stacked capacitors), an exciter transformer, and variable frequency power supply. The amount of reactor modules and capacitors will vary depending on the test voltage. The step-up exciter transformer supplies the AC voltage required for the circuit. The variable frequency power supply provides an adjustable input voltage and adjustable frequency to the exciter (30–300 Hz) and controls the output voltage of the HV reactor. The output of the HV reactor is injected into an entry point on the GIS (either a test bushing or line exit bushing) one phase at a time, until all phases are tested throughout the entire GIS. Variable frequency allows for a smaller HV reactor to be used (allowing for better transportability, more flexibility for on-site testing).

Factory acceptance test is at levels indicated in IEEE C37.122 [1] and IEC 62271.203 [6]. The field test voltage is 80% of the factory test level, due in part to the possibility of uncontrollable environmental factors in the field, such as close proximity to grounded/nongrounded objects, weather and high humidity, corona, sharp metallic edges, etc. In order to pass the power frequency, withstand test, the application voltage needs to be applied for a total of 60 seconds. However, it is important to condition the equipment by slowly increasing the test voltage so it is more likely to pass – this allows the particles (if there are any) to move into areas within the GIS where they are not affected by the electric field. Conditioning increases the likelihood that the equipment will pass the required level of withstand voltage. If there is a flashover during testing, it is important to determine the cause; possibilities include a gas gap flashover, burning up a particle, sharp edge, insulator failure, etc.

5.6.3.2 Pretesting Precautions

The power frequency withstand test presents challenges to the testing personnel and the GIS that need to be identified and accounted for prior to energizing the test equipment. In addition, the following items must be performed and verified prior to commencing with any high-voltage test on a GIS:

- Test Plan: A power frequency withstand test plan is a sequence of tests in accordance with OEM and customer requirements for commissioning a GIS. The test plan would include the testing voltage level, switch configuration, the point of voltage injection, test sequence, if phases are tested separately or together, etc.
- The configuration and assembly of the test equipment is performed satisfactorily according to the testing manufacturer's specifications
- GIS electrical grounds and bonding bars are securely in place
- Station electrical clearance of equipment is properly completed including adjacent equipment isolated and grounded
- Pertinent individuals are briefed on personnel safety regarding HV testing
- Site areas affected by the test have been roped off/barricaded
- Gas zones are indicating at nominal density
- Gas zone gas quality checks and gas leak tests are completed satisfactorily
- All breakers and GIS devices are in the proper configuration for the test
- Current transformers secondary winding has been properly shorted
- Voltage transformers have been electrically disconnected from the circuit under test
- Removable links have been removed in accordance with the test plan

5.6.3.3 Testing Existing and New Equipment

In this section, information is given for the case that exiting and new GIS equipment will be tested with power frequency AC withstand voltage.

New GIS may be extended to add new circuit breakers, new bay positions, additional feeders, etc. Customers have requested guidance on what voltage level the connected existing equipment can withstand, recognizing that this equipment could have lost some of its dielectric withstand capability due to age. Removable links may have been removed between new and existing equipment, to allow the new equipment to be tested to required IEEE C37.122 [1] or IE 62271-203 [6] levels, but when links are reinstalled, the link compartment may need to be tested.

It is recommended that the testing entity contacts the existing original equipment manufacturer for recommendations regarding test levels. If unavailable, test to 110% of the rated maximum voltage (U_r) divided by $\sqrt{3}$ (Justification: IEC 62271-203 clause C.3.2.1) [6].

5.6.3.4 Testing GIS Equipment Including Cable Connections

In this section, information is given for the case that GIS equipment will be tested together with connected high-voltage power cables using power frequency AC withstand voltage when the high-voltage cable cannot be disconnected.

There are instances where the cables have been installed in their final position and cannot be removed. This may be due to many factors such as scheduling, scope of supply, lack of qualified personnel or equipment to remove the GIS-cable link. In other cases, due to the type of test equipment used, power cables and the outdoor termination of the opposite end may be the only feasible access to perform the withstand test on the switchgear.

Typically, GIS designs provide an isolating device to allow separation of the high-voltage power cable from the primary circuit. An isolating device can be an isolating link (removable by hand), or a disconnecting switch. The disconnect switch associated with the power cable must be located in such a way that all the GIS sections are properly tested prior to final commissioning.

Alternatively, if a nonremovable cable is part of the primary circuit, the cable and corresponding air termination may be utilized to test the switchgear; the lesser of the two ratings should be considered for the test voltage. In addition, the following should be considered.

Due to the large capacitance of power cables, a power supply capable of delivering the necessary output current must be considered; therefore, the total capacitance of the test object must be obtained. Given the large power demand, a resonant test system is often utilized. These systems either vary the compensating reactance or vary the test frequency. Since the variable frequency of 20–300 Hz may be utilized for GIS hi pot testing, either system may be utilized. In the event that a voltage transformer is connected to the test circuit, care must be taken to not exceed the saturation voltage at the test frequency.

Care must be taken to not surpass the maximum permissible test potential for the cable system. Different types of cable and accessories are covered under different guides and standards. The manufacturer of the cable and accessories must be consulted for acceptable maximum voltages, frequencies, and durations. Withstand time is often 60 minutes, which is less than the test requirements for GIS.

Cable systems are often designed to be free of any partial discharge at the test voltage of $1.7U_r$, although for lower rated cables, field PD voltage has not been harmonized. This should allow for PD monitoring of the switchgear at the PD test voltage; however, various measurement techniques must be employed to cancel the external noises.

5.6.3.5 Air Electrical Clearances During Testing

Prior to conducting the AC withstand test, proper air electrical clearance distance based on the test voltage between the equipment energized during field testing and workers, adjacent energized, and/or grounded objects must be considered.

WARNING: If possible, it is always recommended to de-energize all equipment in the testing area before performing high-voltage tests.

The equipment energized during the test includes the testing apparatus (voltage divider, reactor, and exciter transformer), electrical lead which connects the testing apparatus to the GIS, and other portions of the GIS that may become energized during the test (such as GIS bushings at another portion of the GIS or direct transformer connections or cable pothead assemblies).

The testing area must be assessed to identify the surrounding areas that pose a risk to field personnel and equipment. High-voltage withstand testing equipment requires multiple reactors and creates an odd-shaped electric field. If the testing apparatus is too close to an adjacent energized/grounded object, it can affect the results of the test, a flashover may damage equipment, or shock and burn nearby workers.

Following are four electrical clearance distances to consider, along with recommended best practices for each:

- Air clearance distance from equipment energized during field testing to workers
  - Only qualified electrical workers may be within the minimum approach distance, all other workers are prohibited from entering the area (refer to NFPA 70E Article 110) [16].
- Air clearance distance from equipment energized during field testing to grounded objects
  - Rule of thumb for test equipment placement to grounded objects: minimum 1× length of the test bushing to all structures as long as there are no sharp objects, as the bushing length is often considered minimum strike distance. Use minimum 1.5× length of the longest bushing in case sharp objects are present.
  - Use insulating blankets made of dielectric materials per ASTM D1048 [17] placed over grounded objects that are very close to the portions energized during testing.
  - Distances within gas-insulated spaces should also be considered, such as within a cable pothead enclosure or transformer bushing enclosure or gas-insulated surge arrester. These compartments will need to have the disconnecting link removed and filled to full rated gas pressure.
- Air clearance distance from equipment energized during field testing to other energized equipment
  - Rule of thumb for test equipment placement to energized/live equipment: minimum 2× highest voltage bushing length of equipment in the testing area. It may be necessary to perform a peak-to-peak calculation of out of sync waveforms to determine the maximum differential voltage. Other risk factors to consider are VFTs, possible lightning events, and odd electric field situations. It may also be beneficial to look for minimum clearances that are corona-free. Refer to Figure 5.11.
- Air clearance distance from equipment energized during field testing to reduce corona or PD (partial discharge) during tests
  - Under certain conditions, the localized electric field near energized components and conductors can produce a tiny electric discharge or corona that causes the surrounding air molecules to undergo a slight localized change of electric charge. The testing equipment shall be located in an area with sufficient electrical clearances to nearby grounded objects in order to reduce corona and PD during the test. Rule of thumb: 610 mm (24 inches) per 100 kV.

5.6.3.6 Testing GIS Equipment Including Cable Connections

In this section, information is given for the case that GIS equipment will be tested together with connected high-voltage instrument transformers, when the high-voltage instrument transformer cannot be disconnected.

Field testing according to C37.122 [1] would typically require a voltage transformer (VT) not rated to the voltage level of the low-frequency AC voltage withstand to be removed from the tested circuit. If the VT is not removed, the field test voltage may never be achieved and may overheat and damage the VT.

GIS VTs are often designed with removable links or integral isolating switches, which when opened will permit the independent testing of the primary circuit. Otherwise, VTs are often physically removed, which can be a laborious task.

Figure 5.11 Dimensional clearances during high-voltage AC withstand tests (Reproduced by permission of IEEE)

Due to many factors such as scheduling, scope of supply, lack of qualified personnel, or equipment, VTs may not be removed for the low-frequency AC voltage withstand test. In this case, the maximum field test voltage will be limited by the values shown on the VT nameplate (usually the phase-to-ground voltage). Always refer to the VT nameplate prior to determining the field test levels. Saturation effects of the VT may be used to limit the maximum voltage induced on the secondary of the VT. For this, the VT manufacturer needs to be contacted.

5.6.3.7 Testing GIS Equipment Including Surge Arresters Connections

In this section, information is given for the case that GIS equipment will be tested together with connected high-voltage surge arresters, when the high-voltage surge arresters cannot be disconnected.

Most high-voltage gas-insulated surge arresters can be disconnected from the circuit under test by a removable link, or an isolating switch, or removing the surge arrester completely and replacing with a high-pressure cover.

The surge arrester must be disconnected from the circuit in order to allow the hi-pot test set to reach the required field test voltage. The surge arrester will begin to conduct current when the voltage exceeds the MCOV (maximum continuous operating voltage) level (usually only 1.05% of the rated phase to ground voltage level), and therefore the circuit under test will never reach the required level.

If the surge arrester cannot be removed, then the test voltage must be limited to the MCOV level, or a portion of the tested circuit that contains the surge arrester could be switched out.

If the surge arrester has a monitoring device (which may count surges and/or provides surge arrester health and recommends condition-based maintenance), then it should be removed or damage to the monitor or future erroneous maintenance requirements may result.

5.6.3.8 Locating an Internal Failure

If a failure occurs during AC high-voltage withstand testing, following are useful suggestions to help locate the physical point of internal failure:

- Utilizing a gas tester, check for decomposed gas in compartments tested during the failure. However, there may be times when the fault power is not enough to produce enough decomposed gas to detect.

As a preventative measure, an accelerometer/piezo/acoustic-based monitoring device can be attached to the GIS enclosure and will provide a readout based on the intensity of the internal flashover in the compartments tested during the failure. When multiple devices are applied to the enclosure in various areas under test, the device that indicates the highest reading will be located closest to the internal flashover.

An other precaution possibility is to connect oscilloscopes to the partial discharge (PD) sensors to measure the time delay between the sensor locations and calculate the arc location.

References

1 IEEE C37.122 (2010) IEEE Standard for Gas-Insulated Substations.
2 IEEE C37.122.1 (2013) Guide for Gas Insulated Substations Rated Above 52 kV.
3 IEC 62271-4 (2013) High-Voltage Switchgear and Controlgear - Part 4: Handling Procedures for Sulphur Hexafluoride (SF_6) and its Mixtures.
4 Kuschel, M., Gerlach, M., Gorablenkow, J., and Kloos, A. (2011) Testing Labarotories for High Voltag Power Equipment – Aspects and Requirements to Ensure Reliable Energy Supply, XVII International Symposium on High Voltage Engineering, Hannover, Germany, 22.-26. August 2011.
5 IEC 62271-1 (2011) High-Voltage switchgear and Controlgear – Part 1: Common Specifications.
6 IEC 62271-203 (2017) Gas-Insulated Metal Enclosed Switchgear for Rated Voltages above 52kV.
7 IEEE C37.09-2018 - IEEE Standard Test Procedures for AC High-Voltage Circuit Breakers with Rated aximum Voltage Above 1000 V.
8 IEC 62271-100:2021, High-voltage switchgear and controlgear - Part 100: Alternating-current circuit-breakers.
9 EN 50052 (2017). High-voltage switchgear and controlgear - Gas-filled cast aluminium alloy enclosures.
10 EN 50064 (2018). High-voltage switchgear and controlgear - Gas-filled wrought aluminium and aluminium alloy enclosures.
11 EN 50068 (2018). High-Voltage Switchgear and Controlgear - Gas-filled wrought steel enclosures.
12 EN 50069 (2018) High-voltage switchgear and controlgear - Gas-filled welded composite enclosures of cast and wrought aluminium alloys.
13 EN 50089 (1994). Cast Resin Partitions for Metal Enclosed Gas-Filled High-Voltage Switchgear and Controlgear.
14 IEEE Std C37.122.7TM 2021, IEEE Guide for Field Testing of Gas-Insulated Substations Rated Above 52 kV.
15 IEEE C37.122.3 - Guide for Sulphur Hexafluoride (SF6) Gas Handling for High-Voltage (over 1,000Vac) Equipment.
16 NFPA 70 (2020) National Electrical Code, National Fire Protection Association.
17 ASTM D1048 series (2019) Standard Specification for Rubber Insulating Blankets.

6

Installation

Authors: 1st edition Hermann Koch, Richard Jones, James Massura
Reviewers: Phil Bolin, John Brunke, Michael Novev, Pravakar Samanta, Devki Sharma, 2nd edition John Brunke, Michael Novev, and Pravakar Samanta

6.1 General

In this chapter, practical information has been collected from hundreds of GIS projects over the last thirty years. The assembly of GIS equipment is complex and requires detailed planning and coordination of the work. With large amounts of equipment involved in the assembly, a well-structured work coordination plan is a fundamental requirement to provide an efficient work flow and a high-quality installation. Safety considerations and helpful assembly tips are provided.

Significant advanced planning is required including detailed engineering and design, preparation of construction specifications, material ordering, identification of material lay down and storage areas, future GIS extensions, other utility interconnections at the remote ends, identification of required permits, energization plans, testing and commissioning needs, documentation control, quality assurance steps, and so forth.

The project planning includes preliminary site surveys, core borings, security assessments, identification of environmentally sensitive areas, determining heavy equipment needs for site grading, civil works, and material handling, developing gas processing procedures, and scheduling special test equipment deliveries, personnel needs and qualifications, training, etc.

Visual inspections, testing for correct wiring of control cables, bus leakage and gas quality checks, main circuit resistance measurement, mechanical switching checks, interlocking checks, grounding measurements, and instrument transformer tests are some of the recommended tests explained in this chapter.

The reader will be provided with an overview of the requirements to plan and execute a successful and efficient GIS installation.

6.2 Installation

6.2.1 Introduction

The following information is provided as guidance to supplement the manufacturers' instructions or to use in the event that GIS modifications are required on older equipment, perhaps no longer supported by the manufacturer. As described below, individuals involved in the installation of GIS

Gas Insulated Substations, Second Edition. Edited by Hermann J. Koch.

should be fully qualified, aware of the high-voltage electrical and mechanical hazards, and the chemical handling aspects of the work. It is generally an accepted practice to have a manufacturer's representative present during the assembly to direct utility crews or contractors. Clearance rules apply before energizing, see Chapter 5 Testing Section 5.5 on-site field testing. Control voltage tagging will apply before energization. Once GIS equipment is energized, the owner's or utility's operations and safety requirements are usually applied, including: minimum approach distances to exposed live parts (e.g., bushing connections) and switching and tagging procedures. In accordance with local or regulatory directives, the owner or utility may specify gas handling practices and SF_6 use records (e.g., the quantity of gas in each compartment).

6.2.2 Safety Considerations and Assembly Tips

1) As in the case of most power equipment, there is a danger of electrical shock and significant injury. GIS equipment also introduces the complexities of working with pressurized compartments and gas handling. When installing new GIS, particularly additions or expansions to energized existing equipment, always follow proper lock-out tag-out procedures and confirm that devices or hardware are properly sectionalized, grounded, and depressurized before accessing a GIS compartment.
2) If the installation involves a cable trench or vaults, basements, or confined spaces below grade or floor level, the space should be tested for oxygen levels. SF_6 is heavier than air and may accumulate in low or confined areas of a facility.
3) GIS equipment has multiple mechanisms in small equipment enclosures and can be operated remotely. To avoid injury, confirm that motor mechanisms and spring/pneumatic and hydraulic operators are discharged and de-energized before any work begins.
4) Use approved tools for the work, including the correct size wrenches, operating handles, and other hardware. Some projects will require metric and English toolsets to work with locally supplied materials and the GIS.
5) Current transformers may be present in the GIS assembly. To avoid damage, the secondaries, if not in use, should be shorted before energization.
6) Avoid excessive force during the assembly process; epoxy insulators can crack with a significant jolt or unusual lateral or axial stresses during alignment.
7) If GIS equipment is provided with view ports to confirm disconnect or ground switch positions, do not directly observe while the device is opening or closing. Electrical arcs may be generated that could damage eyesight.

6.2.3 General Project Planning Outline

A complete installation plan, including considerations for future addition of GIS bays and equipment, is essential for a coordinated, efficient successful project. Generally, the preassembled equipment and bus sections coupled with the manufacturer's instructions will dictate the assembly sequence and follow a series of steps including:

1) Preconstruction meeting between the user and the manufacturer including the installation contractor
2) Site preparation including grading, installation of drainage, foundations and grounding, access roads, and auxiliary power/water/sanitary/waste removal facilities
3) Staging of construction equipment required during the installation
4) Final alignment and leveling of foundations and associated equipment
5) Receiving, unloading, and storing GIS equipment

6) On-site GIS primary assembly
7) Leak testing
8) Protective Relays, SCADA, Communications, Revenue Metering, station service and the associated. Control wiring installation
9) Vacuum processing and gas filling with moisture and gas quality tests
10) Mechanical or operational tests on the switches, instrument transformers, and circuit breakers
11) Primary bus high-voltage dielectric and partial discharge tests. It is also important to identify the test connection point, particularly in the case of indoor equipment where a temporary air to gas test bushing may be needed
12) Commissioning and energization
13) Construction punch list and site cleanup

6.2.4 Future GIS Expansion Considerations

The initial design of a GIS facility should include provisions for expansion and future needs. As an example, if present requirements based upon planning studies determine two transmission lines are required in a breaker-and-a-half scheme at this time, and within the next ten to fifteen years load growth will require additional bus connections, space should be allocated for the future equipment, particularly if a new building is part of the project. In addition to space allocations, it may be a prudent step to install isolation disconnects where the future sections are to be installed. When the future connection is made, SF_6 gas is removed from the disconnect zone and the existing bus on the opposite side is only reduced in pressure, not completely emptied. This step will minimize potential contamination and moisture infiltration in the bus.

6.2.5 Advance Planning and Preliminary Site Evaluation

As part of the preliminary engineering, the various parties (owner, contractor, manufacturer) should jointly develop a *composite project schedule* for the complete project. The schedule should be as detailed as possible with key milestones well defined. Accurate planning and project coordination is an essential element for success. Major elements of a schedule should include:

1) Engineering and regulatory approvals
   a) Management and project organization information (milestone completion dates, key payments)
   b) Document and drawing approval process
   c) One-line, current and voltage drawings and electrical schematic release sequences
   d) Identification and coordinated release of construction drawing packages (civil, structural, electrical primary and GIS installation, secondary wiring and controls)
   e) Installation (regional siting board, conservation, local community) permits and building approvals
   f) Regional transmission authority or public service commission approvals (security, communications, protection systems)
2) Material procurement
   a) *Large long lead-time components:* GIS equipment/circuit breakers, bus, transmission cable and transformer bushing terminations, and other large interconnected elements – cables and transformers
   b) *Routine electrical construction materials:* conduit, fittings, cable tray, control cable
   c) *Civil and structural materials:* steel supports, anchor bolts, rebar, concrete
   d) *Consumables:* cleaning supplies, paint, lubricants

3) Construction sequence
   a) Site preparation: yard grading, fencing and ground mats
   b) Civil works, foundations, and building erection
   c) Electrical primary equipment
   d) Secondary control wiring and cabinets
   e) Miscellaneous: station service power, paging, security, lighting
4) Testing and commissioning elements
   a) Preliminary tests and measurements (cables and point to point)
   b) Vacuum and gas processing
   c) High-voltage conditioning
   d) Energization and in-service measurements
5) Project closure
   a) Construction records/as-builts, maintenance test record turnover
   b) Spare parts and tools
   c) Operations and maintenance manuals and procedures

These steps are part of the project management. The above-mentioned elements should be further broken into specific tasks and have start dates, time durations, estimated completion dates, person/company responsible, percent completed, and so on. Once the tasks are identified by the team, commercially available software can assist with the schedule development.

Project progress should be reported on a regular basis with the report frequency variable dependent on the project stage or criticality. While regulatory and engineering progress may only require monthly updates, construction activities are usually reported on a weekly basis. Both progress and delays should be identified to help make effective changes and avoid impacts on the critical path elements. Subcontractor activities should also be tracked on the composite schedule to avoid potential conflicts. Dependent on the owner's bid process, on some projects multiple companies or firms may be involved in the work. As an example, Firm A may supply the GIS equipment, Firm B the GIS bus, Firm C the local control panels, with Firm D responsible for the overall design and engineering. Included within the advanced planning would be coordination between these firms. Periodic meetings throughout the project will facilitate scheduling, material deliveries, construction in congested areas, safety, and the overall efficiency of the work.

6.2.5.1 Document Control

Document control is another critical tool for project success. A master summary document lists all the required studies, design calculations, preliminary/conceptual drawings, design drawings, material lists, and perhaps other pertinent correspondence (regulatory approvals). The document or drawing number, title, and description are included along with the responsible party preparing and reviewing the information, transmittal, receipt and approval dates, its latest revision number, and date. The drawing issues are also identified, for example, preliminary, design review, and for construction. Drawing or document transmittal sheets are also helpful to track the changes as engineering releases, owner reviews, and released-for-construction drawing issues are made. Transmittal sheets should include in the title: the project name, customer name, customers' contract number, sales order number, bid number, action required (for construction or installation, preliminary review, information only, return with comments), and so on. The body of the transmittal should identify each document or drawing included, the recipient's name and address, and when the document or drawings should be returned.

6.2.5.2 An Initial Site Assessment and Survey

This is generally made before any installation or construction work is initiated. The survey should be a written document that is periodically reviewed and updated. A sketch or layout of the installation is a useful tool to identify hazards, lay down areas, environmentally sensitive sections or wetland boundaries, trailer placement, excavated soil storage, and so on (see Figure 6.1).

Hazards that should be identified include existing energized overhead lines, buried facilities, for example, gas lines, electric cables, subsurface structures, and water or sewer lines. Many utilities provide free underground services' identification or contract third-party firms, who will locate and mark underground utility equipment. In most regions which are densely populated, it is a legal requirement to contact the designated identification authority before any excavation begins. Once the physical hazards are identified, they should be clearly marked with signage, tape, paint, reflective drums or cones, and designated on the site sketch. This information should be included as part of the site brief for installation and construction personnel, and posted in crew are trailers or similar meeting locations. As the work progresses and storage areas are eliminated, traffic patterns change, and equipment energized, construction personnel should be briefed, and the site survey sketch revised.

6.2.5.3 Site Transportation Access

A field survey to identify the preferred shipping route into the site is recommended. Specific concerns may include restricted delivery hours due to crew operations or a residential/urban location, bridge heights or weight limitations, oversized load or vehicle width limitations, noise, and so on. Directions to the site, material receipt times, and a project construction contact should be provided on purchase orders and with each shipment.

6.2.5.4 Security

Local police should be contacted and informed of the construction activities and as necessary after work hours patrols requested. Local authorities may also supply historical information on incidents in the immediate area, which would be helpful to determine the level of security required. Material storage areas are most susceptible to theft and vandalism. The use of lockable storage containers, "conex" boxes (see Figure 6.2), fencing, security lighting, and, if the situation warrants it, surveillance cameras or security guards should be considered. Cable reels, copper ground grid conductors, "salvage" dumpsters with copper scrap, and other high-value materials should be covered or stored out of the public view.

6.2.5.5 Material Storage

Equipment lay-down and storage areas should be defined (see Figures 6.3 and 6.4) that provide heavy equipment access, facilitate large component off-loads, and are secure. Indoor or covered areas are preferred, but if outdoors, a well-drained location should be selected. Crane clearances from energized lines should be reviewed. Material handling equipment certifications should be checked for current status and the latest lift/rating tests confirmed. Operators and rigging licenses should be checked for current status and the certification data recorded.

Some utilities will permit storage in an existing switchyard, in which case GIS bus, equipment crates, and such should be placed at least 10 feet from the perimeter fence. At new "green field" locations, erection of a security fence is recommended, or for indoor locations a secure lockable storage room, basement area, or loading dock bay may suffice.

6.2.5.6 Environment, Health and Safety Plan (EHS)

An environment health and safety plan (HASP) should be developed for the site. The document should include:

1) Scope of work
2) Project personnel: roles and responsibilities, qualifications
3) Hazard identification and risk assessment

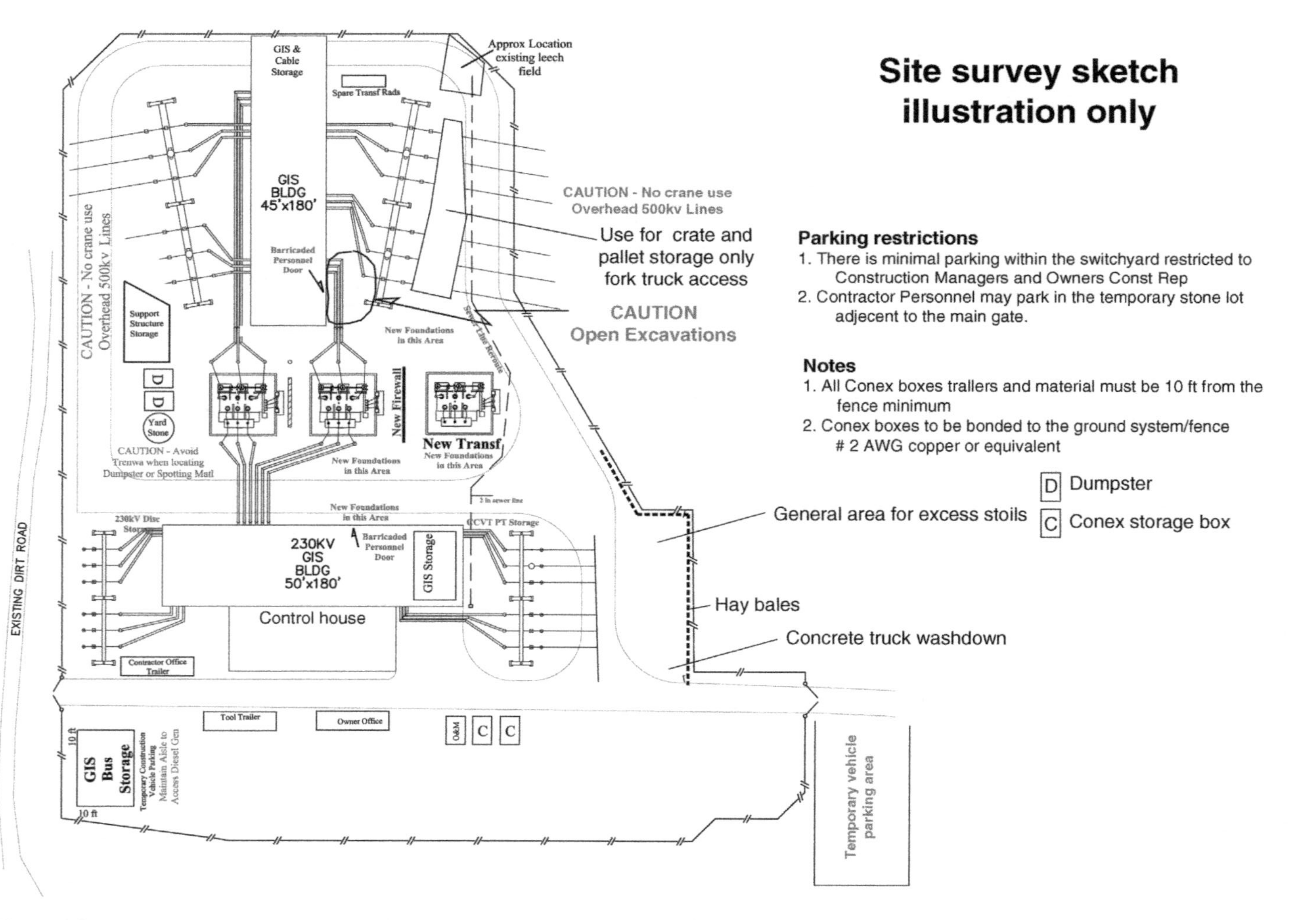

Figure 6.1 Example site survey sketch (for illustration only) (Reproduced by permission of Tech S Corp)

Figure 6.2 Conex box (Reproduced by permission of Tech S Corp)

Figure 6.3 Gas bus off-loading and storage for large amounts of SF_6 (Reproduced by permission of Tech S Corp)

Figure 6.4 GIS crates and bushings storage on site (Reproduced by permission of Tech S Corp)

4) Communications: emergencies, incident reports, safety meeting, and job briefs
5) Technical safety requirements: personal protective equipment (PPE), clearances, tool inspection, rigging equipment, equipment isolation, tagging, fall protection, and traffic control
6) Compliance criteria and requirements: audits, issue identification, and mitigation steps
7) Environmental aspects: guidelines and permits, erosion controls, preventative measures, response steps (spills), and wetland/species protection
8) Site-specific concerns

Before personnel begin installation work, they should be provided with a EHSP briefing. Personnel should also sign a site log to state that they understood the EHSP, emergency procedures, and environmental specific concerns (e.g., wetland boundaries). Site evacuation information including an emergency marshaling area and local hospital information should be posted in conspicuous locations throughout the work site. Individuals with first aid or other emergency medical training should be identified. Emergency supplies should be checked for current condition and quantities including fire extinguishers, first aid kits, emergency eye washes, and environmental spill cleanup kits. Periodic site safety updates and personnel briefings are recommended as the installation progresses. As a supplement to the EHSP, material safety data sheets (MSDSs) should be assembled for every chemical, solvent, cleaner, lubricant, fuel, and so on, used at the installation site and placed in a readily accessible binder.

6.2.5.7 Environmentally Sensitive Areas

These areas, including endangered vegetation, animal nesting, and wetland boundaries, should be marked and documented on the site assessment survey sketch. Any street, roadway, roof, or similar facility with drains leaving the installation/construction area should also be identified and the outfall marked. Spill cleanup kits (plastic drums with absorbent materials) should be placed where there is a possibility of spills, including areas adjacent to construction equipment, in order to contain hydraulic fluid or fuel leaks. As required by the authority having jurisdiction, a site environmental plan and/or a spill prevention, control, and countermeasure (SPCC) plan may be required. These should also be placed in readily accessible areas.

6.2.5.8 Specialized Equipment Requirements

Gas-insulated substation or bus installations require specialized equipment for gas processing and testing. This equipment should be procured and checked well before the actual use.

1) Gas-processing equipment with adequate storage capacity. SF_6 gas is handled using commercially available gas-processing trailers that contain oil-free vacuum pumps, and process equipment, gas storage tanks, compressors, filters, and dryers. The size of the individual gas compartments and the evacuating and storage capacity of the gas-handling equipment are especially important in large stations.
   Suitable evacuation equipment and perhaps an auxiliary vacuum pump may help expedite the gas processing. To fill compartments directly from gas cylinders or gas-handling equipment, heat sources (e.g., wraparound blankets for cylinders) may be needed to counteract the chilling effect of the expanding gas and maximize gas removal from the cylinder.
2) Gas cylinder regulators and an electronic SF_6 gas leakage detector.
3) Electronic analysis equipment to measure SF_6 gas purity and moisture levels.
4) Dry air in sufficient quantity to back-fill all compartments and support construction efforts, specifically for temporary storage during assembly of large bus/equipment sections or temporary work stoppages (holidays).

5) Cleaning or bus protection supplies including:
   a) Commercial-type vacuum cleaner with high efficiency particulate air (HEPA) filters and nonmetallic accessories.
   b) Clean plastic gloves and work clothes. Some utilities, especially if energized equipment is in the project vicinity, may require fire-retardant (FRE) clothing. One important aspect is that the installation crew wear clean clothing that will not introduce contaminates (loose threads, dirt, metal chips) into the bus during assembly.
   c) Lint-free cloths and manufacturer-recommended solvents.
   d) Temporary plastic bags or covers for sealing openings after components have been opened.
6) Specialized tools supplied and recommended by the manufacturer. When tools (manual operator cranks, jigs, alignment templates, measurement devices) not readily available on the open market are required for installation and maintenance of the equipment, one new unused set should be furnished by the supplier before commissioning.
7) Miscellaneous
   a) Ventilating equipment
   b) Welding and metal working tools/equipment
   c) Handling and lifting equipment: nylon slings, cranes
   d) Ladders and platforms as required
8) Special electrical test equipment
   a) Insulation resistance tester
   b) Micro-ohm meter
   c) Circuit breaker stroke measurement and travel and timing test equipment
   d) High-voltage insulation test equipment. High-voltage test equipment (series resonant) is required to check the quality of the insulation system before energization. Supplemental entrance bushings or adapters may be required, particularly where direct cable or transformer connections are made. Termination caps and corona plugs may also be needed for closing the end of an assembly when the entire bus has not been completed.

6.2.6 Training

6.2.6.1 General

GIS training includes three parts: the construction/assembly crew, owner's operations and maintenance (O&M) personnel, and the dispatchers/operators. Where possible, members of the owner's O&M group should participate with the construction/assembly crew and observe the installation, gas handling, and commissioning of the GIS. O&M crews should be encouraged to take progress photographs, inspect assemblies before gas compartments are sealed, particularly where direct transformer or cable connections are in use, operate disconnects and ground switches, observe commissioning tests, record SF_6 gas filling and gas purity/quality tests, and in general understand the equipment operations before the primary circuit is energized. Dispatchers should visit the site after construction completion, and preferably before energizations, to familiarize themselves with the equipment operation, gas zones, tagging procedures, and other operational procedures.

6.2.6.2 Construction and Assembly Crew

The construction/assembly crew should be provided with instructions on the GIS switchgear and bus before the work begins. Color-coded cross section views of the equipment will provide the assembly crew with a better understanding of "what is in the tank or bus assemblies." The training should be in two parts. The first part consists of a classroom setting where a manufacturer's representative can discuss correct component handling and rigging, the "delicate" nature of the equipment and bus, the criticality of component alignment, site and installation cleanliness and dust control, the impacts of moisture on bus and gas processing, overnight or longer term "open" times, the importance of dry air back-fills, regulated gas handling/weighing and environmental reports, leak and vacuum rise tests, high-voltage commissioning tests, and turnover records. The second part of the training consists of field work, inspecting the site or installation area, the equipment or bus to be installed, confirming hardware weights and rigging plans, and reviewing assembly procedures.

During morning "tailboard" meetings, where the day's work plan is discussed, the initial training should be reinforced and crews periodically reminded of the importance of cleanliness in the work area and bus sections, tool accountability, and other technical aspects of the work. Safety topics should also be included in the morning briefing and include rigging, gas handling, required personal protective equipment (PPE), and so on.

Other items to consider during the initial training are to provide the project organization chart, overall schedule, site drawing(s), any required assembly checklists, emergency and security/access procedures, and any special site-specific hazards or environmental restrictions.

Individuals with specific skills should provide copies of their credentials, including, but not limited to, welding certifications, riggers and equipment operator's licenses, first aid/medical/safety training, and so on. If multiple firms are involved in the project, the interface points should be clearly defined for the construction crews. Representatives from the firms should provide in advance of the work drawings that identify the assembly points, including alignment criteria, tolerances, wiring termination points, gas fill responsibilities, and such. It is also important to use a common measurement reference for all parties e.g., survey site marker, top of foundation point, floor level, or final finished grade.

6.2.6.3 Operational and Maintenance Crew

Similar to the construction crews, the owner's O&M personnel training should include both classroom work and practical "hands-on" tasks. Classroom training should include a detailed description of the GIS equipment operation, including internal cross section illustrations, gas handling, safety aspects (emergency fault procedures with arc by-product detection and handling), ground and disconnect switch operations (motorized and manual) and multiple switch position indications (stops, target, viewports, flags, bore scopes, cameras, etc.), circuit breaker maintenance (lubrication, operations counters, interrupter operations), acceptable moisture and gas quality measurements, and so on.

The practical training should include breaker operations, correct ground switch and disconnect for open/close, gas density meter locations, local control cabinet interfaces, and remote-control locations (if used).

6.2.6.4 Dispatch Crew

Dispatch training should focus on the electrical operations of the GIS equipment and couple with the importance of the mechanical aspects, including the gas zone integrity, gas zone boundaries, alarm and trip conditions, interlock logic diagrams, tagging procedures, and so on.

6.2.6.5 Other On-Site Personnel Not Directly Involved with Assembly

For other on-site personnel except perhaps material deliveries, for example, visitors, civil or mechanical contractors installing foundations, HVAC systems, buildings, and so on, electrical hazard awareness training is recommended, so their personnel are aware of the site hazards and/or restricted areas.

6.2.7 Material Receipt and Control

6.2.7.1 General

With the material storage area defined, a storage plan should be developed with materials to be used toward the end of the project, placed toward the rear or less accessible areas to avoid double picks. The plan should identify the material, its location on site, and any required routine checks. Elevated signs identifying the various components may be helpful if the storage is outdoors, where snow accumulation is a concern. Equipment stored outdoors should be on dunnage (wood pallets or framing) a sufficient distance off the ground to avoid water, snow accumulation, and so on, with sufficient room to walk between boxes or hardware, inspect nameplates or shipping labels/component IDs, check storage pressures, and so on.

6.2.7.2 Bus Bar, Disconnect Ground Switch, Circuit Breaker Shipping

GIS bus, disconnects, and ground switches may be shipped with dry air at a slight positive pressure. When this equipment is likely to be stored for extended time frames, the pressures should be checked on a regular, weekly basis and recorded. Similarly, circuit breakers may be shipped with a positive SF_6 pressure, which is "topped off" after installation. The breaker should be equipped with a density gage that can be used to monitor the breaker compartment(s) storage pressure. Control cabinet heaters may also require station service power to minimize accumulation of moisture in the cabinet.

6.2.7.3 Incoming Material Inspection

All materials received on the job site should receive an incoming material inspection. The manufacturer may also have specific inspection criteria to record for critical items. In general, the inspection should check for shipping damage including cracked or damaged containers, broken wood support frames, or other signs the equipment was dropped or damaged by impact. Some devices (e.g., circuit breakers) are shipped with pressure/density gages and a zero-gage reading, due to shipping or handling problems, may indicate a loss of positive pressure, with moisture infiltration into the equipment a possibility. All hardware should be checked against the shipping manifest. The manufacturer and carrier should be advised of any damage or unusual observations. Any impact recorders or similar handling quality or shock detection devices provided should be interrogated for evidence of mishandling. Oversea shipments in particular may be handled and picked several times before reaching the final destination. Global positioning system (GPS) devices, attached to the larger crates or GIS assemblies may help to track shipment deliveries.

Equipment shipping weights should be provided on the manifest documents and verified against the container/crate labels and properly rated material handling equipment provided on site. For assembly purposes, installation crews also require the actual equipment or hardware weights less the crate or packaging materials. If there is any question, the manufacturer should be consulted to confirm the weights before any equipment is lifted. Center of Gravity and load pick points should also be clearly identified.

The GIS bus should be inspected for dents, gouges, cracks, and other indications on the surface that could signal interior damage. Significant dents in the bus may alter the electrical insulation and mechanical strength properties of the bus.

Once materials have been inspected and received, depending on the job size, an inventory control may be required to ensure that materials are installed or that consumables (cleaning supplies, lubricants, sealants) are ordered in a timely manner to avoid delays. In general, a material inventory table consists of: the item description, supplier, part number, quantity, P.O. number, order date, ship date, delivery date, drawing reference, storage location, responsible individual, and notes. Included in the notes would be special instructions like connect control cabinet heaters or check shipping gas pressure.

6.2.7.4 SF_6 Gas Management

SF_6 gas management is another critical element. As a significant greenhouse gas, SF_6 requires careful controls. In some cases, there are government regulations that by law require SF_6 to be weighed upon receipt at a job site, the individual cylinders weighed as gas is extracted and pumped into a bus (see Figure 6.5), and then the gas bottles individually weighed before leaving the installation site. All these weight measurements are recorded and officially transmitted to the appropriate regulatory authorities. Weight record and incoming gas quality record sheets are provided as in Figure 6.6.

Depending on the site conditions, a common gas handling and storage area may be helpful. SF_6 cylinders are color coded with a green top and silver on the bottom with high internal pressures. The standard safety considerations for pressurized bottles should be followed including: not lifting by the valve protection cap, securing bottles not in use, ensuring material handling equipment is available (carts/lifts) to move bottles, and so on. A gas quality measurement should also be made once each gas zone is filled. These data will be used to establish a maintenance baseline or benchmark. For gas handling refer to IEC 62271-4 [1] and IEEE C37.122.3 [2].

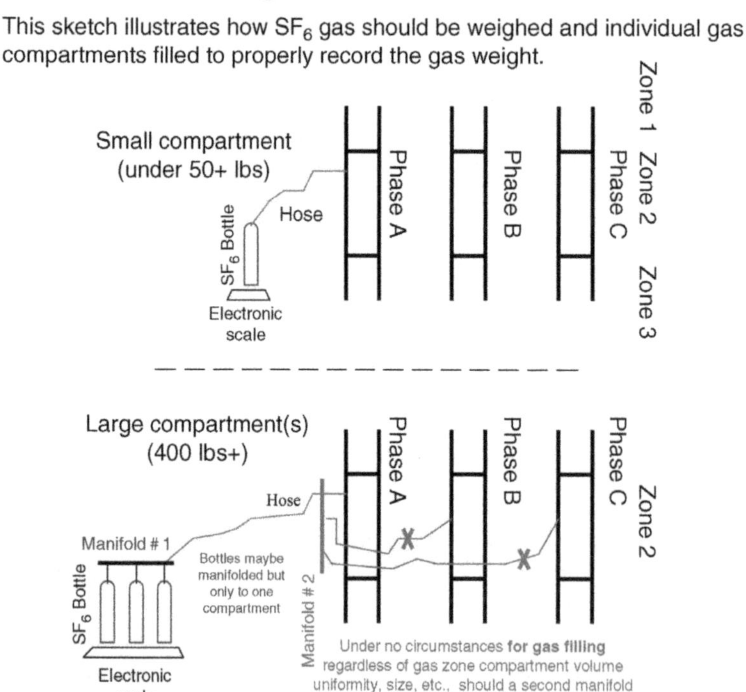

Figure 6.5 Gas weight record and sketch (Reproduced by permission of Tech S Corp)

Memorandum
SF_6 Project Gas Inventory

Date: xx/xx/xxxx
Individual Recording Data: ____________

Subj: **SF_6 Gas quality Measurements (New Bottles Received XX/XX/XXXX)**
Meter: RH Systems SF_6 Analyzer 973 v1.2

Cylinder No	Date/Time	Weight lbs	Vessel Pressure	Dew/ Frost Point 0C@atmP	Volume Ratio ppmv	Volume SF_6 %	Notes
992536							
52364Q							
12568L							
785236							
15487AA							

Figure 6.6 SF_6 project gas inventory (Reproduced by permission of Tech S Corp)

6.2.8 Installation/Assembly Instructions

6.2.8.1 General

As a mechanically bolted or welded assembly, the GIS installation relies on a level, rigid base. The foundation or baseplate level tolerances are critical to the success of the project. Generally, the manufacturers have established specific criteria for the foundation and baseplates. Today's GIS installations may use an imbedded channel where the equipment is initially tack welded and, once the final position is established, a bead weld is installed along the support foot. Other installations may use leveling bolt/nut combinations. During the design phase of the project, the manufacturer and owner should review the installation requirement and agree on the foundation design criteria. During construction and equipment placement, careful, precise surveyed control points will be required to ensure correct equipment locations.

The manufacturer's drawings and assembly steps should be provided to the installation crew well in advance of the work to allow sufficient time for review and site survey work. For more details see IEC 61936-1 [3].

6.2.8.2 Shipping Covers

Shipping covers should not be removed until the equipment or bus is actually ready for connection. Caution: to prevent moisture infiltration, GIS components are shipped with a positive pressure of dry air or nitrogen. Before completely removing the shipping covers, the shipping pressure should be relieved. Interior cleanliness of the bus should be maintained using plastic covers attached to the outer enclosure with tape. The interior compartment should only be open during flange cleaning or the actually connection process. If the bus or equipment "open time" is extended, it can impact on the time required to reach an acceptable vacuum level. If an extended open time is necessary, the vacuum time may be reduced by increasing the dry air or nitrogen back-fill set time [1, 2].

6.2.8.3 Large Pieces

When moving large pieces of equipment or bus, tag lines are recommended to control the load. If the installation is outdoors and high winds are present, the installation crew may opt to delay the work due to the large "sail" area and inability to control the lift. If precipitation or high humidity is present, the crew should avoid opening and exposing the compartment interiors to moisture.

6.2.8.4 Clean Bus and Flange

Materials used to clean the bus or flanges should be nonconductive and lint-free. Cleaning materials should be properly and regularly disposed of, in approved containers to minimize the possibility of a fire. Alcohol or similar chemicals should be stored in approved lockers or spaces with sufficient ventilation. Fire extinguishers should be placed in the vicinity of cleaning solvents stored or in use.

6.2.8.5 Lubricants and Sealants

Only manufacturer's approved lubricants, greases, sealants, and similar compounds should be used in the assembly. Use of other chemicals may have adverse and corrosive impacts on the insulators, mechanism links, or other GIS components.

6.2.8.6 Tools

All tools used in the vicinity of open bus should be subject to an inventory and the count verified before the bus is sealed. An internal bus and equipment inspection should also be made with personnel looking for tools, extraneous cleaning materials, loose hardware, shipping blocks, foam rings, or any other item that might impact the insulation system integrity.

6.2.8.7 Bus Bar Assembly

The assembly of the bus should be carefully directed by a crew leader. Forced fit of the bus or other components should be avoided to prevent insulator or contact damage. While expansion joints may provide some help with component alignment, these joints are not intended to correct major displacements or misalignments of the bus. Once the flanges or GIS sections are secure, bolted connections should be checked for the proper torque (see Table 6.1). It is a good practice to mark the bolt heads after the final torque measurement see Figure 6.7.

Checklists should be completed as the equipment, various sections, or compartments are assembled and include items such as connection tightness (bolt torque), visual inspection

Table 6.1 Approximations of US and international dimensions and torques (Reproduced by permission of Tech S Corp)

US size[a]	Bolt size	Hex head[b] (mm)	Torque (kg cm)	Torque (N m)	Torque (lb ft)
	M6	10	60	5.9	4.3
1/2	M8	13	140	13.7	10.1
11/16	M10	17	280	27.5	20.3
3/4	M12	19	480	47.1	34.7
15/16	M16	24	1200	118	86.8
	M20	30	2200	215	159
1¼	M22	32	3000	294	217
	M24	36	3900	382	282
	M30	49	7700	755	557

[a] US sizes are approximations for comparison only.
[b] Hex head is the wrench size.

Figure 6.7 Bolt head marked as torque checked (Reproduced by permission of Tech S Corp)

for debris or tools, and so on. Are pressure relief vents pointed away from areas where operating activities occur or personnel are present (walkways)? Is adequate maintenance space provided around the equipment? As most breaker interrupter assemblies today slide out of their enclosure to replace arc contacts or nozzles, will structural supports, cable tray, conduit runs, gas monitors, and so on, impede or restrict the access? Ground connections should be verified.

Where responsibilities change either between manufacturers, for example, GIS equipment by supplier 1 and gas bus by supplier 2 or between contractors or subcontractors, a joint supplier interface assembly checklist should be developed and used to confirm the correct dimensional measurements.

Disconnect and ground switches should be inspected after installation to ensure correct operation. Manual operation should be verified and contact penetrations measured. External linkages should be examined for correct alignment as well as proper operation of position indicators, stops, flags, or other similar devices. If temporary power is available, motorized switches should be operated. Caution: generally, circuit breakers should not be operated at below design

Figure 6.8 GIS assembly moved into place by mobile crane (Reproduced by permission of Tech S Corp)

Figure 6.9 GIS breaker moved into place by building bridge crane (Reproduced by permission of Tech S Corp)

pressures since the SF_6 gas may be used to help in dampening the interrupter assembly operation. Figures 6.8 and 6.9 show GIS modules or breakers being moved into position with a mobile crane or a building bridge crane. Figure 6.10 shows a circuit breaker being slid on rollers into its final bus position.

6.2.8.8 SF_6 Gas Handling

After a compartment assembly is complete, generally nitrogen or dry air with low moisture content is used to back-fill the compartment. Manufacturers will vary on the length of time the back-fill is needed. Field experience indicates 24–48 hours usually results in optimum moisture absorption. At the end of the back-fill, the vacuum process using a gas cart with vacuum capabilities (see Figure 6.11) begins to test the compartment integrity. A vacuum manifold is shown in Figure 6.12, which may help facilitate the evacuation process. After a predetermined time of 1 to 2 hours of vacuum, a rise test is conducted, with readings taken typically at 10-minute intervals, over a one-hour time frame. The compartment vacuum levels will initially rise but should eventually level out or flatten for an acceptable test (see Figure 6.13). For more detailed information see IEC 62271-4 [1] and IEEE C37.122.3 [2].

Winter outdoor GIS installations pose some particular challenges. Moisture removal via a vacuum process is very difficult at low ambient temperatures, so provisions need to be made to provide supplementary heat and temporary enclosures. These enclosures can be made from wood and plastic, and should include blankets or similar insulation. Outdoor gas handling and processing may also require supplemental heaters. If possible, gas bottles should be stored in a heated enclosure before use.

The installation crew should maintain records of how long the bus or equipment was open, the time spent with the nitrogen or dry air back-fill, the vacuum rise test, and the total vacuum time to reach accepted manufacturer levels. In the absence of manufacturer's instructions, the bus is usually taken to 300 μ PA and equipment is taken to 1000 μ PA, and then vacuum is continued for an additional hour. These records should be on a per-compartment, per-phase basis (for an iso-phase bus).

As the back-fill, vacuum, and gas process continues, the installation crew should tag the status of each compartment (see Figure 6.14). These tags can include "Under Vacuum," "Partial Pressure," and "Fully Pressurized," and may be color coded. The installation crew training

Figure 6.10 GIS breaker slide into position on roller pads using come-alongs (Reproduced by permission of Tech S Corp)

Figure 6.11 Gas cart for vacuum and filling GIS (Reproduced by permission of Tech S Corp)

Figure 6.12 Typical vacuum manifold (Reproduced by permission of Tech S Corp)

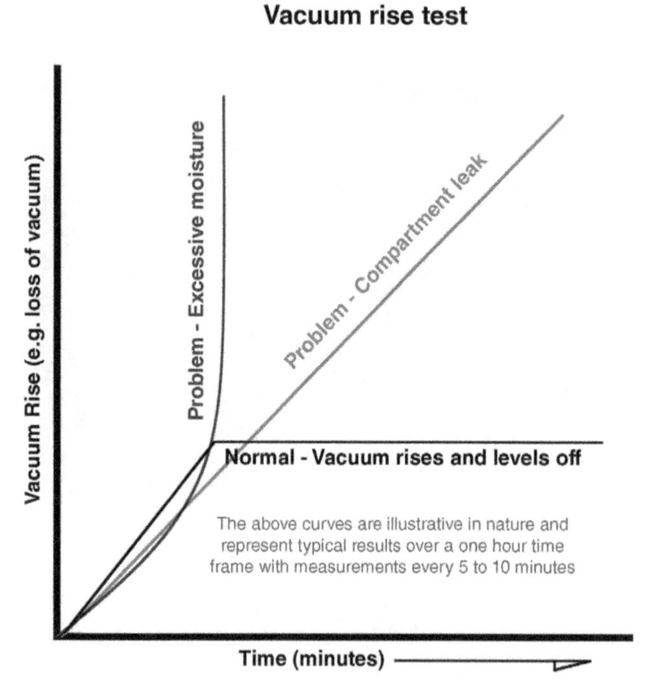

Figure 6.13 Vacuum rise test (Reproduced by permission of Tech S Corp)

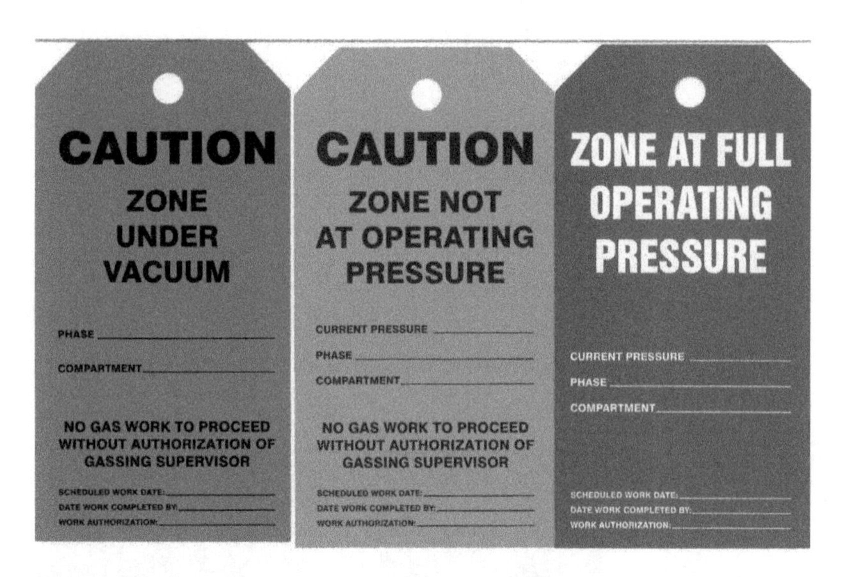

Figure 6.14 Gas zone processing tags (Reproduced by permission of Tech S Corp)

should describe the importance of the tagging system and the color code. Tags should be placed in conspicuous places near gas valves (see Figure 6.15). One or two individuals should be responsible for the entire gas handling process including tagging as the work progresses. A log should be maintained on the job site, particularly on large projects with continual updates provided for each step. Posting overall gas zone diagrams with the status of each zone aids project

Figure 6.15 Example of tagged gas zone adjacent to the valve (Reproduced by permission of Tech S Corp)

Figure 6.16 Assembly of GIS circuit breakers (Reproduced by permission of James Massura)

management, enhances safety, and provides the installation crew with a ready reference of the job status. Figures 6.16, 6.17, 6.18–6.20 show various GIS facilities and components under assembly.

6.2.8.9 Retro-Fit

During retro-fit, it may be necessary for safety reasons to reduce the gas pressure in adjacent gas zones. Personnel should determine the GIS gas zone boundaries by locating the barrier insulators using drawings or if available, the manufacturer's physical bus markings. In some cases, owners will specify a color-coded band using tape or paint on the bus or insulator indicating a barrier/gas zone barrier or boundary (see Figure 6.21). A generally accepted practice is to reduce the

Figure 6.17 Assembly of the bus bar (Reproduced by permission of James Massura)

Figure 6.18 Assembly of voltage transformers (Reproduced by permission of James Massura)

immediately adjacent zone to 2 or 3 PSIG and the zone immediately beyond to 50% nominal operating pressure (see Figure 6.23). Posting a gas zone diagram for the work crew is also a prudent safety measure. The gas zone diagram should show the zones involved with the work and the condition of the remaining zones within the facility. In addition to tagging the gas zone conditions, lock-out tag out procedures are a prudent and in some areas a requirement. Figure 6.22 is one method to lock a mechanism and prevent its operation.

On any project, gasket and O-rings should be properly stored until ready for use in a dry, room temperature space. After a GIS flange is cleaned and inspected, the appropriate sealant should be applied followed by the gasket or O-ring. Before installation, the gasket or O-ring should be inspected for cuts,

Figure 6.19 Cable termination to the GIS (Reproduced by permission of James Massura)

Figure 6.20 GIS connection to the overhead line (Reproduced by permission of James Massura)

nicks, rigidity (dry out), or similar damage that would compromise the gas seal. In the case of retrofits, when flanges are opened, particularly after a long time period, the gasket or O-ring is likely to have a "set," and means it should be replaced. Desiccant should also be replaced in the affected zone.

6.2.9 Gas-Insulated Substation Tests

6.2.9.1 Construction Visual Inspections

These tasks generally include:

1) Check all bonding and grounding conductors are installed, connected, and tight.
2) Check the condition of the gas valves for normal operations.
3) Inspect the appearance and condition of the primary GIS, for example, damaged paint, construction scraps, tight structure support connections, viewport covers, and so on.

Figure 6.21 Example of gas zone barrier insulator location (dark band) (Reproduced by permission of Tech S Corp)

Figure 6.22 One example a GIS disconnect lockout method (Reproduced by permission of Tech S Corp)

4) Cleanliness of the circuit breaker cabinet, disconnect and ground switches, and marshaling and local control cabinets.
5) Equipment labeling including nameplates and device identification plates.
6) Overall job site appearance and cleanliness.
7) Condition of safety equipment, including fire extinguishers, first aid kits, and eyewashes.

6.2.9.2 Control Cable

Typically, control cable tests are in two categories. First, the insulation integrity is tested using a 1000 V source (Megger), which may be manual or motorized. The owner or manufacturer should

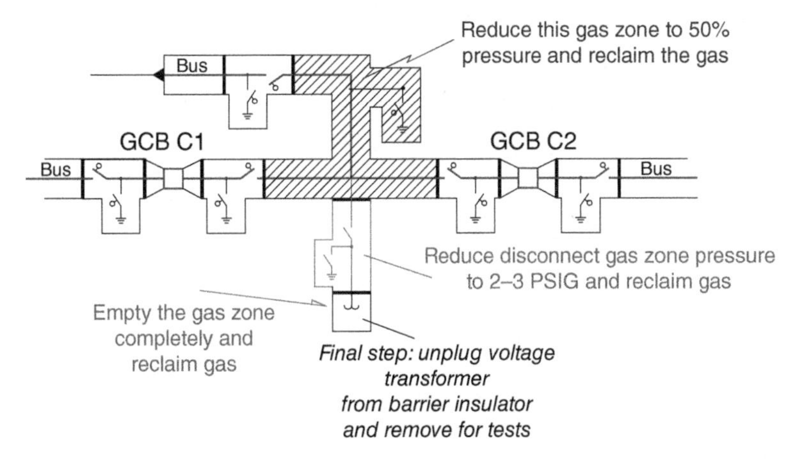
Gas zone pressure reduction
This sketch illustrates a recommended pressure control in the bus as for example when voltage transformers are installed or removed.
Reduce this gas zone to 50% pressure and reclaim the gas
Bus
GCB C1
GCB C2
Bus
Bus
Reduce disconnect gas zone pressure to 2–3 PSIG and reclaim gas
Empty the gas zone completely and reclaim gas
Final step: unplug voltage transformer from barrier insulator and remove for tests

Figure 6.23 One example a GIS disconnect lockout method (Reproduced by permission of Tech S Corp)

be consulted for a site-specific procedure; however, the objective is to confirm that individual conductors are insulated from each other and the cable shield, if one is present. The cable jacket integrity is also tested. The objective is to detect missing or damaged insulation. Successful tests should show high resistive readings.

The second category is a point-to-point continuity test to verify that the conductors are landed on the correct terminals. Installation personnel isolate the connections using slide links and then verify, using sound powered phones, buzzers, or similar equipment, that the connection is correct.

The installation crew should maintain records of the insulation tests and the point-to-point wiring checks. It is also a generally accepted procedure to trace the various protection, control, interlocking circuits, etc., during the point-to-point checks, and "yellow line" the associated electrical wiring or schematic drawings as the checks are made to verify the secondary circuit integrity.

6.2.9.3 Bus Gas Leak Checks

Once the various GIS compartments have been processed and gas-filled, a second leak check may be recommended by the manufacturers, particularly for long bus runs. Each flange is covered by plastic and sealed with duct tape. A small stone is placed in the bottom of the plastic "bag." This plastic "enclosure" remains for 12–24 hours. An SF_6 detection device is then inserted into the bag. If there are gas leaks, as SF_6 is heavier than air, it will settle to the bottom of the plastic bag (see Figures 6.24) and activate the detector. This will only last some hours before the SF_6 will be diffuse in the surrounding air. Figure 6.25 shows examples of two models of SF_6 gas detectors.

6.2.9.4 Gas Density Monitor and Local Alarm Tests

This test involves keeping the gas density monitor wired but isolated electrically with open terminal block slide links or a similar switch mechanism, before mechanically isolating the density switch using appropriate valves. Bleed a small amount of gas off and using an ohmmeter at the terminal block, observe and record the alarm point. Two-way radio or a similar communication may be required for large installations. In a similar manner, slowly introduce some gas and as the

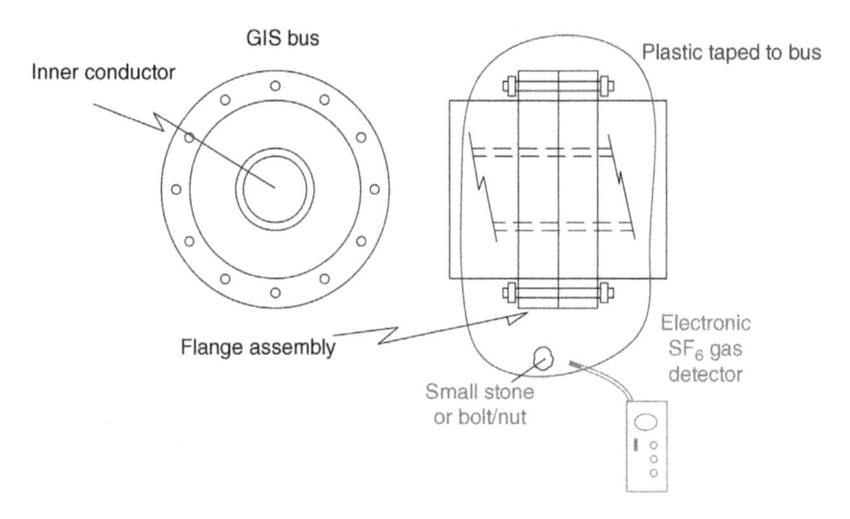
Flange gas leakage test
This sketch illustrates on method to check for the integrity of a gas bus flange for SF_6 leakage
Tape a plastic bag to the bus and insert a small weight, SF_6 is heaver than air so it will sink to the bottom of the plastic. After 24 hours insert an electronic leak detector probe, there should be no reading or gas detection.
GIS bus
Inner conductor
Plastic taped to bus
Flange assembly
Electronic SF_6 gas detector
Small stone or bolt/nut

Figure 6.24 Flange gas leakage test (Reproduced by permission of Tech S Corp)

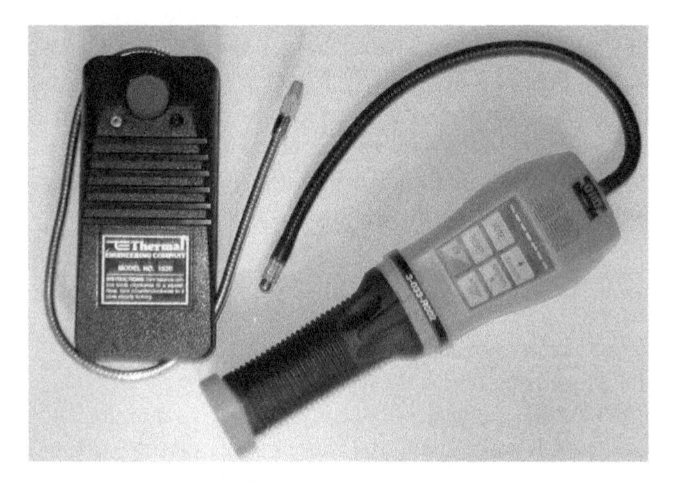
Thermal

Figure 6.25 Gas zone leakage detectors (Reproduced by permission of Tech S Corp)

pressure rises, observe the density switch reset point. The test purpose is to verify the final wiring and actual alarm points.

6.2.9.5 Primary Circuit Resistance

Generally, this test requires a 100 A DC μΩ meter with independent voltage and current sources. Tests are conducted on an assembled GIS section as specified by the manufacturer, isolating the circuit using a ground switch and removing its grounding strap. The various GIS switches in the test circuit are configured to meet the measurement requirements, and with the current lead connected to the "floating" ground strap and the μΩ meter voltage lead connected to the ground switch enclosure, 100 A are injected into the circuit. The readings obtained are compared to engineering calculations, which may include a tolerance specification. An acceptable field measurement agrees with the engineered calculation.

6.2.9.6 SF_6 Gas Quality Tests

Once the gas compartments are processed and filled to the correct pressure/density, gas purity tests are conducted (see Figure 6.26). These tests require a small amount of gas and identify the level of moisture present and the purity of the SF_6. For new GIS equipment installations, acceptable values are moisture levels in the 150–300 ppm range and gas purity in the 99.5% range. Some manufacturers decrease the acceptable parameters in older or reprocessed gas to 500 ppm moisture with a gas purity of 98%. These values should be recorded for maintenance baseline use when gas quality measurements are made in the future. A table similar to Figure 6.6 should be prepared to record the individual gas zone SF_6 moisture content and purity for future maintenance baseline comparison purposes. For more details see IEC 62271-4 [1] and IEEE C37.122.3 [2].

6.2.9.7 Circuit Breaker Tests

Caution: before proceeding with any breaker operations tests, any closing or tripping prevention pins or lock devices should be identified. Depending on the test, the pins or lock devices may or may not be required. The breaker operating pressures should also be verified to confirm and the circuit breaker is at full operating levels. For details see [4, 5, 6, 7].

Mechanism Stroke, Wipe Measurement

These tests are generally performed at the factory and verified in the field to ensure that shipping or the installation procedure has not damaged the breaker mechanism or caused it to deviate from acceptable tolerance levels. Manufacturers will provide the measurement technique and acceptable limits for both the stroke and main contact wipe. This set of measurements is usually done when operating the circuit breaker manually (see Figure 6.27).

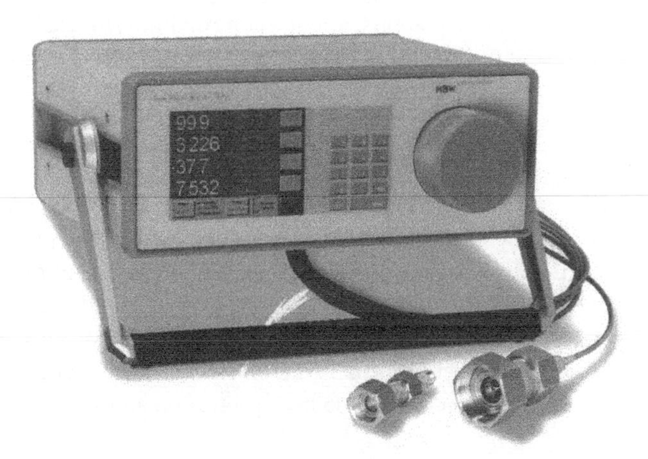

Figure 6.26 SF_6 gas analyzer (Reproduced by permission of Tech S Corp)

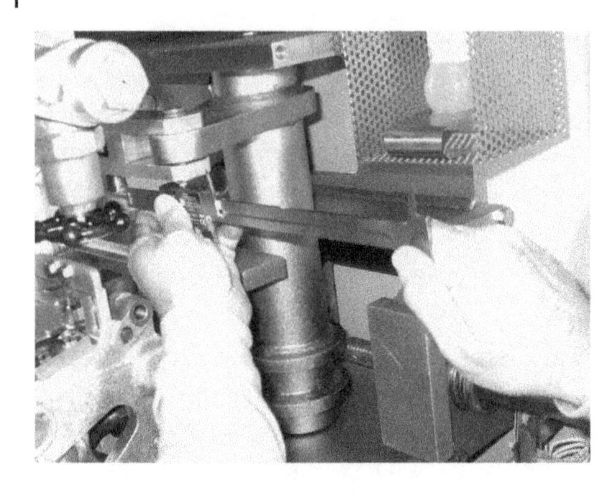

Figure 6.27 Circuit breaker stroke measurement (Reproduced by permission of Tech S Corp)

Open/Close Operation Including Anti-pumping

This test will confirm whether the circuit breaker properly opens and closes. One manual test is performed and then the necessary open and close operations at the operating control voltage, as identified in the specification, for example, open and close "X" times in a specified time. The objective is to observe whether the lights and indicators work properly and the owner's specification requirements are met.

The anti-pumping test involves closing and opening the breaker with a continuous signal on the close coil. If the anti-pumping circuit is properly functioning, the breaker should not re-close until the close coil is de-energized.

Travel and Timing Test

These tests are to confirm the dynamic operating performance of the breaker by measuring the operating times on a per pole basis and identifying any discrepancies or deviations with the phases. Manufacturers will typically provide the timing range and the maximum phase discrepancy limit. Any readings outside the manufacturer's stated values should be investigated and adjustments made as necessary. Figure 6.28 is typical instrumentation used to check circuit breaker contact timing.

Low Gas Tripping and Block Close Operations

This test is on the circuit breaker gas system to prove whether gas leakage detection (multiple alarm levels) is operational and the control limits, for example, serious loss of gas, will block close and prevent the breaker operation.

The manufacturer usually confirms that the tests are acceptable before a circuit breaker is placed in service.

6.2.9.8 Ground and Disconnect Switch and Circuit Breaker Interlock Tests

GIS equipment is generally provided with interlocks to prevent incorrect operations; for example, most disconnects are not designed for load break operation and similarly ground switches cannot be closed without interconnected circuit breakers and disconnects open. The manufacturer's interlock logic diagram should be reviewed to verify the various open and close combinations and confirm that incorrect operations are blocked. These interlock schemes should also be checked for correct integration into the owner's protection and switching logic.

Figure 6.28 Circuit breaker timing equipment (Reproduced by permission of Tech S Corp)

6.2.9.9 High-Voltage Bus and Equipment Conditioning Tests

Before the GIS installations are placed in service, a high-voltage conditioning test is performed with momentary voltages in excess of the equipment rated line-to-ground voltages. The objectives of the test are: to identify any abnormalities in the bus (loose hardware, tools, cleaning material inadvertently left in the bus) that could compromise the internal electrical clearances, identify excessive moisture levels and move conductive and semi-conductive materials to low stress areas or particle traps to prevent insulation flashovers. The test voltage levels and time durations vary between manufacturers and are based upon the equipment voltage class, but for illustration purposes on a 362 kV class, 1050 BIL installation, the test levels may be as shown in Table 6.2.

The high-voltage test on site is usually done by a resonance test procedure. The inductance of the test transformer and the capacitance of the GIS are chosen in such a way that they form a resonance circuit in the range of 50–100 Hz. This allows much smaller test equipment, which is shown in Figure 6.29.

The benefits of using a series resonant test are the inherently limited fault current available, compared to a fault energy present when the GIS equipment is connected to the transmission system. A typical test on 362 kV equipment will take approximately an hour. A three-person team is recommended including a safety observer, a timer, and the test set operator. The substation or switchyard low voltage (120/208 V, 220 V, 380 V, 277/440 V) station service connections and capacity should be identified early in the project. If insufficient capacity is available for the test set, a small portable generator may be required.

The GIS can be tested in multiple ways. To expedite the process, one methodology is to test large bus sections at once, taking into consideration the test set capacity. The disadvantage of this approach is that if a discrepancy is identified, the test section will need to be sectionalized into smaller increments to isolate the problem area. A second approach is to begin with smaller test sections, but this will require more test time.

During normal tests at the higher voltages, corona may be heard around the test set wire connected to the GIS bushing.

The test fails, if the required voltages cannot be attained or the test set trips off before the peak voltage level is reached. Note: if new GIS equipment is retrofitted to older designs and both sections must be tested concurrently, it may be prudent to reduce the test levels to avoid problems with the older equipment.

Table 6.2 Test levels (Reproduced by permission of Tech S Corp)

Step	Step (kV)	Time (min)
1	100	20
2	125	15
3	175	10
4	225	2
5	275	1
6	300	1
7	325	1
8	350	1
9	400	1

Figure 6.29 Resonance high-voltage test equipment (Reproduced by permission of Tech S Corp/Energy Initiatives Group)

After a test failure, given the low energy involved using the test set, repeat tests can be performed to help isolate the problem. If possible, the test section should be reduced with portions of bus systematically removed from the test, until the problem section is located. It may also be helpful to locate individuals at points along a bus to help locate the "ping" or audio noise that may occur when a flashover (test set trips) occurs inside the bus. Personnel should be positioned in safe locations, well clear of the test set and its connections. In case of particles, it is possible that the particle will disappear (destroyed by the arc) and with the next HV test applied, the equipment will pass the test. For this reason, the criteria to pass the test is 15 impulse voltage and maximum 2 fail, if these are not the last two impulse voltages, the test is passed.

6.2.9.10 Instrument Transformer Tests

Current and voltage transformer tests include polarity, ratio, and current transformer saturation curves. These tests require voltage and current injection and are equivalent to testing conducted on similar air-insulated equipment. The results are compared to the manufacturer's published data.

6.2.9.11 Other Tests/Records

AC Station Service Measurements, Heater and Control Cabinet Light Operations

This work involves checking the heater circuits for current with a clamp-on amper meter, observing correct light operations using a door or manual switch, confirming all power receptacles are operational, and checking for tight connections.

Ground fault protection should be provided for receptacles, particularly if the local control or marshaling cabinet is located outdoors. Three-phase 60 or 100 A disconnects with independent power sources located on the exterior of the local control or marshaling cabinets may also be convenient to connect welders or gas cart/processing equipment.

The equipment and control cabinet heater test results should be recorded. The circuit breaker and other heaters in the GIS primary circuit are critical, particularly in cold climates to prevent SF_6 liquefying and loss of gas density. Heaters may also be required at transition points between heated indoor locations and outdoors. Thermostatic control settings (if used) should also be verified and the levels recorded.

Final Circuit Breaker Counter Readings

Before equipment is made available for owner operations personnel, all circuit breaker counters are recorded and the values entered in the commissioning records.

Turnover Gas Zone Density Readings

In a similar manner, all gas zone pressures are recorded including circuit breakers on a compartment, per phase if an iso-phase bus is installed. The bus temperature is also recorded and the pressure reading temperature-compensated. The value obtained is compared to the engineered nominal value with acceptable readings are within the engineered value.

Note: Some owners/operators may prefer gas density gages with numerical readings. In other designs, a color-coded gage may be acceptable with the gage needle in the green area on the gage face indicating satisfactory, the yellow area indicating caution or investigation of a potential leakage situation, and the red area requiring immediate action and potentially a significant problem. The gas density gage design should be discussed during the initial engineering discussions.

6.2.10 Commissioning, Energization, and Outage Plan

Most transmission systems in densely populated areas are heavily loaded and, as a result, commissioning plans may need to be submitted months in advance of the actual ready-for-service date. A typical approach is to split the commissioning into two parts: first, the overall energization plan and, second, the outage plan. The energization plan includes:

1) Use the owner's area or dispatch one-line drawings to identify where the new GIS equipment will interface with existing facilities.
2) Define outage responsibilities between the owner, contractor, test company, control authority, and so on, including personnel and time requirements.
3) Identify the general outage steps, for example, energize transmission line 1, bus sections W-1, W-2, and cable X.
4) Determine points of measurement for in-service readings and phase checks using voltage and current transformers recording voltages and currents as various elements are energized.

5) Establish all pre-energization documentation requirements including: construction punch list resolution, a detailed schedule, final regulatory or permit approvals, regional operations compliance, and so on.
6) Define when the operations ownership transfer will occur.

It is helpful to prepare a checklist with input from all the responsible parties and list the various energization tasks and the individual(s) with responsibility for specific assignments.

Also included in the pre-energization approval process may be protection and controls organizations that will require confirmation that regional security and protection criteria are met including: cyber security, redundant protection schemes, back-up AC and DC power sources, dual and independent circuit breaker trip coils, and so on.

As the pre-energization work progresses, meeting(s) with the owner's dispatch personnel occurs where the switching and tagging protocols, commissioning personnel qualifications, communication methodology, and personal protective equipment requirements are discussed. If the project is sufficiently large, a commissioning group organization chart may be helpful, which also defines the various individual's responsibilities.

The second step in the commissioning process is to develop a detailed outage plan. As part of the commissioning process, a separate group may be required to prepare detailed switching orders. Usually, an outage or cutover plan is drafted by the installation contractor and then revised and approved by the dispatch group. In complex transmission systems, there may be several levels of approvals, including the local authorities, then regional system operators, and, if the interconnection is across borders or regional control areas, perhaps a national authority has jurisdiction. The outage plan identifies:

1) The schedule, timing, and actual devices and their identification numbers (disconnect switches, circuit breakers) to be used as the new GIS equipment is energized.
2) How sections or portions of the new GIS installation will be energized.
3) Hold points in the process as measurements are taken and equipment operating conditions are checked.
4) If applicable, "soak times" where the equipment is energized, but no load is applied.
5) Control or Hold points for all involved personnel to "sign off" as the steps are completed.

If the commissioning process occurs over several months or even years, depending on the interconnection of transmission lines, cables, or generators, a commissioning log may be necessary so that, as the test crews and engineers change, a coherent history of the work and measurements is made.

6.2.11 Maintenance and Turnover Documents

As part of the commissioning process, the owner, manufacturer, and contractor should periodically meet and develop a turnover plan. Some of the items to consider are:

1) What maintenance and test records are required and in what format (hard copy/paper or electronic)?
2) Are special software packages needed, for example, to access cable schedules and termination information in the future?
3) Project documentation and as-built drawing transfers?
4) How will warranty items be addressed?
5) What are the routine maintenance requirements?
6) Emergency manufacturer's contact and recommended spare parts information?
7) Are any spare parts included in the purchase? If so, an inventory should be made and a storage area identified.

Figure 6.30 Tool rack (above framed one line and gas zone drawings) (Reproduced by permission of Tech S Corp)

8) An inventory of special tools should be made including manual disconnect handles and circuit breaker manual jack assemblies and a storage area/rack identified. It may be beneficial to the O&M crews to provide commonly used tool storage points in the vicinity of the GIS equipment including one line and gas zone diagrams (see Figure 6.30).
9) Are all gas zones/boundaries and devices including viewports labeled with the owner's approved nomenclature, particularly disconnect and ground switches and circuit breakers?

A turnover checklist developed by the responsible parties may help ensure that all the documentation (as-built drawings, cable schedules, O&M manuals, test records) tools, spare parts, permits, gas handling records, and so on, are properly addressed.

6.3 Energization: Connecting to the Power Grid

6.3.1 Grid Connection Considerations (Long-Term Planning)

Consideration must be given to the geometry and clearances of the transmission lines and their interconnection to the GIS equipment. The property lines, easements, and rights-of-way will also impact the GIS connections. Gas-insulated equipment and direct cable connections thereto have the desirable characteristics and geometry to place phase spacing in much closer proximity than can be achieved with conventional air-insulated systems.

The sequencing of the transmission, generation, or transformer lines will likely be dictated by the bulk power authority or the sponsoring utility. The project staff must consider the clearances (both electrical and personnel) required to construct and commission the facility including any temporary structures or line supports needed.

For the cutover of the lines to be completed in a timely fashion, the protective relaying at the remote line terminal(s) must be compatible and coordinated with the new gas-insulated equipment. Factors to consider in the design include the philosophy of relaying (current differential, directional comparison blocking, permissive overreach transfer trip, etc.) and the communication system to the remote ends (microwave, fiber optic, dedicated phone line, satellite).

6.3.2 GIS Grounding System

The grounding system for GIS equipment varies markedly from air-insulated substations. Because of the relatively high capacitance value between the center conductor and the concentric housing of GIS equipment, particularly at higher voltages, it is possible to develop significant charges on to the GIS housing during the switching process. For this reason, the grounding system for GIS equipment will typically be multipoint grounded. Calculations are required to determine the touch, step, and transfer potential and design the grounding system to limit the potential levels to safe values.

The project team should consult with the chosen manufacturer during the design of the ground grid and ancillary equipment. Particular attention should be placed on the ground connections within the gas-insulated equipment and their connections to the overall substation ground system.

6.3.3 Gas Zones

The number of gas zones required for GIS equipment will vary significantly as to whether the equipment is manufactured in isolated phase connections or a 3-in-1 design with a common housing enclosure for three phases. The design will vary from manufacturer to manufacturer, but, generally, the higher the voltage class, the greater the likelihood that isolated phase connections will be required.

Prior to manufacture, it is important to establish sufficient gas zones in the equipment to allow for maintenance or repair without jeopardizing or preventing the operation of adjacent GIS equipment. GIS manufacturers may impose limits on adjacent gas zone density differentials, that is, operations may not be allowed to maintain full density on one zone while the adjacent zone is under full vacuum. For this reason, an intermediate zone may be required and should be carefully considered.

Consideration should be given for the use of independent valves for the gas zone density gauges or contacts. The advantage of this valve during the maintenance or repair operations usually offsets the small added cost.

6.3.4 Operational Considerations

Operating personnel of the sponsoring utility or independent power producer (IPP) should be engaged early in the establishment of naming conventions. Some examples of naming conventions or standards include switch or breaker identification, gas zone identification, color coding of signage, position indication method, mimic bus patterns, viewport and camera use, and so on.

Operating personnel, in concert with SCADA personnel and the bulk power authority, should be consulted in the formation of SCADA tabulation standards, point priorities, and data limits. An orderly plan for commissioning the new SCADA points as they become "live" should be reviewed with all affected parties.

References

1 IEC 62271-4:2013 High-voltage switchgear and controlgear - Part 4: Handling procedures for Sulphur hexafluoride (SF6) and its mixtures.
2 IEEE C37.122.3 - Guide for Sulphur Hexafluoride (SF6) Gas Handling for High-Voltage (over 1,000Vac) Equipment.
3 IEC 61936-1:2021 CMV Power installations exceeding 1 kV AC and 1,5 kV DC - Part 1: AC.
4 IEC 62271-1 (2011) High-Voltage switchgear and Controlgear – Part 1: Common Specifications
5 IEC 62271-100, High-Voltage Alternating Current Circuit Breakers.
6 IEC 62271-203 Gas-Insulated Metal Enclosed Switchgear for Rated Voltages above 52kV.
7 IEEE C37.122 (2010) IEEE Standard for Gas-Insulated Substations.

7

Operation and Maintenance

Authors: 1st edition Hermann Koch, Charles L Hand, Arnaud Ficheux, Richard Jones, Ravi Dhara, 2nd edition Richard Jones,
Reviewers: 1st edition Phil Bolin, Noboru Fujimoto, Dave Solhtalab, Richard Jones, Devki Sharma, George Becker, Dick Jones, Hermann Koch, 2nd edition Ryan Stone, Patrick Fitzgerald, Gerd Ottehenning, Coboyo Bodyona and Hermann Koch

7.1 General

The Chapter 7 topics focus on GIS operations and maintenance. A significant difference between conventional air-insulated substations (AIS) and the gas-insulated substations (GIS) is that the GIS-energized electrical components are enclosed within a grounded, pressurized metallic enclosure. One of the major differences between GIS and AIS is concerned with the operation of the switchgear with respect to the circuit breakers, disconnect switches, and grounding switches. This is because in the United States, viewports are required to provide visual viewing of the open or closed disconnect switch gap or the open or closed grounding/earthing switch position. In other countries, instead of the visual control of the open gap of the disconnect switch, the kinematic chain according to IEC 62271-203 requirements is used to prove the correct disconnect [1]. There are other special features and tools related to the operation of a GIS such as, cameras, endoscopes, mitigation of induced currents in the metallic enclosure, gas density alarms, local control cabinets, interlocking/automation, remote control, unique mimic diagrams, gas zone indication markings, and accessibility requirements. These topics are covered in Section 7.2.

Maintenance, Section 7.3, provides practical experiences from existing GIS in use for more than 30 years. It explains the nature and sources of faults including their repair. Common maintenance procedures on measures to define the quality of the insulating gas (SF_6), visual and optical inspections, gas handling records, and leak detection methods are addressed. Specific practices with respect to interlocks, gas zone management, spare parts, and special tools are explained. In one section, experiences and lessons learned are brought together on gas zone identification, secondary interface design, wiring practices, and hardware options. Preconstruction training, monitoring procedures, GIS retrofit, and bus leak repair topics are also explained on the basis of practical experiences.

In Section 7.4, the repair process of GIS is explained. Information is given on the types of expected failures, which require repair of a GIS, how much time is required to effect repairs, and how service continuity is related to the design.

Gas Insulated Substations, Second Edition. Edited by Hermann J. Koch.

Section 7.5, Extensions, explains how a GIS can be extended after it has been installed and operated for many years. This section gives information about the definition of the interface between two GIS of different dimensions. It also shows the work that needs to be done when extension is not anticipated from the initial design stage, how service can continuously be provided during extension work can be provided, and how the new interface can be tested after extension.

In Section 7.7, retrofit or upgrade of GIS is explained. Here, the challenges of old switchgear are discussed, and information is provided to retrofit or upgrade a substation. A specific view is taken on the replacement of circuit breakers in GIS to increase switching capacity.

The case of overloading GIS is explained in Section 7.8 on overloading and thermal limits of the GIS for typical ratings of 1250 A up to 6300 A rated continuous current. The principle of the GIS design for continuous rating current is shown under the thermal and dielectric conditions given in a substation design. The determination of the thermal limits based on IEEE and IEC standards are given. The maximum continuous load current and the short-time overload capacity are explained and equations for calculation of overloads are presented.

7.2 Operation of a Gas-Insulated Substation

7.2.1 General

Both AIS and GIS use circuit breakers, disconnect switches (isolators), and grounding switches, and have various means of indicating their position, either opened or closed. Operation of a gas-insulated substation (GIS) uses most of the same principles as operating an air-insulated substation (AIS) although the various active components are physically configured differently.

The first obvious difference is that the blades of the disconnect switches and the grounding switches used in a GIS are surrounded by grounded metallic enclosures. This enclosure prevents the blades from being readily and easily visible to determine their fully opened or fully closed position. The second obvious difference relates to the bus conductors being located inside grounded metallic enclosures. These enclosures prevent the bus from being grounded with portable personnel grounds except at very discrete locations, such as air-to-gas bushing terminations at transmission lines and transformer banks.

Generally, a GIS requires more extensive electrical interlocking between the circuit breakers, disconnect switches (isolators), and grounding switches. The specific detailed method of operating and interlocking is generally specified by the user of the GIS.

7.2.2 Circuit Breaker

Since one of the main purposes of a circuit breaker is to automatically and rapidly de-energize faulted transmission lines, transformer banks, and buses, an opening operation is initiated by protective relays, which are generally installed remotely in a relay and control room. This operation may also include automatically reclosing of the circuit breaker one or more times. The opening or closing of a circuit breaker can also be initiated by human intervention from several locations, including manually from the circuit breaker mechanism cabinet, electrically from the local control cabinet, from the relay and control panels remote from the circuit breaker, or from supervisory controls remote from the substation.

A circuit breaker, generally, is not prevented (interlocked) from opening and/or closing by the position of either a disconnect switch or a grounding switch. However, a circuit breaker is prevented from operating or is automatically required to operate by parameters that affect its successful function, such as low interrupting and insulating gas density, low mechanism pressure, pole disagreement tripping (in the case of circuit breakers with independent pole operation), or monitoring of the proper position of the circuit breaker before initiating an operation. As an example, a circuit breaker must be in the fully closed position before an open operation can be initiated and, conversely, a circuit breaker must be in the fully open position before a close operation can be initiated.

The indication of the position of the circuit breaker can be monitored at the circuit breaker mechanism cabinet via mechanically operated semaphores, at the local control cabinet, at the relay and control panels remote from the circuit breaker, or from supervisory controls remote from the substation with positions indicating red and green lights or semaphores. Figure 7.1 shows a GIS circuit breaker and Figure 7.2 is a typical mechanism control cabinet. For more details see [7, 8].

7.2.3 Disconnect Switches

Disconnect switches in a GIS are used for the same purpose as those in an air-insulated substation (AIS). They are used to isolate various components of the substation, such as circuit breakers, transmission lines, transformer banks, buses, and voltage transformers. They generally do not have significant interrupting capability except for small quantities of charging current associated with short pieces of bus. These charging currents are in the range of 0.5–2.0 A. For more details see [9, 10].

Disconnect switch operating mechanisms include:

- Motor operated with manual hand crank override
- Manual hand crank operation only
- Single-phase operation or three-phase group operation. Figures 7.3–7.5 show GIS disconnect in several configurations and an motor operated isolation switch.

The disconnect switch position, open or closed, may be determined by one or more of the following:

- An indicating device, such as red and green lights or semaphores, in the local control cabinet
- A mechanical semaphore in the switch operating mechanism cabinet
- The physical position of the linkages that drive the switch blade

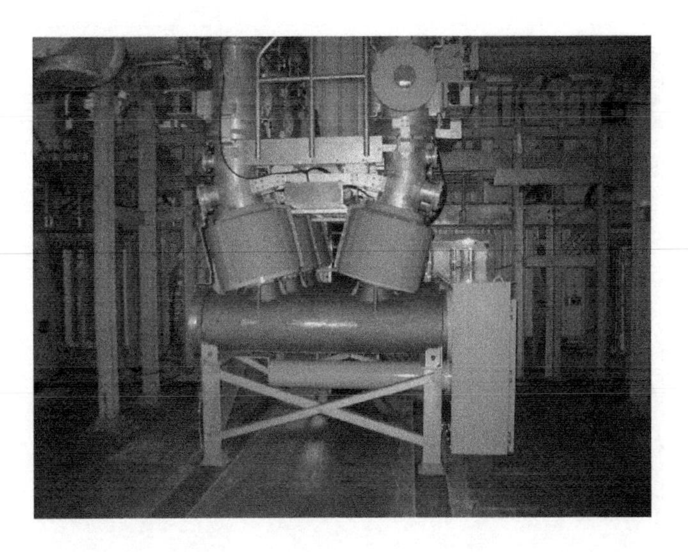

Figure 7.1 GIS circuit breaker (Reproduced by permission of Tech S Corp)

Figure 7.2 Circuit breaker mechanism control cabinet (Reproduced by permission of Tech S Corp)

Figure 7.3 GIS disconnect switches (Reproduced by permission of Tech S Corp)

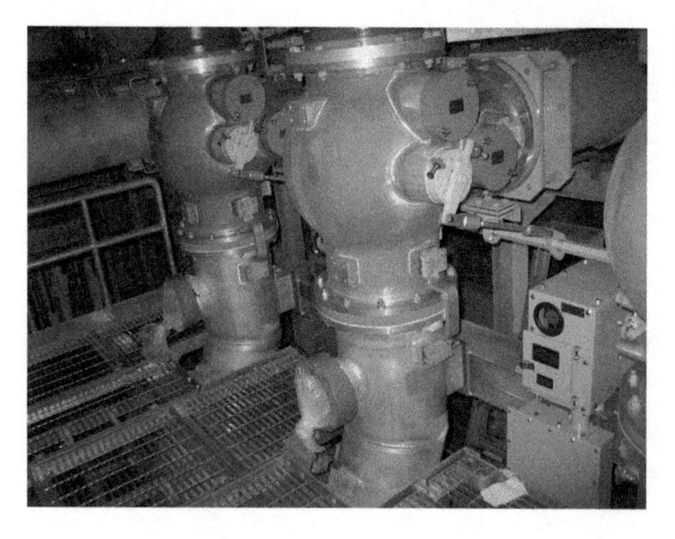

Figure 7.4 GIS ganged disconnect switch (Reproduced by permission of Tech S Corp)

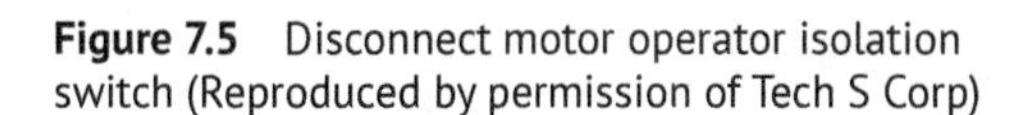
Figure 7.5 Disconnect motor operator isolation switch (Reproduced by permission of Tech S Corp)

- Direct/camera viewing of the position of the switch blade through a viewport in the grounded metallic enclosure. Where motorized disconnect or ground switches are used, some utilities and owners require a knife switch to ensure the motor mechanism is de-energized during maintenance. In addition, the motor mechanism may require decoupling from the disconnect operating arm or it might be required by maintenance rules to mechanically block the motor operated drive. These items should be considered in the initial stages of the facility design. Typically, during the initial stages of a project, a utility or new GIS owner will meet with the GIS supplier to review their operating procedures. Figures 7.6 and 7.7 show different types of semaphore or target position indicators. Figure 7.8 is an example of physical position stops on the linkage.

7.2.4 Nonfault-Initiating (Maintenance) Grounding Switches

Nonfault-initiating (maintenance) grounding switches in a GIS are used for the same personnel protection purpose as those in an air-insulated substation (AIS) similar to a portable personnel grounding connection made with a hook stick. They are used to ground various de-energized components of the substation, such as circuit breakers and voltage transformers. These grounding switches generally do not have fault-closing or induced current-interrupting ability, but are capable of carrying fault current when in the closed position and a small quantity of continuous current for the purpose of testing circuit breakers and current transformers that are out of service. For more details see [9, 10].

Nonfault-initiating (maintenance) grounding switches typically have the same type of operating mechanisms and method of determining switch position as disconnect switches.

External, removable links in the grounding switches are needed to disconnect a grounding switch blade from the external ground. In the closed position, the removable links allow electrical access to the center conductor and facilitate timing tests on circuit breakers, conductivity tests, and current transformer measurements. To assist with operator identification, the outer housing of ground switches may be painted a green, red, or similar color to differentiate from the gray or

Figure 7.6 Switch position indicator – semaphore (Reproduced by permission of Tech S Corp)

Figure 7.7 Switch position indicator – target (Reproduced by permission of Tech S Corp)

aluminum color of the GIS components. Figure 7.9 shows a ganged three phase ground switch and its operator.

7.2.5 High-Speed (Fault-Initiating) Grounding Switches

High-speed (fault-initiating) grounding switches are unique to GIS as they are not typically used in air-insulated substations. Their primary purpose is the same as grounding switches in air-insulated substations and nonfault-initiating grounding switches in GIS. They also serve the same purpose as portable personnel ground connections made with hook sticks. For more details see [11].

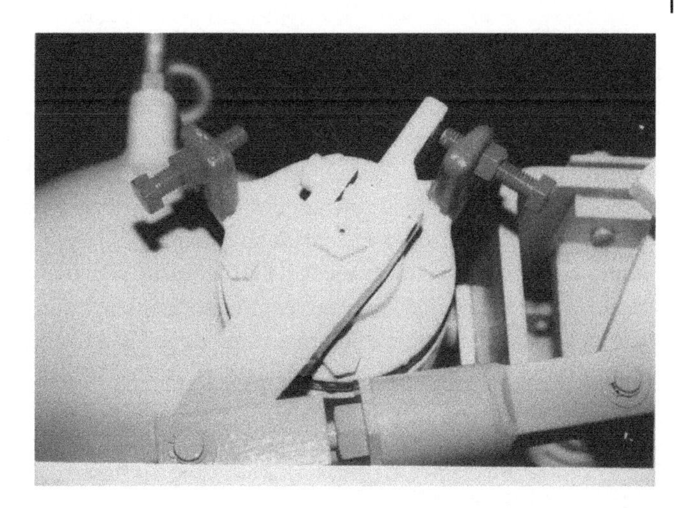

Figure 7.8 Switch position indicator – stops (Reproduced by permission of Tech S Corp)

Figure 7.9 GIS ground switch (Reproduced by permission of Tech S Corp)

High-speed (fault-initiating) grounding switches have the additional capability of closing an energized conductor, creating a short circuit without receiving significant damage to the switch or the enclosure. High-speed (fault-initiating) grounding switches are used to ground various active elements of the substation, such as transmission lines, transformer banks, and main buses. In some GIS facilities, high-speed ground switches are used to initiate protective relay functions. They are not typically used to ground circuit breakers or voltage transformers. High-speed (fault-initiating) grounding switches are also designed and tested to interrupt electrostatically induced capacitive currents and electromagnetically induced inductive currents occurring in de-energized transmission lines in parallel and close proximity to energized transmission lines. They can also remove DC-trapped charges on a transmission line.

High-speed (fault-initiating) grounding switches typically have motor operating mechanisms with spring assists for rapid opening and closing of the switch blade. They typically use the same methods for determining the switch position as disconnect switches.

Depending on the design and customers' maintenance practices, external removable links in the grounding switches may be needed to disconnect a grounding switch blade from the external ground. These removable links are required to facilitate timing tests on circuit breakers, conductivity tests, and current transformer measurements. Figure 7.10 is an example of a high speed of fast acting ground switch connected to a bus.

7.2.6 Three-Position Disconnect/Grounding Switches

At operating voltages in the range of 34.5–161 kV three-phase GIS, a three-position switch is frequently used. This style of switch combines a disconnect switch with a grounding switch. With one operator and one blade, the switch can be placed into the closed position, the open position, or the grounded position.

Three-position switches typically have the same type of operating mechanisms and method of determining switch position as disconnect switches. If three-way switches are used, and the GIS station will be operated remotely, provisions should be made to remotely change the position selector switch (open, closed, or grounded) to avoid a need for operations personnel to travel to the GIS station and change the switch manually.

Similar to the above, external removable links in the grounding switches may be needed to disconnect a grounding switch blade from the external ground for test measurements. For more details see [9, 10].

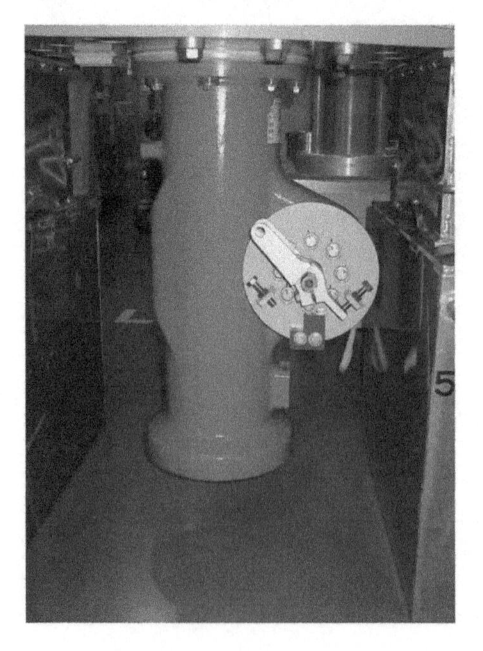

Figure 7.10 GIS fast acting ground switch (Reproduced by permission of Tech S Corp)

7.2.7 Voltage Transformers (VTs)

Similar to AIS stations, GIS equipment employs voltage transformers to reduce the bus high voltage to lower control levels of 120/208 V for protective relays, control and metering and similar functions. Typical turns ratio, insulation integrity, and so forth tests can be made. It is important in the initial stages of design to conduct a ferro-resonance study if the voltage device is a wire wound type. The capacitance of the GIS equipment may interact with the voltage transformer inductance creating a ferro-resonance condition and potential equipment damage. The study should determine if a load can be added to the VT secondary to de-tune the condition. Figure 7.11 is an example of a GIS wire wound voltage transformer. Low power instrument transformers using e.g. capacitive divider are used with digitaization of substations, see section 9.14. For more details see [12, 14].

7.2.8 Current Transformers (CTs)

CTs are generally located in the circuit breaker bushing turrets; however, specific designs involving direct cable or transformer connections may require standalone CTs. As in the case of AIS standard, current injection and ratio tests can be performed. However, to access the main bus as stated earlier, ground switches will need to be isolated and the test currents or voltages injected through the ground switch onto the bus. These tests are conducted with the main bus de-energized and isolated from any operational portions of the remaining GIS switchyard. It is also important to remember that when the tests are completed and before the bus or equipment is energized, the CTs must be connected or shorted, otherwise equipment damage may result. Figure 7.12 is an example of CTs mounted in the circuit breaker bushing turret near the top of the figure. Low power instrument transformers using e.g. Rogowski coil principle are used with digitaization of substations, see section 9.14. For more details see [12, 13].

7.2.9 Switch Viewports

In order to meet the requirement in most utility and owner operating procedures, to visibly "check open" or visibly "check closed" a specific device, viewports are typically provided. A viewport provides the operator via a glass portal method to determine the disconnect or ground switch blade position. Viewports can be a small diameter and require a borescope or camera or something larger in the 50–100 mm range where an operator can use a flashlight and directly

Figure 7.11 GIS inductive voltage transformer (Reproduced by permission of Tech S Corp)

Figure 7.12 GIS inductive current transformers in circuit breaker bushing turrets (Reproduced by permission of Tech S Corp)

observe. Operations cautions include: never look into a viewport while a device is being switched, arcing may occur with the switching operation. All viewports should be clearly labeled with the device number being observed. There should be no question from the operator which device opened and its position. Before a facility is commissioned, personnel should be provided an opportunity to directly observe a switch operation. It is also helpful to provide a video camera for future operators. Since this is one of the unique aspects of GIS, operators should be well trained in switch operations. Further if tools are required to access a viewport, they should be readily available and provided in a common storage locker or adjacent equipment rack. Two examples of viewports are shown in Figures 7.13 and 7.14, with the second example including covers. View port requirements are defined in IEEE C37.122 [15].

7.2.10 Gas Compartments and Zones

A gas compartment is defined as an enclosure that contains gas isolated from the atmosphere and other compartments. In older GIS design, two or more gas compartments may be connected externally with small diameter gas pipes. A gas zone is defined as a section of the GIS that contains one or more gas compartments that have a common gas monitoring system and whose gas density fluctuates in unison. On gas zones refer also to section 2.9 and [6, 15].

Figure 7.13 GIS viewports – A (Reproduced by permission of Tech S Corp)

Figure 7.14 GIS viewports – B with covers (Reproduced by permission of Tech S Corp)

A few principles concerning gas compartments and zones are suggested:

- Generally, each phase of an isolated phase GIS contains its own compartments and zones separate from the compartments and zones of the other two phases.
- Generally, each phase of the circuit breaker is its own zone to eliminate the spread of contaminants created by the operation of the circuit breaker to other compartments.
- Generally, a buffer compartment is installed on each side of each phase of each circuit breaker. In the case of invasive maintenance or repair, the buffer compartment with reduced gas pressure acts as an additional safety barrier against pressurized SF_6.

- The quantity of gas in each compartment is, typically, restricted to the amount of gas that can be conveniently handled by the available gas processing equipment. The SF_6 gas pressure is normally higher in the circuit breaker gas zone than other gas zones due to the switching needs.
- Each gas zone includes an SF_6 gas density monitor, a gas filling valve, and a pressure relief device. Figures 7.15–7.18 provide examples of gas density meters, gas fill valves, and a pressure relief.

7.2.11 Interlocking

Because of the compactness of a gas-insulated substation, and the general difficulty in readily identifying the various active components, it is customary to electrically and/or mechanically

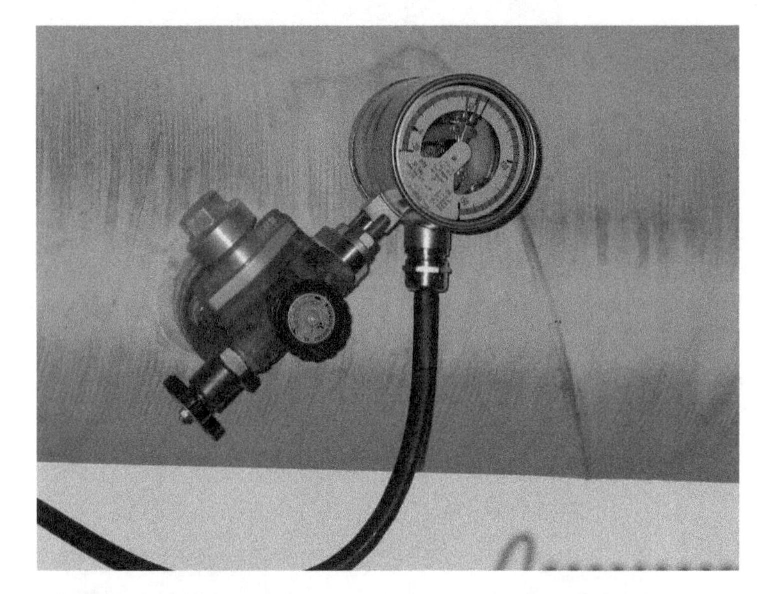

Figure 7.15 Gas density monitor and fill valve (Reproduced by permission of Tech S Corp)

Figure 7.16 GIS gas density monitor (Reproduced by permission of Tech S Corp)

Figure 7.17 GIS gas fill valve (Reproduced by permission of Tech S Corp)

Figure 7.18 GIS pressure relief (Reproduced by permission of Tech S Corp)

interlock circuit breakers, disconnect switches, grounding switches, transmission lines, and transformer banks.

The three main requirements for an interlocking system include:

1) A disconnect switch should be prevented from interrupting or making load current.
2) A disconnect switch should be prevented from closing into a grounded bus.
3) A grounding switch should be prevented from closing on to an energized bus.

A few examples of interlocking are in order. Typical is the circuit breaker position, between disconnect switches on each side of the circuit breaker, and two grounding switches. Each grounding switch is between one of the disconnect switches and the circuit breaker. The main bus and the transmission line are assumed to be energized.

- For the first example, the circuit breaker is closed and is either energized or de-energized. The two grounding switches are open. The two disconnect switches should be prevented from being closed or opened, thus stopping the disconnect switch from either interrupting or making load current. Interlocking requirement 1 is thus satisfied.

- For the second example, assume that the circuit breaker is out of service and de-energized, and that one or both of the two grounding switches are closed. The interlocking prevents either of the two disconnect switches from being closed, preventing them from connecting an energized circuit into a grounded bus. Interlocking requirement 2 is thus satisfied.
- Conversely, for the third example, the circuit breaker is energized (open or closed) and one or both of the disconnect switches are closed. The interlocking prevents either of the two grounding switches from being closed, preventing them from closing on to an energized bus. Interlocking requirement 3 is thus satisfied.
- A fourth example involves switching equipment on opposite voltage sides of a transformer bank. If a high-speed grounding switch on one side of a transformer bank is closed, effectively grounding the transformer bank, the disconnect switches isolating the transformer bank on the opposite side will be prevented from being closed into the grounded transformer. Interlocking requirement 2 is thus satisfied.
- Conversely, a fifth example also involves the transformer bank. If one or both of the isolating disconnect switches on one side of a transformer bank are closed or grounded, the high-speed grounding switch on the opposite side of the transformer bank will be prevented from being closed into the energized transformer. Interlocking requirement 3 is thus satisfied. What about checking to ensure that the line/bank potential is dead?
- A sixth example involves the switching equipment of a transmission line. The most practical way of interlocking is to monitor the voltage on the transmission line and prevent the high-speed grounding switch from being closed if there is system voltage present on the transmission line. Interlocking requirement 3 is thus satisfied.
- A seventh example involves the main bus. If the high-speed grounding switch on the main bus is closed, all of the disconnect switches that are connected to that same main bus are prevented from closing. Interlocking requirement 2 is thus satisfied.
- Conversely, an eighth example also involves the main bus. If one or more of the disconnect switches connected to the main bus are closed, the high-speed grounding switch connected to that main bus is prevented from closing. Interlocking requirement 3 is thus satisfied.

There may be other interlocking requirements for disconnect switches and grounding switches. For example, if the motor of a switch is running, the manual method of operating the switch will be blocked. If the manual method of operating the switch is engaged, energizing the motor will be blocked.

7.2.12 Local Control Cabinets (LCCs)

GIS equipment by virtue of its compact and unique design does not allow the user many options in terms of the components (switches, terminal blocks, indicating lights) selected for use within the equipment. The exception is the local control cabinet (LCC) or marshaling box (MB), which is the generally accepted point of interface between the utility and GIS equipment. The user has the option to install control or monitoring equipment within the cabinet or simply use the enclosure as a wiring marshaling point. As an example, many users design their switchyard facilities with the equipment control panels in close proximity to the protective relay panels. In that case, disconnect and ground switch controls may not be necessary in the LCC and the cabinet would be used primarily for wiring terminations with possibly a push button control for circuit breaker maintenance purposes. Indicating lights may also be provided on a mimic board within the LCC to assist operations personnel. In the event that the LCC is located some distance from the control house or protective relay installation, operating switches could be installed to minimize the travel distance of the operators.

The LCC also provides the user an opportunity to require switches, terminal blocks, heaters, lighting, convenience receptacles to be installed that are in compliance with the user's standards. It is important to most users to provide a consistent representation of the substation or switchyard controls to the operators whether the individuals are located in a GIS or AIS facility.

As a guideline, the dispatch, operations, and maintenance groups should discuss and agree upon the LCC or MB design. Some items to consider:

1) Where is the primary control point (at a remote dispatch location, in a control house, or at the LCC)?
2) If the control will be remote and standard controls are provided locally in the control house, are redundant controls needed in the LCC?
3) How will the breaker be operated for standard de-energized timing tests? Most manufacturers provide a push button operator at the breaker.
4) Will the LCC or MB be a central point for gas or equipment malfunction alarms? One option is to install a PLC that can collect all the alarm point for transmittal to a central processor in the control house where remote alarms signals can be sent and an HMI located. The historical approach is to install standard user-specified annunciators within the LCC.
5) Will the cabinets be installed indoors where standard NEMA 12 [16] cabinets can be used or outdoors in corrosive atmospheres (refineries, fossil power plants) where a NEMA 4X [17] may be better suited?
6) How many LCC/MBs are needed? Generally, one cabinet per circuit breaker is satisfactory with bus and exit runs included in the breaker LCC/MB. Note: each LCC/MB should be provided with a gas zone diagram designating the equipment and bus within its zone and the interfaces to adjacent zones.
7) Where will the LCCs/MBs be manufactured? It is generally not necessary to manufacture the cabinets at the point of the GIS equipment manufacturer. LCCs and MBs while still the responsibility of the GIS manufacturer could be produced at panel shops with experience fabricating cabinets to the user's specifications.
8) How will the control wiring be run from the GIS equipment and the control house? LCCs/MBs can be provided with top side or bottom entry. There should be sufficient space and spare capacity within a design so that the wiring can be grouped and designated to the GIS or to the control house. In addition, the user should specify if a separation of protection circuits is required by a governing body such as NERC or its regional authorities.

Local control cabinets (LCCs) typically contain the following equipment:

- Operating handles for circuit breakers
- Pushbuttons or operating handles for disconnect switches and grounding switches
- Devices such as red and green lights or semaphores to indicate whether a circuit breaker, disconnect switch, or grounding switch is open or closed
- A remote/local permissive switch to block remote operation when local operation is required
- An annunciator to indicate and monitor the status of the GIS, including gas density
- A marshaling cabinet and junction box for control and power cables emanating from the GIS and from the control and relay building and terminating in the LCC
- Electrical interlocking schemes either hard wired or using a logic control IED

- Circuit breaker controls and disconnect switch controls
- A mimic diagram
- Protection relays when specified by the user. Figures 7.19 and 7.20 are examples of LCC cabinets for use indoors and outdoors

  Interposing relays for remote SCADA operation or a real time automation controller IED.

7.2.13 Alarms

In order to effectively operate a gas-insulated substation (GIS), the status of the apparatus needs to be continuously monitored similar to monitoring the apparatus in an air-insulated substation (AIS). However, due to the criticality of the SF_6 insulation system, gas monitoring in a GIS is much more extensive than in an AIS. Typically, the following alarms are utilized:

- Low gas density (approximately 90%) in each gas compartment and zone signifying a gas leak
- Low–low gas density (approximately 80%) in each gas compartment and zone signifying that the dielectric ratings of the apparatus can no longer be met
- Low circuit breaker operating mechanism pressure (pneumatic, hydraulic, or spring)
- Low–low circuit breaker operating mechanism pressure signifying that the circuit breaker can no longer successfully open or close. In this case, the protection scheme may be designed to block any operation. It should be noted that in addition to providing insulation, in most puffer-type circuit breakers the gas also acts as a damper/cushion for the operating mechanism in most buffer-type circuit breakers.
- Loss of voltage to the circuit breaker mechanism's motor supplying operating energy
- Loss of the DC control voltage or voltages to the circuit breakers
- Loss of DC control voltage to the annunciator in the local control cabinet

Figure 7.19 Local control cabinet – outdoor (Reproduced by permission of Tech S Corp)

Figure 7.20 Local control cabinet – indoor (Reproduced by permission of Tech S Corp)

- Pole disagreement operation
- Excessive run-time of the circuit breaker operating mechanism's motor supplying operating energy
- Overcurrent operation of the circuit breaker mechanism's motor protective circuit

7.2.14 Switching a GIS

This section deals with the actual operation of a GIS. The operation is best illustrated by using an example of switching scenario as follows. In this example, the portion of the Koch substation shown is energized. The Hermann line needs to be removed from service for maintenance on the transmission line. An operator has been dispatched to the station to remove the line from service. Although experienced with switching air-insulated substations and receiving training on operating a GIS, this is the first time this operator has actually switched a GIS. The operator has received written and verbal orders and permission to remove the Hermann line from service. These are the steps that he takes, Figure 7.21 is a schematic that may assist the reader.

1) Upon arriving at the substation, the operator verifies on the annunciator that there are no alarms indicating low gas pressure in compartments 1102 through 1112. He also notices that the general appearance of the circuit breakers and disconnect switches is entirely different from what he is used to seeing in an air-insulated substation. Also, he notices that there are no open-air conductors visible in the station, useful for tracing the path of power flow and applying

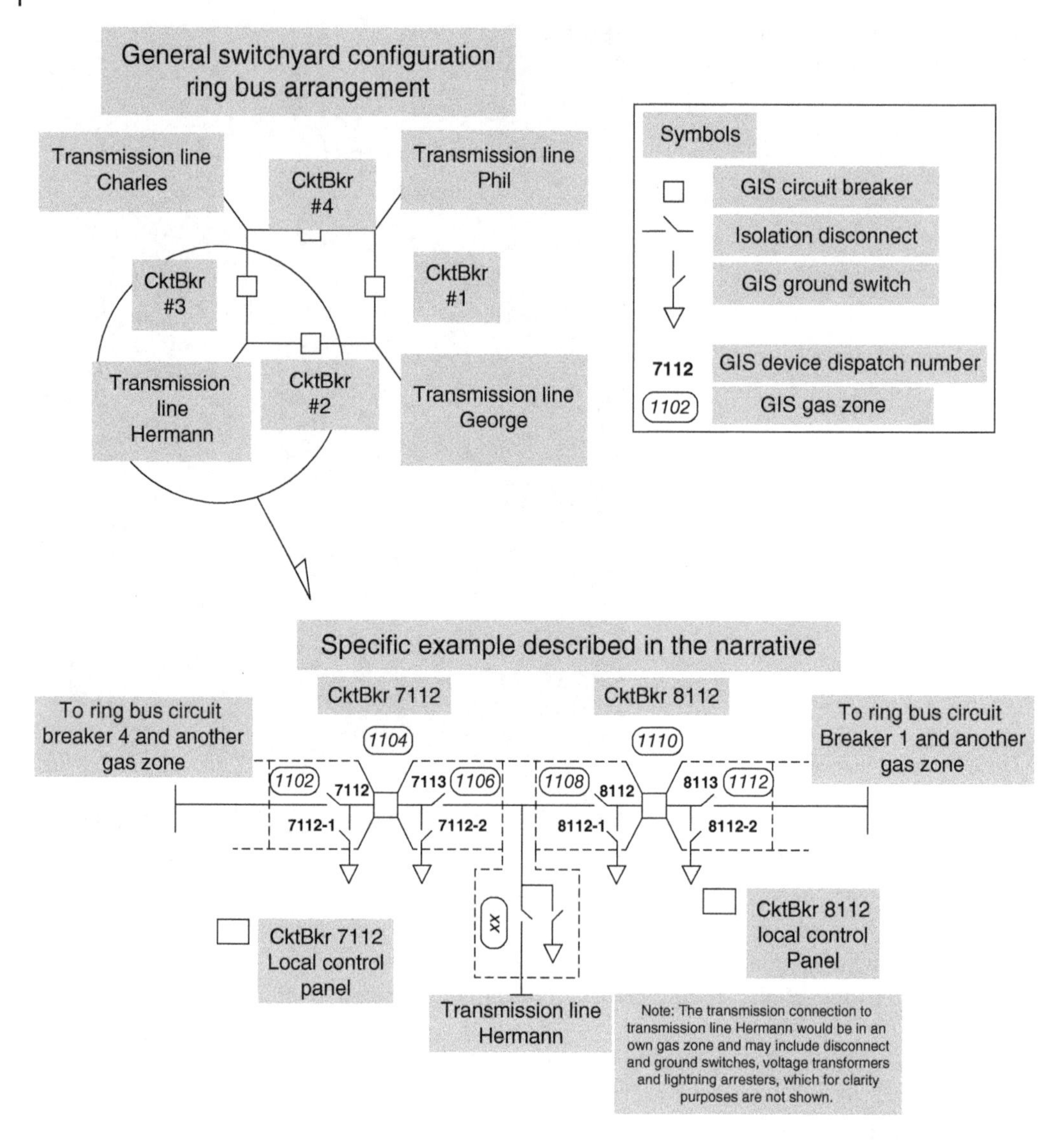

Figure 7.21 Gas zone schematic example (Reproduced by permission of Tech S Corp)

portable grounds. He is highly dependent on accurate signage to provide a guide to operate the correct devices.

2) The operator verifies that the station is normal and all disconnect switches and circuit breakers are closed and all grounding switches are open. The position indicators in the local control cabinets (LCC) indicate the status.
3) In the appropriate LCC, the operator locates the control handle for circuit breaker 7112 in gas compartment 1104 and moves it to the opening position. He hears the circuit breaker bang open and notices that the indicating lights change from red to green, signifying an open-circuit breaker. Note that the color indication might change from user to user. The operator also knows that the circuit breaker could have been opened from the control house at the substation or remote from the regional switching center, but today the orders require him to switch the equipment locally.

4) The control device for disconnect switch 7112 in gas compartment 1102 is then located and manipulated. The noise from the motor driving the disconnect switch open can be heard and the indicating lights change from red to green, signifying an open disconnect switch.
5) The control device for disconnect switch 7113 in gas compartment 1106 is then located and manipulated. Again, the noise from the motor driving the disconnect switch open can be heard and the indicating lights change from red to green, signifying an open disconnect switch.
6) The control devices for grounding switches 7112-1 and 7112-2 in gas compartments 1102 and 1106, respectively, are then located in the LCC and manipulated individually. The noise from the motor driving each grounding switch closed can be heard individually and the indicating lights change from green to red, signifying closed grounding switches.
7) The operator then moves to the adjacent LCC containing circuit breaker 8112 in gas compartment 1110. Inadvertently, the operator manipulates the control device for disconnect switch 8112 in gas compartment 1108 before opening the circuit breaker. Because of the interlocking that prevents a disconnect switch from interrupting load current, nothing happens and no damage occurs. The operator steps back, reviews what he just did, and concludes that he had tried to make an inappropriate switching action. He then proceeds to correctly perform the similar function of first opening the circuit breaker, followed by opening the adjacent disconnect switches, and lastly closing the grounding switches, as was accomplished in the previous steps.
8) The operator's orders require that the blades of the disconnect switches and grounding switches be visually verified for the correct position. In an air-insulated substation, a quick glance at the blades would have completed the verification. Because this is a GIS, the blades are enclosed in an opaque metallic enclosure and are not readily visible. However, since each of the viewing ports has previously been clearly, distinctly, accurately, and uniquely labeled as to phase and switch designation, each viewing port is readily found. With the aid of a handheld flashlight, the positions of the blades are found to be in the correct position.
9) As the operator was verifying the position of the switch blades, he notes that the mechanical semaphores on the operating mechanisms for the circuit breakers and disconnect switches indicate that all phases are open. The mechanical semaphores on the operating mechanisms for the grounding switches indicate that they are closed.
10) The operator now applies tagging as required (physically or electronically or both) by the dispatching authority for the substation and reports to dispatching supervision that the Koch terminal of the Hermann line has been de-energized and verified. He also asks for the status of the switching at the opposite end of the Hermann line and requests permission to proceed to ground the line. The remote end of the Hermann line has been de-energized and he receives the required permission.
11) The operator returns to the LCC that contains the control device for high-speed grounding switch 7114 in gas compartment 1107, locates and manipulates the controls, hears the noise from the motor driving the grounding switch closed, and observes that the indicating lights change from green to red, signifying a closed grounding switch.
12) With the aid of a handheld flashlight, the positions of the three blades of the high-speed grounding switch are determined to be in the correct fully closed position.
13) The operator now reports to supervision that the Koch end of the Hermann line has been grounded and that he is returning to base.

7.2.15 Conclusion

The GIS uses the same type of equipment as the AIS, such as circuit breakers, disconnectors, ground switches, current and voltage transformers, but in a different way. With the metallic enclosure, the high-voltage parts are not readily accessible, but this also provides an excellent safety personnel separation from energized parts. The enclosure also helps to minimize atmospheric contamination and corrosion of the energized devices. The enclosure presents a disadvantage, if operations or union contracts require a visible means of disconnection for the switches, however, this can be corrected with the use of view ports. In some cases, camera systems are attached to the view ports for convenient checks. For grounding/earthing of bus or equipment sections, ground switches need to installed in the GIS and the owner and GIS supplier should review the design to ensure all required grounding operations and maintenance requirements are met. This is different from the AIS where a ground/earth cable can be placed at any section of the air-insulated substation.

The use of ground switches versus portable or personnel safety grounds may be a new concept to operations and maintenance personnel. But with proper training individuals will see there is no difference in the protections provided.

7.3 Maintenance

7.3.1 General

GIS has demonstrated high reliability over the last few decades. GIS manufacturers are promoting a "maintenance-free" concept. This does not mean that maintenance is not required at all, but experience has demonstrated that very minimum maintenance is needed for GIS compared to other technologies. The many reasons for this are detailed in other chapters of this book. This section will focus on various aspects of maintenance and operation of GIS.

7.3.2 Common Maintenance Procedures

GIS manufacturers provide users with recommended maintenance plans. These can vary slightly between manufacturers but the basic guidelines are as follows.

7.3.2.1 Visual Inspection

On a regular basis (ideally several times a year), it is recommended to perform a visual inspection of all GIS elements. The equipment does not need to be de-energized. The purpose of this inspection is to check that there is no sign of unexpected wear or misoperation of the equipment.

Typical operations performed during this inspection are:

- Record and check SF_6 density using meters or installed probes (when available).
- Record switching device operations using the operation counters (when available).
- Check oil pressure and tightness (when hydraulic mechanisms are used). Check compressor run times and proper operation for pneumatic systems. In the case of spring operators, make a visual inspection for any abnormalities.
- Check proper functioning of low voltages devices (indicators, heaters, etc.).

7.3.2.2 Minor Inspection

This inspection can be performed every 5–10 years on GIS components but the inspection can also be dependent on a number of operations of switching devices. The purpose is to check the proper operation of all switching devices. For this, the corresponding equipment needs to be de-energized. Laboratory analysis of the gas may help identify unusual wear, insulator problems, or other issues due to arcing or partial discharge and can be fixed before it degenerates to an unexpected major failure.

This maintenance does not require opening gas compartments. Typical operations performed during this inspection are:

- Check of SF_6 pressures (density)
- Check of SF_6 density relay operations (including the wiring and alarms)
- Check of SF_6 gas purity
- Check of SF_6 by-product and impurity content (SO_2 and moisture, if compartments are not equipped with absorbers)
- Locate any SF_6 leakages (in case of alarms since the last inspection)
- Record and check operating times of circuit breakers (from auxiliary switches). Exercise the circuit breakers and switching devices
- Check of correct operation of pressure switches, in the case of hydraulic mechanism use
- Check of the proper alignment and operation of position indicators
- Check of control and alarm functions

7.3.2.3 Major Inspection

This inspection can be performed every 15–20 years but is strongly dependent on the number of operations of switching devices. It might be decided based on the GIS history not to open gas compartments and extend inspection by another 5–10 years. Major inspections are generally more condition-based than time-based maintenance. Opening of some compartments can be required during such inspections.

In addition to the tasks performed during minor inspections, the typical operations performed during major inspections are:

- Lubrication of various linkages and drives
- Overhaul of the hydraulic mechanism with oil, filter, and switches replacement plus maintenance on the rams and drive mechanisms. Inspection of the circuit breaker interrupter assembly including nozzles and contacts
- Opening and inspection of the switching devices if they have reached the limits recommended by the GIS manufacturers
- Replacement of gaskets and absorbers when compartments are opened
- Record and check of travel curves for circuit breakers

Overhaul of equipment is needed when it has reached its end-of-life. This is usually determined based on the recommendations and experience of the user. However, an overhaul operation will require the expertise of the original equipment supplier, while the other inspections can usually be performed by the user provided that appropriate training has been provided by the GIS manufacturer.

The conditions of the tools and equipment used for maintenance, such as the gas-recovery cart, have also to be regularly and carefully checked and maintained.

7.4 SF_6 Gas Leakage Repair

SF_6 gas leakage is a concern on multiple levels including the environmental impacts, degradation of the GIS insulation system integrity, and cost of the gas. Releases of SF_6 are also becoming reportable incidents in some countries due to the atmospheric greenhouse gas impacts. In GIS equipment, most leaks are identified during the initial assembly and are related to flange misalignment, pinched O-rings or gaskets, dirty or corroded surfaces, and the like. Leaks are also attributed to incorrectly installed by-pass piping, loose flange nuts, poor gas density gauge mounting, and similar instrumentation connections.

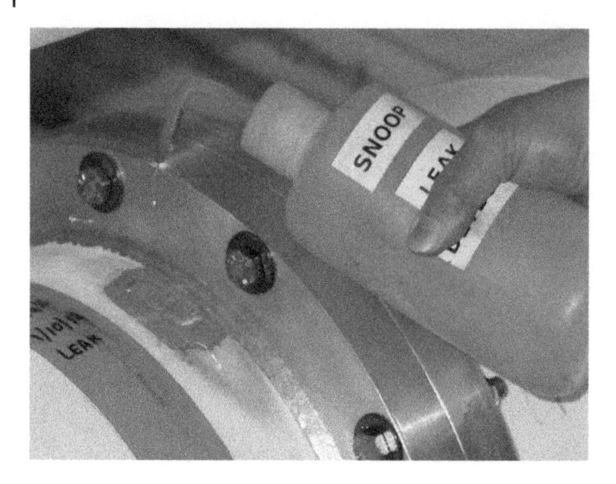

Figure 7.22 GIS leak detection using liquid snoop (Reproduced by permission of Tech S Corp)

Once a gas density meter has alarmed a loss of gas problem, as discussed in Chapter 6, bagging and using an electronic sensor is one method to detect leakage. Other proven methods include: using soap/snoop (see Figure 7.22) a liquid that will produce bubbling and IR cameras are available that "see" the leakage. Once the leak is identified, it should be marked and documented. Simple repairs like tightening a fitting may be done quickly, however, a significant leak as in the case of a pinched O-ring will likely require an outage, gas removal of the impacted zone, disassembly and cleaning the flange, O-ring replacement, reassembly, pressurized testing, bus evacuation, and refilling. Depending on the size of the repair, a high-voltage test may also be a prudent step, to both confirm no foreign materials are in the bus and allow for particle conditioning or movement into low potential traps. Tripping a test set due to a problem is much better than a fault occurring with a transmission line and system connected. All these steps for a large bus section may require several days or a week.

If a leak repair is necessary, maintenance crews should carefully plan the work including:

1) Identify and assembly of all the required parts. If materials are required from the OEM supplier, consider overnight shipments depending on the criticality of the equipment and the leak rate.
2) Assemble and check the required gas storage (confirm sufficient volume is available), vacuum, and gas handling equipment.
3) Have the required cranes, dunnage, temporary bus, or equipment supports available. Confirm the material handling equipment is properly rated and has been recently tested/certified in accordance with local regulations.
4) Schedule crew training – who will do what, when
   a) Assign one experienced individual in charge of vacuum and gas processing. He or she should maintain the gas handling records and weigh the gas removed and replaced including any additional new gas required.
   b) Describe the work gas zone and the adjacent compartment gas pressures; and post a marked-up gas zone and one-line diagram.
   c) Emphasize cleanliness of the work area, control of cleaning materials and hazardous waste disposal (cleaning solvents)
5) Preplan the switching, tagging, and grounding to isolate the leaky section.
6) Inventory all tools to be used for opening and repairing the bus.
7) Before sealing the bus – make a tool count to confirm all hardware is accounted for and have at least two individuals inspected the interior bus work area for foreign materials, cleaning rags, or tools. Check for the replacement of desiccant bags.

The objective should be to minimize the time the bus is open and exposed to contamination and moisture.

In an emergency or if spare parts require a long lead time to manufacturer, there are firms that specialize in SF_6 leakage stops using polyurethane materials. An older equipment for example, a cast bushing fitting may leak and these firms after reducing pressure in the bus, can install an external enclosure, inject a sealant, and remove the cover. The sealant will harden in a few hours but remain elastic to allow for thermal expansion (see Figure 7.23a–c). The disadvantage of sealants is that it make a repair more complicated and should only be used if immediate repair is not possible.

Figure 7.23 Leaky bushing repair (a) – cast bushing fitting (b) – enclosure sealant injected (c) – sealed and ready for paint and conductor connections (Reproduced by permission of Tech S Corp)

7.5 Repair

7.5.1 Nature and Sources of Faults

The latest surveys on reliability of GIS are published by CIGRE brochures TB 509, TB 513, and TB 514 on the reliability of GIS. These reports are part of the Final Report of the 2004–2007 International Enquiry on the Reliability of High Voltage Equipment covering all switching equipment. Brochures TB 509 cover general matter of the survey, how data have been collected and evaluated [2], TB 513 gives information on GIS reliability with statistical data of practical experiences with GIS installed mainly in Japan, which has the largest number of GIS [3] and brochure 514 which is focused on GIS practices, how and when it is used [4]. The reports give very detailed information on the GIS experience over the last decades and the best practices users apply for operation and maintenance of their assets.

Faults in iso-phase bus are typically line to ground and due to bus contamination including moisture. In the case of 3-in-1 bus: In addition to single line to ground faults, double line to ground faults and three phase faults are possible. It is important to note that in the event of faults and likely a pressure relief opening, maintenance personnel should be trained how to approach the equipment. A smell of "rotten eggs" indicates caustic gases and perhaps solid contaminates may be present. Appropriate respiratory and skin exposure protection should be provided before disassembly.

7.5.2 Repair Times and Service Continuity

When GIS equipment fails, the repair time and the service continuity of the GIS is dependent on the level of damage, location of repair facilities, and the availability of spare parts.

Regarding the spare parts, the GIS manufacturers recommend that a minimum set of spare parts covering minor operations should be available on site (e.g., a set of gaskets in case a compartment needs to be opened). For some specific critical applications, it may be recommended that specific GIS components (like a circuit breaker interrupter assembly, gas-barrier insulator, gas pass insulator, or disconnect active parts or a complete spare pole) should be available. The availability of these components on site may help reduce the time to repair.

It should be noted that if spare parts are purchased, it is equally important how the spare parts are stored. Most metal components should be in a dry location, in a sealed box or drum with sufficient desiccants to ensure that the part remains free of moisture accumulation and rust. Alternatively, if complete assemblies are purchased, they could be kept under gas or dry air pressure. Stored parts should be periodically inspected to ensure the storage gas pressure is correct. Material in drums should have the desiccant periodically changed. The purchase of O rings, gaskets, and similar materials should be carefully considered. Rubber, neoprene, and similar materials may dry out, become brittle and unusable after 10 years on the shelf. GIS suppliers should identify the shelf life and owners of critical equipment may want to institute a periodic purchase plan as shelf lives are reached.

The physical arrangement of the GIS also has a major influence on the time to repair. For a circuit breaker repair due to interrupter assembly damage, the time will be shorter if active parts can be removed without dismantling the CB enclosure. However, due to the design requirements for some restricted space projects, other GIS components or structural members may need to be disassembled and removed before the circuit breaker can be accessed. The best time to identify and avoid these difficulties is during the initial design discussions. However, if the limitations are unavoidable, the design documentation and drawings should provide a plan for access. Specifically, to improve the service continuity, some specific features can be implemented directly into the original GIS design (e.g., the addition of removable elements or isolating gaps at the right locations). The user should clearly specify in the original request for a quotation what is expected in terms of service continuity. The repair time and service continuity should also be addressed by the planning engineering in the transmission system design (e.g., what design provides the most reliability for the critical line, bus, or transformer assets a ring bus, breaker and a half, double bus, and so on).

Recent standards and guides have also started to address some requirements for improving the service continuity. This can be found in Annex F of IEC 62271-203 [1] or in the IEEE GIS Guide C37.122.1 [6].

7.5.3 Examples of Repair

GIS equipment problems particularly in older designs, can occur from internal insulator flashovers due to moisture or contamination, component failures like misaligned disconnect or breaker contacts, broken operator insulator arms, dispatch errors including incorrect switching, and other mishaps all of which, could be serious in nature and require a significant amount of time to analyze the failure, access the components, order replacement parts, and install the new hardware. As stated earlier, unlike AIS facility repairs, GIS can require weeks and lengthy outages.

Typically, gas leakage is the most frequent maintenance problem. Unfortunately, small or slow gas leakage may not be identified immediately on a gas density meter. Given the cost and environmental concerns with SF_6 gas, it may be a prudent measure to periodically use an IR camera to inspect for gas leakage. Similar to using thermography to identify conductor hot spots in an AIS facility. If a significant repair is required, personnel from operations, maintenance, and dispatch should prepare a repair plan with one individual "in charge." It is also generally a good approach to involve a manufacturer's representative with the work. See Section 7.4 for additional details.

7.6 Extensions

7.6.1 General

GIS have now more than 50 years of service experience around the world. As for air-insulated-type substations, gas-insulated substations have also been extended during their lifetime. Due to the specific design and arrangement of GIS, most extensions have been implemented by the original equipment supplier. However, some have been performed by a different manufacturer. As the number of GIS installed around the world is increasing, it is expected that more and more GIS will be extended in the foreseeable future.

There are no international standards that deal with GIS interfaces, as there is between GIS and power cables or between GIS and power transformers. However, an IEEE document (a Guide) exists and has recently been revised. IEEE C37.122.6 [17], Recommend Practice for the Interface of New Gas-Insulated Equipment in Existing Gas-Insulated Substations rated above 52 kV. C37.122.6 gives recommendations when a GIS is designed to accommodate a future extension or when an extension is required for a GIS for which an extension was not initially planned. The Guide also covers extensions made by the original manufacturer or by a different manufacturer.

The main challenges when extending a GIS are the knowledge and information (or lack thereof) of the existing equipment and the division of responsibilities of the various parties involved.

The connection parts are usually design-protected by the original equipment supplier and the required design drawings are not always easy to obtain. It is the responsibility of the user to provide the necessary information to the supplier of the extension. Obviously, when the extension is performed by the original equipment supplier, many of the difficulties are avoided. Figure 7.24 gives an example of the interface between two types of GIS from same manufacturer. In cases when the extension is performed by a different manufacturer, the collection of information from the original design can be a more delicate process.

Therefore, it is recommended that any possible extension should be anticipated during the initial design and delivery of the GIS.

7.6.2 Work To Be Done When an Extension Is Planned from Initial Design Stage

When first designing the GIS, it is recommended that any requirement for the future extension should be identified. This includes requirements such as:

- Space in the GIS room or building
- Civil work
- Proper application of buffer gas zones to minimize outage time
- Sizing of auxiliaries
- Integration into the control scheme
- Grounding of equipment
- Site installation requirements of new GIS
- HV testing of newly installed equipment

The IEEE document lists all of these requirements. However, more importantly, it gives some recommendations for the integration of a standard GIS interface that would facilitate future extensions. This interface is made of a rather simple bolted connection and therefore is independent of the original equipment manufacturer technology.

A minimum amount of information must be communicated by the original GIS manufacturer, such as gas densities, diameter of flanges and enclosures, and size of connecting bars. This is well described in Annex A of the IEEE C37.122.6 document.

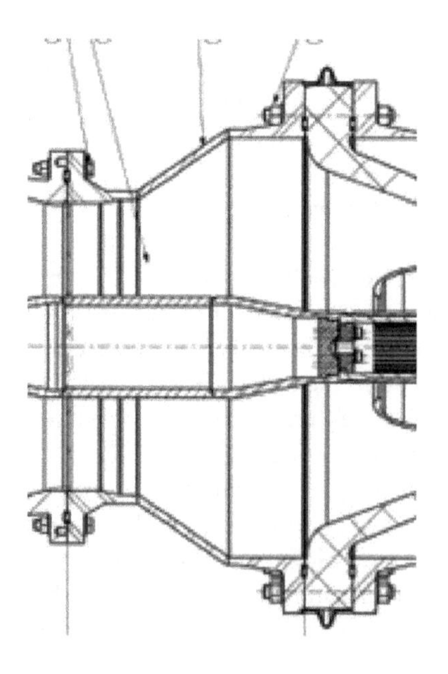

Figure 7.24 Interface between two GIS of different sizes: old GIS (right), new GIS (left), and enlarged diameter of conductor connection from old to new (center) (Reproduced by permission of IEEE)

7.6.3 Work To Be Done When an Extension Is Not Anticipated from Initial Design Stage

When an extension is required, but without an optimum plan for the extension, work can be more complex than with the case described above.

In addition to the previous items discussed above, special attention must be paid to the following issues:

- Obtain detailed drawings of each component where a connection is required (type and size of contacts, detailed design drawings).
- Retrieve technical information about the existing equipment (dielectric ratings, different gas pressures, and alarm levels).
- Get maximum photographs of existing material on site. Collect all nameplate information and measure exterior items like bus diameters and flange bolt patterns.

When all this information is collected, the next step is to perform a detailed design of the interface. This usually requires accurate geometric information of the existing substation (level, altitude, axis, building, civil work reservations, etc.) as this will determine the exact arrangement of the interface connection and position of extension bays. Special attention must be placed on the pressure withstand capability of various components, especially the insulators.

For some projects, the detailed drawings of existing equipment may not be easily available. That will complicate the design of the interface by the supplier of the extension, especially if he was not the original equipment supplier. To cover this special case, some additional work is required, including a site survey involving the opening of compartments where the interface needs to be connected. This work is performed under the strict responsibility of the owner of the GIS. If opening of compartments is not feasible, reverse engineering from existing spare parts is another possibility, as is X-ray analysis of the interface location (see Figure 7.25). With these last two methods, there is a small risk of mismatch between the components.

7.6.4 Service Continuity During Extension Work

When the work is performed on site, some outages will be required to connect the new equipment to the existing equipment. Depending on the existing GIS configuration, some adjacent feeders

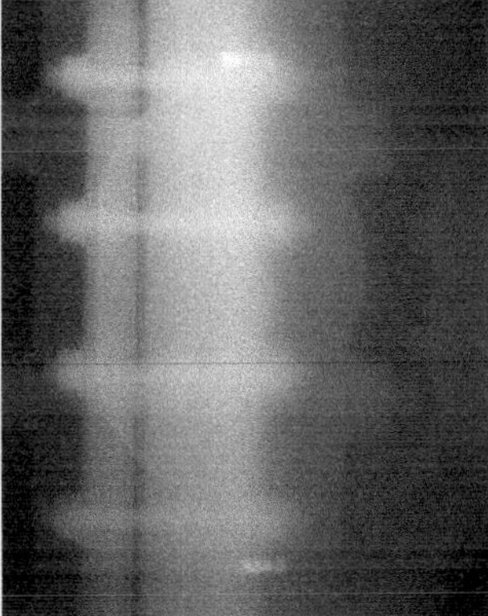

Figure 7.25 X-ray radiography example of an existing GIS at the interface location: (left) X-ray film for exposure, shown in black, and (right) exposed result with nuts and bolts visible (Reproduced by permission of IEEE)

may be switched off during the work in order to assemble the connection or to HV test the connection. Given the heavy loaded transmission systems, this type of work has to be well planned in advance perhaps over a few months or a year.

The IEEE document C37.122.6 gives some detailed guidelines to minimize the outages during the extension work. This can be achieved by adopting specific features during the initial design of the GIS, such as the inclusion of additional gas zones in the bus bar, including maintenance links, adding isolation disconnects, or gapping the bus.

7.6.5 Testing of Interface

After installing the interface and extension bays, the usual on-site tests, as recommended by IEEE or IEC standards, need to be carried out to check the integrity of the newly installed equipment.

The dielectric (HV withstand) test is among the recommended tests. Nevertheless, this test must be fully discussed and agreed to with the owner of the GIS, as the test will put stress on some parts of the existing GIS equipment. The voltage level applied during the test should consider the condition and quantity of the existing equipment to be stressed during the test.

There is no unique answer to the issue of testing and a clear method statement must be established between the parties before performing the test.

7.7 GIS Retrofit or Upgrade

7.7.1 Introduction

Gas-insulated substation technology is about 55 years old with several gas-insulated substations in service for more than 50 years. These substations are approaching the end of their service life. Even though there is no clear definition or method to determine the service life of the switchgear, the performance of the gear will be naturally diminished because of aging. Deterioration may occur in the areas of gas leaks and mechanical wear, requiring frequent maintenance. Typically, the switching components (especially circuit breakers) get worn out toward the end of the service life due to switching operations under load and fault conditions, while the majority of the other GIS components normally remain in relatively good condition. As a result, it may be more economical just to replace the defective/worn-out component instead of the complete gear, thereby extending the service life of the equipment at minimal cost. In case of increasing load and hight rated and short circuit ratting may be required a replacement is recommended.

Figure 7.26 explains a typical trouble/availability life cycle of GIS equipment.

7.7.2 Challenges with Old GIS

Some of the challenges with old GIS include:

- Frequent maintenance resulting in higher maintenance costs
- Longer down-time resulting in loss of revenue
- Higher risk of internal faults due to insulator deterioration, aging, and cracking
- Larger gas compartments
- Higher SF_6 gas leakage at flanges and monitor piping
- Lower safety

An old design from the first generation of GIS is shown in Figure 7.27.

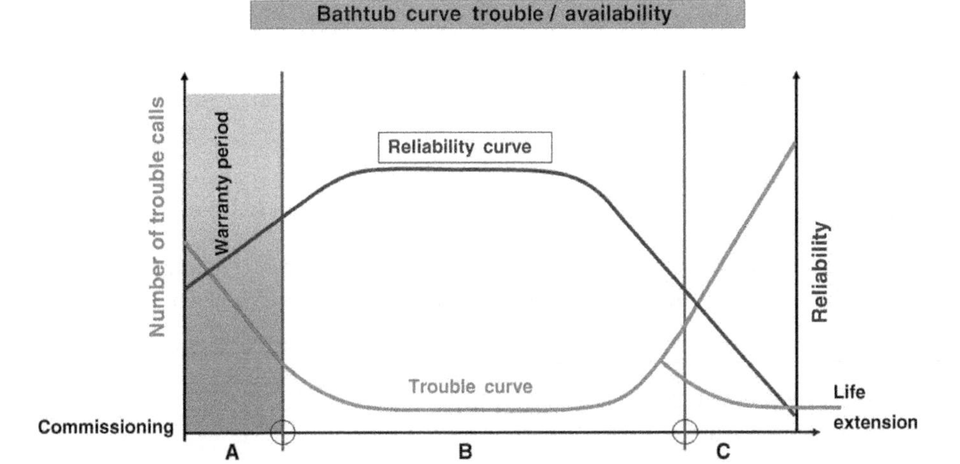

Figure 7.26 Typical life cycle behavior of GIS equipment (Reproduced by permission of ABB)

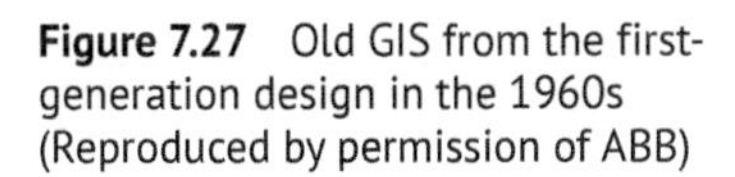
Figure 7.27 Old GIS from the first-generation design in the 1960s (Reproduced by permission of ABB)

7.7.3 Retrofit or Upgrade

Retrofit is a process of replacing or upgrading aged, worn-out, or defective components of switchgear, with latest state-of-the-art components resulting in higher reliability, safety, operational performance, and increased service life of the switchgear with minimal cost. The cost of replacing a component of the GIS is usually lower than replacing the complete GIS. Retrofit also provides the flexibility of replacing one component at a time, so that there may not be any major financial burden (from operational changes) to the owner at any given point in time. During the retrofit, the majority of the substation still remains in service.

The new generation components are also usually more compact than older generation equipment, so no additional space would be required for retrofitting. Interface modules should be used, as required, to connect old components with new components.

Even though it is not common, it is possible to retrofit one manufacturer's GIS with another as long as the interface details are provided by the initial manufacturer/customer.

Another aspect of retrofit could be refurbishment of AIS with GIS to utilize the advantages of this technology. Concepts discussed in Section 7.6 (Extensions) may be useful in interfacing old and new equipment, especially of a different design and manufacturer.

The benefits of retrofit can be summarized as follows:

7.7.3.1 Reliability

- High reliability and availability
- Reduced maintenance
- Long-term availability of spares
- Short lead-time of spare parts

The new-generation design looks much simpler and uses less parts for the same function, which leads to higher reliability of the GIS. A circuit breaker bay enclosure of a GIS is shown in Figure 7.28.

7.7.3.2 Safety

- Reduction in risk of internal faults
- Improved personnel protection

The solid grounded enclosure of the GIS gives maximum safety from touching the high-voltage part in the substation. The strong metallic enclosure also gives maximum safety for the situation of an internal arc. Pressure relief devices prevent the GIS enclosure from bursting (see Figure 7.29).

7.7.3.3 State-of-the-Art Technology

- Latest design (faster operating times, lower mechanism energy requirement, shorter mechanism charging times, lower reaction forces, etc.)
- Higher ratings (continuous and short circuit current, mechanical endurance, etc.)
- Reduced overall size, diameter of enclosures and gas volumes
- Reduced gas leakage rates
- Type testing according to the latest standards
- More compact gear

The latest design of a high-compact GIS with higher ratings, reduced gas volume, high gas tightness, and gas density sensors is shown in Figure 7.30.

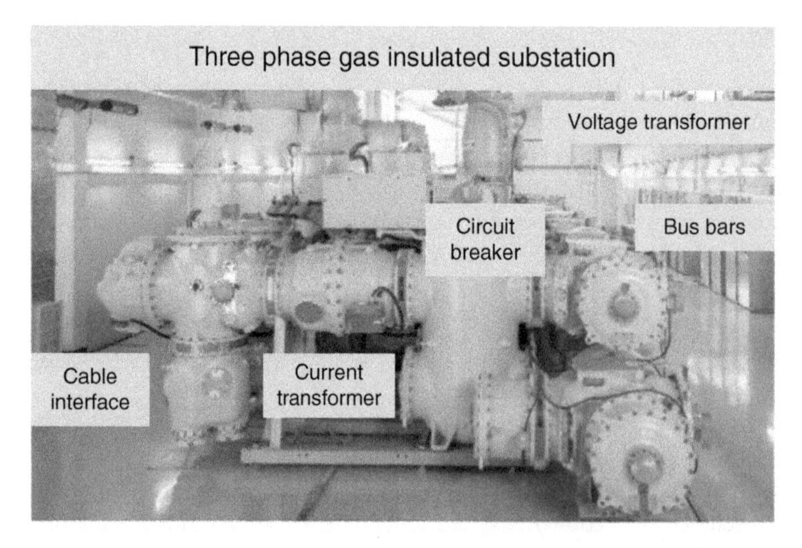

Figure 7.28 New-generation design of today (Reproduced by permission of Siemens AG)

Figure 7.29 Maximum protection from touching high-voltage parts by grounded metallic enclosures (Reproduced by permission of ABB)

Figure 7.30 Latest design of compact design GIS with operation drives and gas sensors attached (Reproduced by permission of ABB)

7.7.3.4 Project Execution

- Short project execution time
- Minimum outage time
- Avoid long land acquisition and permitting process

7.7.3.5 Investment

- Reduced investment
- Minimization of maintenance costs
- Warranty for retrofitted components

Replacement of old breakers and upgrading in ratings is often done in existing substations. As the newer design is smaller in size than the older one, replacement and extensions are no problem. In Figure 7.31, an example is shown for such an extension.

Here are some photographs of retrofit (see Figures 7.32–7.34).

Figure 7.31 Extension of an existing GIS with one three-phase bay of an indoor installation (Reproduced by permission of ABB)

Figure 7.32 Extension of an existing GIS with one three-phase bay of an outdoor installation (Reproduced by permission of ABB)

7.8 Overloading and Thermal Limits

7.8.1 General

The GIS is designed for specific current ratings. These ratings are fixed by the users and are dependent on network requirements. In order to limit the number of possible ratings, the international standards have established a series of ratings that should be selected from the R10 series. Typical ratings are 1250, 2000, 3150, 4000, 5000, and 6300 A. These ratings are given for an ambient temperature, which does not exceed 40 °C, and the average value measured over a period of 24 h, which does not exceed 35°C. The equipment is designed so that the temperature rises do not exceed the limits given in the corresponding standards (Table 3 of IEC 62271-1 or IEEE C37.122.1) [5, 6].

During normal operation, the load current going through the GIS should not exceed these rated continuous currents. However, in some circumstances, it is possible to go beyond these values

Figure 7.33 Old GIS beaker replacement: (left) before and (right) after (Reproduced by permission of ABB)

Figure 7.34 Replacement of old voltage transformers: (left) CCVT and (right) with new ones (PTs) (Reproduced by permission of ABB)

without jeopardizing the integrity of the equipment. These conditions are designated as overload ratings and the methods to calculate them will be explained in this section.

7.8.2 Design for Continuous Rating Current

First, let us explain how GIS is designed to carry the normal rated current. Two major factors have an impact on the dimensioning of the GIS: the dielectric withstand voltage and the rating current.

When designing a GIS, the size of the components is first influenced by the dielectric parameters. These parameters are related to the various voltage conditions on the network and also to the minimum temperature the GIS is designed for. Tables of IEC 62271-1 or IEEE C37.122.1 give dielectric values the GIS must withstand according to different system voltage levels.

Another parameter influencing the size of the GIS is the filling pressure of the dielectric fluid, like SF_6 gas. The higher the pressure, the better is the dielectric withstand. Nevertheless, the maximum pressure is also fixed conveniently according to the minimum temperature the GIS equipment has to support, typically −25°C or −30°C. Indeed, below a certain temperature limit, the SF_6 gas inside the GIS enclosure will condense and the dielectric integrity of the GIS can be at risk.

This dielectric approach often determines the size of the enclosures and internal conductors. However, sizes of GIS have reduced over the years, thanks to optimization of the dielectric designs, and the current rating assigned to the GIS can also influence the dimensions of the GIS. Now that networks are operated at higher current values, the rated current can impact more and more on the design.

Table 7.1 gives some typical dimensions of GIS equipment for different voltage levels.

With the dimensions of conductors, typical current ratings that can be achieved using standard aluminum enclosures and conductors are in the range shown in Table 7.2.

The values shown in Table 7.2 do not mean that higher current ratings cannot be achieved, but the size of the GIS may need to be adapted slightly to meet higher current performances.

7.8.3 Determination of the Limits

IEC and IEEE have adopted the same rules to test and prove the limits of current ratings. They are given by the temperature rise test requirements. The maximum temperature limitations of

Table 7.1 Dielectric required size

Parameter	Voltage rating		(kV rms ϕ-ϕ)	
	145	**242**	**362**	**550**
BIL (impulse, kV, peak, ϕ-G)	650	900	1050	1550
Conductor field (kV/mm, peak, BIL)	15.67	16.78	15.79	17.02
60 Hz max. operating voltage (kV, rms, ϕ-G)	84	141	209	318
Conductor field (kV/mm, rms, 60 Hz operating)	2.02	2.63	3.14	3.49
Ratio of BIL/peak operating voltage	5.47	4.51	3.55	3.45
Standard factory test voltage (kV, 60 Hz, rms, ϕ-G)	310	425i	500	740
Size of conductor OD (mm)	88.9	101.6	127	177.8
Enclosure ID (mm)	226	292	362	495.3

Table 7.2 Typical current ratings

72.5 kV	2500 A
145 kV	3150 A
245 kV	3150 A
362–420 kV	4000 A
550 kV	5000 A
800 kV	6300 A

various parts of the GIS when carrying its rated continuous current are given for an ambient temperature of 40°C.

For GIS equipment, typical values of maximum temperature rise are the following:

- Contacts in SF_6: 65 K
- Enclosures that do not need to be touched during normal operation: 40 K

The total temperature of these parts depends both on the actual load current and the actual ambient temperature. If the temperature is lower than the 40°C, the GIS can be operated continuously at a current higher than the rated continuous current. If the current is lower than the rated continuous current, the maximum ambient temperature of operation of the GIS can be higher than 40°C.

Another case to be considered is the temporary overload. GIS components have a thermal time constant. Until the limit is reached, the equipment can be operated at higher values than it is intended for. Continuous or temporary overload should be established, based on the results obtained from the temperature rise test and test parameters, like rated current, thermal time constant, temperature rise, ambient air temperature, and maximum operating temperatures, as defined in Table 3 of IEC 62271-1 [5, 6].

7.8.4 Maximum Continuous Load Current

Equipment may be assigned an overload capability for higher than rated normal currents based on a lower ambient temperature provided the temperature does not exceed the maximum value temperature specified in Table 3 of IEC 62271-1 [5, 6]. The relationship of the maximum total temperature, the temperature rise due to the I^2R losses, and the ambient temperature are shown in Figure 7.35.

7.8.5 Short-Time Overload Capability

Equipment may be assigned an overload capability for a higher than rated normal current for a temporary period provided the temperature does not exceed the maximum temperature value specified in Table 3 of IEC 62271-1 [5, 6]. To determine the maximum overload values, a pre-load condition should be specified. The relationship between exponential heating and the overload, rated continuous, and pre-load currents is shown in Figure 7.36.

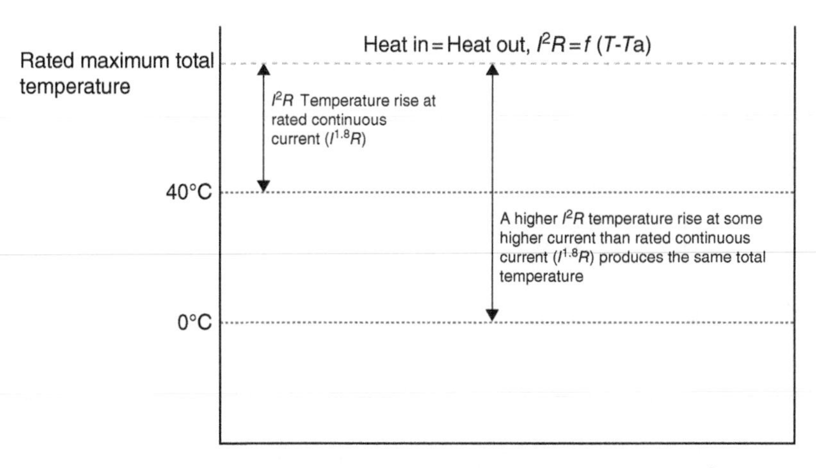

Figure 7.35 Relationship between the rated maximum temperature (T), I^2R losses, and ambient temperatures (T_a) (Reproduced by permission of IEEE)

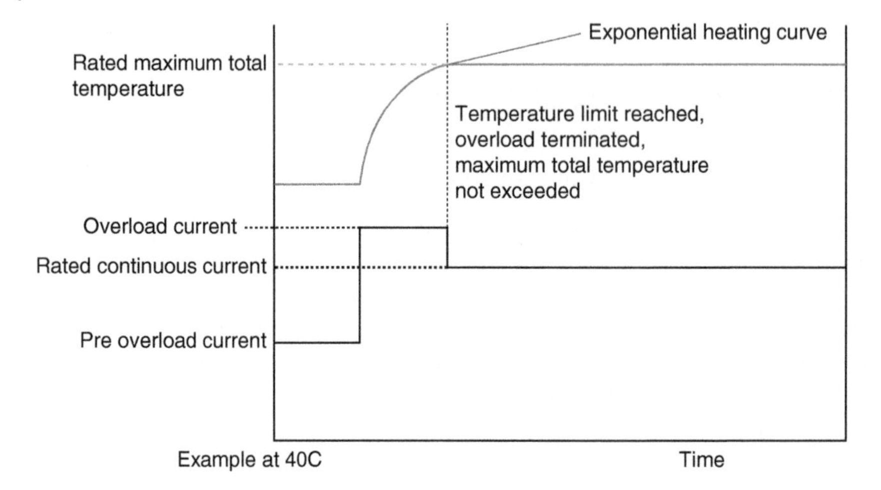

Figure 7.36 Relationship between exponential heating and the overload, rated continuous, and pre-load currents (Reproduced by permission of IEEE)

7.8.6 Equations/Formulae to Calculate Overloads

The formulas to calculate the temperature rise are available using the following parameters:

- The allowable continuous current (I_s) for a given ambient temperature θ_a (Formula 1)
- The operating temperature during overload (Formula 2)

The allowable duration (t_s) of the temporary current I_s after carrying a current I_i (Formula 3)
Formula 1. Ambient temperature effect

$$I_s = I_r \left[\frac{\theta_{max} - \theta_a}{\Delta\theta_r} \right]^{I/n}$$

with I=1
Formula 2. Operating temperature during overload

$$\theta_s = \Delta\theta_r \left(\frac{I_s}{I_r} \right)^n e^{-t/\tau} + \theta_a$$

Formula 3. Allowable duration of temporary current

$$t_s = -\tau \ln \left[1 - \frac{\theta_{max} - Y - \theta_a}{\left(Y \left[\frac{I_s}{I_i} \right]^n - 1 \right)} \right]$$

$$Y = (\theta_{max} - 40) \left[\frac{I_i}{I_r} \right]^n$$

where

θ_{max} = maximum allowable total temperature (°C) according to Table 3 of IEC 62271-1 [5, 6]

θ_a = actual ambient temperature (°C)

$\Delta\theta_r$ = temperature rise at normal current I_r

I_r = rated normal current (A)

τ = thermal time constant (h)

n = overload exponent taking into account material, heat radiation, convection, and so on

I_i = initial current before application of overload current (A)

I_s = overload current (A)

t_s = permissible time (h) that the overload current (I_s) can be carried without exceeding the maximum temperature allowable (θ_{max})

In general, no additional temperature rise tests are required if an exponent $n = 2$ (as a conservative estimate) is used for the determination of the operating temperature during overload or allowable overload duration. An exponent lower than $n = 2$ may be used for the calculation of the overload rating. It has to be demonstrated by calculation from test data.

Note that the time constant corresponds to the time taken to achieve 63% of the final temperature rise after stabilization.

7.9 Maintenance and Operations Pointers

7.9.1 Design Considerations – Direct GIS Connections to Cable and Transformers

One advantage to GIS equipment is to reduce the equipment footprint and space required for the substation installation. In case of dense urban projects or within particular building configurations, direct connections of power cables and transformers maybe required. However, where space is available, making cable and transformer connections using an air gap provides several advantages.

7.9.1.1 Testing

Both transformers and power cable test connections can be made directly without the need to evacuate gas compartments, remove links, test, evacuate, and restore the SF_6 gas. Air gaps also eliminate SF_6 gas processing, which minimizes risks to the GIS including introduction of moisture and contaminates and possible SF_6 releases to the atmosphere.

7.9.1.2 Repairs/Replacement

In the event a transformer fails due to an internal fault, for example, turn-to-turn failure, short-circuit damage, etc., air gaps will facilitate the transformer removal and not require access to the GIS gas zones. If a direct connected transformer fails in addition to loss of the transformer and dependent on the substation configuration, it may impact the overall substation operation, for example, a transformer connected to a supply bus. Prudent spare unit contingency planning should be included where direct connection applications are employed.

Direct connected cable introduces similar operations concerns. Using air gaps faulted cable sections or air terminations can generally be removed and replaced relatively quickly with little or no impact on the GIS substation operations.

7.9.1.3 Spare Power Transformers

Spare transformers employing air bushings are more readily available on the market. Also, in the event a utility or industrial firm has multiple substations using the same transformer design, it will facilitate purchase of a common spare.

As an example, photographs 1 and 2 of Figure 7.37 are a 345 kV/115 kV substation where a direct connected transformer was initially used. A transformer bushing failure occurred with significant debris and damage within the transformer requiring transport to a factory repair shop. Once the transformer was removed, the GIS 345 kV and 115 kV buses were restored using isolation disconnect switches, bus corona plugs, and specially fabricated covers. At significant expense, an used

Figure 7.37 Direct transformer connection photo 1 substation overview, photo 2 air insulated transformer connection, photo 3 gas insulated transformer connection (Reproduced by permission of Tech S Corp)

equivalent transformer with air bushings was located, and temporary buses (air connections) were designed tied directly to the overhead transmission lines.

In later years after load grew sufficiently, a second gas insulated transformer was added with air connections. See photograph 3 of Figure 7.37.

Also note on the top of the air gapped transformer are test terminals/links to avoid disassembly of the main power connection and further facilitate power factor and other routine tests.

7.9.2 Construction Approach

Construction and assembly can be completed by: in-house/owner team, the equipment manufacturer (OEM), or third-party contractor, however, the skill level and experience of the selected team should be carefully evaluated. How many GIS facilities has the team installed? Is the team aware of the GIS installation differences and critical tolerances? What recommendations does the team have to make the GIS installation more efficient, safe, and perhaps less costly? Is the team experienced with SF_6 gas processing and the required documentation?

Prebid meetings (Figure 7.37) are critical, first for the bidders to understand the project scope so accurate and reliable proposals can be prepared and second to address the installation teams' questions. Postaward meetings should be held to discuss assembly details, review construction methodologies, confirm overall schedule, discuss outage coordination, and evaluate problems in a timely manner. This would also be an opportunity to outline crew training and include issues that are specific to GIS work, for example, bus cleanliness, material handling, moisture-level concerns, environmental regulations, and permits.

7.9.3 Construction Planning

Equipment Staging, Material Handling Equipment, Laydown Area, Security – A well-drained area should be selected for material laydown with terrain that will support material handling equipment. The area should be reasonably accessible from the job site taking into account seasonal rain or snow fall. If heavy snow falls are anticipated, all equipment in the laydown area should be inventoried and a reference map created. The materials should also be placed to allow gas bus and equipment pressures to be periodically checked. Security should also be considered if the laydown in an area known for theft or vandalism? Would fencing or a security officer presence be prudent?

Materials Damaged by High or Low Temps – Paints, lubricants, gases, cleaning solvents, and similar materials generally provided by the manufacturer should be protected from extreme temperature variations. Some solvents and cleaning fluids used on GIS equipment may also be flammable and require special isolated storage lockers or facilities.

Control Cabinet Heaters – Most local control cabinets and breaker control cabinets have internal resistive heaters. Generally, the manufacturers require these to be connected during storage to minimize condensation build-up inside the control cabinets.

Periodic Checks on Pressurized Components – GIS bus and other components (circuit breakers) are shipped with a positive pressure of dry air or possibly SF_6. It is important to ensure the pressure is checked on a regular basis and replenished if needed, particularly if the equipment will be stored for any length of time (Figure 7.38).

7.9.4 Installation Crew Training

Training the crews is critical to the success of a GIS project. It is important that they understand the bus is not simply a pipe. Cross-section drawings or photographs of the internal insulators,

Figure 7.38 Project planning (Reproduced by permission of Tech S Corp)

mechanisms, switches, etc., will provide a better understanding of the need for careful handling, cleanliness, and other consideration specific to GIS equipment.

The Gas Zone Diagram should be explained, discussed, and posted in a readily accessible area. Each crew member should be briefed on the importance of the drawing, and understand the gas pressures, processing steps, boundaries (Barrier insulators), etc., for the gas zone where they are working. The diagram should also be revised and updated as the work progresses.

Site Operations and Safety Talks – These are generally called morning tailboards where the crews discuss the days assignments, responsibilities, safety criteria, and other concerns. GIS subjects should be integrated in these talks and include: compressed gases handling, gas cart operations (vacuum, processing, filtration), bus cleanliness, emergency procedures, environmental record keeping, device and gas zone identification methods, substation color codes used, etc.

7.9.5 Quality Assurance

Regular in-process construction inspections should be conducted to ensure specification compliance. Once concrete is poured incorrectly, the rework can be expensive. Shielded control wiring is another concern and should be addressed clearly in the specifications. It is a costly error once the control wring is installed in a bulk power GIS facility to realize unshielded cable was installed and needs to be replaced.

7.9.6 SF_6 Gas Management

Since SF_6 is such a potent greenhouse gas, environmental authorities have placed strict controls on its use and management. Establish a reporting system early in the project and measure/weight all SF_6 as it is delivered, used, and bottles returned. Accountability is very important and may be needed in the event of an on-site audit by authorities. Returned bottles also generally have residual SF_6 that should measure and included in the gas management plan documentation.

7.9.7 Engineering Considerations

Studies should be conducted early in the design and include grounding, overhead shielding, insulation coordination, load current, and fault duty, etc. One study that is sometimes overlooked is ferro-resonance between capacitive elements such as grading capacitors in multi-break circuit breakers and inductive voltage transformers. If the condition occurs, costly thermal damage may result. A study can quickly identify if some type of mitigation is needed. In most cases, a small auxiliary load on the voltage transformer secondary winding will resolve the problem.

Drawing Reviews – Maintenance and Operations involvement in the design process is another important step. O&M personnel can identify needs that the engineering team may inadvertently overlook. The O&M team can check for compliance with their operations procedures, standards, and practices. Among many other topics, concerns of adequate live part clearances, power supply needs for gas carts and test equipment, gas density meter configurations, for example, color (red, yellow, green) bars or numerical density values, union or user's requirements for visible disconnects, etc., can be addressed.

Color Coding Components – Some utilities require ground switches, gas zone barrier insulator boundaries, disconnect switches, circuit breaker cabinets be painted a unique distinct color for operations purposes. This should be included in the purchasing specification to avoid field painting. Particularly, in the case of retrofit projects, the existing color-coding scheme should be continued. The color code system should also be discussed and explained to the installation crew.

Control Panel (LCCs) Similarities – Utility and industrial users generally try to specify the local control cabinets to mirror their existing facilities. This approach gives the operators a common understanding of their substation operations and minimizes confusion (human error). The GIS engineering team should include the user's operating practices and methods in the LCC design.

7.9.8 Spare Parts and Special Tool Accountability

Each GIS project is a bit different with their spare parts and special tools requirements, but invariably at the end of the job, spare parts have been used during commissioning and special tools are missing. Once the spare parts and special tools are delivered to the site, an inventory should be conducted. Tools not needed for the assembly should be given to the user's team using a documented transmittal form. Spare parts should be handled in a similar manner. All spare materials provided to the user should be placed in secure storage. Parts needed for commissioning should be provided, the removal from stock documented, and replacement hardware ordered.

7.9.9 Device Labeling

This is a critical point for all personnel working in GIS facilities. Individuals should learn the various device labeling methods. Each switch, breaker, and other hardware should be labeled clearly in accordance with the drawings or one-line diagram. Note: some users may have multiple device nomenclature systems, specifically a number for engineering and design and a second number for operations and dispatch purposes. Costly mistakes can occur particularly on retrofit projects and during commissioning if personnel do not understand the labeling system or misidentify a device.

7.10 Lessons Learned

7.10.1 Operator/Dispatcher Training

Dispatch personnel should not only be aware of the electrical configuration but also the mechanical or gas zone configuration. A problem occurred several years ago where a major metropolitan city was having voltage support issues. The system operators correctly believed two new pipe-type cables could supply reactive support of approximately 120 MVAR each. The cable ran from the GIS substation to a remote power plant in the city proper about 7 miles away.

Earlier in the day before the low-voltage condition occurred, utility crews removed a voltage transformer for further inspection by the OEM. The VT was making unusual vibrating sounds that were determined later to be uneventful. Since two different utilities were involved and in order to provide a safe work zone, the maintenance order required switches SW1, SW2, SW4, and SW5 be opened and tagged, the cable de-energized and tagged at the remote power plant location. The crew completed their work and left for the day backfilling GZ 3 with low-pressure dry air.

The Dispatchers noted that SW1 and SW2 were opened and tagged, so there was electrical isolation from the work zone. As the voltage problem continued to degrade, the Dispatchers decided to close the switches and energized the 345 kV Cable A from the remote power plant switchyard into the GIS substation with the SW1 and SW2 switches open. There was an immediate fault that cleared within a few cycles.

The electrical configuration before the fault is shown in Figure 7.39.

The information unavailable to the Dispatchers at the time was the gas zone overlay showing the compartment barriers. While SW1 and SW2 were open, they were located in gas zone GZ 3 that was filled with low-pressure dry air. The Fault damage at the time of cable energization is shown in Figure 7.40.

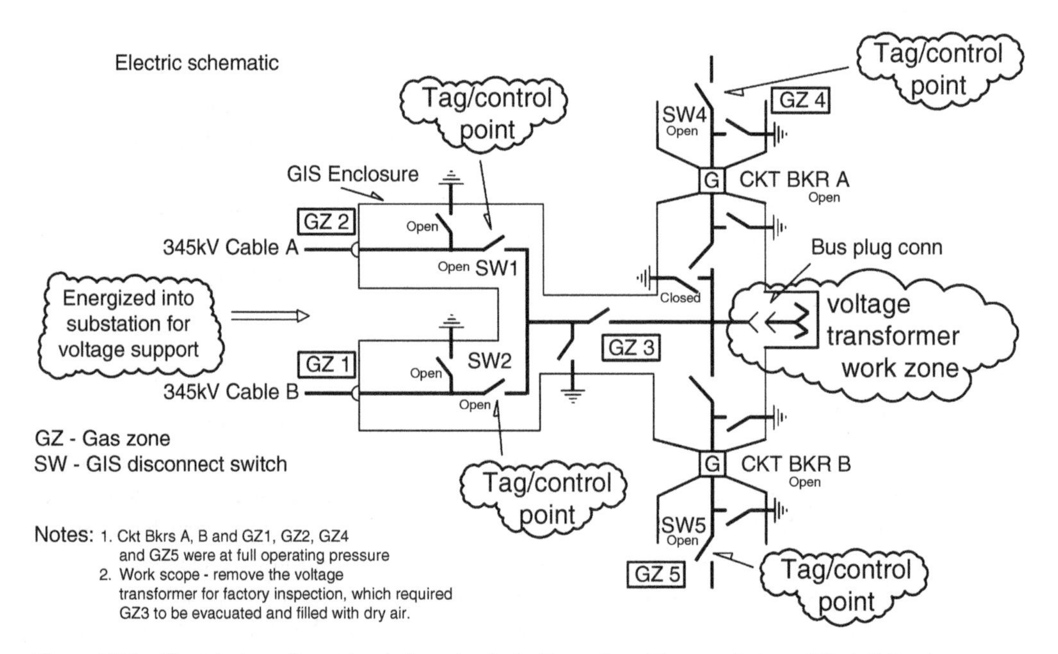

Figure 7.39 Electrical configuration before the fault (Reproduced by permission of Tech S Corp)

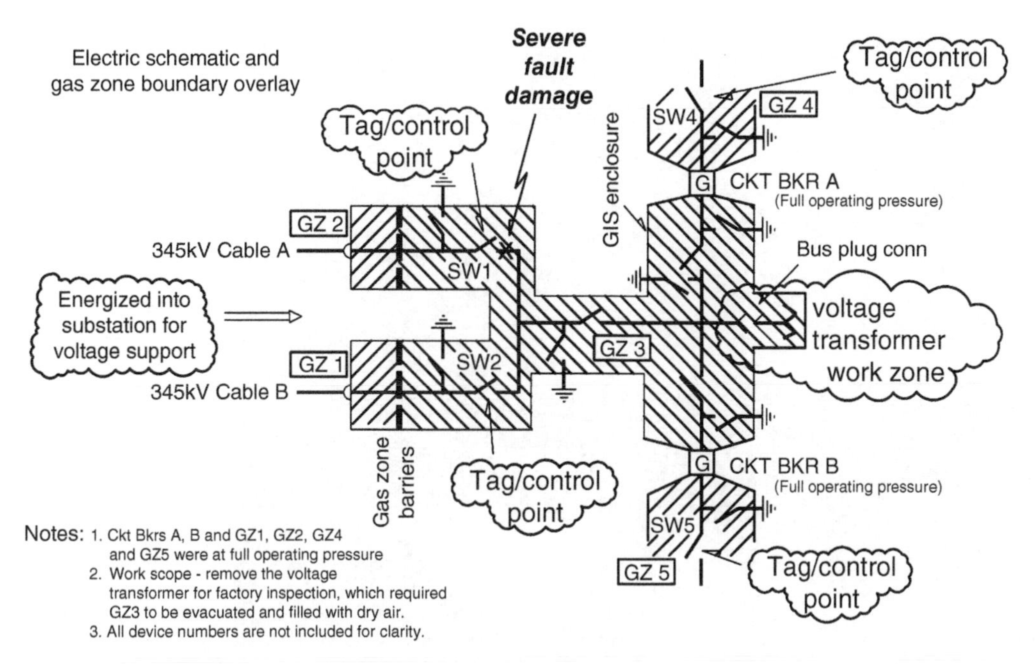

Figure 7.40 Fault damage at the time of cable energization (Reproduced by permission of Tech S Corp)

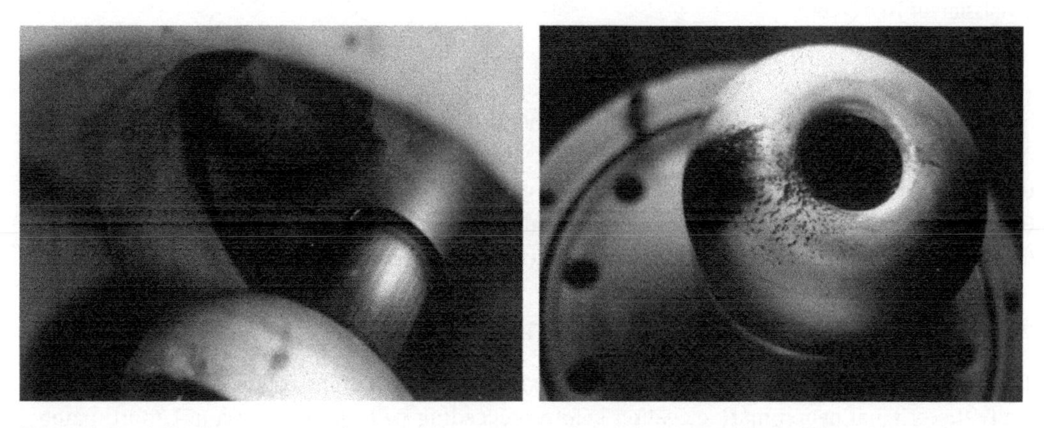

Figure 7.41 Left photo: outer enclosure; right photo: arc damage at the conductor (Reproduced by permission of Tech S Corp)

As noted in Figure 7.41, the bus faulted and the repair work took several months to complete. Lesson: involve the operations personnel in the design and include operations in training programs on all aspects of the GIS equipment.

7.10.2 Improper Rigging – Dropped Bus

In another case, an equipment operator had no understanding of the internal GIS Bus components. On his own initiative to help expedite the work, he rigged and moved a 60-foot length of bus with no tag lines or support personnel. The bus was dropped on a hard surface and the support insulators cracked and damaged. This required return of the bus section to the manufacturer at a cost of $20K. Lesson: Assure safe lifting procedures and train the crews that the bus and other GIS components have internal parts and mechanisms that need to be moved carefully. When large bus

Figure 7.42 Tag line – correct bus rigging using tag lines and support personnel (Reproduced by permission of Tech S Corp)

bar sections are moved, capacity and quality of the lifting gear should also be thoroughly checked. All material handling equipment including cranes should be load tested before the work is begun, to confirm the capability of the complete lifting gear system including slings. Once lifting operations are started, all unnecessary personnel should be removed from the lift area and individuals doing the work should be clear of the drop zone (Figure 7.42).

7.10.3 Retrofit Projects – Bench Marking on Retrofit Projects

This project involved bus extension work in an existing substation. The engineering team failed to verify the original bench mark elevation before proceeding with the new design. Unfortunately, the as-built drawings were not properly documented and updated. The original bench mark elevation was off the mark by approx. 4 in.

In a very similar case, the engineering team also failed to verify the original bench mark elevation of the existing foundations and complicated the error by not fully investigating and resolving conflicting information on the owners' drawings before proceeding with the new design. The original bench mark was off the mark by approx. 1 ft 4 in.

In each case, there was a significant cost to redesign the new bus connection to the existing bus.

7.10.4 Measurements Before Assembly

On another project, a voltage transformer link was installed very loosely and generated significant partial discharge. Measurements were never taken to verify the proper length and fit of the link. Had further investigation of the loose fit been done? corrective action could have been taken before the bus assembly was completed, gas processing finished, and the equipment placed in service.

7.10.5 Bus Assembly Double Checks

It is always a good idea to have a second member of the team double check the bus periodically during the assembly process for tools, cleaning materials, shipping supports, etc. On multiple occasions, personnel have left extraneous materials in the bus. These items generally are detected during the high-voltage qualification tests. It is both time consuming and costly to evacuate the SF_6 and disassemble the bus only to find a shipping block or hand tool was left from the installation.

7.10.6 Circuit Breaker Locking Pins

Most circuit breakers are equipped with locking pins or some other device to manually and mechanically block the open and close operations while technicians or maintenance personnel are working on the equipment. These temporary removable pins or devices are a required and important equipment safety feature to protect the maintenance individuals from severe injury if the circuit breaker was ever inadvertently operated. In this case, a technician was performing routine breaker commissioning tests including travel measurements, completed his work, and left the job site. The locking pins were left in their blocking position. This was unknown to the construction management team or operations group. When the breaker was given the normal close signal, the closing coil overheated and caught fire. Lesson No. 1: once test or maintenance personnel are finished, thoroughly review the job task and inspect the equipment. Lesson No. 2: locking pins or similar devices should be included in the construction lock out tag out procedure and documented in a log book. If possible, a ribbon or flag should also be attached to the locking pin, so personnel can tell at a glance if a pin is in place locking/blocking the circuit breaker mechanism operation.

References

1 IEC 62271-203 (2017). Gas-Insulated Metal Enclosed Switchgear for Rated Voltages above 52kV.

2 CIGRÉ TB 509 (2013). Final Report of the 2004–2007: International Enquiry on Reliability of High Voltage Equipment, Part 1 – Summary and General Matters.

3 CIGRÉ TB 513 (2013). Final Report of the 2004–2007 International Enquiry on Reliability of High Voltage Equipment, Part 5 – Gas Insulated Switchgear (GIS).

4 CIGRÉ TB 514 (2013). Final Report of the 2004–2007: International Enquiry on Reliability of High Voltage Equipment, Part 1 – GIS Practice.

5 IEC 62271-1 IEC 62271-1 (2017). High-voltage switchgear and controlgear - Part 1: Common specifications for alternating current switchgear and controlgear.

6 IEEE C37.122.1 IEEE C37.122.1 (2014). IEEE Guide For Gas-Insulated Substations Rated Above 52 KV.

7 IEC 62271-100 (2021) High-Voltage Alternating Current Circuit Breakers.

8 IEEE C37.09 (2018), IEEE Standard Test Procedures for AC High-Voltage Circuit Breakers with Rated Maximum Voltage Above 1000 V.

9 IEC 62271-102:2018, High-voltage switchgear and controlgear - Part 102: Alternating current isconnectors and earthing switches.

10 IEEE C37.30.5-2018 - IEEE Standard for Definitions for AC High-Voltage Air Switches Rated Above 1000 V.

11 IEC 62271-112:2021 RLV High-voltage switchgear and controlgear - Part 112: Alternating current high-speed earthing switches for secondary arc extinction on transmission lines.

12 IEC 61869-1:2007 Instrument transformers - Part 1: General requirements.
13 IEC 61869-2:2007 Instrument transformers - Part 2: Additional requirements for current ransformers.
14 IEC 61869-3:2007 Instrument transformers - Part 3: Additional requirements for inductive voltage transformers.
15 IEEE C37.122 (2010), IEEE Standard for Gas-Insulated Substations.
16 NEMA 12 Enclosure Rating, 2021, National Electrical Manufacturers Association (NEMA).
17 IEEE C37.122.6-2013 - IEEE Recommended Practice for the Interface of New Gas-Insulated Equipment in Existing Gas-Insulated Substations Rated above 52 kV.
18 NEMA 4x classification for indoor/outdoor enclosures, National Electrical Manufacturers Association (NEMA).

8

Applications

Authors: 1st edition Hermann Koch, William Labos, Peter Grossmann, Arun Arora, and Dave Solhtalab, 2nd edition Hermann Koch, and Dave Mitchell
Reviewer: 1st edition Phil Bolin, Hermann Koch, Devki Sharma, Ewald Warzecha, George Becker, John Brunke, Peter Grossmann, Arnaud Ficheux, Pravakar Samanta, Scott Scharf, Ravi Dhara, and Chuck Hands, 2nd edition Denis Steyn, Petr Rudenko, Stefan Schedl, Scott Scharf, Mark Kuschel, and Bala Kotharu

8.1 General

This chapter is a reflection on various gas-insulated switchgear (GIS) applications are shown based on the last 40 years of a world-wide experiences from the world's leading manufacturers and users. An overview of GIS equipment layouts is given in Section 8.2 for single bus, double bus, ring bus, H-scheme, and breaker and half-scheme arrangements.

In Section 8.3, several GIS project are presented and the reasons for utilizing GIS are explained. Reference projects GIS and several GIS typical applications are shown and it is explained for what reasons GIS was chosen.

In Section 8.4, a case study is presented on how to expand an existing substation with a 500/230 kV GIS in limited space. Air-insulated switchgear could not meet the high-power transmission requirements within the space limitations; therefore, the design of GIS was utilized as a solution.

Section 8.5 refers to Mobile GIS solutions, focusing on containerized and truck mounted GIS solutions for medium voltage and high voltage.

In Section 8.6, examples of a gas-insulated bus are given.

The possibilities of mixed technology substations are explained in Section 8.7. Here, different technical combinations of gas-insulated and air-insulated substation equipment are explained.

In Section 8.8, the expected future developments of GIS are explained. Reduction in size, simpler design, life-cycle evaluations, functional specification, intelligent GIS, integrated electronic devices, Rogowski coil, and capacitive divider are some of the topics discussed.

8.2 Typical GIS Layouts

Operational requirements and reliability of the power system are the major deciding factors used to determine the GIS layout. In addition, future extensions, service, and maintenance as well as investment costs be taken into consideration when deciding on a suitable substation layout. The following arrangements show common layouts used for GIS substations.

Gas Insulated Substations, Second Edition. Edited by Hermann J. Koch.

8.2.1 Single Bus Arrangement

Single bus arrangement GIS has the lowest investment cost to design and built, as such it is used when reliability is not the main priority. This may be the case for small substations in relatively less important loads where additional redundancy is not required. In these cases, the limitations to operational and maintenance activities can be accepted. The substation would be affected by an outage in the case of bus bar failures and service or maintenance activities. A single bus bar arrangement of a three-phase insulated GIS is shown in Figure 8.1.

8.2.2 Double Bus Arrangement

Operation flexibility and reliability is higher by extending the arrangement by a second bus bar when selecting the double bus arrangement. For important system points, the higher investment allows the substation to be operated with two bus bars that are coupled via a tie breaker. Each feeder is connected to the two bus bars. Maintenance can be done on one bus bar while the other bus bar stays in service. A double bus bar arrangement of a three-phase insulated GIS is shown in Figure 8.2.

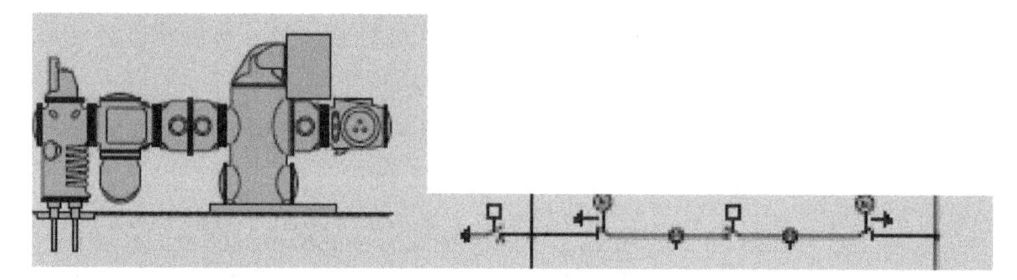

Figure 8.1 Single bus arrangement three-phase insulated (Reproduced by permission of Siemens AG)

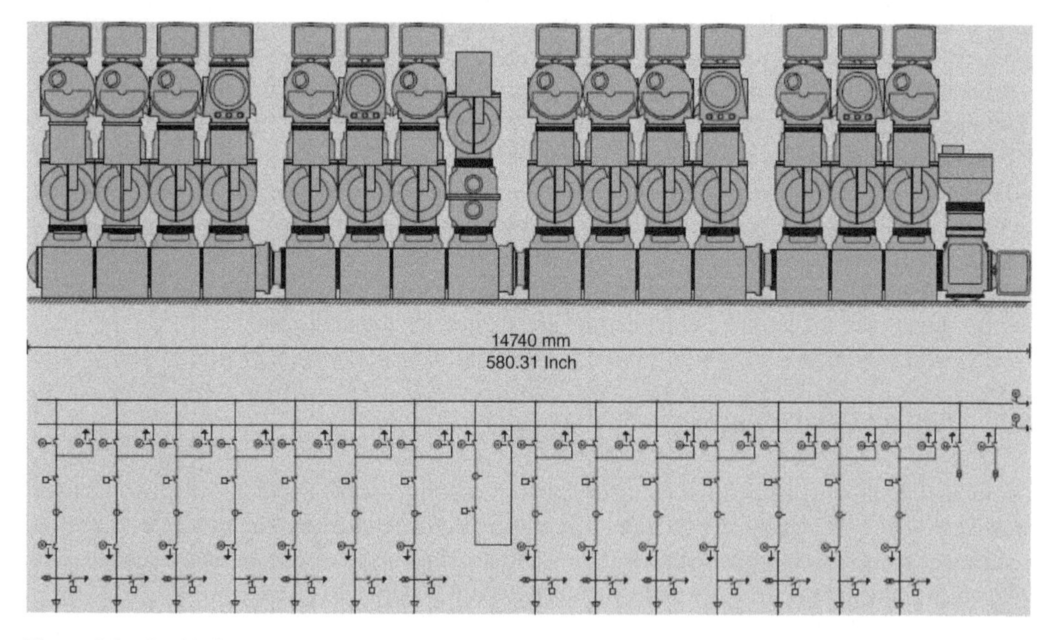

Figure 8.2 Double bus arrangement three-phase insulated (Reproduced by permission of Siemens AG)

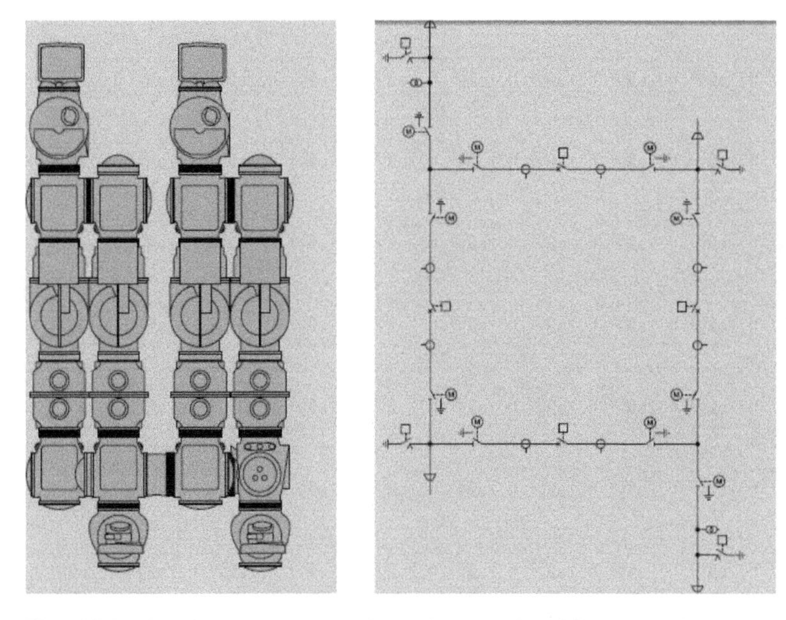

Figure 8.3 Ring bus arrangement three-phase insulated (Reproduced by permission of Siemens AG)

8.2.3 Ring Bus Arrangement

In a ring bus arrangement, the GIS bays are arranged in a ring, which provides good reliability at moderate costs since there is no extra bus bar. In case of a failure in one bus section, only the circuit in that bus section will be affected, and the other circuits can remain in service. Performing maintenance on one GIS breaker can be done by isolating this bus section and keeping the other bus sections in operation. A ring bus bar arrangement of a three-phase insulated GIS is shown in Figure 8.3. The ring arrangement typically involves up to six GIS feeders and is limited for extensions.

8.2.4 H-Scheme Arrangement

The H-scheme can be described as two single bus sections that are coupled by a center circuit breaker. In comparison with the single bus arrangement, the H-scheme provides higher reliability but also higher costs due to additional circuit breakers. In case of a breaker failure, the complete substation would not be out of service. Maintenance of one feeder can be done while the other feeders stay in operation. An H-scheme arrangement of a three-phase insulated GIS is shown in Figure 8.4.

8.2.5 Breaker and a Half Arrangement

With a relatively high investment, the breaker and a half arrangement ensure very high reliability. Even in the event that one bus bar would fail the power supply of feeders will be kept in service. A breaker and a half arrangement of a single-phase insulated GIS is shown in Figure 8.5.

This layout allows flexible operation. Any breaker can be isolated, for maintenance or service, while the other feeders can still be operated. The two bus bars are energized under regular operation conditions. As the name suggest this layout has, one and a half breakers per circuit. Three circuit breakers are designed for two circuits where each circuit shares the circuit breaker in the center.

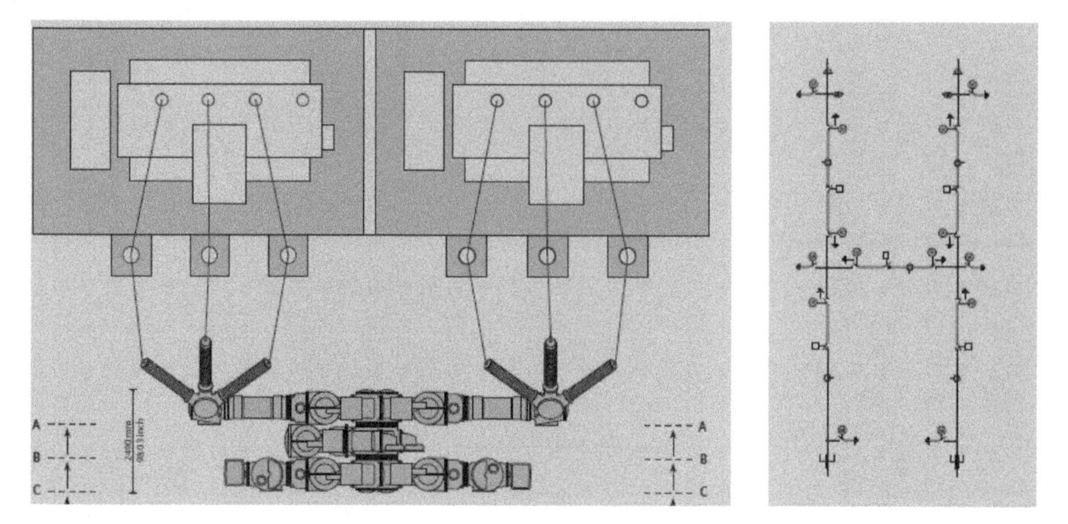

Figure 8.4 Typical H-scheme (H5) arrangement three-phase insulated (Reproduced by permission of Siemens AG)

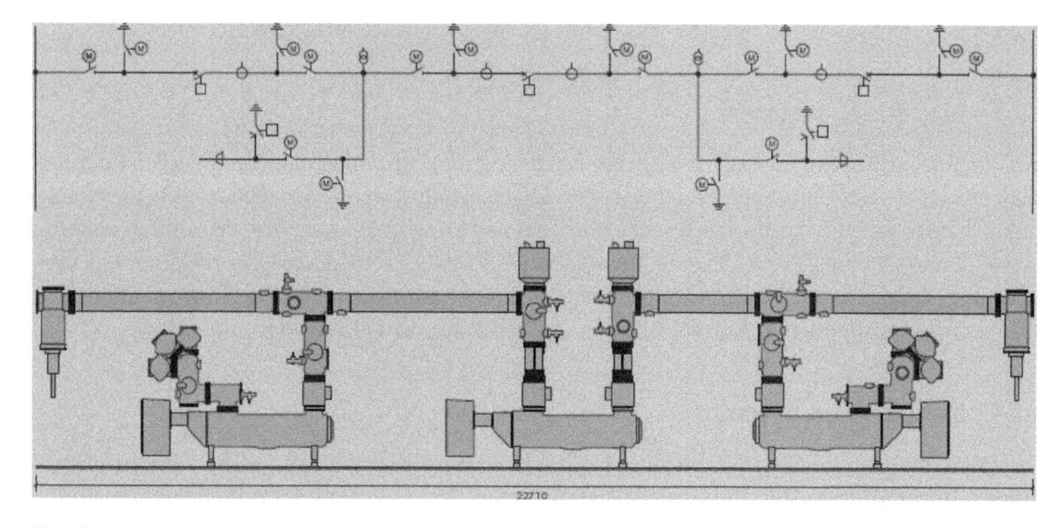

Figure 8.5 Breaker and a half arrangement single-phase insulated (Reproduced by permission of Siemens AG)

8.3 Reference Projects

8.3.1 500 kV Indoor GIS and 115 kV AIS

8.3.1.1 Introduction

The 500 kV/115 kV substation was built as a GIS on the 500 kV transmission side because of the site's proximity to the flood plains. After extensive studies, it was determined that a 500 kV GIS would be more cost effective than an air-insulated conventional 500 kV substation. The studies included the cost of site preparation to increase the grade elevation, structures and structure foundations, site and equipment grounding, and conventional 500 kV equipment, and compared it with a 500 kV GIS that required a smaller footprint. The savings in projected maintenance costs were considered as advantage. The 115 kV distribution side of the substation was chosen as air-insulated substation (AIS) technology [10].

8.3.1.2 Technical Data

The substation has two voltage levels. On the 500 kV side in GIS technology and on the voltage level 115 kV side in air-insulated technology. The current rating is 3150 A at 500 kV and 3000 A at the 115 kV side (see Table 8.1). The basic insulation level (BIL) is according to IEEE C37.122 [2], which is harmonized with IEC 62271-203 [1]. The short-circuit current switching capability is typical for these voltage levels.

The substation was built in 2011 in the short total time of less than a year. The 500 kV GIS is an indoor installation with a compact GIS and control building, as shown in Figure 8.6.

Table 8.1 Technical data of the 500 kV/115 kV GIS/AIS substation

Rated voltage (U_r)	500 kV	115 kV
Rated current (I_r)	3150 A	3000 A
Frequency (f)	60 Hz	60 Hz
BIL (U_{BIL})	1550 kV	550 kV
Short circuit (I_s)	63 kA	40 kA
Date of erection	2011	2011
Number of bays	5	9
Layout scheme	Ring bus (future breaker and half-scheme)	Main and transfer bus (future double bus double breaker)
Type of encapsulation	Single phase	Open air
Number of circuit breakers	4	7

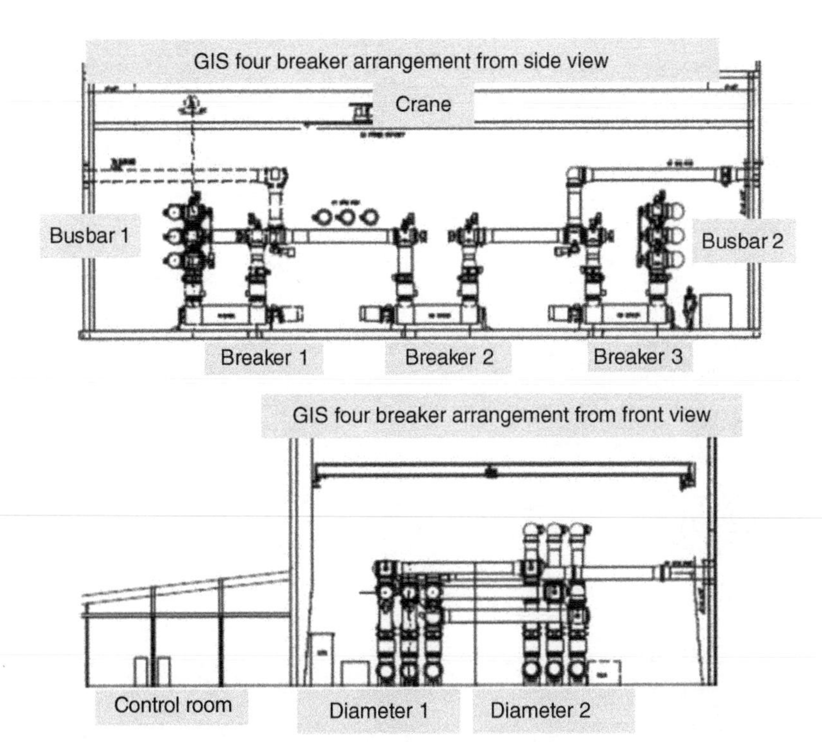

Figure 8.6 Layout of the 500 kV GIS and control building

8.3.1.3 Electrical Layout

The electrical layout of the substation consists of a 500 kV GIS and 115 kV AIS. The 500 kV GIS electrical layout is shown in Figure 8.7. The four-breaker arrangement shows two incoming and two outgoing lines. The breakers are connected in a ring. The current rating of the in- and outgoing lines is 3000 A. Disconnector and ground switches are positioned on each side of each breaker and for each of the circuits.

8.3.1.4 Physical Layout

The drawing of the physical arrangement in Figure 8.8 shows the location of the 500 kV GIS bays and the connection of the ring busses. The compact design of each bay includes three-phase breakers, disconnector switches, ground switches, and voltage transformers. The control room is separated from the GIS.

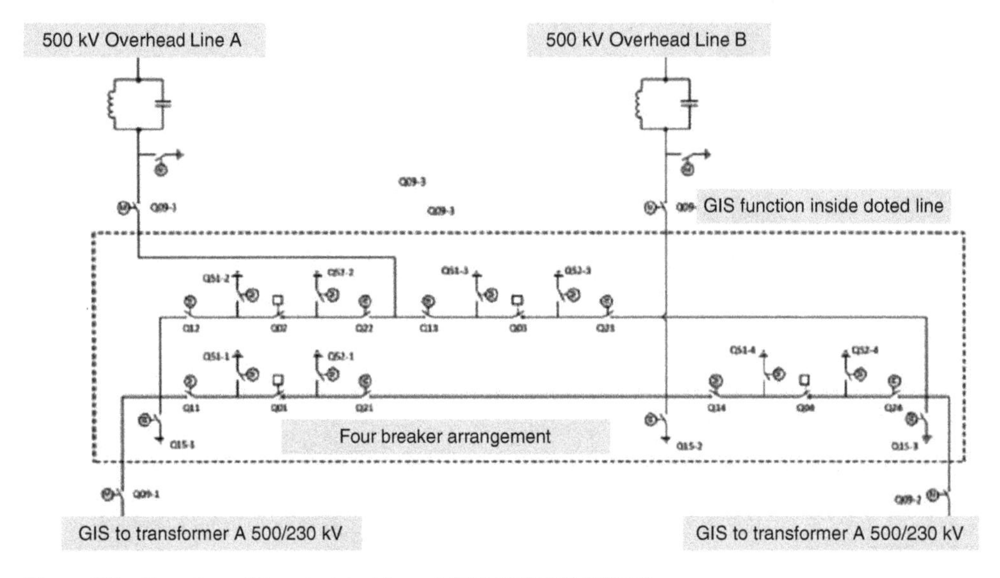

Figure 8.7 Overview of the electrical layout 500 kV GIS/115 kV AIS

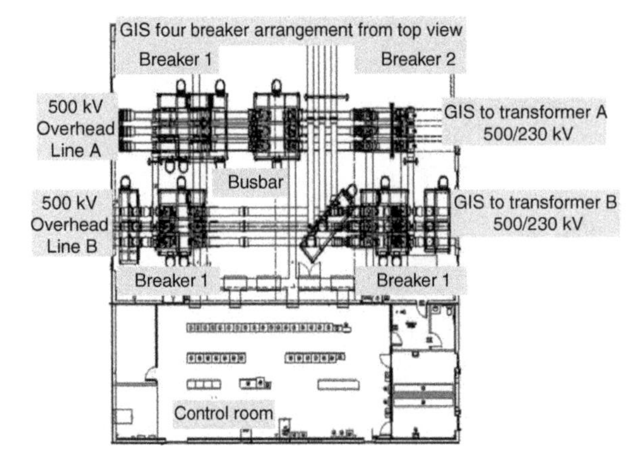

Figure 8.8 Physical arrangement of the 500 kV GIS

8.3.2 115 kV GIS Switching Station

8.3.2.1 Introduction

The 115 kV GIS switching station was as a result of a comprehensive set of reviews, studies, and investigations. Compliance assessments including short-circuit adequacy and compliance with mandatory reliability standards were performed. Also analyzed were contingency overloads based on a 10-year transmission plan of the utility, physical condition assessments of equipment, overall infrastructure condition, adequacy of protection, and control equipment and operational assessments related to outage scheduling and limitations.

The results of the reviews indicated that the existing structure and existing ring bus expansion was not viable, as most equipment was near or beyond fault duty available and the interconnection of proposed generation or future transmission expansion was restricted.

8.3.2.2 Technical Data

The GIS switching station is an indoor installation with a breaker and half scheme. The technical data are shown in Table 8.2.

8.3.2.3 Electrical Layout

The overview of the electrical layout of the substation is shown in Figure 8.9. The breaker and half-scheme arrangement with two bus bars are a typical design found in North America and some locations in Asia. Four incoming and four outgoing lines require 12 circuit breaker bays.

8.3.2.4 Physical Layout

The physical arrangement of the 115 kV breaker and half-scheme arrangement shows the very compact breaker bay arrangement with multilevel bus connections. Long runs of buses connect the four overhead lines in the substation by SF_6 gas-to-air bushings. An overview of the physical layout is given in Figure 8.10 in a simplified drawing.

The overview of the gas zones of the whole substation arrangement of the GIS is shown in Figure 8.11 in a simplified drawing. It is critical for safe operation and maintenance that gas zones are clearly identified. The gastight insulators of GIS are generally identified by colors, for example, yellow on the outside of the insulator or the metallic ring holding the insulator. Each gas zone also has its specific SF_6 gas pressure, as shown in Figure 8.12. The different gas pressure compartments are selected by their function. Circuit breakers have a higher pressure because of their interrupting

Table 8.2 Technical data of the 115 kV GIS switching station

Rated voltage ($\boldsymbol{U}_r$)	115 kV
Rated current (I_r)	4000 A
Frequency (f)	60 Hz
BIL (U_{BIL})	650 kV
Short circuit (I_s)	63 kA
Date of erection	2010
Number of bays	4
Switching scheme	1½ breaker scheme
Type of encapsulation	Single phase
Number of circuit breakers	12

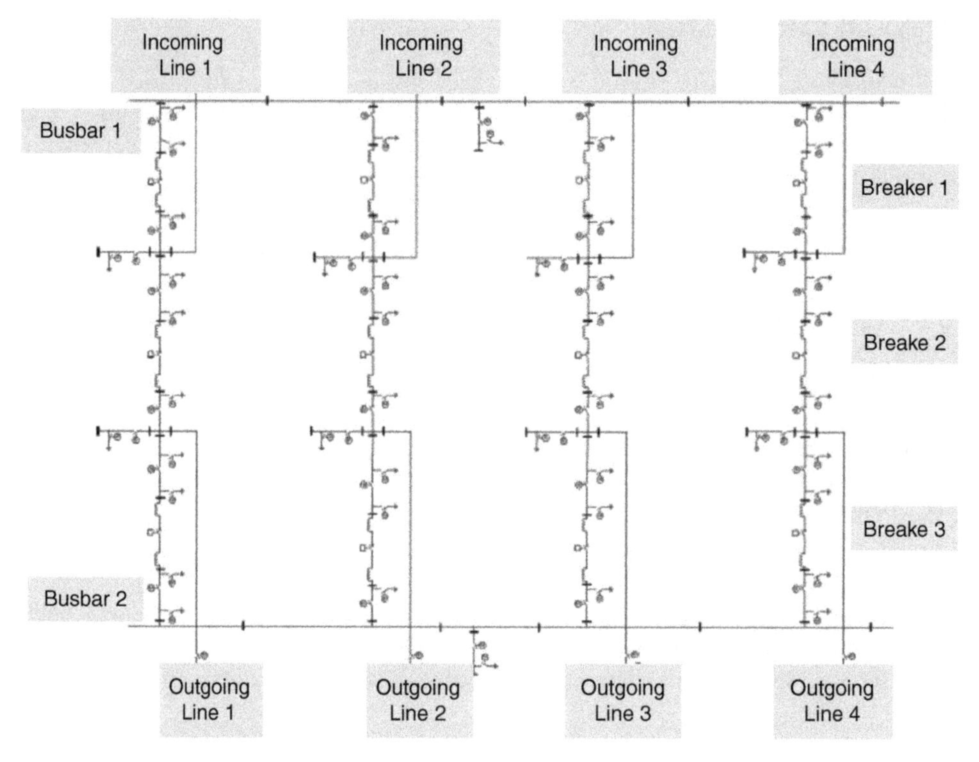

Figure 8.9 Electrical layout of the 115 kV breaker and half arrangement, simplified

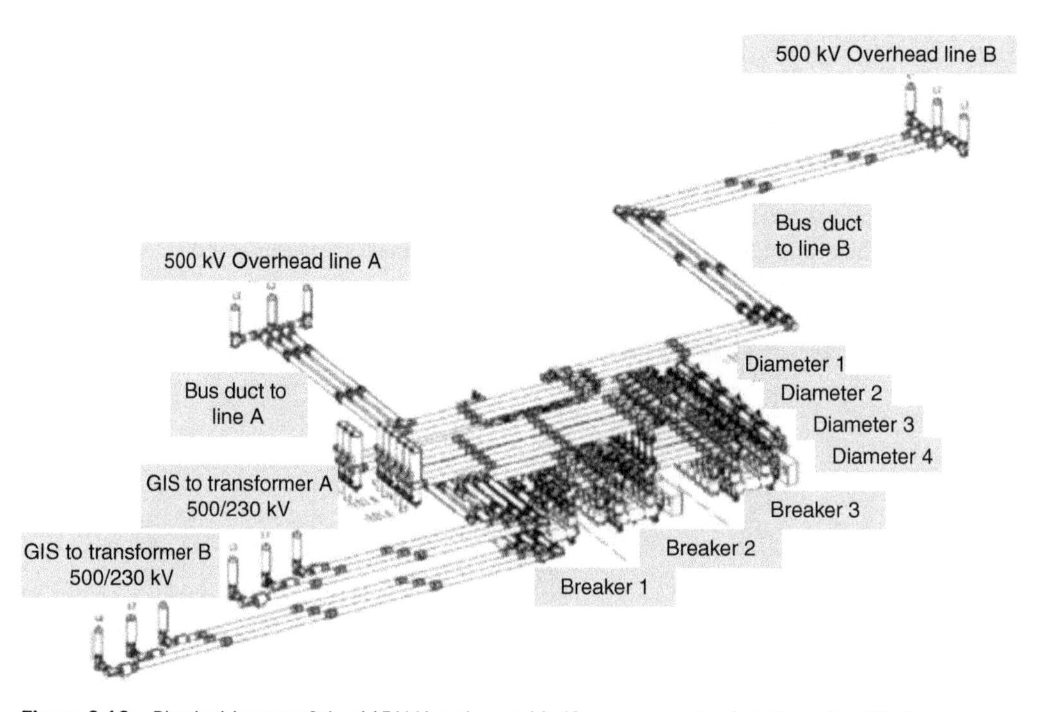

Figure 8.10 Physical layout of the 115 kV breaker and half arrangement substation, simplified

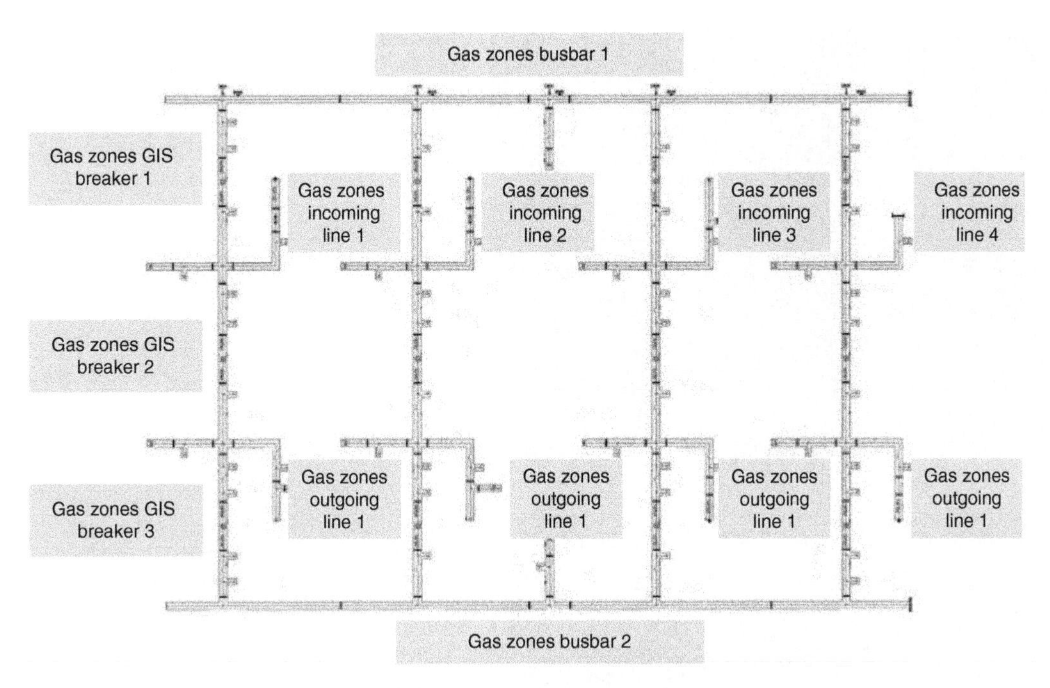

Figure 8.11 Gas zone scheme, simplified

SF_6 gas filling pressure Gauge pressure at 68°F (20°C)		
Circuit breaker (CB)	81.2 psig	(5.6 bar)
Current transformer (CT) at CB	81.2 psig	(5.6 bar)
Current transformer (CT) at feeder	75.4 psig	(5.2 bar)
Other parts of switchgear	75.4 psig	(5.2 bar)
Surge arrester (SA)	62.4 psig	(4.3 bar)

Figure 8.12 Typical gas filling pressure of the different GIS compartments (Reproduced by permission of United Illuminating Company)

function. Current transformers and voltage transformers usually also have a higher gas pressure because of their smaller size and compactness. Other gas compartments like bus bars or surge arresters usually have lower gas pressures.

8.3.2.5 Aerial View

The view into the GIS building in Figure 8.13 shows the side of the breaker and half-scheme arrangement. Behind the stairs on the ground level, three circuit breaker enclosures form the basis of the installation. On the top of the installation, the interconnecting bus bars are shown. The GIS equipment is single-phase encapsulated. The walkway on top of the circuit breaker enclosures allows access to the view ports and switch position indicators of the installation.

The dimension of the lightweight, low-cost GIS building is shown on Figure 8.14. The hall is placed inside the existing air-insulated substation on a small footprint. The basic advantage of the

Figure 8.13 View into the GIS hall with breaker and half scheme (Reproduced by permission of United Illuminating Company)

Figure 8.14 View of the GIS lightweight, low-cost design building (Reproduced by permission of United Illuminating Company)

building around the GIS allows access in any weather condition (rain, storm, snow). Routine maintenance and other works can be carried out independent from the environmental conditions.

In this case, no air conditioning of the hall is needed for operation. There are only natural ventilation openings in the wall to allow adequate cooling.

From the inside, the buses pass through the building wall connecting the GIS with the overhead lines using SF_6 gas-to-air bushings, as shown in Figure 8.15. The bushings are supported on steel structures. A view of wall penetrations for outdoor connection from the GIS to underground cables is shown in Figure 8.16. The case of a connection from the GIS to the overhead line through SF_6 gas-to-air bushings and the wall penetrations are shown in Figure 8.17.

In the case of cable connections, the GIS is connected by wall penetrations to the cable connection housing. The cable potheads are used to interface with GIS and XLPE (cross-linked polyethylene) cables. To protect the cable and GIS against transient overvoltage, surge arresters are located on top of the cable potheads supported by steel structures.

The secondary local control cubicles of the GIS are located in the GIS building on opposite sides of the GIS. Accessible to operators and away from the primary substation equipment, the control panels provide a good overview for safe operation. A view of secondary local control cubicles is shown in Figure 8.18. Each section of the cubicle line represents one bay of the GIS.

The substation control information is available on monitors installed in the GIS building also. The clear structure of the interface monitor and the real-time information of the actual position

Figure 8.15 Wall penetration by buses for connection to overhead lines by SF_6 gas-to-air bushings supported on steel structures (Reproduced by permission of United Illuminating Company)

Figure 8.16 Wall penetrations and cable potheads with surge arresters on the top (Reproduced by permission of United Illuminating Company)

of each circuit breaker, disconnect switch, ground/earth switch, and voltage and current values in the different sections of the substation allow for a quick overview and give detailed information.

8.3.3 345 kV and 4000 A Indoor Expendable Ring Bus GIS

8.3.3.1 Introduction

The existing outdoor GIS was first-generation equipment that developed gas leaks and environmentally friendly and reliable GIS was required.

Figure 8.17 Connection of overhead line to the GIS through SF_6 gas-to-air bushings and wall penetrations (Reproduced by permission of United Illuminating Company)

Figure 8.18 Local control cabinets (Reproduced by permission of United Illuminating Company)

8.3.3.2 Technical Data

The 345 kV and 4000 A indoor expandable ring bus GIS has the technical data shown in Table 8.3.

8.3.3.3 Electrical Layout

The overview of the layout of the substation is shown in Figure 8.19. There are in- and outgoing overhead lines at the 345 kV level. The 345 kV GIS is located indoors.

8.3.3.4 Physical Layout

The physical layout of the 345 kV GIS is shown in Figure 8.20. The four bays are located in the building center in a compact configuration. The buses are connected on top of the bays and the outgoing buses for overhead lines penetrate through the building walls. In Figure 8.19, a bird eye view gives an overview of the substation. The trailer-mounted device covered with a tarp in front of the building are used for the SF_6 gas handling. The vacuum pump device provides vacuum below 2 mbar to extract air or SF_6 from the GIS equipment. This is important to get air out and with the air the moister inside the GIS. After vacuum is reached and held for several minutes, the SF_6 handling device is used for easy filling of SF_6 gas into the GIS gas compartment with the required filling pressure.

Table 8.3 Technical data of the 345 kV indoor GIS

Rated voltage (U_r)	345 kV
Rated current (I_r)	4000 A
Frequency (f)	60 Hz
BIL (U_{BIL})	1550 kV
Short circuit (I_s)	63 kA
Date of erection	2005
Number of CB bays	4
Switching scheme	Ring expandable to 3 bays of 1½ breaker scheme
Type of encapsulation	Single phase
Number of circuit breakers	4

Figure 8.19 Bird's eye view of the 345 kV GIS (Reproduced by permission of United Illuminating Company)

Figure 8.20 View into the 345 kV GIS building (Reproduced by permission of United Illuminating Company)

Figure 8.21 Front view of the 345 kV GIS bays including cubicle of the circuit breaker spring drive operator (Reproduced by permission of United Illuminating Company)

The compactness of the GIS bays is shown in Figure 8.21. Each bay includes a cubicle for the spring-operated drive mechanism of the circuit breaker, which is attached to the front side of the bay. Bus bars and connections to the outside overhead lines are mounted on top of the GIS bays.

The local control cabinets (LCCs) are placed in a separate control room, as shown in Figure 8.22. This option is sometimes selected to minimize the time of operational personnel spent inside the building room of the 345 kV GIS primary equipment. All sensors, control switches, and motor operations are interconnected by cables from the GIS into the LCC, located in the control room.

The internal view of the LCC in Figure 8.23 shows the frame-mounted control devices and cabling. The LCCs include the control switches for remote operation of bay disconnect/ground switches and circuit breaker, monitoring of each SF_6 gas compartment, and motor control for the disconnector and ground switch drives as required for partial discharge monitoring, temperature control, voltage, and current measuring.

Figure 8.22 Local control cabinet (LCC) in a separate control room for the 345 kV GIS (Reproduced by permission of United Illuminating Company)

Figure 8.23 Local control cabinet (LCC) internal for the 345 kV GIS (Reproduced by permission of United Illuminating Company)

8.3.4 69 kV and 3150 A Indoor Double Bus GIS

8.3.4.1 Introduction

The 69 kV and 3150 A indoor double bus GIS had to meet the requirements of compact design for accommodation in the existing building located in the middle of a residential community and with high expectations on esthetic appearance. This substation contains 69 kV GIS, 12 kV GIS, power transformers, protection, and control panels and meets the seismic requirements. Underground dry-type cables with plug-in connectors are used for the 69 kV GIS interface.

8.3.4.2 Technical Data

The eight bays have the following technical data (see Table 8.4).

8.3.4.3 Electrical Layout

The electrical layout of this double bus bar GIS is shown in Figure 8.24 by the gas compartment schematic. The structure of the two bus bars on top and bottom and the cable connections in between are indicated.

Table 8.4 Technical data of the 69 kV indoor GIS

Rated voltage (U_r)	69 kV
Rated current (I_r)	3150 A
Frequency (f)	60 Hz
BIL (U_{BIL})	650 kV
Short circuit (I_s)	40 kA
Date of erection	2009
Number of bays	8
Switching scheme	Double bus bar
Type of encapsulation	Three phases
Number of circuit breakers	8

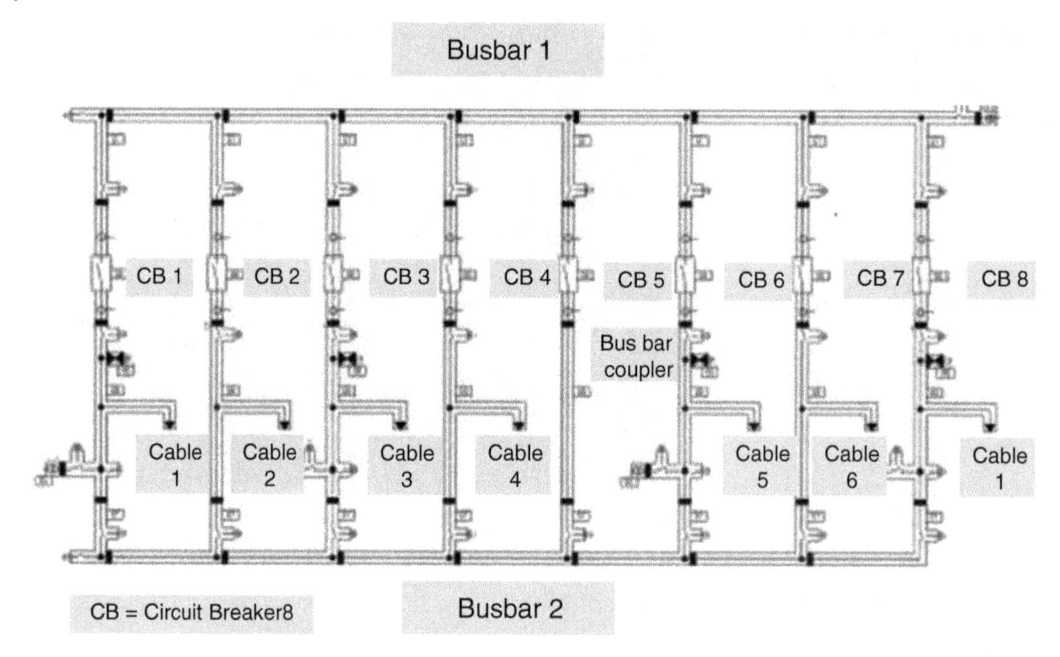

Figure 8.24 Electrical layout shown by the gas compartment schematic, simplified

8.3.4.4 Physical Layout

The compact and systematical physical layout of the 69 kV GIS is shown in Figure 8.25. The three-phase encapsulated bus bars are shown at the upper and lower ends of the bays. In between, are breaker bays with plug-in cable connections and three-phase encapsulation.

The three views are shown in three different bay arrangements in Figure 8.26. The vertical enclosure holds the three-phase circuit breaker with operating mechanism on the top. The upper two bays show high-voltage (HV) cable connections. The right view shows the bay with a voltage transformer on top of the HV cable connection enclosure including a three-position disconnect and a ground switch. The left view shows a bay without a voltage transformer but with a three-position disconnection and a ground switch. In the bottom view, a bus coupling bay is shown.

In Figure 8.25, a plan view of the double bus arrangement is shown with eight circuit breaker bays connected to it. The bay structure is easy to see and understand. For better accessibility, bus bar extension sections are inserted between each bay. Still the total size is small and suitable for underground installations (Figure 8.27).

The local control cabinets are located on the opposite side of the GIS. In Figure 8.28, an example of local control cabinet is shown. A view of the 69 kV GIS building matching the surrounding residential neighborhood is shown in Figure 8.29.

8.3.5 115 kV and 1200 A Container Ring Bus GIS

8.3.5.1 Introduction

This substation was built to enhance the reliability and capacity of the grid, replacing an existing old AIS of lower capacity. This GIS was built in a container because of a very short installation time. The severe climatic conditions at the location of the installation were also an important factor for the containerized solution and for easy access in cold winter periods, as well as the low visual impact due to proximity to a shopping center.

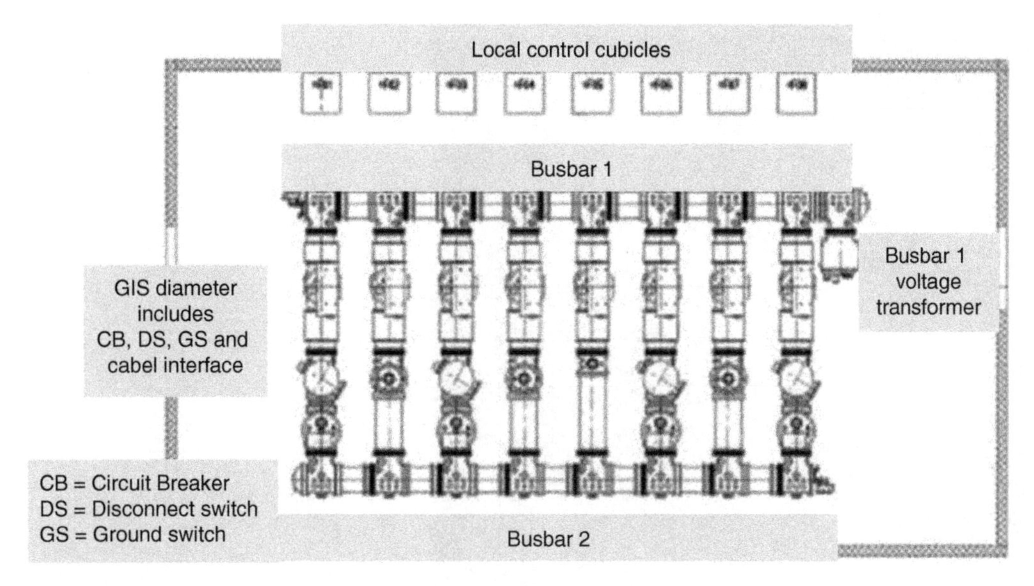

Figure 8.25 Physical layout of the 69 kV GIS equipment, simplified

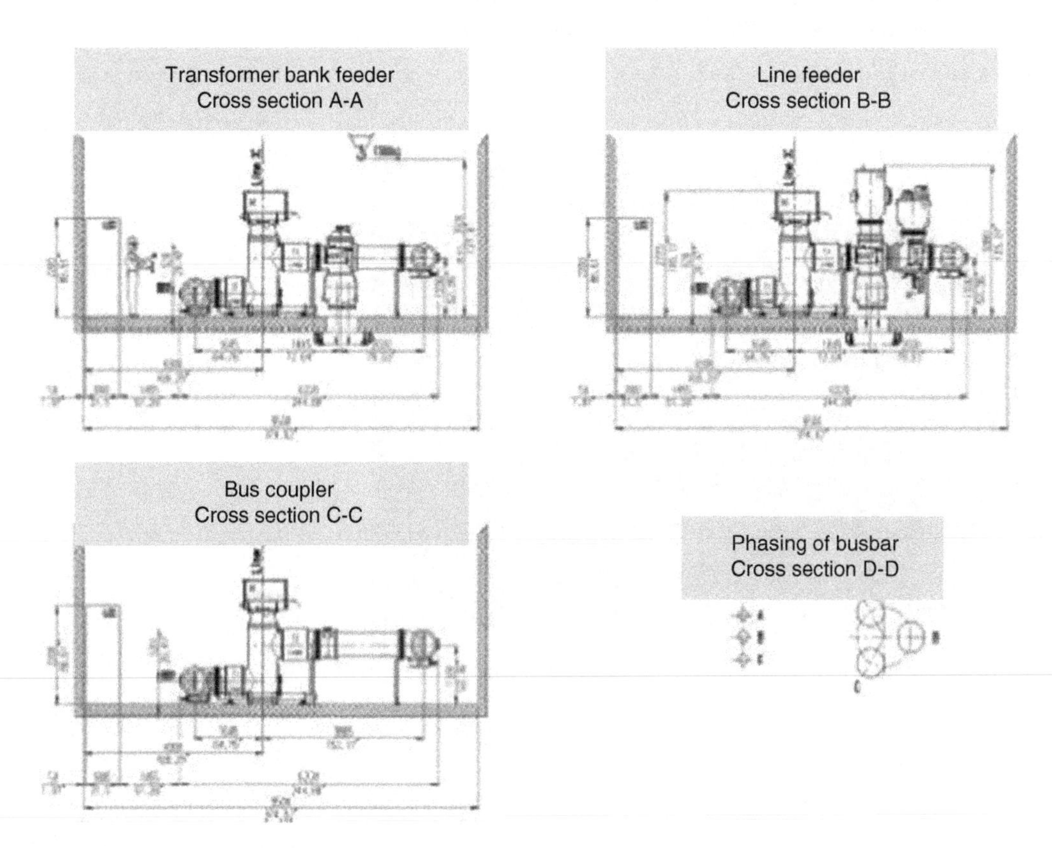

Figure 8.26 Side views of three different bay arrangements, simplified

Figure 8.27 View of the one-bay 69 kV GIS, principle design (Reproduced by permission of Siemens AG)

Figure 8.28 Local control cabinet (Reproduced by permission of Siemens AG)

8.3.5.2 Technical Data

The 115 kV GIS technical data is shown in Table 8.5.

8.3.5.3 One Line Diagram

The electrical layout of this four-circuit breaker ring bus 115 kV GIS is shown in Figure 8.30. All four connections of the GIS are aerial connections. The GIS has a three-phase enclosure design.

Figure 8.29 Substation building for GIS in a residential area (Reproduced by permission of Siemens AG)

Table 8.5 Technical data of the 115 kV container GIS

Rated voltage (U_r)	115 kV
Rated current (I_r)	1200 A
Frequency (f)	60 Hz
BIL (U_{BIL})	550 kV
Short circuit (I_s)	40 kA
Date of erection	2009
Number of bays	4
Switching scheme	Ring bus
Type of encapsulation	Three phases
Number of circuit breakers	4

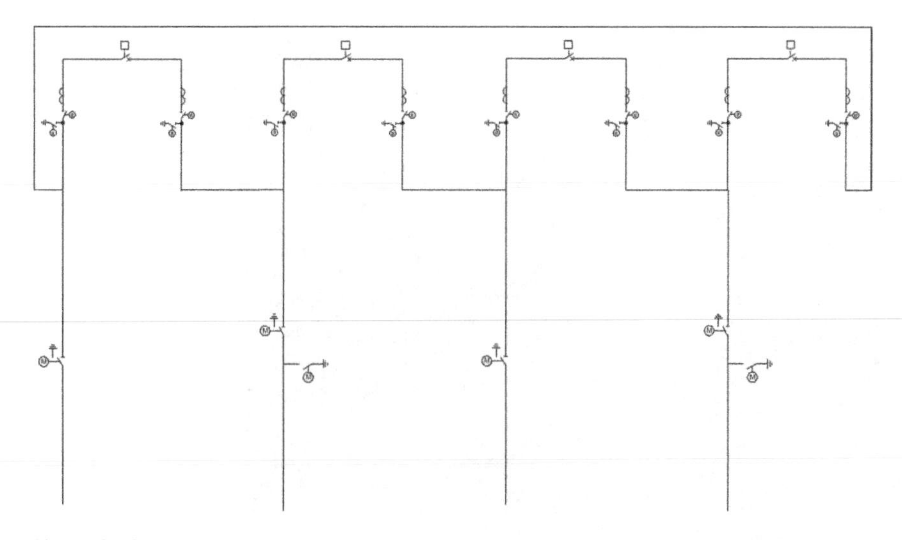

Figure 8.30 One-line schematic showing the electrical ring bus switching scheme

8.3.5.4 Physical Layout

The footprint of the container including the local control cabinets is only about 6 m × 12 m, with a height of about 3 m, which demonstrates the compactness of the whole GIS.

Below are some pictures documenting the ease of installation at the site. Figure 8.31 is a winter view of the old substation that needed to be replaced due to age and an increasing power demand. The delivery and placement of the first fully equipped container is shown in Figure 8.32. A view into the container with the three-phase insulated GIS on the left side and the local control cabinets on the right side is shown in Figure 8.33. A crane hoisted the container and placed it to the right on the prepared foundation.

Figure 8.31 Old 115 kV substation before replacement (Reproduced by permission of ABB)

Figure 8.32 The 115 kV GIS container installation (Reproduced by permission of ABB)

Figure 8.33 View inside the container with the GIS on the left and the LCCs on the right (Reproduced by permission of ABB)

Figure 8.34 View of the 115 kV GIS container and the transformer (Reproduced by permission of ABB)

Figure 8.34 shows the GIS container and the transformer. The transformer is connected by an SF_6 gas-to-air bushing in this case.

An overall view of the new substation with the 115 kV GIS placed container is shown in Figure 8.35. The picture shows the fence and the masonry wall around the GIS with a dead-end structure for transmission line take-off.

8.3.6 115 kV and 2000 A Outdoor Single Bus GIS

8.3.6.1 Introduction

The 115 kV and 2000 A outdoor single bus GIS was built outdoors because of space limitations well as esthetics of the surrounding residential area.

Figure 8.35 View of the 115 kV GIS substation from the street (Reproduced by permission of Hyundai)

Table 8.6 Technical data of the 115 kV outdoor GIS

Rated voltage (U_r)	115 kV
Rated current (I_r)	2000 A
Frequency (f)	60 Hz
BIL (U_{BIL})	550 kV
Short circuit (I_s)	40 kA
Date of erection	2010
Number of bays	4
Layout scheme	Single bus
Type of encapsulation	Three phases
Number of circuit breakers	4

8.3.6.2 Technical Data

The 115 kV outdoor GIS has the technical data given in Table 8.6.

8.3.6.3 Electrical Layout

A view of the layout of the substation is shown in Figure 8.36. A single three-phase bus bar connects the four bays to the overhead lines.

8.3.6.4 Physical Layout

The 115 kV outdoor GIS is designed in a long-extended way as required by the location of the in- and outgoing overhead lines. The control cubicles are located next to the three-phase vertical circuit breaker enclosure. See Figure 8.36.

Figure 8.36 View of 115 kV outdoor GIS (Reproduced by permission of Hyundai)

8.3.7 345 kV and 4000 A Indoor Breaker and Half-Scheme GIS

8.3.7.1 Introduction

The 345 kV and 4000 A indoor breaker and half-scheme GIS was installed as a result of a comprehensive set of transmission planning studies and investigations. A switching station with transformation was required to allow for the connection of a new 345 kV line and to a 115 kV transmission line to a generation station.

8.3.7.2 Technical Data

The 345 kV indoor GIS that was installed is single-phase insulated and follows the breaker and half-scheme. The technical data are given in Table 8.7.

Table 8.7 Technical data of the 345 kV indoor GIS

Rated voltage ($\boldsymbol{U}_r$)	345 kV
Rated current (I_r)	4000 A
Frequency (f)	60 Hz
BIL (U_{BIL})	1550 kV
Short circuit (I_s)	50 kA
Date of erection	2009
Number of bays	16
Switching scheme	Breaker and half-scheme
Type of encapsulation	Single phase
Number of circuit breakers	16

8.3.7.3 Electrical Layout

The 345 kV indoor GIS has a total of 16 bays and follows the breaker and half-scheme assembly, as shown in Figure 8.37. There are 12 in- and outgoing lines and two bus bar systems, which make the scheme relatively complex but satisfies the requirements of the network.

The top view arrangement of the GIS and the transformers are shown in Figure 8.38, showing the compactness of GIS. There are two transformer and two shunt reactor connections in the upper

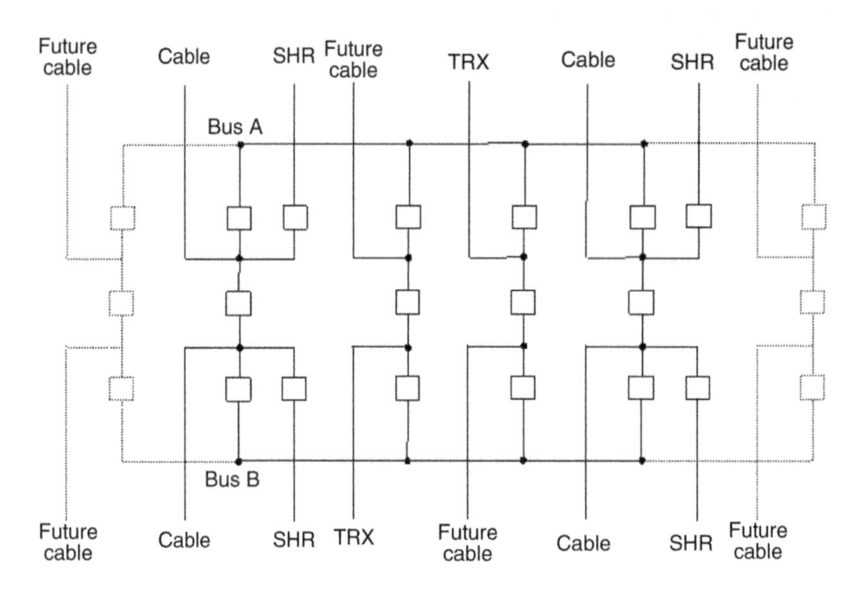

Figure 8.37 Electrical single-line scheme of the 345 kV indoor GIS (Reproduced by permission of United Illuminating Company)

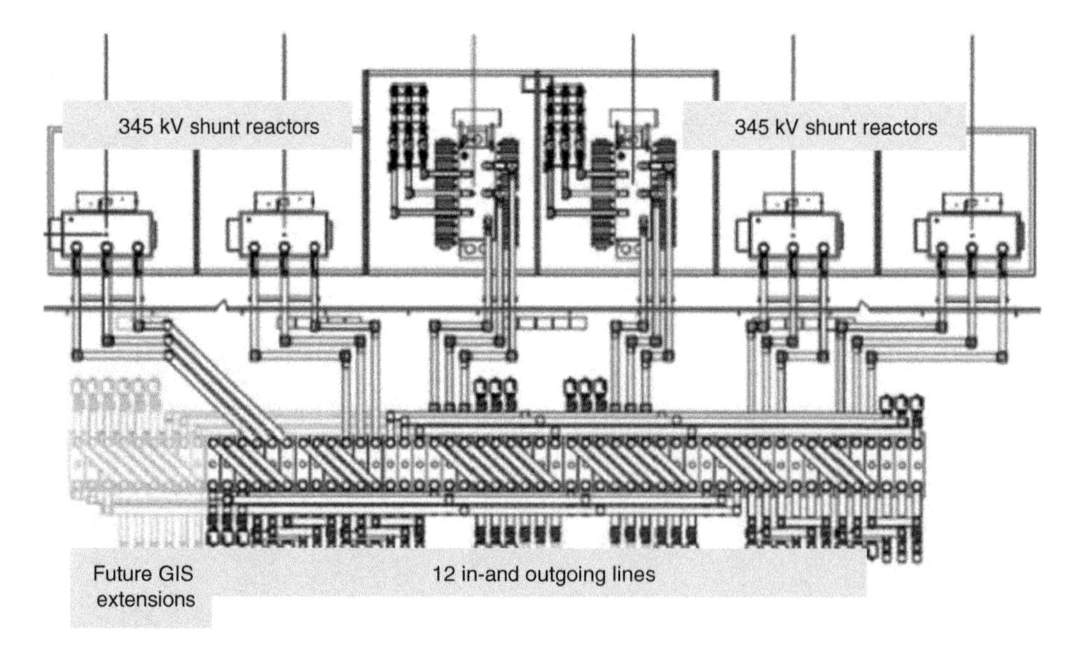

Figure 8.38 Top view of the GIS, transformer, and cable connection arrangement of the 345 kV indoor GIS (Reproduced by permission of United Illuminating Company)

part of the one-line diagram. The lower part shows six cable connections and two spare cables. On the left side of the GIS, space is saved for the full bay of the breaker and half-scheme expansion.

8.3.7.4 Physical Layout

The physical size of the 16 bay GIS is shown in Figure 8.39. The single bays are placed in one row with connections to the overhead lines through the wall with a single-phase gas bus. The buses are connected to the GIS bays on the top.

The wall bushings to the overhead lines outside the building are at about 4 m in height, and so are the bus ducts. Steel support structures are provided to run the buses between the building wall and the GIS (see Figure 8.40).

The local control cubicles are integrated at each bay and are fully accessible from the walkway in front of the GIS. All operational parts or indications can be operated or seen from the walkway at the front (see Figure 8.41).

The rear side of the GIS holds the direct cable connection enclosures to connect the XLPE cables coming from the cable trench in the building floor. The cable connection in accordance with IEC 62271-209 [4], cable connections for gas-insulated metal-enclosed switchgear for rated voltages above 52 kV – Fluid-filled and extruded insulation cables – Fluid-filled and dry type cable terminations (see Figure 8.42).

Figure 8.39 View into the GIS hall (Reproduced by permission of United Illuminating Company)

Figure 8.40 View at wall bushings on the steel support (Reproduced by permission of United Illuminating Company)

Figure 8.41 View of the walkway and local control cabinets in front of the GIS bays (Reproduced by permission of United Illuminating Company)

Figure 8.42 View from the rear side of the GIS with directly connected cables (Reproduced by permission of United Illuminating Company)

The whole GIS is located inside an industrial building. From the outside, the building does not give the impression that it contains a 16 bay 345 kV gas-insulated switchgear with a 4000 A current rating (see Figure 8.43).

8.3.8 115 kV and 3150 A Indoor Ring Bus GIS

8.3.8.1 Introduction

It is hard to imagine an electrical substation that is not only esthetically attractive but also adapted to the local environment and architecture. Such is the case with the new 115 kV GIS.

In operation since 2005, it consists of a two-building facility equipped with four 115 kV GIS bays, several 25 kV switchgears, two 28 MVA transformers, 115 kV and 25 kV duct banks, protection and control equipment, and SCADA and RTU (remote terminal unit) equipment. The substation is designed for an ultimate configuration with six-position 115 kV ring buses and provisions for a total of four power transformers. Its specificity, however, lies in its unique architecture.

Figure 8.43 Outside view of the GIS building (Reproduced by permission of United Illuminating Company)

The buildings are designed so as not to resemble a substation but rather mountain barns, chalets, or other typical buildings in the area. The majority of the local population would probably never imagine that the two barn-like buildings at the bottom of a valley contain a state-of-the-art electric distribution substation.

8.3.8.2 Technical Data

The 115 kV and 3150 A indoor ring bus GIS is three-phase encapsulated and has the following technical data given in Table 8.8.

8.3.8.3 Electrical Layout

The GIS is rated at 115 kV, 40 kA, and 3150 A, an indoor design, and all the connections are through underground dry-type cables. The cables are connected to the GIS through simple plug-in connections.

8.3.8.4 GIS Components

The circuit breaker has three single-phase interrupting elements contained in the same enclosure equipped with a spring-operated mechanism. A combined three-position disconnect and grounding switch is operated by an electrical operating mechanism directly attached to the equipment.

Table 8.8 Technical data of the 115 kV indoor GIS

Rated voltage ($\boldsymbol{U}_r$)	115 kV
Rated current (I_r)	3150 A
Frequency (f)	60 Hz
BIL (U_{BIL})	650 kV
Short circuit (I_s)	40 kA
Date of erection	2005
Number of bays	4
Layout scheme	Ring bus
Type of encapsulation	Single phase
Number of circuit breakers	4

Each of these is equipped with large 14 mm (5.5 inch) diameter view ports located close to the floor level in order to be able to see them without any access platforms or cameras.

A make-proof grounding switch is used for safe grounding of primary circuits, including closing on short-circuit and switching-induced currents.

Telescopic coupling elements with sliding sleeve joints on the conductors are used to allow the removal and installation of components for assembly, extension, and maintenance work without dismantling other switchgear components. The GIS is equipped with customized instrument transformers. The GIS was supplied with local control cubicles fully wired, tested in the factory, and mounted on the GIS.

8.3.8.5 Physical Layout

The physical layout of the 115 kV indoor GIS shows a very compact design with a small footprint of 8 m by 5 m only, as shown in Figure 8.44. The complete 115 kV indoor GIS is located inside a masonry construction type of building and all connections are by underground cables, as shown in Figure 8.45.

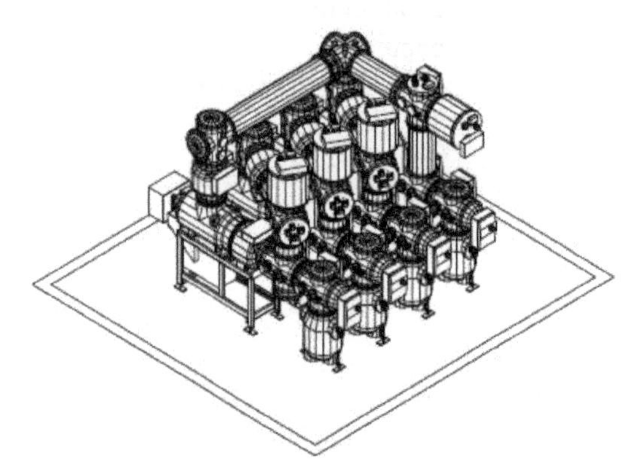

Figure 8.44 The 115 kV indoor GIS layout (Reproduced by permission of Alstom)

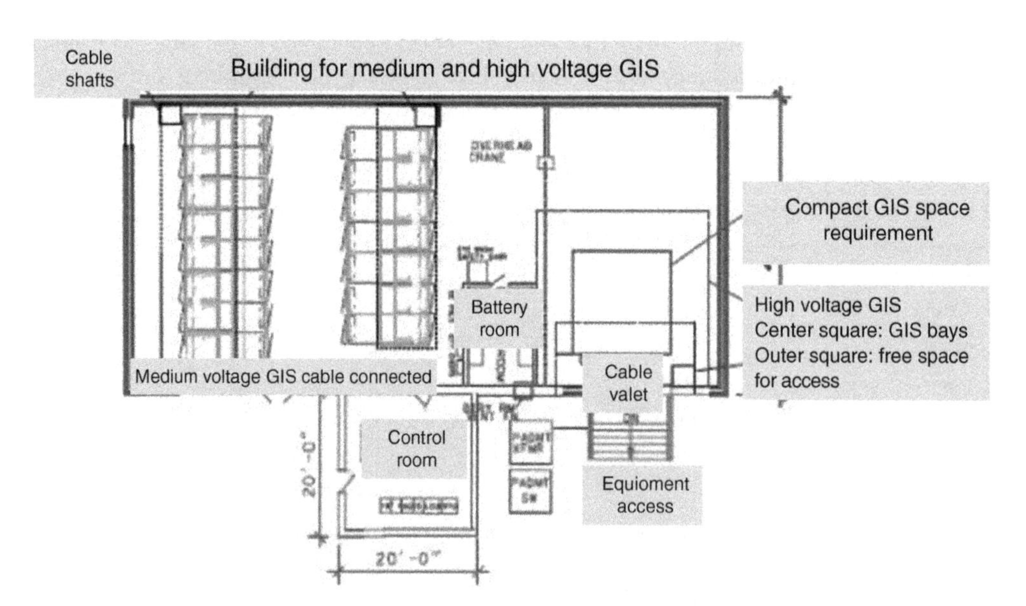

Figure 8.45 Masonry construction type building of the 115 kV GIS (Reproduced by permission of Alstom)

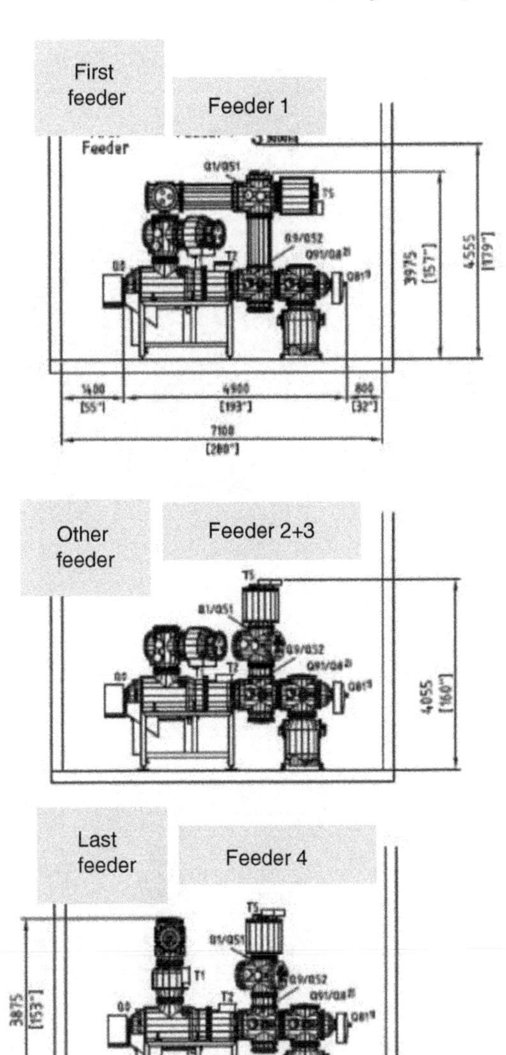

Figure 8.46 Side view of three different bays (Reproduced by permission of Alstom)

In Figure 8.46, three elevation views of different GIS bay arrangements are shown. The upper one shows a vertical circuit breaker enclosure with the bus bar on top and above the bus to close the ring. At the right, the cable connection enclosure is attached to the ground. The cables come in from the basement. The middle view shows the same without the bus bar to close the ring on top, but a voltage transformer on top of the circuit breaker enclosure and a cross bus bar to the next bay. The bottom view shows the same as the middle view but without the cross-bus bar to the next bay.

In Figure 8.47, an interior view of the 115 kV GIS room within the building is shown. When inside the building, the GIS is placed in the center of the room, which provides good access from all sides. For better accessibility to disconnect and ground switches on top, a walkway is built into the GIS assembly (see Figure 8.48).

The medium voltage of the 25 kV GIS and the 115/25 kV transformers are located inside the second building, shown in Figure 8.49. In Figure 8.50, a total view in the landscape of the 115 kV/12 kV indoor GIS is given.

Figure 8.47 View into the 115 kV GIS building (Reproduced by permission of Alstom)

Figure 8.48 View of the 115 kV GIS inside the building (Reproduced by permission of Alstom)

Figure 8.49 View of the medium voltage 25 kV GIS and 115/25 kV transformer building (Reproduced by permission of Alstom)

Figure 8.50 Total view of the 115 kV/12 kV indoor GIS (Reproduced by permission of Alstom)

8.3.9 69 kV and 2000 A, Indoor Ring Bus GIS

8.3.9.1 Introduction

The 69 kV and 2000 A indoor ring bus GIS was built because of space limitations.

8.3.9.2 Technical Data

The 69 kV ring bus GIS has the technical data given in Table 8.9.

8.3.9.3 Electrical Layout

The overview of the electrical layout of the 69 kV indoor GIS is given in the single line in Figure 8.51.

The ring bus is connected to four incoming and four outgoing lines. The incoming lines are protected by surge arresters from transient overvoltage. Disconnect switches and ground switches are located along in- and outgoing lines and disconnectors are located after each circuit breaker on the ring disconnect switches are located to be able to isolate each circuit breaker.

Table 8.9 Technical data of the 69 kV indoor GIS

Rated voltage (U_r)	69 kV
Rated current (I_r)	2000 A
Frequency (f)	60 Hz
BIL (U_{BIL})	325 kV
Short circuit (I_s)	40 kA
Date of erection	2009
Number of bays	8
Switching scheme	Ring bus
Type of encapsulation	Three phases
Number of circuit breakers	8

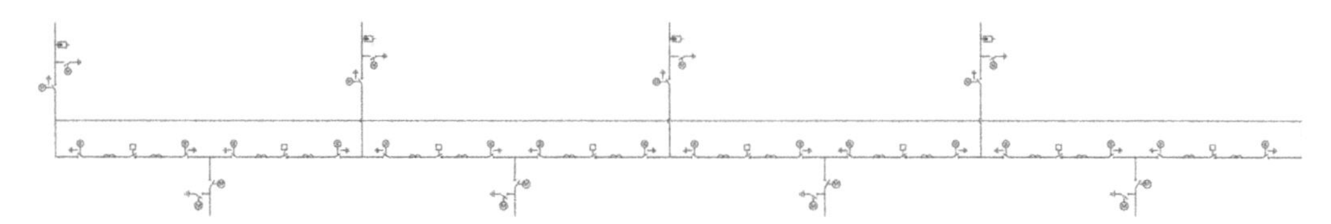

Figure 8.51 Overview of the single-line diagram of the 69 kV indoor GIS

8.3.9.4 Physical Layout

The physical layout of the 69 kV indoor GIS is shown in Figure 8.52 from the rear side with the cable connection on the back part. In Figure 8.53, the front side is shown with the vertical circuit breaker enclosure. The cable connections to the GIS are connected from the basement below the GIS floor. The circuit breakers are in a vertical position with the driving mechanism on top of the enclosure. The circuit breakers are spring operated while the disconnect and ground switches are motor operated.

8.3.9.5 Picture View

The vertical three-phase enclosed circuit breakers are shown in Figure 8.54. The operation mechanism for the circuit breaker is at the top of the circuit breaker enclosure.

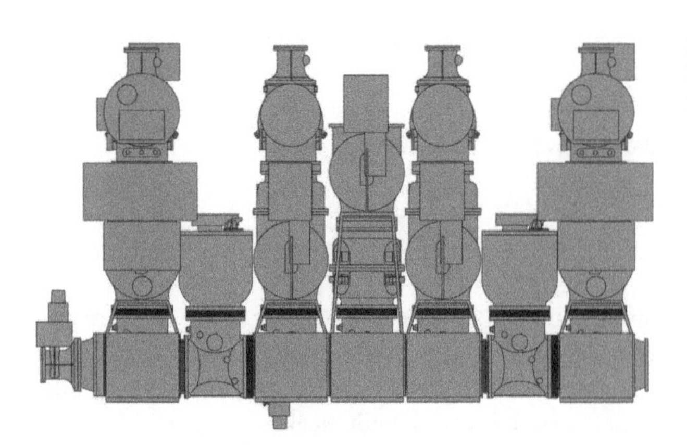

Figure 8.52 Top view of the 69 kV indoor GIS assembly (Reproduced by permission of Siemens AG)

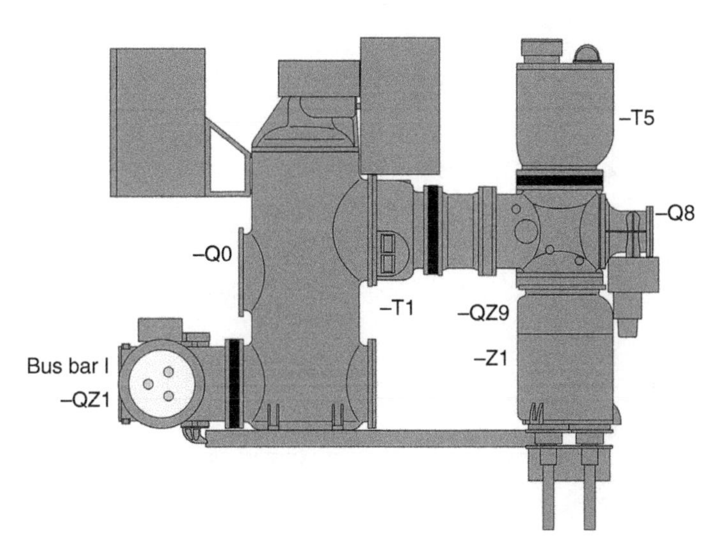

Figure 8.53 Side view of the 69 kV indoor GIS assembly (Reproduced by permission of Siemens AG)

Figure 8.54 Single bay GIS with vertical three-phase circuit breaker enclosures (Reproduced by permission of Siemens AG)

Figure 8.55 View of 69 kV indoor GIS assembly with vertical three-phase insulated circuit breaker (Reproduced by permission of Siemens AG)

Figure 8.55 shows the principle of a three-phase encapsulated GIS assembly with the single bus bar in front and the cable connection at the back side. The local control cubicles, which are also called integrated control units, are attached to the GIS bays in this example.

The cable connecting enclosure is flanged to the vertical circuit breaker enclosure. Figure 8.54 shows a principle set up.

8.3.10 138 kV and 230 kV Outdoor Ring Bus GIS

8.3.10.1 Introduction

The 138 and 230 kV outdoor ring bus GIS was built in a very small space that was available in a narrow strip between highways.

Table 8.10 Technical data of the 138 kV outdoor ring bus GIS

Rated voltage (U_r)	138 kV	230 kV
Rated current (I_r)	2000 A	3000 A
Frequency (f)	60 Hz	60 Hz
BIL (U_{BIL})	650 kV	1050 kV
Short circuit (I_s)	63 kA	63 kA
Date of erection	2009	2009
Number of bays	6	6
Switching scheme	Ring bus	Ring bus
Type of encapsulation	Three phases	Single phase
Number of circuit breakers	6	4

8.3.10.2 Technical Data

The 138 kV and 230 kV outdoor ring bus GIS has the technical data given in Table 8.10.

8.3.10.3 Electrical One-Line Diagram

The one-line diagram of the six-breaker ring connected 138 kV three-phase GIS is shown in Figure 8.56. The six-breaker ring connected 230 kV single-phase GIS is shown in Figure 8.57. The 138 kV GIS has seven underground and three aerial connections. The connections to the transformers and to the overhead lines are made by SF_6 gas-to-air bushings.

The single-phase insulated 230 kV GIS has three incoming and two outgoing lines. The incoming lines are connected by underground XLPE cables from bottom exits to the cable trench. The outgoing line is connected by SF_6 gas-to-air bushings to the transformer.

8.3.10.4 Physical Layout

The physical layout of the 138 kV and 230 kV GIS in Figure 8.58 represents a solution to the space constrains required for this GIS substation. The connections to the transformers are typically made by bushings. The in- and outgoing lines often use cables as shown for the 132 kV GIS in Figure 8.59.

8.3.11 500 kV and 4000 A/8000 A Indoor Breaker and Third GIS

8.3.11.1 Introduction

The 500 kV and 4000 A/8000 A indoor breaker and third GIS were built indoors because of severely cold climate conditions with high winds. The first GIS wear supplied by two different manufacturers, additionally, the high-rated gas bus was manufactured by another GIL/GIB manufacturer. All the buses are drawn below the building grade.

8.3.11.2 Technical Data

The technical data of the 500 kV indoor GIS is shown in Table 8.11.

8.3.11.3 Electrical Layout

The overview of the 500 kV indoor GIS described below by the one-line switching scheme is shown in Figure 8.60. The high voltage and current rating require high reliability and availability of the GIS even in case of maintenance or repair on circuit breakers. That is why a 1⅓-scheme was selected with 14 outgoing lines requiring a total of 19 circuit breakers.

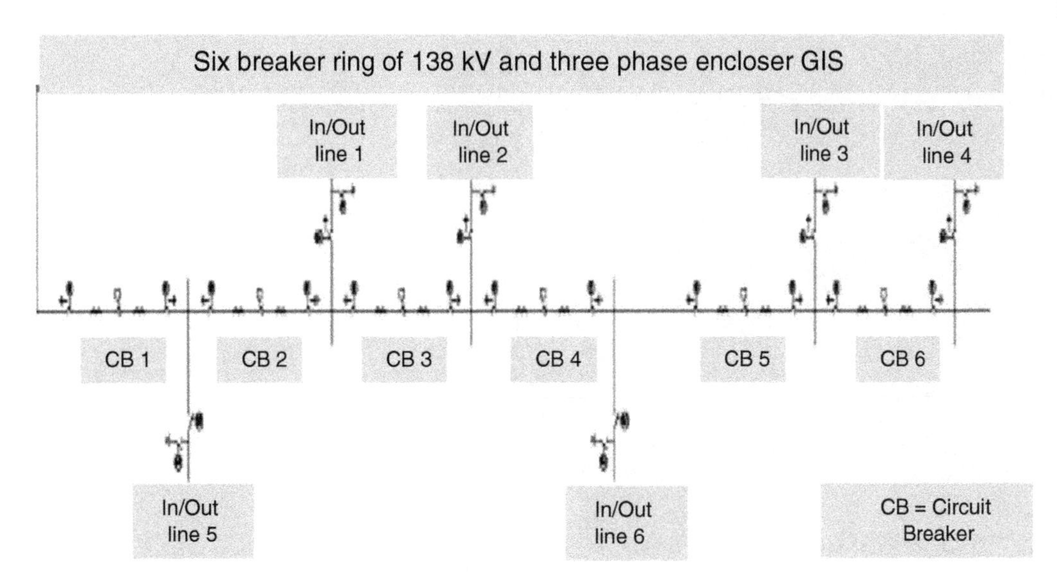

Figure 8.56 One-line diagram of the 138 kV outdoor GIS

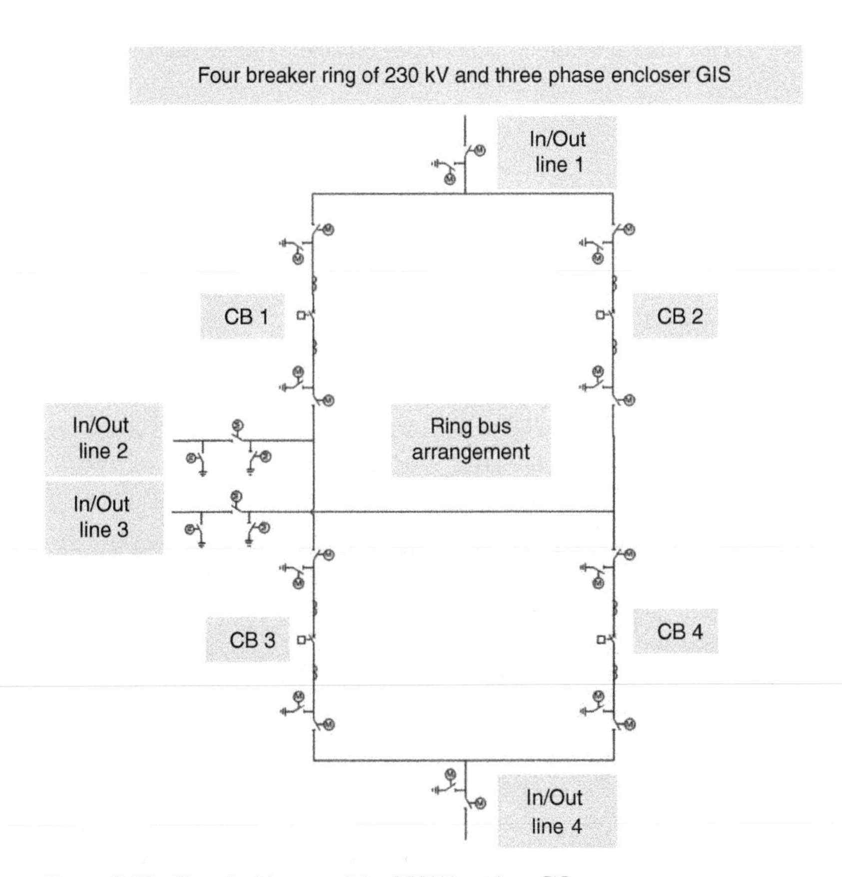

Figure 8.57 Electrical layout of the 230 kV outdoor GIS

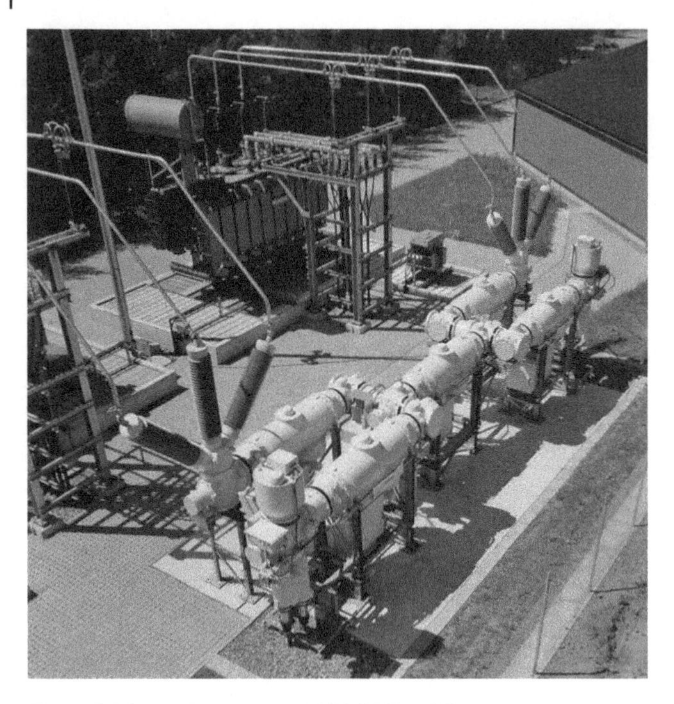

Figure 8.58 Typical layout of 138/230 kV GIS to transformer connection with narrow space requirements (Reproduced by permission of Siemens AG)

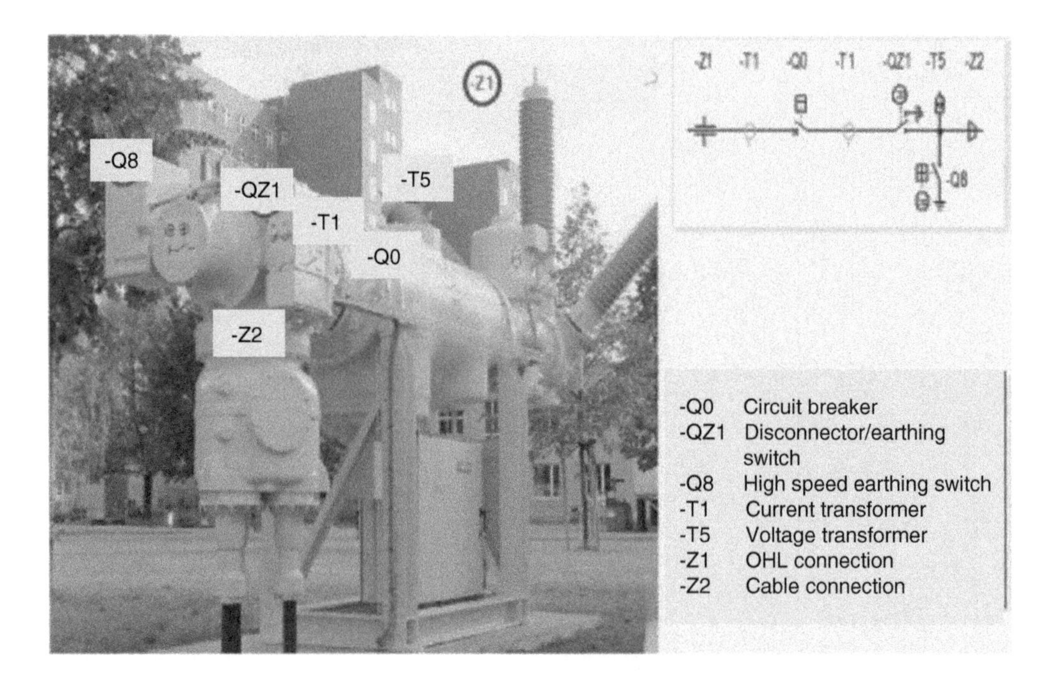

Figure 8.59 Side view of the substation with the 138 kV GIS (Reproduced by permission of Siemens AG)

Table 8.11 Technical data of the 500 kV indoor GIS

Rated voltage (U_r)	500 kV
Rated current (I_r)	4000 A (bus 8000 A)
Frequency (f)	60 Hz
BIL (U_{BIL})	1550 kV
Short circuit (I_s)	80 kA
Date of erection	Different periods through 2010
Number of bays	Five diameters
Switching scheme	Breaker and third
Type of encapsulation	Single phase
Number of circuit breakers	19

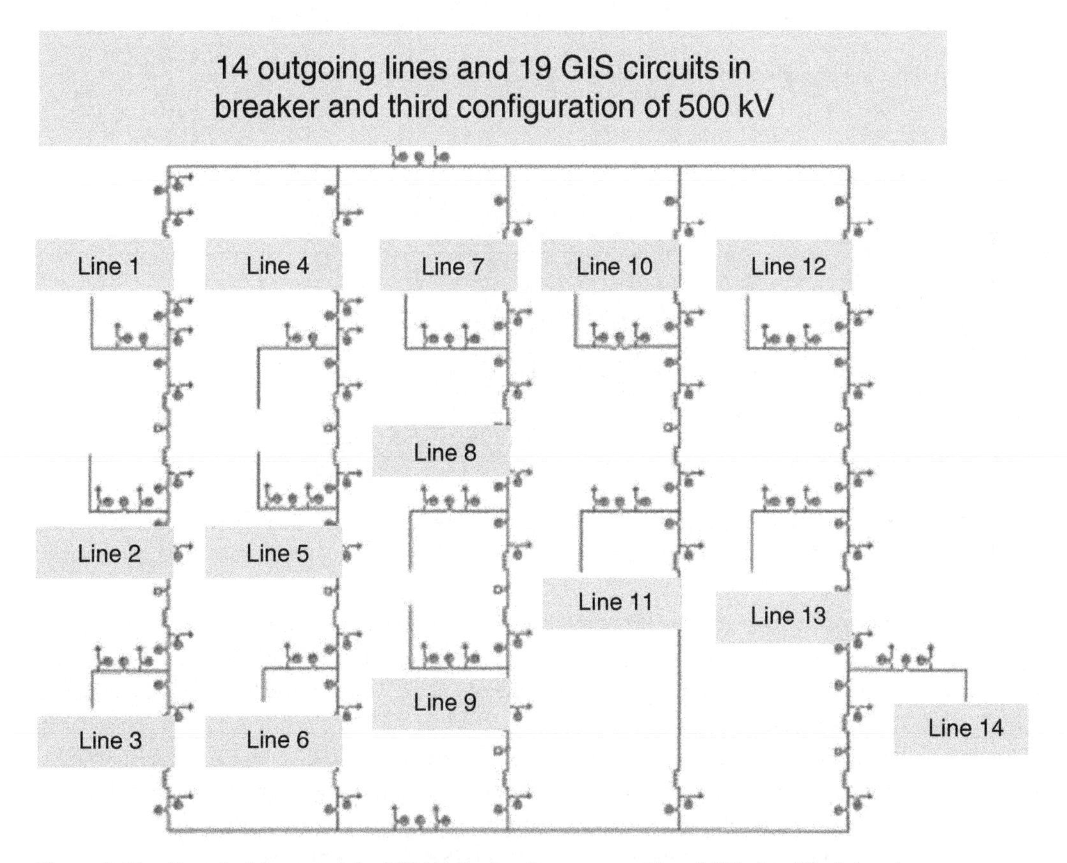

Figure 8.60 Electrical layout of the 500 kV single-phase encapsulated GIS, simplified drawing

8.3.11.4 Physical Layout

The physical layout of this large substation is color coded by manufacturer, with the dark colors indicating the latest extension. The graphic in Figure 8.61 gives an impression of the physical layout of the substation.

The following photographs provide a clear impression of this substation. The vertical mounted circuit breakers in the lighter color have a height of more than 6 m. The new bus section

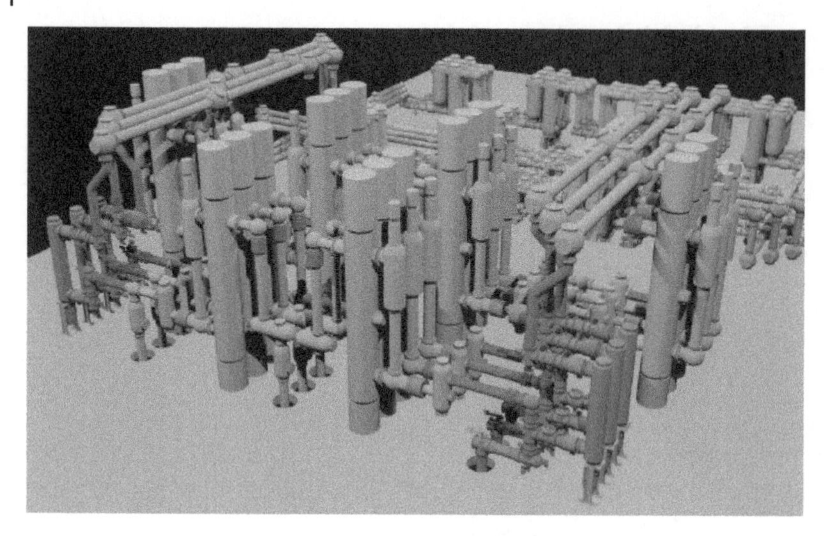

Figure 8.61 Physical layout of the 500 kV indoor GIS (Reproduced by permission of ABB)

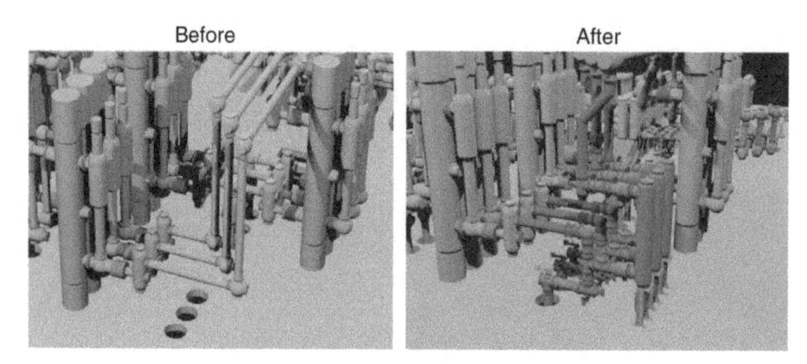

Figure 8.62 Graphical view of the extension of one manufacturer's old-generation GIS with another manufacturer's new-generation GIS (Reproduced by permission of ABB)

connecting to the existing bus and the outgoing bus connecting to the bushings are shown in gray. Figure 8.62 shows the extension of a new GIS bay connected to the existing GIS. The photographs in Figure 8.63 give a good impression of the size of the old 500 kV GIS on the left and the extension with the new GIS design on the right.

The extension of the old-generation GIS with the new-generation GIS from the same manufacturer is commonly used when higher power transmission is required. Well-prepared planning and integration of the new design of the GIS is needed prior to project execution. The following graphics and photographs will provide an example of such a GIS extension process.

Figure 8.64 shows the replacement of the existing bus bar seen on the left with one bay of the 500 kV GIS on the right. Figure 8.65 shows the existing three-phase bus bar before replacement on the left and the three-phase, single-phase insulated 500 kV GIS at the same position connected to the existing substation on the right. Figure 8.66 shows the connection from the GIS to the overhead line outside the GIS building. The GIS bus bars are mounted on steel structures on the ground and then connected via SF_6 gas-to-air bushings to the overhead line above. The SF_6 gas-to-air bushings overvoltage surge arrestors are used to protect the GIS from lightning overvoltage's coming from the overhead line.

Figure 8.63 Photographs of the extension of one manufacturer's old-generation GIS with another manufacturer's new-generation GIS (Reproduced by permission of ABB)

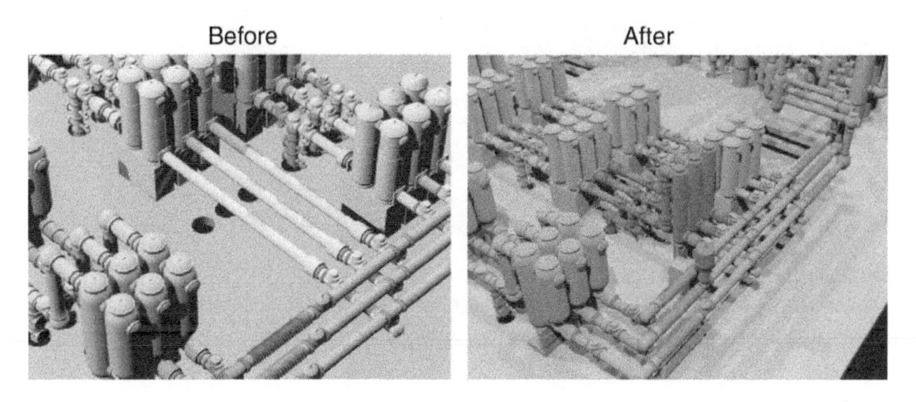

Figure 8.64 Graphical view of the extension of an old-generation GIS with a new-generation GIS, from the same manufacturer (Reproduced by permission of ABB)

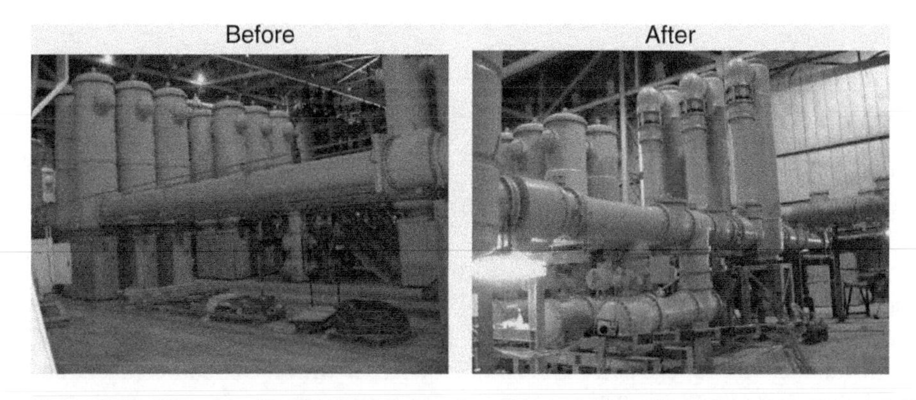

Figure 8.65 Photographs of the extension of an old-generation GIS with a new-generation GIS, from the same manufacturer (Reproduced by permission of ABB)

Figure 8.66 Outgoing buses are shown below the building grade (Reproduced by permission of ABB)

8.3.12 69 kV and 1600 A Outdoor Single Bus GIS

8.3.12.1 Introduction

The 69 kV and 1600 A single bus GIS was built as an outdoor GIS. GIS was used due to space constraints and low maintenance cost and reliability requirements.

8.3.12.2 Technical Data

Table 8.12 shows the technical data for the 69 kV outdoor GIS.

8.3.12.3 Electrical Layout

Figure 8.67 shows the substation one-line diagram demonstrating the two independent parts of the GIS equipment.

8.3.12.4 Physical Layout

The physical layout of this 69 kV substation in the following and elevation drawing views give an impression of the design.

The elevation view in Figure 8.68 shows the SF_6 gas-to-air bushing for connection to the overhead transmission line. A plan view of the two parts of the 69 kV GIS is shown in Figure 8.69. Another elevation view of the 69 kV GIS with an underground cable interface is shown in Figure 8.70, while another view of the 69 kV GIS shows the feeder with the circuit breaker and underground cable interface (see Figure 8.71).

Table 8.12 Technical data of the 69 kV outdoor GIS

Rated voltage (U_r)	69 kV
Rated current (I_r)	1600 A
Frequency (f)	60 Hz
BIL (U_{BIL})	650 kV
Short circuit (I_s)	50 kA
Date of erection	2000
Number of bays	6
Switching scheme	Single bus bar
Type of encapsulation	Three phases
Number of circuit breakers	4

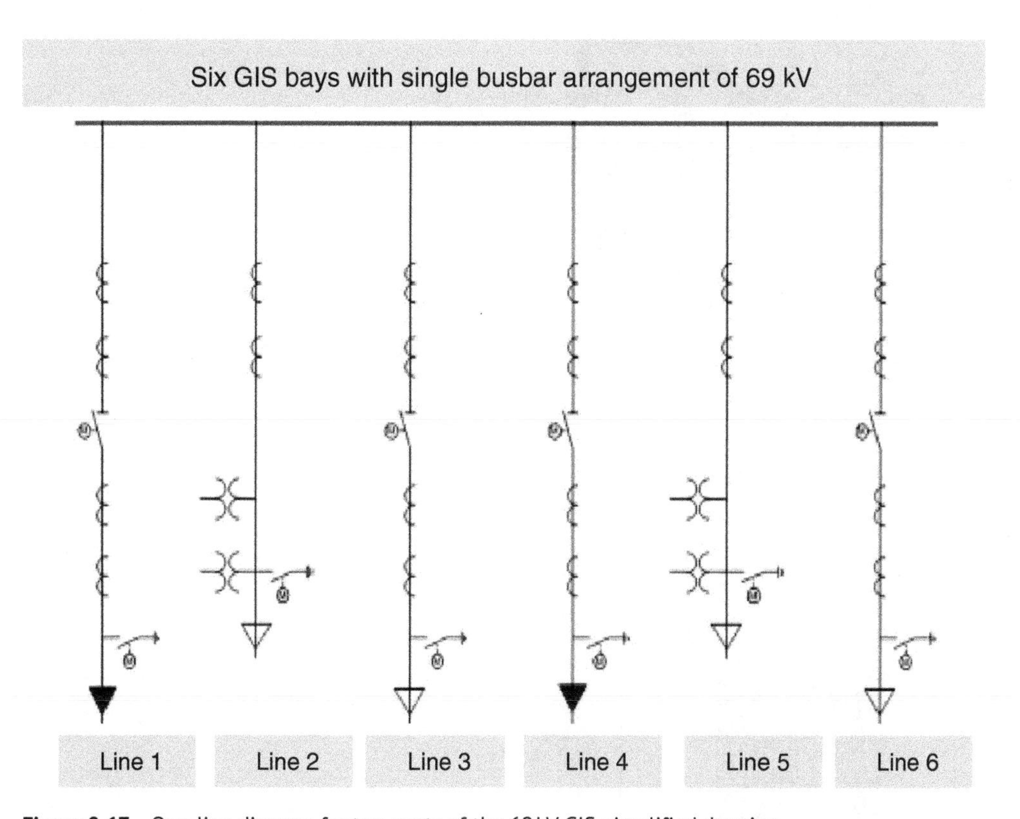

Figure 8.67 One-line diagram for two parts of the 69 kV GIS, simplified drawing

The photograph of Figure 8.72 gives a good impression of the compactness of the GIS. On the left side, the local control cabinet is shown with open doors, while on the right side, the compact design of the three-phase insulated GIS is shown.

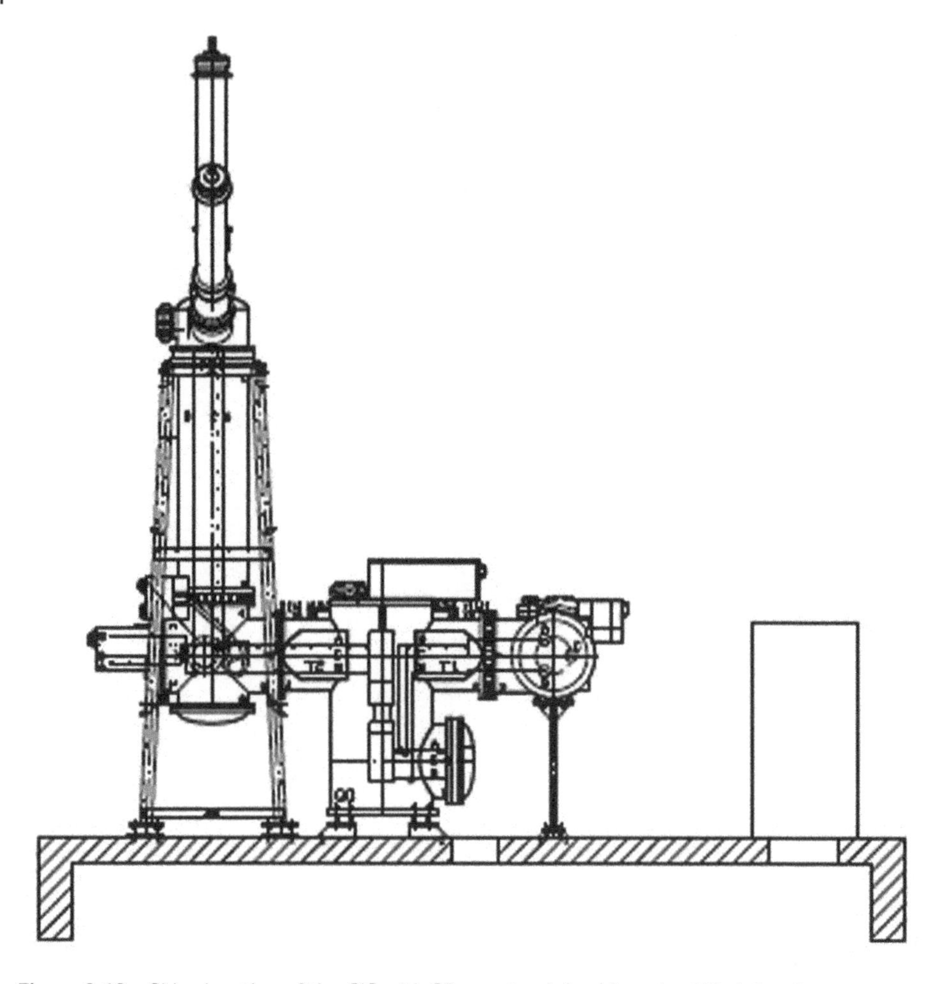

Figure 8.68 Side elevation of the GIS with SF_6 gas-to-air bushing, simplified drawing

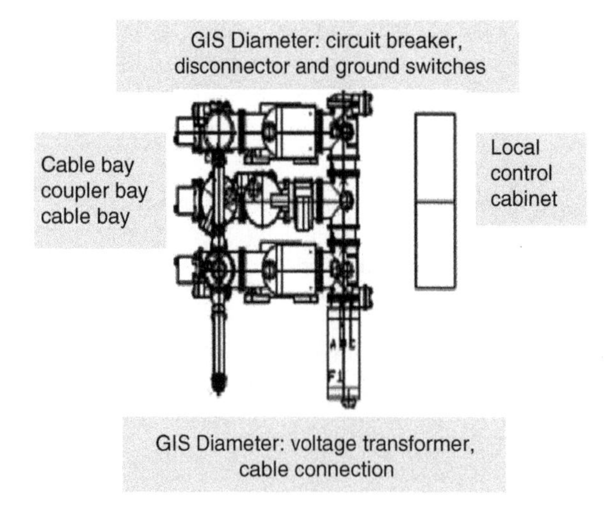

Figure 8.69 Plan view of the 69 kV GIS substation, simplified drawing

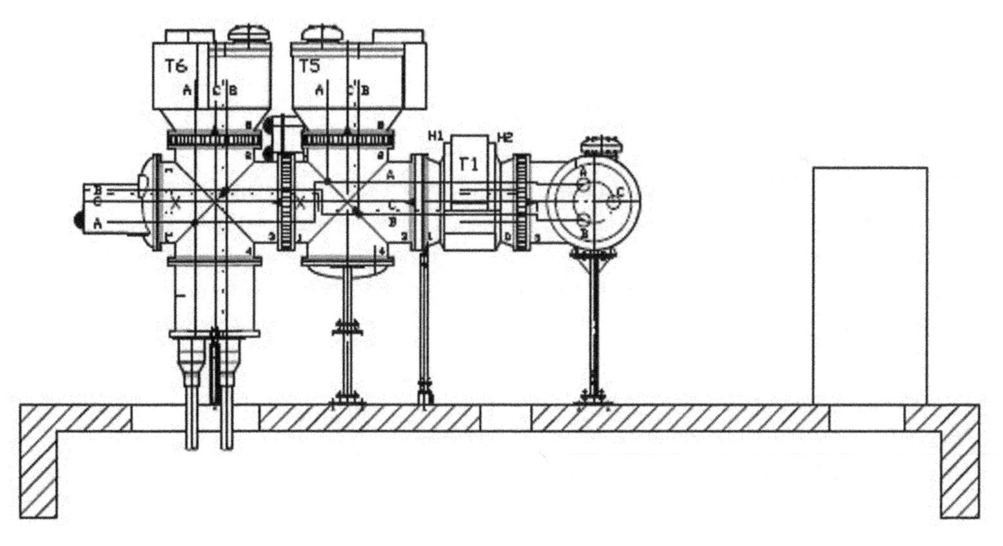

Figure 8.70 Side elevation of the 69 kV GIS with underground cable interface, simplified drawing

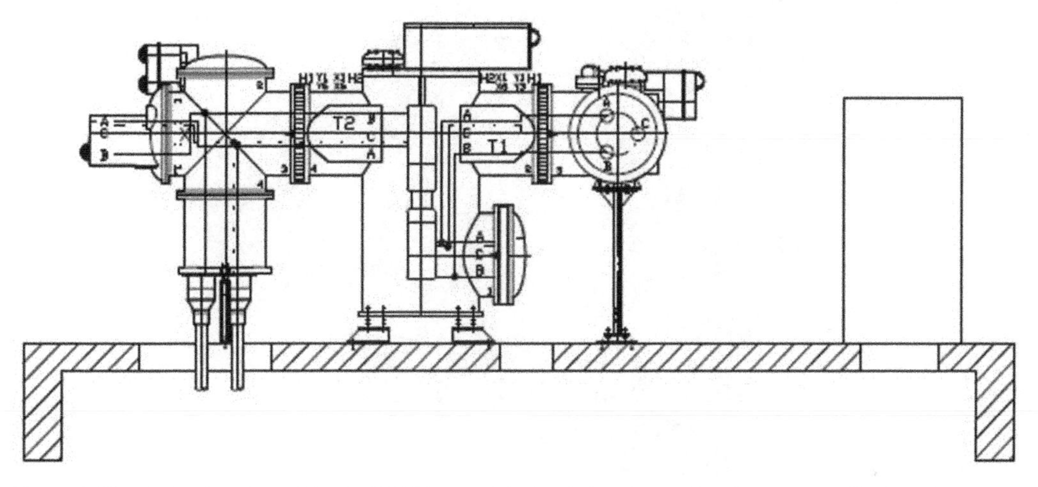

Figure 8.71 Side elevation of the 69 kV GIS bay with circuit breaker and underground cable interface, simplified drawing

Figure 8.72 Compact design of the 69 kV GIS (Reproduced by permission of Siemens AG)

8.3.13 69 kV and 2000 A Underground GIS

8.3.13.1 Introduction

The 69 kV underground GIS was built in close proximity to a residential area in a densely populated community. The park area at a road junction had been chosen for this underground GIS installation. Connected by underground cables the substation is completely invisible to the public, as shown in Figure 8.73. The photograph of Figure 8.76 is a good example of underground GIS in a residential area.

8.3.13.2 Technical Data

The underground GIS is three-phase insulated and follows the double bus bar scheme with an operating and transfer bus. The technical data are given in Table 8.13.

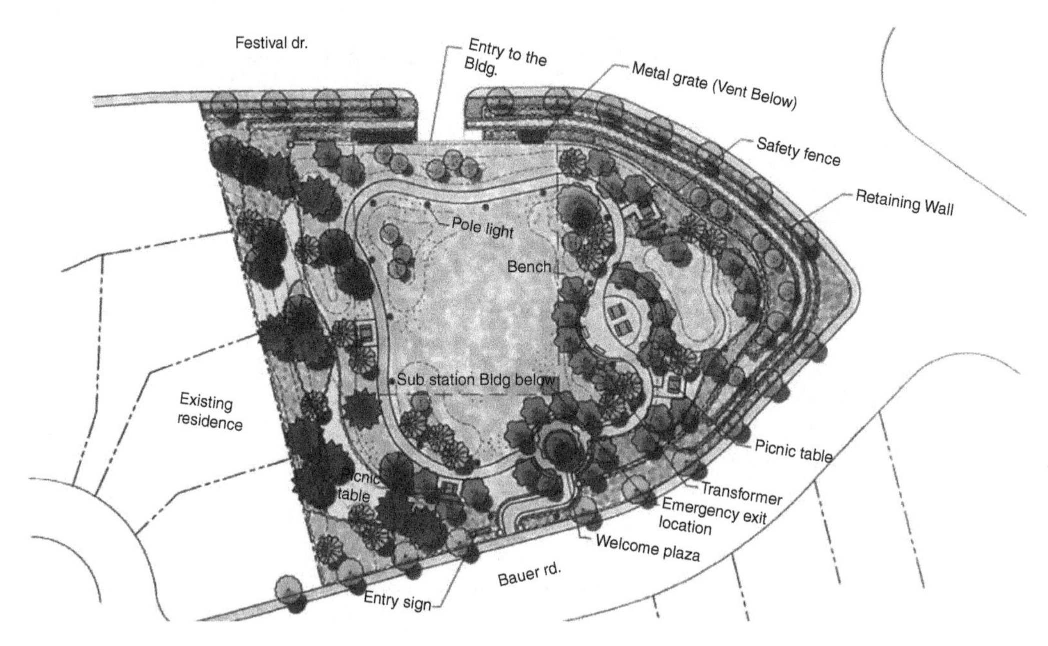

Figure 8.73 Underground substation for 69 kV below a city park (Reproduced by permission of Siemens AG)

Table 8.13 Technical data of the 69 kV underground GIS

Rated voltage (U_r)	69 kV
Rated current (I_r)	2000 A
Frequency (f)	60 Hz
BIL (U_{BIL})	350 kV
Short circuit (I_s)	40 kA
Date of erection	2006
Number of bays	8
Layout scheme	Double bus
Type of encapsulation	Three phases
Number of circuit breakers	8

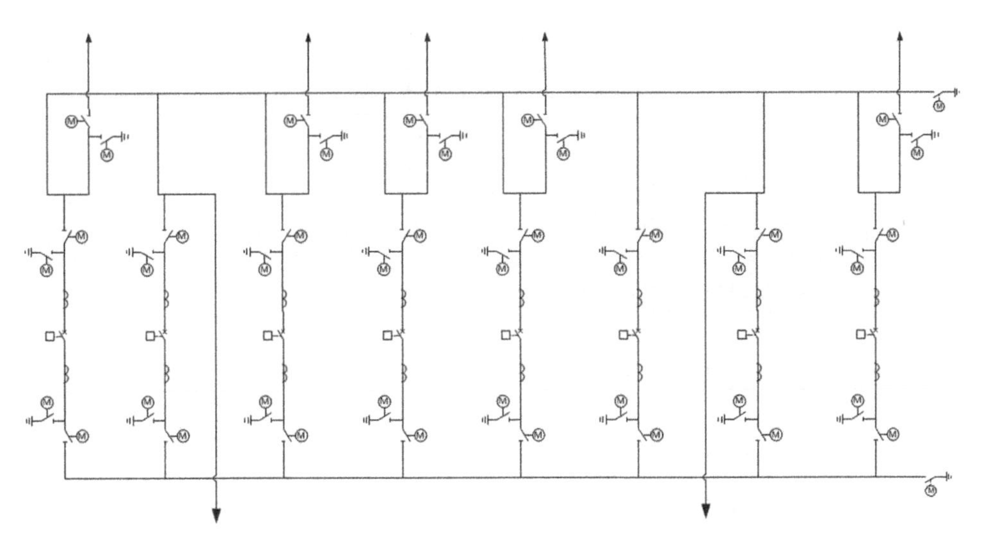

Figure 8.74 Electrical single-line scheme of the 69 kV underground GIS (Reproduced by permission of Siemens AG)

Figure 8.75 Top view of the 69 kV underground GIS (Reproduced by permission of Siemens AG)

Figure 8.76 Side view of the cable bay (Reproduced by permission of Siemens AG)

8.3.13.3 Electrical Layout

The underground GIS has a total of eight bays and follows the double bus bar scheme, as shown in Figure 8.74. There are seven in- and outgoing lines and one coupling bus of the operation and transfer bus.

8.3.13.4 Physical Layout

The layout of the construction site of the underground 69 kV GIS is shown in Figure 8.75. The top view shows the compact design and small footprint of the GIS. The view in Figure 8.76 shows the underground site in an overview and how it is integrated into the surrounding.

8.3.13.5 Photos of the Underground GIS

The high-voltage transmission of 69 kV GIS and the 12 kV medium-voltage distribution GIS are placed in one hall under the park above. In Figure 8.77, the entrance to the underground substation is shown. Standing in the park today, one will give no indication of having a 69 kV/12 kV substation just below your feet. This perfect integration into the urban area is shown in Figure 8.78 with the emergency exit under the pyramid.

Figure 8.77 Entrance to the underground substation (Reproduced by permission of Siemens AG)

Figure 8.78 Emergency exit of the underground substation in the park (Reproduced by permission of Siemens AG)

8.3.14 69 kV GIS under Severe Environment

8.3.14.1 Introduction

Pacific Gas and Electric (PG&E) needed to replace the 60 kV bus structures at Humboldt Bay Switchyard. Located in the North West corner of the State of California, the lattice steel bus structures in the station have been under extreme environmental conditions, subject to ocean salt air and fog. This station is also located in a seismically active area. Due to limited switchyard area available, it was decided to replace the 60 kV bus with a 69 kV compact GIS, located in a pre-engineered building.

8.3.14.2 Technical Data

The technical data for the 69 kV indoor GIS is shown in Table 8.14.

8.3.14.3 Project Details

The Eight Bay 60 kV main and auxiliary bus and structures (four lines, three transformers, and one substitute circuit breaker) were replaced with a 60 kV breaker-and-a-half (BAAH) GIS with four diameters with a spare terminal and room for one additional diameter. The GIS is setup in an inverted BAAH scheme with the two 60 kV buses located on top of each other (Bus 2 on top and Bus 1 below) and the three circuit breakers for each diameter located adjacent to each other.

The GIS equipment is a three phase GIS rated at 72.5 kV, 350 kV BIL, 2,500 Amp. continuous current, and 40 kA interrupting current. The GIS is connected to the existing equipment using three-phase 60 kV gas insulated bus (GIB) and three phase SF6-to-air bushings.

The pre-engineered building, which houses the GIS and its local control cabinets, is 28 m (85 foot) long by 12 m (36 foot) wide. The relay and control equipment are housed in a separate building.

8.3.14.4 Physical Layout

The physical layout of the substation is shown in the following Figures 8.79 to 8.83. A lightweight building was erected to provide protection against the salty air, water and moisture. In Figure 8.79 the outside view shows the large door for delivery of the GIS into the building.

The low center of gravity of the GIS gives good protection against the acceleration related to seismic activity. Figures 8.80 to 8.83 show the GIS layout during the erection period.

Table 8.14 Technical data of the 69 kV indoor GIS under severe environment

Rated voltage (U_r)	69 kV
Rated current (I_r)	2500 A
Frequency (f)	60 Hz
BIL (U_{BIL})	350 kV
Short circuit (I_s)	40 kA
Date of erection	2013
Number of bays	8
Switching scheme	Main and auxiliary bus
Type of encapsulation	Three phases
Number of circuit breakers	8

Figure 8.79 Outside view of the lightweight GIS building (Reproduced by permission of PG&E)

Figure 8.80 Inside view of the 8 bays GIS during erection. Front side (Reproduced by permission of PG&E)

Figure 8.81 Inside view of the 8 bays GIS during erection. Rear side (Reproduced by permission of PG&E)

Figure 8.82 View to vertical circuit breaker enclosures with connected ground and disconnect switches (Reproduced by permission of PSE&G)

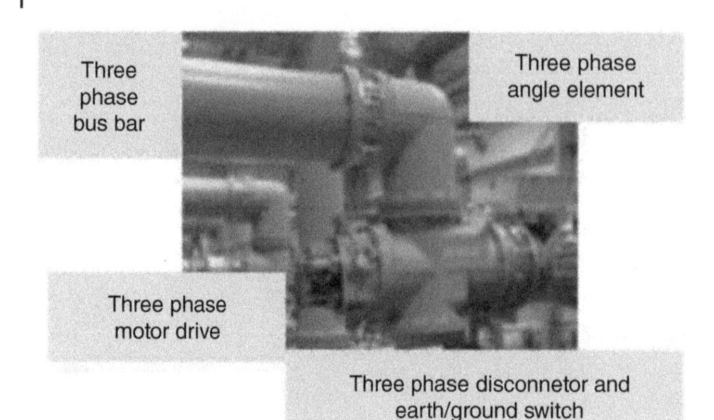

Figure 8.83 View to outgoing three phase bus lines. Rear side (Reproduced by permission of PSE&G)

8.4 GIS Case Studies

Authors: 1st edition William Labos and Peter Grossmann, 2nd edition Hermann Koch
Reviewer: 1st edition Steven Scharf, Devki Sharma and Ewald Warzecha, 2nd edition Petr Rudenko, Stefan Schedl, and Denis Steyn

8.4.1 Public Service Electric and Gas Company – New Jersey

8.4.1.1 230 kV, 80 kA Bergen Switching Station

The Public Service Electric and Gas Company (PSE&G) is New Jersey's largest electric and gas utility and has been in business since 1903. PSE&G serves 2.2 million electric customers and 1.8 million gas customers in a service territory that cuts a diagonal slot from the New York State line in the north to Philadelphia in the southern end, all in the State of New Jersey. The service territory is home to 70% of New Jersey's population and has a peak summer load of nearly 11 000 megawatts. PSE&G has 245 switching stations and substations covering a range of voltages from 500 kV to 4 kV and until 2011 had no gas-insulated substations (GIS) in service. PSE&G has been somewhat reluctant to deploy GIS based on cost, ability to adapt work practices, and some general fear of the unknown. Fortunately for PSE&G, even in New Jersey's highly urbanized landscape, the company was always able to find land to either build new facilities or to expand existing stations without the need to resort to compact technologies like GIS [9].

The story of the Bergen GIS expansion would not be complete without knowing the history of how PSE&G made the decision to go with GIS in the first place. In recent years, PSE&G has been involved with the development of several high-profile transmission projects. The first of these projects was the Susquehanna–Roseland (SR) 500 kV transmission line from Berwick, PA, to Roseland, NJ, a total length of 145 miles. The construction of RS will require a 500 kV breaker station to be built in Hopatcong, NJ, to provide an in/out connection of the new SR line and the sectionalizing of an existing 500 kV transmission line from Branchburg, NJ, to a facility in New York State. An existing 230 kV switching station in Roseland will also need to be expanded to accommodate two 500 kV transformer banks and a three-bay breaker and half 500 kV switchyard. PSE&G System Planners had envisioned the construction of a 500 kV network linking the service territory from north to south and east to west since the late 1960s and land was purchased for future use since those early days to facilitate the construction of planned future facilities. In the late 1960s and early 1970s, PSE&G was able to build a major 500 kV/230 kV station at Branchburg (western New Jersey), three nuclear units and associated 500 kV yards in southern New Jersey, and a 500 kV/230 kV station in central New Jersey with little or no opposition.

Fast forward to the present time, and, like most other utilities, PSE&G no longer has the luxury of building what it wants, where it wants, or when it wants. The first major opposition to large air-insulated substations (AIS) surfaced on the SR project.

PSE&G produced preliminary conceptual designs and renderings for AIS on the SR project. It became very obvious during initial public hearings and zoning board meetings that the citizens and local body politics were opposed to these projects as configured. Large plots of land purchased decades ago could not be fully utilized or utilized at all. Changes in land utilization rules, zoning, wet lands, endangered species, and very vocal public opposition required a change in strategy.

GIS was the answer, providing an extremely compact design, built in a simple but attractive building within the fence lines of existing facilities or, in the case of a new facility, with a very minimal footprint. PSE&G's first GIS placed in service was a three-bay extension of the existing 500 kV yard at Branchburg. After a lengthy and contentious zoning/site planning process, the Branchburg expansion was redesigned to GIS. Since that time, both stations that are associated with SR are being built as GIS. For a company that had no GIS in 2011, they now have seven stations in service with voltages of 500 kV, 230 kV, 138 kV, 69 kV, and even 26 kV medium voltage. An overview of the history of the Bergen switching Station is given in Figure 8.84.

The Bergen switching station rebuild is the largest of PSE&G's GIS switching station projects with 31 circuit breakers. The Bergen switching station was originally constructed in the late 1950s as a 138 kV AIS straight bus station. It was designed to accommodate two138 kV transmission lines, output from a two-unit 650 MW coal-fired generating station (2×325 MW), a 138 kV/26 kV 120 MVA subtransmission station, and a 26 kV/4 kV 30 MVA distribution substation. In the early 1970s, a no interconnected 230 kV two-bay breaker and a half AIS yard was built, which was not interconnected to the original 138 kV yard. This yard was interconnected to the PSE&G 230 kV system via several 230 kV lines and also supplied power to a four-bay 100 MVA 230 kV/13 kV distribution substation. As load in the area grew that station was expanded and interconnected equipment was placed in nonoptimal positions. The transformers shared positions with lines and were connected to main bus sections.

Although the Bergen property is fairly large, it contained a significant number of wetlands, making permitting difficult. Since this is a very active and heavily loaded station, a substantial part of the station would need to be built and energized to permit cutovers from the existing station. Although an important station from the time of its initial energization, its importance has been increasing, with the addition of the interconnection to New York City. This facility will soon need to be classified as a NERC CIP site. The site although large, is in a heavily traveled area, surrounded by major roadways with equipment located relatively close to the fence line. It quickly became

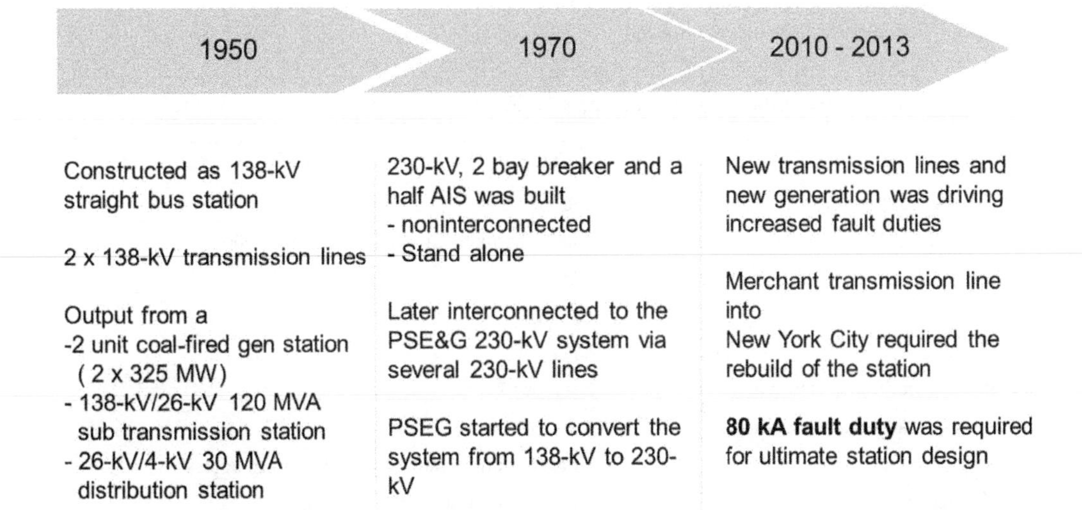

1950
1970
2010 - 2013
Constructed as 138-kV straight bus station
2 x 138-kV transmission lines
Output from a
-2 unit coal-fired gen station (2 x 325 MW)
- 138-kV/26-kV 120 MVA sub transmission station
- 26-kV/4-kV 30 MVA distribution station
230-kV, 2 bay breaker and a half AIS was built
- noninterconnected
- Stand alone
Later interconnected to the PSE&G 230-kV system via several 230-kV lines
PSEG started to convert the system from 138-kV to 230-kV
New transmission lines and new generation was driving increased fault duties
Merchant transmission line into
New York City required the rebuild of the station
80 kA fault duty was required for ultimate station design

Figure 8.84 History of Bergen

apparent that a GIS solution would meet all of the before-mentioned requirements and concerns. The conclusion of the site evaluation is shown in Figure 8.85.

Over time, the PSE&G 138 kV system was becoming less important and was becoming a bottleneck to serving increasing load throughout the territory. PSE&G had already converted several facilities/lines from 138 to 230 kV and it was decided that this conversion would also be required at Bergen. The addition of new transmission lines and new generation in the area was constantly driving an increase in fault duty. In 2010, as part of additional required modifications to support a merchant transmission line into New York City, it was decided to rebuild and reconfigure rather than modify the station. New PSE&G system planning reliability criteria would not permit the existing straight bus design in a new facility as well as the sharing of bus sections with lines and transformers. The station's ultimate design also required the ability to handle an 80 kA fault duty. The technical data are given in Table 8.15.

The electrical layout of the substation is shown in Figure 8.86 and the physical layout of the substation is shown in Figure 8.87.

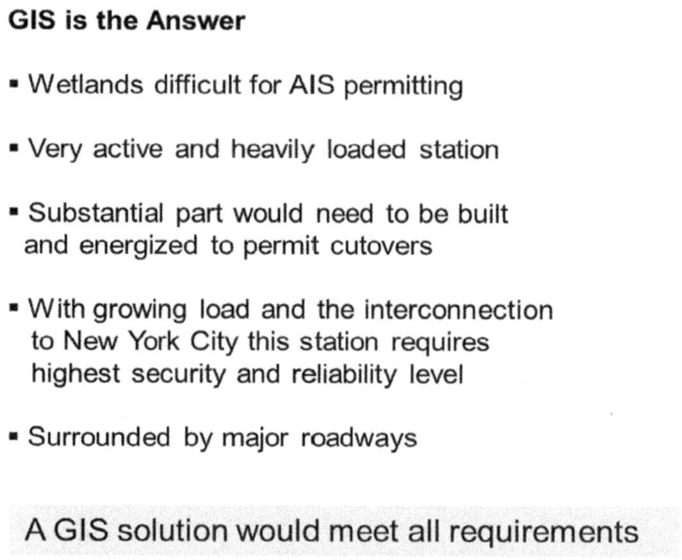

GIS is the Answer

- Wetlands difficult for AIS permitting
- Very active and heavily loaded station
- Substantial part would need to be built and energized to permit cutovers
- With growing load and the interconnection to New York City this station requires highest security and reliability level
- Surrounded by major roadways

A GIS solution would meet all requirements

Figure 8.85 Conclusion of the site evaluation

Table 8.15 Technical data of the Bergen switching station

Rated voltage ($\boldsymbol{U}_r$)	245 kV
Nominal operating voltage	230 kV
Rated current (I_r)	4000 A breakers and bay (main bus 5000 A)
Frequency (f)	60 Hz
BIL (U_{BIL})	1050 kV
Short circuit duty (I_s)	80 kA
Date of construction/service year	2012/2013
Number of circuit breaker positions	31
Switching scheme	9 breaker and a half bays, 4 section breakers
Type of encapsulation	Single phase (isolated phase bus)

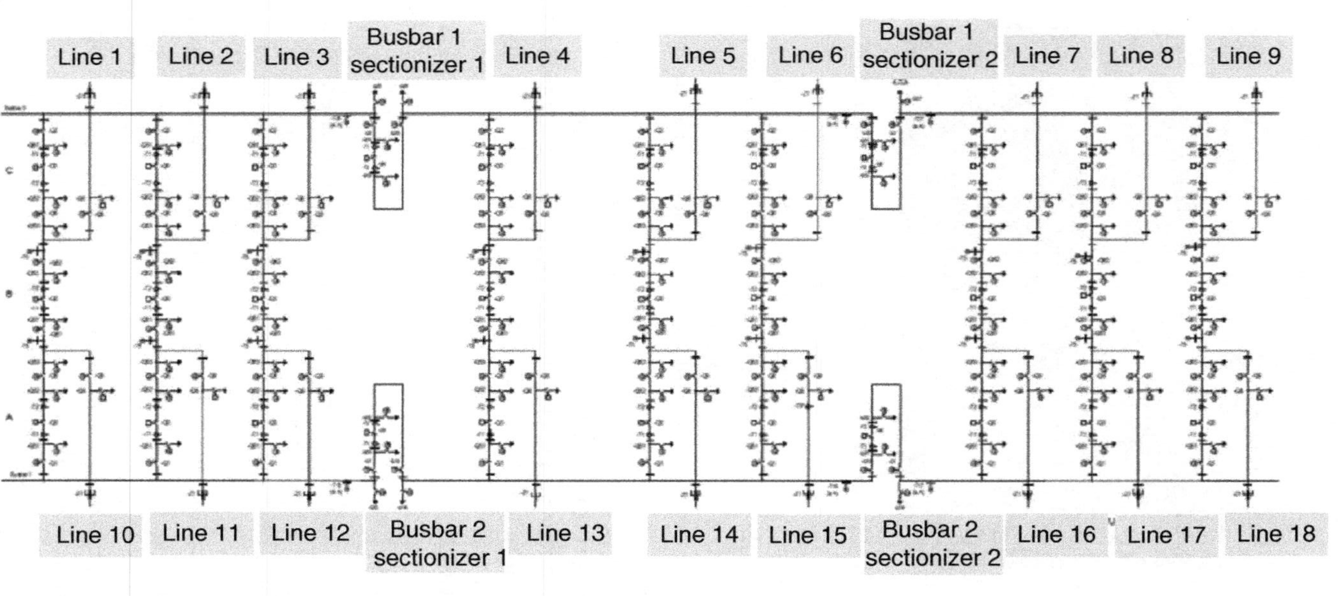

Figure 8.86 Electrical layout of the substation (Reproduced by permission of PSE&G)

Figure 8.87 Physical layout of the substation (Reproduced by permission of PSE&G)

The Bergen Switching station represents the latest state of the art GIS technology. To interrupt short circuits from up to 80 kA, the GIS circuit breakers are equipped with interrupter units that use the principle of the dynamic self-compression system.

Hereby, the energy of the arc is used to build up SF_6 pressure inside the switch unit to interrupt the short-circuit current.

The arc assistance of interrupting the short-circuit current allows operating the circuit breaker with significant less energy than would be needed using puffer systems.

Less energy for the circuit breaker operation results into reduced dynamic forces. Therefore, the spring operating mechanism is an excellent fit for the arc assist interrupting system. Since both closing and opening of the circuit breaker are operated with springs no pressurized hydraulic oil or compressed air is needed.

The Bergen Switching Station has been specified and built to have unique features to optimize operability and maintainability.

a) Sufficient gas zones to minimize impact to the entire system in case of a catastrophic failure, if degassing and or repairs are necessary.
b) Full capability trending gas monitoring systems that tracks the gas leakage performance of all gas zones
c) UHF partial discharge measurement system. Partial discharge sensors inside the GIS compartments act as antennas to detect ultrahigh-frequency signals that occur in the case of partial discharges.
d) Full time permanently installed remote TV monitoring of all disconnects and ground switches, to facilitate rapid verification of device positions without climbing.
e) Extensive use of permanent platforms to eliminate unsafe climbing when accessing equipment.
f) Use of "flat" layouts with extra spacing between equipment to mimic AIS designs and avoid operator confusion.
g) Indoor protective buildings for security and to minimize weather impacts.

One specialty of the Bergen station is the camera monitoring system mentioned under point d. This camera system is used as a safety control for the open and close position of the disconnect and

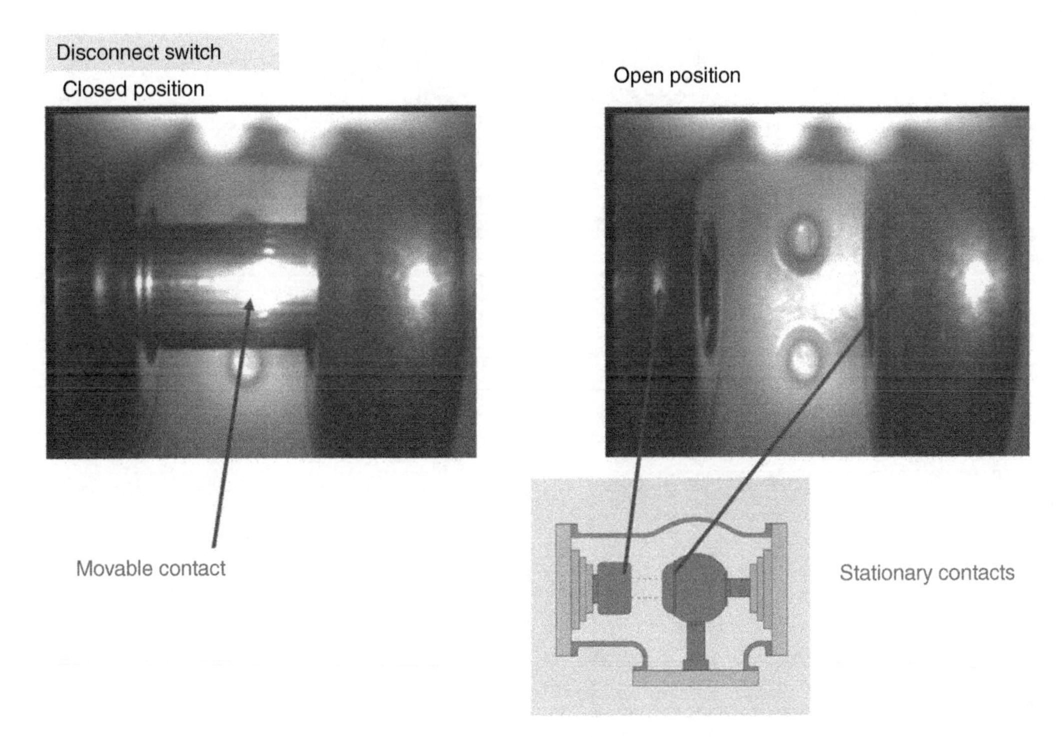

Figure 8.88 Camera system for disconnects and grounding switches (Reproduced by permission of IEEE)

ground/earth switches. In Figure 8.88, the close position (left photo) and open position right photo of the disconnect switch is shown.

The gas density monitoring system of the Bergen Station offers a concentrated information of all gas compartments in the control room. Each gas zone ID can be viewed on the control monitor indicating the actual gas density, the identification of the electric circuit, the related phase of the three-phase system, the nominal density of the gas compartment, the densities of the alarm stages 1 and 2, the installed of filling density, the installed mass of SF_6, and the over pressure alarm level. These technical values are for information. For fast and secure reaction during operation colored information on gas loss, leak rate, previous gas loss, and alarm stages 1 and 2, data are given. The graphical indications in diagrams and gas zone diagrams give quick orientation on the situation. See Figure 8.89.

The UHF Partial Discharge (PD) Monitoring System is installed in the high voltage part of GIS and comes with the factory assembled GIS equipment. The location of the UHF sensors is depending on the electrical voltage signals of high frequency and physical design of the substation. Typically, the UHF sensors are located at each phase of the bus bar and the in- and outgoing lines. The primer uses of the UHF partial discharge monitoring system is for the high-voltage on-site commissioning test to prove the correct assembly of the type and routine-tested GIS installation. In Figure 8.90, the principle of the UHF antenna is shown in a graphic top left. A plate-type antenna receives the partial discharge high-frequency (up to GHz) voltage signals in the μV and an expert system will identify the type of defect measured. The measured signals and evaluation will be used to condition the GIS when first high voltage is applied in voltage step up to the on-site test voltage of 80% of the type test value. Only if the partial discharge signals are below the PD intensity level below equivalent to 5 pC, the GIS will go into operation. The photo on the top right shows the UHF PD signal connection in the termination box. The UHF PD signal cable may be connected to the mobile measuring equipment shown in the photo on the bottom right. It is not recommended to continuously measure PD during the operation of the GIS, because the long-time experience

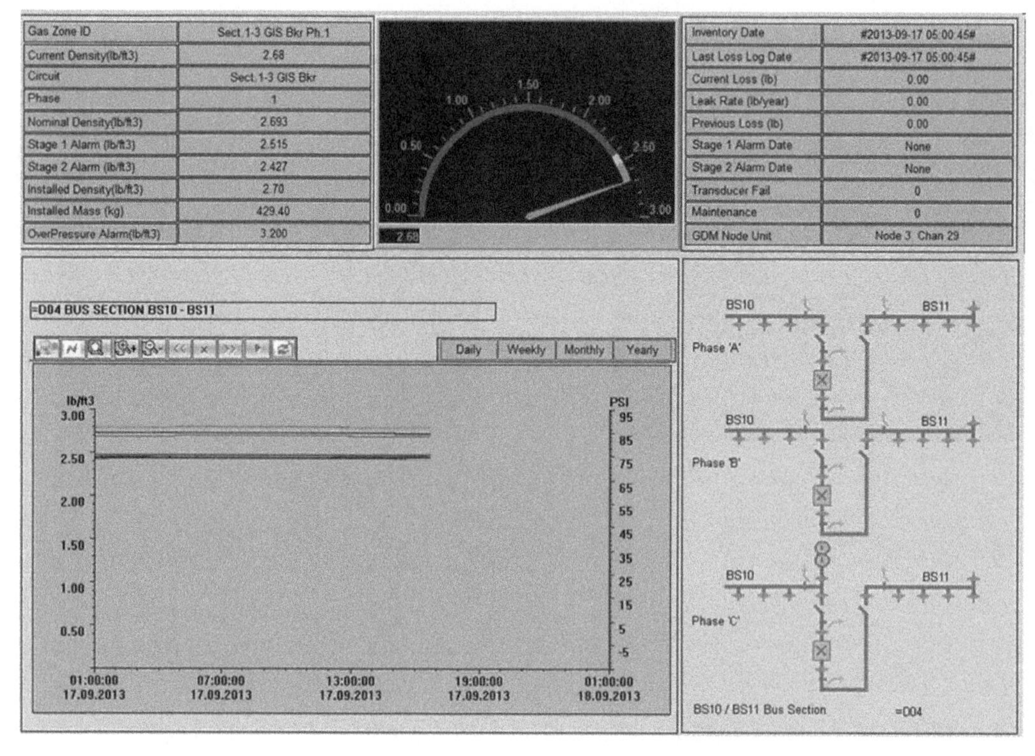

Figure 8.89 Gas density monitoring system (Reproduced by permission of IEEE)

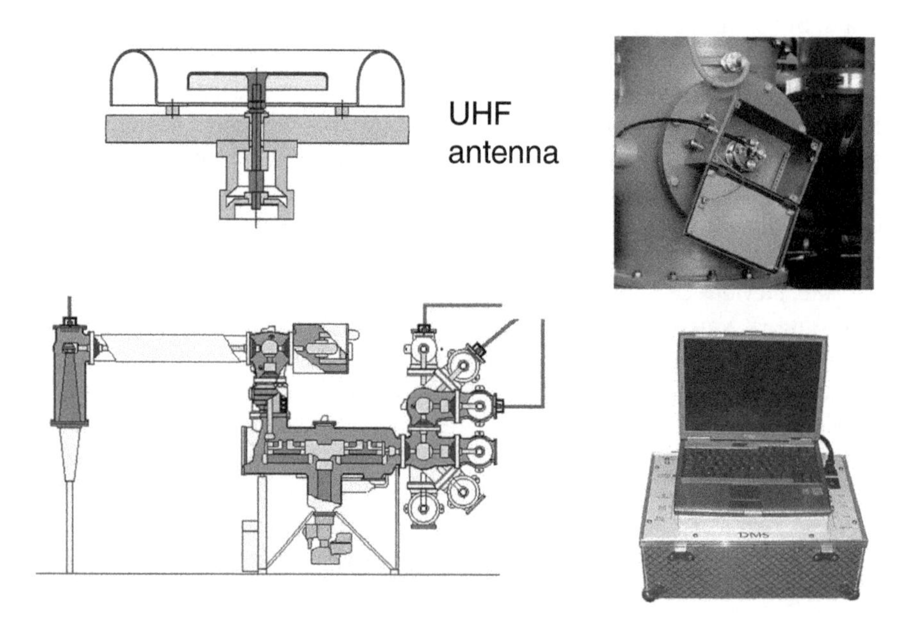

Figure 8.90 UHF partial discharge monitoring systems (Reproduced by permission of IEEE)

shows that GIS in service will not change its PD behavior for life time. Only if the GIS has been opened for maintenance or repair activities, it is recommended to repeat the PD UHV measurement.

So, by the end of 2010, it was decided that Bergen would be nearly totally rebuilt as a nine (9) bay, breaker, and a half station with four (4) main bus sectionalizing breakers (31 breakers total) 230 kV

80 kA with extensive use of gas-insulated bus (GIB) to permit connections to existing remote equipment. All of the transformers would have their own bus sections and the original 138 kV to 26 kV transformers would be replaced with upsized 230 kV to 26 kV units, the original 138 kV yard reduced in size and interconnected with the 230 kV yard via a 230 kV to 138 kV autotransformer. All of the 230 kV, GIS equipment would be enclosed in a heated/ventilated steel-walled structure for both security and weather proofing. Specifications were written, and competitive bids were solicited from three major GIS produces that were able to build equipment capable of handling 80 kA fault duty. The GIS equipment was ordered in June 2011 and delivered in December 2012. Ground was broken in mid-2012 and commissioning and cutover were completed in early 2013. As of the writing of this case study (June 2013), Bergen is the largest GIS installation in North America with 80 kA short-circuit breaking capability.

In Figures 8.90–8.97, the work sequences of the erection of a GIS are shown.

In Figure 8.91, a view to the construction site is presented after the steel works of the foundation is finished and before the concrete is poured in. The foundation is settled on 400 steel piles of 26 m to 33 m (80–100 ft) depth. The cable raceways are embedded in concrete slabs. The foundation concrete is 3′6″ ft thick and has dimension of 73 m by 33 m (220 ft by 100 ft). The control building has a size of 20 m by 26 m (60 ft by 80 ft).

The erected building is from metal frame type and covered with steel panels which are translucent at the upper row, see Figure 8.92. Two cranes were required for the erection of the GIS installation, each crane with lifting force of 5 tons. The building is located in a hurricane area and must resist strong winds, which has been proven by Hurricane Sandy in 2012.

The installation process of GIS is shown on two photos of Figure 8.93. The left photo gives a bird eye view into the building with already installed GIS in the back-part of the photo and the storage area for the preparation of the next GIS sections to be assembled in the front part of the photo. The

Figure 8.91 Providing the foundation (Reproduced by permission of PSE&G)

Figure 8.92 Erection of the GIS building (Reproduced by permission of PSE&G)

Figure 8.93 GIS under construction (Reproduced by permission of PSE&G)

right-side photo gives impression on the storage area, where the container and wooden box deliveries are getting unpacked and prepared for the assembly. This storage area is used for the complete installation process and must be synchronized with the steps of assembling the 31 breaker positions. Two teams were working from building back-side up front. To meet the tight schedule, the preassembly works are key element to shorten the total installation time.

At the end of the installation process, the complexity of the GIS at Bergen Station is shown in Figure 8.94. The left photo shows the middle walk way of the breaker and half installation with horizontal, single-phase insulated circuit breakers on the left side of the walkway and current

Figure 8.94 GIS installation completed (left) and walkway in GIS bays (right) (Reproduced by permission of PSE&G)

Figure 8.95 Bergen switching station, 230 kV, 80 kA with gantry supports (Reproduced by permission of IEEE)

transformers on the right side of the photo. The photo at the right side of Figure 8.94 shows a walkway for operation personal when entering the GIS bays to check, for example, the position of a disconnector or ground switch, or to connect a gas density measuring equipment or to connect a UHF sensor for measurements. These permanent platforms are designed to eliminate unsafe climbing when accessing the equipment.

The view from the outside to the building shows the gantry supports for the connection of the overhead lines entering the Bergen substation is shown in Figure 8.95. The gantries are connected to the building using gas-insulated bus (GIB). Overvoltage protection for transient overvoltage,

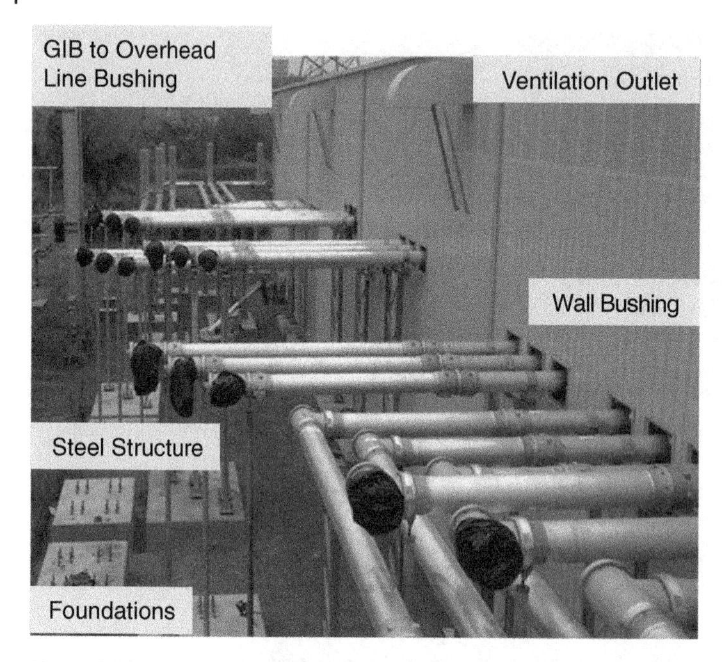

Figure 8.96 Gas-insulated bus section to connect the GIS in the building to the in- and outgoing lines

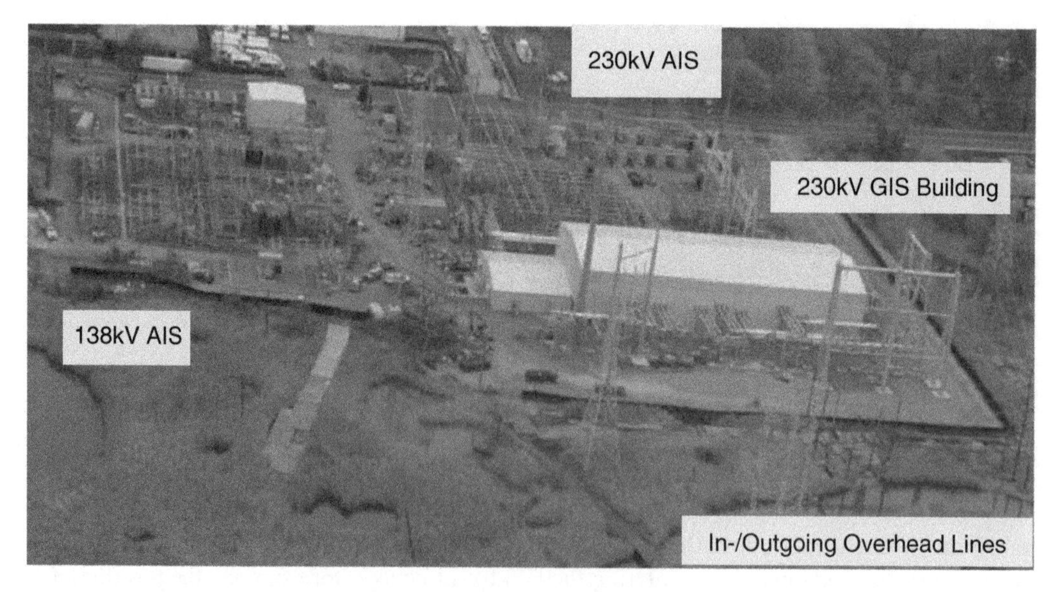

Figure 8.97 Bergen switching station aerial view (Reproduced by permission of PSE&G)

coming from lightning strokes into the overhead line, is given by surge arresters positioned at the entrance to the GIB.

For the connection of the GIS to the in- and outgoing lines a gas-insulated bus (GIB) is used, see Figure 8.96. The GIS three phases are passing through building wall, where the enclosure is ground connected to building. The GIB is single phase isolated using aluminum enclosure pipes which are bolded by flanges. The total single-phase length of the GIB is 2700 m (8200 ft). The use of GIB offers the advantage that line can be easily cross-out to get connected to the relevant overhead or cable line.

The aerial view of the Bergen switching station is show in Figure 8.97. The GIS is inside the white color building, which will replace the air-insulated substation. The performance of the new GIS with 80 kA short-circuit interruption capability is at the 230 kV voltage level outstanding. The substation provides from New Jersey to deliver electrical energy to the New York Metropolitan Area. It is a Milestone in PSE&G history and stands for the innovation of the network. This includes a significant adjustment for people and processes within PSE&G changing from air-insulated substation technology to gas-insulated substation technology. This includes intensive education and training of the people for the operation of GIS. The Bergen Switching Station demonstrates the high performance of GIS capabilities where AIS solutions are limited.

8.4.2 Street House Project Study

8.4.2.1 Metropolitan Area

This comprehensive project study shall give ideas for similar situations in large metropolitan areas around the world. Compact design of gas-insulated substations (GIS) and transmission line technology (GIL) offers solutions for high-power transmission lines within a narrow corridor. Here, the focus is on the use of GIL.

To combine the electric power transmission with public transportation in trains, traffic on streets, and living in apartment houses, basic requirements on safety and electromagnetic field strength need to be fulfilled. The GIL can offer such features as external safety even in cases of an internal arc fault, because the aluminum metallic enclosure keeps the effects of the internal arc inside the GIL. No external impact in case of an internal failure. This allows to combine GIL with public transportation or living.

The second feature of low electromagnetic field strength around the power transmission line is fulfilled by the GIL because of the induced, reverse current in the GIL enclosure, which superimposed to the magnetic field related to the current in the conductor of the GIL. The resulting magnetic field of both currents which are 180° phase shifted below 5% of the full current magnetic field. This allows to bring the GIL close to living houses where low magnetic field values (in some countries only 1 μT) have to be fulfilled. So, GIL offers the basic requirements to bring high-power electric transmission line together with housing and streets.

The density of population of metropolitan areas is very high. Several ten millions of people may live and work in this area. The electric power consumption is high and does further increase in the future with more electric load for computers or vehicles. The rooting for new power lines to feed the city is hard to find in general and available only with some kind of underground solutions.

The traffic situation is in the same way close to its limits and new traffic corridors practically not available. This is valid for individual traffic with cars but also the case for public transportation by metro trains or subways.

A possible solution is to combine both the electric power supply lines with the traffic ways. In Figure 8.98, the left photo shows a top view of the city center with the very limited space for electric power lines or additional traffic streets. The right graphic indicates the possibility to combine electricity and traffic starting at the outer ring highway and following an existing 110 kV overhead

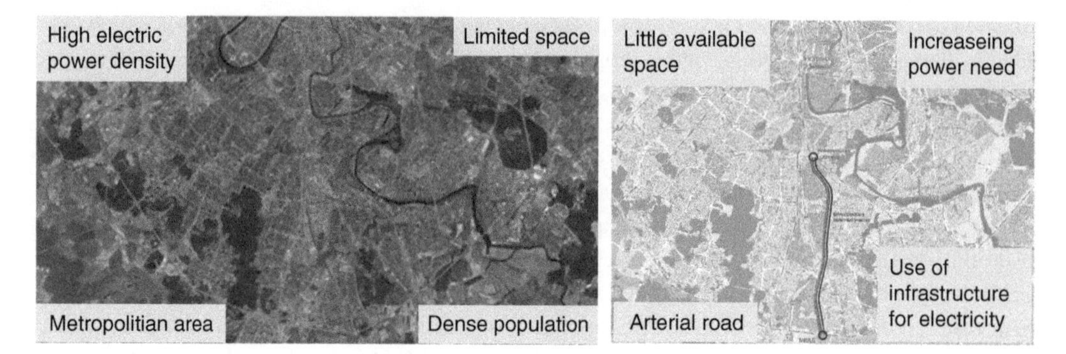

Figure 8.98 Left: top view of city center, right: possible route of electric power line and street (*Source* left: google maps, right: [5–6]).

transmission line root into the city center. The major idea was to connect the line to the outer 500 kV transmission line ring, to upgrade the existing line to 500 kV, and to integrate this in GIL technology into the new street house. The length of this parallel corridor is about 1 km. This solution uses only one route in the city center.

8.4.2.2 GIL Technical Options

GIL technology is based on two aluminum pipes, one for the outer enclosure to keep the pressurized insulating gas (mixture of N_2 and SF_6 or technical air), the other for the inner conductor pipe to carry the high current up to ratings of 4000 A. The typical operation voltages are 400 kV or 500 kV depending on the network requirements with a continuous power transmission capability of up to 3500 MVA for 1 three-phase system. In case of short circuits in the network, the GIL can carry 63 kA for up to 3 s without overheating. The high-voltage type test voltages are depending on the network requirements 1425 kV or 1600 kV lightning impulse voltage. The gas pressure is 0.7 MPa, which is relatively low for a technical system. There are only four functional units of GIL necessarily needed to build a transmission line: straight unit of about 120 m long sections with an elastic bending ability of down to 400 m bending radius, an angle unit to make directional changes up to 90°, a disconnecting unit to separate gas compartment section of up to 1200 m and compensator unit for thermal expansion compensation of the enclosure pipe in case of tunnel laid versions.

The GIL is a completely welded system with a 100% proof of gas tightness by ultrasonic test equipment in an automated process of each weld. The diameter of the encloser pipe is 500 mm and 2 three-phase systems can be laid in round tunnel of 3 m in diameter or in a square tunnel of 2.4 m width and 2.6 m height.

There are four different laying options for GIL: above ground (see Figure 8.99 left), in a trench under a cover (see Figure 8.99 right), in a tunnel (see Figure 8.100 left), or directly buried in the soil (see Figure 8.100 right).

The above-ground laid option (see Figure 8.99 left) is using foundation of concrete or steel to fix the aluminum pipes in distances of about two pipe lengths. This will need a foundation any 25–30 m, depending on the GIL power transmission capability. The GIL pipes are open to the air above and needs no paint or any corrosion protection, because the aluminum oxide layer will protect against weather, rain, and any normal surrounding climate. The trench width is about 5 m.

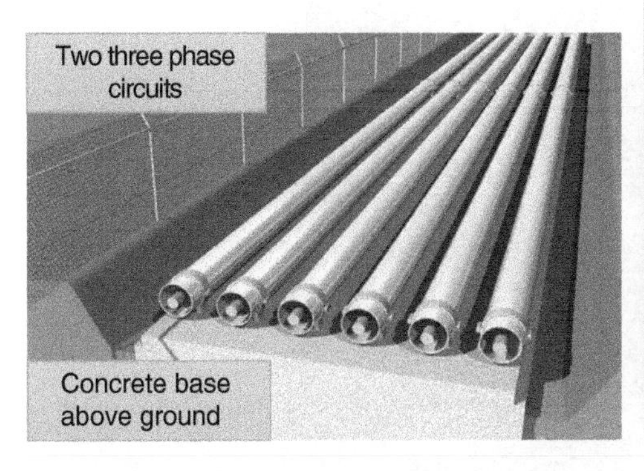

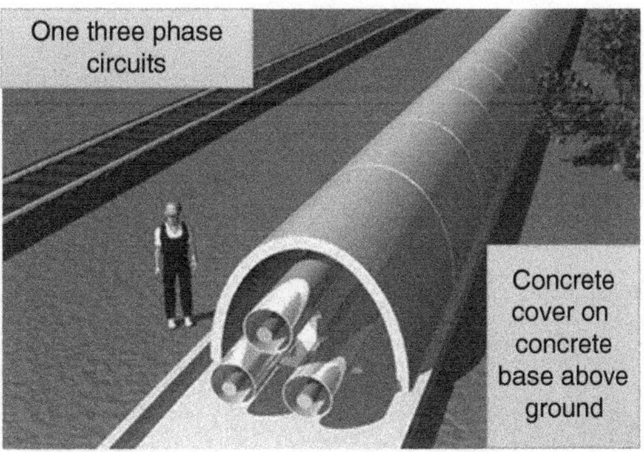

Figure 8.99 Left: GIL laid in an open trench, Right: GIL laid under a concrete cover (*Source* [7, 8])

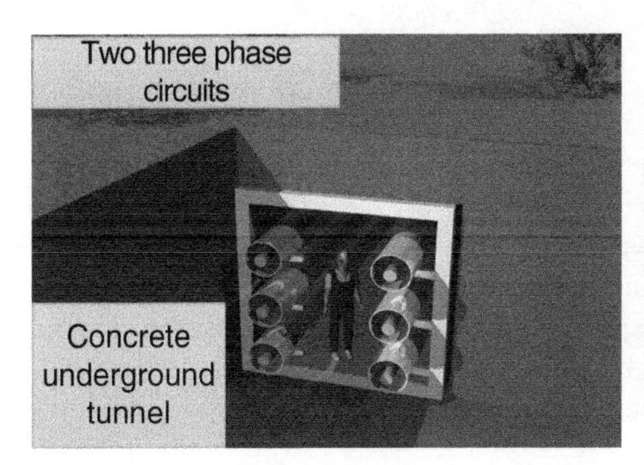

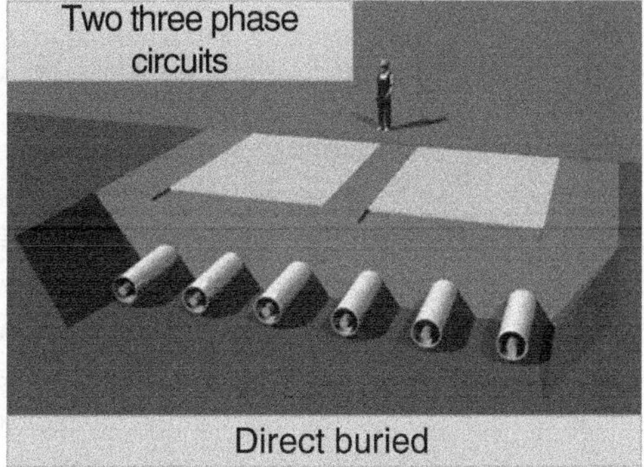

Figure 8.100 Left: GIL laid in a concrete tunnel, Right: GIL laid directly in the soil (*Source* [7, 8])

The trench-laying option with concrete cover (see Figure 8.99 right) may have 1 or 2 three-phase, single-phase-insulated GIL systems. The GIL pipes are arranged in a triangle and steel structures in 25–30 m distance will hold the GIL pipes. The cover is made of sheet steel or concrete structure, which are pre-manufactured in factories and assembled on site. The trench width for 1 three-phase GIL system is about 2 m and for two about 4 m.

The tunnel-laid GIL (see Figure 8.100 left) is using round or squared tunnel design. The round tunnel is using the boring machine and will typical be used in greater depth. The minimum of laying depth of the bored tunnel is 10 m. The squared tunnel needs 2.4 m width and 2.6 m height for 2 three-phase GIL system and is usually made by factory manufactured concrete section of about 5 m length or the sections are made on site in a continuous laying process. The tunnel version will only need a trench width of 3 m.

The directly buried GIL used in open-field areas where a trench can be excavated from the top and kept open for length of 500–1000 m. Directly buried GIL needs a corrosion protection of a PP or PE layer to protect the aluminum pipes in the soil. The trench width for 2 three-phase GIL systems is about 5 m and the depth of laying is about 1–1.5 m. The soil on top of the GIL can be used for agriculture and smaller plant, e.g. a park in the city after construction is ready.

In Figure 8.101, a view in a tunnel of 2.4 m width and 2.6 m height is shown. The single-phase insulated GIL pipes are laid on steel structures fixed to the tunnel wall. Each steel structure is electrical grounded so that the full-induced electric current can flow in the encloser pipe and reduce

Figure 8.101 GIL laid in a squared tunnel of 2.4 m width and 2.6 m height (*Source* [7, 8])

the outer magnetic field. The bending radius shown in the Figure is 700 m. The tunnel is accessible during operation with low risk to the surrounding in case of internal electrical arc because the impact stays inside the solid aluminium enclosure pipe, as proved in type test with using insulating gas mixtures of N_2/SF_6. The magnetic field strength in the tunnel is low at rated current because of the reverse-induced electric current in the enclosure pipe.

8.4.2.3 Street House Concept

The Street House Concept is using the very limited space in a city or metropolitan area to combine the combination of living, public traffic, and electric power supply.

As shown before on the example of the city center (see Figure 8.98), space is very limited in city centers and the concentration of buildings with high electric power consumption is high. To bring electric energy into the city center, high voltage levels of 400 kV or 500 kV is needed in close neighborhood with people working, living, walking, shopping and with the traffic of trains, metro, and cars. For maximum use, the Street House Concept was found, see Figure 8.102, which gives an overview. The Street House is a combination of office buildings and space, living apartments, parking lots, shops, restaurants, walk ways, and little parks with infrastructure of electric power transmission. Figure 8.102 gives a graphical impression of building complex.

The cross-sections shown in Figure 8.102 give information on the interior of the building (left side) and the dimensions of the street on top of the building and the GIL tunnel below (right side). The five levels of the Street House offer parking space on the ground level, office rooms and living apartments at the levels two to four, infrastructure supply systems including the three-phase GIL electric power transmission system at the right side of level five, and on top three lanes of street traffic in one direction. The GIL is connected from the fifth building level to a tunnel to existing overhead line.

The same building is used for the other traffic direction.

The dimensions of the Street House are 9 m wide for the three-lane street and has a total width of 12 m. The tunnel for the GIL has two chambers of each 4.5 m width, which takes the GIL and offers more room for other infrastructure installations like water pipes, rain water collecting system, savage pipe, district heating, low-pressure natural gas pipes, and telecommunication lines. These infrastructure lines are easy to maintain in service with street excavations needed (Figure 8.103).

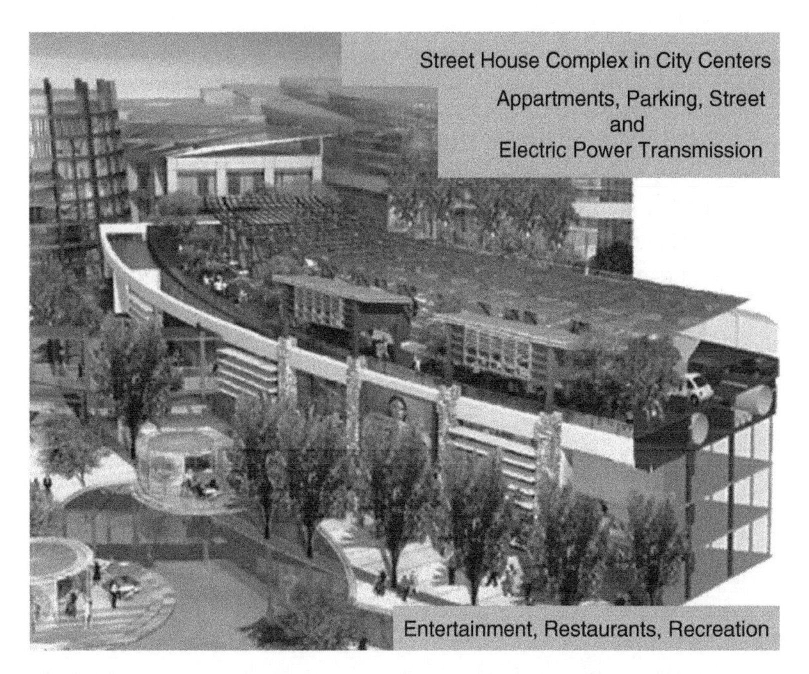

Figure 8.102 Overview Street House Concept (*Source* [5, 6])

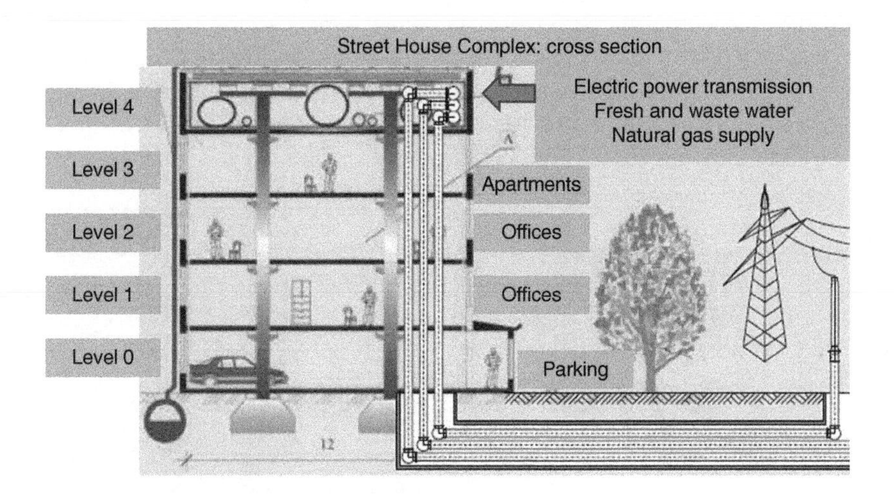

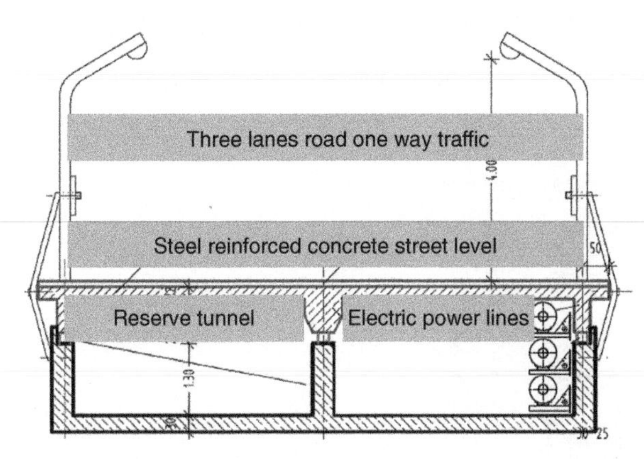

Figure 8.103 Upper: cross-section Street House Concept, lower: detail concrete tunnel for electric power transmission (*Source* [5, 6])

This type of Street House is used along the street between junction points. At the junction points, larger building complexes are used as shown in Figure 8.104.

At each junction point of the street routing from the outer ring highway into the city center (see Figure 8.98), a large building complex is foreseen to give space for offices and apartments. Two building complexes with 12–15 levels will make the junction point to a new area of city live with working place, living apartments, and any facility needed to get shopping or for entertainment like restaurants, cinemas, or clubs.

In Figure 8.104 on the left side, the building is located above the existing street crossing adding new lanes on top of the building and using the large junction area for the two high-rising building complexes. The right side of Figure 8.104 gives another impression of a junction without tall buildings. It shows also the traffic flow of the three lanes on top of the building.

The cross-section of the Street House at a junction point is shown in Figure 8.105. On top of the building, three to four lanes are offered for traffic in each direction. On the side, noise-reducing walls protect the neighborhood and direct the noise to toward the sky.

Bellow the street level, service room is provided for the infrastructure installations including the electric power transmission using GIL. The fourth level is also used to damp the noises coming from the street level from the office rooms and apartments below. At the ground level in the center of the Street House, two lanes for service are provided. These lanes are for buses, taxi, or metro trains for public transportation in the city center.

In Figure 8.106, a section of the Street House is shown where access is given for car traffic to enter street lanes on level 5 from the ground-level streets. Four-elevation buildings at each end and on each side of the Street House section allows car access for up-going and down-coming traffic in both traffic directions (Figure 8.106, left side). Such connections are needed at street crossing

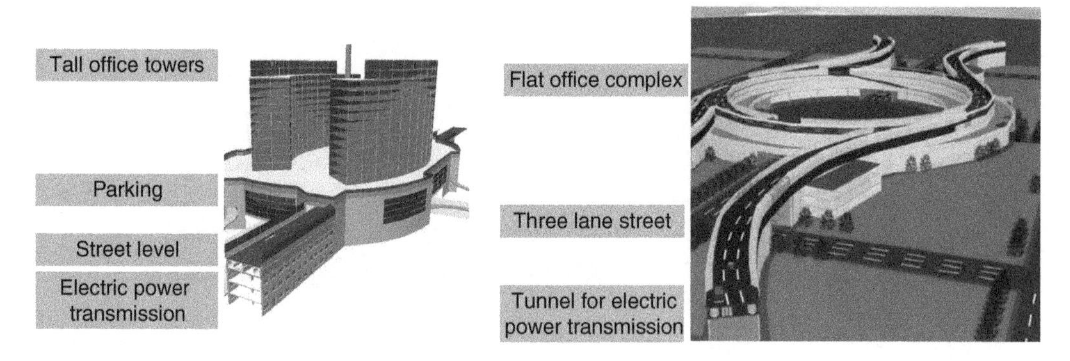

Figure 8.104 Left: Street House Concept with tall office towers, right: Street House Concept without tall buildings (*Source* [5, 6])

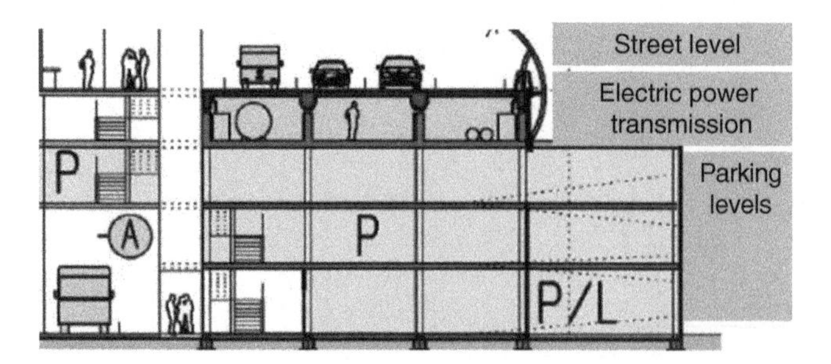

Figure 8.105 Cross-section of Street House at junction point with parking sections (*Source* [5, 6])

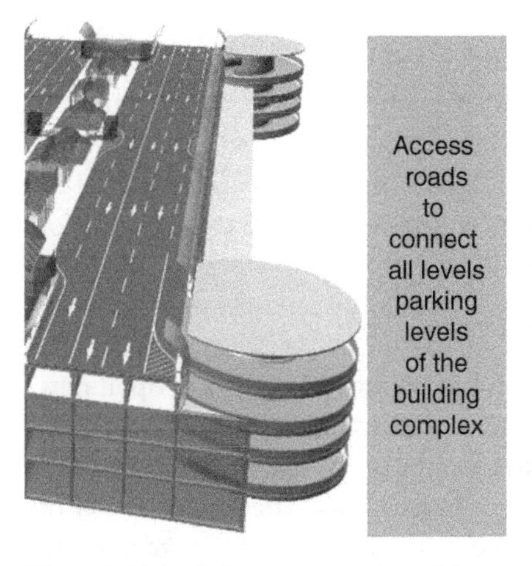

Figure 8.106 Left: access roads and integration of building complex to street junction, right: night impression of street junction building and road complex (*Source* [5, 6])

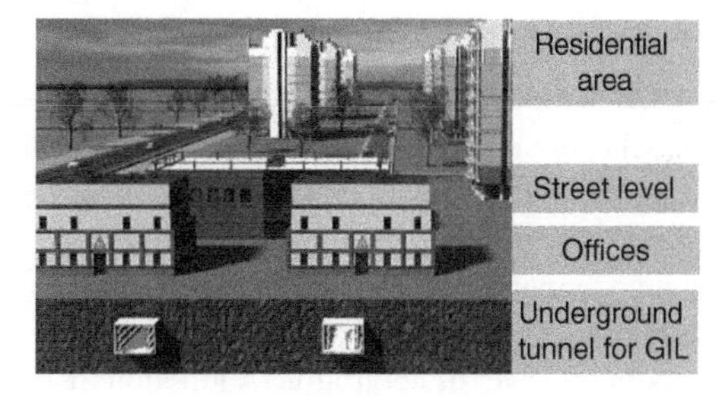

Figure 8.107 Safe electric power transmission lines using GIL below office buildings and residential areas (*Source* [5, 6])

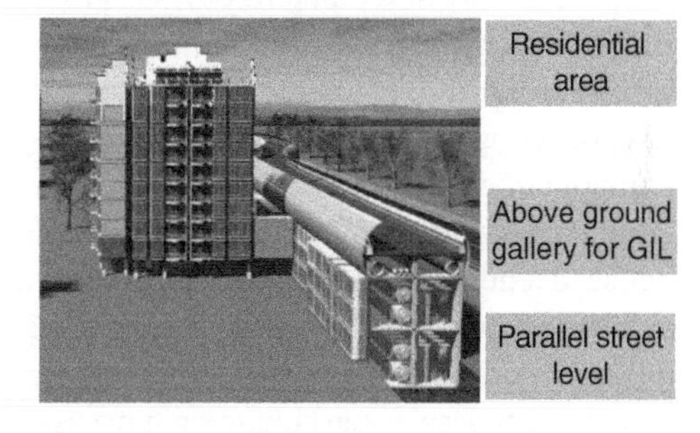

Figure 8.108 Impression of street house complex with electric power transmission lines installed in above ground galleries using GIL (*Source* [5, 6])

points to give the possibility to change the road. The graphic on the right side of Figure 8.107 gives a right impression of such a Street House section.

The Street House can only be built in sections without existing building. In such cases, as shown in Figure 8.107, the GIL can be laid in a tunnel constructed under the buildings. Because of the low outer magnetic field strength around the GIL and the inert safety of the GIL even in the event of an internal failure, the GIL is a safe and reliable solution.

In cases when tunneling is not recommended because of difficult soil conditions, it is possible to use the above-ground installation of GIL in an open trench. In the graphic animation, an open-trench construction made of steel and/or concrete give also space for other infrastructure systems like water, savage system, district heating, or telecommunication (Figure 8.108).

8.4.2.4 Conclusion

The Street House project study has been carried out by the project "Strassenhaus" [5, 6] in cooperation with city planner and experts form electric power transmission technology from Siemens. The explained solutions to solve the need of high-power lines in densely populated areas are using the compactness of gas-insulated high-voltage transmission lines (GIL) and combine its advantages with the need of traffic and other infrastructure requirements. GIL can be used for transmitting electric power in close vicinity to public from people traveling, working, or living because of its low magnetic field strength outside the GIL and the safety even in case of an internal failure.

The time of ever-growing cities will ask for such solutions in future, and GIL can open new ways. Space in cities is rare and a large amount of the space in the city is used for traffic facilities like streets, park houses, parking areas along the streets. At the same time, large cities have a big lack of green. Not enough parks, water, or just green areas. The combination of street and house to the Street House can help to improve the infrastructure and make a worth living city.

8.4.3 City Junction Project Study

8.4.3.1 Introduction

Densely populated areas in cities are subject of the project study to investigate the possibilities and project cost for compact power transmission lines using high-voltage gas-insulated transmission lines (GIL) and high-voltage cables. Cities and metropolitan areas are growing bigger each year with no tendency to stop growing. Such mega cities are all around the world, and this project study stands as an example for solutions with today's GIL and cable technologies to transmit high amount of power underground. A junction in a city center was chosen to show the possible technical solution for GIS, GIL, and cables to underground a high-voltage transmission line and to create free space for city development projects [10, 16, 17].

In this study, a specific routing has been investigated along a highway as part of an outer highway ring around the city center to use crossing points with radial major roads crossing the highway and connecting to the city center. From a city developing point of view, this concept will allow new metropolitan concentration point outside the city center. This trend is seen worldwide, and one example how a city can distribute the concentration away from the center is Atlanta, where with Buckhead a new center was created about 30 km outside the city center connected by a metro line and offering anything for living like work offices, living apartments, hospital, university, restaurants, sports, theatre, and any kind of shopping. To create such concentration spots in a densely populated city area, some changes are necessary. Typically, large city areas have wide corridors for overhead lines in the outskirts of the city, which have been fare away from the city when they were built. But today, after the city grew, they are in the city surrounded by houses, office, shops, and other city infrastructure.

The idea behind this project study is that the wide space of the 550 kV and 230 kV overhead lines are replaced by the compact design of GIL for the 550 kV part and cables for the 230 kV part. The section to be undergrounded is 3–4 km long, depending on the route. Crossings over Highways and some mayor roads have to be solved. The highways are part of the outer ring of Toronto and the main streets are running radial into the city center. GIL and cables will be used to make space for a new city concentration point for city live at this junction point. That's why the project study got the name Toronto Junction.

In this project study, an overview will be given on how such a complex problem can be solved and what are options to realize a new city center. The section around the city center will be explained as an example, and technical information will be given on how to overcome obstacles along the underground transmission line using GIL and cable.

8.4.3.2 City Junction North Line

The City Junction section is on area investigated. An overhead line corridor with 550 kV and 230 kV transmission line circuits is following a highway with two side crossings. The area around the highway is populated with houses, offices, and industrial buildings. The land under the overhead line is free of buildings.

The route for the underground GIL and cables is chosen to follow the highway as close as possible to combine these two infrastructures and make free land available, see Figure 8.109. One big road and a railroad line are crossing the planned routing of the GIL and cables. Two main streets are seen as direct connections into the city center. The existing railroad line into the city center can be used later to operate a metro lines from this new junction point into the city center. A local transformer station needs to be connected to the 550 kV and 230 kV lines for the electric power supply of this neighborhood. A small river needs also to passed by the underground GIL and cable transmission line. The total length of the underground transmission line is 3600 m.

Going underground in a city will bring some surprises of other infrastructure already in the soil. In Figure 8.110, an overview is given about the area at City Junction North Line along the planned route of the GIL and cable. South of the highway, a vital gas pipeline follows in parallel the route of the highway. To undergo or bypass gas pipelines may be expensive because of safety measure to be taken. Gas pipelines are pressurized pipes with a flammable gas inside. Building a tunnel next to it needs to be well planned and safety precaution are taken. The main water supply and waste

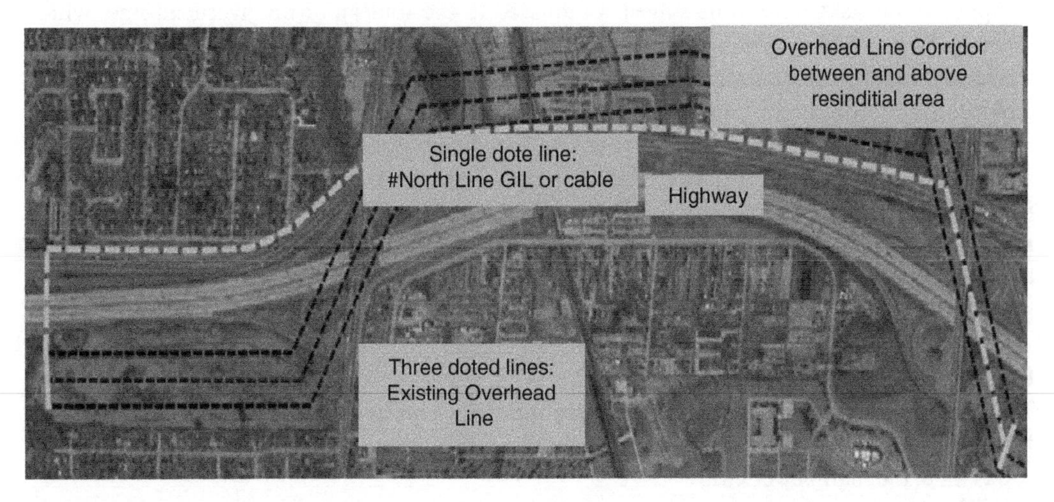

Figure 8.109 City Junction North Line, black line: existing overhead line, orange line: underground routing for GIL and cable (*Source:* Google maps).

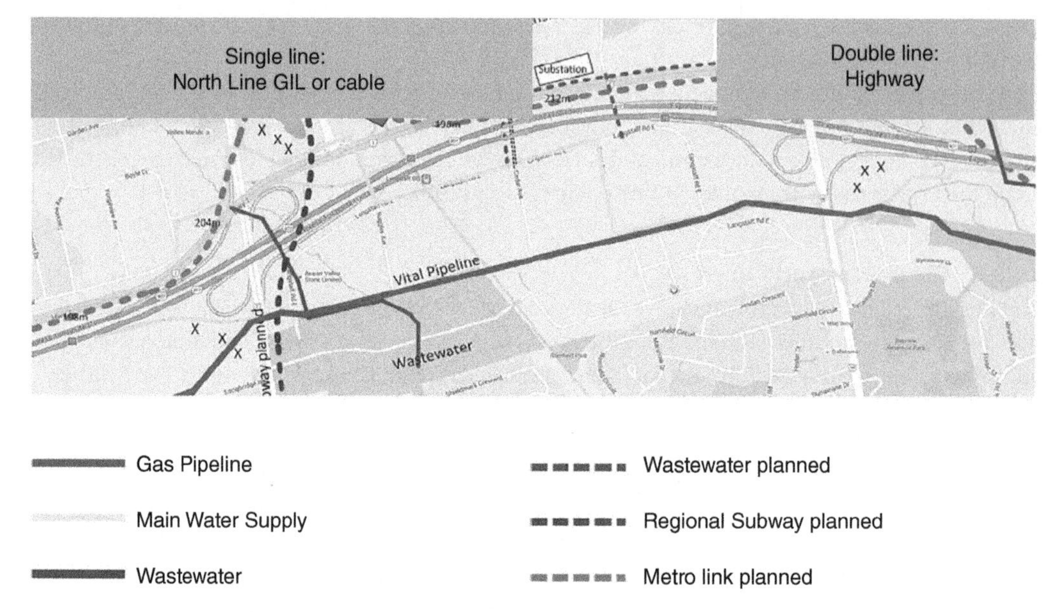

Figure 8.110 Infrastructure at City Junction North Line (*Source:* Google maps)

water pipes follow the highway on the north and south side in parallel with several crossings and branches. Main water supply pipes are easier to handle during construction than gas pipelines. They may be interrupted or bypassed if necessary. In both cases, the GIL and cable tunnel could underpass gas and water pipe by drilling technology for bored tunnel. In addition, many new pipes and lines are planned in future as shown with the dotted lines.

For the public connection of this new city concentration point, a new metro line and a new subway connection are planned for the connection to the city center.

Figures 8.109 and 8.110 show that creating a new city concentration point like the Junction a complex set of tasks have to be solved. Compared to the routing of the overhead line, which seems to be just a more or less straight line with some bends, the route under the ground level has to solve a large number of obstacles. For this, many different technical solutions need to be applied.

The cost for the investment to go underground are higher than the solution above ground, but the new free space in high-price city area offers a much greater value for the investors. The gain of the city is to continue to grow with population without getting into traffic over-load because of too many people need to travel into one city center. The decentralization of working offices, business, living apartments, shopping, restaurant and any other facility of city live, including recreation areas and green within the city, will bring benefit to people living here.

8.4.3.3 GIL Technical Options

To solve the technical requirements to underground, the overhead lines of 550 kV and 230 kV of two technologies may be used: for 230 kV line, the solid-insulated high-power cable and for the 550 kV line, the high-power GIL.

The 230 kV high-voltage line is using cross-linked polyethylene (XLPE) cables. The rate voltage of this cable is 245 kV and was type test by 1050 kV Impulse Withstand Voltage. It can carry 1500 A under typical ambient conditions. The transmitted continuous power of this cable is 600 MVA. A principal design is shown in Figure 8.111.

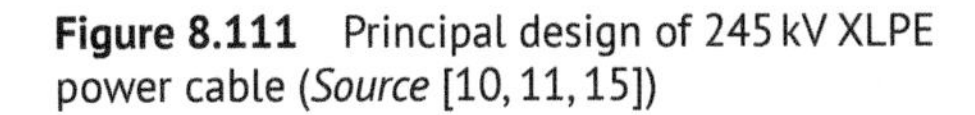

Figure 8.111 Principal design of 245 kV XLPE power cable (*Source* [10, 11, 15])

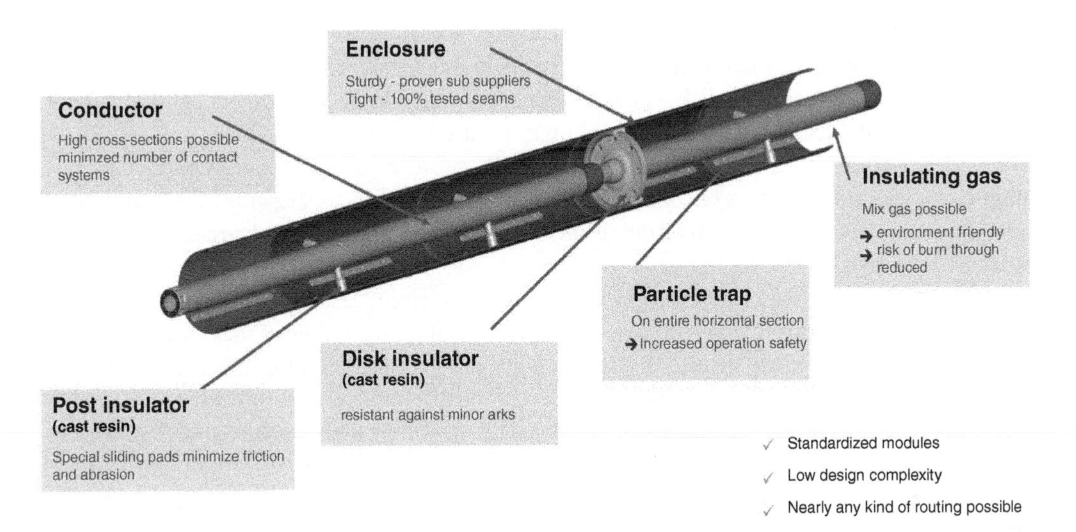

Figure 8.112 Principal design of 550 kV GIL laid in a tunnel [7, 8, 10]

The 550 kV line is undergrounded by the gas-insulated transmission line (GIL). The 550 kV type-tested GIL was applied to 1675 kV Impulse Withstand Voltage and can carry 3675 A continues current under typical ambient conditions. This allows to transmit a total electric power of 4700 MVA. The insulation is made with a mixture of N_2 and SF_6 or technical air at large diameter. The gas pressure is 0.7 MPa, which is relatively low for a technical system. In case of short circuits in the network, the GIL can carry 63 kA for up to 3 s without overheating.

The principal design is of the GIL is shown in Figure 8.112. The conductor pipe with large cross-section of electrical aluminum (typical 6000–8000 mm^2) allows the high current rating of 3675 A. The conductor pipe is held in the center of the enclosure pipe by post-type insulators made of cast resin, which are able to slide inside the enclosure pipe to compensate the thermal expansion of the conductor pipe. Disk type insulators fix the conductor pipe to the enclosure pipe and act as fix point for the thermal expansion of conductor pipe. Particle traps at the bottom of the enclosure pipe will collect any free moving particle inside the encloser pipe and guaranties the high reliability of GIL in service. These standardize module are connected by welding the conductor pipe and

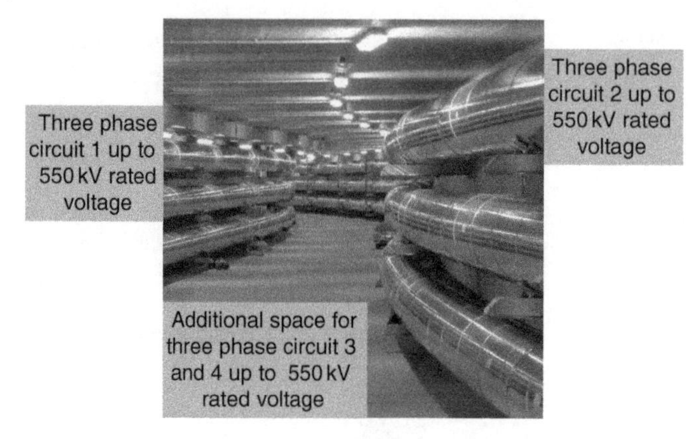

Figure 8.113 Example of 2 three-phase circuits 550 kV in a tunnel with reserve space in the middle for additional 2 three–phase circuits (*Source* [7, 8, 10])

the encloser pipe to a gas-tight compartment for lifetime. With flexibility using elastic bending, the GIL can follow most landscapes with a bending radius down to 400 m. For directional changed up to 90° angle units are used. The disk-type insulators may be gas tight and non-gastight. In case of gas-tight disk insulators, gas compartments of up to 1200 m length are formed.

The GIL is a completely welded system with a 100% proof of gas tightness by ultrasonic test equipment in an automated process of each weld. The diameter of the encloser pipe is 500 mm and 2 three-phase circuits can be laid in round tunnel of 3 m in diameter or in a square tunnel of 2.4 m width and 2.6 m height.

In Figure 8.113, a view in a tunnel with 2 three-phase GIL circuits is shown. The tunnel is made for a total of 4 three-phase circuits with two additional circuit in the middle of the tunnel for future expansions. The single-phase insulated GIL pipes are laid-on steel structure fixed to the tunnel wall. Each steel structure is electrical grounded so that the full induced electric current can flow in the encloser pipe and reduce the outer magnetic field. The bending radius shown in the Figure 8.110 is 400 m. The tunnel is accessible during operation with now risk to the surrounding in case of internal electrical arc because the impact stays inside the solid aluminum enclosure pipe. The magnetic field strength in the tunnel is low at rated current because of the reverse-induced electric current in the enclosure pipe.

8.4.3.4 Project Concept Study

The route study of Junction Project Concept shows in Figure 8.114 various obstacles (yellow line). Starting with crossing the highway on the left of Figure 8.114, the GIL and cable route will have to pass the access road to the highway, then there follows a narrow section with high elevation differences between buildings and housings on the left side of the GIL/cable and highway on the right, and at the right-end side in Figure 8.114, a whole highway crossing needs to be passed to reconnect to the overhead line.

The total length of the route is 3600 m, and for different sections, different technical solutions are needed. The detailed description below will explain the typical solutions.

In Figure 8.115, the typical technical solutions of the GIL for the 550 kV line are indicated in principle symbols. On the left side, an overhead line connection to the GIL is made by a high-voltage bushing. The GIL section is separated in fix point sections of the conductor pipe for thermal expansion indicated by a black triangle. The thermal expansion of the enclosure pipe is using angular (indicated by two horizontal double lines) or axial (indicated by two vertical double lines) expansion units. The road crossings and bridges are indicated by square blocks.

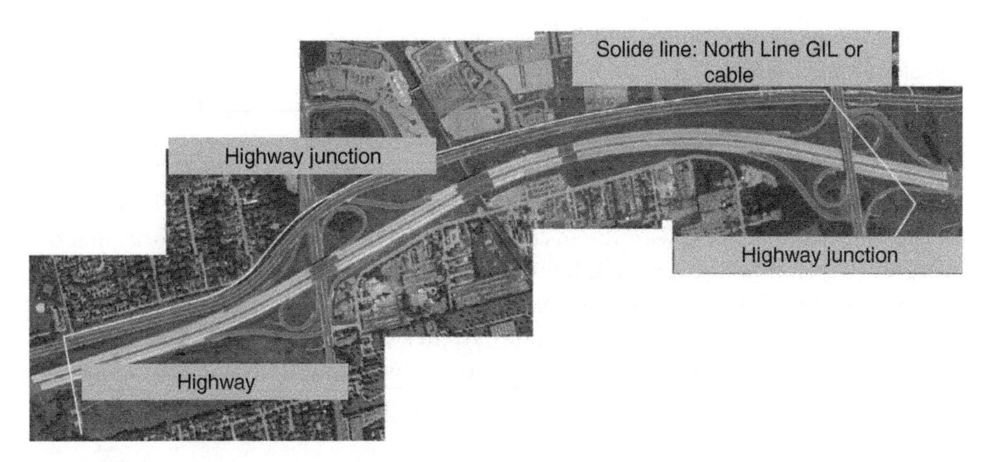

Figure 8.114 Top view of City Junction North, solid line (*Source:* Google maps)

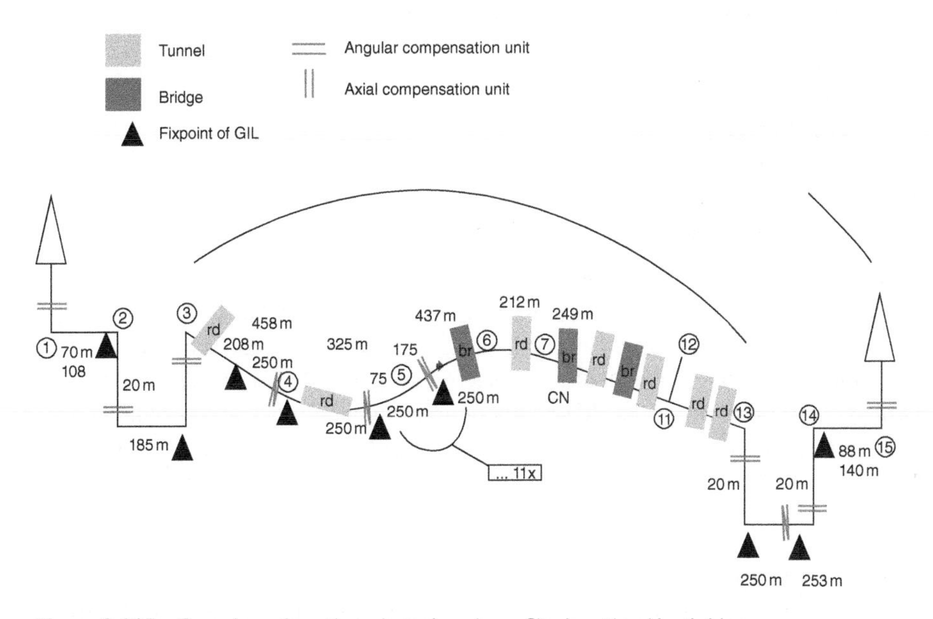

Figure 8.115 Overview of routing obstacles along City Junction North Line

There is a total of 15 obstacles identified. Numbers 1–3 indicate the overhead line tower area with a vertical shaft, a drilled section under the highway, and a vertical shaft on the other side of the highway. Between numbers 4 and 5, a straight section is going into a bend with a radius down to 400 m. From number 5 to 6, the street crossing is indicated as an obstacle. Number 6 indicates the beginning of crossing the access road to highway and from number 7 to 8 the railroad crossing. The numbers 8–12 mark a straight section under a street, an avenue and is connected to the local transformer station. Numbers 12 and 13 marks the bend over two avenues. The last section starting from number 13 with a vertical shaft, a tunnel under the highway, a vertical shaft on the other side of the highway, and the connection to the overhead line by bushings.

For each of these obstacles, specific technical solutions are required. In the following, some typical designs are explained. In a real application, there are always project-specific requirements, which need to be fulfilled. Therefore, the solutions explained for some typical obstacles along the line can only show some design feature, which are recommended when GIL is used as the electric power transmission technology.

In this case of Junction Project, various tunnel-type laying options have been applied for solving the obstacle requirements. The rigid structure of GIL aluminum pipes allows to fix the pipes on steel structures with relative long distances between the fixing points of typical 12–15 m. GIL aluminum pipes also can use angle units to change the direction of the route up to 90° to allow narrow bends when required. GIL with the insulating gas inside can be used in horizontal or vertical laying and any slope in between. The low magnetic field around the GIL allows the use of small tunnel sizes with low impact of the magnetic field radiation of personal in the tunnel. The low temperatures of the GIL even in the case of rate currents will allow several GIL three-phase circuits in one tunnel with only forced ventilation required. Low magnetic field and low temperatures of GIL in operation make also possible to add other infrastructure in the same tunnel (Figure 8.115).

Typical Solutions to Overcome Obstacles

In the following, six different typical obstacles will be explained in detail:

- Transfer section from overhead line to GIL
- Crossing the highway by a tunnel
- Section with high elevation differences
- Section with narrow space
- Section street crossing
- Section with access road underpass

The design for technical solution to overcome obstacles may vary from project to project, because there are many factors that influence the best, most economic, and most reliable solution in one specific project. These typical cases only can give indication on how to find good solutions.

Factors influencing the solution are related to the electric data as voltage level, current rating, transient over-voltages, ambient conditions like temperature, soil conditions, accessibility of the construction site, transportation limitations, maximum weight for transportation, authority requirements, safety during construction and operation, traffic interruptions, and final the cost of solutions.

Transfer Section from Overhead Line to GIL

The transfer from the overhead line to the GIL is needed at both ends of the undergrounded 550 kV section. The end tower of the overhead line is connected to a portal with three single-phase connections to the single-phase isolated GIL using an air/SF_6 bushing, see Figure 8.114. The bushing is mounted on steel structures set right above the vertical shaft to connect the tunnel. The vertical shaft covers a foot area of 2.5 m by 16 m. The distance of the vertical shaft to the overhead line portal is 4 m.

High lightning stroke probability into connected overhead lines or a reduce transient overvoltage level require overvoltage surge arresters in parallel to the air/SF_6 bushing. The surge arresters are not shown in Figure 8.116, it would be placed in parallel to the air/SF_6 bushings.

The clearances between the phases of the overhead line are 8 m, which gives also the distance between the air/SF_6 bushings.

Crossing the Highway Using Small Drilled Tunnels

The section under the highway crossing between numbers 2 and 3 in Figure 8.117 is using a drilled tunnel with an inner diameter of 3 m. The drilled tunnel is about 10 m below the highway. The round tunnel gives space for 2 three-single-phase-insulated GIL pipes. Two are mounted on the

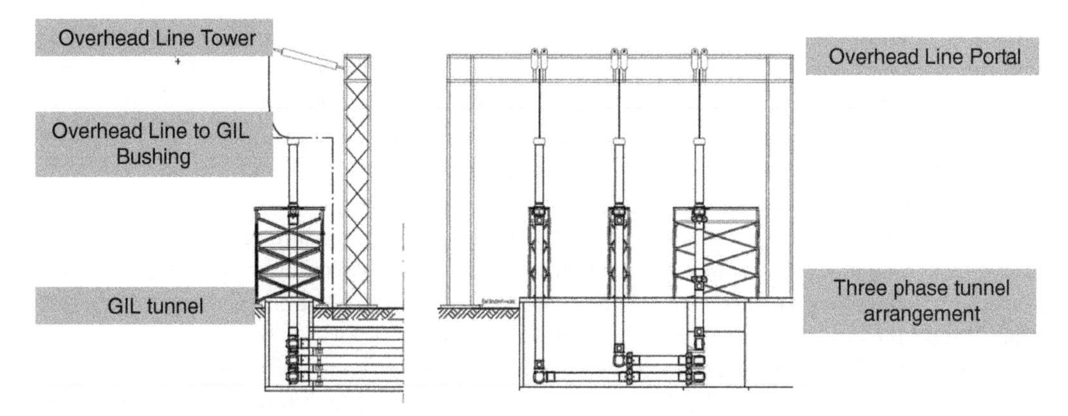

Figure 8.116 Connecting point overhead line (OHL) to GIL in a tunnel (*Source* [10, 12–14])

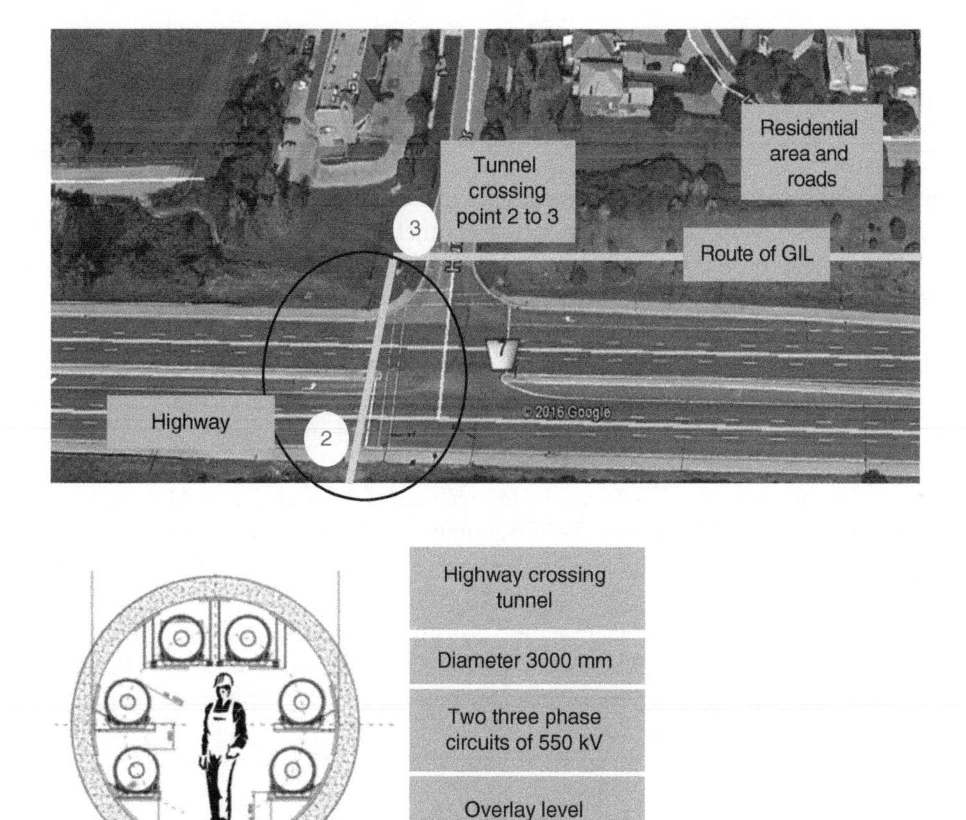

Figure 8.117 Upper part: example of highway crossing tunnel between two shafts, lower part: highway crossing tunnel of 2 three-phase 550 kV GIL circuits (*Source* [10, 12–14])

upper part of the tunnel and two on each side of the round tunnel. This gives space for personal to walk in the middle of the tunnel and to install a rail transportation system in the tunnel during erecting to bring in the GIL segments.

The GIL pipes in the tunnel are hold by steel structures, which give the mechanical strength to hold the weight of the GIL pipes and to be used as solid grounding point for a low-impedance ground connection. This low-ground impedance is needed to allow the GIL the induced reverse current in the enclosure pipe to compensate the magnetic field of the conductor pipe. The steel structures do not fix the GIL but let them slide to allow thermal expansion of the enclosure pipe and to avoid thermal expansion forces to the steel structure, which would be a big force.

At the location of number 3, the other side of the highway, the drilled tunnel coming from under the highway is connected to a square tunnel following the route of the highway. At this corner, a connecting shaft between drilled tunnel and square tunnel is erected. The GIL is using 90° bending units to make the directional change in the connecting shaft.

In total, there are three drilled tunnels of 3 m inner diameter crossing the highway: two tunnels, which hold 2 three-phase 550 kV GIL circuits each and one tunnel for 4 three phase 230 kV cable circuits (Figure 8.117).

The spacing between the tunnels can only be fixed after the geological survey has clarified the soil conditions.

Section with High-Elevation Differences Along Highway Using a Large-square Tunnel with Segments

The route of the GIL and cable tunnel in parallel to highway follows a slope of some meters of high difference toward the neighborhood of a residential area. In Figure 8.118, this section is between mark numbers 3 and 4.

The total available width of the route for laying the GIL and cables is 33 m. In Figure 8.118, the principal of laying is shown without scale. There are two tunnel sections in one tunnel structure. The tunnel structure is built from the top in an open trench. It is separated in one tunnel segment for the 4 three-phase 250 kV cable circuits and a second tunnel segment for 4 three-phase 550 kV GIL circuits.

The cable tunnel segment has 2 m width and 2.7 m height. The GIL tunnel segment has 5 m width and 2.7 m height. Both tunnel segments are separated by a concrete wall. The GIL aluminum pipes are mounted on steel structures, two on the wall on the side of the tunnel and two in the center of the tunnel.

The steel structure is solid connected to the steel reinforcement of the tunnel, and these are connected to ground electrodes. This is to provide the low-ground impedance for the full rated induced reverse current in the GIL enclosure pipe to eliminate the magnetic field of the conductor of the GIL.

Between the GIL and cable tunnel, a concrete vertical wall gives mechanical stability of the slope toward the highway. In Figure 8.118, the principle of this construction is indicated. The specific location of tunnel, vertical wall, and highways varies along the route between mark numbers 3 and 4.

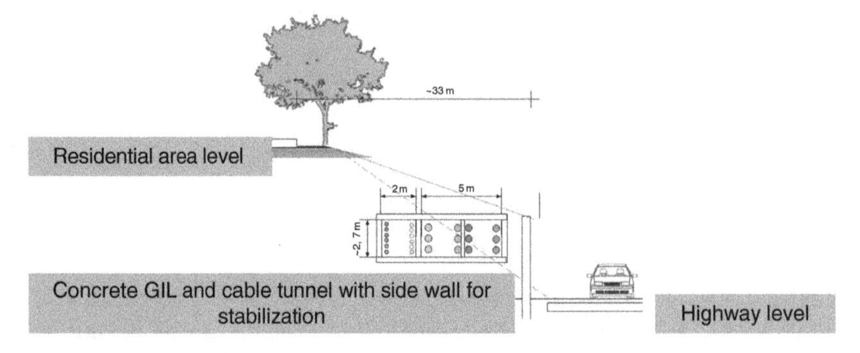

Figure 8.118 Example of height deviation level between residential area and highway (*Source* [10, 12–14])

Section with Narrow Space Using a Large-Square Tunnel with Segments

The route of the GIL and cable tunnel in parallel to the highway between mark numbers 4 and 5 of Figure 8.115 follows the highway with a narrow stripe between a road and a highway as shown in Figure 8.119. The slope between road and residential area is gone, the GIL tunnel and the highway are at the same elevation level.

This section is using a tunnel system with two tunnel segments: one for 4 three-phase 230 kV cable circuits of 2 m width and 4 three-phase 550 kV GIL of 5 m width. The height of the tunnel is 2.7 m. Same tunnel structure as in Figure 8.117.

Between the tunnel and the highway, a noise reduction wall is keeping away the traffic noises of the highway from the living area around residential area road.

The tunnel wall toward the highway is designed as a crash proof wall.

Section Street Crossing Using a Large-Square Tunnel on a Bridge

Between the mark numbers 5 and 6 of Figure 8.115, the highway is crossing a street and is connected to another Express Way. This crossing area is solved by a bridge construction for the GIL and cable route with a tunnel elevated to the height of the highway by using steel enforce concrete pylons as shown in Figure 8.120.

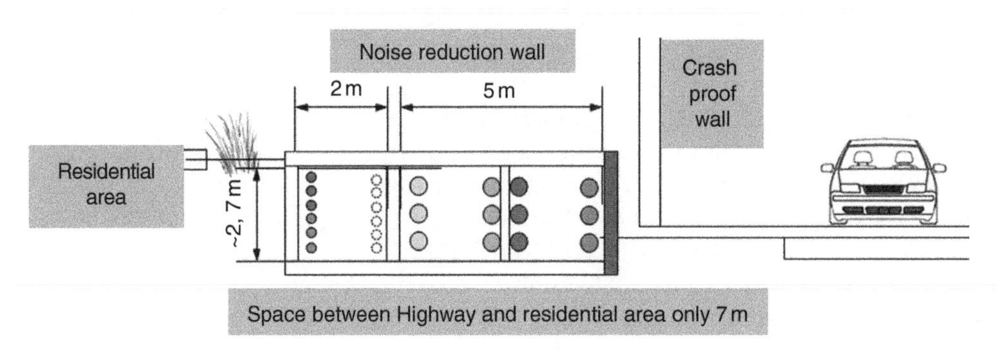

Figure 8.119 Example of narrow space between residential area and highway (*Source* [10, 12–14])

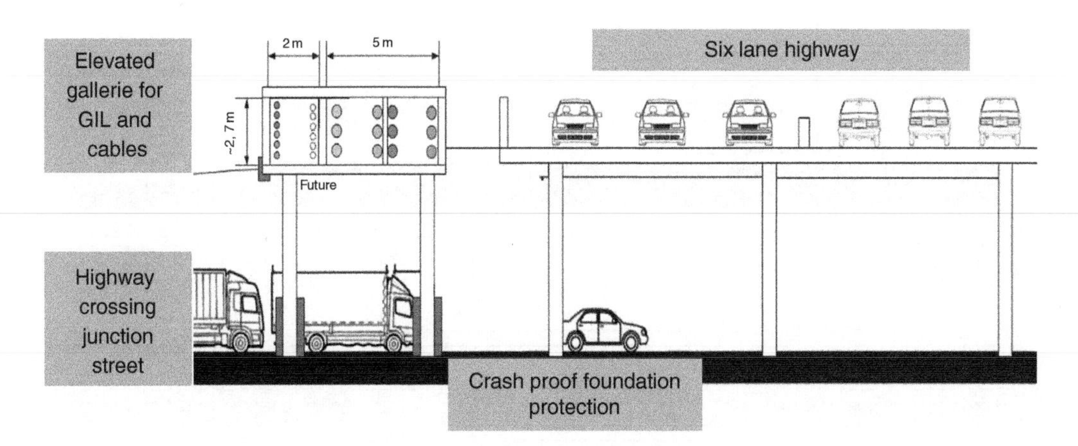

Figure 8.120 Example of bridge above high junction street section (*Source* [10, 12–14])

The tunnel has, as before, two tunnel segments. One tunnel segment of 2 m width holds 4 three-phase 230 kV cable circuits and the other segment of 5 m width holds 4 three-phase 550 kV GIL circuits. The tunnel height is 2.7 m.

The length of the crossing section with a tunnel on a bridge is 170 m. The steel enforce concrete pylons are protected against traffic accidents and are traffic crash proofed.

The same concept is used for the crossing of rail road line (mark number 7 and 8) and the Cedar Avenue (mark numbers 9 and 10).

Section with Access Road Underpass Using a Large-Square Tunnel with Segments

Between the mark numbers 6 and 7 of Figure 8.115, an access road to the highway is crossed by a tunnel under the road. This is a typical solution for cases when the traffic can be de-routed, and the tunnel is built from the top in an open trench as shown in Figure 8.121.

The tunnel structure is the same as before with two tunnel segments: one tunnel segment of 2 m width for 4 three-phase 230 kV cable circuits and the other tunnel segment of 5 m width with 4 three-phase 550 kV GIL circuits.

The length of the tunnel crossing under the access road to highway is 70 m.

For this solution, the geographical soil investigation is needed to avoid problems when constructed. In cases of unstable soil condition or rock, it might be more cost effective to the before explained tunnel on a bridge solution.

In this case of crossing the access road to Highway, the underground tunnel solution is following the underground tunnel of the sections before and after and will not be visible on the surface.

Undergrounding the Total Length by Using a Large Drilled Tunnel

This solution offers the possibility to connect the total length between the remaining overhead line end points by a large drilled tunnel of 3.2 km length. Such a large tunnel will have an inner diameter of 6 m and is separated into four sections. Two sections for 4 three-phase circuits of 230 kV and two sections for four circuits of 550 kV GIL, see Figure 8.122.

The two 230 kV power cable circuits will transmit 2 × 600 MVA of electric power, and two more cable circuits are in spare for future extension. The four 550 kV GIL circuits can transmit 4 × 3500 MVA in two tunnel segments.

Two 550 kV GIL circuits are fixed to the middle wall of the GIL tunnel segment and two 550 kV GIL circuits are fixed to the roof of the GIL tunnel segment, as shown in Figure 8.122.

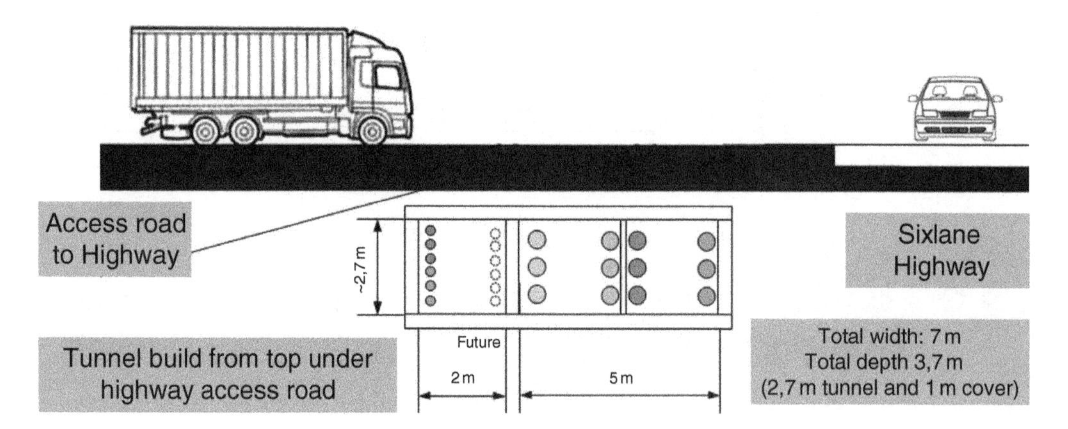

Figure 8.121 Example of underground tunnel section at access road to highway (*Source* [10, 12–14])

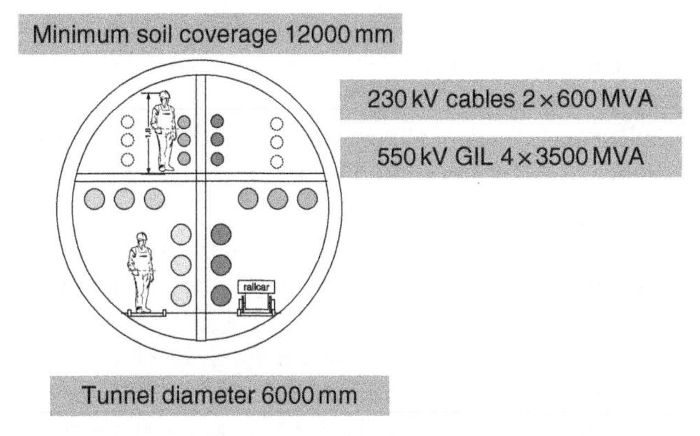

Figure 8.122 Example of large tunnel with four segments for 550 kV GIL and 230 kV cables (*Source* [10, 12–14])

Figure 8.123 Example of tunnel boring machine (*Source:* [18])

The drilled tunnel is laid a minimum of 12 m below the surface to keep the distance of two times the diameter from the tunnel top to the surface. The specific route of the tunnel needs to be studied in geographical study taking into account the soil condition.

The tunnel boring machine technology is developed over the last years to much more economical solutions because of improvement of the accuracy of the tunnel routing during construction. Also, the geological survey of the soil conditions and obstacles is more precise and the investigations bring better results for the drilling and less risk for the tunnel boring process.

In Figure 8.123, a tunnel boring machine is shown in principle. On the left-end side of the graphic, the tunnel boring head is drilling in the soil by using hydraulic forces from the tunnel boring machine from the shaft wall on the right-end side of the graphic. Next to the boring machine head, the concrete tunnel segments are lined along with the progress of drilling toward the left side. The concrete tunnel segments, so called Tübbings, are stored at the tunnel shaft and transported into the tunnel to the boring machine head as the speed of the drilling process requires. For one tunnel section of about 5 m length, about 4–5 Tübbings are needed.

The Tübbing technology is a well proven and worldwide used technology for long tunnel. Tunnel of several 10th of kilometers are already constructed for railroad, vehicle traffic, or electrical transmission. An electric transmission tunnel has been built for 400 kV cables in Berlin, Germany, on a length of 11 km or in Singapore the 18.5 km Gambas to May Road tunnel, and 16.5 km East–West Tunnel, only to mention some.

The soil from the drilling process is transported out of the tunnel as a fluid mixture of lean concrete (Bentonite) by pipes and pumps. On the surface, the Bentonite is separated from the soil and cycled back into the tunnel for more soil transportations.

The efficiency of the tunnel boring technology is much depending on the soil conditions. Hard-soil-like rock will require a different drilling head than soil of sand. The worst is a mixture of different soil conditions. This needs to be studied in geological survey.

The principle laying process of the GIL in the tunnel is shown in Figure 8.124. In the graphic is shown on the left a tunnel close to the surface coming from the overhead line connection point to the shaft of the deep laid drilled large tunnel.

The shaft is used to bring in the tunnel boring machine and will have a diameter of 15–20 m depending on the type of boring machine. This shaft will have a depth of 15–20 m as it is required for the coverage of the large tunnel. The GIL laying process is using this shaft for access to the tunnel. In this example, the diameter of the shaft is 14.5 m and the depth is 20 m.

The total length of the tunnel of about 3.6 km requires per single phase about 300 enclosure and conductor pipes of about 12 m length each. The pipes will brought down the shaft by crane to the welding location of the GIL assembly. The automated assembly process will produce 12 m long GIL segments, which is pulled into the tunnel.

To speed up the laying process, several welding machines can work in parallel and such assembly location can be positioned at each end of the tunnel.

The routing of the GIL from the small tunnel close to the surface, through the shaft, down to the large tunnel will be realized by using angle units of the GIL with 90° angles.

The advantage of the deep-laid large tunnels is that a direct routing can be chosen, in this case 3.2 km for the deep tunnel versus 3.6 km for the surface close tunnel. In addition, when using deep tunnel routing, there are less obstacle as close to the surface, this can save additional cost and can make the deep tunnel the most economical solution. In any case, this needs to be studied in a project study because there are too many parameters to impact each of these solutions.

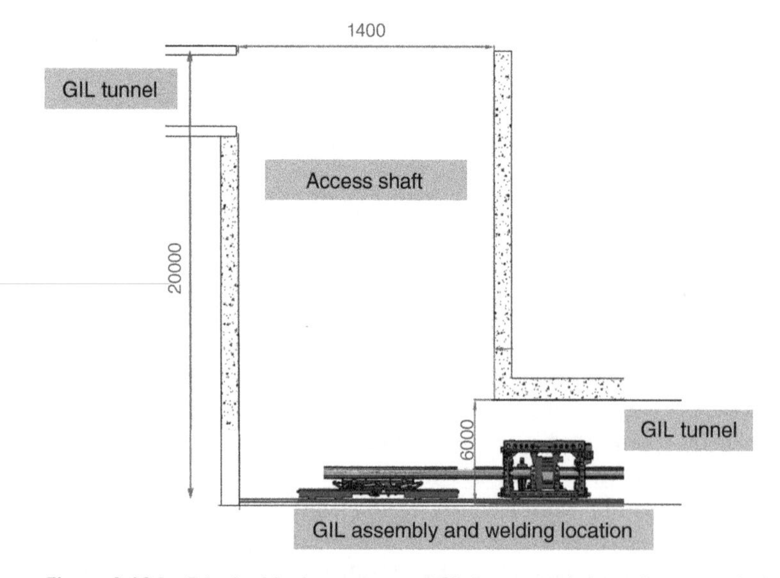

Figure 8.124 Principal laying process of GIL in a tunnel through access shaft (*Source* [10])

8.4.3.5 Technical Design

There are two important technical design features influencing strongly the technical solution of the underground power transmission using GIL and cable. These are forced ventilation and the time for the laying process. These shall be explained in the following.

Forced Ventilation

The electric power transmission creates losses in the conductor, which is generating heat in the tunnel. There are limited temperatures for the safe operation of the GIL and cable, which is at a maximum temperature of about 120–140 °C. Above such temperatures, the insulating material will be damaged and lose its electric insulation capability. The design of the insulation needs to be made in a way to stay below these temperature limits.

In this application 550 kV GIL and high-power transmission capability of 3500 MVA are the dominating heat resource in the tunnel. The 230 kV cables with 600 MVA power transmission will not be at thermal limits.

The following assumptions are taken. The air intake temperature is 35 °C. The maximum permitted temperature in the tunnel is 50 °C. The distance between the shafts for ventilation is 500 m, see Figure 8.125.

There are four 550 kV GIL circuits installed in the tunnel. The aluminum conductor has a cross section of 5340 mm using electrical aluminum. The continuous current rating is 3150 A, and power transmission losses per circuit is 562 W/m.

The tunnel dimensions are 5 m width and 2.7 m height, as shown in Figure 8.125.

The thermal calculation requires an air volume of 57.65 m^3/s to transport the heat out of the tunnel. This requires an air speed of 5.22 m/s by the size of the tunnel. This is a relative low speed of air less than 20 km/h and can be realized by standard ventilation.

Forced Ventilation of a Deep Tunnel

An example of a ventilation shaft is shown in Figure 8.126. Such shafts need to be integrated into the surrounding by visual impact and by noises. Especially in a city or metropolitan area,

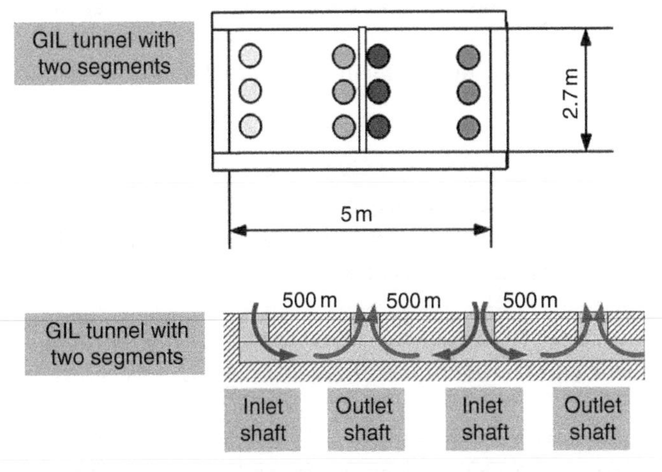

Figure 8.125 Upper part: example of ventilation of four GIL circuits in two tunnel segments, lower part: example of ventilation shafts for tunnel with two segments (*Source* [10])

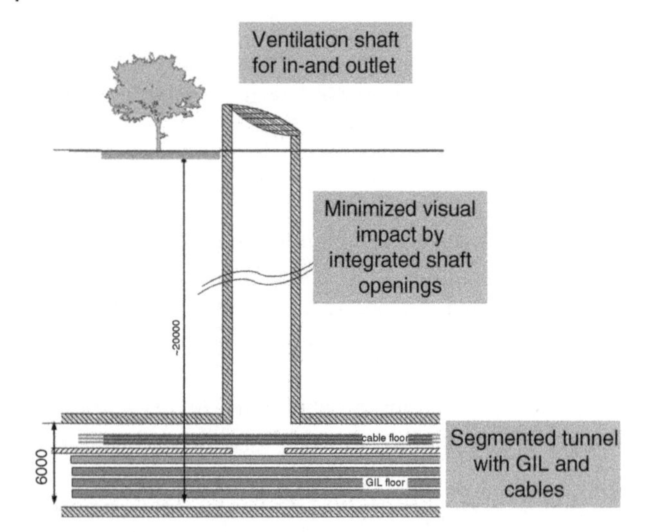

Figure 8.126 Ventilation shaft to segmented tunnel and minimized visual impact (*Source* [10])

consideration needs to be made to integrate the structure into the picture around and to avoid noises, which are mostly heard in the summer when the temperatures are high and the people are living more outdoors.

The shafts also need to fulfill the safety requirements for personal safety. Therefore, different solutions need to be adapted to local requirements. The creativity of architects can help to find attractive solutions like integration in a fountain or spring, or to use the ventilation shaft as an art object.

In Figure 8.126, an example is shown how the ventilation shaft can be connected to the tunnel. The shaft shall have a diameter of 3 m and has a depth of 20 m from the soil surface. The 2–3 m diameter is needed to generate a noiseless airflow at the surface of the ventilation shaft.

The shaft needs to have a cover of trafficable mesh grid to avoid people or animals to fall into the shaft. The mesh grid may be installed vertical, horizontal, or in gradient.

The access from the tunnel to the ventilation shaft is made by openings of the tunnel segments for the cables and the GIL sections.

The fans may be positioned at the tunnel segments to keep them away from the surface exit and reduce the noise coming to the outside.

Forced Ventilation of a Tunnel Closed to the Surface

For ventilation shafts close to the surface, the access is much easier than with deep tunnels. The simple way is to have horizontal cover with a mesh-grid to protect people and animals for falling in and to locate the fan in the tunnel section away from the ventilation opening to avoid noises on the outside.

In some cases, it might be more attractive to add a head house on top of the ventilation shaft to give access to maintenance personal into the tunnel by stairs or a ladder as shown in Figure 8.127.

The size of such a head house would be 5 m wide, 7 m deep, and 4 m high and will have a lockable access door. Such a building would have a bigger visual impact and need to fit into the surrounding.

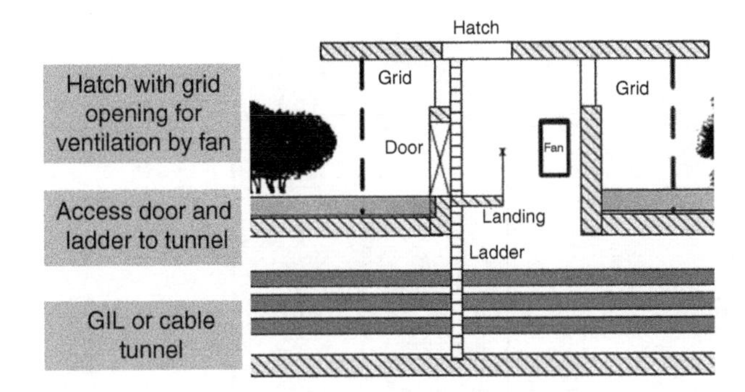

Figure 8.127 Ventilation and access shaft of GIL or cable tunnel integrated to environment (*Source* [10])

The fan could be installed in the tunnel segments or in the head house depending of the type of fan. In the head house, large fan diameters can be used, which produce less noises.

Again, architects are asked to be creative for solutions that can be accepted by public and may be will less visible or seen as an art object.

Underground tunnel projects for electric power transmission in cities or metropolitan areas are large infrastructure measures, which need an overall sensitive project management. It should not end with the project not realize able because of ventilation shafts. This requires to let the public and neighborhood participate with their ideas and wishes early in the project planning. This will help to be sensitive for the public feeling and situation.

Installation Process in a Tunnel

The GIL installation of long length underground transmission lines is using a large amount of single enclosure and conductor pipes. Depending on the length of one pipe which is typically 12–14 m for one kilometer, 80 or 60 pipes are needed. For three phases, these are 240 or 80 aluminum pipes. They all need to be welded on site and pulled into a tunnel or an open trench. This makes the impression that this assembly of GIL takes much more time than with cables where 500–1000 m cable can be delivered on one cable drum.

But this impression is not true. Why? The GIL assembly and laying process is working with parallel working stations at the same time. When doubling the working places, the time half, by four working places, it is a fourth, and so on. This means that the project execution time can be adapted to the requirements at the location.

It is clear that it will be increasing the working places the cost for investment of tools and machinery is going up, and an optimization process is needed to bring time for project execution and cost for equipment to an optimum. In the case of the long tunnel project here, it might be the best solution to have two working locations at each end of the tunnel.

The principle of one working place for GIL assembly is shown in Figure 8.128.

On the right side of the graphic, the 12–15 long enclosure and conductor pipes are brought into the tunnel through the shaft from the ground level. In the shaft at the tunnel entrance, the welding machine of the GIL is located. The welding process is automated and includes the positioning of the pipes before welding, welding the conductor or enclosure pipes, and make a 100% ultrasonic test of the enclosure weld as this need to be gas tight and is a pressurized compartment.

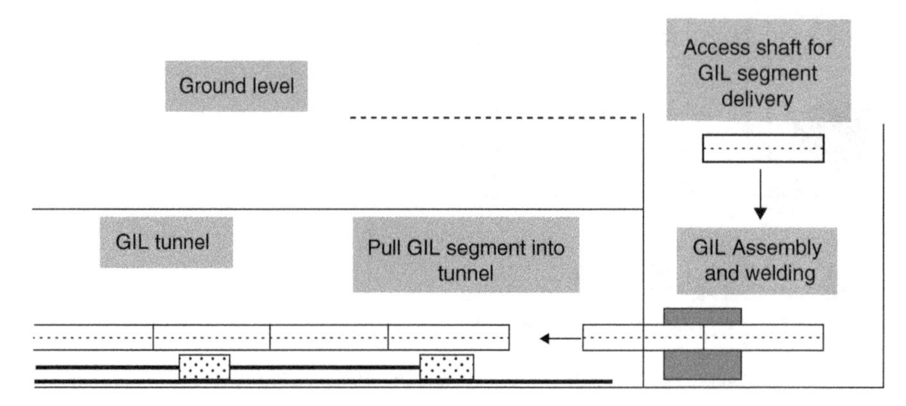

Figure 8.128 Assembly of GIL through tunnel shaft and pull into tunnel (*Source* [10])

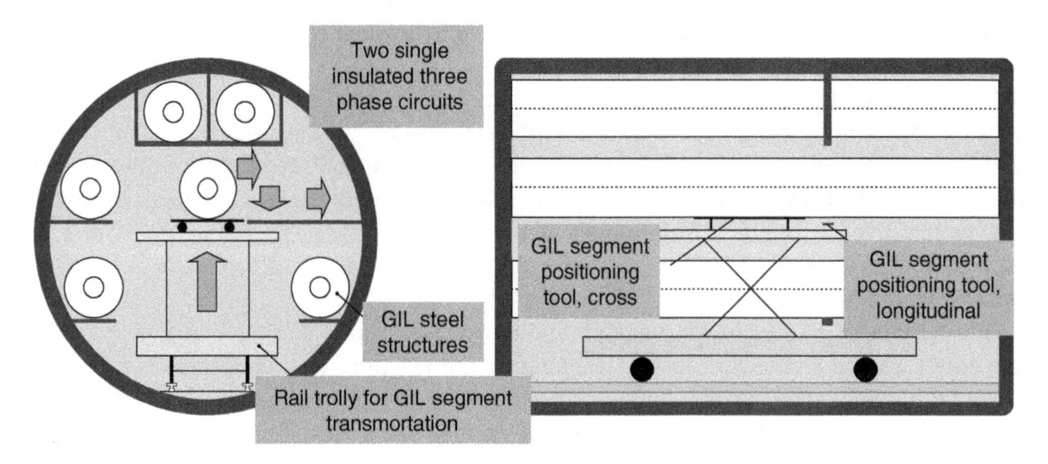

Figure 8.129 Lifting GIL segments to positions at tunnel steel structures (*Source* [10])

The welded GIL segment with enclosure, conductor, and insulator is then pulled onto a roll or sliding mechanism into the tunnel. Typically, 10 GIL segments are welded to one section of 120 m or 150 m length. These GIL sections are transported into the tunnel by a small train system and will be positioned as the final position.

In this working sequence, additional working locations are located inside the tunnel, which will further speed up the laying process of the total length of the GIL.

In Figure 8.129, the positioning of the GIL section in the tunnel at the final position is shown. A rail system with little transportation cars is located in middle of the tunnel and will bring 120 m or 150 m GIL sections into a several kilometer-long tunnel. At the final location, the little rail cars can lift the GIL section to the right level and the GIL section will be welded to the next GIL section. The steel rack to hold the GIL are fixed to the tunnel in about 25 m distances.

8.4.3.6 Conclusion

The City Junction project study investigated the possibility to replace overhead lines of high-power ratings (here 5150 A at 550 kV) in a city area and make space for city development. On a length of about 3.6 km, the overhead line has been replaced by 550 kV GIL and 230 kV cables. Several highways, streets, creek, and a railroad track had to overcome, and different technical solutions like small drilled tunnels, surface close square tunnel build in an open trench, elevated tunnels to a bridge, and a large deep in the ground laid tunnel for GIL and cables. These solutions have been adapted to the different local requirements along the route.

The project study shows that with the obtained space along the route of the existing overhead line, new city development can be realized. Space for office building, apartments for living, entertainment like restaurants, music halls, cinemas, and theatres can be found for new high-value investment in the metropolitan area. Limits of growth are the drivers of such developments when city centers are getting too large, and travelling time in and out take too much time, every day! There are some city developments around the world using this concept to make living and working more attractive to people.

Modern GIL and cable technology offer the technical solutions to bring large amount of electrical energy underground, at acceptable cost, in reasonable time and with liable and long-lasting power transmission.

8.5 Mobile Substations

8.5.1 General

The compact design of a GIS allows technical solutions that are very unconventional as a mobile GIS. This does not mean that the GIS could be moved around while connected to the high-voltage network; the meaning is that for temporary use a GIS including all control equipment could be installed inside a container or on a trailer and can be transported from one temporary use location to the next.

This might be the case when substation extensions or upgrades need to dismantle sections of the substation or in case of a disaster to replace a damaged substation until it is replaced or repaired. Typical voltage levels of such mobile GIS are at the lower high voltages up to 170 kV.

8.5.2 Containerized GIS

For rated high voltages up to 150 kV complete switchgear bays and the required control gear can be installed inside a 40-foot container. Only the SF_6 gas-to-air bushings need to be attached on top to get the bay connected to the substation. A 123 kV solution is shown in Figure 8.130. The container with the GIS bay is mounted onto the concrete basements. The three-phase encapsulated bushing connection is attached to the GIS bay on top of the container. The GIS bay is assembled and routine tested in the factory so that only two weeks were needed to install the containerized GIS in the substation. This fast erection time is a big advantage of containerized GIS and also the easy accessibility during operation under controlled indoor conditions.

To extend the containerized GIS assembly only additional containers need to be added and connected to the substations air insulated bus bar. This is a conventional solution for extending existing air insulated substations, as shown in Figure 8.131.

Figure 8.130 Containerized substation, 123 kV (Reproduced by permission of Siemens AG)

Figure 8.131 Containerized substation, 123 kV – extension is in extremely confined spaces (Reproduced by permission of Siemens AG)

8.5.3 Truck-Mounted GIS

8.5.3.1 General

GIS can be mounted on trucks for temporary use in a substation. Lower voltage ranges are three-phase insulated and several GIS bays are located on one trailer. For higher voltage levels, larger single-phase insulated GIS can be installed on a trailer and assembled on site to a complete bay. The advantages of such truck-mounted GIS are fast installation times and the possibility to move the GIS from one location to another as needed.

8.5.3.2 Truck-Mounted GIS of 72.5 kV

In Table 8.16, the technical data of a three-phase insulated, truck-mounted GIS are shown. At the voltage level of 72.5 kV, the current rating of 4000 A and a short-circuit rating of 31.5 kA is relatively high. The basic insulation level of 325 kV lightning impulse voltage is a standard value of IEEE C37.122 and IEC 62271-203.

The substation layout is shown in the single-line diagram of Figure 8.132 with four in- and outgoing lines and one bus bar. Lines 1 and 2 are connections to overhead lines and lines 3 and 4 are connections to transformers.

The complete GIS substation with four bays and one bus fits on one 40-inch standard trailer and can be transported under normal traffic conditions on the highways. The weight of the GIS is relatively light with its aluminum enclosure and the insulating gas inside. No special weight requirements need to be fulfilled for street transportation.

The control cubicles for the substation bay control are also attached to the GIS installed on the trailer; they have been tested in the factory, and are ready for connection on site. Only integration into the substation control and protection system is required. The electrical connection to the overhead lines, air-insulated bus bars, and the transformers is made by the SF_6 gas-to-air bushing, which can be attached on the top of the bays (see Figure 8.133).

Table 8.16 Technical data of a mobile three-phase insulated 72.5 kV GIS mounted on a truck

U_r	72.5 kV
I_r	4000 A
U_{BIL}	325 kV
I_s	31.5 kA

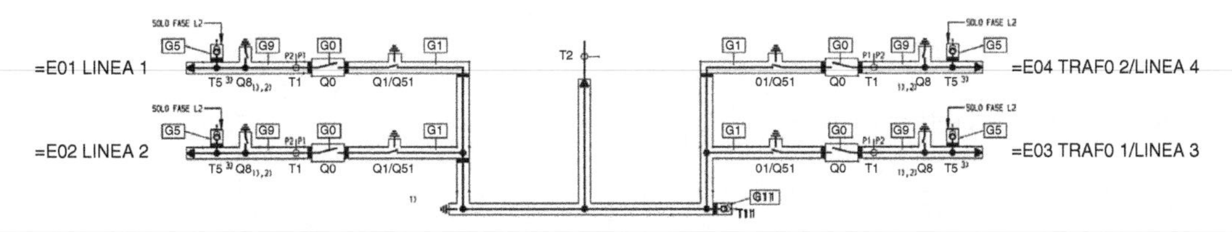

Figure 8.132 Single-line substation layout of a 72.5 kV GIS mounted on a truck trailer – four lines and one bus bar (Reproduced by permission of Siemens AG)

Figure 8.133 Substation layout of a 72.5 kV GIS mounted on a truck trailer – four lines and one bus bar (Reproduced by permission of Siemens AG)

8.5.3.3 Truck-Mounted GIS of 420 kV

The technical data of a 420 kV trailer mounted GIS is shown in Table 8.17. The rated voltage of 420 kV has a standard current rating of 4000 A and a standard short-circuit rating of 62 kA. The basic insulation level of 1420 kV lightning impulse voltage is a standard value of IEEE C37.122 [2] and IEC 62271-203 [1].

The larger size of the 420 kV GIS will not allow installation of a complete three-phase bay on top of one trailer. The single-phase insulated GIS at this voltage level allows only one phase to fit on one standard truck size. As shown in Figure 8.133, the single-phase prefabricated, pretested unit consists of the circuit breaker enclosure in the center and the disconnecting and ground switches connected at each end of the circuit breaker enclosure. This preinstalled single-phase section is then assembled to a three-phase bay on site and connected to the air-insulated bus bar, overhead lines, and transformers. The photograph in Figure 8.134 shows the 420 kV GIS mounted on a truck for transportation to the substations, and in Figure 8.135, the installed single-phase GIS unit is shown in the substation.

The final functional testing can only be made on site where the bay control and protection is connected.

Table 8.17 Technical data of a mobile 420 kV GIS

U_r	420 kV
I_r	4000 A
U_{BIL}	1420 kV
I_s	63 kA

Figure 8.134 Mobile 420 kV GIS mounted on a truck (Reproduced by permission of Siemens AG)

Figure 8.135 Mobile 420 kV GIS connected in the substation (Reproduced by permission of Siemens AG)

8.5.3.4 Mobile High- and Medium-Voltage Substations

The principle of a mobile substation concept is to have a complete substation available in case of emergency or natural disaster to replace high and medium equipments. Therefore, the substation is divided into high- and medium-voltage modules, which are individually installed on a trailer. The trailers are then brought to the location and are interconnected. The following modules (trailers) are chosen, as shown in Figure 8.136.

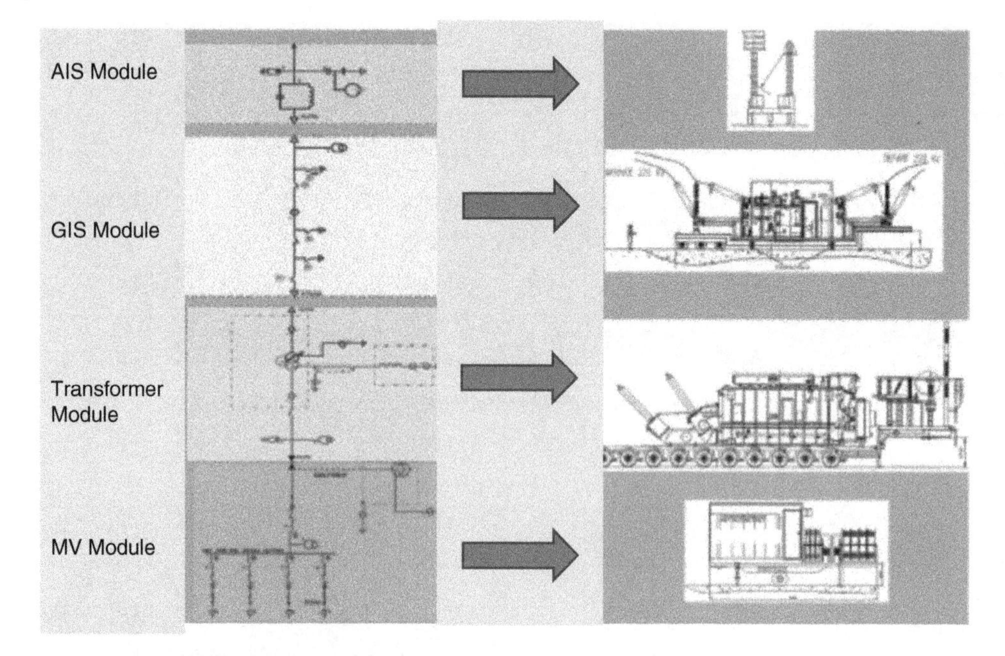

Figure 8.136 Mobile high- and medium-voltage substation

The AIS module holds a voltage instrument transformer, a surge arrester, and a compensation coil and capacitor, if needed. The GIS module holds the complete GIS bay including the circuit breaker, disconnector and ground switches, voltage and current instrument transformers, and at both ends SF_6 gas-to-air bushings.

The transformer module holds the transformer and the transformer-to-air bushings at both ends. The medium-voltage module with medium-voltage GIS is installed in a container including the protection and control equipment.

The trailer with the high-voltage GIS container needs electrical clearances on both sides for the SF_6 gas-to-air bushings, which fulfill the air insulation requirements. The compact design of the GIS does not allow direct connection of the air-insulated lines because of large distances between phases needed in air. Therefore, the three phases are extended to the side to fulfill these criteria, as shown in Figure 8.137.

The high-voltage GIS on the trailer may be connected by cables or by overhead lines, as shown in Figure 8.138. On the left side of the graphic, it is shown how to connect the GIS to a cable and on the right side, it is shown how to make a T-connection to an overhead line or the substation air-insulated bus bar (Figure 8.139).

The medium-voltage GIS is installed inside a container on the trailer and also includes all control, protection, and auxiliaries (see Figure 8.139).

With this modular concept, it is also possible to install more complex substations using multiple trailers. As shown in Figure 8.140, a four-line, one-bay bus bar arrangement can be realized easily using the different trailers.

At 72.5 kV voltage levels, a 15-bay double bus bar configuration including control, protection, and auxiliaries can be placed in one container. Factory assembled and tested; the container is ready to be connected on site. The factory-tested container with 15 bays of a 72.5 kV GIS on the crane at the substation is shown in Figure 8.141.

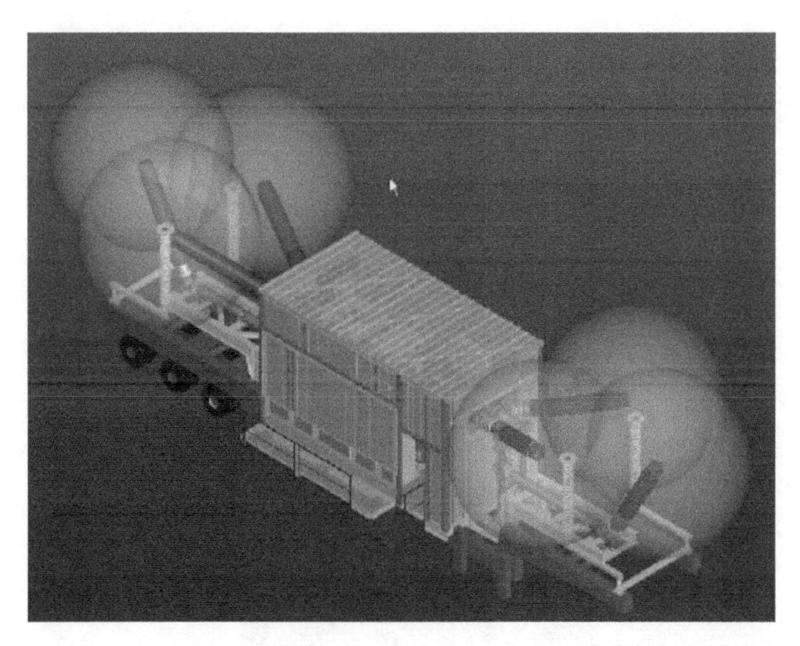

Figure 8.137 Mobile high-voltage GIS mounted on a truck trailer – electrical clearance design in air (Reproduced by permission of Siemens AG)

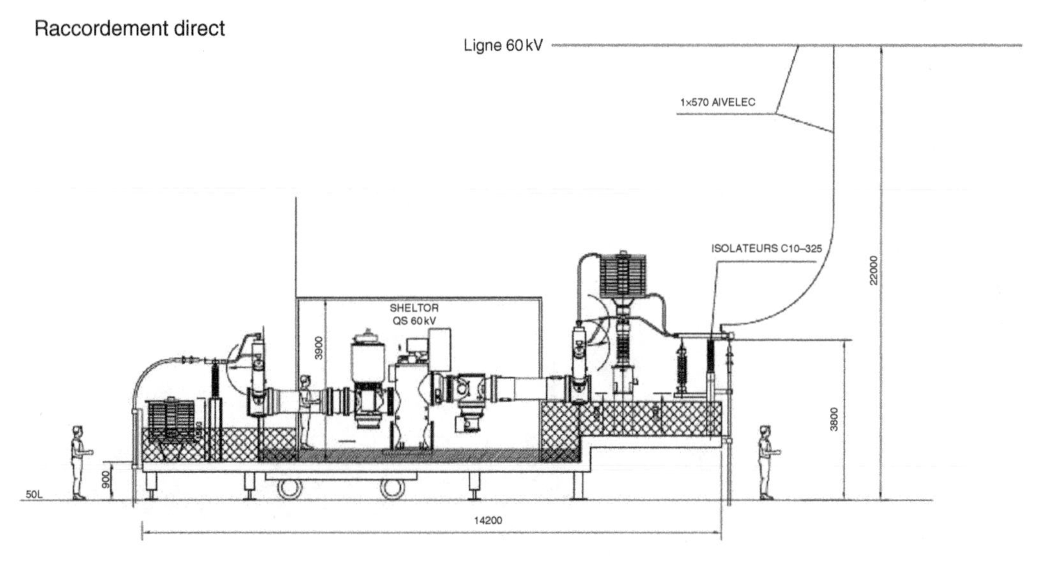

Figure 8.138 Mobile high-voltage GIS mounted on a truck trailer – T-connection to the substation air insulated bus bar (Reproduced by permission of Siemens AG)

8.5.4 Mobile Gas-Insulated Substation

8.5.4.1 General

Electric utilities have been using "Mobile" equipment for many years providing reliable electric service to their customers. Mobile substations offer the means of providing service incorporating power transformers, circuit breakers, and associated devices in the event permanent equipment

Figure 8.139 Mobile medium-voltage GIS mounted on a truck trailer – cable connection to the load side (Reproduced by permission of Siemens AG)

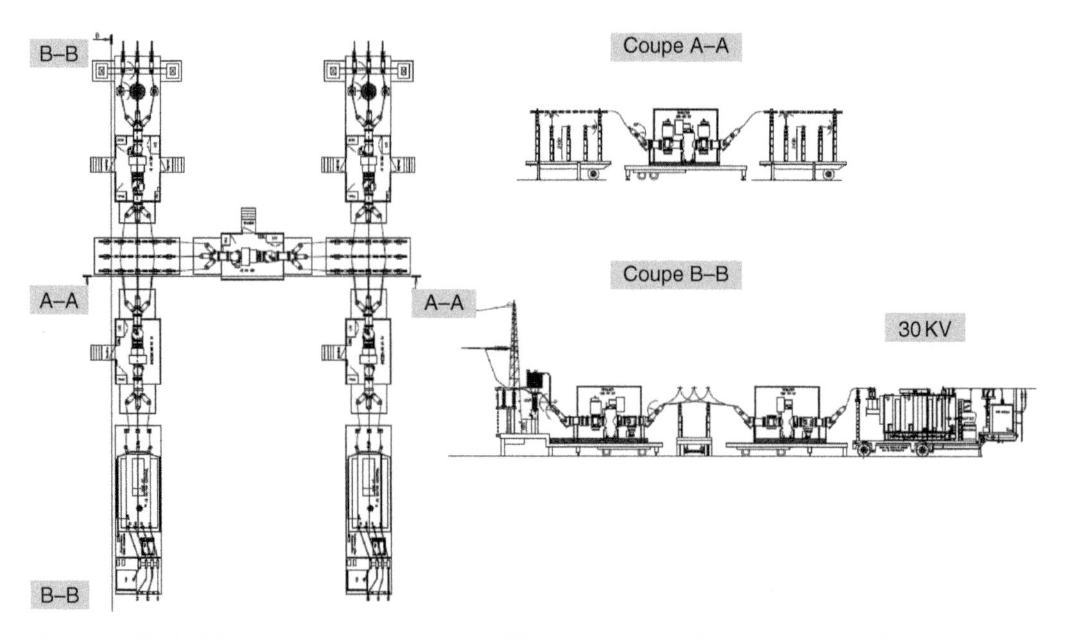

Figure 8.140 Mobile high- and medium-voltage GIS substation – multibay configuration: four lines and one bus bar (Reproduced by permission of Siemens AG)

Figure 8.141 Mobile 72.5 kV voltage GIS – 15 bays double bus bar configuration with control, protection, and auxiliaries in one container (Reproduced by permission of Siemens AG)

becomes unavailable. Currently, the majority of these mobiles are used for distribution circuits since transmission circuits require such larger and heavier equipment.

In this section, our objective is exploring the development and use of a mobile gas-insulated substation. Since this equipment is being used on higher transmission voltages, we will not include power transformers only the switching and protection of higher voltages.

8.5.4.2 Applications

Service Restoration

One of the core objectives for electric utilities is providing reliable electric service. This expectation is more than ever before. The dependence on automated systems for industry, business, and home is crucial today. The availability of a mobile gas-insulated substation provides the utility options in returning the grid to normal in a fraction of the time. These options include equipment replacement due to equipment failures and natural disasters such as flooding and high winds.

Project Support

In the recent years, electric utilities operating in the United States have begun projects to improve and upgrade the grids they operate. To accomplish this scheduling and coordinating of equipment, outages substation and transmission line facilities have become more difficult.

Utility users of mobile gas-insulated substation find options in meeting project schedules. The use of these mobiles can provide temporary service so outage lengths can support project construction. As assumed while utilities are constructing their customers still expect reliable electric service.

Short-Term Installation

Electric utilities have experienced an increase in the number of the projects that fall outside the normal timeline for construction. Previously, three- to five-year project cycle time would fall short for some of today's needs. As third-party developers increase the footprint of renewable generation interconnects demands are placed on the utility to support these efforts. On occasion, developer's schedules don't align with the utilities. Mobile gas-insulated substations offer the utility options that may include providing electric service in a shorter period of time by installing equipment for the short-term while permanent equipment is completely installed.

8.5.4.3 Considerations

Physical Size

Stored Since the majority of the time, the mobile GIS will be in storage adequate thought needs to take place. One immediate question is: Will the mobile GIS be stored under roof or in the open? It is recommended to store the mobile GIS under roof. Limited exposure to the sun and falling weather will minimize maintenance activities and increase service years. Clearly, the dimensions of the mobile will dictate where and how long-term storage is accomplished.

Installed A critical factor for future uses of the mobile gas-insulated substation is the ability to install within substation. A suggestion could be an assessment of your company's substations and determine the maximum size for the mobile. By doing this, you could identify the number of substations the mobile could be used.

Weight

Stored Total mobile weight must be confirmed for identifying a storage location. Example a concrete pad needs to be designed to handle and distribute the trailer and pad access.

Installed In order to transport the mobile across public highway and roads weights must be provided to local highway departments in order for hauling permits. These permits are required in local areas to determine roads and routes suitable for the mobile GIS. Also, this weight is crucial in designing the utilities substation driveways and movement throughout the facility.

Equipment Layout

Adequate thought needs to be carried out on the location of equipment and devices on the trailer. Is should be accepted many associated items will accompany the GIS equipment. These may include storage batteries, gas handling equipment, gas storage bottles, relay control panels, and communication panels. The final locations of these panels such not limit or jeopardize the operation of any equipment.

Equipment Connections

The purpose of this section is addressing how the mobile GIS is going to be connected to the electric network or grid. The available choices are transmission cables or air entrance bushings. Both options have specific requirements. If transmission cables are used then special connections are required in the GIS to accommodate this connection. As for using the air entrance bushing option measures need to be taken to limit mechanical stress on the bushing when connecting to the network.

For each connection, option positive and negative factors need to be considered and addressed. Examples are, but limited to include for cables, the amount of area for bending of the cable.

Compact installations may not be achieved because of cable routing. Regarding air entrance bushing longer set up time would be required since a bushing may need to be installed.

Personal Access

One negative factor with mobile substations is the close arrangement of equipment and the limited access for troubleshooting and maintenance. Since likely the mobile GIS will be mounted in an enclosed trailer additional measures need to take place. One of the key measures is the entry and exit of persons performing activates on the equipment. Thought must include offering the emergency exit of the trailer in the event of fire or loss of breathing air.

8.5.4.4 Mobile Gas-Insulated Substation Applications

Transformer

This application is the protection of the power transformer being provided by the mobile GIS. The mobile GIS could be used for either the high-voltage or low-voltage winding of the power transformer. Clearly, the application depends on the nameplate rating of both the mobile GIS and the power transformer.

Line Protection

This application is the protection of the transmission line provided by the mobile GIS. Clearly the application depends on the rating of the transmission line. Considerations need to be made for surge protection and short line faults. Both of these factors would require on the mobile GIS being able to handle these requirements.

Switch Bypass

This application would use the mobile GIS to bypass a transmission air-break switch. On occasion, these air-break switches can't be removed from service for maintenance or other activities. In order to continue and not disrupt service, the mobile GIS would become the switching device. Clearly, this is an aggressive approach for which this solution is better than taking a line outage.

8.5.4.5 Specifications

Several technical specifications will be involved with the creation of a mobile GIS, GIS Specification, and vehicle – trailer specification. As with a typical GIS specification, several items are included after all it's a total substation hence the name, gas-insulated substation.

GIS

The gas-insulated substation specification should include the following items and equipment.

Power circuit breaker
Instrument transformers – current transformers and voltage transformers
Disconnect switches
Fast acting switches
Ground switches
Surge arrestors
Storage batteries
Local control cabinet
Relays and relay panels
SF_6 gas handling equipment
SF_6 storage

Trailer

The gas-insulated substation specification should include a specification for the trailer that includes these items.

Trailer dimensions
Trailer weight ratings
Trailer weight capacity
Number of doors and locations
Entry locations for air entrance bushings or cables
Entry location for electrical service
HV AC requirements
Electrical requirements
Lighting requirements
Fire detection system
Oxygen monitoring system

8.5.4.6 Installation Plans

The true value of owning a mobile gas-insulated substation is having many opportunities for use. One method to consider to maximize the use is identify the locations and applications in advance where the mobile can benefit operations and construction. Consider looking at substation locations in advance creating a plan for specific equipment in that substation. Identify the location for setting up the mobile and identify a materials list for energizing. This plan could even include the switching order for placing the mobile in service. As you can seem if efforts such as these are completed the user will confirm the "where" and "how" a mobile GIS is utilized.

Another idea for consideration is the creation of "set up" procedures for energizing the mobile GIS. Doing this in advance will identify a step-by-step actions plan that identifies each task but also coordinates all groups and departments performing parallel activities in a small area. This effort not only helps accomplish activities safely but efficiently.

One last idea for consideration is performing a "mock drill" setting up the mobile GIS. Clearly this is a costly suggestion, however, as a former user we saw benefit. The drill would create an opportunity for all groups and departments to carry out the procedures and confirm all previous planning without being driven by the demands of a system emergency.

8.5.4.7 Conclusion

The objective for these writings was to generate thought, suggestions, and questions for considering the purchase of a mobile gas-insulated substation. Mobile substations in general are part of service contingencies and key components for electric service restoration and project execution while maintaining service through construction.

8.6 Mixed Technology Switchgear (MTS)

8.6.1 Introduction

Gas-insulated substations (GIS) have long been known for their compact design and reliability. This is why manufacturers and users have been looking for options to utilize portions of GIS equipment to provide compact, flexible, and reliable solutions in existing air insulated substations (AIS). Mixed technology switchgear (MTS) was born of the idea of fitting this type of compact solution for optimizing existing air-insulated areas that were not able to handle expansion using fully air-insulated equipment. MTS is also known as a hybrid system.

This section of the GIS/GIL handbook discusses the MTS and its applications.

8.6.2 Definition of MTS

Mixed technology switchgear is a compact switchgear assembly consisting of at least one switching device directly connected to or sharing components with one or more other devices so that there is an interaction between the functions of the individual devices. Such assemblies are made up of individual devices that are designed, tested, and supplied for use as a single unit.

The interaction between devices may be due to proximity, sharing of components, or a combination of both. The MTS assembly normally contains components of air and gas insulated substations and may be delivered entirely prefabricated or partially assembled.

Figure 8.142 shows the interaction between and evolution of the technologies. Figure 8.105 shows the space requirement comparison between the technologies. The MTS space requirements can be as little as 30% if the air-insulated equipment is used. Figure 8.143 shows three technical solutions, one for air-insulated substations (AIS), mixed technology substations (MTS), and gas-insulated substations (GIS). The substantial difference in space requirement is obvious and relates to the portion of how much of the substation is SF_6 insulated. In an AIS only, the circuit breakers are SF_6 insulated, while in an MTS the circuit breakers, disconnectors, and ground switches are SF_6 insulated, but not the bus bar. The most compact design is offered by the GIS where all switching devices and the bus bar are SF_6 insulated.

8.6.3 MTS Design Features and Applications

The mixed technology design features mirror those of GIS, while placed in modular units. Those features include compact design, high reliability, integrated functions, modular systems, preassembled and tested transportation units, reduced construction time, and easy exchange of complete modules. It also offers optimum life cycle costs (investment, operation, and maintenance) to users.

The most common MTS application is an assembly installed in an existing AIS. Lack of space in this type of station could result in the need for different types of MTS assembly. MTS is also suitable for applications with high operating frequency (capacitor and reactor). In the next section, examples of different types of MTS applications are discussed.

The most common MTS assembly combines a circuit breaker, circuit breaker disconnects and grounding switches, current transformers, and control unit in a single assembly. Figures 8.144 and 8.145 show examples of this type of MTS assembly.

There has been a more recent application of MTS technology. It is a combination of two circuit breakers and their disconnect switches and current transformers, called the double breaker. The double breaker device has applications in double bus, double breaker, or ring bus arrangements. In Figure 8.146, a double circuit breaker installation is shown in an MTS design for 145 kV rated voltage. The switchgear is mounted on steel structures with a common base frame. The connection to the overhead line is made by an SF_6 gas-to-air bushing on top of the switchgear.

8.6.4 MTS Application Examples

The following six examples show MTS applications from around the world. In Figure 8.147, a cable connected double bay outdoor installation is shown for a 145 kV, 31.5 kA short-circuit current, and 4000 A rated current. Figure 8.148 shows an application of an MTS located on the roof of the

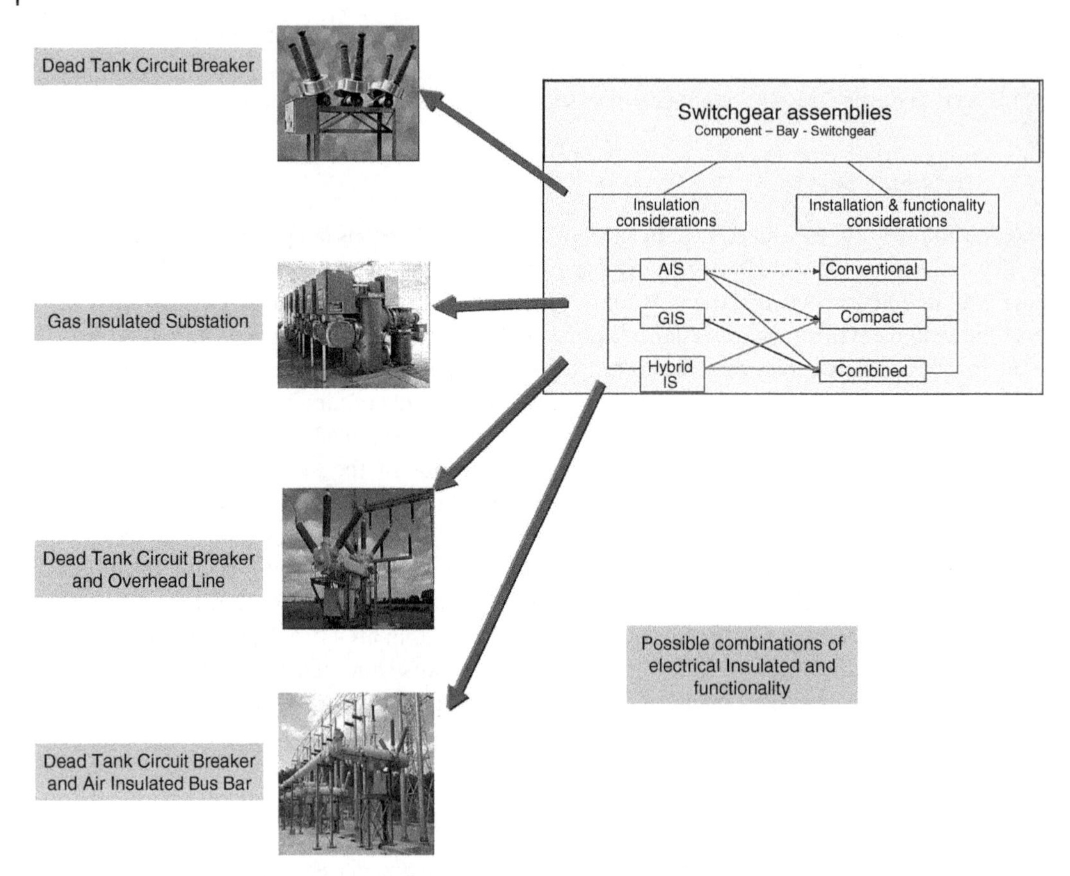

Figure 8.142 Overview of a typical design of mixed technology switchgear (MTS) (Source pubic internet: siemens.com/gis)

substation building. Due to the relative low weight of the MTS equipment such solutions are feasible. In Figure 8.149 a single-phase insulated outdoor bay of 230 kV, 50 kA short-circuit current and 3150 A rated current is shown. Figure 8.150 shows a three-phase insulated outdoor bay to connect two bus bars of 132 kV, 31.5 kA short-circuit current and 3150 A rated current. In Figure 8.151, a three-phase insulated outdoor bay connects a wind farm of 145 kV, 31.5 kA short-circuit current and 4000 A rated current. Figure 8.152 shows a single-phase insulated outdoor bay to sectionalize an air-insulated bus bar in a substation of 132 kV, 31.5 kA short-circuit current and 4000 A rated current.

8.6.5 Conclusion

Mixed technology uses the design features of GIS to create modular compact substation solutions for optimizing existing air-insulated areas that were not able to handle expansion using fully air insulated equipment. The MTS assembly normally contains components of air- and gas-insulated substations and may be delivered entirely prefabricated or partially assembled.

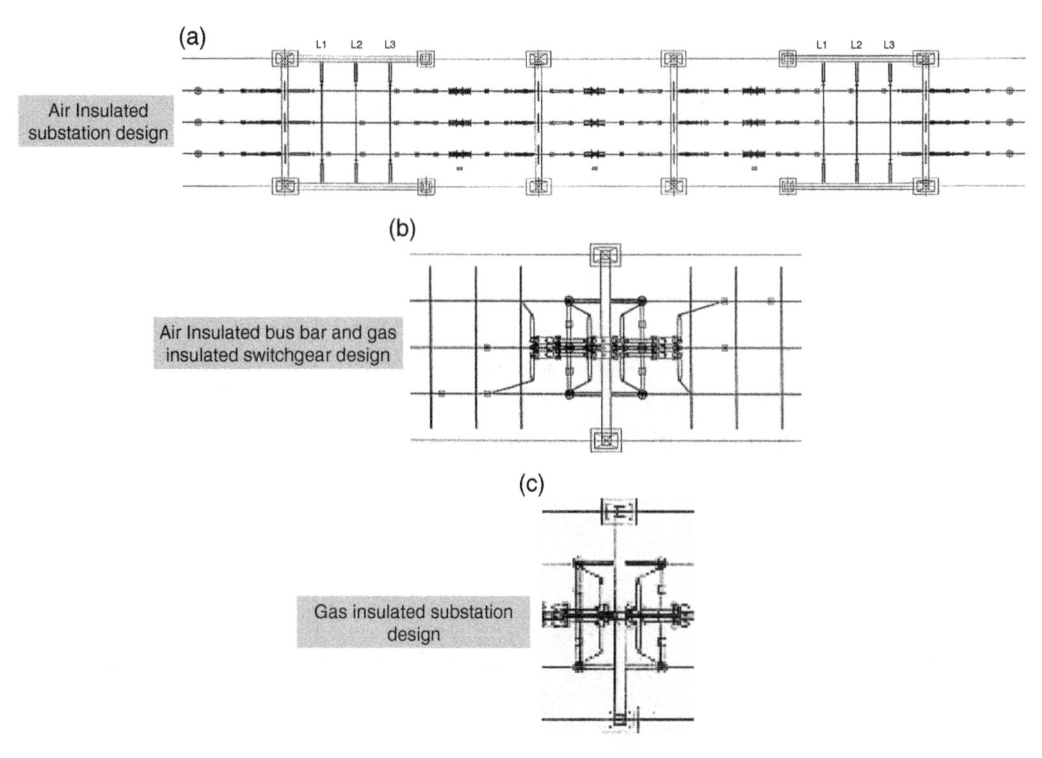

Figure 8.143 Space requirement comparison of AIS, GIS, and MTS circuit breaker bays of 420 kV level. (a) AIS 200 m long. (b) MTS with air insulated bus 60 m long. (c) GIS 20 m long

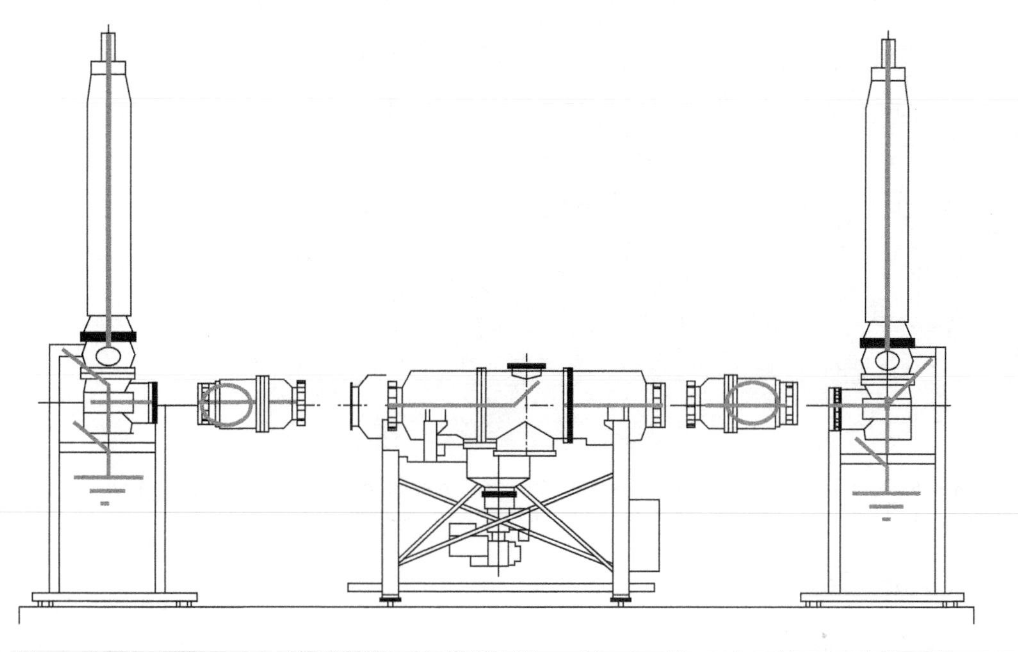

Figure 8.144 MTS module 420/550 kV with CB/SW/CT combination (Reproduced by permission of Siemens AG)

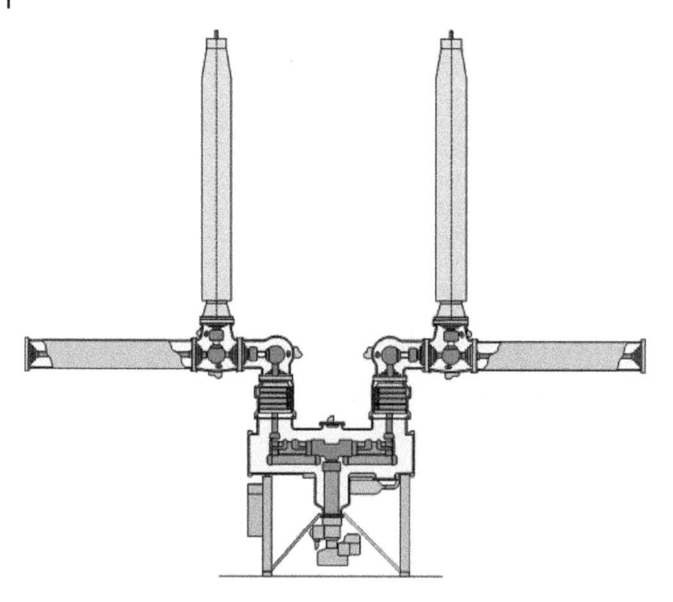

Figure 8.145 MTS 550 kV bay module with CB/SW/CT combination and vertical bushings (Reproduced by permission of Siemens AG)

Figure 8.146 MTS 145 kV double circuit breaker with circuit breaker, disconnector, and ground switch combination and vertical bushings (Reproduced by permission of Siemens AG)

8.7 Future Developments

8.7.1 Reduction of Size

The size of the GIS is directly related to the cost. The bigger the size, the more material, more space, larger buildings, and more shipments are required. The development evaluation of GIS has already reduced the size of GIS by 70–80% over the last four decades, as shown in Figure 8.153.

Figure 8.147 Compact substation – double outdoor cable bay of 145 kV, 31.5 kA short-circuit current and 4000 A rated current (Reproduced by permission of Siemens AG)

Figure 8.148 Compact substation – double bay on the roof of a substation building of 145 kV, 31.5 kA short-circuit current and 4000 A rated current (Reproduced by permission of Siemens AG)

Figure 8.149 Compact substation – single-phase insulated bay of 230 kV, 50 kA, 3150 A (Reproduced by permission of Siemens AG)

Figure 8.150 Compact substation – three-phase insulated outdoor bay of 132 kV, 31.5 kA short-circuit current and 3150 A rated current (Reproduced by permission of Siemens AG)

Figure 8.151 Three-phase insulated outdoor bay to connect a wind farm of 145 kV, 31.5 kA short-circuit current and 4000 A rated current (Reproduced by permission of Siemens AG)

Figure 8.152 Single-phase insulated outdoor bay used to sectionalize an air-insulated bus bar in a substation of 132 kV, 31.5 kA short-circuit current and 4000 A rated current (Reproduced by permission of Siemens AG)

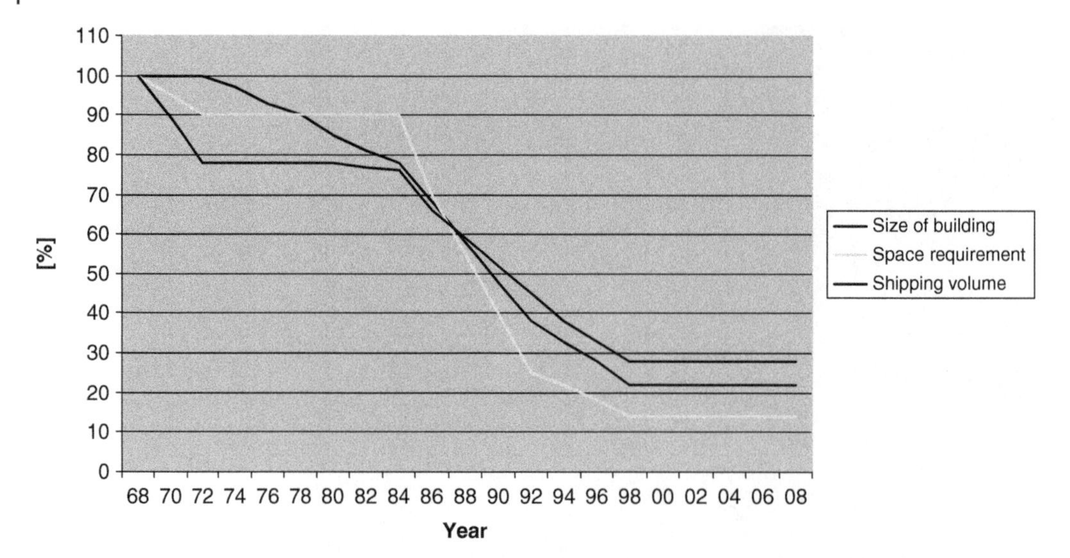

Figure 8.153 Progress in size reduction, example of the 145 kV GIS (Reproduced by permission of Siemens AG)

Figure 8.153 shows the progress in reduction of the building size, space requirement, and shipping volume since 1968 when the first GIS was built to present. The reduction was directly related to the increased knowledge of high-voltage gas-insulated technology. The size decrease between 1980 and 2000 has reached a compactness of GIS that will probably not be repeated again. The limits in compactness are given by needs for accessibility and the possibility to repair or exchange parts of the GIS. Only at higher voltage levels such as 400 kV, 500 kV, and higher might some reduction in size be possible. At the lower voltage levels, the design has reached the limits of compactness due to accessibility restrictions.

8.7.2 Simpler Design

Technical developments in manufacturing processes and new materials for metal parts as well as for insulating parts will further drive GIS design to simpler technical solutions. This will reduce material and manufacturing costs and because of simplicity, it will increase the reliability further. When a design is simpler, it usually extends the lifetime of the equipment and will reduce maintenance.

Simpler design in general is using new materials or manufacturing processes to reduce manufacturing costs without reducing functionality and reliability. The number of different parts to be assembled in a 145 kV GIS bay have been reduced from around 20 000 parts in the first-generation GIS in the 1970s to the present range of 4000 parts. At the same time, the performance of switching capabilities and the reliability went up from about 1000 A rated current to about 3000 A. The number of type-tested mechanical switching operations went up from 2000 to 20 000. All of this relates to simpler design.

8.7.3 Life Cycle Cost Evaluations

The traditional way of substation planning was based on using the same single-line diagrams for air-insulated and gas-insulated substations. This does not reflect the opportunities given by the modular structure of GIS.

Today and in the future, much more attention is given to evaluate not only the investment cost but also the total life cycle cost of the substation. This also includes the optimization of the single-line diagram taking into account the higher reliability of the equipment. Breaker and half-schemes can be replaced by ring bus or double bus bar arrangements to take care of the higher reliability of the circuit breaker and to reduce the installation cost.

The life cycle cost does not only cover the initial investment cost but would also cover cost impact coming from operation, outage times, reliability, maintenance, repair cost, and time, electrical transmission losses for lifetime, dismantling cost, and the whole decommissioning cost including waste disposal or reuse of materials. This sophisticated life cycle assessment evaluation will have a strong impact on substation solutions and will bring advantages to GIS in general because of the high reliability and availability, as found in CIGRE Study. To reach this optimization, close cooperation between the manufacturer and user is needed.

8.7.4 Functional Specification

Functional specifications offer a cost reduction potential on the total substation cost. The specific GIS solutions (e.g., double bus bar, ring bus) to fulfill the requirements of the functional specification will have a standardization effect on GIS solutions. Standardized GIS can be produced on a larger market share with higher numbers of standard design to be delivered. The key is that standards for functions need to be created on a substation level not limited to today's existing company-specific solutions.

System-related standardization will bring a large-step cost reduction when the user can define standard functional values for complete substations. Manufacturers then can offer standardized products with fixed ratings. The standardization will lead to less variation and the most cost reductions will be made with simplified project engineering, operation, and maintenance.

8.7.5 Intelligent GIS

What is an intelligent GIS? This is a good question.

It only can be answered in the context of future use and tasks of a GIS in a regenerative transmission network with fluctuating energy generation and quickly changing directions of energy transmission. Power generation of a wind farm can change within half an hour from 100% power generation to 0%, solar power generation can change from 100% power generation to 20% in 1 min by a single large cloud. The requirement for a GIS from such an operation scenario is that besides the switching operations much more sensitive measurements of current, voltage frequency, and phase angle are necessary and need to be integrated into the GIS. This cannot be solved by classic voltage transformers and current transformers on a magnetic basis with an iron core and a 100 V or 5 A measurement signal. This equipment is too expensive for the need of an intelligent GIS.

Integrated nonconventional sensors for voltage current, gas pressure, temperature, and other monitoring functions will be connected to information technology (IT) systems and provide more information to the network operator. This allows new protection and control functions in relation to actual network conditions. The required devices are already available today as optical sensors for current and voltage measurements, capacitive dividers for voltage measurement, or a low power Rogowski coil, and are explained in the following sections.

8.7.6 Integrated Electronic Devices

Intelligent electronic devices (IEDs) will bring intelligence of primary and secondary equipment directly to the bay level. Digital optical IT connections are made directly to the circuit breaker, disconnector, and ground switch drives in each bay, as also to the current and voltage instrument transformer of conventional inductive or nonconventional types, also in the bay level.

The IED will introduce the bus system down to the single operational or measuring device in the GIS. Control cabling with copper conductors will be replaced by one optical fiber connection serving for any exchange of information or for sending commands. The basic communication protocols for standardized data exchange are available in the IEC 61850 series of standards.

8.7.7 Rogowski Coil

The Rogowski coil provides highly accurate secondary output without saturation because there is no iron core involved. The air core makes the measurement linear but delivers only a small voltage, for example, 1 V instead of 100 V. It enables a compact equipment arrangement and can be integrated into the GIS with little space requirement because there is no iron core. The Rogowski coil has high immunity to noises/surge voltage because the coupling factor is much lower than for a coil with an iron core. Typical ratings are shown in Table 8.18.

A design example of the Rogowski coil is shown in Figure 8.154, where the coil is placed inside the GIS at the grounded enclosure.

The primary conductor in the GIS enclosure acts as the primary winding for the Rogowski coil. The secondary winding of the coil has an air core and is connected to an analog/digital converter

Table 8.18 Typical ratings for the Rogowski coil

Rated primary current	4000–2000 A
Accuracy	0.2%
Rogowski output	1 V

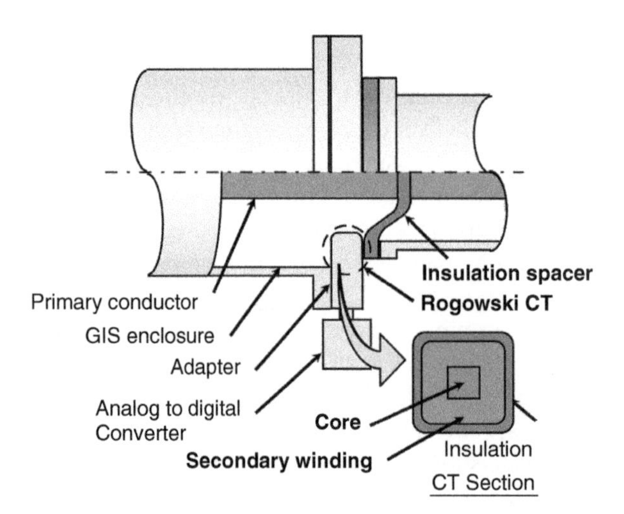

Figure 8.154 Construction of a Rogowski coil (Reproduced by permission of Alstom)

as IED and then is connected by optical bus to the protection and control system. The primary conductor has a high-voltage potential and carries the current. It acts as the primary winding of the Rogowski coil.

The secondary windings are at ground potential at the GIS enclosure to transform the current in the primary conductor into an induced voltage of the secondary windings of the Rogowski coil. An electronic card then converts the measured value into a digital signal to be connected to the protection, control, and measurement system (see Figure 8.155).

In Figure 8.156, the application of a Rogowski coil in a GIS is shown. The size of the Rogowski coil is much smaller than the conventional current transformer. Only one air coil covers all measurement requirements for protection, control, and current measurement down to 0.2% accuracy. This is possible because of the linearity of the coil and is realized by different settings of the converter. Conventional current transformers need different iron cores.

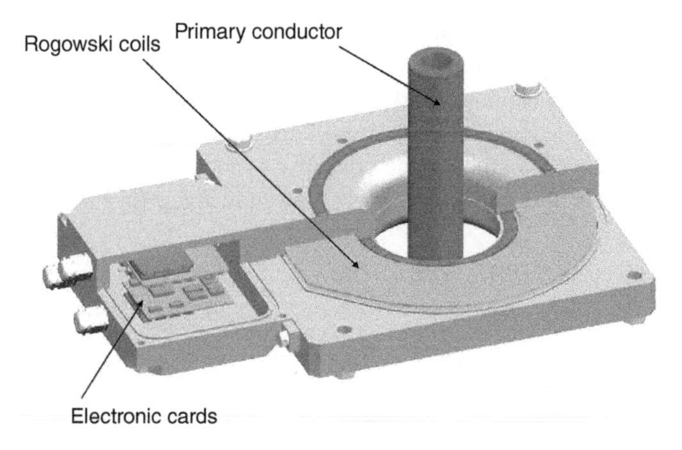

Figure 8.155 Principle of the Rogowski coil (Reproduced by permission of Alstom)

Figure 8.156 Application of a Rogowski coil (Reproduced by permission of Alstom)

8.7.8 Capacitive Divider

Capacitive dividers provide highly accurate secondary output signals. This technology has been available for many years. Developed in the 1980s with electronic amplifiers, their size and cost are strongly reduced using the integrated circuits of today. They enable compact equipment arrangement for GIS applications and can be integrated with an intermediate electrode as the electric field sensor on the inside of the GIS enclosure.

The secondary voltage is low and needs to be amplified and temperature compensated by the analog/digital converter (IED). In Table 8.19, typical values are given for a 550 kV GIS. The capacitive dividers show high immunity to noise/surge voltages.

In Figure 8.157, the primary conductor inside the GIS enclosure forms a capacitive field. The intermediate electrode is the sensor, which is related to a field potential at its location of design. The capacitive field distribution is known in the cylindrical setup of a tubular conductor inside a tubular enclosure. With this knowledge, the actual voltage of the conductor can be identified and measured. The analog/digital converter transforms the measured value according to IEC 61850 into a digital protocol to be sent by the integrated electronic device (IED) to the bay controller. The capacitive sensor is very small and can be located at almost any location inside the GIS.

In Figure 8.158, the principle of a capacitive divider is shown in a 3D graphic. The primary electrode is the conductor. The secondary electrode is the isolated secondary electrode, which measures the voltage.

The conductor of the GIS acts as the primary electrode of the capacitive divider. The secondary electrode is a metallic device, for example, an aluminum foil around the conductor, placed inside

Table 8.19 Typical ratings for the 550 kV GIS

Rated primary voltage	550/3 kV
Accuracy	0.2%
Secondary output	1 V

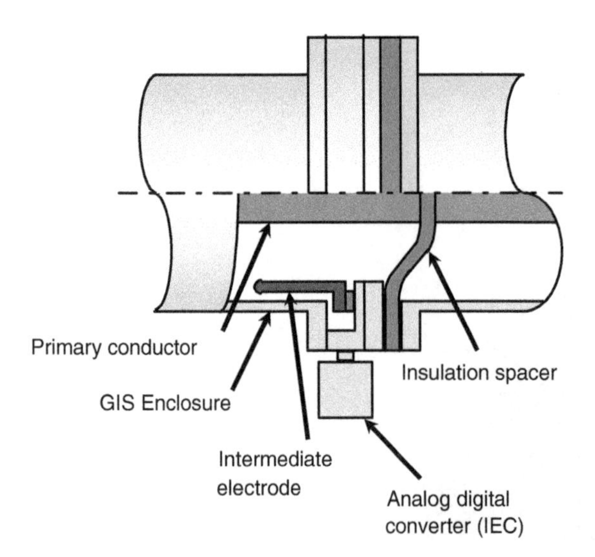

Figure 8.157 Construction of a capacitive divider (Reproduced by permission of Alstom)

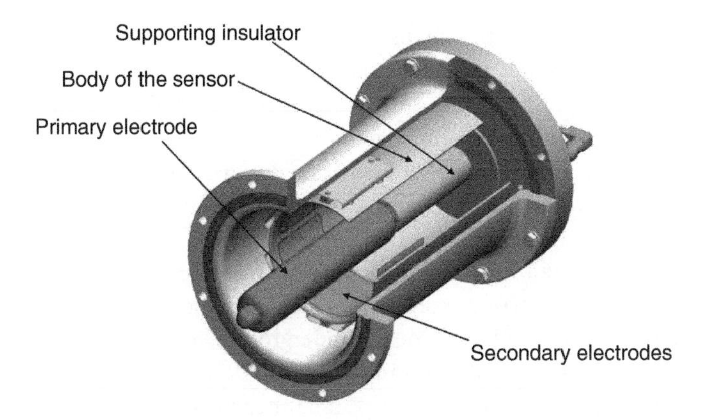

Figure 8.158 Principle of a capacitive divider (Reproduced by permission of Alstom)

the GIS enclosure but is electrically insulated toward the GIS enclosure, which is grounded. The secondary electrode is then connected to the amplifier and analog/digital converter, which is then connected to the protection and control system. The amplifier, analog/digital converter, on-board computer processor, and the optical connector form an IED.

8.8 Underground Substations

Author: Hermann Koch
Reviewer: Chiranjeevi B., Kotharu, Mark Kuschel

8.8.1 Introduction

According to a report from the UN published in 2018 [19], two out of every three people are likely to be living in cities or other urban centers. This growing density of population combined with the scarcity of usable real-estate in the cities and metropolitans needs creative solutions by the utility operators. One of such solutions is underground substations.

Underground substations as the name suggests are substations built completely below grade, i.e. under the ground. One would find such substations in metropolitans, cities, and densely populated areas with limited availability of space where safety and functionality are of high importance. The gas-insulated high-voltage technology helps meet criteria above by offering compact substation solutions.

Certain design aspects shall be considered when designing a high-voltage substation built to operate fully or partly underground. These include soil and ground water conditions, fire protection, noise, public safety, and maintenance requirements.

In this chapter, we will cover some examples and concept underground substation designs built and operated in different places.

8.8.2 Critical Aspects of Designing Underground Substations

8.8.2.1 Soil and Ground Water

The soil at the project location has the biggest impact on the feasibility of an underground substation. A proper site survey and geo technical investigation with soil and rock samples is necessary to study the feasibility of selected site. A good mixture of compact soil is a good site for such substations while shallow bedrock makes it a challenging site for construction, swamps are a strict no for such substation solutions. In cases where sandy soils exist, the cost of preparing the site for building an underground substation increases dramatically as it requires unique civil works to make the site usable such as deep concrete or steel piles.

Surrounding soil pressure is considered as a parameter for the structural design of underground substation walls. Depending on this, the building walls and ceilings need to be constructed to withstand the forces.

Shallow ground water table is another issue encountered in cities close to the ocean. In few cases, the ground water table was observed to have started as shallow as 1 m below top grade. While water treatment and proper drainage systems are required for all underground substation constructions, shallow ground water table increases the challenges involved with designing such systems. This water pumping and drainage systems are required both at the time of construction and later for the time of operation of the substation with an expected life time of at least 50 years before a need for revising the calculations.

During construction, the building pit is left open, and the accumulated groundwater is pumped out as long as the construction work continuous keeping the pit accessible. This is one way to prepare the construction site, while another way is to work under water to construct the building ground floor and the walls with water proof concrete. This work is carried out by underwater divers and is preferable only if large amount of water needs to be pumped. Figure 8.159, lower photo, gives a visual impression of water pumping system.

The example in Figure 8.159 on the upper part shows the level of groundwater of the underground building at the middle of building height. This means that the building walls need to be of

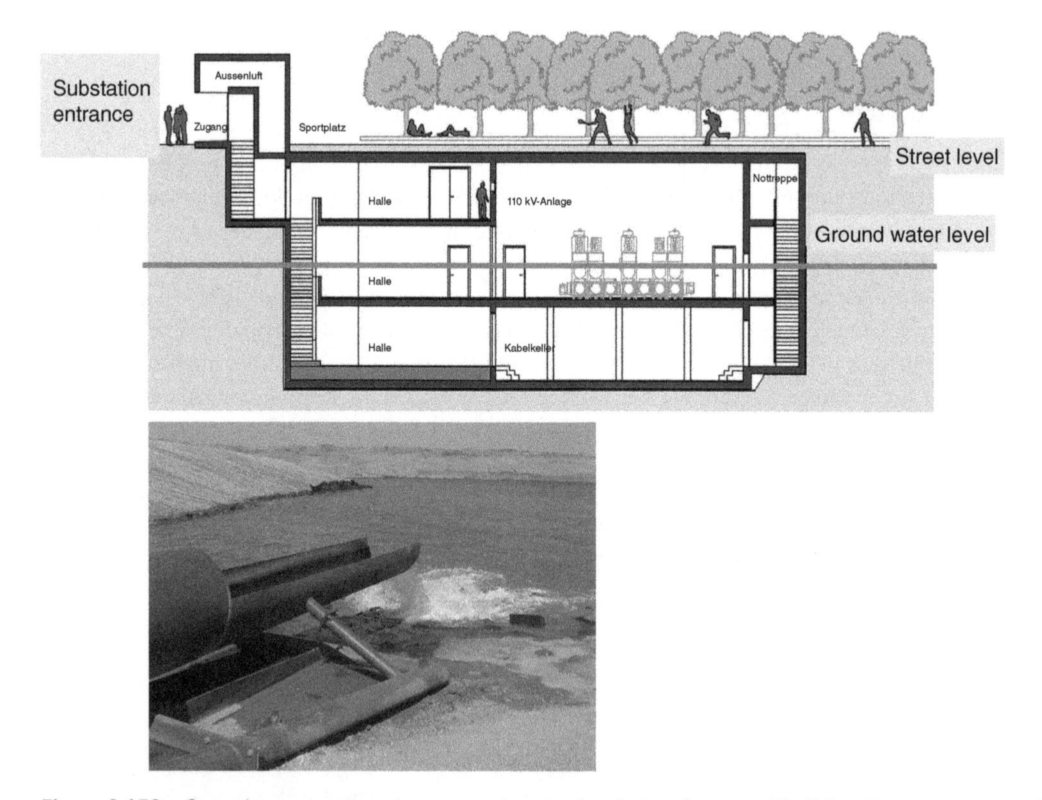

Figure 8.159 Groundwater treatment, upper part: water level at underground building, lower part: ground water pumping system (*Source* [20])

water proof concrete to avoid water seepage into the building. In addition, the interior of the building needs drainage system to collect water if any inside the building and a water pumping system to pump out this collected water above the groundwater level.

Though the costs associated with installing and construction of such systems are considered a onetime investment, the operational costs shall be considered for throughout the lifetime of the substation.

The above discussed drainage systems and water pumps require sensors and monitoring systems for normal, storm, and heavy rain to prevent the underground substation from flooding.

8.8.2.2 Fire Protection

Fire would cause the most server damage to an underground substation. If not extinguished in short time, it leads to loss of the electric supply, permanent damage to equipment and building. Fire if any in an underground substation should be avoided. Necessary fire prevention systems should be installed to limit the damage to the equipment in case of an emergency. Mineral oil, typically used in transformers, is flammable and has a high heating value, which results of high temperature of more than 2000 F, enough to destroy a concrete building structure to accommodate for high temperatures, K-type transformers are selected for such applications to withstand higher operating temperatures. Figure 8.160 gives an impression on fire in substations by oil.

Gas-insulated switchgear with SF6 gas or vacuum used as arc quenching medium do not necessarily be the feed the fire in emergencies; however, they do create high-temperature arc faults in case of an internal failure of the equipment. Type test series are conducted to prove that the external faults do not have an impact on the gad and vacuum chambers of these equipment. Hence, nonflammable materials such as stainless steel with flame-resistant coating are used as part of the construction, e.g. steel structures, concrete walls, and steel beams for framing.

Preventive fire-protection systems are used to protect the equipment from fire hazards in underground substations. The fire mitigation and prevention measures in an underground substation are classified as active and passive.

Passive firefighting measures include using nonflammable materials for installation inside the underground substation to reduce fire load in the building, special ventilation systems to reduce

Figure 8.160 High-temperature fire of burning oil of transformer (*Source* [21])

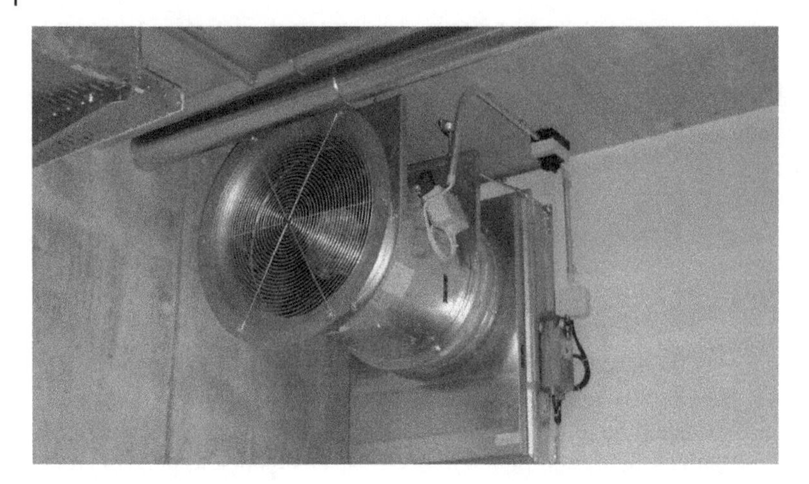

Figure 8.161 Separated ventilation for transformer and building (*Source* [22])

the risk of fire due to heat and gas trapped inside the building. Such ventilation management systems are used to control the oxygen brought into the building and to extract the smoke from building. The smoke extraction system will only be used after the fire is extinguished by the fire extinguishing device. Activation control for the fire extinguishing device and ventilation system are installed at the entrance of the building, for safe operation, see Figure 8.161.

The mineral-oil-filled power transformers are usually the biggest risk for fire in substations. One passive firefighting method used in such cases is to replace the mineral oil by ester oil, which has a much lower fire load.[1] Ester and mineral oil transformers are similar in design and their power transmission capabilities. A further step is to use SF_6 gas-insulated transformers, which is a non-flammable insulating gas. In this case, the transformer foot print/volume is larger than a typical transformer of same MVA of an oil-insulated transformer. In addition to that, a separate large facility for cleaning and cooling of the SF_6 gas is required.

Active fire-protection is installed to put out a fire in case of an emergency. Automatic Fire extinguishing devices such as sprinkler systems using water, CO_2, halon, nitrogen, inergen, or a combination of water-based solutions for foam production are installed in active fire protection systems. The goal of these systems is to prevent oxygen from the burning material, e.g. the oil of the transformer. Therefore, the gaseous, fluid, or foam fire extinguisher are released into the room by a sprinkler system.

Sprinkler systems are installed on the ceilings of the underground substations with heat-resistant steel pipes to pump the fire extinguishing material within a short period of time to keep limit the overall flames and temperature inside the building.

When using fluid fire extinguishing system, a drainage system with fire barriers are required to prevent spread of fire to other areas of the building. For environmental reasons, the fluid fire extinguishing material shouldn't come in contact with the groundwater outside the building. Therefore, a special retention tank with a capacity to hold the volume of transformer oil and other fire extinguishing fluids is required.

Figure 8.162 shows a sprinkler system on the ceiling of an underground substation in the left photo and a covered and sealed drainage system in the ground is shown in the right photo.

1 Fire load is defined as a way of establishing the potential severity of a future fire. It is the heat output per unit floor area, often in kJ/m^2, calculated from the calorific value of the materials present.

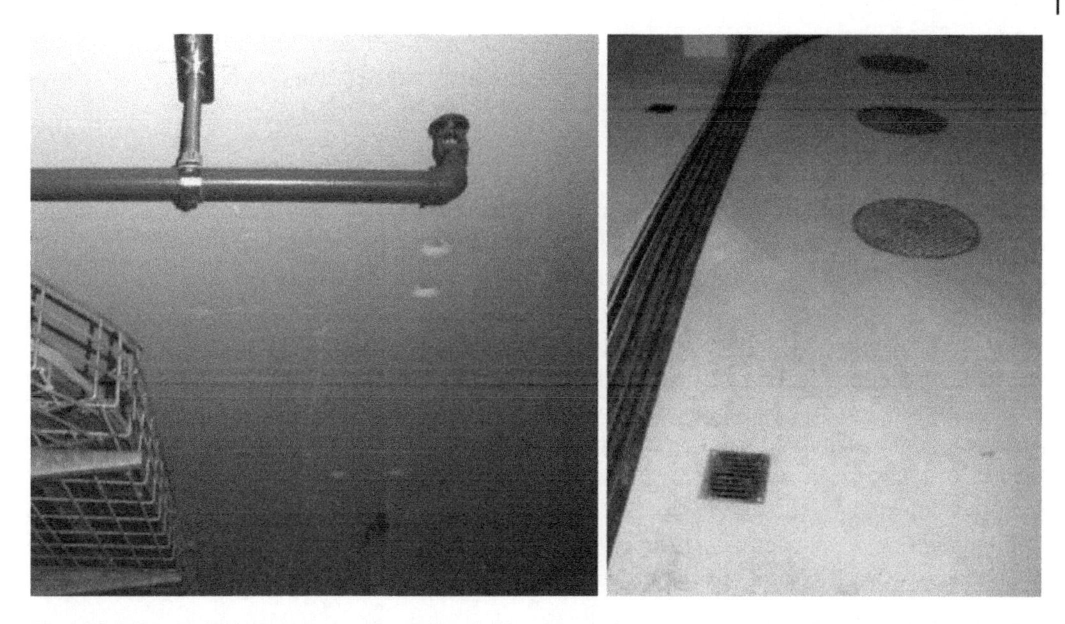

Figure 8.162 Sprinkler system on the ceiling (left), water drainage system on the flour (right) (*Source* left [23] and right [24])

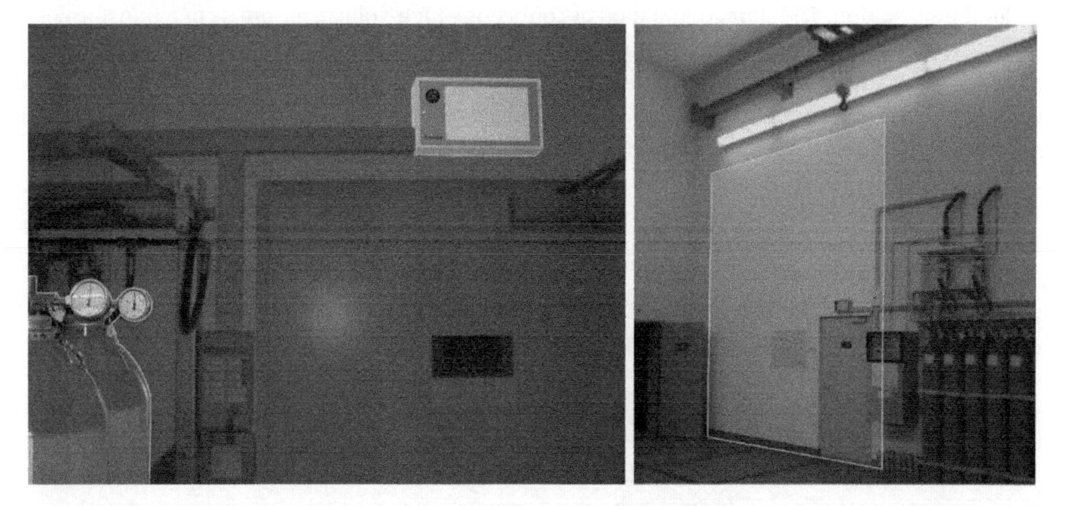

Figure 8.163 On the left photo, the fire protection control system is shown, and on the right photo, fire-resistant doors to the transformer room are shown (*Source* left [25, 26] and right [26])

The design and construction of sprinkler and drainage system is depending from fire load, volume of oil, and the room volume for available oxygen [5, 6].

Figure 8.163 shows the storage cylinders for fire suppression gases (nitrogen or CO_2) is shown in gray bottles on the right. The stored volume of these gas cylinders must sufficiently fill the room to replace the ambient air completely. The firefighting gas needs to stay in the room until the fire is completely extinguished. To help the fire extinguishing, the high-voltage protection system is designed to trip off the electric power supply within 300 ms. After the fire is extinguished, the ventilation is used to replace the firefighting gas by ambient air.

The human safety and environmental (HSE) regulations require emergency exits to reach the outside of the building on a safe way. Doors from the switchgear room and the transformer room need to be fire-resistant for 120 minutes, following international standard IEC 61936-1 [27]. The doors must have a fire-protection control system to automatically close in case of fire such that the active fire protection can work effectively.

8.8.2.3 Dust and Fresh Air Protection

Underground installations need to be hermetically sealed to the outside world. However, fresh air is needed for personnel entering the underground substation for installation and maintenance activities. A ventilation system is installed to pump and circulate fresh air into the building as needed.

In addition to pumping fresh air, this ventilation system is also used for filtering the air from dust and other particles. The dry indoor conditions collect dust on devices over the time, which can cause failures or even total shut down of the entire substation. Cleaning is very challenging when it comes to high-voltage systems; hence, a shutdown of the substation might be required in narrow layouts. For this reason, the fresh air-protection system is used to filter dust and other particles from air in the underground substation.

In Figure 8.164, on the left photo, a fresh air outlet is shown below the transformer, and on the right photo, the fresh air outlet is under the ceiling.

8.8.2.4 Transformer Termination

Connection of the high-voltage terminal of a power transformer is a measure of safety and reliability for the substation. The standard method of connecting high-voltage terminals of a transformer to the switchgear is by using bushings and conductors with terminal pads. This solution requires more space, which is a challenge in underground substations. For this reason, encapsulated terminations are used at the transformer terminals, and insulated HV cables are used. This increases the safety while reducing the overall space required.

While the typical underground high-voltage applications are at 110 kV, 132 kV, or 145 kV and in some cases 330 kV, 400 kV, or 500 kV. The use of encapsulated transformer terminations reduces

Figure 8.164 Fresh air outlet and dust cleaner, left: under the transformer, right: on the ceiling (*Source* [28])

Figure 8.165 Left: transformer-encapsulated termination with XLPE at 110 kV, right: GIL connection transformer to GIS, 800 kV (*Source* left [29] and right [30])

the required electrical clearances need for air-insulated connection by more than the factor of ten (10). This space reduction will compensate for the additional cost of the encapsulated transformer termination by providing savings on size of the building.

Today solid cross-linked polyethylene cables such as XLPE are used for such applications as seen on Figure 8.165, left photo.

For higher current ratings of up to 6000 A and usually at 400 kV, 500 kV, 800 kV, and 1100 kV, gas-insulated transmission lines (GIL) are used for electrical connections. These installations are often used in hydropower plants to connect the HV side of the transformer inside the cavern of the hydropower plant to the overhead transmission line outside the plant, Figure 8.165, right photo, gives an impression of an 800 kV GIL in a tunnel.

8.8.2.5 Ventilation and Humidity

Ventilation of an underground substation is a complex task. As discussed in above sections, ventilation systems in underground substations are used in fire prevention and fresh air circulation. Typical power transformers, switchgear units, and other power equipment generate heat. A proper ventilation system is required to pump out the hot air and replenish it with cooler fresh air to prevent overheating, which could lead to fire accidents. The operating and ambient temperature limits in an enclosed building are stricter than for that equipment installed in open-air setup. A room temperature of 40 °C is too warm for people to work inside, while switchgear and transformer could reach 70 °C in operation. To design such ventilation systems, volume of air that needs to be exchanged keeping in mind the thermal losses of the substation equipment is taken into account.

The heat exerted from the equipment is the losses through energy conversion or transformation. Power transformers typically exert the heat in such substations; and hence, they are designed to be inside a separate room within the station with necessary vents and cooling systems.

With limited availability of space in city centers for installing ventilation outlets and in order to keep the noise level to the neighborhood in acceptable limits, the ventilation shaft is guided through the building to the roof as shown in Figure 8.166.

The volume of air that needs to be pumped out of the underground building is proportional to the size of vent ducts. While it is inversely proportional to the air speed. Lower the volume of ventilated air, the higher the air speed is required for same amount of heat to transported out of the building basement where the substation is located. These are two main optimization criteria while designing air vent systems for underground substations. High volume of air needs large size of ventilation shafts, which is usually limited by space limitations. High speed of ventilated air produces loud air flow noises and is limited by noise limitation in the neighborhood. The result is a compromise of air volume and air speed.

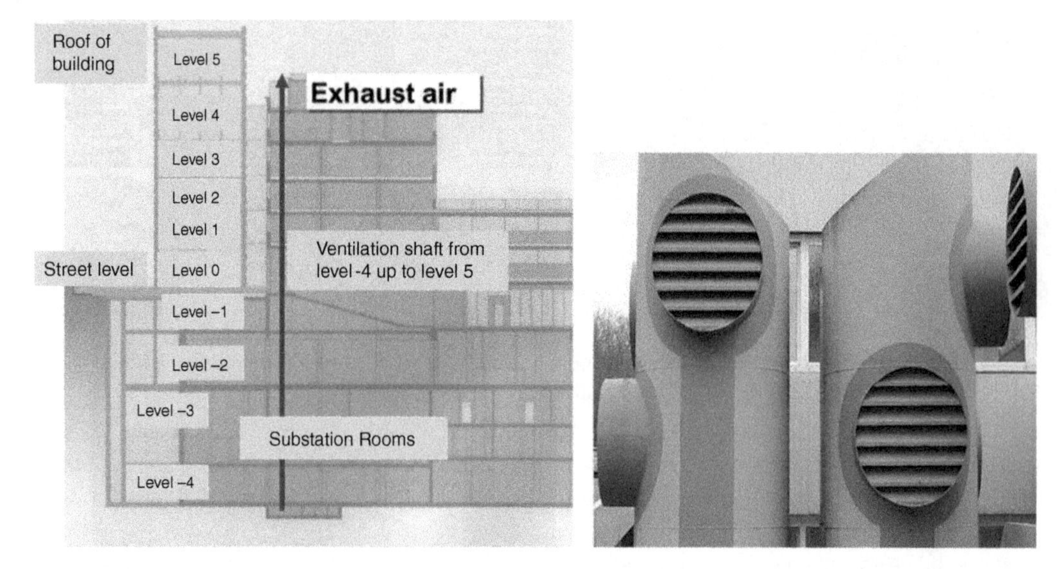

Figure 8.166 Left: substation ventilation shaft from the basement of the building to the roof, right: ventilation outlet on the roof (*Source* [22])

Humidity is another important aspect for the design of the substation ventilation concept. If the ambient air has a high humidity and is ventilated into the underground substation building large amount of water will condensate in the building creating a "rainfall" and lots of water. This needs to be avoided. Such situation may happen in summer with warm and very humid air. In such cases, the ventilation system needs to be interrupted and a moister control system will take the control at that point.

One way to avoid humidity coming in large amounts into the underground substation building is to minimize the volume of air exchange. For this, air conditioners are installed in the underground room to take out heat and reduce air speed for ventilation, helping distribute the humidity better while reducing the air humidity as shown in the Figure 8.167.

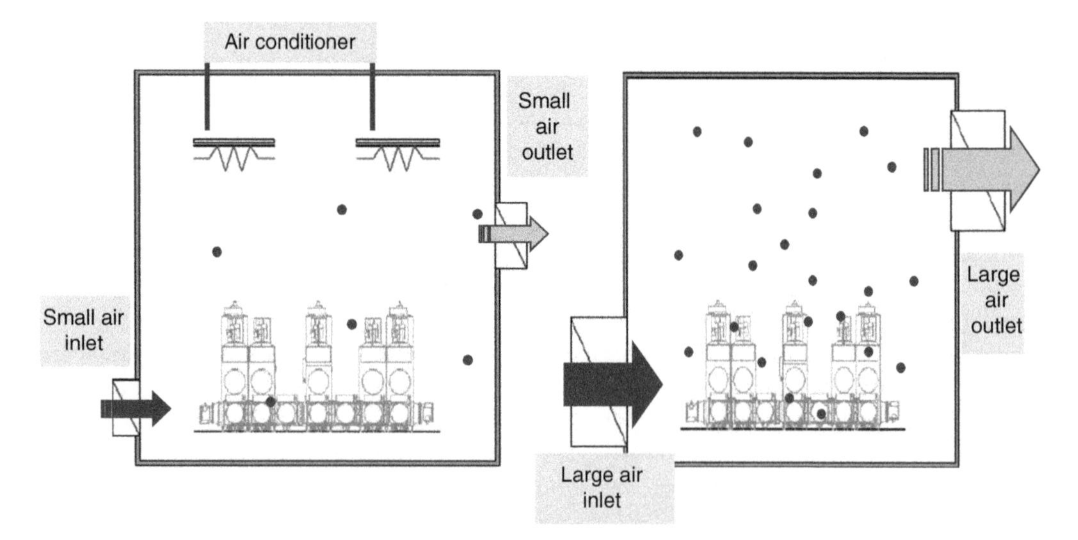

Figure 8.167 Reduced air exchange by using air conditioning, left: switchgear room with air conditioner and small ventilation, right: switchgear room without air conditioner and large ventilation [10]

Figure 8.168 Typical air conditioner system to compensate the heat released by underground substation buildings (*Source* [31])

In Figure 8.168, an air conditioner unit is shown, which is used to reduce the air volume needed from the outside for cooling the underground substation.

As discussed before, power transformers are one of the main sources of heat in an underground substation. The heat generated by the transformer can be used by using a heat exchanger between the transformer oil circulation and the heating system of the building. This reduces the heat energy needed for the building heating and heats the water supply in the winter and in the summer as needed. Excess heat if any can be released from the building. Figure 8.169 shows the heat exchanger assembly to the transformer [32].

8.8.2.6 Noise Emissions

Underground substations are usually built in densely populated areas where noise pollution can be the reason for public appeal and in some cases require costly measures to reduce noise levels. Some common sources of the noises at an underground substation are power transformers producing 100 Hz vibration noise, the noise of ventilation to high air speed or the noise of corona in case of air-insulated high-voltage devices installed, e.g. high-voltage bushings to air for transformer connections.

The high-voltage switchgear does not produce loud noises during operation of its switches; however, operating high dynamic forces, mechanical switch opening in less than 0.1 sec causes high dynamic noises. Operation of circuit breakers or switches is relatively seldom, and therefore, these dynamic noises of switching are not that critical for an underground substation. Modern spring-operated design produces much less dynamic noise than the older designs.

Figure 8.169 Heating exchanger from transformer oil to the building heater (*Source* [32])

It is important to design the underground substation with low noise levels and installing noise damping solutions is required when building in an urban setup. With each noise damping solution, a check of the thermal impact is required, because most of the noise damping measure are also reducing the cooling capabilities.

Commonly used noise reduction methods for underground substations are as follows:

- Choosing a low-noise transformer design. Such transformers use very precise steel sheet assembly methods and high-quality steel materials, which reduces the 100 Hz vibration noise. The cost of such transformers is higher, but cost for noise damping measure could be much more expensive.
- High-quality and low-loss transformers produce less heat, which needs much less ventilation and cooling. The higher cost of the one-time equipment installation is balanced by the savings on need for installing a cooling and ventilation system.
- Installing sound absorbing materials on ceiling and walls of the building, as shown in Figure 8.170, left. This will not impact the cooling of the substation devices.
- Sound absorber material installed in the ventilation shaft are shown in Figure 8.170, right. This solution will not allow high air speed for heat exchange and therefore limit the cooling effect.

Another unique way of tackling the noise pollution from underground substation installations is to produce other noises that general public would expect in a public area. "Noise masking" should be approved by local permitting authorities prior to installations. Acoustic solutions/noise masking methods are to use natural noises, which are more accepted to make technical noises of ventilation inaudible

Figure 8.170 Left: sound absorbing cover on the ceiling, right: sound absorber in the ventilation shaft heater (*Source* [33])

When underground substations are located under a public park or an inner city complex open to public, it is challenging to find a suitable location for the ventilation shaft. The noise from such ventilation systems is counter intuitive for the parks or other public access area built around.

One popular noise masking method is to install waterfalls at the ventilation shafts. The sound of water falls/fountains would mask the noise from the air vents, and the public who visit the area would expect the sound of natural water falls making it a win–win situation. This idea is picked up by city planner and architects to find a location in the city for the need of electric power supply with an underground substation solution and combine this technical requirement with the need of having more appealing natural ambience in the city.

The waterfall covers the entrance of an underground substation ventilation shaft. Behind the waterfall air is absorbed from the outside, ventilated into the substation building, and through an exhaust shaft, the warm air is blown out of the underground substation. This requires a large air volume to reduce the air speed and the noise frequency.

There has been a very positive feedback on waterfall noise-masking technique as it is installed in many cities so far. People gather around these waterfalls for pleasure while noise of the falling water dominates the noises produced by ventilation system and flowing air from underground substations.

In Figure 8.171, two solutions of water falls are shown. The left photo shows a water fall where the water flows down steps and produces typical noises of the rivers. The solution on the right shows a water curtain falling over an edge. In this case, the noise produced is similar to rain.

8.8.2.7 Electromagnetic Field

Underground substations also need to be protected from HV transient over-voltages such as lightning and switching impulses as incoming disturbances and the surroundings of the stations need to be shielded from the electromagnetic field (EMF) radiations from the substation equipment.

The incoming over-voltages are relatively easy to cover by over-voltage surge arresters at the line terminations. These lines are typically connected by cables so that direct lightning strokes to the substation equipment are diverted to ground via surge arresters. See Figure 8.172, left.

The shielding from electromagnetic fields is more complex. For the shielding from electric fields, the substations need a solidly grounded system with low impedance with multiple connections of the substation devices with the grounding system. The substation grounding should be connected

Figure 8.171 Acoustic solutions for underground substation ventilation (*Source* [34])

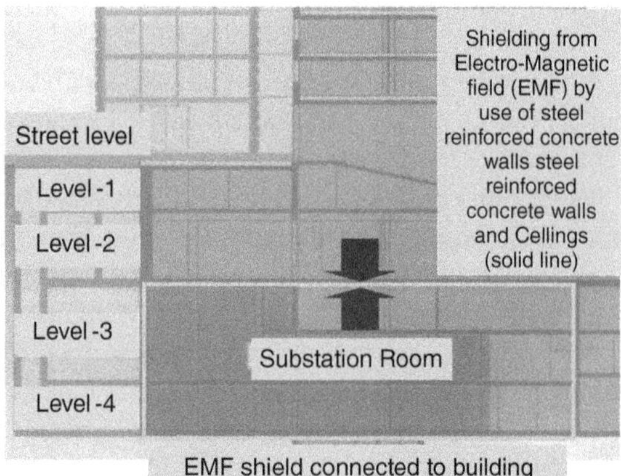

Figure 8.172 Use of surge arresters to protect from lightning strokes (left), EMF shielding concept for underground substation in buildings (*Source* left [35])

to the building grounding system and the ground grid should be designed and installed before constructing the building. The design of a low impedance grounding grid is dependent on factors such as type of soil, length and distance of the ground mesh, and cross-section of the grounding conductor used.

The magnetic field produced by 50/60 Hz power frequency can only be mitigated by distance and right positioning of the equipment grounds. Gas-insulated technologies like GIS and GIL shall be solid grounded to reduce outer magnetic field by 90% inclusive of reverse current of enclosure. The transformer needs a precise design, like for the low-noise transformers, to reduce the outer magnetic field.

For the planning of the EMF requirements at an underground substation, a study is required to with EMF protection zones. Depending on the devices to be installed, their electric discharge voltage and current contribute to the protection zone in which specific maximum electromagnetic field values are given.

In Figure 8.172, the graphic on the right shows an EMF protection zone concept. The substation in the basement of the building is restricted for public access. Only qualified personal is allowed to enter the substation for a limited time.

Outside the marked area, public has access to the rooms of the building or outside the building. People can stay in this area for longer duration or even live there.

Shielding for electric fields and choosing the right location for the magnetic fields are the two most import measure to meet the EMF requirements. It is important to simulate the EMF for the field values before the substation is built. Differences in measurements after the substation is built could be very costly.

8.8.2.8 Control Room

Underground substations are built where there isn't much space available, which means that all elements such as switchgear, transformers, cables, protection, and control equipment shall be designed and installed in close vicinity. This may cause unwanted interferences with the protection and control system. The location of the control room should be out of the direct contact to high-voltage circuit breakers, ground, and disconnecting switches as they produce very fast transient over-voltages. These transients may cause disturbances by electric field radiations or by control wires connected to the sensitive protection and control elements such as relays.

Planning and design of a control room is therefore an important aspect as any changes to the layout post construction are expensive. Main factor to ensure the safe functioning of the underground substation is to choose the right location, laying the control and power cables in separate trenches, shielding control cables, grounding the control cable shield at both ends, and to make sure that the control cubicle doors are closed during operation.

The photo in Figure 8.173 shows a control room right next to high-voltage gas-insulated switchgear in the underground substation. As viewed through the windows, the distance from sensible electronic devices to high-voltage operating switchgear is very close but safely separated.

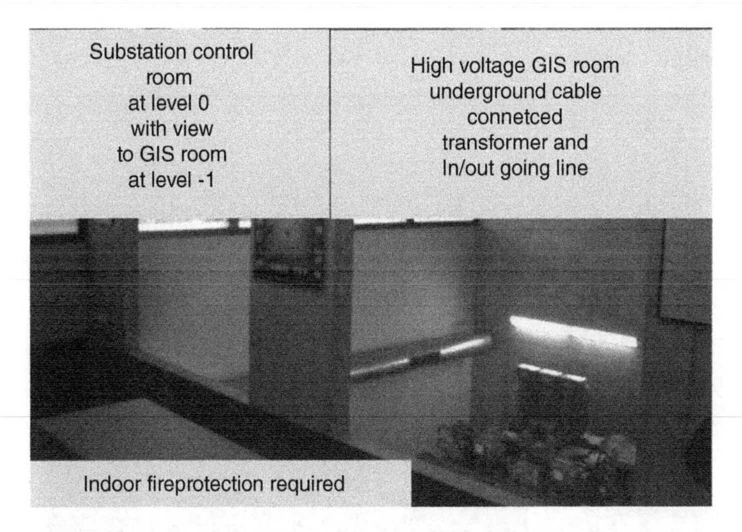

Figure 8.173 Substation control room next to the high-voltage switchgear of an underground substation (*Source* [36])

8.8.2.9 People Protection

The protection of public is the primary criteria of consideration while designing underground substations because of their proximity to the equipment. The dangers of operating high-voltage underground substations are the high-voltage levels which require safe separation/ isolation, dangers due to an internal failure resulting in electric arc, high short-circuit currents, high-magnitude transient over-voltages, loss of pressure from the gas-insulated equipment, electromagnetic fields, excess heat, SF_6 leakage to the surroundings, and the risk of fire.

These are a few aspects that should be considered during the planning and design of an underground substation and need to be solved in a way that the design wouldn't cause any harm to the public and people around substation.

In Figure 8.174 on the left, an underground substation is shown, which is located under a publicly accessible city park. People are in the vicinity of a few meters to the high-voltage equipment in the underground substation. This picture shows the emphasis of safe and reliable design needed for building such substations. Proper signage is required in cases where the basement unit and entrances are close to public access.

The photo on the right of Figure 8.174 shows a playground outside the entrance of an underground substation. Wide and high concrete walls are built behind which the high-voltage equipment is installed; this kind of barriers limits the risk to people and public around even in the case of an internal arc or failure.

IEC 61936-1 [27] details the requirements for the safe installation of underground substations.

8.8.3 Risk and Common Issues

Underground substations are very useful and handy when it comes to resolving the problem of delivering the electricity to densely populated cities with rapidly increase consumers of electricity. However, as discussed in the earlier sections of this chapter, there are a few issues associated with proximity to public such as follows:

- Exposure to EMF/ELF
- Risk of fire due to equipment failure
- Noise pollution
- Environmental issues in event of gas/oil spills

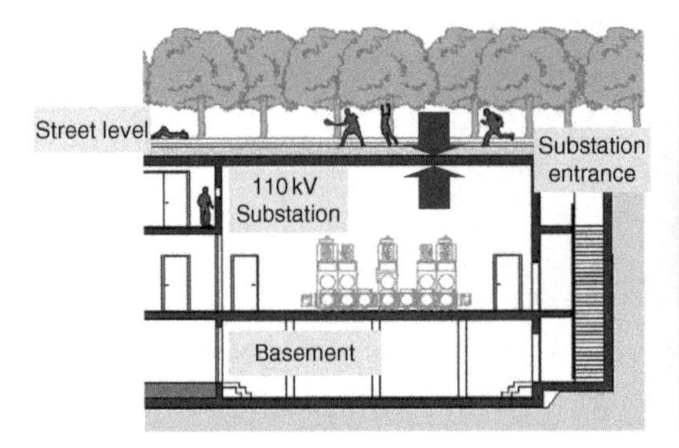

Figure 8.174 Vicinity of people and public, left: at underground substations, right: at a playground (*Source* right [37])

Use of proper mitigation measures and installing necessary preventive systems increases the safety and prevents any loss/damage during the lifetime of such equipment.

8.8.4 Architectural Concepts

8.8.4.1 Introduction

The principal premise for underground substations is the compact design of equipment as underground substations are chosen when space is rare and expensive. The high-voltage power supply is located in the basement to use the space above for prime real estate of commercial value such as office buildings, living apartments, shops or restaurants, or a mixture of all of them.

In Figure 8.175 on the left, a GIS unit is shown for voltages up to 145 kV, rated currents of 3150 A, and short circuit switching capability of 40 kV. The right photo shows a 11-kV medium voltage panel for 2000 A rated current and short circuit interruption capability of 25 kA. These pictures are shown to give a perspective of the space required for GIS equipment. Similar arrangement for an air-insulated substation would be at least 20 times that of a GIS equipment depending on voltage level.

High switching capability is required in city centers to cover the load requirement of the tall buildings with large electric consumption.

The second important aspect of the equipment to be used in underground substations is higher reliability and longer operational life. Based on field data of equipment from the technology leading manufacturers, GIS equipment of voltage ranges of up to 145 kV for the high-voltage transmission network and 11 kV of the distribution network have a statistical life of more than 1000 years for the mean time between failures (MTFB). This has been investigated by CIGRE and is published in CIGRE brochures TB 509-514 [38–43].

Beside switchgear, the transformer needs to be compact design too, to allow the high-power transmission capability and safe operation close to public. The specification of the transformer needs to cover the power losses, noises, and EMF to make sure to reach the limiting values required.

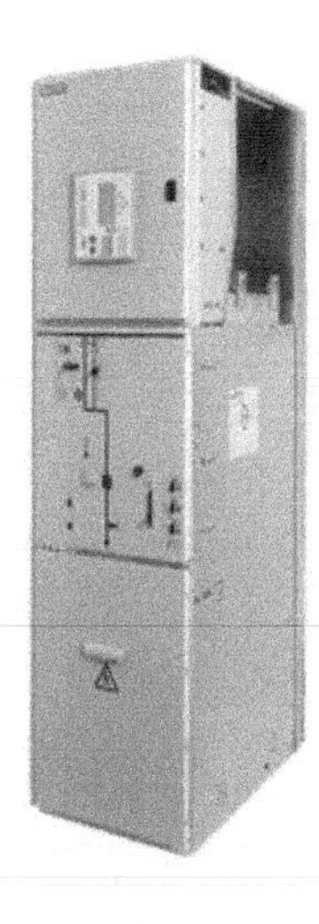

Figure 8.175 Left: 145 kV and 3150 A high-voltage GIS, right: 11 kV and 2000 A medium-voltage GIS (*Source* [44, 57])

Building on the concept of compact design, the GIS technology has been used to design underground substations for various applications as discussed in the below subsections.

8.8.4.2 Principal Solutions

There are three common solutions for installing substations in cities:

- Completely underground substation
- Semi-underground substation
- On-top building substation

The completely underground substation solution is the costliest as the entire station footprint of the substation needs to be excavated and expensive surveys and groundwater holding and stability in the ground are required. This concept is chosen if the space at the location is high priced and the space above the substation can be used for city infrastructure such as a building complex, a city park to provide some green and a quiet space in the city center for public use. In Figure 8.176, the concept of underground substation and the public use of a park on the top is shown.

The semi-underground substation as shown in Figure 8.177 is using the natural geographical slope between the street on the left and park and residential area on the right. The difference in

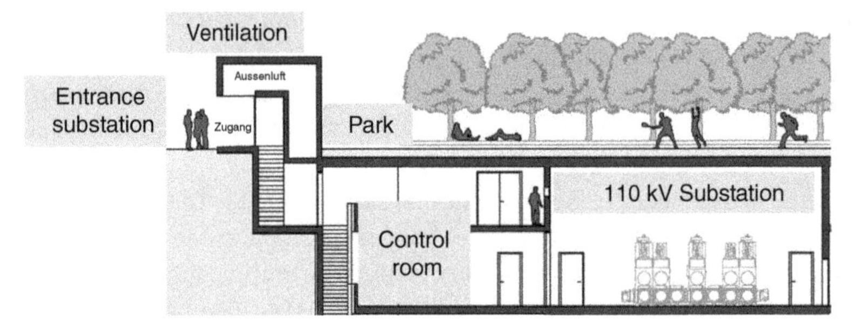

Figure 8.176 Completely underground substation with a public park on the top [45].

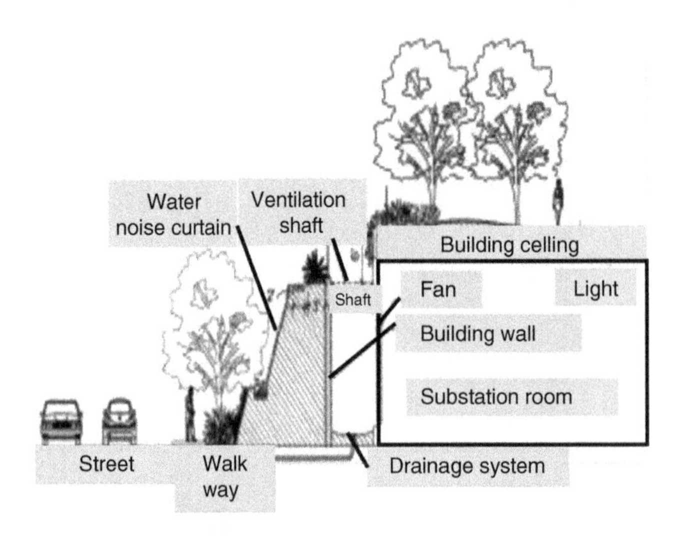

Figure 8.177 Semi-underground substation in a sloop between street, public park, and living area [45].

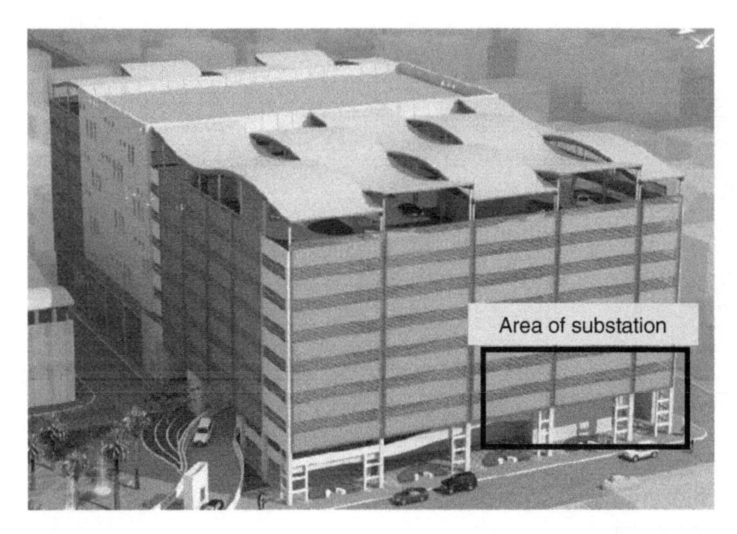

Figure 8.178 Underground substation with a public park on the top (*Source* [46])

height between the street and the residential area provides a visual cover for the substation. The station's access and ventilation are toward the street making it a perfect spot for bringing in the transformers, the heaviest equipment of the substation.

The on-top building substation solution is where the ground floor at the surface level is used to install the substation equipment, while the floor above is used for other infrastructural use. In example shown in Figure 8.178 below, the floors above are used for car parking in the city center. Such solutions for substation are integrated into the city design and architecture. After completion, the high-voltage substation for electric power supply of the city center is not visible anymore.

The choice between underground, semi-underground, and on-top building solution is dependent on the permitting requirements, available real estate, and needs of the city development authorities. City developers around the world are confronted with the task on how to shape the future cities. Following the trend that cities are still continuing to grow [1] and the living standards in the city include quiet living apartments, green parks, close distance to offices, car parking, space recreational activities make it a stronger case for choosing the underground substation solutions.

In the following subsections, we will discuss some exiting solutions of underground substations to give an idea of what is possible in the realm of underground substation application.

8.8.4.3 Architectural Solutions

Discussed below are a few architectural solutions for high-voltage substations in cities.

Substation in large cities is often located in places where they are surrounded by buildings, located in parks, green areas/recreational areas in the city.

Substations in city squares or plazas require extensive architectural design to integrate with the surroundings.

In Figure 8.179, a substation is shown, which is integrated into the architectural design of its surrounding. Modern office buildings, apartment building for city living, stores for daily needs, and restaurants are in the close vicinity of this substation. The substation building is not visible for public and is masked by a modern architectural façade to give an appearance of an office space or a mall.

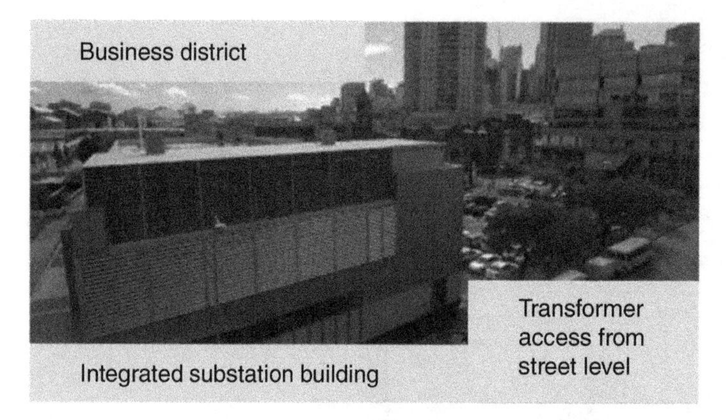

Figure 8.179 Substation building above ground in a city center (*Source* [47])

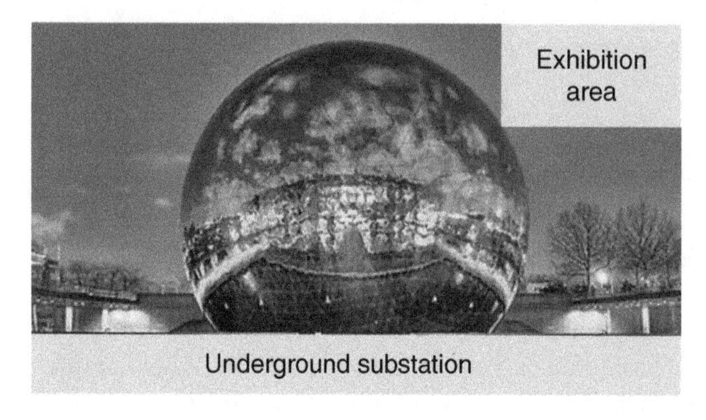

Figure 8.180 Substation in an exhibition area installed underground (*Source* [48, 55])

In Figure 8.180, the substation in build under an exhibition center. People around the area would have no idea that there is a high-voltage substation right in the heart of the city helping meet the city's electricity demand. The underground substation helps meet all the T&D requirements of the neighborhood. Compactness of the gas-insulated equipment not only opens such creative solutions for inner city applications but also meet the safety and reliability requirements.

Underground substations can provide the much-needed space to build parks or other green spaces in city centers, see Figure 8.181.

In Figure 8.182 are glimpses of underground substations as they appear to the public. The substation is completely under the square or plaza. The entrance may be used for entering the substation to bring in equipment and to be used as ventilation. The architectural design ideas are adapted to the buildings in the surrounding.

In Figure 8.183, below, two architectural solutions are shown for substations in the city center. One (on the left) shows an architectural solution of an underground substation that allows public to look into the substation through the glass arrangement. This 132-kV gas-insulated underground substation is designed with various colors and lightning to enhance visual impression.

The solution on the right of Figure 8.183 shows a concept with only a cubicle visible to the public. The cubicle allows access for operational personal and is used for the ventilation to cool the substation (Figure 8.183).

Public city park
Underground substation below pleasure ground

Figure 8.181 Substation under a park or green field in the city center (*Source* [45])

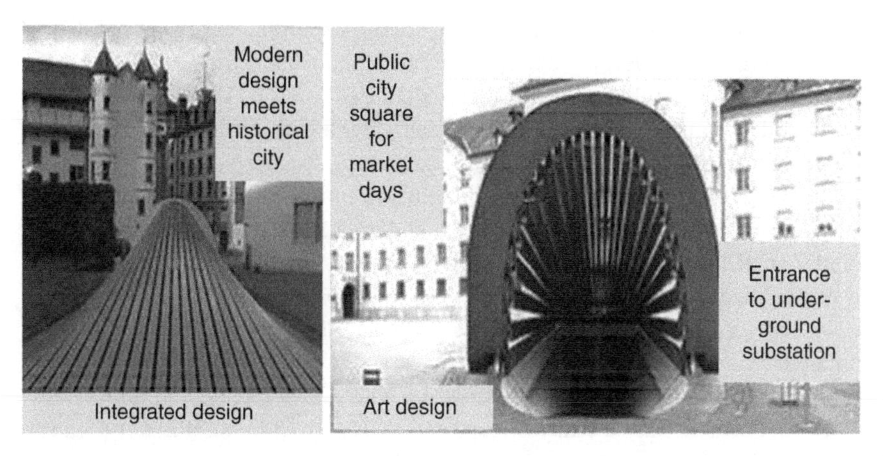

Figure 8.182 Substation under a square or plaza in the city center (*Source* [49])

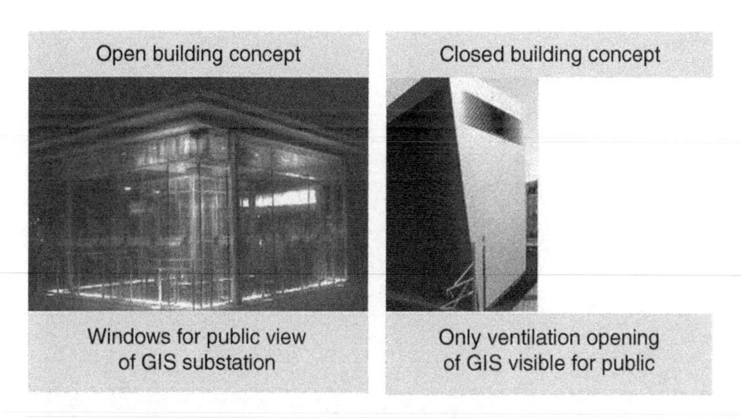

Figure 8.183 Architectural solutions for substation in the city center, left: open-building concept with view into the underground substation, right: closed-building concept for access and ventilation (*Source* [50])

8.8.4.4 Model City Development Solution

City development has to consider numerous aspects when planning the future layout of the city. The goals vary strongly from location to location while the local public opinions, investment plans, political discussions, and forecasted electrical energy demand are some of key elements to be evaluated and decided before approving the underground substation solutions.

As one example, development plan of a city will be discussed in the following. A capital city with about half a million inhabitants in 1980, which grew to more than one million in 2014 and still is growing. The city transformed from a city with low-rise buildings on a wide spread area to a more concentrated city with high-rise buildings in the city center. This implies higher demand for electricity in the city center area.

The business development of the future beyond the oil and gas exports is the driver for development of the financial business, sport, recreation, tourism, and education. For this reason, in 2005, a financial center was founded for governmental and private investments into the city to create new infrastructure development projects. The convention center tower, metro, the new living quarter, and a luxury island are some examples for the city development, see Figure 8.184 [51].

The city developers shall review the development potential of the city with that of regional structure reserving space for certain type of development projects with financial focus. The 2016's bird's eye view of the city center, as shown in Figure 8.185, it is evident that some areas are more densely populated than others and as you get closer to the city center, the concentration of buildings increases. With such concentration of buildings and infrastructure, the need for electric power supply increases. In Figure 8.185 are also three new proposed substations marked as pins, i.e. one at the convention center, another at the Financial District, and another at residential area. High-voltage power transmission of 145 kV brings high about 500–1000 MVA to these substations, which will be distributed to the nearby neighborhoods. The transmission line will use underground cables to connect the new substations, which will be built as either underground or semi-underground substations.

These substations shall be located where the transmission line cables can be laid and where heavy equipment like transformers can be easily transported to. Heavy equipment like the transformers is typically installed on the ground floor for easy installation.

Unlike large power transformers, the gas-insulated switchgear doesn't need any special accommodations for transport as it can be fragmented into shippable units.

Figure 8.184 City center development, left: city view in 1980s, right: city view in 2014 (*Source* [51])

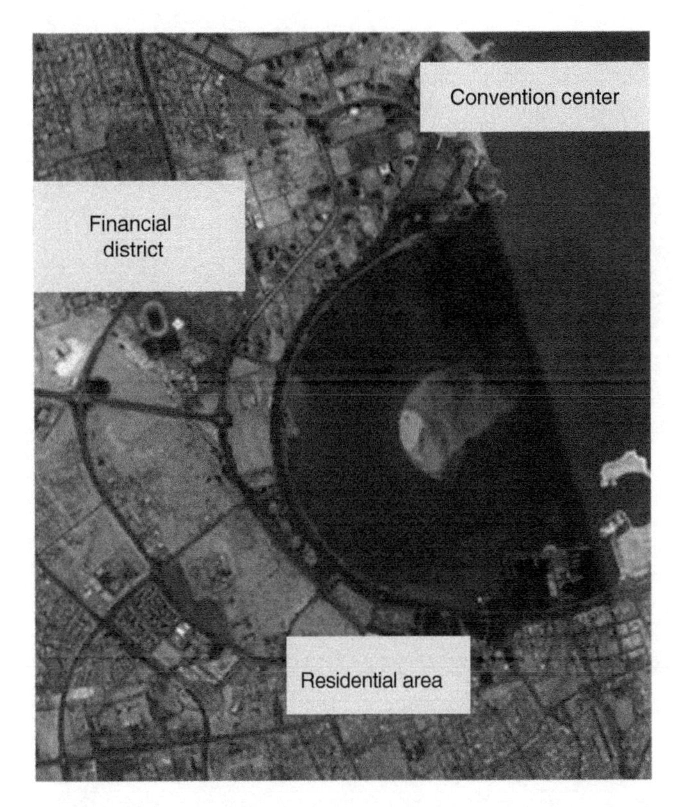

Figure 8.185 City development with three new underground substations (*Source* [52])

The substation planned at the convention center may be placed at a street junction close to the city center as show in Figure 8.186. The area marked is enough to bring in the high- and medium-voltage switchgear equipment, the transformer, and the control and protection equipment. The high-voltage cables for the incoming transmission line and the outgoing cables to the new buildings around the substation will also use the streets.

The marked area gives indication for the space required for underground substation. Depending on the purpose of the construction discussed above, this substation may be part of the building or a shopping mall. The substation may be placed into the basement in case the area above will be used for a park or recreation area or it might be a semi-underground substation with the transformers at ground level and the switchgear below. The solution might also be an above-ground substation installation with the transformer at ground level and the medium- and high-voltage switchgear above.

The photo in Figure 8.186 shows that the city planners considered the new substation at the convention center district and necessary real-estate is left in place as necessary. Timing has an important impact on the success of such projects. It is in the nature of the planning process that a clear picture of what the development will look like at the end of the process.

The project execution plan will have a relative small-time window for the erection of the substation. GIS offers short manufacturing times and fast erection on site because of factory preassembled and pretested modules usually form of one or more bays in one delivery unit. See also Section 8.8.6.

The substation in the Barwa Financial District is located close to a street junction. The area around the marked substation location will be fully developed for the purpose of the financial

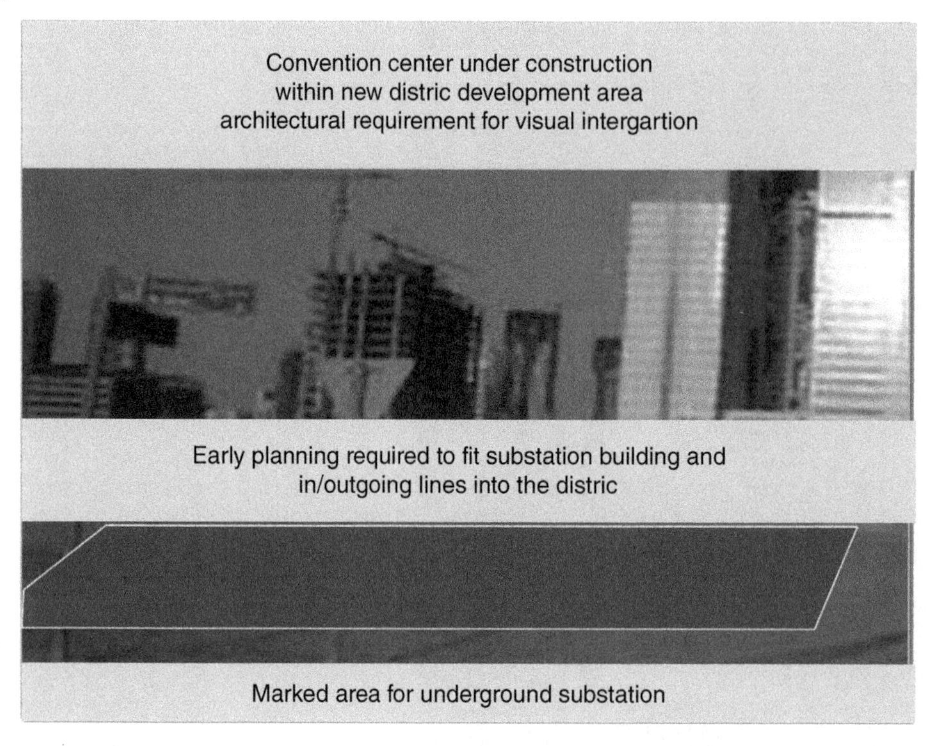

Figure 8.186 Substation integrated at the convention center (*Source* [48])

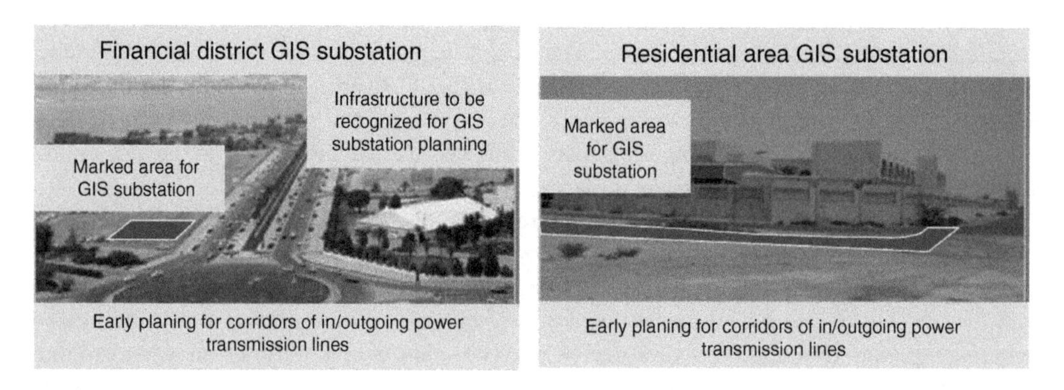

Figure 8.187 Substation integrated in Financial District (left) and city center (right) (*Source* [48])

district. Also, in this case, the location is chosen close to the street for the underground cable terminations. For the incoming transmission line at 145 kV and also for the outgoing distribution lines at 11 kV, see Figure 8.187.

The substation is planned close to a closed housing complex surrounded by a wall. Here, the requirement is not to go under the housing complex and to hide the substation completely underground. No building will be erected at this location. The cables shall run parallel to the wall and shall not go under the housing complex. Noises and EMF are two important requirements for this substation location.

This example of the city center development with three new substations shows the flexibility of GIS substation solution underground, semi-underground, or above ground integrated into the building complex.

Specific requirements coming from the compact design of the substation can be met with GIS at a reasonable price allowing the new buildings and infrastructure in the city center of Doha.

8.8.4.5 Industrial Aesthetic Solution

Industrial buildings can give a unique feeling for the city. One of those cities that likes the technical aspect of its architecture since the beginning of the industrial development introduced a GIS building with large windows for public view. The city technical history goes far back into the 1920s. At this time, the city was one of the world leading producer of electro-technical equipment. The number of products at this time was much smaller as today, but the technical understanding goes deep in public memory and the recognition of technical aesthetics, even today.

In 1989, the requirement for new electrical transmission lines from the outskirts of the city into the city center at 400 kV voltage level was asking for new solutions.

The planning work included underground 400 kV transmission lines into the city center using two circuits of XLPE cables. Two new high-voltage substations in the city center and two extensions existing 400/110 kV substations were required to connect west and east of the city with a tunnel. The tunnel was built about 40 m under the city to be out of obstacles and to go the direct way from substation to substation as the shortest route. An electric power tunnel for the reliable supply of the city center at 400 kV voltage level. Starting in the western part of the city with the first substation, one in the center and two in the eastern part of the city. The electric power link connects to the 400 kV overhead transmission line network.

One substation in the eastern part of the city is close to a wide inner-city boulevard. A busy street with many shops, restaurants, offices, and apartment buildings and many people on the street side walk. The architectural idea is to bring this billion Euro project to people recognition by showing the 400 kV GIS units through a large window on street view side. This view is shown in Figure 8.188.

Figure 8.188 420 kV GIS substation for industrial aesthetics (*Source* [53, 54])

Illuminated at night, this 400 kV GIS substation impresses the bystanders by its color and size. People passing by enjoy the view and are impressed by the equipment used to bring the electricity into their homes and offices. It remembers the people of Berlin about the importance of electric power supply.

Bringing industrial aesthetics to showcase to the people in the city is an interesting approach of city development authorities.

8.8.5 Economics of Underground Substations

In the following section, the principal of cost comparison of an underground substation compared to conventional above-ground substation design is given. For this exercise is assumed that both substations are using gas-insulated equipment (GIS) for the 69-kV high-voltage transmission level of 13 bays and 11 kV for the distribution level of 28 bays. With three transformers installed, each of 40 MVA power transmission capability, the high-voltage cable of 69 kV and 11 kV, low-voltage cables, earthing and lightning protection, civil and contractual works, mechanical and pumping works, and all services are assumed for both solutions. Additionally, excavation and landscaping cost are evaluated for the underground option.

As seen in Table 8.20, the cumulative % cost difference for an underground GIS vs above-ground GIS substation is about 12%. The additional 12% cost of underground option goes toward site civil and structural works such as excavations, landscaping, ventilation architectural design. Additional expenses toward testing and commissioning are also expected for an underground option.

Table 8.20 Cost impact of conventional substation versus underground substation.

Cost comparison of an above-ground and below ground substation					
Unit description	**Quantity**	**Units**	**Assumed % unit pricing for AG option**	**Assumed % unit pricing for BG option**	**Difference**
66 kV GIS switchgear	13	Bays	14	14	0
11 kV GIS switchgear	28	Bays	4	4	0
Transformers – 40 MVA units	3	Nos	15	15	0
HV cables and terminations	1	Lot	2	2	0
LV service equipment	1	Lot	6	6	0
Protection & control systems	1	Lot	16	16	0
LV cables and terminations	1	Lot	8	8	0
Earthing and lightning protection	1	Lot	2	2	0
Excavation work	1	Lot	0.5	3	2.5
Civil and structural works	1	Lot	14.5	20	5.5
Landscaping and architectural works	1	Lot	1	3	2
Mechanical, electrical, and plumbing	1	Lot	7	9	2
Engineering, commissioning, and testing	1	Lot	10	10	0
Totals			**100**	**112**	**12%**

8.8.6 Typical Project Schedule

The timeline of such projects is a key issue for the overall performance of the investment. When large city development projects such as office buildings, apartments, metros, and parking lots are planned, the ability to supply electric power to these new constructions becomes an important issue.

Most of the times, the critical details such as substation load data and allocated real-estate information is made available at a later stage of the planning process, which could make it challenging to design, build, and bring the substation online. Gas-insulated switchgear has an advantage in this aspect given the major parts are manufactured and pretested in the factories and delivered to the construction site to optimize the time needed for installation.

An example project shows the project starting in June 2008 and completed by February 2010, a total of only 20 months. The engineering works on the substation started in June 2008 and were finalized by January 2009 i.e., within six months. The procurement for manufacturing, inspection, and logistics started in July 2008 and was finished in April 2009, within 10 months. The civil works started in August 2008 with excavation, building construction, and finalized installations completed by July 2009, i.e. within one year. The electrical installation of the GIS lasted for only seven months from April 2009 until October 2009. Testing and integration of these systems started in May 2009 and ended in February 2010. The commissioning process of this underground substation lasted from January to February 2010.

A typical underground substation requires a detailed project planning and thorough construction practices. All equipment must be made available when needed as dictated by the construction schedule. The critical aspects of such projects are developing the detailed specifications for the equipment, and the integration into the local electrical network. Often some crucial information is missed out during the initial phases of the project causing last-minute changes and delays in the delivery of the equipment. Hence, proper engagement from all stakeholders is expected from the inception of such projects (Figure 8.189).

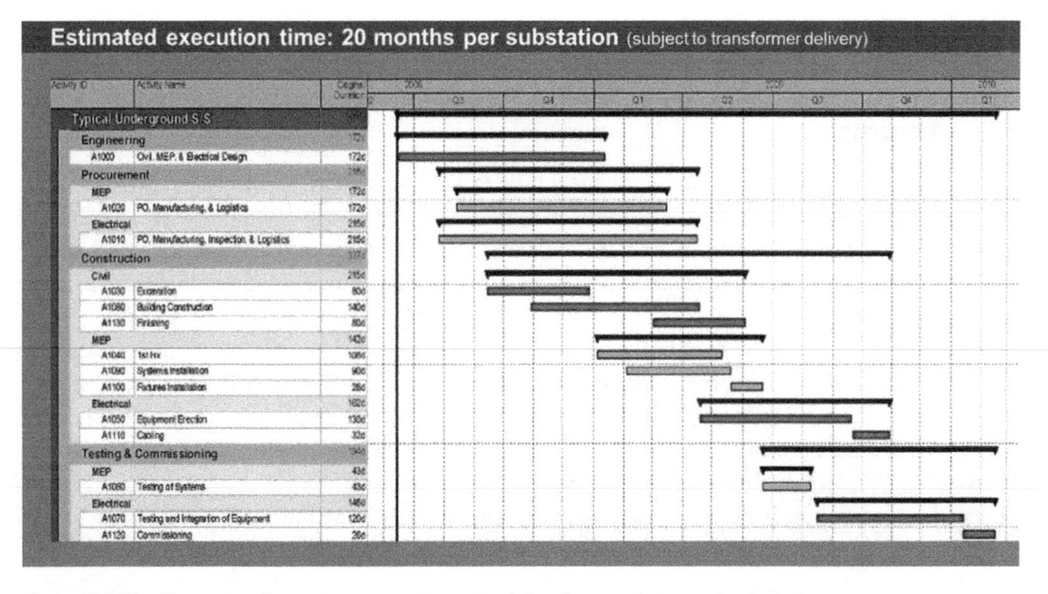

Figure 8.189 Example of a project execution schedule of an underground substation

8.8.7 Conclusion

In this chapter on underground substations, we discussed different concepts of design and installation of underground substations. Through the various examples and case projects, it is evident that underground substations are becoming a popular choice for city planners from around the world where available real-estate is scarce.

We discussed various aspects of underground substations that drive the cost, safety, and reliability of such installations. It is evident through this chapter that underground solutions offer new ways to deliver the electricity to densely populated city centers while not damaging the aesthetics of the city.

The high costs associated with going completely underground are often offset by the value of the land that is now freed up for revenue generating real-estate investments.

Architectural solutions for substations in the city centers such as open-building concepts with glass walls to allow on lookers to see what is inside of these underground substations or closed-building concept with architecturally designed cubicle for access and ventilation and semi-underground and above-ground indoor substations with part basement and part ground floor arrangements are discussed in this chapter.

Commercial aspects of underground substations were discussed in this chapter with an example of a 69-kV transmission and 11-kV distribution substation with a cost difference of only 12% increase compared with a more traditional above-grade GIS installation.

We also discussed project execution schedule of a model underground substation, which shows the entire project only takes about 20 months from inception to bringing the station online to feed the loads.

This chapter is put together as an informative document for all the engineers, planners, and developers looking to meet critical load forecasts/energy demands in growing metropolitans around the world.

8.9 Special Substation Buildings

Author: Hermann Koch
Reviewer: Mark Kuschel, George Becker

8.9.1 Introduction

The compact design of gas-insulated switchgear (GIS) allows substation solutions in special buildings. Worldwide, this possibility is used to fit in substations into living or business areas without notice from the outside. If people do not know that there is a building behind this wall, they wouldn't think of. In some architectonical solutions, the GIS building for a downtown substation is a real eye opener.

There are many different types of buildings used for GIS depending on the need of the substation and the aesthetical aspect. In some cases, for low budget, light weight, light construction buildings, a steel frame with steel blade walls are chosen. In other cases, concrete constructions in special architectonical design are creating real art works or traditional brick stone buildings are used to fit into an existing environmental community.

Investment cost aspect is one view to make the right choice for the building, but in some cases the public, the community neighborhood, or the authorities request low visual impact of the building or low noises of equipment of the substation submitted to the surrounding or protection from possible failure inside the substation to the area around.

The requirements coming from the substation operation may be influenced by ambient atmospheric conditions of high temperatures, high humidity, high sun radiation, or severe climatic conditions with dust and salt in the air. In some cases, the accessibility of the GIS under any weather condition is a reason to choose a building to allow operational and maintenance personal safe access at any day of the year.

In city centers, GIS building are used to fit the substations into the existing infrastructure and to be able to realize the substations function at the given space. Space is used for office buildings, and the basement will host the substation including underground GIS and transformers. Hydro power plants for electric power generation are often located in mountain areas at very remote locations.

The resources of hydropower plant in the mountains are often used for power generation at very remote locations. Hydro power plants for electric generation which often located in mountains at very remote locations. These connections inside the mountains are of made with gas-insulated transmission line (GIL) for high power lines of 5000 A. Constructions inside the mountains are used for GIS [7, 8, 10].

Safety from terrorist attacks or criminal damage of substations may require a building around the substation which includes the GIS and the transformers. A substation behind walls follows the rule: out of sight, out of mind. The building gives protection of the equipment of the substation against bullets or other firearm projects [10, 45–50].

When buildings are chosen for GIS with or without transformers to realize a substation, the original project cost is not the only criteria for the decision of choice. Many views and sights are part of the decision given by public, authorities, national reliability council, or local action teams. In the following, some examples are given from project around the world [53, 55, 71, 74, 75].

8.9.2 Examples

8.9.2.1 110/10 kV GIS Substation in City Centre

In a city center, a new substation of 110/10 kV was planned for the electric power supply of the city center in the 1980s and executed in the 1990s. The concentration of tall business buildings and apartment buildings in the city increase the power consumption so that a power supply at distribution voltage level of 10 kV was not an economic solution. The use of 110 kV would offer a much better and reliable power supply to the city but does require most space. A technical solution with air-insulated equipment could not be applied because of the required space. So, the solution found was GIS from the perspective of space. The space in the city is expensive, and the best use is a building which can hold the substation, including office space and apartments for rent.

As shown in Figure 8.190, the substation is integrated in a building complex and installed underground in Level 3 and Level 4, substation rooms of the building basement. More than 250 t of materials and equipment had to be craned in to a point 15 m below ground level.

The high-voltage substation has six bays of GIS 110 kV and two power transformers 110/10 kV with 40 MVA rated power. The distribution voltage side of the substation has 17 bays of 10 kV GIS to connect the building and the surrounding to the electric power supply.

The 110 kV GIS is using single-phase insulated equipment with a three-phase insulated busbar. The bay control cubicle is directly attached to the bay, and also the hydraulic circuit

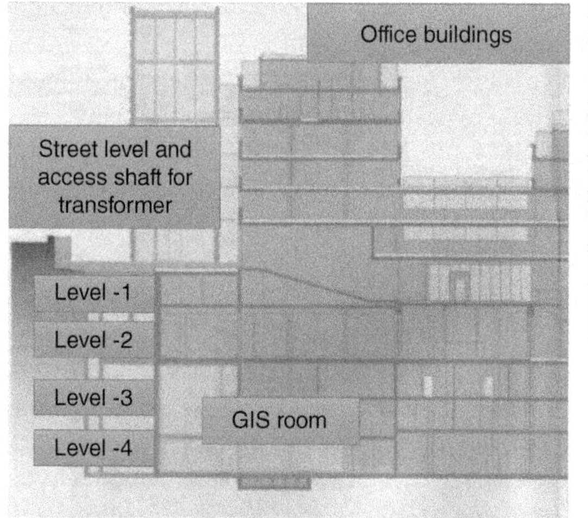

Figure 8.190 Left 110/10 kV GIS Substation in the basement of a building complex, right: 110 kV GIS, 10 bays (*Source* [44, 58–60]).

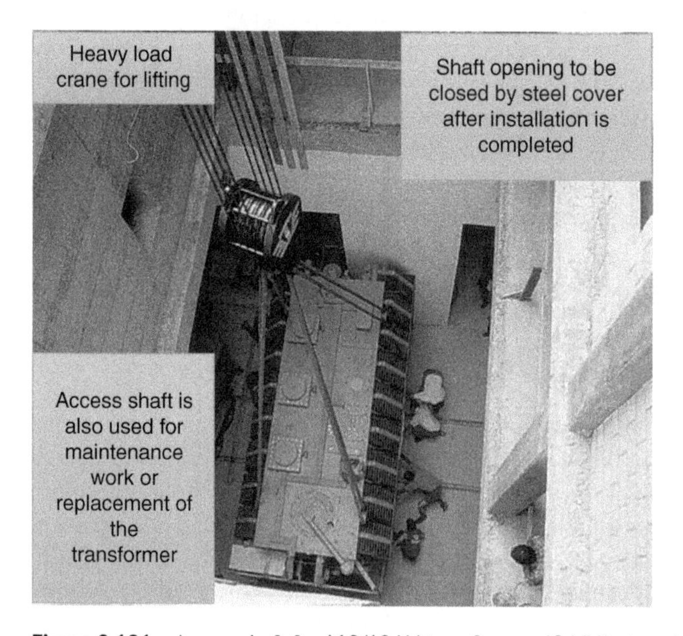

Figure 8.191 Access shaft for 110/10 kV transformer 40 MVA at underground level 4 (*Source* [61])

breaker drive is at the front of the bay, see Figure 8.190. Left figure shows the location of the substation in the building complex and right figure shows a 10 bay 110 kV GIS double busbar assembly

To bring in the transformer, a shaft has been positioned under the street in front of the building. The shaft was open during construction time and the transformer has been lifted down through the shaft into the building basement, see Figure 8.191. After the two transformers have been positioned, the shaft of the street has been closed with a steel cover to allow traffic flow. In case of a damage or repair of the transformer, the shaft can be opened again and access to the transformer in the building.

Beside the access requirement to the transformer and GIS, it is required to install an effective fire existinguishing system in the basement of the building. For this, a nitrogen fire distinguishing system was chosen. In case of a fire, the room will be flooded with nitrogen to replace the air in the basement and with this to distinguish the fire. For this, pressurized storage tanks are installed in the basement of the building.

8.9.2.2 110/10 kV GIS Substation at Four Building Levels

In a four-level building in a dense city area, a 110/10 kV GIS Substation supplies the area around with electric power and gives high reliability of power supply.

The substation has eight bays of GIS which are located at the third floor in a multi-storage building in the row between other buildings. This is one out of several substations in multi-storage buildings located in the city center. At the third level, the 10 kV distribution GIS is located, and a control room at the fourth level. The transformer 110/10 kV is located at the ground level and at the first floor, the compensation coils are located, see Figure 8.192. The building is historically protected and any change on building on the outside appearance is not allowed.

The specific on this project is that the time for erection on site was very limited and the complete installation needs to be in service before the main tourist time of the years starts. The substation needs to be fully integrated into the existing building of historical protection without any change made from the outside. Electrical connections in and out of the building are made by underground cables.

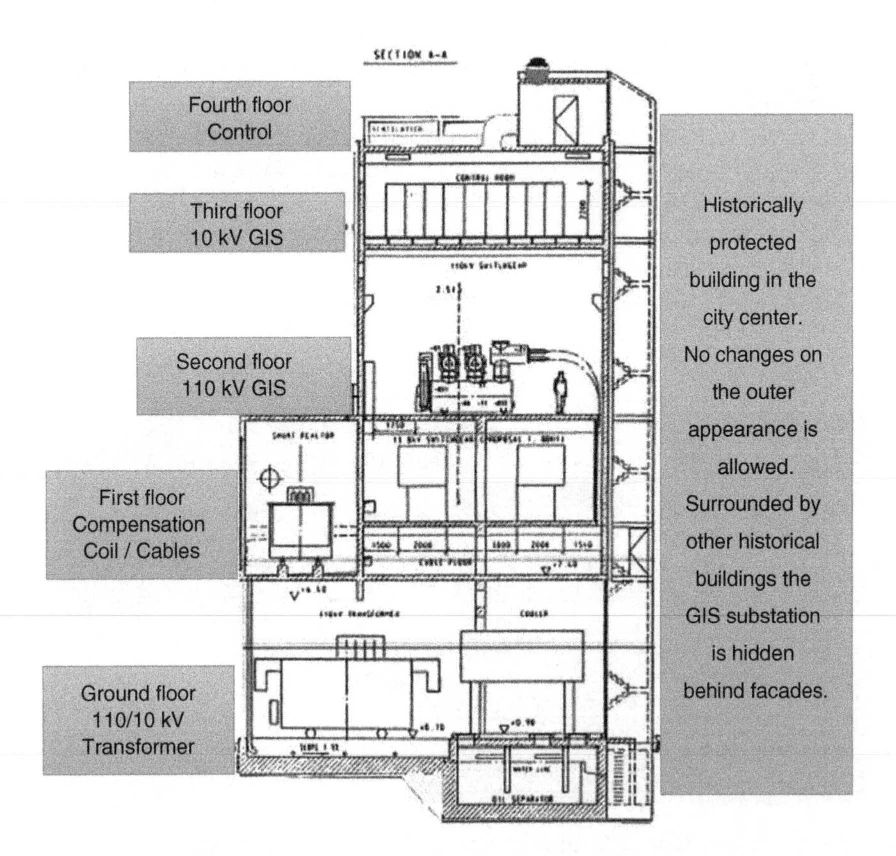

Figure 8.192 Access shaft for 110/10 kV transformer 40 MVA at underground level 4 (*Source* [60, 62])

The access to the construction site was very limited, and high security rules have to be managed. Only registered experts were allowed to enter the construction site, and this includes all workers.

8.9.2.3 Downtown Business District

New high-voltage transmission lines for the electric power supply at the downtown business district in a growing city center were needed; 330 kV underground cables are feeding the city center to connect the new Downtown Business District substation to the electric network

23 bays of 132 kV GIS on the city distribution side and four bays of 330 kV GIS for the connection to the transmission cable are installed. The 330/132 kV transformer connects the voltage levels and was designed in gas-insulated technology using SF_6 to reduce the fire load of the substation building.

The Downtown Business District substation is integrated in a shopping and business complex in the city center. The substation is installed at Level 0 and Level 1, see Figure 8.193. At Level 0, which is below street level, the gas-insulated transformers, 132 kV GIS and 330 kV GIS are located. At Level 0, control and protection equipment and the SF_6 cooling and gas treatment facility are located.

In this underground installation, gas-insulated transformers have been chosen to reduce the fire risk in the building complex. Gas-insulated transformer uses instead of oil the insulating gas SF_6 for dielectric insulation. This reduces the fire risk of tons of flammable insulating oil, but needs the infrastructure facilities to cool and clean the SF_6 gas during operation. For these reasons, a temperature exchanger and gas-filter are installed on the floor above the transformers in the basement.

To avoid the release of SF_6 to the atmosphere, because of the large volume of SF_6 used in the switchgear and transformers at level 0, below this floor at Level −1, a gas storage room is installed without ventilation access to the outside of the building to be able to capture the complete volume of SF_6 used. In case of SF_6 release, because of the molecular weight of SF_6, it will fall through floor openings down to Level −1 and is captured in the room. In such a case, the SF_6 can be re-captured by pumps and stored in containers, see Figure 8.193 left side. At Level +1, air conditioning and ventilation are handling the substation cooling requirements. At ground level, the control room is located, with the SF_6 cooling and cleaning facility.

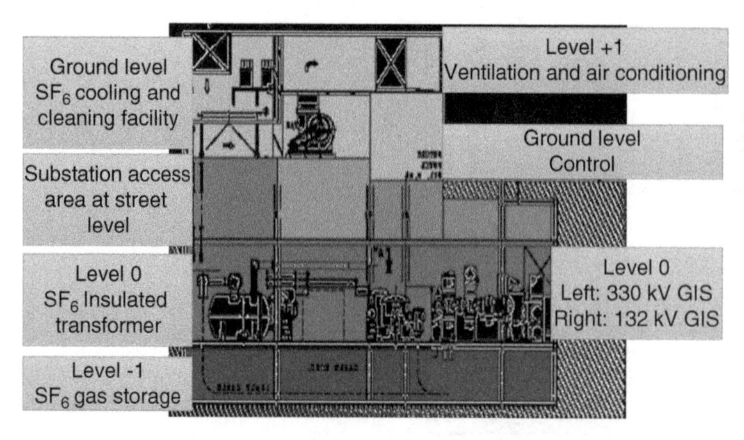

Figure 8.193 330/132 kV Downtown Business District substation, left: location in the building, right: 23 bays of 132 kV GIS (*Source* [60, 62, 63])

The 23 bays of 132 kV GIS feeders connect the downtown area to the incoming electric energy of the 330 kV cables. The 132 kV maximum operating voltage level is using 145 kV GIS equipment of three-phase-insulated enclosures in a double busbar scheme. The feeders are connected by high-voltage cables to the distribution points in downtown area. Figure 8.193 right photo shows the compact design of the 145 kV GIS switchgear assembly.

The four GIS bays of 330 kV transmission voltage connects the Downtown Business District substation in the city center to the in-feeding double system, three-phase cable circuit of about 20 km length. The high-voltage cable is connected to the 330 kV GIS and then to the 330/132 kV power transformer installed at level 0 in the substation building, see Figure 8.194 left photo.

The Downtown Business District substation building architecture is adapted to the central business district of downtown. With large office buildings in bright color and using materials as concrete and steel and glass, the substation building reflects this design. The modern form and clear front design of the substation building give the impression of a modern, functional building complex in a downtown area. The electric lines are all underground and therefore not visible, the equipment is indoor and underground to avoid noises to be noticed in the neighborhood. Electromagnetic influences are minimized by using a solid, low-impedance grounding system and by keeping distance of the switching equipment, the transformers, and the underground cable toward public accessible areas.

8.9.2.4 Hydropower Plant

The GIS cavern substation of the hydropower plant in a mountainous area connects the electric power generated in the dam with the machine transformers and by a 420 kV transmission line with GIL to the outside 420 kV high-voltage overhead line of the transmission grid. The cavern inside

330 kV GIS to connect double three-phase cable system for infeed to city center

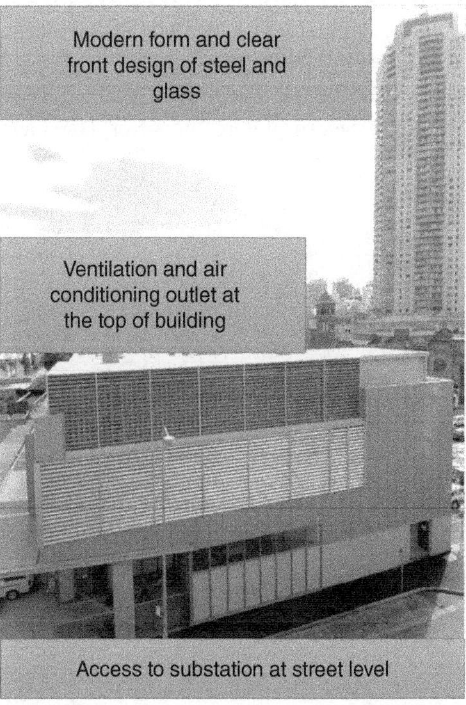

Figure 8.194 330/132 kV Downtown Business District substation, left: four bays of 330 kV GIS, right: building design adapted to business district (*Source* [50, 54, 59, 60])

the large dam holds the hydro turbines, electric power generators, machine transformers, and 420 kV GIS. Space inside the dam and accessibility to the cavern in the dam are very limited and need specific compact solutions.

Hydro dam constructions are large infrastructure projects usually in very remote areas. Difficult to reach, hard to transport equipment, and personal. This requires onsite assembly work and commissioning process for the high-voltage substation in a short period of time and with less work of assembly. GIS offers here the best option as most part of the substation is preassembled and tested in the factory before transportation to the construction site. Figure 8.195.

The seven bays of 420 kV GIS of the cavern substation designed in a double busbar scheme to connect 2 three-phase circuits of GIL for the connection of the machine transformer in cavern of the dam to the overhead line on the outside, as shown in Figure 8.196.

The photo in Figure 8.197 gives an impression on the remote area in which the dam was build. Small access roads in a very mountainous region give the only street access to the Hydropower plant. This gives very specific requirements to the project execution plan and to choose technical equipment to solve the needs of the substation to connect the hydro generator to the overhead line 420 kV transmission line.

8.9.2.5 Hydropower Storage Plant

The hydropower storage plant is connecting the hydropower generation in the cavern by five bays of 245 kV GIS to the high-voltage overhead line outside the cavern.

The access to the hydropower storage plant substation is difficult to reach at an altitude of 2000 m. The only way was to use an aerial tramway, cable car. No roads could be used to bring equipment up in this remote region of the mountains, see Figure 8.198.

The substation of the hydropower storage plant is located in an underground cavern, accessible through a tunnel, to connect the hydropower plant from the machine transformer to 245 kV overhead line outside the cavern. Two GIS bays could be transported on the platform of the cable car at one time, see Figure 8.198.

Each bay of the GIS is factory assembled, routine tested, and transported through the access tunnel into the cavern, where the seven GIS bays are assembled. The substation design is using a

Figure 8.195 Hydropower plant, left: Damn in a mountain valley, right: access tunnel to cavern inside the dam (*Source* [64])

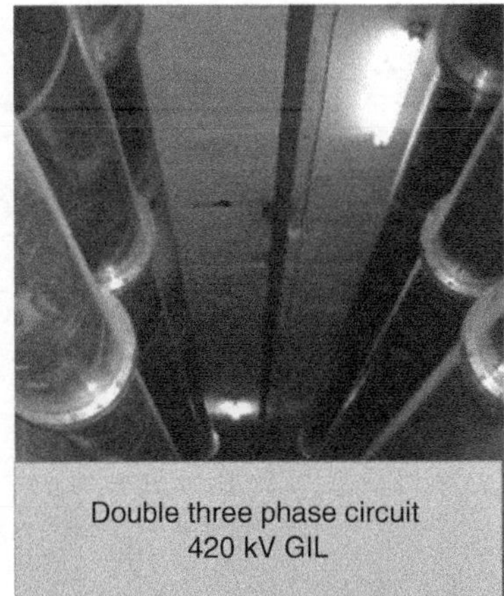

Figure 8.196 Hydropower plant, left: Seven bays of 420 kV GIS, double busbar, right: double three-phase circuit 420 kV GIL (*Source* [54, 60, 65, 66])

Figure 8.197 Hydropower plant, dam overview in a remote mountainous area difficult to get there (*Source* [20])

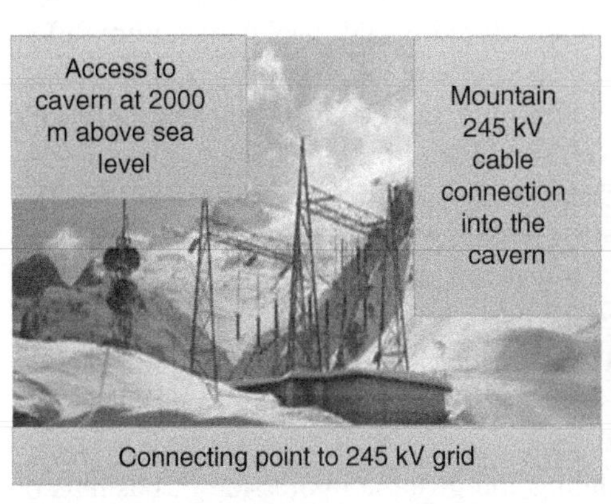

Figure 8.198 Hydropower storage plant, left: connection to 245 kV power grid, right: lifting the four bays of 245 kV GIS by cable car to the substation cavern at 2000 m (*Source* [65, 68])

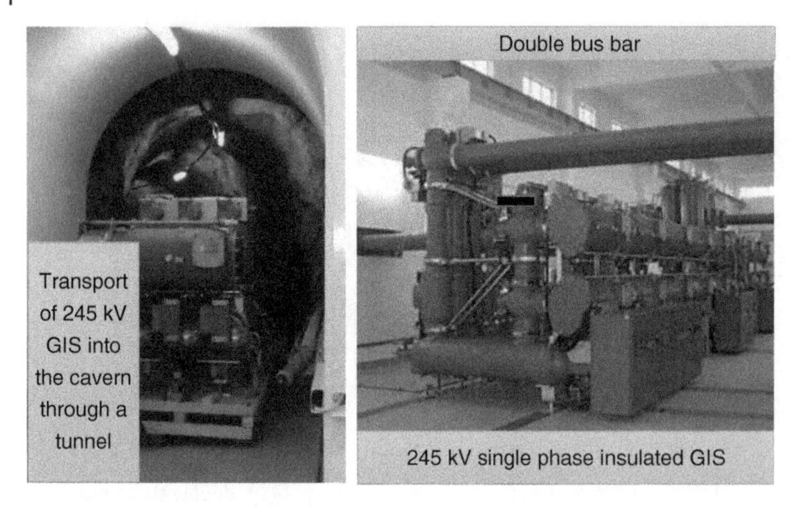

Figure 8.199 Hydropower storage plant, left: entering the cavern through a transport tunnel, right: seven bays GIS double busbar scheme (*Source* [54, 66–68])

double busbar scheme for the connection of two machine transformer from the generator side to the overhead line outside the cavern. A busbar coupling bay allows to couple or separate the two busbars for additional switching options and reliability of the power supply from the hydro power station. See Figure 8.199.

The execution of this project is adapted to the needs at high elevation in the mountains with long winter and short accessible summer time. During installation time, access to the substation is needed at any weather condition. The ambient conditions in the cavern take away extreme weather conditions, storms, ice, and very low temperature to give a more constant condition for reliable operation of the substation. GIS gives the best solution for such application.

8.9.2.6 Substation Building at Desert Conditions

132/11 kV Substation

The substation at warm ambient conditions is realized with GIS inside a substation building adapted to the neighborhood. It meets very warm ambient conditions and the impact from high dust level coming from desert storms together with high humidity and huge temperature differences between day and night.

The substation combines eight bays of 132 kV GIS on the transmission side, three 132/11 kV transformers of 50 MVA, and 60 panels of 11 kV GIS on the distribution side.

The plot size was limited as the location of the substation is in an urban area of Dubai City. The appearance of the building had to fit to the surrounding buildings.

This project is typical for substation solutions in a downtown city area. Existing architecture in the neighborhood gives visual requirements to the possible design solutions for the substation. Connecting the 132 kV transmission lines and the 11 kV distribution lines by underground cables allows to enter the building underground in the basement. No visual impact is given by the electrical transmission and distribution lines, see Figure 8.200.

The building takes the beige color of the surrounding and give the impression of business or manufacturing building. It does not appear as an electrical substation and people who don't know it would not think of this is an electric power substation.

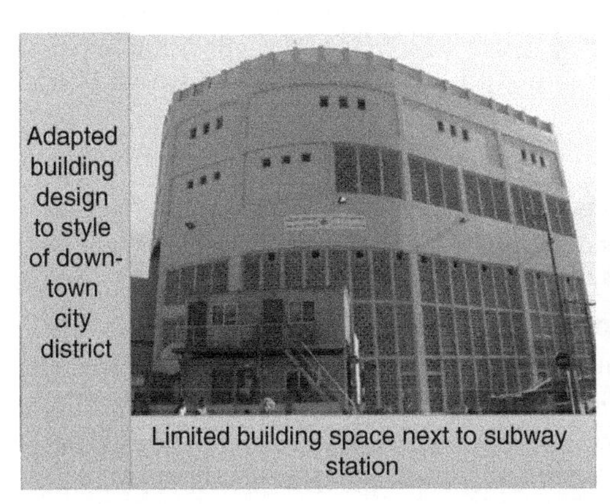

SIEMENS
Eight bays of 132 kV GIS at street level

Figure 8.200 Substation building at desert conditions 132/11 kV, left: substation building adapted to the neighborhood, right: seven bays of 132 kV GIS (*Source* left: [47], right [44, 69])

8.9.2.7 Residential Area City Substation

The residential area city substation of 132/11 kV was built in the city center to provide reliable electrical energy supply to the downtown area. Space was limited and requirements came from the close residential neighborhood around the substation building.

The substation has six bays of GIS on the transmission side to connect two incoming cable lines with the transformer to the 11 kV distribution side. The 31.5 MVA power transformers feed 25 units of 11 kV medium voltage switchgear, see Figure 8.201.

A number of special challenges had to be met in the residential city substation. Because the substation is located close to the homes of many people in the city center, the building has to be particularly pleasing in appearance and the transformer particularly quiet. Switchgear and transformers are inside the two-level building with two levels of basement. The transformers are in a separate room with noise reducing air exhaust shaft on the top of the building to cool the transformers and transport the thermal heat losses out of the building. The cool air is inserted to the transformer room from the basement floor entering through a ground floor shaft beside the

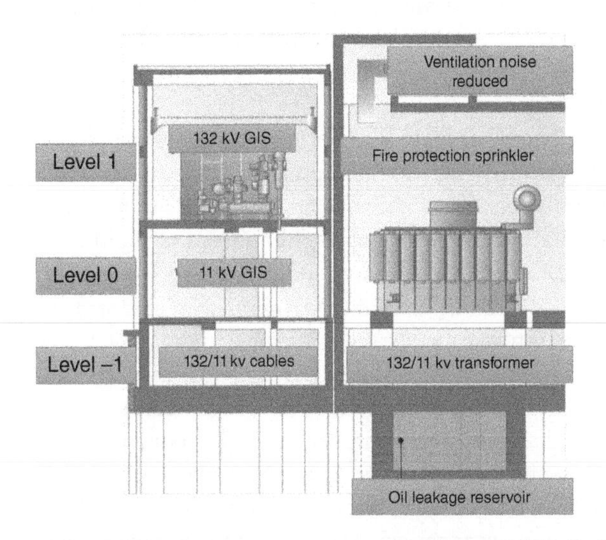

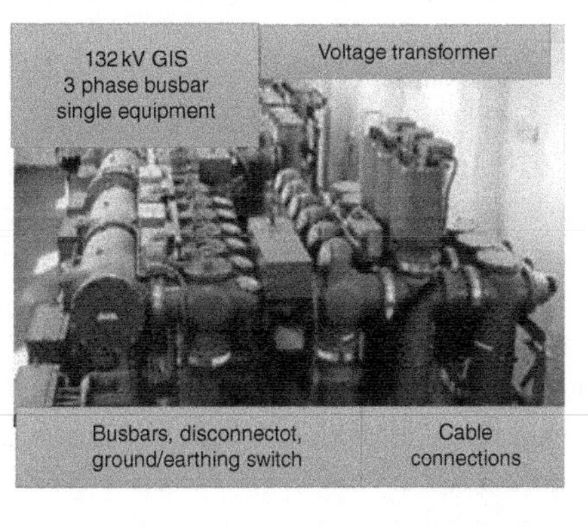

Figure 8.201 Residential area city substation 132/11 kV, left: overview of substation building structure, right: six bays of 132 kV GIS (*Source* [44, 69])

building. Below the transformer at the second basement level, an oil collecting space is installed to hold in case of an oil leakage entering the soil. The transformer room is equipped with a fire distinguishing facility to protect the building from a fire hazard coming from the oil in the transformer. The transformer is connected by 132 kV cables to the GIS on the transmission side and with 11 kV cables on the distribution side.

The 132 kV GIS on the transmission side is located at the first floor of the building and the 11 kV medium voltage GIS panels are located at the ground floor. The incoming 132 kV cables are entering the cable basement and then connected to the 132 kV GIS at the first floor. The outgoing 11 kV cables are leaving the building from the cable basement. See Figure 8.201.

The gas-insulated equipment of the transmission side with the 132 kV GIS and on the distribution side with the 11 kV GIS substation building allows a very compact substation solution. This reduces the space required by the building and gives more option to fit the design into the existing neighborhood architecture. The 132 kV high-voltage GIS is located at the first floor of the building because of its relative low weight and the few numbers of cables of the six GIS bays to be connected to the substation scheme. The transformer is located at the ground floor because of its high weight and the need to separate from the other equipment to reduce fire risk. The 25 bays of medium voltage 11 kV distribution connection are also located at the ground level because of the many cables to be connected.

8.9.2.8 Three-Voltage-Level substation

The three-voltage-level substation covers 220/66/11 kV. There are two substations as part of the city development program and need to fit into modern architecture of this new business district. The high rating of electric power supply for the new business district is provided by a 220 kV transmission line infeed and the distribution by 66 kV and 11 kV voltage levels.

Three-Voltage-Level Substation 1

The three-voltage-level substation project is using 28 bays of 220 kV GIS in a double busbar configuration to connect to the network and the 220/66 kV transformers in the substation. At the 69 kV distribution voltage level 51 bays GIS in a double busbar configuration are used to connect the 11 kV voltage level GIS bays in a separate building with a total of 87 medium voltage GIS bays. There are six 200 MVA transformers 220/66 kV and nine 40 MVA transformers 66/11 kV located in the substations building. In addition, two shunt reactors of 60 MVAr are connected to the 220 kV network and also located in the substation building.

To bring the complete substation into one building, special solutions are required. The substations building is designed in a futuristic architecture, which does integrate aspects of desert tent and business building design features of glass, steel, and concrete. The colors are bright, sand tone to reflect the landscape around the substation. The building is illuminated at night and appears as a light and sky open structure, which does not give any indication of its technical purpose of holding a complete substation of 220/66/11 kV and all the equipment required. The power line connections to the building are underground cables and not visible to the public, see Figure 8.202.

The multilevel substation building has different rooms at different levels to provide space for the substation equipment. In Figure 8.202, on the right photo, a view into the 69 kV GIS room is shown with 20 bays of a double busbar scheme. The compact design of the three-phase encapsulation of 66 kV and single-phase encapsulation of the 220 kV GIS allows to bring all equipment into building in the new district of the city development area, and the appearance needs to fit to the other buildings there. See Figure 8.202.

Architectural style adapted to new city development district

Figure 8.202 Three-voltage-level substation 1 of 11/66/220 kV, left: substations building, right: 66 kV GIS with 20 bays of double busbar scheme (*Source* [50] right: [44, 69, 70])

Figure 8.203 Three-voltage-level substation 2 building of 11/66/220 kV includes parking, left: multilevel building for substation and parking, right: 11 kV GIS double busbar scheme (*Source* [46] right: [44, 69, 71])

Three-Voltage-Level Substation 2

The three-voltage-level substation 2is also located in a multi-storey high-rise building. The substation is similar as three-voltage-level substation 1 and covers the same voltage levels of 220/66/11 kV. The substation building is part of a building assembly with a multilevel parking building and is integrated into the design of the area, see Figure 8.203.

8.9.2.9 Stadium Substation

The stadium substation is a very special project with four bays of 110 kV GIS and 21 bays of 10 kV medium-voltage air-insulated GIS. Two 110/10 kV transformers of 25 MVA deliver the energy required for the stadium operation.

The stadium substation is located in the city center. It was a big challenge to coordinate the implementation of the substation with the ongoing work on site. To bring the electric energy to the stadium GIS offers a solution for underground installation of the 110/10 kV substation with high reliability and low transmission losses. With increasing power supply of large infrastructure

Figure 8.204 Stadium substation left: stadium overview, with 110/10 kV GIS below, right: 110 kV GIS under the stadium (*Source*, left: [72, 74], right: [44])

installations like a stadium in the city center, high voltage level of 110 kV plays an increasing role for power supply in cities. In Figure 8.204, an overview is given about the stadium complex and the four bays of 110 kV GIS located under the stadium [44].

The transformers are brought into the substation under the stadium grass floor using openings, which are closed after the installation. The prefabricated concrete slab can be removed with reasonable effort to get access to the transformers and the substation equipment in case it is needed to be replaced.

High reliability of GIS and transformers designed for a service live of expected 50 years allows underground installations. The substation building under the stadium grass floor allows maintenance during operation of the substation and even in case of failing equipment can be replaced with reasonable effort to open the grass floor and get access to the substation (Figure 8.205).

Figure 8.205 Prefabricated concrete slab under the grass floor (*Source* [73])

8.9.2.10 Monument Substation

Monument substation is located in city park below a city monument using 132 kV GIS to bring electric power into the city center in the middle of a commercial district.

Six bays of 132 kV GIS are connected to two power transformers of 100 MVA are installed in the famous park in Monterrey with the tall monument in the middle of commercial district, as shown in Figure 8.206.

The substation is hidden in the park. An access ramp gives access to the underground substation without notice to the public. Nobody will expect a high-voltage substation at this location without having knowledge that there is a substation. The air-conditioning openings are hidden and give the required air circulation to cool the transformer and switchgear equipment. Stairs are hidden behind landscaping structures to allow operational and maintenance personal to get access to the substation.

The compact design of gas-insulated high-voltage equipment with a reduced space requirement and a connection by underground cable make the 132 kV substation with the ability to deliver two times 100 MVA of electric power into the city center forms this unique technical solution.

The high reliability of GIS and transformers and the low maintenance requirement are granting the long-time use of the substation without the need for replacement of equipment. This balances the higher initial cost for the installation by a higher use of the equipment for a long life time.

The Monument Substation is an example for technical solution possible with GIS equipment in areas with less space availability and the need of high energy consumption. The compact design of GIS and its high-power transmission and short-circuit interruption capabilities is the basis for such substation designs.

Entrance to GIS substation
through monument
Monument city park location
at city center

Figure 8.206 Monument Park substation of 132 kV GIS, left: access to substation through monument, right: substation location (arrow) within a densely populated city area (*Source* [75])

8.9.2.11 City Center Transmission Substation

The city center transmission substation has 12 bays of GIS for 245 kV rated voltage installed underground in the basement of a building. This is one of three underground substations in multi-storey buildings in the city center. The concept is to use the basement under building for the substation including the 245 kV transmission voltage to bring in the electric energy into the city center at low-power transmission losses and to offer a high reliability.

Special caution needs to be taken in such cases for the limitation of electromagnetic fields around the substation, the ventilation, or cooling and the fire hazard mainly with the large amount of oil of the transformers, which is flammable in case of internal failures.

The electric field of the GIS and transformer is shielded to the surrounding by a solid grounding system of the substation. The magnetic field of the GIS is low because of the induced, reverse current in the enclosure, which eliminates the magnetic field of the conductor to the surrounding.

The transmission losses in the GIS and the transformer losses need to be transported out of the substation basement by ventilation or if needed by air conditioning. For this, ventilation shafts are needed at the substation, which need to be designed in way to reduce air steaming noises in the neighborhood by slow air flow speed.

The fire distinguishing system for underground substation is a fundamental requirement to the functioning and safety of the substation. For this, protection for several fire distinguishing systems are available. Some are based on nitrogen to flood the room and displace the oxygen. Others are using water and foam to create the same effect. Such systems are working well in the case of an internal failure with arc faults to kill the fire.

The access to the underground substation is through an open able duct, see Figure 8.207. The GIS bays have been brought in through this shaft by a crane and were assembled bay by bay in the basement of the building. After installation, the shaft is closed but can be opened again in case of replacing or repairing of the 245 kV GIS with a reasonable effort.

The high reliability of GIS is the economic basis for such technical solutions. With a mean time between failure (MTFB) of about 1000 years, the probability of technical failure which needs a replacement of the GIS is very low. The reliability of the operational functionality of GIS is high, and most failure, which may occur during operation time, can be repaired on site, often without

Figure 8.207 City center transmission voltage substation 245 kV GIS, left: shaft opening for GIS installation in the building basement, right: power transmission cable connection to GIS (*Source* [67, 69])

power interruption. This is depending on the substation electrical design and the possibility to clear connections and transfer the power flow to bypass lines. In most cases, a double busbar scheme is used to allow such bypass operations. In some cases, with very high requirements of reliability and non-interrupter power supply, a three-busbar scheme is used to have most switching options. This decision has to be evaluated with the importance of the substation in the network and the additional cost of more equipment needed to make the substation more flexible in switching operations.

8.9.2.12 Schoolyard Substation

The schoolyard substation is using 110/10 kV GIS substation below the yard of a school located in the city center. The transmission voltage side has five bays of 110 kV GIS.

This schoolyard substation has to fulfil special requirements, because it is located directly under schoolyard. The schoolyard is a public location, and in Switzerland high requirements are given for electromagnetic fields around the substation. The electric field shielding is relatively easy to realize by a low-impedance grounding system. The magnetic field needs a special low magnetic field design for the transformer. The GIS is using the induced current in the enclosure to superimpose the magnetic field of the conductor. This results in a 90% reduced outer magnetic field of the GIS. The limiting values for magnetic flux density for new installations at public assessable areas is only 1 μT. The distance from the transformer and GIS from the underground substation room to the floor of the schoolyard above is only some meter.

In Figure 8.208, the lift down of the 110/10 kV transformer is shown. The shaft has been closed after the transformer is installed. In case of a replacement of the transformer, it will be lifted out through the same shaft.

The five bays of 110 kV GIS installed in the underground substation is shown in Figure 8.209. The design is using a single busbar scheme to connect the transformer and the in-feeding cable. The single busbar does not allow many options of switching. Only the disconnecting busbar in two sections will allow to separate one side of the busbar and operate the remaining busbar section in case a failure occurs in one section. This is a decision of the operator based on the importance of this substation to feed the city center, or if other network sections can take over the power supply if needed.

The schoolyard is shown on left of Figure 8.209 and reflects the open space and accessibility to public. Nobody would expect to find a 110/10 kV substation under the schoolyard. And nobody

Lifting the transformer through the access shaft to building basement

Heavy load crane for lifting the transformers weight of about 50 tons

Figure 8.208 Schoolyard substation in a building basement, left: lifting shaft to basement level for substation equipment, right: opening for transformer lift beside building under a square (*Source* left: [76] right: [60, 61])

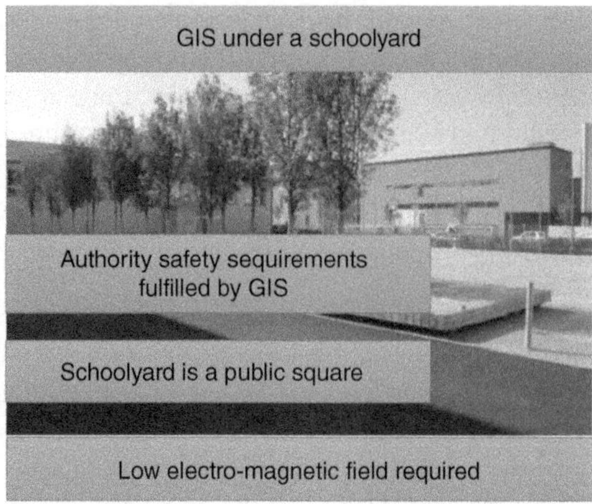

Figure 8.209 Schoolyard substation in a building basement, left: five 110 kV GIS single busbar, right: sketch schoolyard at public city square (*Source* left: [44, 61] right: [72, 74])

needs to be afraid to sit on top of the substation eating lunch or talking with friends. Even in a case of failure in the substation, the impact to the surrounding will be very limited, in most cases nobody will notice the failure when it occurs.

The combination of high-voltage substations and public is coming more often with the growth of cities and the need of electric power supply. GIS technology together with sensible protection system which can clear failure in less than 300 ms and precautions taken for internal arc faults of the equipment allows such solutions.

Engineering studies for electromagnetic field (EMF), transient over-voltage, low-impedance ground system for power frequency and high frequency, thermal design with ventilation or air conditioning, fire extinguishing systems, and handling repair and maintenance are some recommended studies to carry out before executing the underground substation.

8.9.2.13 Remote Area Substation

The remote area substation connects a hydroelectric power plant at 362 kV transmission voltage level with the grid. The remote area substation has nine bays of 362 kV GIS and connects the underground substation in cavern of the hydropower plant with the overhead line on top of the mountain by a 115 m long, vertical shaft.

The remote area substation is located in a very remote area close to the national border with border conflicts. Political unrest and riots during the time of installation required special safety rules for workers and personal on site. Therefore, the total erection time for the substation was from great interest to have as short as possible people in this remote area, which needs personal protection. GIS with its prefabricated bays and bay sections allows short installation times of some month only. An argument for the decision to choose GIS. Figure 8.210 gives an impression on the remote area.

The dam is seen on the right upper part of Figure 8.210 in the back. At this location, the water is guided through a tunnel to the river bed in the front of the picture. The height gained for hydropower generation is 115 m from the turbines to the water dam. The turbines drive the electric generators in the cavern in the mountain, and the generators are connected to machine transformer to transform the generator voltage to a high voltage of 362 kV for the energy transmission out of the cavern by GIL. The shortest way to connect the hydro-generated energy to the overhead transmission line is to go vertical from the cavern to surface where the overhead line towers are connected.

Figure 8.210 Remote area substation with 362 kV GIS located in a mountain cavern of the hydropower plant (*Source* [77])

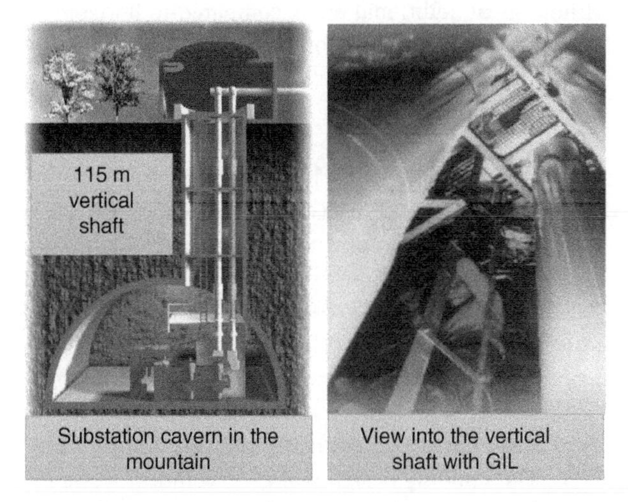

Figure 8.211 Remote area substation, left: sketch of GIS in a cavern and a GIL in a 115 m vertical shaft, right: view into the GIL shaft (*Source* [77])

The nine bays of 362 kV GIS are located in a cavern in the mountain under the dam. The substation and hydro-generation equipment were brought in through an access tunnel. The GIS equipment was transported in segments of a bay, typical units are circuit breaker, disconnect and ground switch, busbar, connection to cable, and GIL. See sketch in Figure 8.211.

The vertical tunnel of 115 m length could not be solved easy by cables. The typical oil cables at the time of installation in 1977 would require a high oil pressure of more than 10 bar, which would not be easy to handle. In addition, the oil would bring a high risk of fire into the tunnel. Solid-insulated cables have not been available at this time and at this voltage level using XLPE insulating material. But also, solid insulated cables in a vertical installation need multiple fixing point to the tunnel wall to avoid long-time floating of the insulating material and a damage of the high-voltage insulation.

The gas insulated transmission line offers vertical solutions by using a construction kit of modular segments. Each GIL segment of 12 m length is fixed to the tunnel wall by steel structures to keep the weight of one segment. This adds up to 10 segments for the total length and is easy to handle. The insulating gas is not depending on the vertical orientation of the GIL because gas molecules can move free inside the GIL enclosure and distribute equal in the gas compartment. The differences of gas density from the bottom to the top can be neglected.

The photo on the right of Figure 8.211 gives an impression of the vertical tunnel and how the GIL has been fixed to the walls of the tunnel. In the middle of the tunnel, a ladder is installed to allow maintenance and operation personal inspection of the GIL in the tunnel. Such inspections are recommended in longer time distances, e.g. once year to check if no corrosion takes place on the steel structures or any changes on the GIL may be noticed. Corrosion might come if water is penetrating the tunnel and when continuous water penetration come to the surface of the steel structures or GIS, some corrosion protection measure is required. This could be corrosion protection coating or to install steel sheets to re-direct the water away from the steel structure and GIL.

The remote area substation is typical GIS/GIL application to solve the space and building requirements in this remote area. The building, in this case a cavern in the mountain, needs to give space for the power station equipment and in case of the GIS, dry ambient conditions and a solid and even floor for GIS installation of bay by bay.

The GIL offers the vertical orientation without great additional effort compared to horizontal laying. The principle uses of 12 m GIL units to form vertical construction sections hold and fixed to tunnel wall. See Figure 8.211.

8.9.2.14 Futuristic Harbor Substation

The futuristic harbor substation is connected to the electric network by 123 kV cables coming in underground to the futuristic building in the harbor area. The 123 kV GIS has seven bays of a double busbar configuration including a busbar section disconnector and coupler bay.

The medium voltage substation of the futuristic harbor substation has nine units for the connection of AC and DC distribution lines. The AC distribution lines are for the harbor infrastructure area, and the DC distribution lines are used for the connection of ships in the harbor to deliver electric power instead using the ship diesel engines. The building gives also room for the control and protection equipment.

The building concept of the substation is to create an eye-catching architecture to bring the importance of electric power supply into focus. The building design will give technical views to high-voltage equipment and bring it into a positive context of modern living. Big glass windows allow the view of the public to look inside the substation and notice the technical equipment, which brings electric energy into every body's life.

It is not to hide a substation behind walls, how it is done in most indoor applications of electric power supply substations. Following the idea: out of sight, out of mind, less problems. Here, with the futuristic harbor substation concept, the message is different: look here what we are doing for you, high-quality equipment brings you the electric energy, see how nice it looks.

There are some projects around the world following this model to get into close contact to the public with technical power supply solutions, e.g. in Berlin and Hamburg, Germany or in Zurich, Switzerland. The experiences made are different. Most people see is positive and like such open views. Others are starting criticism on it in general. As usually in life, there are always some opponents.

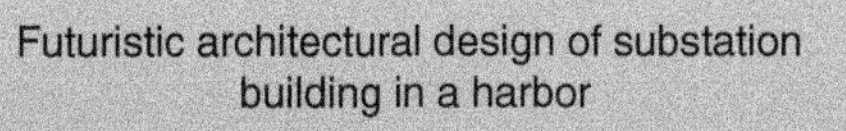

Figure 8.212 Futuristic harbor substation, left: multilevel substation building in the harbor area, right: seven 123 kV GIS bays double busbar scheme (*Source* left: [78] and right: [44, 61])

The GIS project of futuristic harbor substation gives a clear indication for innovative solutions of substation design is possible when using GIS technology. In these cases, the substations in cities harbor district with limited space, connections only be underground cables, architectonical requirements from the surrounding, and the wish to have a representative building to present the importance of high-voltage substations for the common delivery of electric energy to the public and industry (Figure 8.212).

8.9.3 Conclusion

This chapter gives examples of innovative solutions using Gas insulated switchgear (GIS) and gas insulated transmission Lines (GIL) technology. This created solutions of a wide range of special substation building concepts. Nine examples around the world are given to explain certain situations that have led for special substation buildings: located in the basement of a business building in the city center, fit into multilevel building without any change on the outer building appearance, underground in a shopping complex, caverns in the mountains with hydropower plants, urban area building adapted to the neighborhood, under a football stadium, under a schoolyard, under a memorial park, and as an architectonic eye-catcher in the harbor district.

All the innovative ideas in realizing high-voltage substations were only possible be using GIS for the substation transmission and distribution voltage side. In some cases, GIL is used to connect electric energy of the hydropower generation in a cavern in the mountain or a dam to the outside for the overhead transmission line.

The compact design and high reliability of GIS and GIL design is the basis for such applications. Higher cost for equipment and building are compensated by acceptable solutions for the public and authorities. This makes a project more likely to get permission and at the end allows the city development in this area where electric power increase is the consequence of growth.

The high concentration of loads in city center requires high-transmission currents of up to 5000 A and high short-circuit interruption capabilities of up to 80 kA. This can be performed by high-voltage GIS on the transmission voltage side.

At the medium-voltage GIS on the distribution voltage side, the compact design of the switching panels allows many bays at a small space required. Medium-voltage GIS may be air insulated or SF_6 gas insulated.

The transformers are usually using insulating oil and effective fire distinguishing systems. In some cases, SF_6-insulated transformers are used if the power transmission is not so high. Both types of transformers show a high expected life time if operated in the given operation limits of voltage, current, and temperature.

The future development of special substation buildings is seen as a growing market. While GIS technology is gaining market share over the last decades and this will continue because of growing cities and city development. With this, the need for innovative solutions for substations and special building design will develop. Modern architecture in cities will use the technical appearance of electric power supply to bring in more into the mind of the public. Eye-catching forms and design features to integrate cultural symbols, e.g. desert tent or exhibition hall design, will be used more often. GIS and GIL technology will give the required technical features to realize such concepts.

References

1 IEC 62271-203:2011 High-voltage switchgear and controlgear - Part 203: Gas-insulated metal-enclosed switchgear for rated voltages above 52 kV

2 IEEE C37.122-2010 - IEEE Standard for High Voltage Gas-Insulated Substations Rated Above 52 kV

3 Mc Donald, John: Electric Power Substations Engineering, Second Edition, Clause 3: Bio, Michael, Air-Insulated Substations – Bus/Switching Configurations, CRC Press, 2007.

4 IEC 62271-209:2019 RLV High-voltage switchgear and controlgear - Part 209: Cable connections for gas-insulated metal-enclosed switchgear for rated voltages above 52 kV - Fluid-filled and extruded insulation cables - Fluid-filled and dry-type cable-terminations

5 Strassenhaus, Prof. Lipp, https://www.baulinks.de/firmen/strassenhaus.com 2008

6 Ein transdisziplinäres Konzept für die Stadt des 21sten Jahrhunderts (2010), LIFIS ONLINE, www.leiniz-institute.de

7 Gas Insulated Transmission Line (2005), Siemens Technical Brochure, www.siemens.com/gil

8 Gas Insulated Transmission Line (2007), Siemens Technical Brochure, www.siemens.com/gil

9 Labos, William; Grossmann, Peter: Case Study – 80 kA Gas Insulated Substation Bergen Switching Station – New Jersey, T&D Conference 12.-17. April 2014, Chicago

10 Koch, Hermann. 2014. GIS Gas Insulated Substations. United Kingdom: John Wiley and Sons Ltd.

11 Peschke, E. von Olshausen, R. Cable Systemes for High Voltage and Extra High Voltages,Pirelli, Publicis, Erlangen, München, 1999

12 Design guidelines for high voltage lines, 2020, National Grid public internet, www.national grid.com/sites

13 Hydro One Networks Inc. application to reconductor a high-voltage transmission line, Hydro One Networks, Public Internet www.oeb.ce/practicate/applications

14 Report Integrated Regional Resource Plan, Toronto Region, ieso, public internet www.hydroone.com/abouthydroone 2021

15 Public internet, www.shop.bruggcables.com

16 Hermann Koch, Petr Rudenko, Bhaskar Roy, Siemens AG, Germany, Siemens Ltd., Energy Management Division, India, ,Applications of Gas Insulated Lines as Alternative to Conventional Power Transmission Solutions, CBIP/CIGRE Symposium, 2017-07, Delhi, India

17 Technical Brochure Gas Insulated Transmission Line (GIL) 2010, Siemens

18 Public internet (2020), www.schildknecht.com/boringmashine

19 United Nations Report World Population 2018, www.un.org/development/ desa/ en/ news/ population/2018-world-urbanization-prospects

20 Public domain internet waste water, 2020
21 Public domain internet high temperature fire, 2020
22 Public domain internet ventilation, 2020
23 Public domain internet sprinkler installation, 2020
24 Public domain internet drainage system, 2020
25 Public domain internet fire protection sensor, 2020
26 Public domain internet fire resistant doors, 2020
27 IEC 61936-1: Power installations exceeding 1 kV AC – Part 1: Common rules, Edition 2.0 2010
28 Public domain internet air cleaner, 2020
29 Public domain internet transformer termination, 2020
30 Public domain internet transformer and GIL, 2020
31 Air Conditioner Unit https://en.wikipedia.org/wiki/Heating,_ventilation,_and_air_conditioning
32 Public domain internet heat exchanger, 2020
33 Public domain internet acoustic sound absorber, 2020
34 Public domain internet forced ventilation, 2020
35 Public domain internet lightning, 2020
36 Public domain internet substation control room, 2020
37 Public domain internet playground, 2020
38 CIGRE TB 509: Final Report of the 2004 – 2007, International Enquiry on Reliability of High Voltage Equipment, Part 1 - Summary and General Matters, Working Group A3.06, October 2012.
39 CIGRE TB 510: Final Report of the 2004 – 2007, International Enquiry on Reliability of High Voltage Equipment, Part 2 - Reliability of High Voltage SF6 Circuit Breakers, Working Group A3.06, October 2012.
40 CIGRE TB 511: Final Report of the 2004 – 2007, International Enquiry on Reliability of High Voltage Equipment, Part 3 - Disconnectors and Earthing Switches, Working Group A3.06, October 2012.
41 CIGRE TB 512: Final Report of the 2004 – 2007, International Enquiry on Reliability of High Voltage Equipment, Part 4 - Instrument Transformers, Working Group A3.06, October 2012.
42 CIGRE TB 513: Final Report of the 2004 – 2007, International Enquiry on Reliability of High Voltage Equipment, Part 5 - Gas Insulated Switchgear (GIS), Working Group A3.06, October 2012.
43 CIGRE TB 514: Final Report of the 2004 - 2007 International Enquiry on Reliability of High Voltage Equipment, Part 6 – Gas Insulated Switchgear (GIS) Practices, Working Group A3.06, October 2012.
44 Siemens: Gas-Insulated Switchgear up to 145 kV, 40 kA, 3150 A, Type Series 8DN8, 2010, Order Number E50001-G620-A 122-X-4A00
45 Public domain internet park, 2020
46 Public domain internet parking lot, 2020
47 Public domain internet city center, 2020
48 Public domain internet convention center, 2020
49 Public domain internet city square, 2020
50 Public domain internet building concepts city, 2020
51 City center development, Doha, public internet www.wikipedia.org/Doha 1980/2014
52 Public Google maps, Doha, Emirates, 2020
53 Public domain internet industrial building, 2020
54 Siemens: Gas-Insulated Switchgear up to 420 kV, 63 kA, 5000 A, Type Series 8DQ1, 2013, Order Number E50001-G630-A 239-V1-4A00.
55 Public domain internet exhibition area, 2020
56 Public domain internet fire protection sensor, 2020

57 Public domain internet siemens/products/GIS, 2020
58 Koch, Hermann. 2014. GIS Gas Insulated Substations. United Kingdom: John Wiley and Sons Ltd.
59 High Voltage GIS design, public internet www.siemens/gis, 2020
60 High Voltage GIS, public internet www.siemens-energy/gis 2021
61 High Voltage Transformer, public internet, www.siemens-energy.com/products
62 Gas-insulated switchgear (GIS), Applications, Buildings, public internet www.hitachienergy.com/product-and-systems, 2021
63 Gas-Insulated Solutions, GIS, GIL, Hybrid Substations, Mobile GIS, public internet www.gegridsolutions.com/hvmv_equipmen/GIS, 2021
64 Hydropower Plant, Mountain Valley, public interenet https://en.wikipedia.org/wiki/Tehri/Dam 2020
65 Gas-Insulated Transmission Line (GIL), Siemens Brochure, Power Transmission and Distribution, 2007
66 Gas Insulated Line, Siemens Reference List, Power Transmission and Distribution High Voltage Division, 2009
67 Siemens Technical Brochure Gas-insulated switchgear up to 245 kV, 50 kA, 4000 A Type 8DN9, 2014
68 Gas Insulated Substations Applications, www.wikipedia.org/ Gasisolierte _ Schaltanlagen 2020
69 Brochure GIS Substation References, Installation in Special Building Arrangements, Siemens, 2016
70 Gas Insulated Substation Indoor, public internet www.westbay/substation/qatar 2020
71 GIS in Parking Building, public internet www.wikipedia.org/Fahidi-substation 2020
72 GIS below Football Stadium, public internet www.google/football/stadium 2020
73 Prefabricated concrete, public internet prefabricated/concrete/slab/und/grass/floor 2020
74 Football Stadium public internet football/stadium/swiss 2020
75 GIS in Monument Park, public internet monterry/substation/Mexico 2020
76 Power transformer applications, public internet www.siemens/transformers 2020
77 Hydro Power Cavern Plant , public internet www.wikipedia.org/Ruacana/Hydroelectric_Power_Station 2020
78 Harbor Substations Special Building, www.sheher/harbor/building, 2020

9

Advanced Technologies

Authors: 1st edition Hermann Koch, Venkatesh Minisandram, Arnaud Ficheux, George Becker, Noboru Fujimoto, and Jorge Márquez-Sánchez
2nd Edition Hermann Koch, Maria Kosse, George Becker, George Becker, Mark Kuschel, Aron Heck, Dirk Helbig, Uwe Riechert
Reviewers: 1st Edition George Becker, Devki Sharma, Noboru Fujimoto, Venkatesh Minisandram, Phil Bolin, Pravakar Samanta, Hermann Koch, Linda Zhao, Xi Zhu, John Brunke, Dick Jones, Linda Zhao, David Lin, Devki Sharma, and Patrick Fitzgerald, 2nd Edition Michael Novev, Pablo Gonzales Toza, George Becker, Hermann Koch, Dick Jones, Bala Kotharu, Johne Brunke, James Massura, Dirk Helbig, Mark Kuschel, Arnaud Ficheux, Robert Lüscher, and Aron Heck

9.1 General

This chapter provides information related to practical experiences and the latest developments and innovations of substation design using GIS. This does not mean that these topics are less important. Topics such as the Environmental Life Cycle Assessment in Section 9.2 are increasingly important in view of total life cycle assessment and the environmental impact assessment of the product during its entire life cycle, the commissioning and decommissioning processes, and with consultations with authorities. Life cycle cost analysis in Section 9.3 is more of relevance to initial purchasing when searching for the best and most economical solution [1].

In Section 9.4, the effects and influences of insulation coordination is explained and the reasons for studies are given. The impact of overvoltage and the grounding system are explained. Very fast transient (VFT) voltages and transient enclosure voltages (TEVs) are explained for GIS in Section 9.5. Simulation of the VFT phenomena related overvoltage and impact to insulations are explained and related failures are explained, based on the generation and propagation of transient enclosure voltages of GIS. Shock hazard and its preventions are also explained in this section. Information on induced voltages to control wires is given.

The planning issues for the project scope development are discussed in Section 9.6. These issues include planning for the installation, site preparations, insulation of new GIS extensions of existing GIS, and equipment access. Risk-based management is the topic of Section 9.7.

In Section 9.8, the health and safety impact of GIS is explained. This covers risk of burning, risk of breathing toxic gases, risk of touching toxic materials, and risk of electric shock.

The electromagnetic field topics are covered in Section 9.9. Here, the electric field in operation and during short circuit of GIS is explained. Information is given for magnetic fields of GIS in operation and during short circuit.

Gas Insulated Substations, Second Edition. Edited by Hermann J. Koch.

In section 9.10 decomposition by-products of SF6 are explained and handling procedure described. The evaluation procedures for GIS condition assessment are given in 9.12.

Substation of advanced resiliences are explained in 9.12. The new developments of vacuum switch applied to high voltages above 52 kV are shown in section 9.14 including the physical background of arc interruption. Low power instrument transformers of section 9.14 are basis for future digital substations and provide new sensoring possibilities in a network with high level of regenerative power generation. Their function and applications are explained. Digital twin procedures for GIS and GIL and their application are presented in 9.15 to provide a life long digital documentation.

The changes coming with the digitalization of substations, which is required to manage and control a power network with a large number of generation units (wind or solar) and fast reaction time on volatile availability of generated electric energy.

9.2 Environment

9.2.1 Environmental Life Cycle Assessment

9.2.1.1 Introduction

Decisions on procuring products have generally relied on technical compliance, initial and total life cycle costs, and targeting a balance between functional needs, cost, and reliability. With increasing environmental awareness and focus, regulators, manufacturers, users, and customers are tending to consider the environmental impact of the equipment or project in business decisions. One tool resorted to is the "Environmental Life Cycle Assessment – Life Cycle Assessment" (normally referred to as the Life Cycle Assessment – LCA), which considers the entire life cycle of a product. The importance of LCA application in new product development and corporate strategy is gaining prominence.

9.2.1.2 LCA Process

The Life Cycle Assessment (LCA) is a process to evaluate the environmental burdens associated with a product, process, or an activity by identifying and quantifying energy and materials used and wastes released to the environment; and to identify and evaluate opportunities to affect environmental improvements. The assessment includes the entire life cycle of the product, process, or activity; encompassing, extracting, and processing raw materials; manufacturing, transportation, and distribution; use, reuse, and maintenance; and recycling and final disposal [2]. These stages are shown in Figure 9.1.

The following International Standards Organization (ISO) environmental management standards describe the procedures and methods for the LCA:

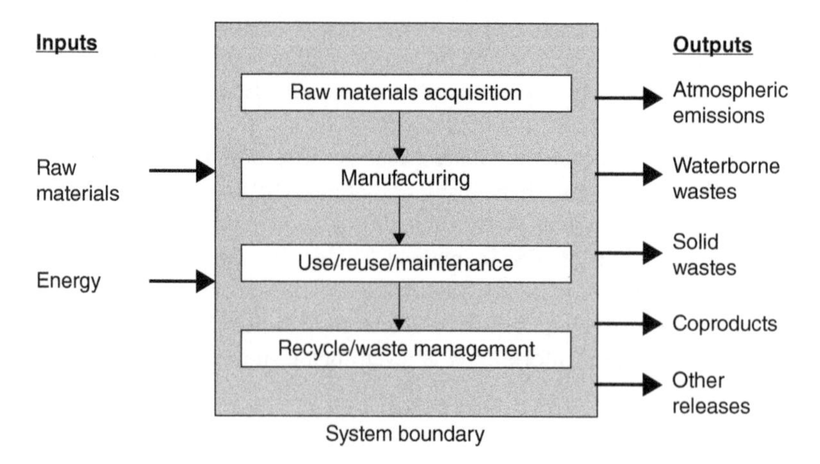

Figure 9.1 Life cycle stages (Reproduced by permission of Siemens AG from EPA, 1993)

- ISO 14 040 Environmental management – Life cycle assessment – Principles and framework [24]
- ISO 14 044 Environmental management – Life cycle assessment – Requirements and guidelines [25].

9.2.1.3 LCA Relevance to GIS

According to the EPA, roughly 80% of all SF_6 produced worldwide is used by electric power industry. The GIS technology utilizes SF_6 as an insulating and arc-quenching medium. *With a global warming potential (GWP) 22,800 times greater than CO_2 and an atmospheric life of 3200 years, one pound of SF_6 has the same global warming impact as 11.4 tons of CO_2.*

When it comes to utilization of electrical equipment that uses SF_6, invariably the focus is on the global warming potential while evaluating the alternate options during project planning or permitting and licensing deliberations. Without a global view of all the environmental issues associated with each of the options and the overall environmental impact assessment, any decision based on a narrow environmental impact review on a single issue could be detrimental from a life cycle assessment perspective.

As noted in PAS 2050:2011 [3], the specification focus is on a single environmental issue [i.e., greenhouse gas (GHG) emissions and their contribution to climate change], but this is only one of a range of possible environmental impacts from specific goods or services. The relative importance of those impacts can vary significantly from product to product, and it is important to be aware that decisions taken on the basis of a "single issue" assessment could be detrimental to other environmental impacts potentially arising from the provision and use of the same product.

9.2.1.4 Industry Response to Environmental Impacts

The high global warming potential of SF_6 was identified in 1995 and was listed as a greenhouse gas in the Kyoto Protocol [4]. This resulted in enhanced efforts to monitor and reduce the overall emissions and usage of the SF_6, institute voluntary emission reduction programs, provide training on usage and handling, and promote improvements in manufacturing process.

Since the first generation of GIS installations in the 1960s, technological advancements and environmental needs have resulted in a significant reduction of GIS dimensions (by approximately 80% for a 145 kV GIS), smaller gas compartments, improvements to design, manufacturing, and testing methods, leading to a positive impact on the environment. The mass of SF_6 in the GIS has been significantly reduced, as reflected in Figure 9.2 [5].

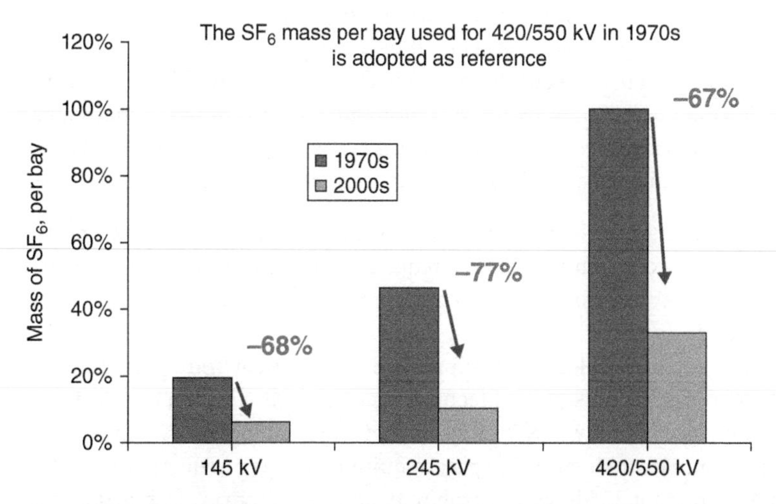

Figure 9.2 Graph showing the significant reduction in SF_6 mass since the 1970s (Reproduced by permission of Siemens AG)

The gas leakage rate of equipment has been on the decrease over the years and manufacturers have been targeting a design leak rate of less than 0.1% per year. The standards [7, 10] have accordingly been updated and currently the standards specify that the leakage rate from a gas compartment should not exceed 0.5% per year. The validation of this is seen during factory design and production tests. The assurance of these low leakage rates requires laboratory conditions and cannot be measured on site. The focus on site is on proper gas handling, proactive gas monitoring for advanced gas leak detection, and gas reuse/recycling.

Gas handling is one area where the release of SF_6 gas to the atmosphere can be prevented. To help achieve this goal, the standards organizations such as IEEE and IEC have published SF_6 gas handling guides [6, 7].

While, Programs, such as EPAs SF_6 Emission Reduction Partnership for Electric Power Systems has resulted in increased awareness and focus on SF_6 emissions and handling, the European Union requires certification of anyone involved in gas handling.

Most of the GIS equipment is shipped from the factory assembled and filled with positive SF_6 gas pressure or dry nitrogen. These require SF_6 topping up or filling on site to the rated gas pressure. Once the equipment is placed in service, it is expected that gas handling will be needed only twice during the equipment lifetime (one equipment overhaul mid-life and at the end of life during decommissioning). Gas handling units are now available that are capable of recovering SF_6 gas to a higher degree prior to opening a gas compartment for internal inspection.

The equipment manufacturers have also started offering online gas monitoring systems in order to continuously monitor and provide early warning of gas leaks. The adaptation of transducers with integrated pressure and temperature sensors has enabled remote SF_6 gas density monitoring, leak rate determination, and proactive corrective actions. This has been featured as an advancement of the conventional practice of utilizing temperature compensated pressure gages/switches with alarm and design limit thresholds. With conventional gage settings of about 10% below the rated gas pressure, the alarm trigger does not render itself for an early action to identify and mitigate the gas emission.

The availability of compact gas view cameras has also facilitated identification of the source of minute gas leaks quickly, avoiding the need to take the equipment out of service to determine the leak location and take prompt corrective action.

9.2.1.5 Life Cycle Assessment (LCA) Studies

As noted previously, life cycle assessment requires a global view of the environmental impacts. LCA studies have been done to cover transmission and distribution electric equipment.

One study was undertaken for a comparative LCA using different switchgear technologies with and without the use of SF_6 evaluated at the level of the switch bays and electric supply grid to a city [8]. The LCA study was carried out in compliance with the requirements of the international standard DIN EN ISO 14 040 [24] and accompanied and evaluated by an external independent audit from the TÜV NORD [26] compliance testing agency. The study produced the following results:

(a) Level of switch bays. Use of SF_6 technology provided an advantage in four out of five criteria of the study: for primary energy consumption, space requirements, acidification, and nutrification potential. The total global warming potential can be reduced subject to low SF_6 losses and in bays with high loading.
(b) Level of power grid. Use of GIS technology lowered all five potential environmental impacts studied. The layout of the power grid using GIS technology compared with the layout of the same grid with AIS technology (with lower SF_6), approximately providing a 27% reduction in primary energy consumption, 86% reduction in space requirements, 21% reduction in global warming potential, 21% decrease in acidification potential, and 29% decrease in nutrification potential.

The reasons for the environmental impact reductions included:

- Stations and equipment in the GIS option can be made with less material and energy than those of the AIS alternative due to substantially better insulation and arc-extinguishing properties of SF_6 in comparison to air.
- The compact construction of GIS lends itself to locating substations near inner city load centers, thereby leading to the reduction of losses by deploying high voltage transmission lines to power the substation and serving customers by using short medium voltage lines.

The AIS alternative instead required the location of the substations in the outskirts of the city, powering the substations via a HV/MV transmission line, and serving customers using longer medium voltage lines (medium voltage lines have larger transmission losses than high voltage lines because of higher currents).

A study undertaken to perform a preliminary assessment of the life cycle carbon emissions of the transmission network in Great Britain [9] reveals that transmission losses alone account for 85% of total carbon emissions of CO_2 while SF_6 losses are around 12%. It suggests credible opportunities to deliver carbon benefits through lower transmission losses.

Life cycle environmental impacts of various substation equipment have been characterized [10]. This facilitates modeling the impacts of individual power grid components and also assembling the different components to model a specific system in order to calculate the overall environmental impacts for that system.

9.2.1.6 End-of-Life Recycling

The end-of-life stage relates to the impacts associated with the demolition and recycling or disposal of materials. Being compact and of modular design, a GIS lends itself to easy dismantling. The initial phase includes evacuation of SF_6; efficient gas-handling units are capable of recovering SF_6 to the highest degree. Reclaiming and reuse of SF_6 strategies have evolved, which include reuse of gas in new equipment and refeed of used gas into the SF_6 production process for converting it into new gas or safe incineration at the gas manufacturing facility. With new developments in contaminated gas processing, incineration is a rarity. Detailed gas-handling instructions are given in [27, 28].

The GIS is mainly composed of metals (aluminum, copper, and steel), which can be effectively recycled. Depending on the GIS population of the same vintage in the system, it may be desirable to retain some of the components as spares.

Many manufacturers have now developed patented technologies to recycle SF_6 gas. However, we cannot recover and recycle 100% of the SF_6 gas. The % of gas that cannot be reused shall be neutralized for disposal.

9.2.2 Eco Design

9.2.2.1 Introduction

Energy demand has been increasing on a regular basis over the last few decades. This growth trend is expected to continue in the coming years. Due to the significant impact of human activity on the environment, the delivery of electricity must take a sustainable approach. As a result, GIS manufacturers continue to integrate eco-design techniques throughout the development of their GIS, leading to a reduction in the overall impact to the environment over the life cycle of the equipment.

9.2.2.2 Eco-Design Approach

The GIS manufacturers now follow the environmentally conscious design (ECD) process, described in IEC 62430 [29]. This process involves the identification and evaluation of environmental aspects, the integration of eco-design concepts into the existing design process (creation of significant environmental parameters, the fixing of environmental targets, and development of improvement strategies), the review and continual improvement of products, and, finally, communication of progress made. It is now common practice to integrate eco-design engineers into the development teams to bring their expertise to the design, manufacturing, transport, installation, operation, and decommissioning of the GIS.

To evaluate the environmental impact of GIS entire life cycle needs to be covered: from rough material extraction, to manufacturing, then distribution (transport and delivery), use (operation), and, finally, end of life, as shown in Figure 9.3.

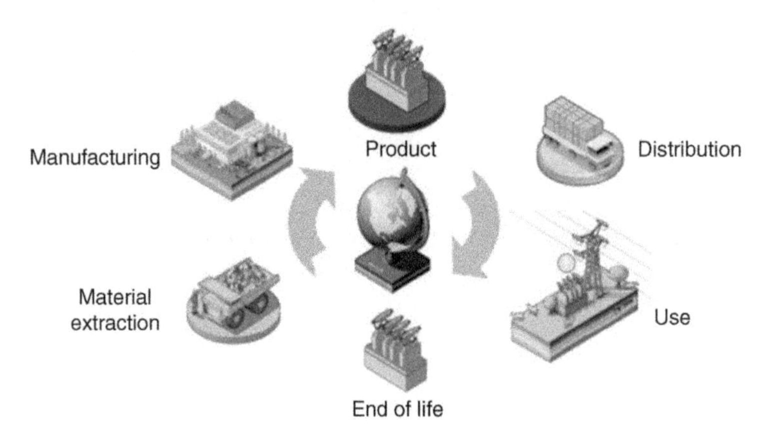

Figure 9.3 Life cycle concept (Reproduced by permission of GE's Grid Solutions now GE)

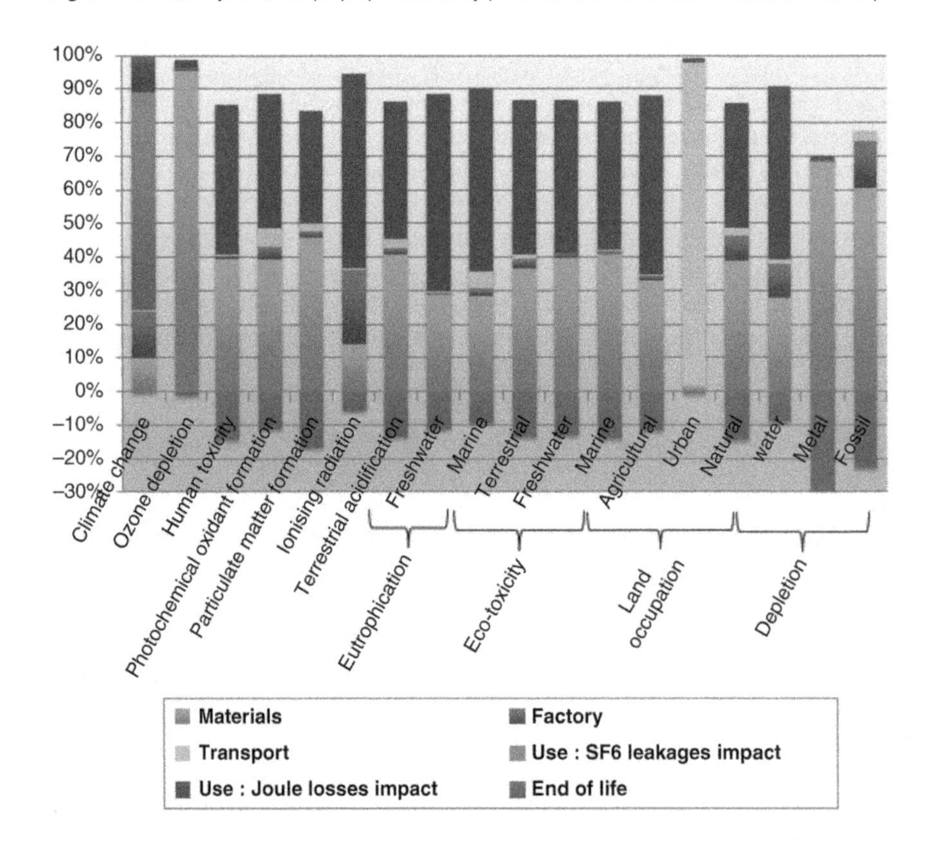

Figure 9.4 Example of LCA results for a GIS 420 kV GIS (Reproduced by permission of GE's Grid Solutions)

Table 9.1 Recycling process for GIS/GIL materials

Material	Process
SF_6	To be pumped out of the GIS/GIL and stored in container Reused in other GIS/GIL High reuse value
Aluminum enclosure	Cut in sections and reused in aluminum production High reuse value.
Aluminum conductor	Cut-in sections are reused in aluminum production High reuse value
Resin insulators	Separated and reused for degraded materials Lower reuse value
Sliding contacts	Separated and reused metals Silver separated high reuse value
Compensator enclosures	Separated and reused metals High reuse value
Cast aluminum enclosures	Separated and reused in aluminum production High reuse value

9.2.2.3 Life Cycle Analysis

A life cycle analysis (LCA) is done using specific environmental software. The typical LCA result for a 420 kV GIS is shown in Figure 9.4. This includes environmental impact of the following factors: material, transport, use/Joule losses, factory, use/SF_6 leakage, and end of life. These factors are applied to 18 environmental processes which are given in Figure 9.4. In Table 9.1, the recycling processes for GIS and GIL are explained. Almost all materials used for building GIS and GIL can be recycled and reused. They are easy to separate as the main components are no mixed materials: SF6 gas, aluminum enclosure and conductor, resin insulators, sliding contact of copper and silver, compensator enclosures of aluminum or stainless steel and cast aluminum enclosures. This covers about 99.5% of the used materials. In Figure 9.4, the vertical axis gives reference of the different impacts listed on the horizontal axis as a percentage related to the first column of climate change.

The LCA can also give indications where the main impacts of the product can be. We can simplify the LCA results as a pie chart, as shown in Figure 9.5.

Three interesting and important results can be deduced from the above example:

Figure 9.5 GIS 420 kV environmental impact sources (Reproduced by permission of GE's Grid Solutions)

- The main environmental impact of the GIS product is the energy (joule) loss during the use phase (calculation made assuming 40 years of operation).
- The SF_6 leakages during the use phase are of concern with respect to global warming.
- The optimization of types of materials and their quantities also represents major ways of reducing the environmental impact of GIS.

The above example is given for a typical 420 kV GIS, but the trends are also valid for any other type of GIS at any voltage level.

9.2.2.4 Examples of Eco-Design Improvements in GIS

Management of Hazardous Substances

GIS must be safe for people and for the environment. To meet this requirement, some substances should be avoided. Manufacturers have adapted their R&D and production activities to avoid the use of certain specific substances (hexavalent chromium or lead) in manufacturing process (like surface treatment or welding).

Reduction of Joule Losses

As shown previously, Joule losses have a high environmental impact. Designers must select materials and contacts with a very low resistance and optimize on the quantity of conductors used to transmit electricity. As an example, for the same 420 kV GIS, two technical solutions can be available: one with a conductor diameter optimized for nominal current and cost and another one with a conductor of larger diameter optimized to reduce the joule losses. Although the second solution is more expensive at the initial investment phase, the differences is totally offset when the cost of joule losses are accumulated over the product's 40-year life cycle. It also reduces the total environmental impact of the product by 8% (average of all the environmental indicators), as can be seen in Figure 9.6.

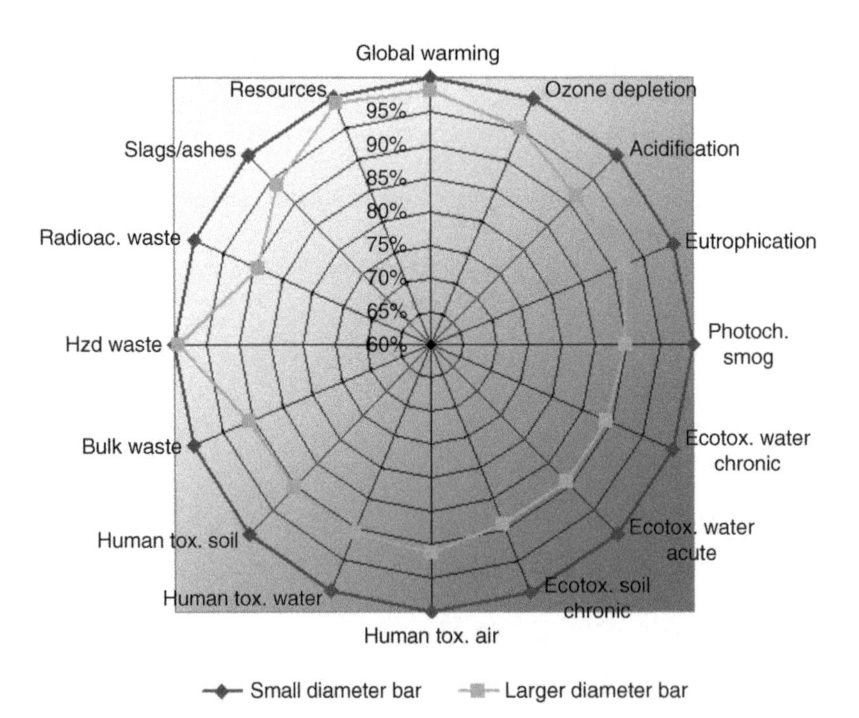

Figure 9.6 420 kV GIS with different diameters of the bar (Reproduced by permission of GE's Grid Solutions)

SF_6 Gas Management

SF_6 is used in GIS due to its excellent dielectric properties. It is also chemically inert, nontoxic, and has very good arc-switching characteristics. Today, it is recognized that there are several alternatives to SF_6 for high voltage equipment are under investigation. This covers flour ketone, flour nitrile, and technical air, see clause 2.9. However, this gas is a potent greenhouse gas and has been identified in the Kyoto Protocol. As a consequence, GIS manufacturers and GIS users have put in place specific policies to control and reduce SF_6 emissions. For instance, during the manufacturing phase, many practices and procedures are already in place to limit SF_6 emissions. The inventory methodology to quantify SF_6 emissions has been defined by CAPIEL [30] (the manufacturers' association) and is now largely used by GIS manufacturers: SF_6 emission = SF_6 input – SF_6 output ± delta stock.

SF_6 gas detectors (see Figure 9.7) are commonly installed in GIS manufacturing premises. They analyze the air, indicate the SF_6 concentrations, and activate an alarm in the case of leaks, allowing the operator to immediately shut down and determine the causes of the emissions. All the leaks are recorded via the acquisition unit, and corrective actions can be put in place in order to avoid repetition of these leakages. During the use phase, gas tightness and optimization of equipment volume are key elements in controlling the emissions of SF_6.

Manufacturers continue to reduce their leakage rates during operation – presently less than 0.5% per compartment per year according to IEC 62271-203 [28] or IEEE C37.122 [31]. On the latest generation of GIS, this low rate of leakages is achieved thanks to the proven and widely used EPDM (ethylene propylene diene monomer)-type gasket solution. It offers high resistance to extreme temperatures and also to decomposition by-products sometimes produced in the SF_6 gas. The goal is to continually reduce SF_6 emissions.

Some activities on site, such as the decommissioning of equipment, requires recovery of SF_6. A recycling process usually involves a company that specializes in the recovery and treatment of SF_6, so that nearly 99% of SF_6 gas is recycled and reused (according to the CIGRE guide, "SF_6 Recycling Guide: Reuse of SF_6 gas in electrical power equipment and final disposal") [32].

End-of-Life Management

The end of life is also a very import phase of the product life cycle. The design phase offers a high degree of liberty to optimize the product's end-of-life impact and to increase the recyclability rate. Solutions using recyclable materials are preferred to nonrecyclable ones.

When there is no recycling solution, such as in the case of epoxy resin insulators, GIS designers tend to reduce the quantities used. However, use of recyclable materials to facilitate end of life can be effective only if the different materials can be dismantled and separated. For example, copper cannot be recycled with aluminum or steel as it pollutes the recycled material and decreases its welding or elongation properties.

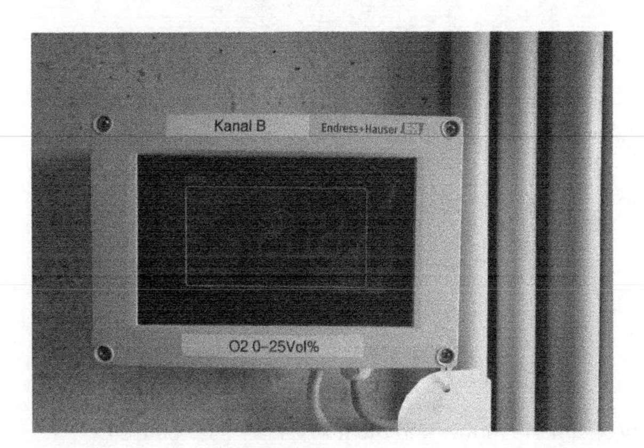

Figure 9.7 SF_6 gas leakage detector (Reproduced by permission of Siemens)

In order to facilitate recycling, manufacturers need to support the users to manage the end of life of their equipment. Recycling instruction manuals, detailing how to dismantle the equipment, and how to sort out the different families of materials, are now becoming a common deliverable by the GIS manufacturer.

Example of LCA between Two Generations of GIS

At the end of product development, a life cycle assessment of the new product is commonly performed. It provides a way to compare the results with the previous generation. Example results are illustrated in Figure 9.8. This calculation has been performed with SimaPro software and ReCiPe 2008 methodology.

We can see that the environmental impact of the latest 245 kV GIS has been significantly reduced compared to the previous generation; on average, on the 18 indicators this represents a reduction of 16%. This reduction has been achieved by the reduction of the joule losses during the use phase, by reducing the number of electrical contacts, by reducing the active parts lengths, and the increase in active parts sections.

On the "urban land occupation" indicator, the environmental impact has been reduced by more than 60% by reducing the wooden packing material by 65% because of less material used.

9.2.3 Environmental Impact Case Study: Transmission Project Using GIL or XLPE Cables

9.2.3.1 General

The environmental impact of technical systems has an increasing importance world-wide. In industrial countries in Europe, North America, and Japan, strong laws are in place that can cause high cost through operation, for example, when damaging environment or at the lifetime end when the technical systems need to be dismantled. Contaminations of soil would lead to soil treatment and exchange. Besides these toxic hazards to the environment, for several years the impact to global warming and the CO_2 equivalent contribution of building and operating technical transmission

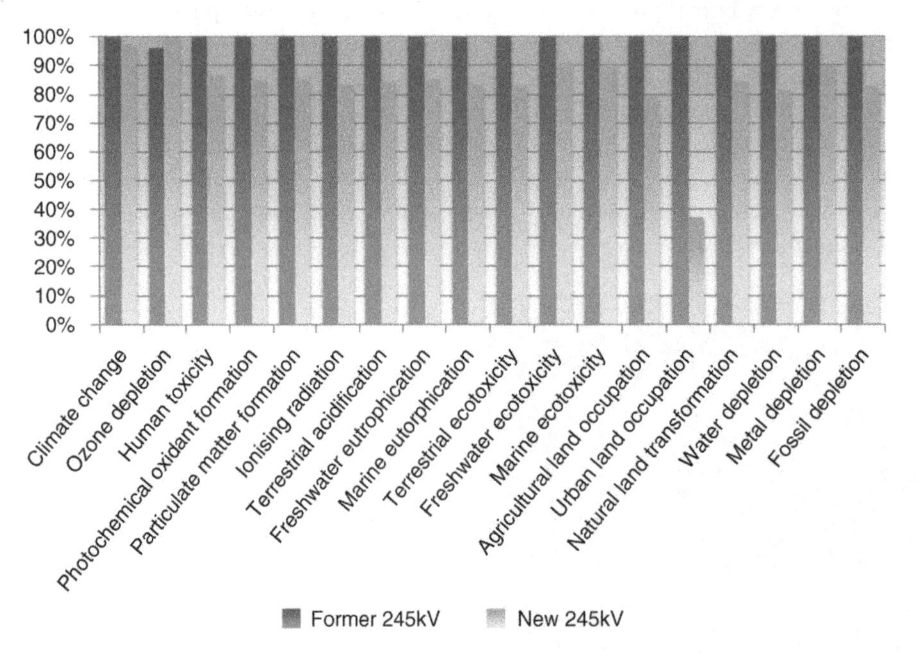

Figure 9.8 Comparative life cycle assessment between the former and the new 245 kV (Reproduced by permission of GE's Grid Solutions)

systems is part of the total project evaluation and will add cost, for, example, to the need for CO_2 contribution shares. This leads to a higher evaluation of transmission losses during operation.

In emerging countries like China, India, and Brazil the importance of environmental impact is strongly increasing with the growth of the industry and business and the increasing number of environmental hazards known from these countries, for example, air pollution and water pollution. Legislators in these countries are working hard to close legal gaps for environmental hazards, and in some cases, the limiting values are sharper than those in industrial counties, for example, EMF exposure values.

In this section, the environmental impact of a transmission project using GIL or XLPE cables will be presented.

9.2.3.2 Toxic Materials

Both GIL and XLPE cables do not use toxic materials in normal operation mode. In the case of GIL, the main toxic material comes from the use of SF_6 and its by-products [33, 34]. Among these byproducts SO_2 F_2, S_2F_{10}, and SF_4 are known to be highly toxic. The use of SF_6 has to follow handling procedures, defined in IEC 62271-4 [35] and C37.122.3 [27, 33], as a mandatory standard requirement to keep SF_6 always in closed compartments and to avoid exposure to the atmosphere. SF_6 itself is not toxic. However, in the case of internal arcing the high temperatures of the arc can produce SF_6 by-products, which can be toxic. In this case of internal arcing, the SF_6 and its toxic by-products stay inside the metallic enclosure and will be reclaimed by skilled experts, as defined in the above-mentioned IEC and IEEE standards. No toxic material is spilled into the surroundings, even in the case of an internal arc of the GIL.

In the case of XLPE cables no toxic material is related to the installation and operation process. Only in the case of a ground/earth fault of a single-phase cable, which is typically used for high power transmission at extra high voltages (e.g., 400 kV or 500 kV), will the arc melt the XLPE and burn it at the fault location for a length of some centimeters. The burned materials will be exposed to the surrounding soil and the atmosphere. The quantity of contaminated soil is relatively small and can be replaced during the repair process. The toxic gases in a tunnel need to be ventilated out of the tunnel before personnel can enter.

9.2.3.3 Thermal Impact

The thermal impact of GIL and XLPE cables depends on the type of laying (in a tunnel or directly buried) and the current rating. The higher the current rating, the higher are the thermal losses of the transmission system and the accepted maximum temperatures of the materials used. The ambient temperatures and the soil temperatures give the limits for the maximum power supply.

For GIL maximum current ratings of 4000 A for the tunnel laid and also for directly buried GIL systems have been approved. The maximum current rating installed above ground is 8000 A. For XLPE cables, maximum current ratings of 2500 A have been laid in tunnels and 2000 A for directly buried versions for 2500 mm^2 copper conductor.

9.2.3.4 Maximum Enclosure Temperature

The enclosure temperature is limited in tunnels to 60 °C according to IEC standards in order to avoid burning personnel when they touched equipment during the operation. For higher temperatures, the enclosure surface must be protected against touching by personnel with their hands. Nevertheless, before such high enclosure temperatures can be accepted it is necessary to approve any internal temperature that does not exceed limiting temperature values of materials. This could cause thermal aging and a destruction of the transmission system.

9.2.3.5 Recycling of Materials

After the end-of-life time, both the GIL and XLPE cable systems need to be dismantled. Used materials shall be brought into the recycling process for reuse, as far as possible. The recycling process of the GIL follows the steps in Table 9.1.

The recycling process of XLPE cables is different for the cable materials and for the cable joints. The cables are typically shredded and the materials are separated into their main components copper, steel, aluminum, and XLPE. For this process, a special machinery is necessary. The joint materials can be separated into steel structure materials and cable materials for shredding.

9.2.3.6 Fire Risk

The fire risk of GIL and XLPE cables is relatively low. In both cases, nonflammable materials under normal operation conditions are used. The flammability temperature of aluminum is in the range of 350–550 °C depending on the type of alloy. The XLPE flammability temperature is above 300 °C. Such high temperatures are only possible in the case of an internal fault caused by short-circuit currents of several tens of thousands of amperes, for example, 50 or 63 kA. These short-circuit ratings are only available for less than 0.5 s and therefore the fire exposure is very much limited to the arc location area. In the case of cables, an outer paint with calcium is used to prevent them from burnings after the arc fault is cleared by the protection system by opening the related circuit breakers.

9.2.3.7 Smoke and Toxic Gases

Smoke and toxic gases only exist for GIL and XLPE cables in the case of a ground/earth fault under short-circuit rated currents.

In the case of the GIL no external impact occurs in the case of a gas mixture filled GIL using 20% SF_6 and 80% N_2 and large gas compartments of several hundred meters in length. Typical gas compartment lengths of the GIL are up to 1 km. In this case, the internal arc will not develop a high pressure so the pressure relieve device will not open. Because of the gas mixture the footprint of the internal arc does not generate a high enough temperature to melt the aluminum alloy enclosure of the GIL. Even in the case of an internal arc there is no external impact of smoke and toxic gases.

In the case of XLPE cables, a ground/earth fault under the condition of short-circuit currents will cause a section of the cable to be destroyed and generate smoke and toxic gases. The area of the cable destroyed is restricted to some centimeters of length. The volume of toxic gases is not very large and for directly buried XLPE cables no high risk to personnel is seen. However, for tunnel installations, when high pressure waves are released to the tunnel by the exploding cable section, the smoke and toxic gases in the tunnel need to be removed by ventilation before personnel can enter the tunnel.

Cable tunnels are designed as non-accessible tunnels for personnel in order to minimize the number and times for personnel to be in the tunnel, thus reducing the risk of harm.

9.2.3.8 Noises and Vibrations

Both transmission technologies for GIL and XLPE cables are passive elements and do not generate noise in operation. Caused by the electric current of 50/60 Hz, mechanical vibrations of 100/120 Hz are generated. These vibrations are of small amplitudes and any fixing point or structures for GIL and XLPE cables need to withstand those vibrations. Any kind of mechanical resonances need to be avoided.

In the case of tunnels and bridges, vibrations may come from traffic and the structures and fixing points for GIL and XLPE cables need to withstand these low-frequency vibrations (typically 1–10 Hz) and, again any resonances must be avoided.

9.2.3.9 Cooling and Ventilation Systems

Depending on the transmission losses and the ambient thermal condition, cooling, and/or ventilation might be necessary. In the case of short-time high loading of the GIL or XLPE cable, it might be necessary to ventilate or cool a tunnel or a trench by adding water pipes. Ventilation of a tunnel is effective under certain conditions, such as maximum ambient temperatures, maximum air speed in the tunnel, and distances of ventilation shafts. Cooling systems are more effective but also much more costly and only a few projects use such technology. The installation costs are high and the operation costs including maintenance works are expensive for the complete lifetime of the transmission system.

The first choice is to avoid ventilation and cooling systems by strong enough transmission systems for the required power transmission.

9.3 Life Cycle Cost Analysis

9.3.1 Introduction

Life cycle cost analysis (LCCA) is an economic methodology for assessing the total cost of ownership over the entire life cycle. It can be used to evaluate the cost of a full range of projects, from an entire substation, modifications to existing substations, and/or to a specific transmission or distribution system component. While the initial cost is a factor in the decision-making process, it is not the only factor. The principle of life cycle cost analysis is applied taking into consideration the cost of acquisition and installation, operating costs, maintenance costs, and the cost of renewal/disposal in their selections [11].

LCCA employs well-established principles of economic analyses to evaluate long-term performance of competing investment options/alternatives. The LCCA process is performed by summing up the discounted monetary equivalency of all benefits and costs that are expected to be incurred in each option/alternative. The investment option/alternative that yields the maximum gains is considered the optimal option/alternative. LCCA can be used as a support system for making informed and conversant choices in equipment selection.

The objective of LCCA is to choose the most cost-effective approach from a series of alternatives to achieve the lowest long-term cost of ownership. It is good engineering and good utility practice to employ the total owning cost methodology to assets using the LCCA method. This methodology should be applied in evaluating major distribution and transmission system equipment and components to ensure safe operation, maintain acceptable reliability levels, optimize design efficiency, and provide a legacy of equipment that reduces O&M costs over the life of the asset.

9.3.2 Scope

Efficient and reliable operation is of particular importance with respect to gas-insulated substation (GIS) equipment. They must execute their function reliably, preferably lifelong, and must be as economical as possible over their complete lifetime. This guideline is based on the principles of LCCA, which has been used for quite a long time to assess single units like circuit breakers, transformers, switches, and so on. The user should apply this guideline in the evaluation of GIS equipment in comparison to air-insulated substation (AIS) equipment to take into account not only the acquisition cost but also to consider the land acquisition, operational, renewal/disposal, and residual costs. This analysis can also be extended to evaluate GIS equipment offerings from different manufacturers.

9.3.3 Coordination

An effective cost breakdown structure and cost chain are essential for successful LCCA and the calculation of the net present value (NPV) of a project, equipment, asset, or alternative (Figure 9.9).

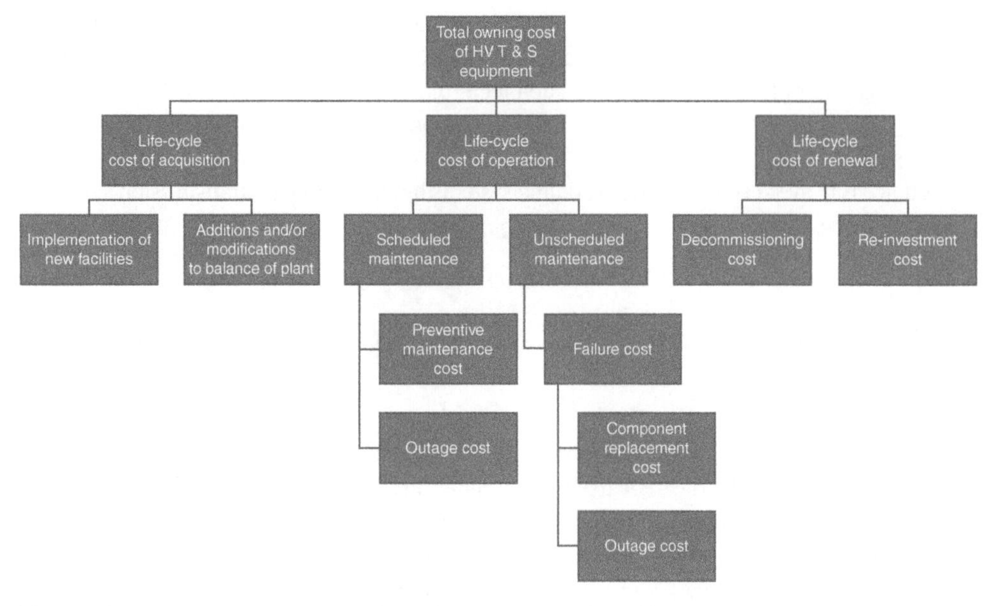

Figure 9.9 Total owning cost string (Reproduced by permission of United Illuminating Company)

Coordination and support of all departments in an organization will be required to establish costs of installation, operation, renewal/disposal, residual value of GIS equipment, discount rate, and escalation/inflation rate.

9.3.4 Methodology

This guideline sets forth a total owning cost evaluation methodology that calculates the life cycle costs (LCCs) for the following:

- Cost of acquisition (CA) and installation
- Cost of operation (CO)
- Cost of renewal (CR)/disposal
- Residual value (RV)

Defining the exact costs of each category can be somewhat difficult since, at the time of the LCC study, nearly all costs were unknown. However, through experience, the use of reasonable, consistent, and well-documented assumptions, a credible LCCA can be prepared.

If costs in a particular cost category are equal in all project alternatives, they can be documented as such and removed from consideration in the LCC comparison.

9.3.4.1 Cost of Acquisition and Installation (CA)

The first step in the completion of the LCCA of a project alternative is to define the cost of acquisition. It is the sum of all the initial investment costs of the alternatives. Initial investment costs are costs that will be incurred prior to the operational commissioning of the equipment. All initial costs including the installation cost are to be added to the LCCA total at their full value (Figure 9.10).

9.3.4.2 Cost of Operation (CO)

The second step in the completion of the LCCA of a project alternative is to define all the future operation costs of the alternatives. The operation costs are annual costs of a sustainable operation of the equipment. Among other things these costs include the cost of staff training and complete

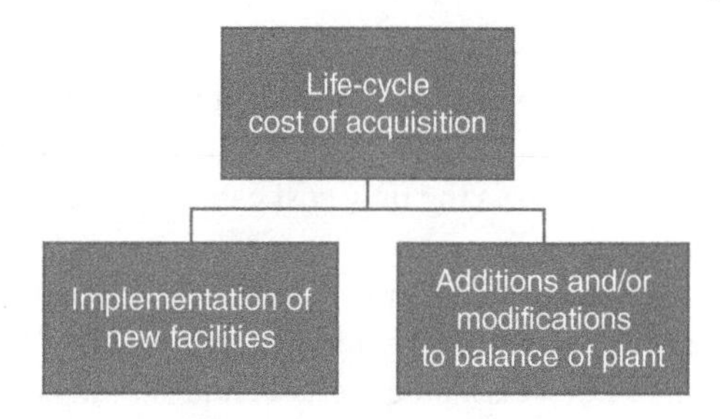

Figure 9.10 Cost of Acquisition (Reproduced by permission of United Illuminating Company)

maintenance expenditures due to different strategies, for example, time-based, condition-based, or corrective. All operation costs are to be discounted to their present value prior to addition to the LCCA total (Figure 9.11).

9.3.4.3 Cost of Renewal/Disposal (CR)

The third step in the completion of the LCCA is to define all the future costs for work, materials, and disposal in conjunction with the rebuild of the existing equipment and possible profits in the disposal of steel and copper. These costs are unanticipated expenditures that are required to prolong the life of equipment without replacing it.

Renewal/disposal costs are by definition unforeseen so it is impossible to predict when they will occur. For simplicity, repair costs should be treated as annual costs. All renewal disposal costs are to be discounted to their present value prior to addition to the LCCA total (Figure 9.12).

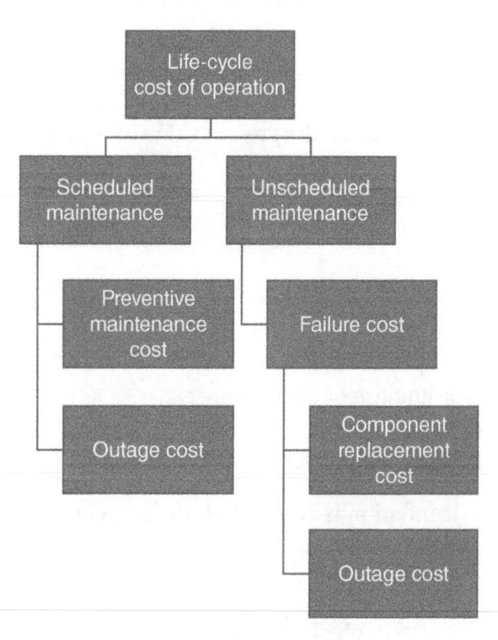

Figure 9.11 Cost of Operation (Reproduced by permission of United Illuminating Company)

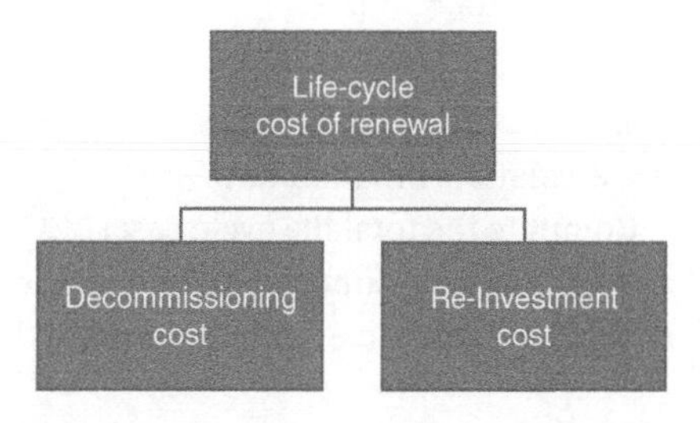

Figure 9.12 Cost of renewal (Reproduced by permission of United Illuminating Company)

9.3.4.4 Residual Value

The fourth step in the completion of the LCCA of an alternative is to define the residual value of the alternative. Residual value is the net worth of the equipment at the end of the LCCA study period. This is the only cost category in an LCCA where a negative value, one that reduces cost, is acceptable.

Since an LCC is a summation of costs, a negative residual value indicates that there is value associated with the equipment at the end of the study period. Whatever the reason for the remaining value, it is a tangible asset and should be included in the LCCA.

A positive residual value indicates that there are disposal costs associated with the equipment at the end of the study period. Perhaps the costs are related to abatement of hazardous material or demolition of the structure. Whatever the cause, these are costs of equipment ownership and should be included in the LCCA.

A zero residual value indicates that there is no value or cost associated with the equipment at the end of the study period. This rare instance occurs if the intended use of the equipment terminates concurrent to the end of the study period, equipment has no value, and it can be abandoned at no expense.

9.3.5 Procedure

Life cycle costing methodology involves adding up all the costs of equipment over the term of the evaluation, with the costs in any one year being discounted back to the base period. The discounting process seeks to reflect the time value for money and reduce all future sums of money to an equivalent sum of money in the base period (e.g., in today's dollars or dollars at the time of procurement). This discounting process estimates the present value (PV) of future costs.

To determine the present value of future costs the following formula is used:

$$\mathrm{PV} = \sum_{t=1}^{T} A_t x \frac{1}{(1+d)^t}$$

where

PV = present value
A_t = amount of cost at time t
d = discount rate
t = time (expressed as number of years)

If inflation is considered, the present value formula then includes the effects of both inflation and discount rate, and becomes a function of d, a, and t:

$$\mathrm{PV} = \sum_{t=1}^{T} A_t x \frac{(1+a)^t}{(1+d)^t}$$

where
a = escalation/inflation rate

Compute the total life cycle cost (LCC), which is the sum of present values of cost of acquisition and installation, annual cost of operations, costs of renewal, and residual value for each alternative. The alternative with the lowest LCC will normally be considered the optimum design.

However, this method is not sensitive to budget and equipment delivery time constraints, which must be considered in the final evaluation:

$$\text{LCC} = \text{NPV} = \text{PVCA} + \text{PVCO} + \text{PVCR} - \text{PVRV}$$

where

NPV = net present value
PVCA = present value of cost of acquisition and installation
PVCO = present value of annual cost of operation over the life cycle study period
PVCR = present value of costs of renewal over the life cycle study period
PVRV = present value of the residual value at the end-of-life cycle study period

9.3.5.1 Present Value

To combine the initial cost with future expenses accurately, the present value of all costs and expenses must first be determined. Present value is "the time equivalent value of past, present, or future cash flows as of the beginning of the base year." The present value calculation uses the discount rate, escalation/inflation rate, and the time a cost was or will be incurred to establish the present value of the cost in the base year of the study period. Since most initial expenses occur at about the same time, initial expenses are considered to occur during the base year of the study period. Thus, there is no need to calculate the present value of these initial expenses because their present value is equal to their actual cost. The determination of the present value of future costs is time-dependent. The time period is the difference between the time of initial costs and the time of future costs. Initial costs are incurred at the beginning of the study period at Year 0, the base year. Future costs can be incurred any time between Year 1 and the final year of the study period. The present value calculation is the equalizer that allows the summation of initial and future costs.

Along with time, the discount rate also dictates the present value of future costs. Because the current discount rate is a positive value, future expenses will have a present value less than their cost at the time they are incurred. To consider the effect of inflation the present worth factor includes the effects of both inflation and discount rate.

Future costs can be broken down into two categories: one-time costs and recurring costs. Recurring costs are costs that occur every year over the span of the study period. Most operating and maintenance costs are recurring costs. One-time costs are costs that do not occur every year over the span of the study period. Most replacement costs are one-time costs.

To simplify the LCCA, all recurring costs are expressed as annual expenses incurred at the end of each year and one-time costs as incurred at the end of the year in which they occur.

9.3.5.2 Study Period

The second component of the LCC equation is time. The study period is the period of time over which ownership and operation expenses are to be evaluated. Typically, the study period can range from twenty to forty years, depending on the owner's preferences, the stability of the user's program, and the intended overall life of the facility. While the length of the study period is often a reflection of the intended life of a facility, the study period is usually shorter than the intended life of the facility.

9.3.5.3 Discount Rate

The third component in the LCC equation is the discount rate. The discount rate is defined as "the rate of interest reflecting the investor's time value of money." Basically, it is the interest rate that would make an investor indifferent as to whether he received a payment now or a greater payment at some time in the future. Obviously, as the economics of the world around us change, so too does the discount rate.

9.3.5.4 Escalation/Inflation Rate

The fourth component in the LCC equation is the escalation/inflation rate. Inflation reduces the value or purchasing power of money over time. It is a result of the gradual increase in the cost of goods and services due to economic activity. The escalation/inflation rate is the annual percentage increase in the price of goods and services.

9.3.6 Finalized LCCA

Once all pertinent costs have been established and discounted to their present value, the costs can be summed to generate the total life cycle cost of the alternatives. After this has been done for all the alternatives, a summary of the results should be prepared. This summary will compare the total life cycle costs of the cost of acquisition and installation, operational costs, cost of renewal, and residual value of all the alternatives.

The LCCA needs only to address the cost categories that are pertinent to specific equipment. However, to ensure an accurate comparison of alternatives, all LCCA evaluations of the alternatives must incorporate the same cost categories. The LCCA of each alternative should include:

- A brief description of the alternative (GIS or AIS)
- A brief explanation of the assumptions made during the LCCA
- A site plan showing the integration of the proposed equipment on the site and necessary site improvements (additions or new construction to accommodate proposed equipment)
- A summary table that compares the total life cycle costs of the initial investment, operations, maintenance and repair, replacement, and residual value of all the alternatives, as shown on an LCC spreadsheet
- Budget, equipment delivery time, or other constraints that must be considered in the final evaluation

The equipment alternative with the lowest overall life cycle cost and satisfying specific constraints, if any, should be selected for procurement.

Consider the following example, a comparison between a 145 kV AIS and GIS that includes a five-circuit breaker H-type arrangement with two feeder lines and two transformer positions (Figure 9.13).

The LCCA calculation considers a turnkey solution and comprises any equipment in between the incoming lines and outgoing power transformers. The cost chain analysis is completed and the costs are documented. Some of the major costs analyzed and assumptions are:

Permitting and licensing
Land acquisition
Site development, fencing, and grading
Primary and secondary equipment
Control building/GIS building
Erection and commissioning
System engineering
Structures, grounding, cable trays, and so on
Direct operating and maintenance costs

De-commissioning
Net present value calculation (6% interest, 2% inflation rate)

The chart in Figure 9.14 shows the LCCA calculation of costs for each alternative normalized to 100%, which is the cost of the more expensive alternative. Another way of looking at the costs is the impact over the life of the assets (Figure 9.15).

By looking at the cost impact over the life of the assets, one gets a true picture of the total owning cost of the compared alternatives. The asset manager can then choose the equipment alternative with the lowest overall life cycle cost while satisfying that specific constraints, if any, should be selected for procurement.

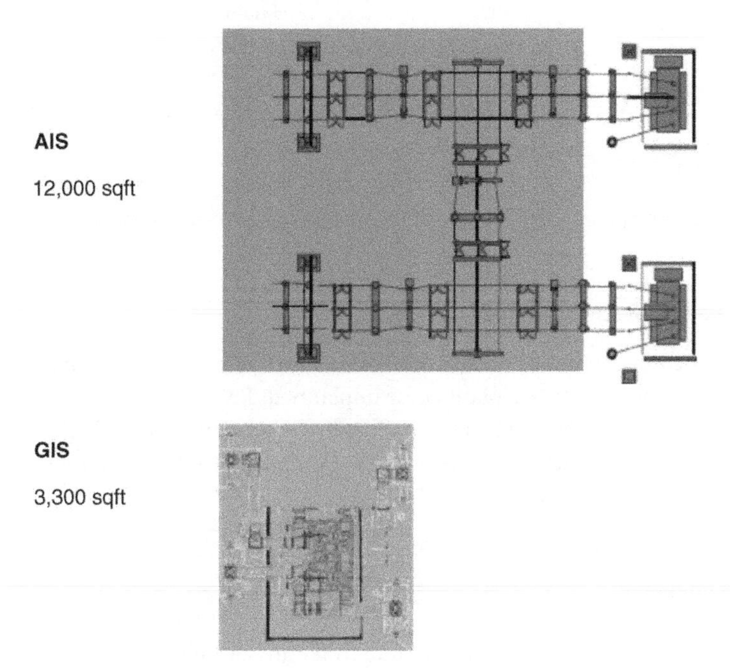

Figure 9.13 AIS and GIS arrangements (Reproduced by permission of United Illuminating Company)

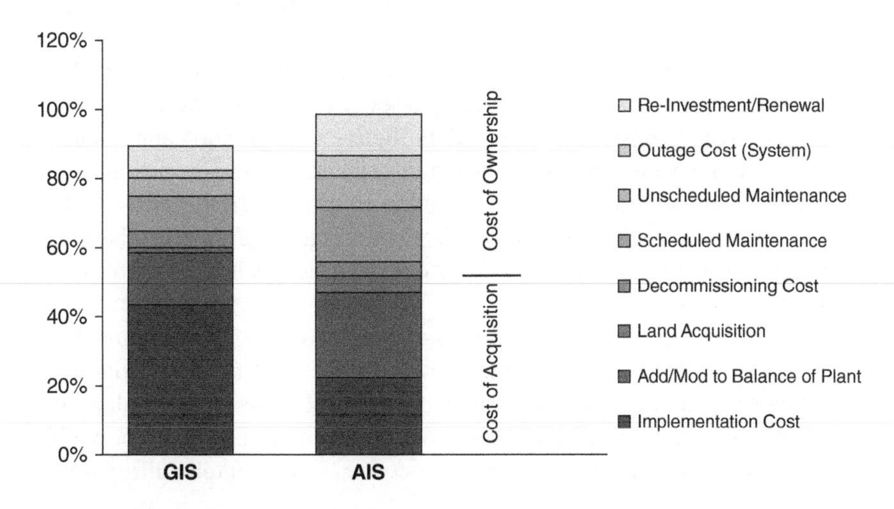

Figure 9.14 Normalized cost comparison (Reproduced by permission of United Illuminating Company)

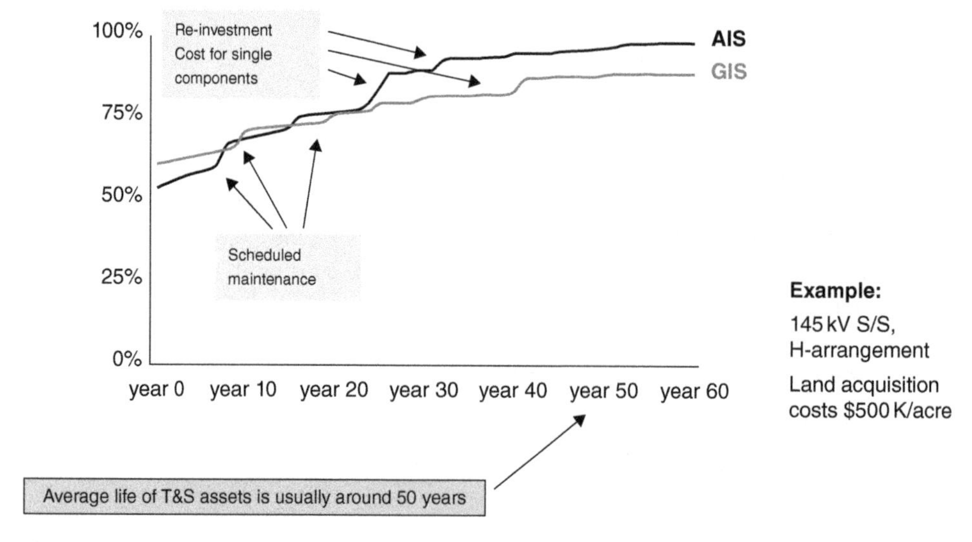

Figure 9.15 Comparative life-cycle cost comparison over the life of the assets (Reproduced by permission of United Illuminating Company)

9.3.7 Reliability, Availability, Failure Rates, and Cost

The reliability of a GIS is represented by a complex factor determined by the total duration and frequency of outages, regardless of whether they are planned or unplanned. Reliability is usually expressed by the availability of the GIS and is defined as the probability that at any point of time the GIS circuit breaker bay is operating satisfactorily or is ready to be replaced. GIS experience surveys show that the average availability factor (F):

$$F = \frac{1 - \text{outage time}\left(\text{planned and unplanned}\right) \times 100\%}{\text{total operating time}}$$

is about 99.8%. The remaining 0.2% consists of unavailability due to failure and repair time and planned/unplanned maintenance. The user should calculate the average availability factor (F) for all alternatives and determine the weight of this factor in relation to the results of the LCCA calculation.

In general, the factors that have a major influence on the combination of costs and availability are normally decided upon during the concept and design phases of the GIS life cycle. The most important factor is the initial cost including land acquisition; the costs of operation and maintenance are generally given a lesser weight. Life cycle cost analysis can be used to demonstrate that in the case of critical installations and important locations in the network, the initial costs can be offset by the benefits of reduced outage times. Improved data collection and evaluation methodologies along with increased focus on the ultimate energy customer will yield a better balance between cost and reliability.

The various phases of the life cycle of a GIS are: planning, concept, engineering, acquisition of equipment, installation, commissioning, operation, maintenance, and finally decommissioning and disposal. Data collection during the various phases of the life of a GIS plays an integral part in the development of an asset management strategy that minimizes cost while maximizing availability. Figure 9.16 illustrates the relation between costs and reliability (availability) [19].

The conclusion that can be drawn from this figure is that the overall total owning cost becomes a minimum at point A, while maintaining the highest possible reliability.

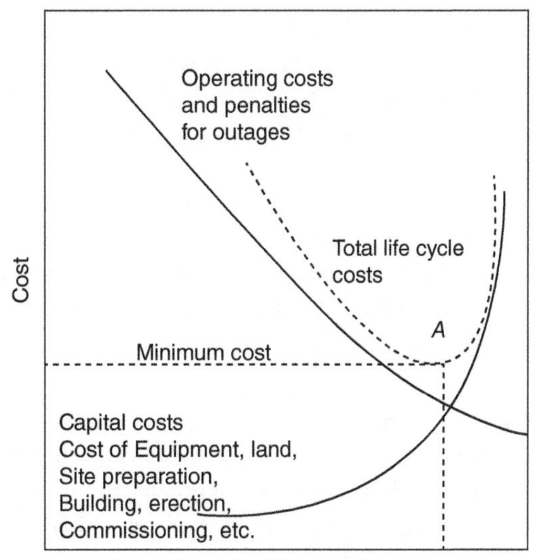

Figure 9.16 Reliability versus life-cycle trade-off (Reproduced by permission of United Illuminating Company)

9.4 Insulation Coordination Study

9.4.1 General

Insulation coordination is the basic criteria for the reliability of the GIS. The process of insulation coordination covers the physical properties of the used insulating gas, the manufacturing process including the tolerances and quality deviations, the overvoltage of the electrical power transmission system, and the experiences made with the GIS in service for more than 40 years. This all leads to rules for the design, the overvoltage protection, and the need of network studies for GIS projects to cover the transient overvoltage possible in the system.

From an operational point of view some main topics are of importance. The major failure rate (MFR) is important for the operator to evaluate the impact of power delivery interruptions. This has been studied by CIGRE and is published in their Technical Brochure (TB), numbers 510, 511 and 512 [2–4]. From a technical point of view the voltage levels of the dielectric-type test, routine test, and on-site tests of the equipment and overvoltage levels of the network are to be considered.

The GIS standards IEC 62271-203 [28] and C37.122 [31] offer voltage levels for each rated voltage, which is the higher value of the proposed voltage levels of IEC 60071 [42]. The reason for this is the high equipment cost and the high design cost of GIS, which need to cover a large world market of different overvoltage levels.

9.4.2 Overvoltage

Lightning overvoltage on transmission systems result from three possible causes. They are listed in the order of increasing severity below:

- Induced voltages
- Shielding failures
- Backflashes

Induced voltages have a maximum magnitude of approximately 200 kV and are of importance typically on low voltage systems. These voltages arise when the lightning strikes an object near to the line but not the line itself.

A shielding failure occurs when the lighting stroke terminates directly on the phase conductor of the transmission line. The shield wires on top of the transmission towers limit the occurrence of direct strikes to the lines. The shielding failures that do occur statistically are of a much lower current magnitude than the average lighting surge and as a result do not generate very high voltages in the substation.

Lightning striking the shield wires or transmission tower can flash over the insulator to the energized phase conductors; this phenomenon is called a "backflash." It is this backflash that can create the high transient voltage surges into a substation. The backflash has the fastest rate of rise of any external transient overvoltage coming into the GIS, and it is the most important phenomena for insulation coordination.

9.4.2.1 Backflash

A flashover from the shielding or tower structure to the phase conductor is commonly referred to as a backflash. It occurs when the lightning surge terminates on a tower or a shield wire. This produces a voltage wave at the tower top which travels down the tower. The wave, which is reflected back from the bottom of the tower, increases the voltage at the tower top; with low tower footing resistance the reflection is reduced. If the footing resistance is not low enough, or the lightning current is high enough, the voltage at the tower top can increase until an insulation failure occurs across the insulator string, injecting the surge onto a phase wire. A backflash creates a surge with a faster rate of rise than is created by a direct lightning strike on a conductor.

The insulator voltage withstand level with regards to lightning backflash is called the "positive polarity impulse critical flashover" (CFO). This is the voltage that causes flashover 50% of the time when the insulator is subjected to a standard $1.2 \times 50\,\mu s$ voltage wave.

The magnitude of a voltage wave injected on to a phase wire by a backflash can be somewhat higher than the rated insulator CFO. One reason is that the insulation strength is a probability function.

More important is that, depending on the rate of rise of the lightning surge current, the voltage impulse at the tower top can be steeper than the $1.2 \times 50\,\mu s$ test wave. The insulator flashover level is very dependent on the steepness of the voltage stress. Adjusting the CFO for steeper voltage waves is very complicated. The *IEEE Guide for the Application of Insulation Coordination* (IEEE Std. 1313.2–1999) [43] recommends 20% as a conservative margin to be added to the CFO to account for both of these effects.

9.4.2.2 Detailed Steps Adopted for Calculation of Various Parameters

The tower footing resistance (R1) was assumed to be less than $20\,\Omega$ for the towers nearest the terminal station and $10\,\Omega$ was used for the study as the footing resistance decreases with higher currents. Based on the definitions of BIL and the critical flashover (CFO) the following relationship was used:

$$\mathrm{CFO} = \mathrm{BIL} / 0.9616$$

The steepness of the surge voltage at the point of backflash is essentially infinite. As the surge propagates along the phase wire, the steepness is reduced by the corona and other factors. However, because the distance from the flashover point to the protected equipment is short, only the corona needs to be considered.

The corona is the partial ionization of air around a conductor that occurs when the conductor voltage produces an electric field equal to the breakdown strength of air. The field strength is uniform from the surface of the conductor to the edge of the ionization envelope. Therefore, the corona can be thought of as suddenly and dramatically increasing the diameter (and therefore the capacitance) of the conductor. It is this capacitance that reduces the severity of the overvoltage impinging on the protected equipment. To account for the corona, IEEE Std. 1313.2 [43] provides the following estimate for the steepness of the voltage surge entering the substation:

$$S = \frac{K_S}{d} \text{kV} / \mu s$$

where $K_S = 700\,\text{kV km/µs}$ for transmission lines with a single conductor and d is in km. The distance from the struck towers to the cable pothead locations varied from circuit to circuit.

The voltage surge decays exponentially with a time constant τ:

$$\tau = \frac{Z_s}{R_i} T_{sp}$$

where Z_s is the surge impedance of ground wire and T_{sp} is the travel time for one line span, R_i is the tower footing resistance, and Z_s is estimated from the output of the transient calculation program.

9.4.2.3 Effect of the Circuit Configuration on the Rise Times at the Station Entrance

For a station directly connected to an overhead line, the rise time at the station entrance is similar to the rise time at the point of flashover but with the rate of rise reduced. This rate of rise is also affected by the capacitances connected to the line at the station entrance, such as wound PTs and CCVTs. The higher the capacitance of the equipment, the more the rate of rise is reduced.

The traveling wave of a lightning stroke will reach the substation and finally the GIS. With reflections at an open switch or breaker the voltage level can double and needs to be withstood by the insulation of the GIS. This covers the insulators and the insulating gas. The insulating gas can be SF_6 or in some cases a gas mixture of N_2 and SF_6. The use of gas mixtures will cause reduced breakdown voltage according to the percentage of SF_6, as shown in Figure 9.17.

Overvoltage limiting devices as surge arresters are used to protect the GIS when connected to overhead lines. The voltage levels of the surge arresters are part of the insulation coordination study. They might be connected to the overhead line by air bushings or in some cases be integrated into of the GIS. When integrated in the GIS light frequency transient overvoltage can be limited due to the low impedance connection to the GIS conductor.

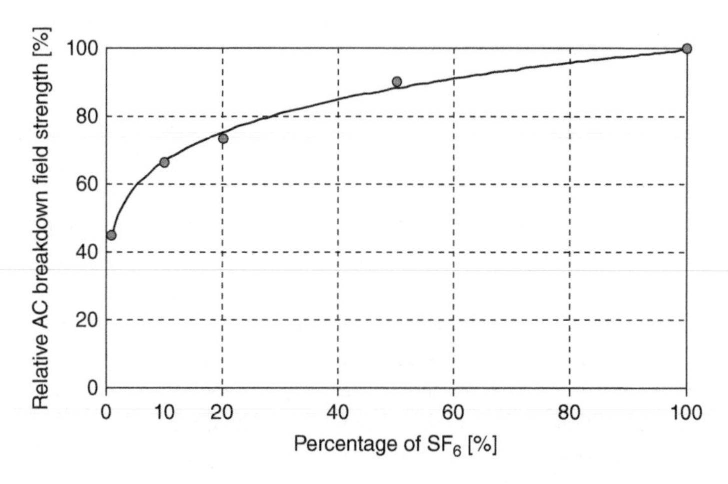

Figure 9.17 Relative AC breakdown field strength of the SF_6–N_2 gas mixture (Reproduced by permission of Siemens AG)

GIS switching operations also cause high frequent transient overvoltage, which can damage the power transformer/reactor when directly connected to the GIS. In such a case, gas-insulated surge arrester are also recommended. Surge arresters can reduce the maximum overvoltage of the GIS connections. The limit voltage values are related to the surge arrester characteristics and the maximum power frequency overvoltage of the system and the superposed transient lightning and switching impulse voltages.

If the GIS is connected to a line and a transient overvoltage is entering the GIS when the disconnector is open a 100% reflection (see Figure 9.18) will double the voltage value at the GIS connection point. This can cause too high voltage values and a flashover might occur. Therefore, this situation needs to be covered by calculations of the insulation coordination study and a surge arrester to limit the voltage level might be recommended.

9.4.3 Grounding

The grounding of the GIS is the most important protection against overvoltage stress. Low resistance and low impedance ground connections are needed to cover the requirements of the power frequency grounding and high-frequency transient overvoltage grounding. Grounding connection of the GIS needs to have multiple connections to the building and substation ground grid to also be effective for high frequencies. The cross section of the ground wire and the distance of the ground mesh are important factors of the effective grounding. The ground mesh is part of the insulation coordination study. The safety of persons and rise of touch voltage levels and step voltages also in case of a transient voltage rise from lightning stroke or switching operations need to be considered. For more detailed information see [44, 45].

9.5 Very Fast Transients

9.5.1 General

Very fast transient (VFT) voltages and transient enclosure voltages (TEV) are two physical phenomena related to GIS. Because of the use of SF_6 and its very fast arc distinguishing properties in the case of switching high voltages and due to the compact design of GIS, VFT, and TEV are important to be recognized and the proper design and grounding are required. The phenomena and the technical solutions are explained in the following sections.

9.5.2 Very Fast Transients in GIS

One of the unique phenomena associated with GIS is the existence of very fast transients (VFTs). VFTs are transient phenomena generated by breakdown of the SF_6 gas, which occur as a result of insulation failure during service or test conditions and the breakdown of the gas between contacts of a switch (disconnector or breaker) during operation. The breakdown of SF_6 gas occurs rapidly, usually within a few nanoseconds. As a result, the transients generated have very short rise times and inherently a high-frequency content. The transients lead to the development of very fast transient overvoltage (VFTO), high voltage transients on external surfaces of grounded GIS enclosures or other grounded components (transient enclosure voltage, or TEV), and transient interference voltages on control wiring and auxiliary equipment.

Although flashovers and insulation breakdown will cause transients in any type of equipment, whether it be GIS or not, GIS transients will have a much higher-frequency content, leading to a

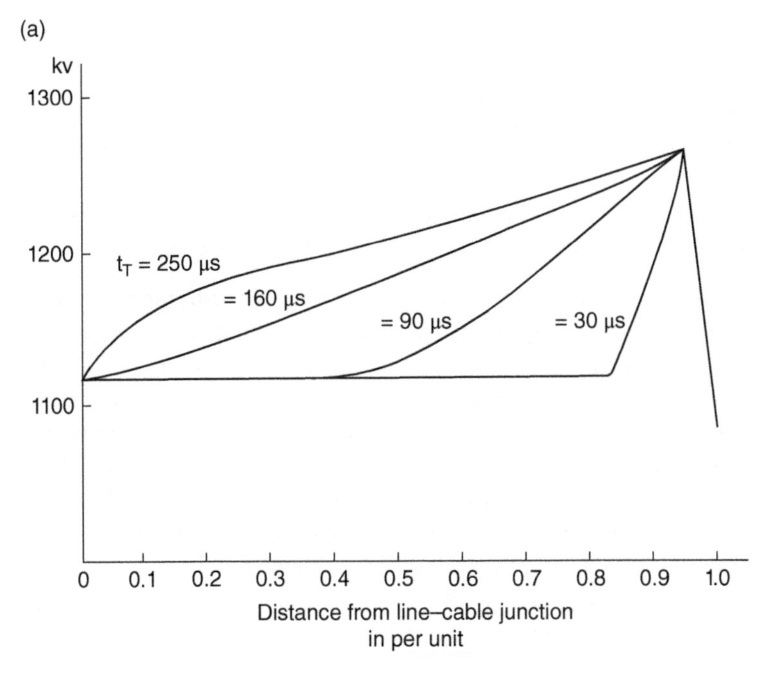

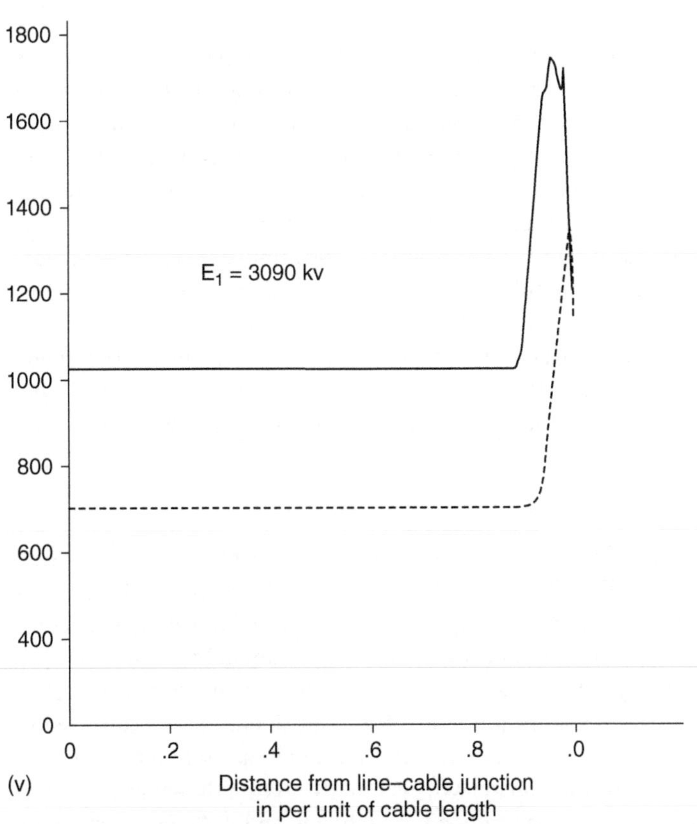

Figure 9.18 Voltage profile along a connected line to a GIS with the reflection effect at the GIS with an open disconnector (Reproduced by permission of Siemens AG)

number of issues quite distinct from those experienced in conventional stations. These stem from two characteristics:

1) Dielectric breakdown of the SF_6 occurs very rapidly in GIS, as indicated above. Typically, the collapse of voltage occurs within a few nanoseconds, leading to short rise-time traveling wave transients.
2) The GIS bus work, which can be considered as a network of short coaxial transmission lines, will support the lossless propagation of high-frequency signals and traveling waves.

The rapid collapse of voltage due to breakdown of the SF_6 gas generates steep-fronted traveling waves, which propagate from the source in both directions. These traveling waves propagate throughout the GIS with little loss, attenuation, and distortion. The front of the traveling wave typically will have risen times as short as 3–5 ns, implying a frequency content extending to a few hundred megahertz. When these waves encounter junctions within the GIS, terminations, or other discontinuities, the waves are reflected or refracted according to transmission line theory. The resulting transient waveform at any one point in the GIS is therefore a superposition of all of the various traveling wave components at that location. In addition, the various components will combine differently at different points within the GIS depending on the origin and direction of travel of the various components. Although the overall transient wave shape may vary, transients will maintain the same initial fast front characteristics and high-frequency content.

9.5.2.1 Simulation of VFT Phenomena

The propagation mechanisms of VFT in GIS also make it amenable to simulations. Simulations can be used to evaluate and assess the transient environment in an installation. The source can be modeled simply with a "step wave" of about 5 ns rise time. A flashover to ground can be considered as a step source with zero impedance. An intercontact flashover caused by the operation of a switch can be considered as a high impedance step source feeding the two sections of the bus in either direction from the switch and with opposite polarity. Sections of the GIS appear as coaxial transmission lines characterized by a surge (characteristic) impedance and electrical length. The value of surge impedance depends of the ratio of the conductor and enclosure diameters, and is typically about 60 Ω. Overall, the GIS can be viewed as an interconnected network of these short transmission line sections. Variations introduced by other components (breakers, arresters, etc.) are usually addressed by the addition of small lumped capacitances or short sections of transmission line of different impedance. References found in the literature can be used to provide further guidance [12].

9.5.2.2 VFT Overvoltage (VFTO)

As discussed above, the VFTs discussed above consist of a number of traveling wave transients propagating throughout the GIS. The GIS will appear as a network of interconnected transmission lines characterized by a surge impedance and length. These individual traveling wave components travel at (effectively) the speed of light, or 3×10^8 m/s (approximately 30 cm or 1 ft per nanosecond). The magnitude, polarity, and direction of travel of these components are determined by changes in impedance at connections, junctions, tees, and so on, in the transmission line network, and the traveling waves undergo reflection and refraction according to transmission line theory. The composite transient wave shape will eventually take on the same characteristics as could be determined by considering the lumped capacitance and inductance values of the various components, especially as the higher-frequency components attenuate and are lost over time. However, the initial portions will maintain the fast-front nature of the traveling waves. Again, depending on the capacitance values, the VFT "envelope" tends to be of an oscillatory

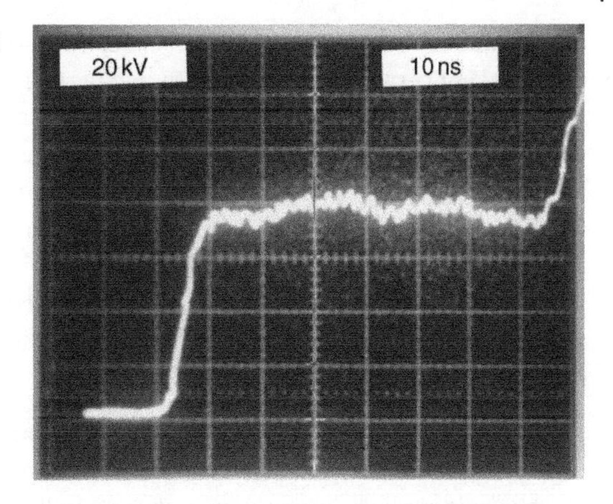

Figure 9.19 Oscillograph trace of the initial traveling wave portion of a VFT waveform (Reproduced by permission of Kineectrics Inc.)

nature with a ringing frequency of a few MHz Eventually, the entire transient wave will die off as well. VFT will generally have a duration not exceeding a few microseconds. Typical waveforms are shown in Figures 9.19 and 9.20.

The specific voltage wave shape at any one point within the GIS is the superposition of all of the component waves arriving at and traveling through that point. Depending on the exact location within the GIS, the overall voltage wave shape can represent an overvoltage. These overvoltage are collectively known as very fast transient overvoltage (VFTO). In typical GIS arrangements, the VFTO magnitudes are modest, rarely exceeding 1.5 times the peak normal AC stress to ground (1.5 per unit). However, the magnitude is dictated by not only the traveling wave characteristics of the station but also the magnitude of the initial step wave generating the VFT. For switching operations, the magnitude varies as there can be a number of "pre- or restrikes" that occur across the contacts during the switch operation; each "strike" will generate its own VFT or VFTO. The highest values of VFTO are seen when the intercontact strike voltage is high, due to trapped charges on the contacts or when switches are operated in phase opposition.

In cases of line-to-ground flashovers, the highest VFTOs are usually created with the test voltage or when failures occur due to some other type of overstress. However, flashover-generated VFTOs create only one set of transients (unless there are multiple flashovers).

9.5.2.3 VFTO and Insulation

VFTO magnitudes are usually not very high. As stated earlier, magnitudes will usually be no higher than 1.5 per unit, although values in the 2 per unit range can occur. However, VFTOs have

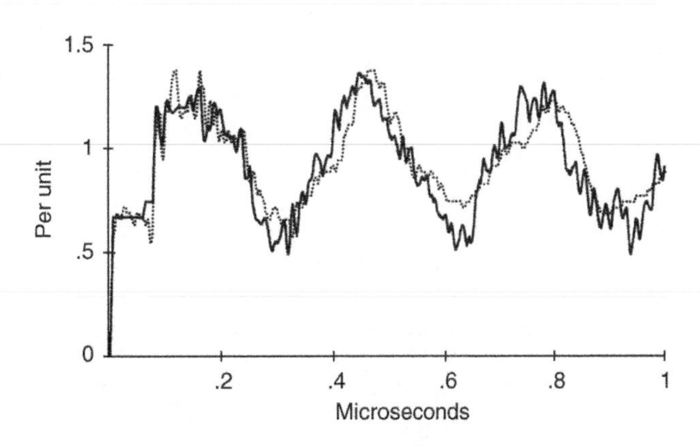

Figure 9.20 Typical overall VFT waveform, shown on a longer time scale. The solid line represents the result of a computer simulation. The dotted line represents an actual measurement (Reproduced by permission of Kinectrics Inc.)

high-frequency components and very steep rates-of-rise. As a result, VFTOs can be, in some cases, of special concern from an insulation perspective.

Traditionally, insulation behavior for steep-fronted voltage stresses has been characterized by a voltage–time (v–t) curve (Figure 9.21). In most cases, the curve predicts that insulation breakdown will occur at higher voltages for short times and for lower voltages when the time increases. This characteristic is generally true for most insulation systems and suggests that insulation problems are a non-issue with respect to VFTO. However, it has been suggested that the v–t curve could possibly "dip" to very low values for short times (in the range of VFTO). Behavior associated with such a dip is not always observed, making its study difficult. Two possible phenomena are sometimes used to help explain this behavior:

1) *Corona stabilization effects.* Corona stabilization can occur in the presence of very sharp metallic objects, such as metallic burrs, particles, and other defects in the insulation system. When high voltage is applied, corona-like discharges will occur around these defects. The charge cloud created by the corona "shields" the metallic protrusion and reduces local stresses at the microscopic level, limiting further discharge activity. In effect, the corona cloud makes the defect look "less sharp." However, for very fast (short-time) stresses, such as VFTO, there might be insufficient time for a charge cloud to develop. As a result, the microscopic stresses are not effectively shielded and breakdown can occur at a lower voltage level than for the corona-stabilized defect.
2) *Statistical time lag.* For an electrical discharge to occur, two necessary conditions are required. First, the electrical stresses must be sufficiently high in a volume of space sufficiently large for a discharge to develop. The second condition is the availability of a free electron within the stressed volume that can trigger the occurrence of the discharge. Free electrons will occur naturally through natural background radiation and through electron emission and molecular collision mechanisms. The time required for such an electron to become available after the first condition for discharge is met is called the "statistical time lag." As the name implies, the time lag has a random element to it – suitable electrons could appear very quickly or after a significant period of time. When a large enough volume is stressed electrically, the (average) statistical time lag is extremely short and generally ignored in insulation studies. However, in the presence of minor defects, such as discussed above, and relatively low overvoltage, the stressed volume can be very small. The statistical time lag can be substantially longer and in the range of many seconds to minutes. Consequently, VFTO stresses will not necessarily consistently

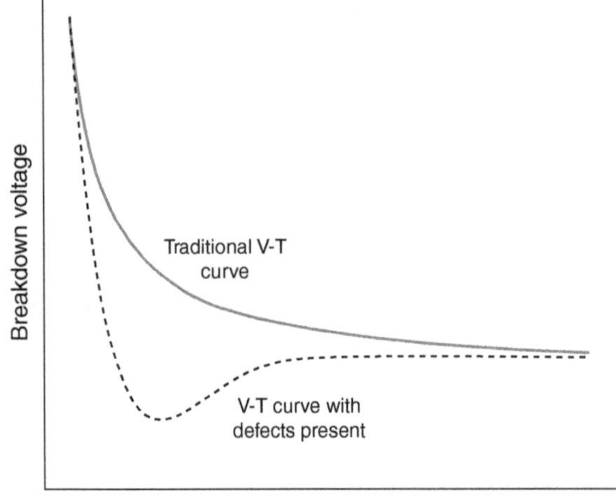

Figure 9.21 Typical v–t characteristic for insulation breakdown. The dotted curve represents a characteristic that occurs under some cases where certain types of defects (sharp protrusions and particles) are present (Reproduced by permission of Kinectrics Inc.)

cause insulation breakdown, even in the presence of minor defects. Any possible "dip" in the V–t characteristic could easily be missed experimentally. VFTO-induced failures are often very random and difficult to identify.

9.5.2.4 VFTO-Related Failures

Insulation Failure

As indicated above, proving that a failure was caused by VFTO is very difficult as such failures, if they were to occur, would be "low probability" events. It is possible that those "unexplained" failures occurring during switching events could have been caused by VFTO but definitive evidence is hard to come by. However, because of concerns, some GIS stations were tested on site using lightning impulse waveforms. Although such impulses are still "slower" than VFTO, their moderately rapid rate of rise combined with a high voltage level is better able to detect the presence of sharp metal protrusions and dangerous particles. Note that particles that are free to move can be readily detected at power frequency using acoustic methods, but particles "stuck" on solid insulator surfaces may not be detected and behave similarly to fixed protrusions. In practice, the impulse method has been successful in finding such defects, which were not previously detected using a full power-frequency withstand test (Figure 9.22).

The on-site impulse test, despite its successes, has not found widespread application, mostly due to its high cost. Alternatively, reliance on high quality of manufacture, assembly (factory and onsite), and more easily applied test methods have proven to be adequate. Power frequency withstand tests are often complemented by acoustic particle detection and sometimes with advanced partial discharge detection methods.

Disconnector Failure [13]

A number of early failures of disconnectors during operation were blamed on VFTO. In these cases, the intercontact arc that occurs during operation was found not to be well-centered and was

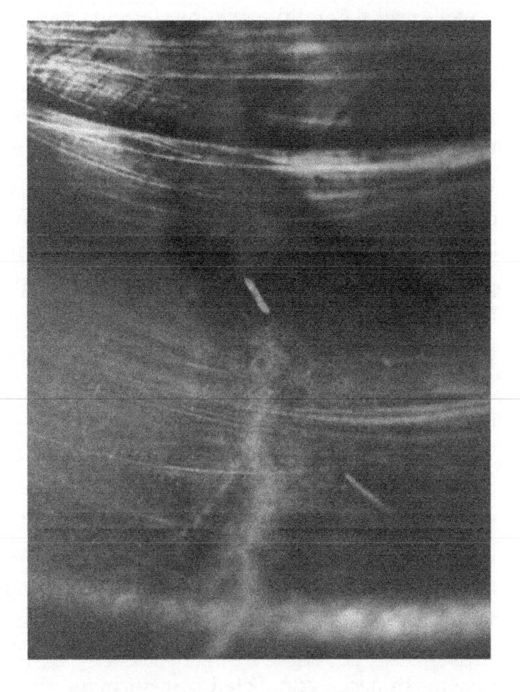

Figure 9.22 Photograph showing a portion of a flashover mark that occurred during impulse testing of a GIS. An impression of a particle (approximately 4 mm long) is clearly visible within the track. The particle was presumed to have been adhered to the insulator surface and was responsible for the flashover. This defect was not detected during the power frequency withstand test, which was applied prior to the impulse test (Reproduced by permission of Kinectrics Inc.)

prone to “wander” and branch to the grounded enclosure as a result of the self-generated VFTO stresses at the switch location. Investigations of the failure mechanisms have led to two distinct developments. First, major manufacturers of GIS equipment have modified their designs to control the behavior of the intercontact arcing better and to minimize the influence of VFTO. Second, with better understanding of the mechanisms of VFTO, geometric parameters that affect the VFTO magnitude were integrated into type test requirements for GIS disconnectors, which are now described in the relevant standards. In any case, failures of this type in modern equipment are rare and are no longer a major issue.

Insulated Flanges and Cable Terminations [14]

In most cases, direct cable-to-GIS connections interface via an insulated flange. The insulation is used to isolate the cable grounding system from the GIS ground which (1) allows the application of cathodic corrosion protection on the cable system and/or (2) controls circulating currents that result when the two grounding systems are directly connected. To maintain ground potentials, the GIS and cable ground systems are physically connected via a separate ground lead, which is designed to take into account the cathodic protection, latent circulating currents and fault currents. The consequence of this type of arrangement is that the separate ground lead is seen as a high impedance path, especially to high-frequency content signals, such as VFTO.

When VFTO is generated, the high impedance of the intended ground connection results in a buildup of voltages across the insulating flange. In many cases, the electrical withstand of the flange is exceeded and flashovers occur. These flashovers are often observed and may be alarming to those unfamiliar with the phenomena. If the only energy dissipated in the flashover is from the originating VFTO, very little damage results, although portions of the insulation may suffer from long-term damage. If such flashovers are known to happen, periodic inspection of the insulating parts (especially insulating washers and bolt sleeves, if such a design is employed) is recommended.

Of greater concern are cases where high circulating currents are possible or there could be internal GIS failure. In the first case, the VFTO-induced flashover creates a conductive path which, under certain conditions, allows circulating currents to flow (unintended) across the insulating flange. Depending on the magnitude of the current, significant damage to the flange could result. A similar situation can also occur in the case of a GIS insulation failure. The internal breakdown in the GIS will create a VFTO, which, as in the other cases, might cause a flashover across the insulating flange. The flashover, as in the previous case, creates a conductive path across the flange, which, depending on the circumstances of the failure, allows the power frequency return fault current to flow across the flange (instead of flowing within the intended ground system). The fault current can create significant damage to the flange. If the damage is extensive, seals could be compromised, leading to a loss of SF_6 gas, cable oil, or both. In the case of cable oil loss, a significant fire hazard exists from electrical arcing at the flange location. One way to avoid flashover at the flange is to use a conductive metallic ring to bridge the isolated flange.

Impact of VFTO on Other Power Apparatus [15]

The VFTO generated internally within the GIS could become incident on externally connected apparatus. Although VFTO magnitudes may not be excessively high, the VFTO waveforms consist of steep fronts and high-frequency components. Other power system components connected to the GIS must be capable of withstanding such stresses. Of particular concern is that of power transformers. Because of the steep front, it is possible that the full VFTO voltage could develop within the first few turns of a transformer winding. The situation is most acute when transformers are directly connected to the GIS, as VFTO is more efficiently coupled into the transformer system. In aerial-connected transformers, the impact of the bushings and the air link will have the effect of reducing the steep-fronted portions of the VFTO, making the situation less onerous. It is difficult to know to what extent this phenomenon has been a problem as transformer failures may be misdiagnosed due to a general lack of knowledge of this phenomenon. Concerned users should consider applying a chopped-wave or other steep-fronted voltage test to transformers used in this application.

9.5.3 Transient Enclosure Voltage

Transient enclosure voltage (TEV) in GIS is a phenomenon related to VFT and VFTO. TEV (also known in the literature as transient ground rise (TGR) or transient ground potential rise (TGPR) [16, 17]) results from VFT, originally generated internally to the GIS, which migrates externally on to the grounded enclosure and other grounded components. TEV generally consists of short-lived, high-frequency transient voltages that appear on grounded parts. TEV can reach high magnitudes (tens to hundreds of kV in some cases) but with durations of only a few microseconds. The main concerns of TEV are (1) direct electric shock hazard, (2) indirect safety concerns such as unexpected “startle” hazards, and (3) as a source of electromagnetic interference on control wiring.

9.5.3.1 TEV Generation and Propagation

Under ideal conditions, VFT generated internally to the GIS does not propagate to external parts. The metallic enclosure of the GIS behaves as a “Faraday cage,” which effectively shields the external components from internal disturbances. However, the GIS enclosure system is not a perfect shield and has a number of electrical “apertures” from which transients can emerge. The largest such opening is usually found at the air terminations. VFTs that propagate to the terminations propagate externally and distribute according to the surge impedances of various components. GIS air terminations can be considered to be a junction of three “transmission lines,” including the internal GIS system, the bushing/overhead connection, and the transmission line formed by the GIS enclosure and the ground (Figure 9.23). Simple application of transmission line theory and with estimated values of surge (characteristic) impedance of the three lines can provide an estimate of, at least, the initial magnitude of the VFT that is converted into TEV. For example, the characteristic impedance of the GIS, overhead line, and enclosure-to-ground systems are typically 60, 300, and 150 Ω respectively. Using these values, we can see that about 60% of the incident VFT magnitude can be converted into TEV:

$$\text{TEV traveling wave magnitude} = \text{incident VFT magnitude} = \frac{2 \times Z_e}{Z_i + Z_e + Z_t}$$

Z_e = characteristic impedance of enclosure-to-ground transmission line (~ 500 Ω)
Z_i = characteristic impedance of internal GIS (~60 Ω)
Z_t = characteristic impedance of externally connected air connection (~300 Ω)

In reality, in the presence of grounded supports, ground leads will act to reduce the TEV magnitude. However, as a result of the short nanosecond rise times (implies high-frequency content) of the TEV, ground leads with lengths of a few feet or more are not effective grounds at short times and behave more as transmission lines as well (Figure 9.24). As such, the impact on the initial portion of the TEV wave is that caused by the impedance discontinuity of the ground lead connection. As single ground wire connection leads have high transmission line characteristic impedances, the TEV will be reduced (typically by 70–80%), but not as would be expected for an effective ground

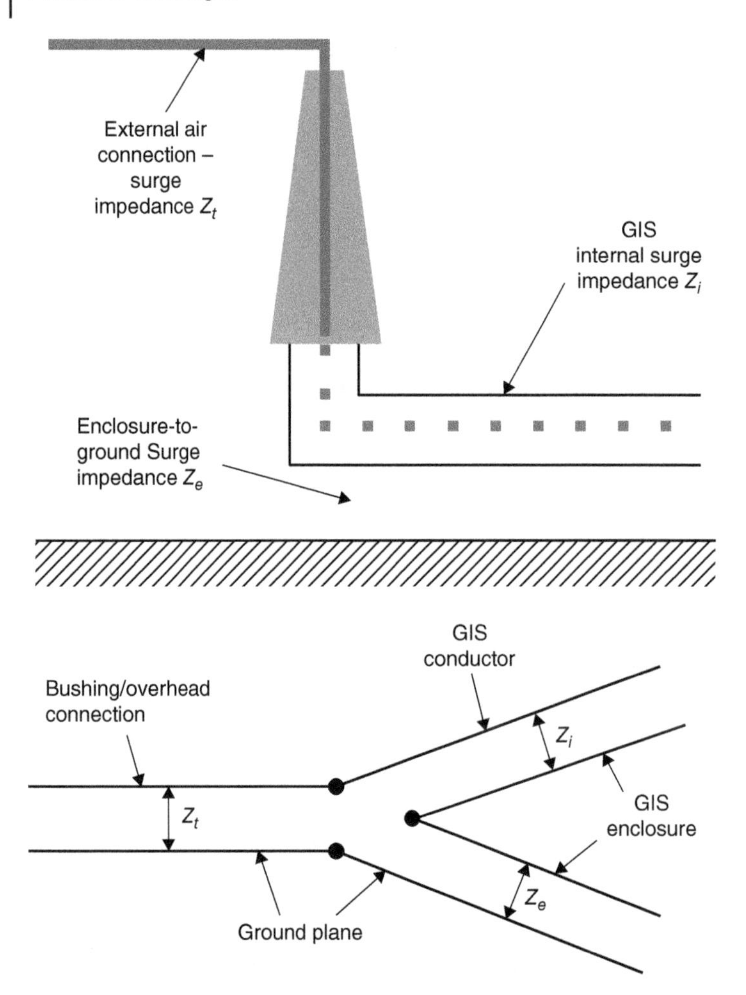

Figure 9.23 GIS termination modeled as a junction of three transmission lines. As a result of the high-frequency content of the VFT waveform, the inside and outside surfaces of the GIS enclosure behave as separate, distinct conductors. The figure explains the mechanism of how internal VFT waveforms emerge externally to become TEV via coupled transmission line segments (Reproduced by permission of Kinectrics Inc.)

with multiple ground wire connections. Such ground leads only become effective when the portion of the transient that propagates along the ground lead has sufficient time to reflect at the ground and propagate back. As a result, grounding connections, especially long ones, will sometimes have less effect on reducing initial TEV magnitudes.

The above analysis is correct for very short times (in the nanosecond range). As with the VFT, the various traveling wave components eventually merge and form an overall characteristic related to the discharge of larger capacitances in the system. As a result, the overall wave shape will be oscillatory (in the MHz range) but will reduce quickly as the system grounds become more effective as time passes. Usually, TEV waveforms persist for only a few microseconds.

The propagation of TEV waves along GIS enclosures is somewhat lossy and TEV magnitudes will be attenuated as the waves travel from the termination back into the station. The consequence of the mechanisms of TEV generation and propagation is that TEV magnitudes are highest at the point where TEV emerges from an internal VFT and is lower as distance from the origin increases. In many GIS configurations, this means that TEV is highest close to the air terminations but is reduced significantly within the main parts of the station.

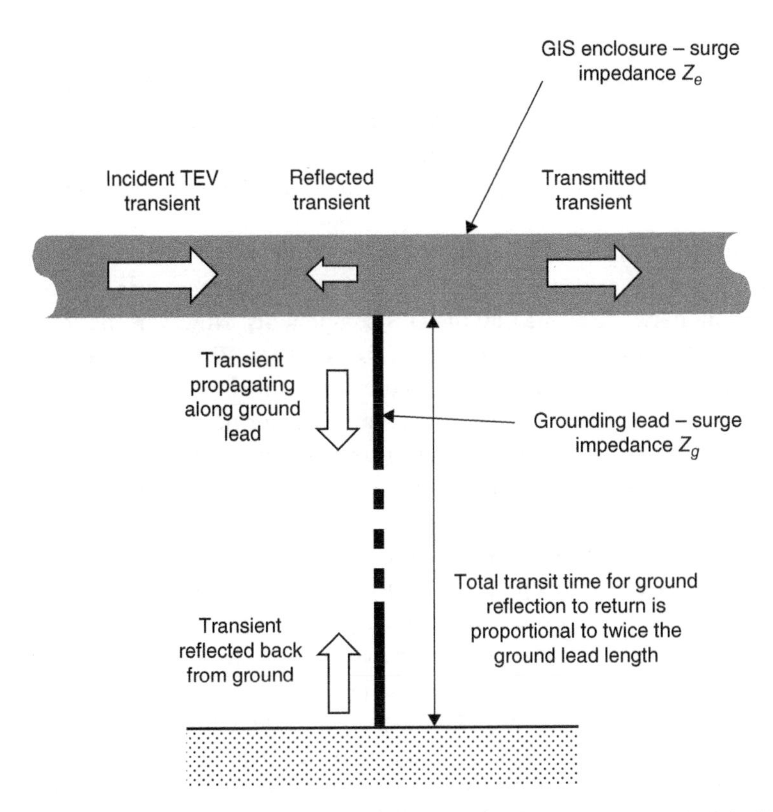

Figure 9.24 As a result of the fast rise-time (high-frequency content) of the TEV waveform, ground lead connections on the GIS enclosure need to be treated as transmission lines. Transients propagating on the ground leads will only start becoming effective after twice the electrical transit time along the lead. Consequently "short" leads are much more effective in reducing TEV than longer ones (Reproduced by permission of Kinectrics Inc.)

The value of the initial TEV magnitude depends on the characteristic impedance of the transmission line formed by the GIS enclosure and ground. In the previous example, this was estimated at 150 Ω. If this value were higher, the initial TEV magnitude would also be higher. Conversely, lower values mean lower TEV values. Lower impedances occur when the GIS enclosure is close to the ground. Low enclosures have the added benefit of making the ground leads shorter and, therefore, more effective. Table 9.2 describes some general station design configurations with respect to the bus-to-air terminations and their impact on the TEV.

Table 9.2 Factors influencing TEV magnitude in a GIS

Lower TEV in station	Higher TEV in station
Enclosure low to ground	Enclosure high above ground
Frequent, short ground connections	Infrequent, long ground connections
Long distance to station	Short distance to station

TEV can also emerge from other electrical "openings" in the station. In many GIS designs, the size of the openings (such as at viewports, etc.) are small and the TEV emerging from these points is also small. However, designs that use an exposed insulator at flanges will tend to allow TEV to emerge. Although these flanges will use shorting straps to maintain the enclosure ground, the straps are too few to effectively contain the high-frequency VFT and some TEVs will emerge from these points.

TEV magnitudes also depend on the magnitude of the VFT that is incident on the electrical openings in the GIS. As discussed previously, VFT magnitudes will depend on geometric considerations of the transmission line network formed by the GIS components. However, the VFT magnitudes will also depend on the originating disturbance. For instance, VFTs generated by disconnector operation depend on the voltage across the disconnector contacts just prior to the formation of the intercontact spark. As disconnectors are operated under normal conditions, this voltage is, at most, 1 per unit (source side of switch) plus the trapped charge level on the load side. Most disconnectors operate relatively slowly so the trapped charge rarely exceeds 0.3 per unit, although a 1 per unit trapped charge is theoretically possible.

However, there are usually many "strikes" of varying magnitude across the disconnector contacts during its operation. Consequently, a disconnector operation will generate a series of VFTs (and therefore TEV transients) for each operation. Breaker operation is similar to disconnector operation but, due to the speed of operation, the intercontact voltage prior to the spark is more random although the possible range of values would be similar. However, a breaker operation, unlike that of a disconnector, will usually generate very few (and possibly only a single) transients. Line-to-ground failures in the GIS will also generate a VFT (and TEV). In this case, the VFT magnitude will depend on the instantaneous voltage at the time of failure. This value can be higher than 1 per unit if some sort of overvoltage was responsible for the failure. Of particular interest are flashovers that occur during on-site high voltage testing. As test voltages can be very high, TEV values will be higher as well.

9.5.3.2 Shock Hazard

Shock hazard due to TEV is difficult to evaluate. Propagation of high-frequency currents through the body and their impact is not well understood. Some have attempted to establish a shock energy criterion (~1 joule) based on an assumed impedance of a human body, but such an analysis has far too many uncertainties to have any validity [18]. On the other hand, there have only been a few cases of electrical shocks reported due to TEV and no incidents of serious injury. The anecdotal evidence would suggest that the shock hazard is insignificant in most cases. Nonetheless, it is prudent to take some precautions. For instance, casual contact with the GIS enclosure should be discouraged, especially in the vicinity of the air terminations, where TEV values could be highest.

Of greater concern to a direct shock hazard is, perhaps, the indirect hazard or "startle effect". Workers may experience minor shocks or view other visual manifestations of TEV. If unprepared, the worker may fall or behave in another manner that could lead to an accident or injury. As a result, some utilities have instituted certain precautions including:

- Education and training. Explain to workers how TEV manifests itself so they are aware of the phenomenon.
- Warning alarms prior to planned switching events. Workers will know that disconnector or breaker operations are about to occur, allowing them time to "stand clear" until an "all clear" is indicated. While this does not cover unplanned events and failures, planned operations would occur far more frequently and cover the majority of the TEV incidents.
- Warnings and exclusion zones. Exclusion zones around key areas such as air terminations, general warnings of transients, and discouraging casual contact.

9.5.3.3 TEV-Induced Interference and Control Wire Transients

In addition to the possible impact on personnel, TEV could also find its way into control wiring and associated equipment. Because of the high-frequency content of the TEV signals, the problems are exacerbated in GIS as compared to conventional installations. The issue is also of more interest as more and more sensitive electronic equipment is used in station environments.

Most of the TEV-related issues can be addressed through adequate shielding of wiring and other components. Perfect shielding protects all sensitive parts by enclosing them in a Faraday cage, which TEV cannot penetrate. However, full and complete quality shielding might not be feasible or practical. Consequently, the following guidelines are usually recommended to minimize the effect:

- All cable should be shielded. Quality foil or solid shields are better than braided shields.
- Cable shields must be grounded at either end. If circulating currents on the shields become a problem, other provisions (such as a separate ground lead) and careful design must be considered.
- Pigtail ground leads at cable ends should be avoided. If pigtails must be used, the leads should be as short as possible and resemble, electrically, a coaxially grounded connector.
- Cables should be routed as close to metallic structures as possible, to minimize induced currents.

9.6 Project Scope Development

9.6.1 Engineering Planning

Prior to the purchase and the installation of a gas-insulated substation, the end user needs to determine present and projected future configurations of the station. During this process, electrical and physical parameters and all constraints dictated by the location of the station should be considered as well. Each user needs to review their operating and maintenance procedures to determine whether revisions will be required when transitioning from an air-insulated substation to a gas-insulated substation. These determinations should be documented in specifications and drawings so that potential suppliers can furnish detailed technical and commercial proposals for the project. Some of the items that need to be considered follow:

(a) Internal user meetings with all responsible departments should be held to define and establish all requirements and constraints of a new substation. Several of the arrangement constraints are identified in IEEE Std. C37.122.1 [46].

(b) The type of site where the substation will be located should be determined and evaluated. Will the GIS site be in an existing building, a new space enclosure, a shared building, underground, in an unusually shaped area, on the crest of a dam, inside the median strip of a freeway, or other site?

(c) Development of the electrical parameters of the station including:
   1) Rated maximum voltage including rated insulation level
   2) Rated short- circuit current
   3) Rated continuous current of each bus, line exits, transformer connections, and bay positions
   4) Current and voltage transformer requirements
   5) Circuit breaker and disconnect and grounding switch control and interlocking requirements
   6) Surge arrester requirements and their locations
   7) Special purpose requirements [SVC (static var compensation), capacitor banks, reactors]

(d) Development of gas zones and monitoring arrangement requirements and gas schematic diagram.

(e) Development of required single line for the station including:
   1) Quantity and location of circuit breakers, disconnect, and grounding switches
   2) Quantity and location of transmission lines terminating at each voltage level
   3) Quantity and location of current and voltage transformers
   4) The type of terminations into GIS
   5) Quantity and size of the power transformer banks that will be installed
   6) Future bays or diameters
   7) Special purpose requirements

(f) Development of required general arrangement:
   1) Location of major equipment
   2) The method that the transmission lines will terminate at the station including air-insulated, gas-insulated bus, solid dielectric cable, and/or oil-filled cable terminations
   3) The method that the transformer banks will terminate at the GIS including oil-to-gas bushing or air-insulated connections
   4) Future bays or diameters
   5) Location of local control cabinets

(g) Determination of the scope and intent of the manufacturer's responsibilities including:
   1) Supply of gas-insulated substation apparatus only
   2) Delivery, installation, labor, testing, and commissioning of gas-insulated substation
   3) Supply of SF_6 gas, gas handling equipment, and labor
   4) Supply, delivery, installation, labor, and commissioning of ancillary equipment, such as foundations, support structures, galleries, and walkways, GIS space enclosure, cranes, TRV capacitors, power supplies, spare parts, and so on, needs to be determined
   5) Arrangement and supply of GIS terminal apparatus in relation to arrangement and supply of air-insulated terminal equipment
   6) Physical clearance requirements for performing high voltage testing, particularly when testing one transmission line or transformer bank while an adjacent line is energized from the system
   7) Supply of auxiliary power for circuit breakers, disconnect, and grounding switches
   8) Extent of the control wiring installation responsibility and interface between manufacturer and user-supplied wiring
   9) Determination of control cable conduit or trench interface between user and manufacturer
   10) Extent of grounding and bonding requirements and interface between manufacturer-supplied grounding and user-supplied grounding
   11) Access to control cabinets, circuit breaker, disconnect, and grounding switch operating mechanisms, gas sampling and fill valve, gas density monitors, view ports, and circuit breaker interrupter removal requirements

(h) A review of the user's maintenance and electrical clearance requirements should be performed in detail and any deviations between gas-insulated and air-insulated substations need to be addressed and resolved.

(i) The site preparation work needs to be defined including core borings, grading (fill and cut), drainage, access roads, lay-down areas, control buildings, maintenance buildings, duct banks, auxiliary power supplies, and perimeter fencing. Areas available to GIS manufacturer should be identified.

See Section 7.6 for details on information that should be considered for extending an existing GIS.

9.6.2 Planning the GIS Project Construction and Installation

A deliberate and complete installation plan, including the future addition of similar equipment, is essential so that all aspects of construction can be reviewed. The preassembled sections of the equipment and the manufacturer's instructions dictate the assembly sequence and, in most instances, follow a series of steps categorized as follows:

(a) Preconstruction meeting between user, equipment installer, and manufacturer
(b) Site preparation including grading, installation of drainage, foundations, and ground grid, access roads, and auxiliary power
(c) Staging of construction equipment required during the installation
(d) Final alignment and leveling of foundation support
(e) Receiving, unloading, and storing GIS equipment
(f) On-site assembly and lay-down area
(g) Grounding of GIS equipment to ground grid
(h) Local control cabinet installation
(i) Connection of control wires
(j) Evacuation and filling with insulating SF_6 gas
(k) Leak testing
(l) Mechanical or operational testing
(m) Dielectric testing of primary circuits with conditioning steps
(n) Cleanup in accordance with applicable regulations
(o) Testing between GIS equipment and balance of plant for system integration
(p) Energization

Other planning considerations are as follows:

- A schedule for work crews should be prepared to provide for more economical use of manpower and to minimize conflicts caused by limited space. Scheduling may also result in the release of specialized skills in the shortest possible time.
- On-site or nearby preassembly areas should be planned when practical so that specialized equipment can be set up and repetitive assembly tasks can be performed under controlled conditions.
- A site layout designating erection equipment locations should be prepared to allow maximum use of the equipment with minimum movement. The layout should include details for each phase of installation so that orderly movement of the equipment can be maintained.
- The capacity of cranes, hoists, gas-handling equipment, welding equipment, and so on, should be considered to ensure that the proper sized equipment is available for the job.
- Electric power, heat, water, and so on, should be available at the appropriate time in the installation sequence.
- Cleanliness, in accordance with the manufacturer's instructions, should be observed at all times.
- Material safety data sheets (MSDSs) and other health and safety information should be readily available to the work crews.

9.6.3 Site Preparation

Regardless of indoor or outdoor installation, the GIS foundation or space enclosure should be complete and all preparations in place prior to the start of erection. Project scheduling should ensure that inappropriate tasks (e.g., civil works modifications) are not planned for the same installation period. The keyword is cleanliness. The long-term reliability of the GIS equipment depends greatly on the level of cleanliness maintained during the installation process. This can be achieved by the provision of a defined clean working area. Additional preparation measures to be taken include the following:

- The manufacturer should specify any local working condition limitations that should be imposed on the erection of the GIS to avoid contamination by particles, dust, water, or ice. Temporary measures in the form of shelters, barriers, or heaters may be necessary to achieve this condition, especially during outdoor installation.
- The party responsible for the on-site erection of the GIS should ensure the availability of the contractually agreed installation tools and accessories (e.g., lifting equipment, tools, and power supply) throughout the full installation period.
- The manufacturer should specify the quantity and qualifications of the personnel needed to complete the installation. The foundation (floor) should be cleared to allow for the layout of the GIS and the concrete sealed to prevent unnecessary dust.
- The unpacking and if necessary general cleaning of the components should be performed away from the final clean assembly area.

9.6.4 Installation of the New GIS

The overall installation process for GIS may encompass many months, during which time other activities associated with the project should continue. Coordination of activities among the project's responsible parties is a necessity, especially with regard to the interface with the HV power transformer and HV cable connections. Time spent in these coordination processes will help to ensure the minimum number of disruptions during the installation process. Disruptions will nevertheless occur and a certain degree of flexibility on the part of all parties is essential. Specific installation procedures are tailored for each manufacturer's GIS requirements, but a typical sequence for the installation of new GIS could be as follows:

- The anchoring/support system is installed and leveled to accommodate civil works tolerances.
- Complete bays and single- or three-phase bay components are installed on their respective supports.
- Interbay connecting elements are installed and bus-coupled.
- GIS equipment is grounded to a ground grid.
- Installation of local control cabinets and interconnecting cables.
- Commencement of SF_6 gas vacuum-filling process.
- Gas-insulated buses, including SF_6 gas-to-air bushings to outgoing power transformers or line positions, are installed.
- Interface components are installed (e.g., GIS to HV cable or power transformers), but bus links remain uncoupled.
- Site commissioning tests are completed, including local control cabinets.
- GIS is subjected to the high voltage withstand tests (refer to Section 9.5).
- Ancillary GIS devices (e.g., voltage transformers and surge arresters) are installed and bus links to high voltage cables and/or transformers are coupled.
- Means of dust control during installation should be taken into account.

To accelerate the overall program some tasks can be done in parallel if the overall standard of the assembly practices is not compromised.

9.6.5 Installation of GIS Extensions

The installation of an extension to an existing GIS substation imposes special conditions on both the manufacturer and plant operator that do not normally apply for the installation of new GIS, which is covered under IEEE Guide C37.122.6 [47] in more detail.

9.6.6 Equipment Access

Structural supports, access platforms, ladders, stairs, cable raceways, conduits, and other auxiliary equipment required for operation and maintenance, as furnished by the manufacturer, should be incorporated into the design. See IEEE Std. C37.122 [48] for more details.

9.7 Risk-Based Asset Management of Gas-Insulated Substations and Equipment

9.7.1 Introduction

Best practice asset management decisions for maximizing performance and minimizing equipment life cycle costs of gas-insulated substations are based upon risks associated with actual equipment condition and historical performance. There are four key steps involved: understanding existing performance, understanding required performance, projecting future performance, and understanding how to bridge gaps with risk assessment. Ongoing risk-based asset management efforts are focused on developing condition assessment algorithms to understand existing performance, project future performance for gas-insulated substation equipment, and provide diagnostics to bridge the performance gaps [19].

9.7.2 Scope

The scope of a risk-based asset management program includes the establishment of goals and objectives to maximize equipment performance and life extension [20].

The primary goal is to provide continually improved risk-based decision–support methodologies for substation equipment asset managers. It envisions that the development will lead to an integrated framework for asset risk assessment, mitigation, and performance improvement.

The primary objective is to develop a methodology that includes the following:

- Ensuring Operation of the Switchgear within its Thermal Capabilities
- Identify Possible Failure Modes
- Observed Degradation Mechanisms Affecting the Switchgear
- Identify Diagnostic Methods
- Document and Implement Preventive Maintenance Assessments
- Document and Implement Major Maintenance Assessment
- Document Periodic Condition Assessment
- Institute a Comprehensive Life Cycle Management Program

9.7.3 Methodology

9.7.3.1 Existing and Required Performance

Figure 9.25 depicts the relationship between the aforementioned elements that formulate a comprehensive risk-based asset management methodology. The overall methodology takes into account test data inputs, operational inputs, and predictable degradation/failure modes to map to

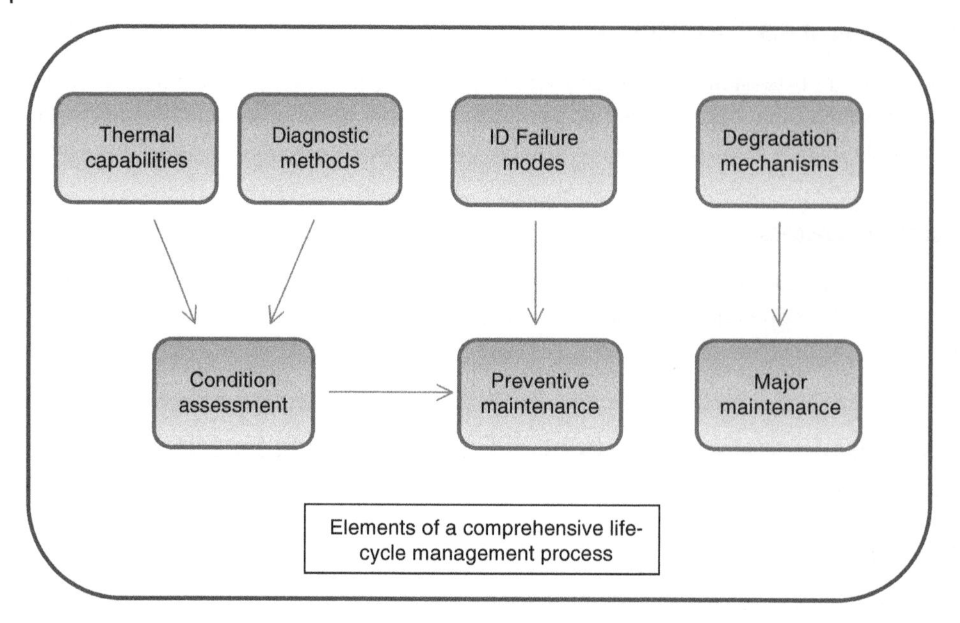

Figure 9.25 Comprehensive life-cycle management process (Reproduced by permission of United Illuminating Company)

the most effective maintenance program(s). These programs then become part of the overall comprehensive life cycle management process, taking into account acceptable levels of risk.

Thermal Capabilities

The switchgear must be operated within the calculated thermal ratings based on IEEE or IEC Standards and regional reliability criteria of the system operating entity.

Preloads for any ratings, if required, should be considered and agreed to by the user and the system operating entity. The thermal rating of the switchgear should be determined by the most limiting applicable element of the switchgear main current path. The scope of the switchgear ratings should include, as a minimum, both normal and emergency ratings. Ratings should consider design criteria, ambient conditions, and operating limitations imposed by the system the switchgear is connected to.

In order to provide an input to an integrated framework for asset risk assessment, mitigation, and performance improvement, continuous monitoring and trending of switchgear loading in relation to calculated thermal capabilities must be accomplished. Any operating time frames that exceed the normal and emergency ratings of the switchgear should be avoided. They should be, as a minimum, documented in value and time, trended, and analyzed as part of the overall view. The cumulative I^2t or the measure of the energy content of an overload transient for the switchgear should be the primary input to the overall analysis for potential degradation of insulation or connections. Operating the GIS outside of calculated thermal ratings will prompt a condition assessment.

Failure Modes

The possible failure modes of the entire gas-insulated substation must be identified. The failure modes must be ranked according to criticality with respect to the operation of the electric system, cost to remedy failures, and their impact on other assets and projects that are directly or indirectly associated with the gas-insulated substation in question.

The failure modes can be categorized in three major ways:

- Minor faults (Type 1)
- Minor faults (Type 2)
- Major faults (Type 3)

Minor Faults (Type 1)

These faults are specific to the SF_6 environment that is outside the high voltage encapsulation, that is, piping, valves, density monitors, fittings, and so on. This type of fault does not generally cause protective relay operation, but they require planned shutdown of parts of the GIS.

Minor Faults (Type 2)

These types of faults occur within the high voltage encapsulation and involve dismantling of the inner parts to remedy. They require replacement of internal high voltage components or insulation. This type of fault may cause protective relay operation, but they require planned shutdown of parts of the GIS.

Major Faults (Type 3)

These types of faults occur within the high voltage encapsulation. They cause protective tripping and generally cause damage to the components of the GIS. This type of fault requires dismantling of many sections of the GIS and requires replacement of the damaged parts. This type of fault may require prolonged system outages to effect repairs.

Internally and Externally Caused Types of Faults

Each of these types of faults can be classified as internal and external. Some examples are now given.

(a) *Internally Caused Faults*
    Moisture
    Particles
    Loss of electrical contact
    High electrical stresses
    Poor contact of current-carrying parts
    Poor quality assurance testing (production and field)

(b) *Externally Caused Faults*
    Accidental damage
    Improper operation of the GIS
    Excessive surge voltages

Identification of possible failure modes will lead to preventive maintenance actions and the establishment of preventive maintenance guidelines.

Degradation Mechanisms Affecting the Switchgear

In general, GIS is essentially immune to many of the degradation mechanisms that affect air-insulated equipment. However, there are certain degradation mechanisms that the user should be aware of, such as whether the GIS is exposed to harsh environments or there were quality control issues during production and any inherent design weaknesses.

The following is a partial list of some of the major degradation mechanisms that should be factored into the life cycle management process to minimize major maintenance activities:

- Frequent operation above thermal limits
- Frequent high electrical stresses due to surges
- Particles in the high voltage encapsulation from production and/or installation
- Excessive moisture in the SF_6 gas
- Impurities in the SF_6 gas
- Contaminants entering flanges and affecting O-rings
- Mechanical wear/aging
- Poor assembly techniques
- Poor management of decomposition products due to switching or interruption of currents

Periodic Condition Assessment and Diagnostic Methods

The feasibility of implementing extensive condition monitoring and diagnostic methods in GIS has been widely studied [21]. GIS is now a highly developed technology and any future developments should be directed towards evolving methods to optimize its operational service and minimize failure mechanisms. Therefore, various aspects of future trends in asset management that will evolve should include partial discharge detection, internal flashover detection and location, SF_6 gas condition monitoring, and circuit breaker diagnostics. Continuous condition monitoring units use remote data collection and logging systems, thereby reducing manpower costs and making it possible to rectify faults and failure modes before they develop into major breakdowns. Figure 9.26

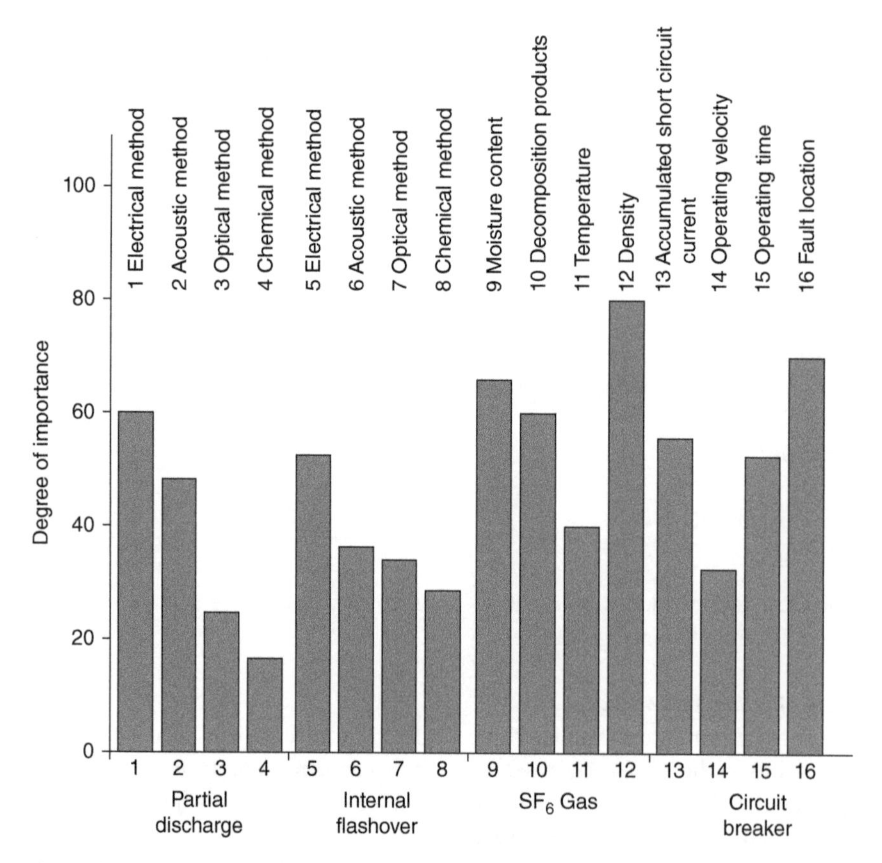

Figure 9.26 Distribution of degree of importance assigned by users for failure modes (Reproduced by permission of United Illuminating Company)

shows the four most important *risk areas* to GIS users and the degree of importance assigned by users of GIS to new technological developments in diagnostic methods that offer improved reliability.

Preventive Maintenance Assessments

This type of maintenance is traditionally applied to most substation equipment and usually has two parts.

The first part is a routine inspection that includes visual checks for the presence of abnormalities. The routine inspections are generally carried out on an annual basis, without the requirement for outages.

The second part may require the detailed strip-down and inspection of components to check for signs of wear or impending failure. Intervals are less frequent and are determined by the number of mechanical operations and severity of switching duties.

Major Maintenance Assessment

This type of maintenance is carried out at intervals of 10–20 years or after accumulating the permissible number of switching operations for circuit breakers and disconnect switches. Circuit breaker bay outages range from 5 to 7 days and could last as long as 14 days, depending on the type of maintenance and whether multiple gas compartments need to be evacuated and refilled.

Comprehensive Life Cycle Management Program

A successful comprehensive life cycle management program includes the management of the cost chains associated with the life of the GIS. The following is an example of a cost chain hierarchy that can be used to formulate a program (see Figure 9.27).

9.7.4 Assessing Risk

9.7.4.1 Projecting Future Performance and Bridging Gaps with Risk Assessment

In its simplest form, risk assessment is the process of enumerating risks, determining their classifications, assigning probability and impact scores, and associating controls with each risk.

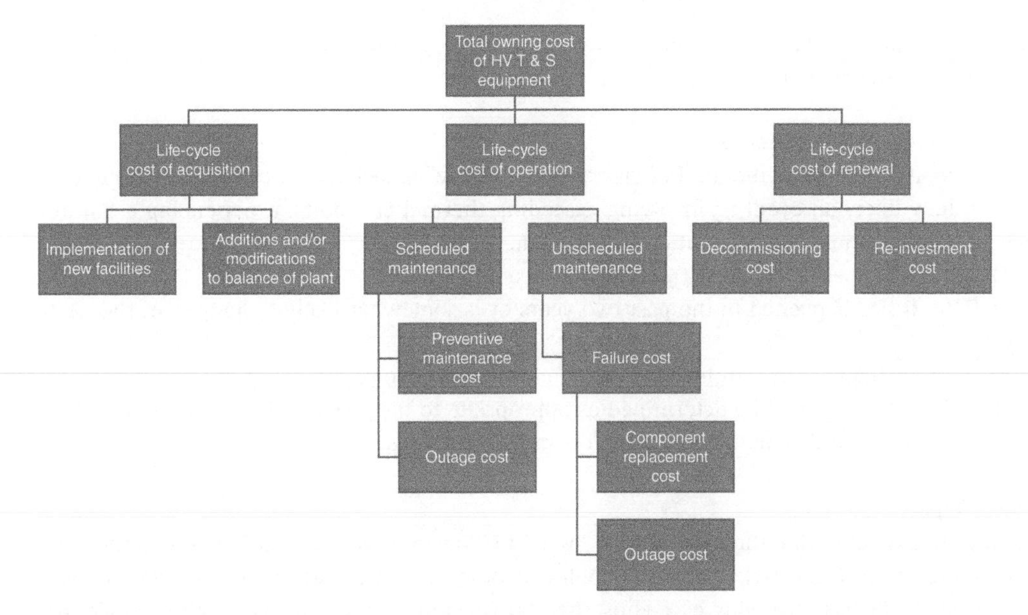

Figure 9.27 Total owning cost string (Reproduced by permission of United Illuminating Company)

9.7.4.2 Types of Risks

Qualitative: This is risk that is measured in terms like "high," "medium," and "low" for probability and impact, looking at the relative value of risk.
Quantitative: This is risk that is measured in dollars and statistical/probabilistic formulae.
Risk assessments: These measure the risk, the potential loss, and the probability that the loss will occur.

The simple formula is $R = L \times P$, or Risk (R) = Loss value (L) $\times$ Probability (P).

For a GIS system, the formula could be represented by the following:

Risk (R) = L (Availability $\times$ Impact) $\times$ P (Probability of occurrence).

9.7.4.3 Assessing Risk Levels

Defining the risk levels is the first step in the assessment process. The levels of risk should be defined with respect to some numerical value continuum such as:

High = 5
Medium high = 4
Medium =3
Medium low = 2
Low = 1

9.7.4.4 Availability

The user should determine the level of risk "high" to "low" associated with the GIS equipment not being available for service because of exceeding thermal capability, experiencing a failure mode, or manifestation of a degradation mechanism.

HIGH. If components of the GIS are rendered unavailable it causes large impacts to the transmission system, impairs business operations, affects customer service, or makes system operations difficult.

MODERATE. If components of the GIS are rendered unavailable it causes productivity loss but does not interrupt customer service and does not severely impact the transmission system or system operations.

LOW. If components of the GIS are rendered unavailable it does not severely impact business operations or system operations.

9.7.4.5 Probability of Occurrence

The user should determine the level of risk "high" to "low" associated with the GIS equipment based on how likely an event is: involving exceeding thermal capability, experiencing a failure mode, or manifestation of a degradation mechanism.

HIGH. It has happened in the past year or is happening now.

MEDIUM. It has happened in the past two years or is somewhat likely to happen in the next two years.

LOW. It rarely happens or is unlikely to happen in the next two years.

Probability ratings should be determined as appropriate to the goals of the assessment and in accordance with the user's maintenance and operating standards.

9.7.4.6 Impact

The user should determine the level of risk "high" to "low" associated with the GIS equipment based on the impact (financial, operational, political, other infrastructure, etc.) to the electric system that an event has: involving exceeding thermal capability, experiencing a failure mode, or manifestation of a degradation mechanism.

HIGH IMPACT. Will cost a significant amount of yearly budget, will cause widespread system operating issues, will significantly damage reputation, or will consume large amounts of resources affecting other work or projects.

MEDIUM IMPACT. Will cost some amount of yearly budget, will cause small or minimal system operating issues, will damage reputation somewhat, or will consume some amount of resources affecting other work or projects.

LOW IMPACT. Negligible effect or cost.

9.7.4.7 Risk Assessment Procedure (A Simple Algorithm Example)

There are six basic steps in the risk assessment procedure:

1) Classify the risk areas.
2) Rate the availability.
3) Rate the probability of occurrence.
4) Rate the impact.
5) Calculate the qualitative value of the risk.
6) Suggest controls to minimize the occurrence and impact.

The rigor that is applied to these steps will be different based on the GIS user's asset management strategy.

The first step in the risk assessment procedure is to classify the risk areas that are most important to the user of GIS equipment. If we return to Figure 9.26, we can easily classify the four most important risk areas:

1) Partial discharge
2) Internal flashover
3) SF_6 gas low density
4) Circuit breaker dielectric failure

The second step in the risk assessment procedure is to rate the level of risk associated with the GIS equipment not being available for service, based on the classified risk areas (see Table 9.3).

HIGH. If components of the GIS are rendered unavailable it causes large impacts to the transmission system, impairs business operations, affects customer service, or makes system operations difficult.

MEDIUM. If components of the GIS are rendered unavailable it causes productivity loss but does not interrupt customer service and does not severely impact the transmission system or system operations.

LOW. If components of the GIS are rendered unavailable it does not severely impact business operations or system operations.

The third step in the risk assessment procedure is to rate the probability of occurrence of the classified risk areas. The following rating method is based on many factors such as the age of

Table 9.3 Rating the level of risk for the four most important risk areas

	High (5)	Medium high (4)	Medium (3)	Medium low (2)	Low (1)
Partial discharge				×	
Internal flashover	×				
SF_6 gas low density					×
Circuit breaker dielectric failure	×				

Table 9.4 Rating the probability of occurrence in the classified risk areas

	High (5)	Medium high (4)	Medium (3)	Medium low (2)	Low (1)
Partial discharge			×		
Internal flashover				×	
SF_6 gas low density		×			
Circuit breaker dielectric failure					×

Table 9.5 Rating the impact of the classified risk areas

	High (5)	Medium high (4)	Medium (3)	Medium low (2)	Low (1)
Partial discharge					×
Internal flashover	×				
SF_6 gas low density				×	
Circuit breaker dielectric failure	×				

equipment, maintenance frequency, observed operating duty, and so on. The rating of probability of occurrence will be different for each user based on the user's application of the equipment and maintenance/diagnostics for the equipment (see Table 9.4).

HIGH. It has happened in the past year or is happening now.

MEDIUM. It has happened in the past two years or is somewhat likely to happen in the next two years.

LOW. It rarely happens or is unlikely to happen in the next two years.

The fourth step in the risk assessment procedure is to rate the impact of the classified risk areas (see Table 9.5).

HIGH IMPACT. Will cost a significant amount of yearly budget, will cause widespread system operating issues, will significantly damage reputation. or will consume large amounts of resources affecting other work or projects.

MEDIUM IMPACT. Will cost some amount of yearly budget, will cause small or minimal system operating issues, will damage reputation somewhat, or will consume some amount of resources affecting other work or projects.

LOW IMPACT. Negligible effect or cost.

The fifth step in the risk assessment procedure is to calculate the qualitative value of the risk based on our simple formula (see Table 9.6):

$$\text{Risk}(R) = \text{Loss value}(L) \times \text{Probability}(P)$$

$$\text{Risk}(R) = L(\text{Availability} \times \text{Impact}) \times P(\text{Probability of occurrence})$$

What this demonstrates is that the user may want to employ controls to mitigate the risk areas in the following order based purely on the qualitative risks:

1) Internal flashover
2) Circuit breaker dielectric failure
3) SF_6 gas low density
4) Partial discharge

Table 9.6 Calculating the qualitative value of the risk

	L (Availability)	*L* (Impact)	*P* (Occurrence)	*R* (Risk)
Partial discharge	2	1	3	6
Internal flashover	5	5	2	50
SF_6 gas low density	1	2	4	8
Circuit breaker dielectric failure	5	5	1	25

Table 9.7 Some of the controls used to minimize the occurrence and impact of the classified risk areas

Risk area	Mitigation control
Internal flashover	High potential electrical testing
	Acoustic partial discharge monitoring
	Optical arc sensor monitoring
	Detection of chemical arc by-products
Circuit breaker dielectric failure	Accumulated short-circuit current data
	Measure mechanism/interrupter operating velocity
	Measure circuit breaker operating time
	High potential electrical testing
	Dynamic resistance measurement
	Static resistance measurement
SF_6 gas low density	Measure moisture content
	Monitor temperature
	Monitor density and trend
	Detection of chemical arc by-products
Partial discharge	High potential electrical testing
	Acoustic partial discharge monitoring
	Optical arc sensor monitoring
	Detection of chemical arc by-products

A quantitative analysis of executing the mitigating controls with respect to the first cost of the controls will need to be accomplished.

The sixth step in the risk assessment procedure is to suggest controls to minimize the occurrence and impact of the classified risk areas. Again, this will be different for each user based on the user's application of the equipment, maintenance/diagnostics for the equipment, and financial drivers and is highly dependent on the risk tolerance of the user.

Table 9.7 shows some of the controls used to minimize the occurrence and impact of the classified risk areas. This is not an all-inclusive list, but it does provide the user with the most widely used controls to mitigate the risks associated with the identified risk areas. The user should evaluate the different diagnostic techniques and determine which approaches are in line with the GIS equipment asset management goals of the user's organization.

The user should seek to employ controls that mitigate multiple risks to maximize performance and value. For example, the same controls can be applied to mitigate the risk areas of internal flashover and partial discharge.

9.8 Health and Safety Impact

9.8.1 General

Health and safety are important aspects for both personnel and the general public during installation and operation of the gas-insulated switchgear. The risks of fire with burning materials producing airborne toxic gases or toxic material that is accessible by touch need to be reduced or appropriate measures need to be set in place to avoid harm to personnel and the public.

Local codes, requirements by authorities, fire department or police, and standards will need to be met for all stages of a GIS installation. Precautions are required during all phases of the GIS life cycle, including erection, installation, operation, and dismantling.

9.8.2 Risk of Burning

Burn hazards may be caused by contact with hot surfaces or burning materials. For GIS, temperature values of components accessible to touch during operation are restricted to a maximum of 80 °C. If any portions can reach higher temperatures then some form of touch protection is required.

The burning risk through materials on fire is avoided by the use of nonflammable materials for GIS, such as aluminum, resins, or copper. These materials are nonflammable under normal operating conditions.

9.8.3 Risk of Breathing Toxic Gases

In normal operation, GIS is free of toxic gases. However, in cases of an internal failure, for example, an earth/ground fault with short-circuit currents, and in compartments with interrupting or switching capabilities (breakers, disconnecting, and grounding switches), toxic gases may be present in the corresponding gas compartments of the GIS. These gases result from the decomposition of the SF_6 gas in the presence of electrical arcs. The power system's protection and control system (and also other possible sensing and measuring systems) will indicate the occurrence of an earth/ground fault and give an indication of a failure in a GIS. In some instances, following a failure, the GIS enclosure may become compromised (either via a breach in the enclosure or by operation of a pressure relief device), leading to toxic gases released into the ambient. If GIS buildings are used, oxygen sensors indicate the loss of oxygen and smoke and other detectors sense other gases inside the GIS building.

In the above scenario, personnel access to buildings is restricted and will only be given to trained personal with appropriate protection. Investigations must be taken prior to allowing personnel to enter the GIS room without danger of breathing toxic gases. Similarly, personnel entering a tunnel or a vault when faults have occurred will also have to use special protective clothing and breathing equipment. Details of the procedures and recommendations are available in the IEC (IEC 62271-4) [34] standard.

9.8.4 Risk of Touching Toxic Materials

Under normal circumstances, GIS does not use toxic materials that are accessible to touch. As discussed above for the case of an earth/ground fault, toxic solid compounds could be generated (in addition to toxic gas – see Section 9.10) and precautions must be taken to protect personnel prior to repair work.

9.8.5 Risk of Electric Shock

GIS is a completely enclosed and solidly grounded electric system. In normal operation, no direct contact to high-voltage parts is possible. For GIS, the outer metallic enclosure is solidly grounded and the grounding system is designed such that no dangerous touch voltage occurs in normal operation and in cases of ground fault currents in the enclosure.

The multiple grounding connections of GIS with the ground grid also generally provide a low impedance path for high-frequency transient overvoltage as they occur during switching operations. In this case, the outer enclosure must be grounded with a low impedance connection to avoid the risk of an electric shock. In some cases, the transient enclosure voltages (TEVs) can become significant and some incidents of shock are occasionally reported. A discussion on this phenomenon can be found in Section 9.5.

9.9 Electromagnetic Field

9.9.1 General

The electric and magnetic field of GIS and XLPE cables has an important impact on the cost of the total project in the case of given limitations. Electric fields can be shielded with the GIS enclosure or the XLPE cable shield. Magnetic fields of power frequency can be compensated by reverse currents in the GIS enclosure or in the XLPE cable shield. Additional shielding by magnetic materials, for example, steel plates, are very costly. External magnetic field strength of GIS is usually low due to the induced revers current in the enclosure.

9.9.2 Electric Field in Operation

When a GIS is in service, the electric field of the power frequency voltage is shielded by the solid grounding system of GIS and XLPE cables. In the case of transient voltages caused by switching operations, the high-frequency transient voltages need a low impedance ground connection to avoid high electric field intensity. This low impedance grounding is usually done by multiple, parallel ground connections to the grounding grid underneath the GIS. For fast transient voltages see section 9.15.

9.9.3 Electric Field during Short Circuit

In the case of a short-circuit current through the GIS or XLPE cable system the ground connections need to be designed higher in order to handle the rated short-circuit currents, for example, 50 kA or 63 kA, and the instantaneous peak current, which could be 2.6–2.7 times the rated short-circuit current.

9.9.4 Magnetic Field in Operation

To shield the surrounding of the GIS or XLPE cable from frequent power magnetic fields it is necessary to compensate the magnetic field of the conductor by an induced reverse current of the outer enclosure pipe of the GIS or by shielding the XLPE cable. The higher the induced reverse current, the lower is the remaining magnetic field around the GIS or XLPE cable. In the case of the GIS, the typical reverse current is in the range of 90–95% of the current in the conductor. In the case of an XLPE, the reverse current is in the range of 30–40%.

The technical possibility of using magnetic materials for outer shielding of the magnetic field of the conductor current will require large cross sections of steel plates to generate a shielding effect at 50 Hz or 60 Hz power frequency. This generates a high material cost of the steel plates and for civil works of the shielding construction.

9.9.5 Magnetic Field during Short Circuit

In the case of short-circuit current ratings the cross section for the reverse current needs to be high enough to limit the temperature rise due to the high short-circuit current rating.

In the case of the GIS the cross section of the enclosure pipe is by typical design large enough (cross sections have 12 000–20 000 mm^2 depending on the enclosure wall thickness). In the case of XLPE cables, the short-circuit currents may need to be limited to prevent them from overheating. In the case of lower current ratings for the cable shielding, the remaining active magnetic field surrounding the cable is higher due to the lower inverse current in the cable shield.

9.10 SF_6 Decomposition Byproducts

9.10.1 General

Since Pure SF_6 gas is nontoxic, the primary issue for safety is displacing air, especially in low-lying regions where heavier-than-air SF_6 can collect. However, when the SF_6 gas is exposed to electrical discharges, as indicated above, SF_6 will dissociate and the resulting byproducts can be toxic. Under certain conditions (such as in a switching compartment or breaker), many of the dissociated components will recombine and the SF_6 gas properties will be restored. However, the process is imperfect and byproducts can also be produced. Most of these byproducts are toxic. There are several situations where such byproducts can be generated

- In switchgear compartments (breakers and switches) – electrical discharges and arcs can form between the contacts of switching devices during operation. This is considered part of "normal operation" – byproduct generation is minimized through careful engineering of the switchgear contacts and through the use of appropriate materials. Nonetheless, byproducts can be generated – switching compartments typically include some absorbent material to reduce and control the concentrations in the gas.
- When partial discharges are present – partial discharges (PD) and corona are low-energy discharges often present when defects are present in the equipment – PD are often used as a marker for deterioration of the insulation system. This situation will generate byproducts at relatively low levels. This situation is considered abnormal.
- When gas-insulated equipment fails electrically, a large, high energy arc is formed between the electrodes generating large amounts of gaseous and solid byproducts. In many cases, the faulted gas is contained within the equipment. However, in some cases, the fault arc can burn through the equipment's enclosure or, sufficient internal pressure is generated to cause operation of a pressure relief device. Both of these cases will result in the direct contamination of the immediate area with byproducts.

9.10.2 Byproducts and Toxicity

Many byproducts' species are generated. The most commonly reported and relevant byproducts are listed in Table 9.8 (based on [22, 23]).

Table 9.8 List of relevant SF_6 decomposition byproducts. TLV (Threshold Limit Values) are often used to determine safe limits but, the exact value and interpretation will depend on specific jurisdictions. Typical TLV are presented here to provide an indication of relative toxicity. Local authorities should be consulted for specific regulations.

Byproduct	Chemical Name	Comments	End Products	TLV (ppm)
SF_4	Sulfur tetrafluoride	Rapid decomposition	HF, SO_2, SOF_2, SO_2F_2	0.1
SOF_2	Thionyl fluoride (or difluoride)	slow decomposition	HF, SO_2	2.5 mg/m^3
SO_2F_2	Sulfuryl fluoride	stable	SO_2F_2	5
HF	Hydrogen fluoride	Stable, but soluble in water	HF	3
SO_2	Sulfur dioxide	stable	SO_2	2
S_2F_{10}	Disulfuric decafluoride	Rapid decomposition in high heat	SF_6, SF_4	.01
AlF_3	Aluminum fluoride (solid)	Stable – provides surface area for adsorbed gases, including toxic byproducts	AlF_3	

Generally, byproducts can be categorized as follows:

1) Short-lived gaseous byproducts – byproducts such as SF_4 and others can be generated following an internal fault. These gases are usually extremely toxic. However, these byproducts tend to react internally with trace amounts of moisture and converts rapidly to secondary byproducts, such as SOF_2. This process is usually rapid and occurs in the timeframe of minutes to 10s of minutes
2) Stable gaseous byproducts – Byproducts (primarily SO_2, HF, SO_2F_2 and SOF_2) can be generated directly by the fault, but also form from the further decomposition of more reactive species, such as SF_4. Byproducts in this category tend to be stable internally to the GIS. SOF_2 and SO_2F_2 can undergo further hydrolysis to form SO_2, but this process takes much longer, usually in the order of days or longer.
3) Byproducts exposed to the atmosphere – Byproducts released directly into the atmosphere will undergo further hydrolysis (reacts with abundant moisture in the ambient) and be converted to SO_2 and HF.
4) Solid byproducts – depending on the nature of a fault, a large quantity of solid "powder" is produced. These are typically fine particles of metallic fluorides. Such powders can be an irritant and many of the toxic gaseous species can be found adsorbed on the surfaces (Table 9.8).

Other byproduct compounds are sometimes reported but these will usually fall into the above categories and share similar properties to others in the same category. One exception is carbon tetrafluoride (CF_4) which is often found when arcs interact with polymeric insulators used internally to the equipment. CF_4 is, however, nontoxic and does not impact on the gas quality in small quantities.

9.10.3 Impact on Worker Protection

Knowledge of the byproducts produced needs to be considered when developing a protocol for worker protection. For example:

1) Short-lived gaseous byproducts can be an issue if there is a sudden release of gas following a fault. This can occur if a pressure-relief device operates or if the GIS enclosure ruptures or burns through. These are rare events and are usually managed through controlled access to the equipment.

2) Stable byproducts (second group) might be a problem if accidental release occurs during gas handling following a fault. As this could occur as a result of a gas-handling error, some utilities will require some sort of respiratory protection as a precaution.
3) Gaseous byproducts can be released into atmospheric air. Normally, faulted compartments are fully evacuated, and faulted gas processed prior to opening for repair. However, workers could be exposed to trace quantities of gaseous byproducts and some form of protection is required. Respiratory protection is usually based on SO_2 levels. HF might also be present but, as HF is highly soluble, HF is more likely to be adhered to surfaces along with trace moisture. Gloves and other forms of protection are usually used. Large quantities of byproducts could also be expelled into the atmosphere in the case of pressure relief operation or a rupture in the GIS enclosure. In this case, procedures usually call for an exclusion zone until the hazard can be assessed and cleaned up. Forced ventilation is also used in the case of indoor installations.
4) Solid byproducts are almost always present to one degree or another. The fine powders are a respiratory irritant and will have toxic gaseous byproducts and HF adsorbed on the powder's collective surface area. Appropriate respiratory protection suitable for filtering fine (μm range) particles is usually required. Cleaning processes (vacuum) should use High Efficiency Particulate Air (HEPA) filtration.

In general, health and safety requirements for are usually based on anticipated SO_2 levels (most of the gaseous byproducts will hydrolyze in atmospheric air to form SO_2 and HF). Depending on circumstances, protect may range from full, supplied-air respiratory protection to use of cartridge respirators. Overall, these procedures appear to have provided adequate protection, provided that the rules and procedures are respected. Reports of injuries are usually the result of improper gas-handling and error, possibly related to a lack of training. Guidance on procedures for gas handling and worker protection are provided in Section 2.8.

9.10.4 Gas Analysis

SF_6 can be sampled and analyzed. SF_6 gas samples can be collected in specially-prepared stainless steel sampling bottles and sent to a laboratory for analysis. Typical analysis may include N_2, O_2, CF_4, SO_2, SO_2F_2, and SOF_2. In some cases, portable instruments are used. These instruments will generally respond to SO_2 and/or HF and may also indicate SF_6 purity. Separate instruments are used to measure moisture – typically, "dew point" hygrometers are used for this purpose. The portable instruments have the advantage of immediacy – measurement results are immediately available, but less detail is available in the result. Laboratory testing offers a more comprehensive analysis but with the inevitable delays and possibility of sample contamination and degradation routed to the laboratory.

Gas analysis is typically done to confirm the absence or presence of decomposition byproducts. Samples are often taken prior to invasive work to positively identify the faulted compartment and to assess the hazard for worker protection. Byproduct levels in a faulted compartment will typically be in the few hundred to a few thousand ppm range. The dominant byproduct following a fault will be SOF_2 although other species will be present. As SOF_2 will be slowly converted into SO_2 (internal to the equipment), some will use the relative quantities to judge the elapsed time since the failure.

S_2F_{10}, an extremely toxic species, is normally not found in gas samples. Any S_2F_{10} produced is usually assumed to decompose into other species. However, it is possible that in certain types of low-level ("cold") discharges, S_2F_{10} could be created in significant quantity – appropriate precautions should be taken.

The byproduct CF_4 is also produced but is usually not removed by filtration or other processes that "scrub" faulted gas. New SF_6 gas will inherently have low levels of CF_4 but "scrubbed" used gas (fit for reuse according to IEC 60480) may have CF_4 concentrations as high as 200–300 ppm. As it is not harmful, CF_4 is treated in the same manner as air contamination.

9.11 Condition Assessment

Author: Hermann Koch
Reviewer: James Massura

9.11.1 Introduction

In this chapter, recommendations are given for condition assessment based on industry experience. The recommended cycles of checks, maintenance, or repair may change for the specific condition of the GIS. This may require more frequent checks, maintenance and repair, and needs to be investigated for each GIS installation. The values given here can only be general recommendations.

Condition assessment of gas-insulated substation is getting more important with increased aging of the installed equipment. There are multiple aspects to evaluate the status of the GIS and to decide if maintenance, repair, up-grade or replacement with new GIS will be the next step.

The manufacturer's instruction manual for the GIS defines the required maintenance typically at 5, 10 and 25 years of operation. In addition, specific requirements of the substation location and connection to the electric network impact the functionality of the GIS and may require different maintenance schedules or replacements.

International standards are referred to the product related standards as listed in References 1 to 13.

In the following clauses, the various aspects of GIS operation and its impact on the life cycle are explained and proposals are made to assess the condition of the GIS.

This chapter combines the information collected by expert engineers of the IEEE PES Substation GIS Subcommittee on the design, development, and operational experiences of more than 50 years of GIS applications in the network.

The information provided can only assist evaluations on GIS condition assessment for a specific GIS installation at a specific location. The parameters influencing the condition of the GIS vary widely and need to be evaluated with engineering knowledge for each specific GIS installation.

The recommendations in the following are based on typical experiences made with GIS substations. It may be that the conditions at the evaluated project differ from these typical conditions and inspections might be needed in shorter time sequences.

Also it is from impotance for a correct evaluation of the condition of the GIS substation to take into account possible changes on the design or electrical circuit of the GIS substation.

9.11.2 Visual Inspection

9.11.2.1 Introduction

Visual inspections are basic maintenance activities in a substation to check the condition and to recognize possible maintenance issues. Most operational practices of users define a time schedule to visit the substation and to check specific features of the GIS. In the following section, information is given which features shall be checked and an explanation is given on the technical background for the visual inspections.

9.11.2.2 Mechanical Inspection and Stress

Visual mechanical inspections are made to find material defects caused by mechanical stresses in the GIS installation. There are different sources for mechanical defects: (1) stress coming from thermal expansion of the GIS enclosures caused by missing or incorrect functioning of thermal expansion compensators, (2) stress coming from external sources like vibrations in the building, (3) impact from street or rail road traffic, (4) impact from seismic mechanical forces applied to the GIS and its steel structures.

This mechanical stress can be identified by deformations or cracking of the steel structures or the GIS enclosures and flanges. Thermal expansion forces are very strong and can deform any steel structure. So, if the visual check indicated a deformed steel beam there must be a reason for this mechanical deformation and the source of this reason must be found. The cause of this deformation must be determined and corrected or more serious failures with impact to the high voltage part of the GIS and possible gas losses by leaking O-ring seals may occur.

Sources of mechanical stress can come from non-functioning slide bars of longitudinal or angle thermal-expansion compensation systems. There could be a blockage of the sliding parts or a high slide resistance which can cause such high mechanical forces.

Vibration or oscillation can be produced by directly connected transformers with aluminum bus duct pipes without mechanical decoupling, rotating machines in the building, by heavy traffic close to the substation foundation, or by smaller or larger earthquakes. In any case, the vibrations may impact the mechanical drive systems of the GIS switches and breakers, or apply permanent forces to the conductor and enclosures which can be damaging in case of resulting mechanical resonance.

Visual Inspection Time Schedule for Mechanical Stress

A visual inspection for mechanical stress at the GIS can identify failures before they may cause more significant damage. Most substations are unmanned and remote controlled. Usually nobody is visiting the substation regularly, except if there is a time-based visual inspection schedule. Such visual inspections shall be done more frequently after the initial commission of the GIS. Initially, a weekly inspection is recommended, which later will turn to a monthly visual mechanical inspection. When the substation and the GIS is in normal operation such inspections can be scheduled on an annual basis.

Some substations are very remote and difficult to reach. In such cases, remote visual inspection tools may be installed like video cameras. This will help to identify changing conditions at an early stage and prevent the GIS from failing and power supply interruption.

Checklist for Visual Inspection Time Schedule for Mechanical Stress

To guide through the mechanical visual inspection, it is recommended to have a checklist made for the specific GIS installation. This checklist may have the following items:

- Deformations at the steel structure
- Deformation at the GIS footings
- Deformations at the GIS enclosure flanges
- Deformations at the slide bars of bus ducts
- Deformations at the slide bars between bays
- Deformations at the thermal expansion compensators
- Damage of the color paint of GIS enclosures
- Damage of color paint of steel beams

9.11.2.3 Corrosion

Corrosion reduces the lifetime of the GIS when it occurs. Both indoor and outdoor GIS need to have corrosion protection. The aluminum enclosures do not need paint to protect the surface, because aluminum is self-protecting by the oxide layer formed on the surface.

Only in the case when water containing chlorides is penetrating the aluminum surface for a long time the oxide layer will be destroyed and corrosion takes place. This corrosion will lead to a pit hole and finally gas losses. To prevent from such corrosion effects the visual inspection shall include to check for water contamination of the GIS surface.

Such water penetration will shorten the lifetime of the GIS. A leaking roof may be the cause that water is frequently contacting exterior of the GIS. Water can also come from ground water or the ground water piping system. This is important when GIS is installed in basements of buildings.

The steel structures and steel compartments for the control cabinet need to have passive corrosion protection for indoor or outdoor use. Steel structures and steel beams are using hot-dip galvanized steel in most cases. The local control cabinets are using color paint.

The inside of steel cabinets for outdoor equipment needs a heater to protect from moisture inside which also causes corrosion.

Most critical are connecting areas of steel materials and moisture stored in the gap between the materials. This remains for a long time and specially in the case when chlorides are entering the gap the corrosion can be accelerated.

Flanges of the GIS may corrode in case of improper installation of the equipment by using the correct grease to avoid water from entering the flange gap. Gap corrosion will damage the flange surface and causes gas leakage at the flange. The visual check of the flange will prevent gas losses and a time planned repair can be scheduled.

For indoor GIS in case of high humidity, to prevent corrosion it might be recommended to air condition the building.

For outdoor GIS, the control and mechanical drive boxes need to be designed to be waterproof. External gas piping should be avoided to prevent from pipe corrosion and decentralized density monitoring devices directly connected to GIS gas compartments are preferred. When the GIS equipment is installed in the proximity of a chemical, steel, or aluminum manufacturing plant, it is particularly important to do the corrosion inspections on a regular basis.

Visual Inspection Time Schedule for Corrosion

The corrosion process is slow for indoor or outdoor GIS and the inspection follows a longer time schedule. After the commissioning is finalized and all checks have been made on the GIS installation, the first corrosion check can be scheduled after one year of operation.

In the following visual inspection checks on corrosion can be scheduled in 3–5 year sequences.

Checklist for Visual Inspection Time Schedule for Mechanical Stress

The checklist for the visual inspection of the GIS may have the following items:

- No permanent water penetration to the GIS
- Inspect the building for leaks from the roof
- Check building for ground water penetration
- Check the steel structure and beams at the connection area
- Check the inside of the local control cabinet
- Check the gap of the GIS enclosure flanges
- Check the humidity inside the building
- Check the control and mechanical drive boxes in the GIS bay

9.11.2.4 Ground or Earthing Grid Connections

The ground and earth connections to the grid are essentials for personal safety of limited touch voltages. In this case of a high voltage ground fault, a short-circuit current needs to be grounded with low impedance. The correct function of the substation protection system also requires a defined and constant ground impedance for the years of operation of the substation.

Changing and increasing ground impedances may have different reasons. Loose connection of the ground wire terminal is one. This could come from the low force of the terminal screw after wrong assembling works or connector or it could be caused by corrosion.

To check the ground or earthing grid connection a visual check might not be enough to make sure the impedance is low enough for correct function. To prove this, a resistance measurement of the ground grid termination is recommended.

At the GIS the ground or earthing grid connection is usually done by connecting the foot point of the GIS bay with the ground or earthing grid of the building or substation. The terminations need to have clear markings with the ground/earth sign.

Visual Checks of the Grounding or Earthing Grid Connections

The visual inspection time schedule for corrosion of the grounding or earthing grid connections need to be checked frequently. The main aspect for the visual check is to see if anything has been changed after the GIS has been installed and commissioning checks have been carried out. This might happen if new devices have been added or others replaced. A visual check of the substation for correct ground or earthing grid connections are recommended on a one-year sequence.

Checks for the correct impedance of the ground or earth grid connection are recommended in a longer time sequence of 5–10 years. Changes of the impedance is related to slow processes like corrosion, mechanical, or material stress or external vibrations. Here, it is recommended to check the conductivity by an electric resistance measurement.

Checklist for Ground or Earth Grid Connections

Checks on an annual basis:

- Check for any change made on the ground or earth grid connection
- Check for clear markings of the ground or earth grid connection

Check of 5–10-year basis:

- Check electric resistance of the connector terminals by measurement
- Check for material overload indications by visual changes (color, form)
- Check for mechanical stress identifications by deformations
- Check for external vibrations to the ground or earth grid connections

9.11.2.5 Cross Bonding and Bonding Connections

Cross bonding and bonding connections of GIS are used to allow the induced current of the enclosure to circulate. The effects are low outer magnetic field strength and low electro-magnetic forces to GIS insulators. The induced current in the enclosure of the GIS is in the range of 80–90% of the current in the GIS conductor. In normal operation, this is about the rated current of 2000, 3000 or 4000 A in most cases. For the short-circuit situation, the current can be 40, 50, 63 kA or even 80 kA for 0.3 or 0.5 s in most cases. The cross bonding or bonding conductors and its terminations need to able to carry such high currents for the given times.

The GIS bay design covers these requirements with copper conductors between the three phases connected to the GIS enclosure flange, the connecting cable or bushing terminations, or for the bus duct for direct transformer terminations. The cross section and the number of nuts and bolts is dependent on the current rating and timing, and will be designed by the manufacturer.

The GIS should be installed, including these cross bonding and bonding elements, and the correct function shall be checked during the commissioning process.

Changes in the substation design may change the cross bonding and need to be checked frequently.

Material stress, over-load and material degradation, external impact to the contact system of the cross bonding or bonding connections need to be visually checked.

Visual Checks of the Cross Bonding and Bonding Connections

Visual inspections of the cross bonding and bonding connections of the GIS need to be carried out frequently to maintain the proper function of the substation. More frequent visual checks of the cross bonding and bonding connections need to follow any change of the substation assembly. Removed bonding conductor bars may interrupt the induced current flow for outer magnetic field strength and for inner electromagnetic field forces during short-circuit events.

Visual checks for changes in the GIS bays are recommend every year.

Assembly failure, corrosion effects, increasing temperatures at the bonding connectors or deformations of the bonding connector bars and screws are slower and infrared measurements, check on deformations are recommended in 5–10-year time sequences.

Checklist for Cross Bonding and Bonding Connections

Checks on an annual basis:

- Check for any changes made on the cross bonding and bonding connections

Check of 5–10-year basis:

- Check by infra-red thermal measurement the cross bonding and bonding connector bars and terminals for high temperature because of increased resistance.
- Check the cross bonding and bonding connector bars and terminals for material overload indications by visual changes (color, form).
- Check the cross bonding and bonding connector bars and terminals for mechanical stress identifications by deformations.

9.11.2.6 Pressure Relief Device

Pressure relief devices in GIS are the mechanical weak points of the pressurized enclosure and work as a safety opening in case of over-pressure. Such over-pressure can occur during the gas filling process to reach the designed filling pressure. In case of wrong settings or nonfunction of the pressure measurement and control during filling process, the design pressure of the GIS enclosure could be reached with the risk of enclosure rupture. In such a case, the pressure relief device will open or break and the gas will be released before any danger of the GIS enclosure rupturing.

The second reason for the pressure relief devices is the situation of an internal arc fault of the GIS. In this case, the gas pressure in the GIS gas compartment will rise quickly, within milliseconds, and could reach values when the GIS enclosure could rupture. To avoid this, the maximum pressure of the GIS gas compartment will be limited by setting the rupture pressure of the pressure relief device to a value less than 10% above the maximum operational gas pressure. The rupture pressure of the GIS enclosers are design for more than 250% over-pressure.

There are different types of pressure relief devices in use: spring operated valves, metallic rupture discs, or ceramic rupture discs. Each design needs to follow the gas tightness, corrosion, and aging requirement for GIS applications.

The safe function of the pressure relief devices requires one for each gas compartment and need to have gas stream control hubs to avoid hot gas penetrating areas where control and maintenance personnel may be. Usually, the gas stream control hubs are directed away from the walk areas. This is part of the GIS substation bay design.

Visual Checks of the Pressure Relief Device

After the commission checks, but before the GIS is energized, the correct design and assembly of the pressure relief device is checked and proven correct. Visuals checks during operation are recommended at frequent intervals. It is very unlikely that any change will be made at the pressure relief devices, because this would require gas works and a de-energization of the GIS and a recommissioning process. Therefore, yearly visual checks of changes made on the pressure relief devices are not recommended.

The impact of gas leakages and corrosion effect of the pressure relief devices are recommended in a long-time sequence of 5–10 years. Gas leakage and corrosion can be visually inspected by changes of color, form, or surfaces. Changes of the form or color of the rupture discs can indicate aging effect and will require a replacement, depending on the design.

Checklist for the Pressure Relief Device

Check of 5–10-year basis:

- Check for gas leakage at the pressure relief device sealing in case of gas leakage of the connected gas compartment.
- Check the form of the metallic rupture discs for deformations

9.11.2.7 Bushing Terminations

The bushing terminations are the contact point from gas insulated to air insulated. This transitions from the dielectric requirement of shorter distances required for pressurized insulating gas with the longer distances required at ambient air dielectric insulation.

The gas pressure side of the bushing is connected to the gas pressure of the GIS. In most cases, a gas-tight insulator allows the dismantling of the bushing from the GIS without opening the connected gas compartment of the GIS.

The bushing may be protected by over-voltage surge arresters to avoid transient over-voltage from entering the GIS.

For indoor GIS, the bushing is installed at the building wall and is exposed to the ambient air conditions of wind, rain, ice, dust, salt, and anything else contaminating the surface of the bushing insulator.

For outdoor GIS, both the bushing and the GIS are under the same ambient conditions.

Bushings are grounded or earthed at the flange connection of the GIS, or in case of a wall passage, the grounding and earthing is part of the building wall.

Visual Checks of the Bushing Termination

Visual checks of the connected bushing termination have to follow two aspects: one is related to safety and grounding/earthing and the other is related to contaminations and aging. The correct ground and earthing are checked with the commissioning process of the GIS and short time checks are only required in case of changes made on the GIS design and circuit. Any change needs to verify the correct grounding and earthing.

The long-time sequence visual check is related to the impact of ambient conditions, which goes in line with the substation equipment under the same ambient conditions. Insulator surface conditions of contaminations may influence the insulating capability, and cleaning or exchange might be needed.

The bushing termination to the overhead line can increase its resistance and produce more heat. This is a long-time effect of the terminations and needs to be measured by infra-red measurements to identify necessary replacements.

Checklist for the Bushing Terminations

Checks of yearly basis:

- Check for any change made on the bushing termination.
- Check the correct conductivity of the grounding and earthing.

Check of 5–10-year basis:

- Check the contaminations of the insulator surface.
- Check the correct connection of the surge arrester, if any.
- Check for high temperatures at the bushing termination by infra-red camera measurements.

9.11.2.8 Local Control Cabinets

The local control cabinets combine circuit breaker (CB), ground or earthing switches (GS) and disconnecting switches (DS) and their connection to the control center and protection devices. Relays and control switches provide the electric signals to the GIS and substation control and protection. Gaskets for gas density measurements, sometime located in the local control cabinet need to be gas tight according to the international standards with less than 0.5% gas loss per gas compartment per year. The general condition of the control equipment is important for the maintenance schedule of the GIS and substation. Controls and power cables bring energy to the motors and drives and link the signaling to the control and protection system. Heaters are used in local control cabinets to avoid inner corrosion caused by condensation moisture.

Relays/Control Switches

Relays and control switches are crucial elements of the substation control. Without well functionality of these the substation cannot work correctly. Mechanical or electronic relays and control switches show a long lifetime when used in the proper way as designed. Changing conditions may cause failing or provide aging effects. These changing conditions might come from too high or too low temperature, to many operations, to high currents because of faults in circuit or dust and salt with moisture causing fault currents.

These impacts are slow in time and may be checked in 5–10-year cycles.

Visual Checks of the Relays/Control Switches

Relays and control switches can be visual checked for any indication of aging by changing colors or deformations. Overload of circuits might be visible. Moisture may cause corrosion of the contacts or switch assembly.

Checklist for the Relays/Control Switches

Check of 5–10-year basis:

- Check the relays and control switches for changed color indicating overload.
- Check the relays and control switches for deformations.
- Check for corrosion indication on the switch contacts or the assembly.

Gaskets

Gaskets are required to keep the gas density in the gas compartment for dielectric safe function of the GIS. Some GIS design is using gas density meters to be installed inside the local control cabinet. In such cases, the gaskets of the pipes to connect the gas density meter with the related gas compartment of the GIS need to be gas tight.

Wrong assembly of the gas pipes and the gas density meters can cause gas leakages.

Vibrations and temperature changes might age the gaskets and cause leakages.

Moisture in the local control cabinet may cause corrosion at the gaskets and soldered pipe connections which will lead to gas leakages.

Gaskets in the local control cabinet have been avoided by new design because such gas leakages of the gas pipes and the gaskets of the gas density meters are difficult to find and are often reason for gas losses in GIS substation.

Visual Checks of the Relays/Control Switches

After the gaskets have been proven as gas tight with the commissioning process, checks for leaking gaskets and pipe will need some time to develop. For this reason, a 3–5-year sequence for visual checks of gaskets and gas pipes is recommended.

Checklist for the Relays/Control Switches

Check of 5–10-year basis:

- Check the gasket for changed color and surfaces, indicating corrosion.
- Check the gaskets and pipes for deformations.

General Condition

The general condition of devices used for the local control is much related to its age and the availability of spare parts. While the GIS has an expected lifetime of >50 years, most control equipment's lifetime is much shorter. Many electronic devices often have only 10–20 years available for spare parts.

Visual Checks of the General Condition

Visual checks of the general condition of the local control cabinet are recommended in longer time sequences. By checking the general appearance of the local control cabinet and its devices, the functionality of the complete substation is enhanced. With missing function of control and protection, the GIS cannot work. Visual checks shall include the availability of spare parts.

Checklist for the General Condition

Check of 5–10-year basis:

- Check the local control cabinet on its general appearance.
- Check the devices of the local control cabinet for available spare parts.

Control and Power Cables

Control and power cables installed in the substation to connect the GIS to control signals or electric power supply will age, depending on their operating conditions.

Control cable may get corrosion at their terminal contacts which might bring contact problems and missing signal transmission.

Power cable may see thermal aging over time if transmitted current is close to its rating.

Visual Checks of the Control and Power Cables

Visual checks of the control and power cables used in the GIS and local control cabinet can give indication of the cable terminations on corrosion on the contacts, or overload occasions of the power cables.

Failing terminations of control cables are difficult to identify, as inner corrosion or loose contacts cannot be seen. For this, electrical resistance measurement is required. But such tests and checks are too extensive and not required to do. If the aging of the control and power cables show such indications, it is recommended to replace the complete cabling including terminations. In practice, this decision needs to be evaluated with the lifetime of the complete GIS or if any up-grade or replacement would be the better and more economical solution.

Check List for the Control and Power Cables

Check of 5–10-year basis:

- Check the local control and power cables for indication of corrosion.
- Check the local control and power cables for indication of over-load conditions (changed color).

Heaters

Heaters are required to control moisture inside the control cabinets. Heaters may fail and moisture may create corrosion inside the cabinet. Especially in regions with strong changes of temperature, this effect can cause failure within a short period of time.

In deserts, for outdoor equipment high day temperature changes to low night temperature. Moisture condensates on the local cabinet wall and produces free water. With the dust and salt in the air, corrosion starts immediately. This will affect any element inside the local control cabinet.

The heater of the local control cabinet needs to be designed for producing enough heat to avoid the condensation of moisture. This requires different heater systems depending on the ambient climatic conditions.

Visual Checks of the Heaters

The visual checks of the heater need to follow these requirements. First is the check of the correct function of the heater at its location and ambient condition. This check needs to be done early in the operation time and with short sequences of time at the beginning. The second check is related to the heater function itself and longer periods are recommended.

Checklist for the Heaters

Check of 1–3-year basis:

- Check for sufficient heat production at the location and its ambient conditions.
- Check the heater temperature setting

Check of 5–10-year basis:

- Check the heater function.
- Check the heater temperature setting.

9.11.2.9 Breaker Operation Inspection

The visual inspection of the circuit breaker operation drive system has to cover three principal types of operation systems: compressed air, hydraulic systems and spring systems. There is a different focus on the visual inspection for each of them.

Compressed Air Operational Systems

Compressed air operational systems are using a centralized compressed air storage tank and compressed air pipes to each of the circuit breaker drives. The whole system is under pressure and leakages in pressure losses may disable the function and need to be found and repaired. The compressor operation sequences are a first indication of pressure loss. If the compressor has to refill the storage tank too often there might be leaks. The second check is with the air pressure pipe and joints to identify possible leaks.

Corrosion on the pipe system and its joints for compressed air may cause leakage of air. The corrosion can be from the outside environment and also from the inside, when the compressed air contains moisture. Moisture filters on the compressor might be ineffective and need to be changed.

In principle, the compressed air system is losing pressure and the compressor needs to pump new air into the storage tank. The time sequences for the compressor are defined in the operational handbook and are typically in the range of some tens of minutes.

Compressed air operational systems for the circuit breaker have been used through the 1960s and later have been replaced by hydraulic operation systems in the 1980s and spring operated systems in the 1990s. There are still substations in operation using compressed air.

Visual Checks of the Compressed Air Operational Systems

The first check is on the sequence of operation of the compressor to refill the compressed air tank. The time sequence should be in the limits given by the operation manual.

The second check is for the pipes and their joints for leaks and corrosion.

Leaks of the compressed air system can occur in short time sequences and it is recommended to check the compressed air operational systems on a yearly basis. Also, the compressed air system is essential for the functionality of the circuit breaker.

Checklist for the Compressed Air Operational Systems

Checks of 1-year basis:

- Check the time sequence of the operation of the compressor.
- Check the compressed air pipes for leaks.
- Check the joints of the compressed air pipes for leaks.
- Check for corrosion from the outside to the pipes and joints.
- Check the moisture content of the compressed air to prevent inner corrosion.

Hydraulic Operational Systems

Hydraulic operational systems for the circuit breakers are using Nitrogen (N_2) high pressure storage and hydraulic oil for the operational mechanism to transmit the mechanical force. The pressure of N_2 tank is in the range of 30 MPa and is directly connected to hydraulic drive mechanism of the circuit breaker. Such hydraulic drive systems are assembled in the factory and routine test before they are assembled to the GIS bay. The probability of failures is low, compared to the on-site assembled compressed air system with the extensive pipe system.

Hydraulic operational systems are equipped with pressure sensors and will indicate loss of pressure.

The sealing system of this high-pressure system may get leakages on the N_2 pressure tank or at the pressurized hydraulic oil system, which will be indicated by oil leaks.

Visual Checks of the Hydraulic Operational Systems

Visual checks on the hydraulic operational system are recommended for the N_2 storage tank and the sealing of the oil filled hydraulic drive mechanism.

Oil outside the hydraulic drive may indicated leakages in the hydraulic part of the mechanism. The N_2 high pressure pump should only be in operation when the circuit breaker has operated. The N_2 storage should not have any gas loss, which will be indicated by frequent operation of the N_2 pump.

Corrosion of the pipes and joints may cause leaks and need to be checked.

Leakages of hydraulic systems are not very frequent after the system has been passed the commissioning check after installation. Also, there is a pressure sensor for indicating pressure loss with each hydraulic drive. For this reason, visual checks are recommended in 1–3-year sequences for oil or N_2 leaks and for 5–10-year sequences for corrosion checks.

Checklist for the Hydraulic Operational Systems

Checks of 1–3-year basis:

- Check the time sequence of the operation of the N_2 compressor.
- Check the compressed N_2 pipes and tank for leaks.
- Check the sealings of the hydraulic oil for leaks.

Checks of 5–10-year basis:

- Check for corrosion from the outside to the N_2 pipes and joints.
- Check for corrosion from the outside of the hydraulic system.

Spring Operational Systems

The spring operation systems for circuit breakers has been introduce to GIS in the 1990s and has the least maintenance requirements. Spring operation systems are assembled and routine tested in the factory and come installed in the GIS bay on-site for substation assembly. No compressed air or oil can leak. The spring is mechanically loaded and maintenance free for normal operation for its lifetime.

The spring of teller or spiral type is loaded by an electric motor and the mechanism is triggered by electric activators from the control and protection devices.

The force of the loaded spring is monitored by sensors and indication is given when the spring is losing force.

Corrosion or high temperature may cause an impact to the spring operation system of the circuit breaker which may require maintenance and repair. The time sequence of such impact is slow.

Visual Checks of the Spring Operational Systems

The visual check of the spring operational system should cover the ambient operation conditions including moisture impact and maximum temperatures. This check is recommended in a 1–3-year time sequence.

The electrical system should be checked for well-connected wires and contacts. Corrosion at the contact system might lead to interruption and need to be checked. These corrosion checks are recommended in a 5–10-year sequence.

Checklist for the Spring Operational Systems

Checks of 1–3-year basis:

- Check for maximal temperatures of the spring operational system.
- Check for moisture impact of the spring operational system.

Checks of 5–10-year basis:

- Check for corrosion from the outside to the spring operational system.
- Check for corrosion of the electrical wires and contacts.

9.11.2.10 Operating Mechanism and Linkage

Operating mechanism and linkages are used to connect the disconnect switches and ground/earth switch with the motor drives. These devices are integrated part of the GIS bay and have been tested in the routine tests at the factory and with the on-site commissioning tests. In normal operation, no visual check is required and maintenance shall follow the recommendation of the operational manual.

Impact to the functionality of the operating mechanism and linkages may come from constructional change at the GIS bay. This should then be checked.

The ambient condition may also change for the GIS installation by water penetration from the roof, high moisture levels, dust, and moisture or higher temperatures which may cause corrosion. Corrosion at the operating mechanism and linkages may cause sliding or motion problems which ends in malfunction of the switches. These effects are developing slow in time but need to be checked.

In case of blocks of operation mechanisms and linkages, connecting rods and bars may get deformed by the force of the motor drives. This indicates possible corrosion of sliding contacts and need to be repaired. The check of such deformations develops slowly and it is recommended to check for deformations in a longer time sequence. Depending on the visual check of deformations, repair should be scheduled.

Visual Checks of the Operating Mechanism and Linkage

Changes on the design and assembly of the GIS bays are coming quite often in the lifetime of the GIS. Installing of additional sensors for monitoring reasons are used also in older GIS to improve its management. Such additional devices need to be installed in a way not to compromise the functionality of the operating mechanism and linkages of switches. This should be checked in short time sequences of 1–3 years.

The ambient conditions of the location of the GIS may change over time and should be checked in longer time sequences of 5–10 years.

Checklist for the Operating Mechanism and Linkage

Checks of 1–3-year basis:

- Check for design changes of the GIS bays and impact to the operating mechanism and linkage.
- Check for change of the ambient condition of the GIS bays.
- Check for deformations at the operating mechanisms and linkages.

Checks of 5–10-year basis:

- Check for corrosion at the operating mechanism and linkage.
- Check for deformations at the operating mechanism and linkage.

9.11.2.11 Gas Piping and Decentralized Gas Density Sensors

First generation GIS was using a centralized SF_6 monitoring system with gas density sensors of each gas compartment located in the local control cabinet. This required large amount of gas piping in each GIS bay. Each joint of the pipe are potential causes for gas leakage, and the soldering method of joints shows aging effect, causing gas leaks. This explains the technical development concerning gas pipes in GIS bays. Nevertheless, there are still many GIS bays in operation using gas piping.

Today gas density sensors are connected to each gas compartment and the gas density signal is transmitted by electric wires to the local control cabinet. Next generation of GIS will use an Integrated Electronic Device (IED) direct connected to the GIS gas compartment to measure temperature and pressure, and to calculate the gas density. The data connection to the local and central control is done by fiber-optic cables. The gas density sensor connected to each GIS gas compartment are using O-ring seals which perform well for the lifetime of the equipment. Gas leakages may occur if corrosion at the O-ring sealing takes place; this may be caused by changing ambient conditions. Water penetration of high moisture level including dust may be the reason for the corrosion. These changing ambient conditions need to be checked.

Visual Checks of the Gas Piping and Decentralized Gas Density Sensors

Visual checks of gas piping are required for GIS installations using central gas monitoring in the local control cabinet and gas pipes for the connection with each gas compartment of the GIS bay. The checks need to include the pipe and the joints which are usually tin-soldered. When gas loss is indicated, a check of the gas piping is recommended. Deformations of gas pipes and its joints may also be the reason for gas leakages. This check on deformations is recommended in short time sequences of 1–3-years. The aging of soldered joints and corrosion are slower processes and may be checked in 5–10-years' time sequences.

For decentralized gas density sensors, the gas density is provided by the proven technology of O-ring sealing. Gas leaks may occur when corrosion takes place at the sealing. This can come from changing environmental conditions of the GIS bays with water penetration or high moisture with dust.

Checks on deformations of the decentralized gas density sensors may be related to works carried out at the bay and are recommended to be checked in short time sequences.

Gas leakages of O-ring sealings are seldom and develop slow by corrosion. Checks for corrosion are recommended in longer time sequences of 5–10-years.

Checklist for the Gas Piping

Checks of 1–3-year basis:

- Check gas pipes for deformation.
- Check gas pipe joints for deformation.

Checks of 5–10-year basis:

- Check the joints and gas pipes for corrosion.
- Check the joints and gas pipes for aging effects.

Check-list for the Decentral Gas Density Sensors

Checks of 1–3-year basis:

- Check decentral gas density sensor for deformation.

Checks of 5–10-year basis:

- Check the O-ring sealing of the gas density sensor for corrosion.

9.11.2.12 Gas Density Monitoring Equipment

Gas density monitoring equipment is needed for the control and calibration of the accuracy of sensors. As recommended by the gas sensor manufacturer a calibration tool is needed to control the gas density sensor. This is necessary in longer time sequences of 5–10-years.

Visual checks of the gas density monitoring equipment are recommended to follow the changes made on the assembly of the GIS bay. Access to the decentralized gas density sensors, which may be at higher physical levels of the GIS, is required.

Visual Checks of the Gas Density Monitoring Equipment

The visual check of the gas monitoring equipment is recommended to prove accessibility to the decentralized sensors within the GIS bay. This check should indicate any changes of the GIS bay assembly which would hinder the connection of calibration tools.

Checklist for the Gas Density Monitoring Equipment

Checks of 1–3-year basis:

- Check gas density monitoring equipment for accessibility of calibration tools.

Checks of 5–10-year basis:

- Check for calibration of gas density sensors according to operational manual of manufacturer.

9.11.2.13 Foundation Settlement

Settlement of foundations of GIS installations may cause mechanical and gas tightness problems. GIS bays are flange connected at the bus bar bay-by-bay. The accuracy of the connection allows about 20–30 mm of tolerance when the GIS gets installed. This is a difficult requirement for GIS installations which can have several tens or hundred meters of length.

In case the foundation settles, mechanical forces are brought to the flange of the GIS. This may cause gas leakages at the O-ring sealings.

The reason for foundation settlement can be caused by the static weight of the GIS and to soft ground construction, or it can be related to dynamic forces of the circuit breaker operation, which causes high forces and may cause material fatigue.

Visual checks of the GIS on foundation are recommended in shorter time sequences by looking for indications of settlements. This could be indicated by changes of the foot points of the GIS, the anchor system in the ground or deformations on the bus bar. This check may be done in 1–3-year time sequences.

In case of a direct transformer connection to GIS bus bars, the settlements may come from the transformer foundation. The connection from GIS to transformer requires an angle compensation unit to compensate such settlings. It is recommended to check the tolerances given by the compensation unit. This is typically a slow process of settling and it is recommended to check in 5–10-year sequences.

Visual Checks of the Foundation Settlement

Visual check of the settlement of the foundation of GIS is recommended in longer time sequences of 5–10-years. Special attention shall be given to direct connections of GIS and transformers by bus duct systems. Here, settlements may be faster and a visual check in 1–3 years sequence is recommended.

Checklist for the Foundation Settlement

Checks of 1–3-year basis:

- Check tolerances of settlements between GIS and direct connected transformers.
- Check for deformations of the GIS bus bar.
- Check for deformations of the GIS foot points.

Checks of 5–10-year basis:

- Check for settlements of GIS bay foot points.
- Check for settlement of GIS anchor system at the ground or floor.
- Check for deformations or sagging of the GIS bus duct.

9.11.2.14 Supporting Structure

The supporting structure of the GIS bay gives mechanical stability for operation under normal or seismic conditions. Supporting structures are designed for each GIS bay to cover the static and dynamic forces. Supporting structures usually use hot-dipped galvanized steel.

Mechanical and dynamic forces from the GIS to the steel structure may increase when sliding mechanisms for compensation of thermal expansion are not working correctly and mechanical forces to supporting structures increase. This may cause deformations of the supporting structures.

In case of seismic requirements, the supporting structure may also get higher dynamic mechanical forces which may deform the support structure.

Changing environmental condition of the GIS installation may create corrosion when water penetrates the steel structure or if the level of moisture is high combined with dust and salt. Corrosion starts at the joint areas at the flanges or foot point for fixing the support structures.

Visual Checks of the Supporting Structure

Visual checks of the supporting structure are recommended in shorter time sequences to check for deformations and for longer time sequences to check for corrosion.

Check for changes at the ambient condition of GIS support structure for water penetration and high moisture and dust.

Checklist for the Supporting Structure

Checks of 1–3-year basis:

- Check for deformations of the supporting structure.
- Check for change of ambient conditions with water penetration or high moisture and dust levels.

Checks of 5–10-year basis:

- Check for corrosion at the support structure.

9.11.2.15 Instrument Transformer Connections

Instrument transformers connections are required for the safe and correct function of the GIS substation. In case of any fault event, the voltage and current transformer signal will trigger protection systems for circuit breaking and disconnecting. In case of normal operation, the voltage and current transformer signals are needed to operate the GIS within the limitation of maximum voltage and current ratings.

Conventional instrument transformers are using the electro-magnetic principle of measurement and status record of voltage and current. Current transformers need to have a defined burden to produce a correct measurement signal. In case of a CT secondary measurement cable open circuit, high voltage may cause damage and malfunction of the system. The cables of voltage and current transformers may be subject to interference with transient over-voltages induced from the electronic side of the control and protection system of the GIS. For this reason, shielded cables are used. These instrument transformer connections are tested with the commissioning test and proven for correct function.

Changes could come from damages to the secondary cables by mechanical forces, or by the corrosion of the cable terminals at the instrument transformer side or in the control cabinet. This includes measurement signal wire and the correct connection of the cable shield.

Visual Checks of the Instrument Transformer Connections

Visual checks of the instrument transformer connections should cover any damages of the secondary cables or the cable rack. Such damages may come from work in or around the GIS installation and need to be checked in short time sequences.

Changes of environmental conditions may cause corrosion to the secondary cable terminals of the instrument transformer and in the control cabinet. These changes are from slower development and are recommended to be checked in longer time sequences.

Checklist for the Instrument Transformer Connections

Checks of 1–3-year basis:

- Check for damages of the secondary cables.
- Check for damages of cable racks.
- Check for changes environmental conditions.

Checks of 5–10-year basis:

- Check for corrosion at the instrument transformer cable terminals.
- Check for corrosion at the control cabinet cable terminals.

9.11.3 Testing

9.11.3.1 Introduction

There are three ways to test for the condition assessment of the GIS: In-service testing, out of service testing and data collection.

The in-service testing requires access to signals and measurements while the GIS is operating under high voltage conditions. There are several such possibilities like gas leakages, gas density monitoring, gas density alarms, partial discharges (PD), radiology and contact resistance measurements.

The out of service testing requires the disconnection of the GIS from the high voltage sources. It will provide circuit breaker timing, grading capacitors capacitance, I^2T analysis, secondary cable Megger, circuit breaker main circuit resistance test, instrument transformer testing, set point of hydraulic system, set point of compressed air system and interlocking checks.

Data collection of the operation of GIS offers the opportunity to evaluate status and condition. The following data can be collected: spare parts availability of manufacture obsoleted design, age of components, identification of critical parts, circuit breaker drive details, health index and probability of failure, known issues with equipment from manufacturer, maintenance practices and history, electrical loading and rating, mechanical stresses, environmental conditions, operating practices, load growth, other information from user, SF_6 nameplate volume verification and circuit breaker, and high-speed ground switch fault clearing data.

9.11.3.2 In Service Testing

For in-service testing, the GIS is operating under high voltage and rated current. It is important to limit in service testing to such information which is relevant for the safe and reliable operation of the GIS. It is not the right way to measure and sense as much as possible because with each secondary system attached to the GIS the possibility of wrong information or failures increase.

First relevance is for the high-voltage dielectric system of the GIS, the insulators and the insulating gas. For this, the partial-discharge measurement, the SF_6 gas density, leakage rate, by-products and moisture monitoring are relevant.

Second relevance is for safe power transmission capability by carrying the rated current without reaching thermal limits. For this, the contact resistance and Dynamic Resistance Measurement on Circuit Breakers are relevant.

For high voltage vacuum circuit breaker, the radiology is relevant to be monitored.

Partial Discharge (PD) Measurement

The partial discharge (PD) measurement gives information about the dielectric condition of the GIS. It will provide information on possible types of dielectric instabilities like spikes at the conductor, metallic particles at the insulators, void in the insulators, loose metallic parts not grounded at free potentials, or free moving metallic particles. For each of such dielectric instability, PD patterns have been identified.

PD measurements are typically carried out on the GIS bay with the routine test at the factory to prove the correct and free of dielectric failures assembling of the GIS. For leading manufacturers, the GIS will not leave the factory without PD free high voltage condition. The measurements are recorded with each GIS bay.

After assembly on site, it is recommended to carry out PD measurements or to apply impulse voltages to prove the assembly of the GIS installation free of failure. The on-site commission tests will be recorded with commissioning documentation.

In operation, it is not recommended to measure and monitor continuously the PD patterns of the GIS, because for the enclosed high voltage compartments, no external impact from ambient climatic conditions will change the dielectric condition. Long-time experiences with GIS show that high voltage failures occur, if any, right at the beginning of the operation, usually within the first year. After this first operation time, very seldom dielectric failures are reported of GIS. See CIGRE reports [1–6].

The PD measurements of the GIS in the factory and on-site are using higher power or resonance frequency test voltages which are 80% of the type test values, which depending of voltage level of the equipment, about 1.5–2.5 times the operation voltage level.

For in-service PD measurements, the applied voltage is the rated voltage of operation. This is a much lower value than with the routine and on-site PD tests.

It is recommended to measure the PD of the GIS in case of any doubt of dielectric instabilities or for quality control of the insulating system after a longer time of operation and depending of the operational history. In case of long-time high-power load or more than normal switching operations, or after some short-circuit interruptions, it may be recommended to get a profile of actual status of PD of the GIS.

Also, transient overvoltage conditions at operational frequency and fast transient over-voltage may influence the dielectric stability if these situations occur often and at high level during operation of the GIS.

PD Measurement Recommendation

It is not recommended to measure continuously PD after the GIS has been commissioned as PD-free.

It is recommended to measure PD of the GIS after longer times of operation depending on the operational history: time and level of transient over-voltages, current level over the operation time, short-circuit interruptions and above normal number of switching operations.

The PD measurements be compared with the PD data available with the routine test in the factory and the on-site commission tests.

SF_6 Gas Density Monitoring

SF_6 gas density provides the dielectric insulation of the GIS and needs to be maintained for the complete time of operation. The GIS will be filled with the filling pressure at the local ambient temperature. The filling pressure is above the minimum operation pressure of the GIS, which provides for the safe operation of the GIS, including switching and circuit breaker interruption.

The gas pressure in each gas compartment of a GIS bay may be different related to the function of the gas compartment. Usually, the gas compartment of the circuit breaker holds the highest gas pressure typical 0.7–0.8 MPa. The lowest gas pressure usually is held in the bus duct without any switching device; inside the pressure may be 0.3–0.4 MPa.

The SF_6 gas monitoring provides for each gas compartment the indication of the required gas density. The gas density is direct indicated by a temperature compensated gas pressure measurement. Depending on the operation requirements of the GIS in the substation, the gas density meter will provide 2 or 3 alarm levels.

After the GIS has been filled with the filling pressure, the first alarm level is be set about 0.01 MPa below the filling pressure and will provide an electric contact signal indicating gas loss. In some operation schemes, there will be a second alarm level another 0.005 or 0.01 MPa lower to provide an electric contact signal for gas loss. The final level of electric contact signal will block the switching function of the GIS bay for disconnector or earth/ground switches and for the circuit breaker another 0.005 or 0.01.

The gas density monitoring is comparing the condition of each gas compartment to the interactive gas schematic screen at the control center. Each alarm and the actual gas density and gas pressure is indicated in the gas schematic. The gas schematic gives an overview on all gas compartments of the GIS installation and will show alarms of gas loss in color.

The gas density meter needs to be calibrated according to the operational manual of the manufacturer. For this purpose, the gas density meter needs to be dismantled from the GIS gas compartment and connected to a calibration tool. The gas density meter is connected to the GIS gas compartment by a ball valve which avoids gas losses during dismantling.

SF_6 Gas Density Monitoring Recommendation

SF_6 gas density monitoring is recommended to check correct function, density and pressure indication, and calibration procedure in sequences as given by the manufacturer.

SF6 Gas Content

SF6 content of GIS installation of several bays and many gas compartments can reach a total weight of some tons. SF_6 is seen as a strong global warming gas and regulations from authorities require to record the SF_6 content for the complete fleet of GIS installed, the losses and the SF_6 weight for refill.

To keep track of the SF_6 content, it is necessary to record and document each gas filling with the value of SF_6 weight.

If gas leakages of the gas compartments need an SF_6 refill to keep the gas pressure above the minimum gas pressure, then the weight of SF_6 gas needs to be measured and recorded in the documentation.

SF_6 Gas Content Recommendation

It is recommended to keep track on each weight of SF_6 filled into the gas compartments of the GIS and record the measurements in the documentation. This starts with the initial gas filling and covers any additional refill, if required.

SF_6 By-products

Using SF_6 in high voltage equipment will produce chemical by-products under the decomposition of SF_6. In operation the SF_6 molecule may get split into fractions, by high electric field strength or switching operations with an arc. In cases of other, molecules are available inside the gas compartment so called by-products are generated.

Most by products are produced inside the circuit breaker gas compartment, less in switch gas compartments and little in passive gas compartments without switching, for example, a bus duct.

The SF_6 by-products indicate the inner condition of the GIS gas compartment by its concentration. A gas probe taken through a ball valve can by analyzed in the laboratory and the result can give indication of aging effects. This may lead to a maintenance activity on a certain gas compartment or even may give an indication of repair or exchange of parts.

SF_6 Gas By-products Recommendation

SF_6 gas by-products analysis may give information on the inner condition of a gas compartment of the GIS. In cases of abnormal operation at high current ratings, long standing over voltages, or high number of switching or interruption activities, it is recommended to check the concentration of the SF_6 gas by-products.

SF_6 Leakage Rate

The international standards require gas-tightness of 0.5% maximum volume loss of SF_6 per gas compartment per annum. Leading GIS manufacturers todays provide gas tightness of less than 0.1% in operational service measure over a long period of time.

In cases of leakage ratings above these values, failures of the sealing system may be the reason. After identification of the leaking gas compartment, measurement to identify the source of the leak are required. There are some repair methods available to be attached from the outside for a temporary solution. In most cases, it will be required to exchange the sealing or the complete GIS module.

To identify the leaking gas compartment the following methods are helpful:

Alarm History

Each alarm of gas loss is recorded in the control system by identifying the gas compartment and the exact time when the electric contact alarm was set and at what alarm level. This indication gives information about the speed of gas losses and if one or more gas compartments are involved.

Online Gas Density Monitoring

Online gas density monitoring systems can measure the gas pressure and temperature over the time and provide information on gas losses. The online gas monitoring system overviews the complete GIS installation with any involved gas compartment.

Trending Data

Trending data for SF_6 gas leakages is calculating the trend for gas loss by analysis of the pressure, temperature, and time data of the gas density monitoring system. To calculate trends long time data collection is necessary as the leakage rate of gas compartment are small and cannot be seen directly because of the impact of the changing temperatures. It may take month for the trend analysis to detect a gas leakage at a gas compartment and long before the first alarm will be trigger. This allows early reaction and planning for repair.

SF_6 Gas Leakage Rate Recommendation

The SF_6 gas leakage rates in GIS are usually small and need a long time to analyze the seals and gas compartments. For this reason, it is recommended to use the information from the alarm history, the online gas density monitoring data and to calculate trending data.

After identification, first repair concepts are recommended to close the leak. In a second step, the exchange of sealings or complete GIS modules is recommended.

SF_6 Gas Moisture Measurement

SF_6 gas moisture measurement is required to keep the performance of the insulation. Moisture in SF_6 needs to be at very low ratings, dew point below −20 °C, for the complete service time. Moisture stored in the cast resin insulators might contribute moisture to the insulating gas which reduces the insulating capabilities and could cause condensation on the surface of insulators.

The low level of gas moisture in GIS and GIL is one of the physical basics for the exception of no temporary pressure test during the operational time of the equipment, because the low moisture content of the SF_6 gas prevents from inner corrosion the pressurized enclosure pipe or housing.

Also, low gas moisture is required to perform the dielectric insulation required for safe operation. Moisture inside the high voltage system may cause condensation at the surface of the insulators and with this it could initiate an electric discharge and arc.

Manufacturers usually add moisture drying material in form of aluminum oxide (AlO_3) to capture the rest of the moisture which may evaporate from the cast resin insulators after the gas compartment has been dried during the commissioning process of the GIS and GIL.

SF_6 Gas Moisture Measurement Recommendation

It is recommended to measure the gas moisture in longer periods of time in case no gas leakage is found at the gas compartment. For this situation, gas moisture measurements may be carried out in 10–20-year time periods in agreement with the recommendations of the GIS manufacturer.

In case of leaking gas compartments, it is recommended to repair the leakage and measure the gas moisture with the recommissioning process.

Radiology

GIS in service needs to follow the IEC [7, 8] and IEEE [9, 10]. standards requirements on emitting x-ray radiation in case when vacuum circuit breakers are used.

X-ray radiation is a physical phenomenon of vacuum switchgear technology. The level of radiation depends on the design of the equipment and has to pass type tests defined in the IEC and IEEE standards.

X-ray radiation limiting values are safety requirements for personal work or stay at the GIS installation. Measurements are required to prove that the limiting values are held.

Changes of x-ray radiation may occur after longer periods of time so that a repeat of x-ray measurement may be recommended.

Radiology Measurement Recommendation

It is recommended to measure x-ray radiation of the vacuum interruption unit in longer time periods of 10–20 years, in agreement with the manufacturer recommendations.

Contact Resistance by Thermal Imaging

Contact resistance may increase over the service time due to contact surface corrosion and increasing contact resistance. This leads to higher contact temperatures and a thermal imaging of the contact system can help to detect failing contacts before they fail.

Using infra-red cameras, thermal hot spot locations of the GIS installation can be made visible. The sensitivity of infra-red measurements from the outside is an easy-to-handle process. The infra-red camera is pointed to the surface of the GIS enclosure and locations of higher temperature indicate that the contact resistance has been increased.

Changing contact resistance is related to the performance of the equipment. The higher the continuous and peak load and current the more likely are changes of the contact resistance. Also, the number of switching operations of the equipment may indicated a higher risk of increasing contact resistance.

The increase of contact resistance is usually a slow process but needs to be repaired because by the time of operation, the conditions are getting more worse. The thermal imaging is a method during operation of the GIS to find locations of increasing resistance and initiate a repair before it may fail.

Contact Resistance by Thermal Imaging Recommendation

It is recommended to measure the contact resistance by thermal imaging in longer periods of time. After the installation and commissioning checks the resistance of main contacts has been measured and documented so that the GIS went into operation with a correct contact resistance. For this reason, a thermal imaging to find increasing contact resistances of the GIS or GIL is recommended after 10–20 years of operation in agreement with the recommendation of the manufacturer.

Dynamic Resistance Measurement on Circuit Breakers

Dynamic resistance measurement on circuit breakers is using an impulse voltage to check the circuit breaker contacts. The impulse voltage may be generated by switching operation of the circuit breaker.

During operation the high voltage circuit breaker contacts with each switching or breaking function the contact system will see degradation. The number of switching or breaking operations is defined by the type test requirements of the GIS. To get an impression on the current status of the circuit breaker the dynamic resistance measurement can be used. With this method an analysis of parameters extracted during the switching operation of the circuit breaker give information on the degradation of the circuit breaker. This is a non-standardized method to analyze the condition of the circuit breaker contact system and results should be evaluated together with the GIS manufacturer.

The degradation of the circuit breaker has been proven by the type test requirements of the GIS and is documented in the type test record. Therefore, dynamic resistance measurement of the circuit breaker is only recommended in case of abnormal operation conditions with high current interruptions, high temperature, high number of operations or any other circumstance for higher stress to the circuit breaker. In such cases, the dynamic resistance measurement may help to find an increased resistance of the circuit breaker main or arcing contact and allows repair before failing.

Dynamic Resistance Measurement on Circuit Breakers Recommendation

Dynamic resistance measurements on circuit breaker main and arcing contacts are recommended in longer time periods and in cases of abnormal, higher load or stress conditions.

For this reason, dynamic resistance measurements on circuit breakers of the GIS are recommended when abnormal operation conditions are applied to the GIS and in time sequences of 5–10 years of operation, in agreement with the recommendation of the manufacturer.

9.11.3.3 Out of Service Testing

For out-of-service testing, the GIS needs to be disconnected from the high voltage network and de-energized.

Out of service testing is recommended with the operational handbook of the manufacturer. Today these out of service tests are not only related to the time of operation, but also take into account the operational history of the GIS. This covers the environmental conditions around the GIS, the ambient temperature, the level of power load, the maximum current, the duration of high load operation, static, and dynamic over-voltages, the number of switching operations, the number of circuit breaker interruptions, the short-circuit current level and the presents of partial discharges. All of these indicators need to be evaluated and a decision will be made if and when the GIS needs to go off-line for maintenance inspections. This evaluation process is recommended to be done together with the GIS manufacturer. Such condition-based maintenance needs the technical data for the complete operation time stored and documented.

The digital substation (see Sections 9.15 and 9.18) of the future is using this in-operational information in a data base linked to the GIS and provides analytical evaluation on the condition of the GIS, including recommendations for maintenance, repair, up-grade or replacement of the existing GIS.

Circuit Breaker Timing and Travel Curve

The circuit breaker timing and travel curve is measuring the absolute time for operation according to the type test record requirements and gives a curve for the travel time of the contact system. These data are measured and documented with the routine tests of each GIS bay and states the mechanical condition of the circuit breaker after leaving the factory in its new condition.

To evaluate the correct function of the circuit breaker driving mechanism after several years of operation, the circuit breaker timing and travel curve measured is compared to the data recorded in new condition.

Slower movement and longer operation times of the circuit breaker indicate aging effect of the mechanical components and replacement or repair may be needed. To evaluate the circuit breaker timing and travel curve the GIS manufacturer should be contacted. Testing equipment is available from the manufacturer.

This measuring method is usually linked to the maintenance cycle recommendation s of the manufacturer and may be carried out with the first recommended maintenance after 25 years as recommended by the leading GIS manufacturers. In case, the condition-based maintenance evaluation results in continuing of operation for some more years (typical 5–10 years), this measurement should also be shifted.

Circuit Breaker Timing and Travel Curve Recommendation

The circuit breaker timing and travel curve measurement is recommended in the time sequences recommended by manufacturer for the first maintenance cycle after 25 years of operation.

Grading Capacitors

Grading capacitors are connected to multiple circuit breaker interrupter units for equal distribution of the rated high voltage in the open position of interruption units. This grading capacitors are mechanically direct-connected to the circuit breaker interruption unit and have to withstand the mechanical and electrical stresses for the operation time of the GIS.

In case of loss of these grading capacitors, the equal voltage distribution across the multiple interruption units may not function correctly and single interruption units might be over-stressed. To avoid failing interruption units, it is recommended to measure the capacitance and correct function of the grading capacitors.

Grading Capacitors Recommendation

It is recommended to measure the capacitance of the grading capacitors when the GIS is de-energized for the, maintenance to prove the capacity value and the correct function. According to the manufacturer operational handbook, the first maintenance is usually recommended after 25 years. In case of longer times for the maintenance, based on the evaluation of the condition assessment of the GIS, this grading capacitor check should also be shifted.

I^2t Analysis

I^2t analysis provides information about the transmission losses of the GIS. With this analysis a defined current over a defined time will produce power losses which will increase temperature according to the local resistance of the conducting system. The temperature rises includes the conductors, joints, and contacts systems.

Based on the thermal constant of the GIS bay and the absolute temperatures measured at the critical point during the type tests, the I^2t analysis gives information on the correctness of the conducting system of the GIS. In case of loose flange or bolted contacts, the resistance will increase and so the temperature. This might help to find aging effects of the conductor system before it may fail.

This analysis may be restricted to critical applications as it is an extensive measurement. Critical applications are those of high continuous current ratings and may indicate higher losses in operation, meaning higher temperatures.

I^2t Analysis Recommendation

The I^2t Analysis is recommended for critical GIS applications of high continuous currents ratings to be applied after the recommended maintenance and with information of higher power losses.

Secondary Cable Megger

The secondary cable megger provides information about the insulation capability of the secondary cables. For this, the insulation resistance is measure by using a high voltage insulation test of 1.9 or 2 kV depending on the applied standards.

Secondary cable may lose insulation capability during operation. The signals of secondary cables are required for the safe function of the high voltage GIS. Secondary cables are selected according its ambient conditions, temperature and current rating. In cases, when condition is close to the cable performance limits, it is required to check the insulation capability to prevent failing in operation.

Checking secondary cables by high voltage test after a long time of operation has the disadvantage of hurting the cable insulation capability with the test and initiate a cable failure in future. Also, such cable high voltage test is extensive in time and it might be a better solution to exchange secondary cables in cases when aging of the secondary cable insulation is supposed.

Analysis of the specific condition of the GIS at its location of operation and the history of environmental and technical operation will help to find the best solution for replacing or testing the secondary cables.

Secondary Cable Megger Recommendation

It is recommended to test secondary cable with Megger to prove the soundness of the insulation to avoid failing with the maintenance recommended by the manufacturer and when indications of aging of the cable insulation are seen. An evaluation and cost estimation should be done before carrying out the Megger if a replacement of secondary cable would be the better solution.

Circuit Breaker Main Circuit Resistance Test

The circuit breaker main circuit resistance test provides information about the conductivity of the main circuit contact system which is a reading for the transmission losses in service. This value is measured during routine test and changes of the resistance will bring information on the health of the main circuit of the circuit breaker.

The circuit breaker main circuit resistance test can be carried out by using isolated ground switches before and after the circuit breaker unit. A defined current (typical 100 A) is inserted to the main contact and the voltage drop will give information on the resistance. The measured value can be compared with the values measured during the routine test on the equipment done in the factory and documented in the routine test protocol. A comparison of the resistance measurement of the new state values with the value measured after years of operation will give indication if the circuit breaker contacts need to be exchanged.

To carry out this resistance measurement of the main contact of circuit breaker is easy to accomplish if an isolated ground switch has been installed with the bay design. If this access is not possible, the measurement may include more than one contacts in the GIS bay and the accuracy of the measurement of main contact of the circuit breaker is lower.

Circuit Breaker Main Circuit Resistance Test Recommendation

It us recommended to use isolated ground switches for the measurement of the circuit breaker main circuit resistance with the maintenance recommended by the GIS manufacturer usually after 25 years or later depending on the result of a condition assessment for the GIS in operation.

Instrument Transformer Testing

The instrument transformer testing of a de-energized GIS allows the correct functionality of the instrument transformers. In the case of instrument transformers for accounting, calibration is important.

The conventional instrument transformers using the electromagnetic measuring principle are important for the safe function of the GIS and the substation. The instrument transformer rated values of voltage or current transmission ratio and the magnetic characteristics of saturation may change in time of operation. According to the instruction manual of the manufacturer, test and calibration are needed.

This test includes also the check of grounding/earthing of the instrument transformers to avoid corrosion or loose contacts of the ground wires.

Instrument Transformer Testing Recommendation

It is recommended to test the instrument transformer correct transmission ratio and saturation with the maintenance recommendation of the GIS, usually after 25 years or as recommended by the instrument transformer manufacturer.

Set Point of Compressed Air System

The set point of the compressed air systems is important for the correct functionality of the circuit breaker operational drive.

Compressed air had been used in the 1960s and 1970s as a centralized operational drive system in substations. The compressed air is stored in central high-pressure storage tank and connected by pipes to each single drive on the GIS.

The set point of the compressed air system is chosen to a pressure value which provides the required switching operations, for example, C–O–C to be carried out.

To check the set point of the compressed air system a calibration pressure sensor is needed to prove the correctness of the setting.

Set Point of Compressed Air System Recommendation

It is recommended to test the set point of the compressed air system with a calibration pressure sensor during the maintenance recommendation of the GIS, usually after 25 years or as recommended.

Set Point of Hydraulic System

The set point of the hydraulic systems is important for the correct functionality of circuit breaker operational drive.

Hydraulic systems have been used since the 1970s as a de-centralized operational drive system in substations. The hydraulic pressure is stored in a N_2 pressure bottle and provides the required energy for one three-phase or three single-phase operation mechanism. The hydraulic system is installed for each GIS bay and has been routine tested in the factory.

The set point of the hydraulic system is chosen to a pressure value which provides the required switching operations, for example, C–O–C to be carried out.

To check the set point of the hydraulic system a calibration pressure sensor is needed to prove the correctness of the setting.

Set Point of Hydraulic System Recommendation

It is recommended to test the set point of the hydraulic system a calibration pressure sensor with the maintenance recommendation of the GIS, usually after 25 years or as recommended.

Set Point of Spring-Operated System

The set point of the spring-operated systems is important for the correct functionality of the circuit breaker operational drive.

Spring-operated systems are used since the 1990s as a de-centralized operational drive system in substations. The energy for the spring-operated system is stored in spiral or plate spring and provides the required energy for one three-phase or three single-phase operation mechanism. The spring-operated system is installed for GIS bay and has been routine tested in the factory.

The set point of the spring system is chosen to a pressure value which provides the required switching operations, for example, C–O–C to be carried out.

To check the set point of the spring-operated system a calibration mechanical-force sensor is needed to prove the correctness of the setting. Some manufacture provides integrated mechanical-force sensors for measuring the energy stored in the spring.

Set Point of Spring-Operated System Recommendation

It is recommended to test the set point of the spring-operated system a calibration mechanical-force sensor with the maintenance recommendation of the GIS, usually after 25 years or as recommended.

Interlocking Checks

Interlocking checks of the GIS bays are for preventing false and unsafe operations.

Interlocks are required for the safe operation of the GIS bay to avoid switching failures which could cause an internal arc and damage the GIS with great danger for personnel. The interlocks are set and checked with the commissioning process of the GIS.

Many GIS installations are under continuous changes and it might happen that the interlocks need to be changed. To check the interlocks, the GIS bay needs to be de-energized.

It is recommended to be checked with the maintenance of the GIS, usually after 25 years of operation or later depending on the result of the condition assessment of the GIS.

Interlock Checks Recommendation

Interlock checks are recommended to be carried out with the recommended maintenance of GIS by the manufacturer to prove the correctness of the settings.

9.11.4 Data Collection

The data collection for the GIS bays installed in the substation will give an overview of information related to the specifics of the installation. It provides the view on what information could be important in case of maintenance, repair, or upcoming failures.

Data collection begins with the manufacturing and routine testing of the GIS in the factory; it collects information during erection and commission tests and operation time. The data collection is the basis for evaluation about the condition assessment of the GIS. It shows the history of the GIS and allows predictions about the upcoming operations. Decisions about maintenance, repair, or replacement strategies are using the collected data of the GIS.

In the following, some aspects are presented on information, strategy, and evaluations based on data collection are given.

9.11.4.1 Spare Parts Availability of Manufacturer Obsoleted Design

The availability of spare parts is important for service of the GIS. Not available spare parts of obsoleted design need to be manufactured as single piece which could be very costly.

GIS installations have been in operation for decades. The first GIS system installed in 1968 is still in operation. The availability aspect of spare parts needs to be seen under the aspect that the type of GIS in operation is no longer manufactured or even the manufacturer of the GIS is no longer in business.

Not all types of spare parts are recommended to keep in storage for such a long time. O-rings or cast resin insulators may change their physical structure and may not be usable after 20 or 30 years. Mechanical parts may be reproducible in case the original production drawings are available, but at high cost.

The most changes are related to the secondary system which usually show lifetimes of less than 10 years. It might be a better way to replace the secondary system instead of storing spare parts.

The condition of the primary part of the GIS is much dependent on the network requirements of power transmission levels or number of switching operations. Because of the high-cost GIS functional units, it is not recommended to store complete, gas-filled GIS units in case for replacement. These GIS functional units need to be maintained as the GIS itself, which causes additional cost.

The spare part aspect is complex and need to be discussed and decided in agreement between user and manufacturer. This agreement may change over time with technology and management changes of the user and manufacturer. Time-based maintenance is more and more replaced by condition-based maintenance or no-maintenance with early replacement of GIS.

Spare part handling and storage is costly and difficult to plan. Beside the material of the spare parts, it is also necessary to have skilled workers to carry out the replacement of GIS parts. This requires the availability of special tools, which may not be available anymore. The older the GIS, the more serious are these aspects. It is like with an old clock: It's nice to have it on the wall, but for repair a specialist needs be available who understands the technology, can reproduce parts and assemble these with special tools. If this is economic or not depends on the view of the owner.

Spare Parts Availability of Manufacturer Obsoleted Design Recommendation

It is recommended for the spare part strategy to be prepared with information right from the beginning of the GIS production. An agreement with the manufacturer for drawings for reproduction of parts of the GIS should be sought. Spare parts which can be stored should be identified in agreement with the manufacturer and ways to reproduce spare part of the GIS by using the manufacturer's production line.

9.11.4.2 Age of Components

The age of components can indicate the quality and expected remaining lifetime of the GIS. The impact of aging is very different for the components used in GIS.

Gas-insulated systems, as GIS, do not show aging effect when operated within the defined technical limitations of voltage and temperature [11]. This reflects the design of electrical insulation capability of the insulators at the required gas density, low moisture content and free of particles.

Cast resin insulators within the GIS are operating far below their maximum electrical field strength. The maximum electric field strength of the cast resin insulator is about the factor 3–4× higher than in gas. For this reason, no electric aging of the insulation is expected.

The operation temperature of the GIS allows maximum temperatures up to 120 °C given by the material limitations of silver-plated contacts and the cast resin of the insulators. The type tested maximum temperature for the GIS current rating is usually between 80 °C and 90 °C. This means, that in case of operating the GIS in its given limit of ambient temperature and rated current no thermal aging can be expected.

The more than 50 years of experience with GIS installed and in service reflects these physical features of gas-insulated systems. In case of dielectric failures, they typical occur at the beginning of operation usually within the first year. In most cases, the reason for these failures is in the assembly process on site or with quality problems during production. When GIS is in operation under normal conditions, usually no dielectric failures are reported.

Metallic enclosures made of cast aluminum are designed as pressurized compartment with a safety margin of factor 2.6 according to standards and do not show any aging effect.

Aging effect are mostly related to secondary equipment and mechanical drives of switches and circuit breakers. CIGRE investigations [6] have shown that failures in GIS are linked to failing motors, mechanical relay, electronic switches, loose cable, contacts, or contact corrosion.

The reasons for these failures are linked to corrosion, overload conditions with too much heat, impact of pollution with moisture and dirt, mechanical stress from the surrounding area with vibrations of machines, traffic, or railroad, impact from resonances in the building or structure.

To get an overview about the age of the different part of the GIS installation, the year of erection and replacement or repair should be listed.

Age of Components Recommendation

It is recommended to list the year of erection, replacement, or repair of each component used in the GIS installation.

9.11.4.3 Identification of Critical Parts

The identification of critical parts will help to be prepared for a failure and the availability of the failed part.

Critical parts of GIS are depending on the importance of their function in the network. To connect a large power plant to the network for delivery of base load energy has a higher power delivery requirement than a remote housing area of low load concentration.

The scheme of the GIS installation at these substations shall reflect the importance of power delivery. In low-level availability, a single bus bar scheme will provide an adequate power supply, while for high-level availability, double or triple bus bar schemes or breaker and half schemes are used to provide more flexibility in cases of failure and provide a high level of power supply.

To choose the right the substation scheme for the GIS, the first step is to avoid power interruptions. Critical parts of GIS can be bypassed by redundancy in the substation scheme and isolate the critical part of a GIS bay for repair.

To identify critical parts of GIS is specific for the manufacturer's design and may be identified by manufacturer's experiences and data collected from GIS in operation.

For the measures coming from identified critical part in a GIS substation, a concept of provision, storage, replacement, and repair needs to be agreed with the manufacturer.

Identification of Critical Parts Recommendation

For the identification of critical parts in substation design, it is recommended to evaluate the importance and scheme of the substation. The need for GIS units and parts to be stored or being produced at the factory on demand is recommended to discuss and agree with the manufacturer.

9.11.4.4 Circuit Breaker Drive Details

Circuit breaker drive details including the circuit breaker number of operations will give information about the requirements for maintenance or exchange.

Circuit breaker electrical and mechanical operations are defined by design and type tested according to IEC or IEEE standards [12, 13]. The standard requires 2000 electrical operations at rated current, 5–10 operations depending on short-circuit current value and the extended number of 10 000 mechanical operations. In practice, under normal network operation, the GIS operates much less than type tested quantities. For special switching conditions, inductive or capacitive compensation units, or in regenerative networks with frequent power flow changes or for industrial applications, for example, in steel mills the number of operations might be higher and can get close to the type tested number. The data of switching operations should be monitored and collected for evaluation.

For circuit breakers operating infrequently, the condition of the driving mechanism may be contaminated by corrosion, dust, moisture, and aging, or degraded parts may cause faulty operations in case when used. For this, the data collection should cover the ambient condition of the driving mechanism to identify changes.

The experience of the manufacturer of possible weak points in the circuit breaker drive system can help to identify parts to be maintained or replaced in time before they may cause a fault function.

Circuit Breaker Drive Details Recommendation

It is recommended to monitor and document the number of circuit breaker operations during lifetime to get a picture of the frequency of switching operations. In relation to the type tested operations, a maintenance or repair plan can be built up using the collected data.

The data should include the ambient conditions of the circuit breaker drive to identify changes and evaluate the need of maintenance, exchange, or repair of parts.

9.11.4.5 Health Index and Probability of Failure

The health index of the GIS gives information about the probability of failures and will guide maintenance activities before failures occur.

This data collection gives statistical information on the health index and probability of failure of the GIS. Information coming from the fleet of GIS equipment of the same design or information from literature or from the manufacturer can build a data base to identify the general health index and the probability of failure of the GIS.

The health index of GIS will give guidance on maintenance and replacement on a general condition of the equipment. This health index will cover the year of erection (age), the intensity of usage (number of switching operations and level of current carrying), the ambient conditions (temperature, moisture) and the experiences recorded with this GIS design (known failures).

This health index will help to identify the need for maintenance or replacement in view of the total fleet of GIS in operation. This is an important part of information to coordinate the network development with an aging set of GIS equipment. GIS has been in operation more than 50 years, and no real aging effect has been identified thus far. In most cases of required replacement, the reason is an upgrade of the network for higher voltage level and/or higher rated or short-circuit levels.

In some cases, secondary effects of non-delivery of spare parts, non-availability of aged protection and control equipment or ambient changes at the substation installation (climatic, temperatures) are the reason for a GIS replacement.

The health index generated by data from GIS operation will give information on the failure probability and with this a maintenance or replacement schedule for the GIS fleet can be established.

Health Index and Probability of Failure Recommendation

It is recommended to collect data from the GIS during the operation time using age, number of switching operations, level of current, ambient conditions and experiences of failures to build a statistical evaluation number for each GIS in the fleet. This number will give indication on when which GIS needs to be maintained, repair, or replaced.

9.11.4.6 Known Issues with Equipment from Manufacturer

Known issues with equipment from manufacturer will help to identify a weak point of the GIS installation and give special attention on this subject.

GIS technology developed along the different voltage levels starting at about 100, 200, 300, 400, 500, 800 and 1100 kV during the last 50 years. Each step of development brought new design for the insulators, the enclosures, the operating mechanism, the control and protection equipment, and the ever-increasing power transmission capabilities from 1000 A up to 8000 A today.

Each manufacturer has made its own experiences in technical solution and knows the weak points best. In agreement with the manufacturer, the known issues should be discussed and measures for the GIS in operation should be agreed upon.

GIS issues haven been found with gas tightness, flange corrosion, high-frequency transient overvoltages, capacitive, or inductive switching, ferro-resonances, particles, partial discharge of insulators, increase of contact resistance, mechanical vibrations by electromagnetic forces, grounding/earthing high-frequency impedances or contact corrosion, to name some.

This strategy is in favor for the manufacturer and user of the GIS because it may help prevent losses from failing equipment and power delivery interruptions. It is recommended to get in contact.

Known Issues with Equipment from Manufacturer Recommendation

It is recommended to the user to get in contact with the GIS manufacturer to discuss known issues with the specific design of GIS and to agree on measures to avoid.

9.11.4.7 Maintenance Practices & History

The maintenance practices of the operator's schedule should be applied to the GIS installation according to the operational history.

Maintenance practices are changing from operator to operator and reflect the philosophy of the owner.

The least maintenance activity is following the procedure to repair only case of a failure has been occurred. This philosophy takes into account that GIS is a very reliable technology and failure are very seldom. The CIGRE proven value for a 100 kV type GIS, which is the most often used voltage level, has a mean time between failure of more than 1000 years before a power delivery interruption will occur. It is understandable that under such conditions, each onsite maintenance will add new risks for failures and may not improve the GIS condition.

In some maintenance practices, the time used is the only measure (time-based maintenance). According to the manufacturer's recommendations of the operational manual, certain maintenance activities are recommended in yearly, 3-year, 5-year, 10-year or 25-year cycles.

Today most user of GIS use the information about the history of the GIS to set up maintenance, repair of replacement schedules. In this condition-based maintenance, the history of the GIS is evaluated together with the manufacturer and timing is giving for maintenance works or re-evaluation in some years (typical 3–5 years). This condition-based maintenance strategy is using the data collection of each GIS to identify its actual condition and make predictions for possible failures. This is the basis for a maintain, repair, or replacement schedule.

The future will bring many more sensors used in GIS installation (see Sections 9.15 and 9.18). With these sensors connected to the evaluation intelligence in the cloud, each parameter coming from the GIS will be used to recommend maintenance, repair, or replacement.

Maintenance Practices & History Recommendation

The maintenance practice for GIS operators needs to be developed in strategic concept for the complete fleet. Three choices are given: maintenance only in case of failure, time-based maintenance or condition-based maintenance.

In future the digital substation with GIS and GIL will be monitored by the evaluation software of the cloud and recommendations for maintenance will be provided.

9.11.4.8 Electrical Loading & Rating

Electrical loading and rating give information about the aging status of the GIS.

The GIS installation is chosen on the basis of data available at the time for the project planning. It is typical to choose the GIS rating at a higher level than required at the time of installation, assuming that the power transmission requirement will increase over the years by typical 2–3%. This would be 20–30% in 10 years or 40–60% in 20 years. This leads to high margin of the rating related to the actual load.

Pressure from the limitations of financial investment budgets may lead to choose GIS with lower power transmission margins or the expected load growth at the substation of the GIS location will be higher than expected, and limits of the GIS design will be reached soon.

It is recommended to collect data on this loading and rating situation for each substation and its GIS for the time of operation to identify early limits of ratings. The limits of rating may come from the transmitted current reaching the rated value or the short-circuit current will increase by more power generation in the network exceeding the rated value of the GIS. Both parameters are fast moving targets, especially in a regenerative network with new installations of photovoltaic or wind energies.

The level of continuous current transmission of the GIS has an impact on aging effects of the contact system and the insulators. For both, contacts and insulators, the maximum reached temperature is the influencing value. With temperatures above about 120 °C, aging will occur depending on the time how long this hot situation will last. It will be important to collect the electric load of the GIS in conjunction with the ambient temperature. Based on the thermal constant measure of the GIS measure and documented with type test records, the maximum temperatures can be calculated. Specific for electric power networks with high contributions of wind and solar energy is that high continuous currents are transmitted for long periods of time.

Electrical Loading & Rating Recommendation

The electric load of the GIS over the time of operation related to GIS rating will give information of aging effects of the equipment. Transmission data of current, short-circuit current, time duration, and ambient temperature are information to be collected for evaluation and decision for maintenance, repair, or replacement.

9.11.4.9 Mechanical Stresses

Mechanical stress to the GIS enclosure needs to be avoided. The GIS enclosure is a pressurized compartment and no permanent mechanical stress not part of the design and type test (pressure, switching) should be avoided.

Mechanical stresses of the GIS might have different sources. Depending on the installation and its location there might be impact from thermal forces, vibrations coming from traffic, railroad, or machinery in buildings with machinery installed. Settlements of foundations of the GIS bays can result in high forces to the flanges of the bus bar. Direct connection of transformers may bring oscillation to the GIS bay involved. In case of earthquake, activities should be monitored and the accelerations be measured.

Thermal forces may stress the GIS in case if the thermal expansion compensation is not working correctly. The thermal expansion is compensated in longitudinal bellow compensators or with angle compensation. In both cases, a sliding mechanism is providing the thermal expansion movement of the GIS enclosure. If the sliding mechanism is not working correctly, high mechanical forces may bring damage to the GIS steel structures or the GIS enclosure. Thermal expansion forces are very high and can reach 160 tons of force. It is recommended to collect data on the correct function of the sliding mechanism of GIS to cover thermal expansions.

Vibrations from traffic, railroad, or machinery in building may influence the functionality of the GIS. Vibration has impact to any drive mechanism of the switches or circuit breaker or for the free moving particle inside the high voltage system of the GIS. The data collection should cover changes around the GIS installation and identify vibrations.

The GIS foundation needs to balance the height-level differences from bay to bay when installed. In case of foundation settlement, the forces to the flanges could reach high values and cause gas leakages over time. It is recommended to collect the data on the GIS foundation during the time of operation to avoid stress to the bus bar flanges.

Special care needs to be taken when the GIS is directly connected to the transformer. Here, a mechanical weak enclosure section in form of a bellow compensator is needed to separate the 100 or 120 Hz vibration from the transformer to the GIS and the high dynamic forces of circuit breaker operation to the transformer. Both sides are sensitive for such mechanical impact. The data collection should cover the direct transformer connection and its correct function.

Mechanical Stresses Recommendation

To identify mechanical stress to the GIS enclosure and steel structures, the data collection should procure data for correct function of the thermal expansion sliding mechanism, external vibrations, foundation settlements, and the correct function of the mechanical decoupling of the direct transformer connection.

9.11.4.10 Environmental Conditions

Environmental conditions might change during the service time of the GIS and should be monitored at the location of the substation.

The environmental conditions of a GIS substations are evaluated for the project design and execution. Local conditions of temperature, moisture, dust, sun radiation, heavy rainfall, soil drying, flooding, wind, snow, ice, and pollution are taken into account for the GIS, the building and indoor or outdoor installation.

Over the time of operation, these parameters may change, which may have impact on the functionality of the GIS. For this reason, it is recommended to collect data about changing environmental conditions at the site of the GIS substation.

In a wider sense, environmental conditions are also related to substation audio-noises or electromagnetic field radiation. Substation audio-noises may come from an increase of the number of switching operations and the opposition coming from neighbors of the substation side. Audio-noises may also come from external high voltage parts with partial discharge at the contacts (corona noises) due to higher moisture in the air or degraded electromagnetic shielding. In this aspect, the radio interferences may also play role.

To recognize changes of the environmental conditions data collection of the related parameters from the substation site will help to identify possible future problems and their sources. In case of environmental changes, action can be taken to avoid failing equipment or problems with neighbors.

Environmental Conditions Recommendation

For changes of the local environmental conditions, it is recommended to collect data on temperature, moisture, dust, sun radiation, heavy rainfall, soil drying, flooding, wind, snow, ice, pollution, audio noises or radio-interferences data for the GIS substation as a basis for the evaluation of counter measures.

9.11.4.11 Operating Practices

Operating practices of GIS in service may change over the years. Switching operations could be more often or current rating could be higher than at the beginning of operation.

The function of the substation in the network defines the way how it will get operated. In classical networks with low impact of renewable energy generation connected, substation operation is relatively seldom and linked to the load changes, failures in the network or planned maintenance works. Such substations may have switching operations once a month or even less.

In networks with high renewable power generation the power flow may change within hours or even minutes. This requires operation under very different load conditions, full load or no load. This load changes in addition requires power factor compensation with switching inductive coil or capacitor banks.

Power flow optimization in an interconnected power transmission or distribution network requires more often redispatchment switching operations. This involves several substations and may need several switching interim steps before the new optimized power is reached. With growing renewable power generation, such redispatchment operations are needed more often.

Operating Practices Recommendation

The operational practice of the GIS substation may change over the time within the power network. Related to higher levels of renewable power generation, redispatchment operations are more frequent, switching conditions of full load or no load are more often, inductive and capacitive power factor compensation units are switched more often, and these operational changes need to be monitored. The data collected for the different operational mode and the frequency of operation will help to evaluate the GIS substation operation and possible maintenance, repair, or replacement activities.

9.11.4.12 Load Growth

Load growth is most often the reason for replacements of GIS or upgrading existing installations. Load growth is the normal for substations, the only question is "how fast will it grow"? The reason behind this is the transfer from oil and gas as primary energy more and more to electric energy. Heating with renewable energy will be one future driver of this tendency as well as the electric vehicles entering the market in the next few years.

When a GIS substation is planned at a specific location in the network the expected load will be estimated for the time of erection and the following years of operation. Operation time for GIS substations is usually in the range of more than 50 years and a precise estimation of the load growth is not possible. In general, there are scenarios with 1, 2, 3 or 5% load growth depending on the expectation in the area.

The reason for load growth may have different source. One source is development of housing in the area, or industry, in some other cases the concentrations of apartments and office buildings will increase load requirement in the area. Also, the share of electrification that is replacing oil or gas for heating by electricity will change the load required. All these aspects cannot be foreseen for such a long period of time of 50 years or more. It is recommended to monitor these parameters and collect data for evaluation of measure for replacement or upgrading the substation.

Replacements and up-grades may use the same voltage level at higher current ratings and additional transmission lines and GIS bays. It is also possible to change the voltage level to the next higher voltage rating. This will require a complete replacement of the substations and more space.

Replacement and upgrade could also use an existing air-insulated substation to install a higher voltage level GIS at the same location, which is possible because of the smaller size of GIS. Also, an outdoor substation can be turned into a small size compact GIS building.

To prepare the future decision of the GIS substation under the condition of load growth, data need to be collected on how the load growth and what measure can be taken in the network.

Load Growth Recommendation

Load growth of the GIS substation will require measures to follow the substation requirements. Data should be collected on what load margin is designed into the GIS when erected, how fast the load is growing at the location of the GIS in the network and what upgrades are feasible. Additional GIS bays, higher current rating, higher voltage rating, higher short-circuit current rating are some examples.

9.11.4.13 Other Information from User

Other information from user concerning the service and operation of the GIS will help to identify possible maintenance or replacement strategies.

Within an electric power network each activity in and on the network will change the electric conditions. Impact is given on current flow when new lines get in operation, when addition generation units are added or some are disconnected. This information should be available for GIS substation evaluation and preparation of future measures.

New technical developments may help to find new solutions based on network changes. User information will help the manufacturer with a more efficient technical solutions.

Other Information from User Recommendation

The user should stay in contact with the GIS manufacturer to inform and discuss changes in the network with consequences to the GIS in operation.

9.11.4.14 SF_6 Nameplate Volume Verification

SF_6 nameplate volume verification may be necessary to record the volume of SF_6 used in the GIS.

The global warming potential of SF_6 requires, in most countries and according to international standards [12, 13] the identification of the volume of SF_6 used in the GIS. This requirement will be fulfilled for new installation with the gas filling process. But for older GIS these data might not be available correctly.

To generate the correct data of gas fill, it is necessary to evacuate and refill the GIS which requires a de-energization and is a strong impact into the GIS operation. A GIS with no failure should not be opened to avoid particles to enter under the onsite working conditions. It also takes quite a long time to refill any of the gas compartments of a GIS installation.

SF6 Nameplate Volume Verification Recommendation

The verification of the SF_6 content of the GIS should be combined with maintenance work which requires gas works.

9.11.4.15 Circuit Breaker/High Speed Ground Switch Fault Clearing Data

The circuit breaker and high-speed ground switch fault clearing data is important to evaluate the need of maintenance.

The impact of fault clearing for circuit breakers and high-speed ground switches is dependent on the fault current level and the time for interruption. Substation protection will record the data during the short-circuit failure and is available in the recordings.

Circuit Breaker/High Speed Ground Switch Fault Clearing Data Recommendation

Get the fault curve data from the protection recording to better evaluate the need for maintenance of the circuit breaker or high-speed ground switch contacts.

9.11.5 Recommendations

9.11.5.1 General

Condition assessment of GIS substations is a complex evaluation process which has aspects in the functionality of the GIS, the substation and the electric power network. Depending on the viewpoint, different measures are recommended to improve the condition of the GIS substation.

From the GIS perspective the equipment and functional parts of the GIS are in focus to identify possible weak points and start replacement before it fails. From the substation perspective, the functionality of the GIS needs to fulfill its task in conjunction with the primary task of the substation as a circuit knot in the network. From the network perspective, the GIS in the substation has to fulfill the power flow without any congestions related to overload of network sections and to allow fast isolation of failure with recovery of the remaining healthy part of the network.

9.11.5.2 Preventative Maintenance

Preventive maintenance is focused on early identification of weak parts or units of the GIS substation to be repaired or replaced before it may fail by a scheduled repair process plan. This concept is getting more and more used by operators of GIS and it requires the availability of data to evaluate the condition of the GIS substation. This includes visual inspections (Section 9.11.2), in and off service test (Section 9.11.3) or data collection (Section 9.11.4). Based on the information available at any time of the GIS operation, evaluation can be made to find solutions for improving the condition of the GIS, substation, or network.

Some of the required information comes with the type test records for the GIS basic design criteria, with the routine test records from the factory tests on each bay and with the onsite commissioning test for the assembly and tests before energization. After the GIS is in operation, more information is coming from the current flow, voltage and overvoltage occasions, switching, and interruption activities, gas density monitoring, circuit breaker monitoring, partial discharge monitoring or operational drive monitoring.

A third part of the information comes from the environmental conditions on changing ambient climate with high temperature, more moisture or pollution in the air, ice load, heavy rain flood, seismic events, mechanical vibrations or mechanical stresses.

From the network, changes may come from increasing power flow, increasing number of switching operations, change to inductive or capacitive switching during full load or no-load conditions or reactor switching, increased short-circuit ratings through additional power generation units mainly through renewable solar and wind power units connected on a decentralized basis in the network.

The availability of this information is in form of test records accessible for condition evaluations. Most of the data can be generated by automated measurements and made available in an evaluation software as it is seen for the future digital substations and digital twin concept (see Sections 9.15 and 9.18).

Gas monitoring data from decentralized gas density meters and electric signal connection to the control center will provide actual data on each GIS gas compartment on any gas leakage. This data can be automatically collected and provided for the GIS condition assessment and possible repair needs for replacing gas seals or as done in most cases, replacement of complete GIS units if they show gas leaks.

Today, most data for the condition assessment and evaluation process have to be collected manually from different sources. Current and voltage from the substation operation, short-circuit currents and over-voltages from the protection system and partial discharge (pd) information from offline operated pd sensors and measuring system are available.

Visual inspections will deliver data on the GIS condition and needs to be collected manually with defined criteria for evaluation. This includes the changes of environmental conditions, detection of corrosion, detection of aging effects on secondary cables and parts,

Tests during service or when the GIS is de-energized are scheduled on occasion if there is any indication of a malfunction, for example, increasing contact resistance. These data are collected and evaluated for the condition assessment of the GIS.

Data coming from the network will indicate different requirements to the GIS in the form of higher load, higher number of switching operations, or even the need for higher voltage levels to meet the power transmission requirements of the future. This data is part of the general evaluation of the GIS condition and possible upgrades or retrofits.

9.11.5.3 Upgrades & Retrofits

Upgrades and retrofits of GIS are often seen after longer operation times and when the GIS is getting closer to its physical limits. With a general increase of electric energy for use in household and industry, the current flow is increasing which may require an upgraded circuit breaker or switch. New switching requirements may need a controlled switching mechanism and higher circuit breaker operations may need a new operational drive.

The data collection for GIS on gas density, gas moisture or partial discharge monitoring may require new equipment installed to existing GIS.

Circuit Breaker Upgrade

The circuit breaker design allows defined values for rated current and short-circuit currents at a given voltage level. The design of the circuit breaker interruption unit may be upgraded to higher values by using more contact fingers. This would require the exchange of the circuit breaker interruption unit and a de-energization of the circuit breaker bay. Most GIS design allow such replacements.

The cost efficiency of such an upgrade needs to be evaluated in the total economic aspect of the GIS substation and its future use.

Circuit Breaker Controlled Switching

Requirements from network operation, the limitations of transient switching over-voltages or the increase of power switching capability can be reached by using a circuit breaker with controlled switching instead of an uncontrolled circuit breaker. This controlled switching devices can be added to existing GIS. Additional voltage and current measurement signals may be necessary for the circuit breaker-controlled switching.

The cost efficiency of such an upgrade needs to be evaluated in the total aspect of the GIS substation and its future use.

Circuit Breaker Operational Drive

During the lifetime of the GIS, the frequency of switching operations of the circuit breaker may come from changes in the function of the GIS substation in the network. The operational drive mechanism of the circuit breaker may reach the limits of the number of switch operation and may be exchanged. For this change usually no opening of the circuit breaker gas compartment is required, but the GIS bay needs to be de-energized.

The cost efficiency of such an upgrade needs to be evaluated in the total aspect of the GIS substation and its future use.

Gas Density Monitoring

The monitoring of the gas density is required for safe operation of the GIS, because the correct gas density provides the proper dielectric insulation. To control gas density of each gas compartment, todays GIS from leading manufacturers is using decentralized gas density meters with an electric signal to the GIS substation control center for gas density ok, gas leakage detected and operation lock for the circuit breaker or switches. With this decentralized and automated monitoring of the gas density, at any time the information on each gas compartment is available in the control center.

Older GIS may not have such an automated and decentralized gas monitoring which requires a manual check of the gas meter at the GIS bays.

It is recommended to upgrade the GIS with decentralized and automated gas density meters.

Online Moisture Monitoring

The low moisture content of the insulating gas is important for the safe function of the GIS. During the commissioning process the low moisture content of the insulating gas is provided usually by a dew point of −20 °C. Cast resin insulators contain moisture which is evaporating to the GIS gas compartment over the time. To capture this moisture in the gas aluminum oxide (AlO_3) is used as a drying agent. In some cases, the moisture content may increase if the stored volume in the insulator is too high or if the gas sealing allows moisture to evaporate into the gas compartment. This happens usually with gas compartments which show gas leakages.

Gas moisture may be measured by taking gas samples or by automated sensors connected to gas filling valves.

In case of any doubt on the correct moisture content of the gas compartment sensor can be added to the gas filling valves without the need of opening the GIS gas compartment.

Partial Discharge System

The condition of the high voltage system of the GIS can be evaluated by partial discharge (pd) measurements. The pd sensors used today are ultrahigh-frequency (UHF) antennas which are installed within the GIS. The antennas are used for the commissioning process to prove that the GIS gas compartments are free of particles, damages, or voids of the insulators, damages on the conductors, or any loose part in the GIS not grounded and producing partial discharges as a free potential part. After the pd measurement proves the correct assembly of the GIS on site, it will not be further used during operation.

In case of any doubt on the integrity of the dielectric system of the GIS, the intergraded sensors can be used for measurements without interrupting the power flow.

9.11.5.4 Recommendations from All Items Above

The condition assessment for GIS is a life-long process of the installation. It starts with the project planning and the GIS substation design and electric circuit. Information collection and future maintenance, upgrades, replacements, or extension planning start here. It is important for the user of the GIS to communicate these aspects closely with the manufacturer, because the maintenance assessment is closely related to the design of the GIS.

The evaluation process of condition assessment during the operational time of GIS should be carried out in close cooperation with the manufacturers. Inviting the experts of the manufacturers in a team for such evaluations is well invested money from the operator to find efficient solutions of the substation development into the future. It may prevent failing GIS and the high cost of power interruption and repair, and it may provide the right decision for the GIS future development of upgrade, retrofit, maintain, replace, or repair.

The technical development of GIS design over the last 5 decades brought big progress on how GIS is used and how the performance has been improved. And this process is not at the end. In this chapter, the focus is on SF_6 type of GIS equipment. Today first installation of alterative gases is in operation collecting data for their future use in the network (see Section 2.9). Vacuum circuit breakers of up to 245 kV for one interruption unit are developed and type tested in 2019. With technical air (O_2/N_2) and vacuum circuit breaker climate neutral high voltage GIS equipment is entering the market (see Section 9.13).

The organization and widely automation of data collection related to the evaluation of the GIS substation condition is a central task and the key factor for a successful evaluation process. Today most data are collected manually by very different sources and this take much time and a large engineering effort because the quality of data collected is essential for the right decision made at the end of the condition assessment process.

With increasing participation of digital equipment in the GIS substation and with new sensors, this data collection will be automated and collected in a cloud-based data base. Analytical software for data evaluation will generate recommendations to the operator of the GIS substation. In some cases, even action will be arranged automatically, for example, with the maintenance personnel.

It is seen that the future development of high voltage substations in the distribution and transmission level will change with digital development in sensor, equipment, and available software. Analytical methods and engineered algorithm will help the GIS substation designer to choose the right technical solution, the operator to find the right decision for safe and failure free operation, and the investor to optimize the overall cost of the power network and its substations. For more details on digitalization of substations see section 9.18.

9.11.6 Life Extension Options

9.11.6.1 General

The expected lifetime in practice of GIS is more than 50 years. This is based on the experiences made since the first GIS installations of the 1960s. No end of lifetime for GIS is indicated by the experience of today in a way that GIS could not be used any longer.

This is linked to the physical aspect of a non-aging gaseous high voltage insulation and under the permission that operation of the GIS is within its given physical limits. In most cases, the location of the GIS substation in the network requires very few switching operations and even less short-circuit interruptions. Also, most GIS installations operate far below the current rating for continuous power transmission.

Reasons for life extension options are often linked to changes in the operation of the GIS substation with higher load, more frequent switching operations, changing environmental conditions of higher ambient temperature, higher moisture in the air or higher air pollution and corrosion effects. These changes may result in different life extension options as follows: new GIS, new components, extended maintenance, retrofit options or additional sensors and data evaluation.

9.11.6.2 New GIS

The way for a new GIS may be chosen in case when the increase of load to be managed in the substation will require a higher voltage level. In this case, the existing GIS may not be able to get operated at the higher voltage rating.

The load increase at the GIS substation may also be covered by higher current ratings which may not be managed by changing parts or modules of the GIS. Current ratings of GIS installation have increase over the last decades from typical 1000–2000 A applications in the past to 3000–4000 A applications of today. Such big increases cannot be realized with the exchange of a few parts of the old GIS, like the number of contact fingers of the circuit breaker.

Beside the load increase, the local short-circuit rating of the GIS installation may grow and may reach short-circuit rating values of 63 kA or even 80 kA. This high short-circuit current will require new type tested equipment and can only be fulfilled with new GIS.

Depending on the total age of the GIS installation and the non-availability of spare parts in reasonable time when needed, may require the installation of a new GIS. This is very dependent on the importance of the substation in the network and the impact of failure with power delivery interruption.

The secondary protection and control equipment are an integral part of the GIS installation and is physically and electrically linked to the primary part of sensors (voltage and current instrument transformers) and actuators (motor drives and relays). In this case, when old secondary equipment cannot provide spare parts for any failing element, it might be necessary to replace the GIS with a new installation.

A new GIS may also be the better option in case of changed ambient or climatic conditions with higher temperatures, higher moisture and pollution or a general aging effect on secondary equipment in total. If the GIS installation shows aged secondary cables, corrosion on motors and drive mechanism or gas leakage from flange corrosion effects it might be the better way and overall, less costly to replace the GIS with a new installation.

The decision for a new GIS needs to evaluate the total condition of the existing GIS installation and the expected requirements for the near future. This evaluation will find the right time to change to a new GIS or it may find that components may be changed, extended maintenance is proposed or a retrofit of parts of the GIS may be best way for the next few years of operation.

9.11.6.3 New Components

The modular design of GIS allows also the exchange of single modules after years of operation. This could be the case if the number of switching operations is higher at the GIS substation as expected and the maximum number of types tested switching operations is coming close. In case, the load is not increasing and any other operational parameter is constant for future substation development, new components can extend the life-time of the GIS installation.

The most likely part to be exchanged would be the circuit breaker unit, the ground- and earth-switch unit or the disconnecting unit. These switching elements have aging effects and need maintenance or even a replacement.

For the replacement of GIS modules, the manufacturer has working schedules which indicate which bay or section of the substation needs to be de-energized for the replacement of modules. The work includes gas-works, mechanical works and, before re-energizing, high voltage tests. The complete process will require some days or weeks of work at the substation depending on the number of modules to be changed.

New components replacing aged GIS modules will provide several more years of operation under the same voltage and current ratings of the substation as it was designed for. Modules of GIS without any moving parts do not show aging effects and there are not limited in lifetime if the operational conditions are not changing.

Changing components in a GIS installation needs to be evaluated with a general view of the future substation development. Load increase, higher number of switching operations, higher short-circuit current ratings at the location of the substation or changing ambient or climatic conditions with higher temperatures, moisture, pollution, and corrosion effects may result a different option for life extension as a new substation, extended maintenance or retrofit.

9.11.6.4 Extended Maintenance

Maintenance cycle time and recommendations for functional parts of the GIS are made by the manufacturer to cover the effects of use and aging of operation in service. The measures recommended reflect the manufacturer design of the functional unit to prevent from failing and provide safe operation for the next maintenance cycle time.

Typical maintenance recommendations are in 5, 10 or 25 years of cycle time. The impact to the operation of the GIS installation is depending on the measures taken for maintenance. Checks of non-high voltage parts may be done without de-energization of the GIS. This includes gas quality checks by using the filling valves, corrosion checks on the enclosers, flanges, steel structures, operational drives or secondary equipment. Other maintenance measures which need access to the high voltage system will need de-energization of the GIS and safe disconnecting from the high voltage sources. This includes the checks of the contact system of the circuit breaker, disconnect, and ground switches. For such check, endoscope technology can be used by opening a manhole to get access. This requires much more impact to the GIS including gas works.

In case of detailed contact inspection of the circuit breaker interruption unit, it might be necessary to dismantle the circuit breaker unit of the GIS. In case of disconnecting or ground switch, such detailed investigation of contacts may be done through a manhole opening.

It is recommended to evaluate the overall condition of the GIS installation at any point of the operation time which maintenance measure are required and which additional maintenance could be combined to provide additional years of safe operation. This decision is based on the actual condition of the GIS, the level of continuous load, switching operation frequency, climatic condition changes with higher ambient temperatures or increasing moisture and pollution causing more corrosion.

The GIS needs evaluation of any of the visual checks, in- and out service tests and the data collection for the operation conditions of the GIS installation. This will bring the required information on possible maintenance needs and should be discussed together with GIS manufacturer and its experience and knowledge on the efficiency of his design.

The goal of extended maintenance is to prevent the GIS substation from failing during regular operation and possible power supply interruption. Each unplanned interruption and the following repair create high cost for the operator of the GIS and can be avoided with suitable maintenance measures.

9.11.6.5 Retrofit Options

The retrofit option of GIS installations may be the best solution in cases when parts of the GIS see changing operational conditions in form of increased load or higher switching frequencies. This may lead to exchange functional parts of the GIS with new modules to meet the new requirements and being able to continue safe operation for the complete GIS substation.

This may be the case for the circuit breaker module, the disconnector or ground switch module or the high-speed ground switch module. For this decision, the GIS installation needs to be evaluated in total about the condition of all functional parts and the operate ability for future use.

Retrofit options are partly renewing the GIS installation with keeping most of the original equipment in place. This requires a larger impact to the GIS substation and a de-energization of parts or even the complete substation.

It is clear that most of the GIS will stay as original units and only as critical parts are renewed. This has to be evaluated with the cost and future performance of a complete replacement of the GIS.

9.11.6.6 Additional Sensors and Data Evaluation

Additional sensors connected to the GIS may help to get more detailed information on the operation conditions. This may include sensors for temperature, gas pressure and density, switching operation data of current, overvoltage, movement of operational drives with time diagram, or transient overvoltage with partial discharge measurement.

Based on such data, evaluations on the condition of the GIS can be calculated and recommendations for maintenance, repair, retrofit, or replacement of the GIS can be drawn.

Sensors are available from the GIS manufacturer and may be installed external to the GIS. The data evaluation needs to be done together with the manufacturer to get correct information on the design of the equipment.

The additional information will help to manage the safe and reliable operation into the future and to decide maintenance measures more accurate to the GIS condition. Unnecessary maintenance work can be avoided and maintenance cycles may be shifted into the future if no need is seen according to the actual condition of the GIS.

9.11.7 Condition-Based Analytics

Condition-based analysis of the GIS installation is a complex process with many influencing parameters. It is of great importance to analyze the GIS based on technical data specific the manufacturer design. For this design data from the type tests, the routine test and onsite commission test are required.

The operational condition data from the substation at its local conditions are needed for the complete lifetime of the GIS to generate a picture of the GIS condition.

The analytics on the GIS condition at each point of time allows one to predict the requirement for maintenance and changes of the GIS substation in context with the changes coming from electric power network.

GIS is a long living product and it is a relative costly installation. This requires a good maintenance concept to operate the GIS safely and reliably for a long time. Experience collected for the last 50 years gives good indications on what to focus the data investigation. Switching part, gas quality and tightness of the compartments, partial discharge analysis of insulators and the high voltage gas compartment integrity, thermal indications on contacts for higher temperature and increasing contact resistance and visual inspections on any material degradation or corrosion effects.

It is important not to follow the principle to measure as much as possible. This increases the equipment cost and may not provide useable information. The way should be to measure and sense such data which allows a benefit for the GIS condition assessment and prevent it from failing. Discussions with the manufacturer of the GIS and following failure statistics of CIGRE [5] to identify relevant data to be sensor and measure for a condition-based analysis of the GIS.

9.11.8 Conclusion

Condition assessment of GIS substation is an effective measure to operate the GIS safely and reliably over a long period of time. GIS is a very reliable technology with low failure rates, but in case of failures high cost for non-delivered electric power and repair of the equipment will follow.

It is recommended to develop strategies for the condition assessment. This includes the easy to perform visual inspections explained in Section 9.11.2. Visual inspections of the GIS installation do not require de-energization of the GIS and provide good information on the GIS condition. Visual inspection shall follow an inspection plan with technical indications on what the inspection focus shall be.

There are several testing measures for in- or out of service test explained in Section 9.11.3 to identify weak points of the GIS installation. These tests shall be discussed with the manufacturer and should be applied in case of a doubt of correct function of the GIS. The principle here is not to test as much as possible but to evaluate indications from the operational manual of the GIS equipment and the technical use of the GIS at the specific substation conditions. Here, certain indications as high current rating, high number of switching operations, changing climatic conditions at the substation with higher ambient temperatures, higher moisture or more pollution may indicate the requirement of a test.

The data collection on GIS should include spare part availability, indication of critical parts, operational drives, failure statistics, load history, mechanical stress, environmental conditions, operating practice of the substation, growing load conditions, gas leaks and fault clearing data. This will help to get a picture of condition of the GIS and the function of the substation in the network. The data collection will start with the erection of the GIS and be continuous for the total operational time as explained in Section 9.11.4.

In Section 9.11.5 recommendations are discussed on the different maintenance methods and their best application. It covers the preventive maintenance, upgrades, and retrofit methods and gives information under which condition these methods may be applied.

The life extension options of Section 9.11.6 give information on the replacement with new GIS, the change of components of the GIS, the use of extended maintenance, the retrofit of modules of the GIS and the improvement which can be made with additional sensors and data evaluation.

Section 9.11.7 gives information on how to set up condition-based analysis and what to focus on. These analyses need to be set up in close cooperation with GIS equipment manufacturer and the experiences made with GIS in operation. A principle for valuable analysis is to focus the data measurements and sensors to those values which contribute to the efficient improvement of the GIS lifelong safe and reliable operation.

The collection of topics above has been drawn from many experts in a working group of the IEEE PES Substations Subcommittee for GIS and GIL. It gives information to substation engineers and personal involved in GIS substation business to find the right decision for each specific GIS application in the network. The big number of substation design and GIS schematics with the variations of influence parameters and local impacts leads to complex evaluations of the condition of the GIS. A wide range of possible maintenance measures are discussed and recommended to be applied. The recommendations can only help to find good solution with your own investigations on the condition assessment of GIS substations. To find the right methods, a discussion with the GIS manufacturer, the GIS operator, and the network operator are required to find best solutions for the GIS substation.

References Section 9.11

1 CIGRE TB 509, Final Report of the 2004 – 2007 International Enquiry on Reliability of High Voltage Equipment, Part 1 – Summary and General Matters, Working Group A3.06, October 2012.
2 CIGRE TB 510, Final Report of the 2004 – 2007 International Enquiry on Reliability of High Voltage Equipment, Part 2 - Reliability of High Voltage SF6 Circuit Breakers, Working Group A3.06, October 2012.
3 CIGRE TB 511, Final Report of the 2004 – 2007 International Enquiry on Reliability of High Voltage Equipment, Part 3 - Disconnectors and Earthing Switches, Working Group A3.06, October 2012.
4 CIGRE TB 512, Final Report of the 2004 – 2007 International Enquiry on Reliability of High Voltage Equipment, Part 4 - Instrument Transformers, Working Group A3.06, October 2012.
5 CIGRE TB 513, Final Report of the 2004 – 2007 International Enquiry on Reliability of High Voltage Equipment, Part 5 - Gas Insulated Switchgear (GIS), Working Group A3.06, October 2012.
6 CIGRE TB 514, Final Report of the 2004 – 2007 International Enquiry on Reliability of High Voltage Equipment, Part 6 – Gas Insulated Switchgear (GIS) Practices, Working Group A3.06, October 2012.
7 IEC 62271-1:2017, High-voltage switchgear and controlgear - Part 1: Common specifications for alternating current switchgear and controlgear.
8 IEC 62271-100:2008+AMD1:2012+AMD2:2017 CSV Consolidated version, High-voltage switchgear and controlgear - Part 100: Alternating-current circuit-breakers.
9 IEEE C37.100.1-2018, IEEE Standard Of Common Requirements For High Voltage Power Switchgear Rated Above 1000 V.
10 IEEE C37.100.2-2018, IEEE Standard For Common Requirements For Testing Of AC Capacitive Current Switching Devices Over 1000 V.
11 CIGRE, LONG-TERM PERFORMANCE OF SF6 INSULATED SYSTEMS, by Task Force 15.03.07* of Working Group 15.03**, On behalf of Study Committee 15, Paris CIGRE Session 2000.
12 IEC 62271-203:2011, High-voltage switchgear and controlgear - Part 203: Gas-insulated metal-enclosed switchgear for rated voltages above 52 kV.
13 IEEE C37.122-2010 Standard for High Voltage Gas-Insulated Substations Rated Above 52 kV.

9.12 Gas-Insulated Substations for Enhanced Resiliency

Author: George Becker
Reviewer: Hermann Koch

9.12.1 Introduction

Gas-Insulated Substations (GIS) can enhance substation resiliency.

Resiliency in general is defined as "a means to withstand and to rapidly recover from disruption." Resiliency with respect to critical infrastructure is "the ability of a facility or an asset to anticipate, resist, absorb, respond to, adapt to, and recover from a disturbance." It is impossible to design a substation that is 100% immune to naturally and human caused physical threats. However, electrical substations must be resilient and operate safely and reliably irrespective of the environments in which they reside. Grid strengthening efforts should improve substation reliability, inherently enhance emergency response and improve recovery strategies related to naturally or human-caused physical events that damage electric substations [1, 2].

Natural events such as severe weather, are a primary cause of unplanned power outages. There are a range of natural events that can cause high impact consequential damage to an electric substation, and include storms, naturally occurring environmental conditions, earthquakes, floods, wildlife etc. The severity and impact of recent weather events such as superstorms, large scale tornado outbreaks, coastal flooding and inland river flooding and severe thunderstorms accompanied by strong straight-line winds have prompted substation owners to examine design strategies to limit potential damage to substations due to natural events.

Human-caused physical attacks and criminal activity directed at electric substations can be high impact events that cause consequential damage and have long lasting effects. As a result of more frequent criminal attacks on critical substations, many utilities are examining design strategies to mitigate the effects of criminal activity directed at substations in conjunction with limiting potential damage to substations due to natural events (see Figure 9.28).

Historically, substation designs for the most part have been based on standardized engineering practices and traditional air-insulated technology. Given the challenges associated with the threats posed by high impact naturally and human-caused physical events that can damage substations, many utilities are turning to gas-insulated technologies as a more resilient solution to harden their substation infrastructure.

9.12.2 Historical Physical Threat Events – Natural and Human-Caused

Electrical substations, especially transmission substations, generation interconnecting substations or substations serving critical loads function as part of critical infrastructure that is essential to the well-being and safety of the public. Ideally, these substations should therefore be designed to be immune to the effects of natural catastrophic threats and nefarious human-caused physical threats. It is impossible to design a substation that is 100% immune to naturally and human caused physical threats. However, well planned substation designs can be engineered and constructed, so that substations serving critical infrastructure are sufficiently hardened to provide the necessary reliability for increasing public demands, and more critically minimize the impact of damage and facilitate fast emergency response. It is possible to make the damage less severe, prolong service and restore service more quickly with a sufficiently hardened facility and a layered approach to physical security [3].

The aftermath of storms such as Superstorm Sandy and Hurricane Katrina, as well as the physical attack on the Pacific Gas and Electric Company's Metcalf Transmission Substation has increased pressure on utilities and municipalities to limit electric system outage frequencies and their durations. In addition, there is an overall heightened awareness among governmental agencies and utilities to harden critical infrastructure, thereby improving grid system reliability in preparation for major storm events and potential nefarious human-caused physical threats.

Conventional substation solutions, based on standardized air-insulated designs, may not provide the best approach for critical substations located in areas susceptible to storms, naturally occurring environmental conditions, earthquakes, floods, wildlife, or other natural threats, or physical attacks and criminal activities.

For example, Superstorm Sandy caused a potentially catastrophic storm surge and inundation of coastal flood waters that posed significant challenges to emergency responders and disaster relief efforts. Power outages caused by deficiencies in some traditional substation designs, and sometimes due to aging infrastructure, further compound these challenges and can delay emergency response. In addition, with growing populations in coastal areas, there is an increased risk of property damage and loss of life.

Figure 9.28 Examples of Natural and Human-Caused Physical Threats, from top, left to right: bird nest in HV, hurricane, copper theft in substations, severe environmental conditions, transformer fire, tornado, terroristic attack to substation and water flooding (various public internet outlets)

The US National Oceanic & Atmospheric Administration (NOAA) [14] identifies the following potential exposure to extreme weather:

- From 1990 to 2008, population density increased by 32% in Gulf coastal counties, 17% in Atlantic coastal counties, and 16% in Hawaii (U.S. Census Bureau 2010)
- Much of the United States' densely populated Atlantic and Gulf coastlines lie less than 10 feet above mean sea level
- Over half of the nation's economic productivity is located within coastal zones
- 72% of ports, 27% of major roads, and 9% of rail lines within the Gulf Coast region are at or below 4-foot elevation
- A storm surge of 23 feet can inundate 67% of interstates, 57% of arterials, almost 50% of rail miles, 29% of airports, and almost all ports in the Gulf Coast

For example, along the U.S. northeast coast line NOAA places the return period for a Category 3 storm between 68 and 70 years on the southern coastline of New England. Likewise, NOAA places the return period for Category 3 storms between 84 and 110 years along the Mid-Atlantic coastline, and between 85 and 170 years for the Northern New England/Maine coastline.

Another aspect to evaluate is the potential for recurring storms of specific severity to a specific area. For example, the NOAA has maintained records that show all hurricanes making landfall within 75 nautical miles of New Haven, CT since 1842. Within this period, nine hurricanes made landfall in New York and Connecticut (see Figure 9.29):

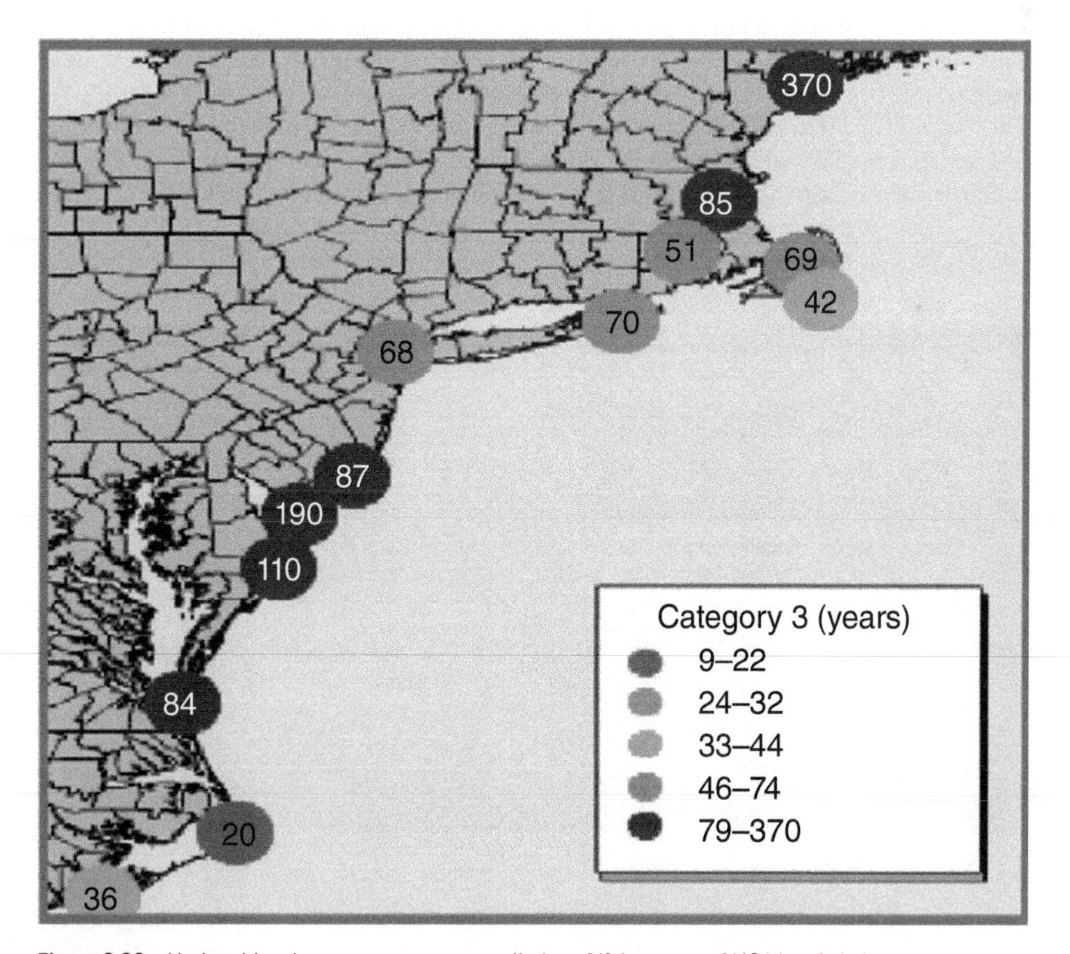

Figure 9.29 National hurricane center return predictions [4] (courtesy of NOAA website)

Likewise, a high voltage substations sniper attack was a 'sophisticated' assault on a substations connenction power transmission line in California. The assault happened on 16th April 2012, in which gunmen fired on 17 high power electrical transmission transformers. The attack resulted in over $15 million worth of damage.

Seventeen transformers were seriously damaged. To avert a black-out of the critical "Silicon Valley Area", energy grid officials were forced to reroute power from nearby power plants to the Silicon Valley Area. The damage at the transmission substation and the action of reconfiguring the transmission system to continue to serve critical loads put the grid in a first contingency outage scenario for an extended period [5–7] (see Figure 9.30).

9.12.3 Resilience Methodology Using Gas-Insulated Substations

The methodology of prepare, prevent using GIS, respond, and recover with respect to the use of gas-insulated technologies to harden substations against natural catastrophic threats and human-caused physical threats is characterized by the following [10–13]:

1) The "prepare" actions consist of identification of threats, identification of the vulnerabilities with respect to the identified threat, assessment of the impact of the threats and maintaining an ongoing awareness of new risks and evolving threats.
2) The "prevent using GIS" actions consist of utilizing substation designs that are sufficiently hardened against natural catastrophic threats, utilizing substation designs that are based on a layered approach to physical security and integrating enhanced shielding measures. The effectiveness of these actions is greatly enhanced using gas-insulated substation designs.
3) The "respond and recover" actions consist of damage assessment, securing the substation area and perimeter and initiating the process of repairing or replacing substation components that may have incurred damage and returning the substation to full operating condition.

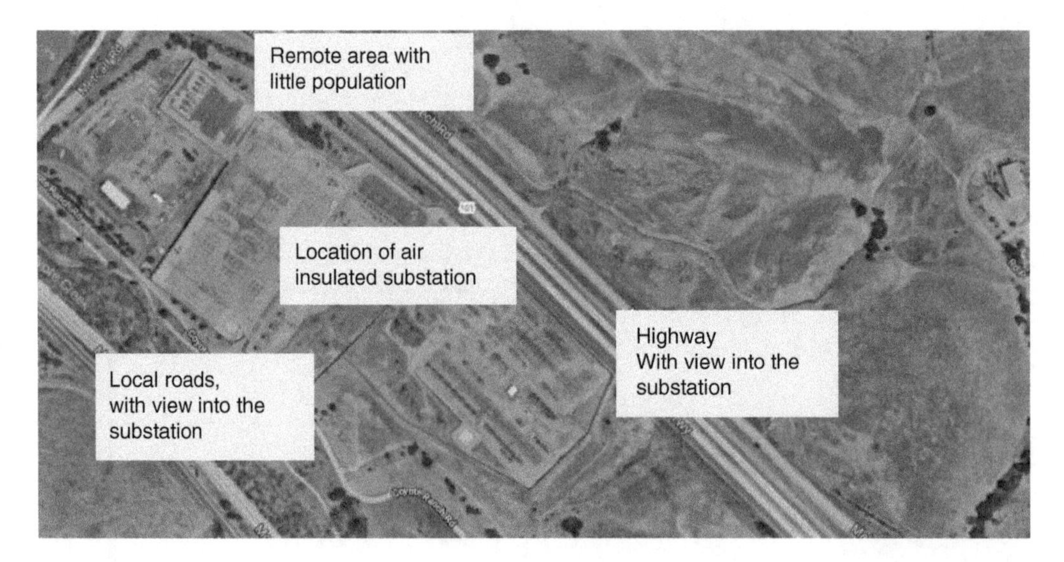

Figure 9.30 Location of transmission substation attached by sniffer gunmen (Google maps website) [5]

9.12.4 Prepare

The definition of resiliency in general is "a means to withstand and to rapidly recover from disruption." In the context of critical infrastructure, resiliency is "the ability of a facility or an asset to anticipate, resist, absorb, respond to, adapt to, and recover from a disturbance."

Preparation in anticipation of a threat(s) or disturbance(s) is the first step to designing a truly resilient substation. Substation owners must "Know the Threats, Vulnerabilities and Impacts." The specific threats to a substation may be different based on the location of the substations and the climate it resides in. The assessment of threats and vulnerabilities of a substation site and vulnerabilities of specific assets is the initial step in fully characterizing the threats [14–16].

To effectively prepare for potential threats to a substation the following actions need to be accomplished:

- Identification of the major physical threats to the substation (naturally and human-caused) and their characteristics
- Identification of the vulnerabilities with respect to the identified threats
- Assessment of the impact of the threats
- Maintaining an ongoing awareness of new risks and evolving threats

The following are short descriptions of the major naturally and human-caused physical threats to electric substations and potential impacts. These threats do not encompass all possible threats, but they are the major threats that are recognized as being potentially damaging to substation infrastructure.

9.12.4.1 Tropical Cyclone Winds

Strong winds from a tropical cyclone can damage or destroy vehicles, buildings, bridges, personal property and other outside objects, turning loose debris into deadly flying projectiles. In the United States, major hurricanes comprise just 21% of all land falling tropical cyclones, but account for 83% of all damage. These are long duration events.

Tropical cyclones can cause serious damage to air-insulated electric substation equipment because the winds can produce flying projectiles that damage insulators, switches, and bushings as well as other critical equipment in electric substations.

9.12.4.2 Tropical Cyclone Heavy Rains

The thunderstorm activity in a tropical cyclone produces intense rainfall, potentially resulting in flooding, mudslides, and landslides. Electric substations that are particularly vulnerable to freshwater flooding may be at a high risk for damages associated with these threats [8, 9].

The wet environment in the aftermath of a tropical cyclone, combined with the heavy salt contamination of the air-insulated electric substation equipment can cause both short term and long-term damage.

9.12.4.3 Storm Surge and Coastal Flooding

A storm surge, storm flood or storm tide is a coastal flood or tsunami-like phenomenon of rising water commonly associated with low pressure weather systems such as tropical cyclones and strong extratropical cyclones. The severity of the storm surge is affected by the shallowness and orientation of the water body relative to storm path, as well as the timing of tides.

Since much of the United States' densely populated Atlantic and Gulf coastlines lie less than 10 ft above mean sea level, storm surge is a very credible threat to electric substations in these areas. During a storm surge or coastal flooding event, access to a substation can be hindered and substation equipment can be submerged. Storm surge can produce large floating objects that can damage substation equipment.

9.12.4.4 Inland River and Flash Flooding

Inland river flooding can cause damage to electric substations very similar to the damage caused by storm surge or coastal flooding. Flash floods can also cause similar and usually more serve damage, except that flash floods occur in a much shorter time frame, and they involve large volumes of water travelling at great speeds and with tremendous force. Flash floods usually always carry large floating objects that can damage substation equipment.

9.12.4.5 Severe Cold, Heavy Snow and Ice

Most electrical equipment sold for outdoor use is designed and built to international standards that require the equipment to operate within design parameters with no loss of performance between −30 °C (−22 °F) and +40 °C (+104 °F). However, a significant number of utilities operate in territories that are subject to temperatures below −30 °C (−22 °F) or even 50 °C (−58 °F). Equipment subjected to such low temperatures can be affected in a number of ways, depending on its construction, insulating medium, lubricants used for its moving parts, gaskets, and seals and materials used for fabrication,

High levels of snow and ice storms affect air-insulated substation equipment in different but equally damaging ways, often leading to equipment failures, alarms, and occasional blackouts if multiple pieces of equipment are affected. Heavy snow not only affects access to equipment in the substation, but also the safety of personnel and the operation of the substation.

The build-up of ice on electrical conductors and steel frame power structures in air-insulated substations can reach unsustainable weights. Built-up ice can accumulate on porcelain bushings and insulators, eventually causing line-to-ground flashovers, taking the equipment out of service.

9.12.4.6 Tornados

Tornadoes have a shorter and more targeted impact than other storms such as tropical cyclones, or ice storms. Tornadic activity may accompany tropical cyclones and severe thunderstorms. The devastating wind energy from tornadoes makes hardening air-insulated designs expensive and improbable with respect to the prevention of outages. The intensity, frequency, and change in expected location cause significant outages for utilities that they may not be prepared for with air-insulated substation designs.

9.12.4.7 Severe Thunderstorm Micro Bursts

A microburst is an intense small-scale downdraft produced by a thunderstorm or rain shower. A microburst often has high winds that can knock over fully grown trees. They usually last for seconds to minutes. Microbursts can also damage substation equipment be putting extreme downforces and moments on cantilevered equipment and strain buses.

9.12.4.8 High Straight-Line Winds

High straight-line winds associated with a land-based, fast-moving group of severe thunderstorms also known as "derecho," can cause hurricane-force winds. Unlike other thunderstorms, which typically can be heard in the distance when approaching, a derecho seems to strike suddenly. A derecho moves through quickly but can do much damage in a short time to air-insulated substation equipment.

9.12.4.9 Earthquakes

Air-insulated substations contain structures to support electrical equipment at significant heights above ground level. These structures are mounted on footings, the anchorage to the footings supporting the equipment is frequently inadequate to prevent overturning or slipping during earthquakes due to the large forces and moments on the structures. Therefore, earthquakes can cause major damage to air-insulated substations.

9.12.4.10 Severe Humidity

Elevated humidity levels will provide environmental conditions that promote increased fungal and microbial growth. Fungal and microbial growth can have detrimental effects on electrical utility equipment. This increased prevalence of fungus and mold on air-insulated substation electrical equipment can cause premature outages, power failures, increased maintenance, and equipment failure.

9.12.4.11 Desert and Severe Heat

Heat is a prime enemy of electrical equipment. Elevated heat levels, above nameplate temperatures, can cause premature failure of air-insulated substation equipment increasing maintenance and operational costs to the utility. Heat and drought environments create large amounts of dust and drying out of the soil that the substation is built on. The dust can cause mechanical operation issues and the dry soil can affect ground conductivity of the substation ground grid. Ultra-violet (UV) deterioration of materials can also be an issue.

9.12.4.12 Human-Caused Physical Attack

Air-insulated substations are vulnerable to human-caused nefarious physical attack unless they are significantly fortified at their perimeter and protected from above. Fortification of large air-insulated substations is extremely expensive. Air-insulated substations like the Metcalf Substation can be targeted with projectile weapons, bombs, air assault vehicles (i.e., drones) and physical attack using hand tools and machine tools by intruders. They are particularly susceptible to physical attack because of the area they encompass, their line-of-sight exposure and their large perimeter.

9.12.4.13 Human-Caused Criminal Theft

Air-insulated substations are also more susceptible to criminal theft for the same reasons they are vulnerable to physical attack, large area and large perimeter. Intruders can find multiple weaknesses in the perimeter protection of the substation to enter and steal equipment and materials and possibly cause outages.

9.12.5 Know the Threats, Vulnerabilities and Impacts

The key to designing a truly resilient substation is to methodically assess what could happen and the vulnerabilities within the substation that need to be addressed. While the reliability requirements are focused on addressing a physical attack, substations fail for other reasons as well. Most of the utilities approach the assessment and subsequent planning from a more holistic perspective that includes physical attack, natural disasters or simply age-related failures.

In Table 9.9 an example for a tropical cyclone is given for characterization of a major natural physical threat to an electric substation, the tabulated vulnerably of equipment and the potential impacts of the threats to the substation and electrical system.

In Table 9.10 an example for Terrorist Physical Attack characterization of a major human-caused physical threat to an electric substation, the tabulated vulnerably of equipment and the potential impacts of the threats to the substation and electrical system.

9.12.6 Prevent

Effective prevention of most catastrophic damage to substations can be achieved by using gas-insulated switchgear as part of the design of substations that are critical to the electric grid and vulnerable to major threats.

Gas-insulated switchgear provides the substation owner with the following advantages with respect to substation design and cost while enhancing resiliency:

- A smaller substation footprint (area and perimeter) that can be located inside a hardened building, underground structure or elevated structure
- Expansion of the security perimeter of the substation
- Reduced construction time and schedule risks during construction

Table 9.9 Tropical Cyclone characteristics, vulnerability, and threat

Identified threat and characteristics	Vulnerability of personnel and equipment	Perceived impact of threat
Tropical Cyclone		
Storm Surge and Coastal Flooding	Flooding – submerged electrical equipment, access/egress problems and dangers for personnel, large floating objects	Requirement to de-energize the substation to minimize potential for major equipment damage due to faults.
	Salt contamination – insulators, bushings, connectors, control/power /termination cabinets	Long term restoration of equipment
	Long term corrosion – steel, cables, termination points	Critical power flows and critical load interrupted, national security risks and public safety risks

Table 9.10 Terrorist Physical Attack characteristics, vulnerability, and threat [10]

Identified threat and characteristics	Vulnerability of personnel and equipment	Perceived impact of threat
Terrorist Physical Attack		
Sniper Attack with Heavy Small Arms Ammunition	Personnel at the substation can be targeted	Threats to human life of workers
	Puncture and rapid leakage of insulating fluids	Long term restoration of equipment
	Destruction of support insulators for HV equipment	Critical power flows and critical load interrupted, national security risks and public safety risks
	Destruction of bushings on HV equipment	

- More easily located near load centers and critical infrastructure
- More easily disguised and aesthetically pleasing with the opportunity for no overhead line entry, "out of sight, out of mind."
- Approximately 15 times the reliability of an air-insulated substation performing the same duty

9.12.6.1 The Tropical Cyclone, Storm Surge and Coastal Flooding Example

Elevating an entire transmission substation is challenging due to the amount of space required for increased electrical clearances at higher voltages. The corresponding amount of earth work, additional concrete for raising foundations, and increased steel requirements to elevate substation structures may not be feasible. Even if AIS (air-insulated substation) equipment is raised out of projected storm surge or flood levels, the substation will remain "outdoors" and is still vulnerable to other elements of the tropical cyclone.

Based on prior installations and case studies, elevating transmission substations with "indoor" GIS (gas-insulated switchgear) has proven to be an excellent solution to flood-prone substations, especially in coastal areas. GIS provides many unique benefits, such as a significantly reduced-footprint which often allows construction in small vacant areas of existing switchyards. This construction technique can solve permitting issues and allow existing substations to remain operational during GIS construction, thus providing reduced outages and shorter cutover durations (see Figure 9.31).

By elevating a GIS substation above predicted storm surge and flood levels, risk of flooding equipment areas is virtually eliminated. In fact, according to the "Economic Benefits of Increasing Electric Grid Resilience to Weather Outages" paper issued by the White House; "Common hardening activities to protect against flood damage include elevating substations and relocating facilities to areas less prone to flooding. Unlike petroleum facilities, distributed utility T&D assets are not usually protected by berms or levees [11]. Replacing a T&D facility is far less expensive than building and maintaining flood protection. Other common hardening activities include strengthening existing buildings that contain vulnerable equipment and moving equipment to upper floors where it will not be damaged in the event of a flood."

In addition, by installing GIS inside an elevated substation, wind, and atmospheric contaminates (e.g., dust, salt, industrial pollutants, etc.) cannot impact the substation's operation. Since GIS equipment has all primary electrical components enclosed and in insulated with SF_6 gas, reliability improves significantly due to no external influences. Therefore, the power grid realizes benefits of improved reliability, as well as safety, and many times improved life cycle costs.

These concepts of using enclosed GIS in hardened buildings or structures are applicable to all major natural physical threats to substations. GIS significantly reduces the physical area and perimeter of the substation. This enables the substation owner to employ the following construction methods for substations to reduce the impact of Naturally Caused Physical Threats, catastrophic events and enhance resiliency, see Table 9.11.

GIS provides enhanced resiliency for each of the identified threats/impacts individually. But when the benefits of the GIS Enabled Construction Methods are combined, the result is a truly resilient substation that is largely immune to most, if not all, naturally caused physical threats.

9.12.6.2 The Terrorist Physical Attack Example

The successful protection of substation equipment from human-caused nefarious physical attack requires a layered security approach. Figure 9.32 shows the widely accepted asset protection basic philosophy which is best approached through three steps:

- The application of concentric rings of security measures
- Inner most layers of security protect the most critical assets
- A structured process for evaluating protection based on impact and risk

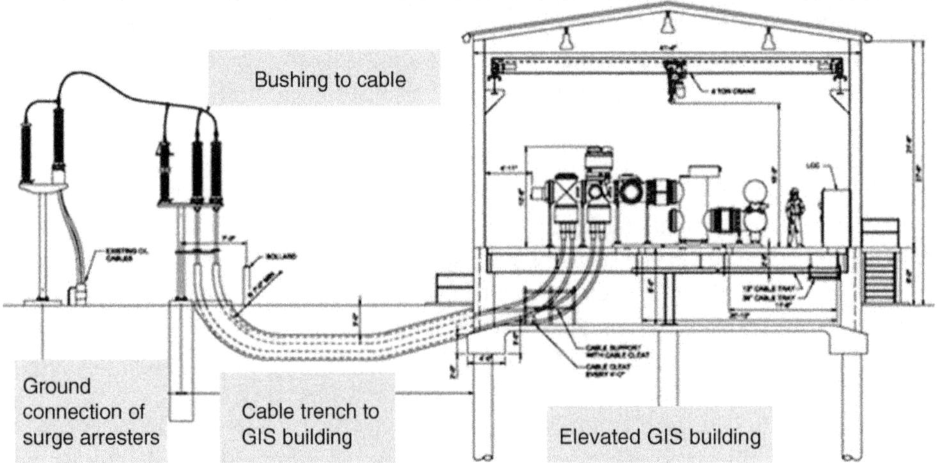

Figure 9.31 Elevated GIS substation concept (Courtesy of Power Engineers and ABB)

The successful protection of substation equipment also requires some form of threat characterization and classification of substations. Threat characterization is accomplished by conducting a threat assessment of each substation site, assessing the risk (probability) of each threat manifesting, including prioritization of assets at the site and the identification of potential consequences and assessing the vulnerability of the site to the risks identified.

For example, The North American Electric Reliability Corporation (NERC) classifies critical substations under its critical infrastructure protection plan (CIP) as; NERC/CIP Critical Substations – substations that if significantly damaged will cause a major decrease in transmission system reliability.

In Figure 9.33 an example of a Threat Characterization Matrix for a Critical Transmission Substation is shown.

Each substation owner can perform a threat analysis and ranking of NERC/CIP Critical Substations and determine the impact of a catastrophic physical attack on those substations.

Table 9.11 Terrorist Physical Attack characteristics, vulnerability, and threat

Identified threat/impact	GIS enabled construction method
Tropical Cyclone Winds	GIS can be installed in a low-profile enclosure/structure that is rated for tropical cyclone force winds. Incoming and exiting transmission and/or distribution lines can be routed underground to protect line terminations.
Tropical Cyclone Heavy Rains	GIS can be installed in a low-profile enclosure/structure that is rated for tropical cyclone force winds which consequently mitigates the salt contamination issues associated with tropical system rains.
Storm Surge and Coastal Flooding	GIS can be installed in a low-profile enclosure/structure that is rated for tropical cyclone force winds and is elevated above the maximum predicted storm surge level.
Inland River and Flash Flooding	GIS can be installed in many locations that a similar AIS cannot. This increases the options of constructing the substation away from the impacts of rivers and areas that are prone to flash floods. If relocation is not an option, then the GIS can be installed in a low-profile enclosure/structure that is elevated above the 500-year inland flood level and on piles that are not affected by surface flash floods.
Severe Cold, Heavy Snow and Ice	GIS can be installed in a low-profile enclosure/structure that is rated for snow and ice loads. The GIS enclosure /structure can be climate controlled to mitigate the effects of low temperatures on the SF_6 filled equipment and all electrical equipment.
Tornados	GIS can be installed in a low-profile enclosure/structure that is all or partially underground. Incoming and exiting transmission and/or distribution lines can be routed underground to minimize the impacts to line terminals.
Severe Thunderstorm Micro Bursts	GIS can be installed in a low-profile enclosure/structure that is all or partially underground. Incoming and exiting transmission and/or distribution lines can be routed underground to minimize the impacts to line terminals.
High Straight-Line Winds	GIS can be installed in a low-profile enclosure/structure that is rated for tropical cyclone force winds and for straight line "derecho" wind forces as well.
Earthquakes	GIS is inherently resilient with respect to the forces of an earthquake. The GIS has a low center of gravity, minimized structure heights and extensions which reduce the effects of external forces and moments. GIS is a highly standardized modular building block design that is seismically qualified as a complete system capable of withstanding at least 0.2 times the equipment weight applied in one horizontal direction, combined with 0.16 times the weight applied in the vertical direction at the center of gravity of the equipment and supporting structure.
Severe Humidity	GIS can be installed in a low-profile enclosure/structure that is climate controlled to mitigate the effects of high dew points and minimizes fungal and microbial growth.
Desert and Severe Heat	GIS can be installed in a low-profile enclosure/structure that is climate controlled to mitigate the effects of high temperatures on equipment ratings, UV and minimize dust impacts on mechanical seals. Also, since the GIS grounding grid system is less reliant on soil moisture within the substation, ground conductivity issues can be minimized.

Once this threat analysis and ranking is accomplished, the substation owner will generally try to identify the substation design, perimeter protection, security equipment/protocols and counter measures to effectively protect the substation from all perceived human-caused physical attacks.

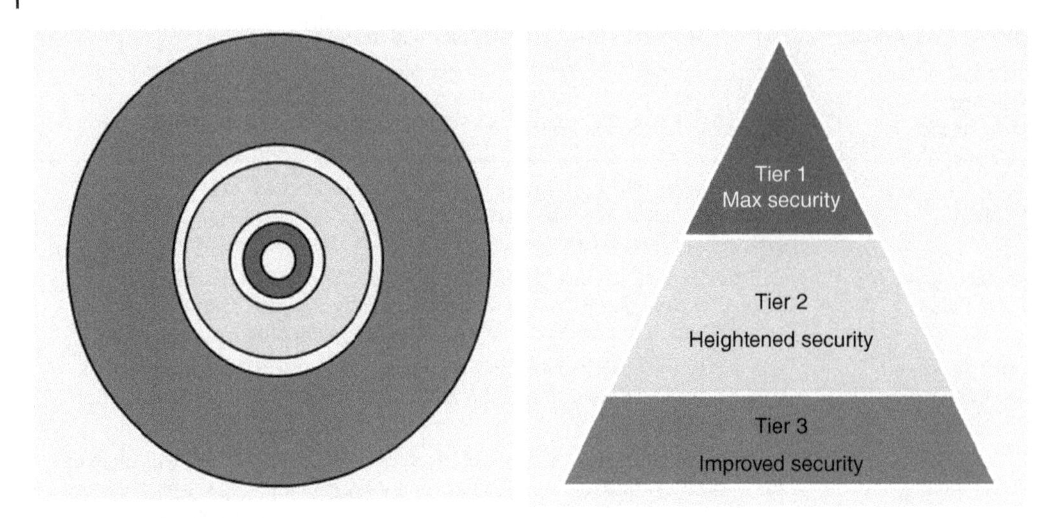

Figure 9.32 Asset protection basics

Security Evaluation Prcoess - example - EAST SHORE - 7.8

Assault Team

Severity \ Exposure	1	2	3	4	5
1	2	3	4	5	6
2	3	4	5	6	7
3	4	5	6	7	8
4	5	6	7	8	9
5	6	7	8	9	10

SCORE: 7

Moderate Severity, Moderate Exsposure. Active overnight area

Maritime

Severity \ Exposure	1	2	3	4	5
1	2	3	4	5	6
2	3	4	5	6	7
3	4	5	6	7	8
4	5	6	7	8	9
5	6	7	8	9	10

SCORE: 10

High severity, High exposure due to Harbour traffic

Sabotage

Severity \ Exposure	1	2	3	4	5
1	2	3	4	5	6
2	3	4	5	6	7
3	4	5	6	7	8
4	5	6	7	8	9
5	6	7	8	9	10

SCORE: 7

Moderate Severity, Moderate exposure. Side Fence adjacent to public road

Standoff

Severity \ Exposure	1	2	3	4	5
1	2	3	4	5	6
2	3	4	5	6	7
3	4	5	6	7	8
4	5	6	7	8	9
5	6	7	8	9	10

SCORE: 7

Moderate Severity, moderate exposure, Large Yard mutilpe buildings

Theft/Diversion

Severity \ Exposure	1	2	3	4	5
1	2	3	4	5	6
2	3	4	5	6	7
3	4	5	6	7	8
4	5	6	7	8	9
5	6	7	8	9	10

SCORE: 7

Moderate Severity, moderate exposure

VBIED

Severity \ Exposure	1	2	3	4	5
1	2	3	4	5	6
2	3	4	5	6	7
3	4	5	6	7	8
4	5	6	7	8	9
5	6	7	8	9	10

SCORE: 9

High Severity, High Exposure due fence line close to distance Yard and Side Fence adjacent to public road

Figure 9.33 Threat characterization matrix for a critical transmission substation

GIS provides the substation owner with a pre-engineered substation design that enables the substation owner to employ construction methods for substations that reduce the impact of Human-Caused Physical Threats and enhance resiliency. GIS designs can expand the outermost Tier 3 layer of security by adding buffer distance in addition to strong wall/

barriers. GIS designs also allow for more concentrated mid-layer Tier 2 and maximum innermost Tier 1 security, see Figure 9.34.

Designing a substation that is truly immune to most human-caused physical threats is challenging. GIS designs in concert with GIS Enabled Construction Methods are an effective method to protect a substation and its associated electrical equipment from severe damage from a nefarious human-caused physical intentional attack. GIS designs allow for the following substation design attributes with respect to protection from nefarious human-caused physical intentional attack:

- Reduced area of the substation, flexible compact arrangements
- Reduced exposed perimeter of the substation, significantly reduced cost for perimeter protection
- Expanded Tier 3 security buffer, intruder neutralization devices can be more easily installed
  - More concentrated fixed cameras with analytics at the perimeter
  - More effective pan-tilt-zoom camera coverage with analytics
- Concentrated Tier 2 and Tier 1 security layers with less potential for penetration
  - More concentrated thermal imaging cameras with analytics
- "Out of sight, out of mind" to blend into surroundings and hidden from aerial imagery
- No line-of-site to equipment for projectile weapons
- Switchgear is completely enclosed with fire protection
- Transformers and reactors can be enclosed and covered with grating overhead to minimize air assault
- Relay control rooms can be located at the interior of the complex to enhance protection
- Interfaces to transmission and distribution system can be more easily concealed underground

Figure 9.35 shows a comparison of an open-air-insulated substation behind a fence and the enhanced resilience of an underground gas-insulated substation.

Figure 9.35 shows an indoor gas-insulated substation located out of the city center in a bird eye view and the inside of the substation building protection the substation against natural and human caused physical threats (see Figure 9.36).

Figure 9.37 shows an indoor gas-insulated substation located in the city center in a bird eye view at night and location drawing of the substation area for protection the substation against natural and human caused physical threats.

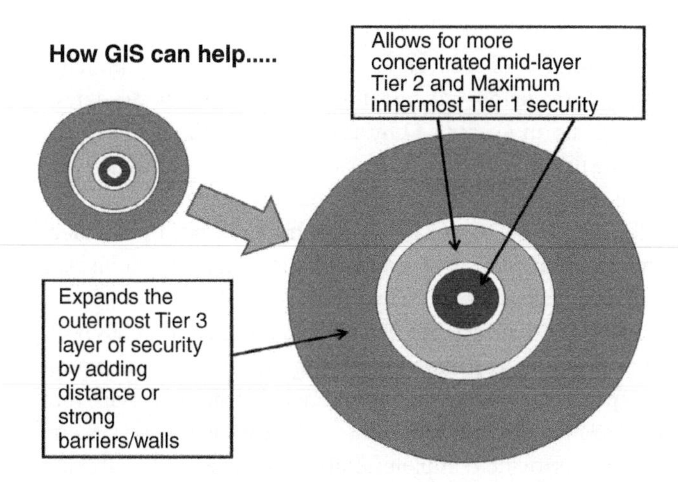
How GIS can help.....
Allows for more concentrated mid-layer Tier 2 and Maximum innermost Tier 1 security
Expands the outermost Tier 3 layer of security by adding distance or strong barriers/walls

Figure 9.34 GIS enhanced Tier 2 and Tier 3 security

Figure 9.35 Traditional AIS versus resiliency enhanced GIS (*source* [12])

Figure 9.36 GIS substation, resiliency enhanced against both naturally and human-caused physical threats located out of the city center (Courtesy of IEEE)

The use of gas-insulated substation designs is a cost-effective strategy that results in enhanced resiliency and higher reliability due to a reduction in substation area and exposure to external threats.

9.12.7 Respond and Recover

Damage assessment and the rapid implementation of recovery steps are essential to minimizing the impact to the electric system caused by damage to a critical substation. Innovative substation solutions can reduce restoration times after storm events or physical attacks. GIS solutions can provide options for substation owners during the process of repairing substations that are damaged.

Mobile GIS designs offer a wide range of switchgear and bus configurations at all operating voltages. These mobile GIS options can be designed to be located on a single trailer or multiple trailers and connected via power cables in any required line or bus configuration. Electric circuit see Figure 9.38 and a mobile GIS on a trailer see Figure 9.39.

The installation of mobile GIS to temporarily replace existing switching equipment that has been damaged can significantly expedite the process of repairing and exchanging damaged AIS equipment or GIS components/modules. The rapid reestablishment "fast re-energization" of service that mobile GIS provides, further enhances the resiliency of the substation and the transmission system.

Mobile GIS equipment can also be used to provide complete temporary substation switching arrangements as a proactive solution for the following scenarios:

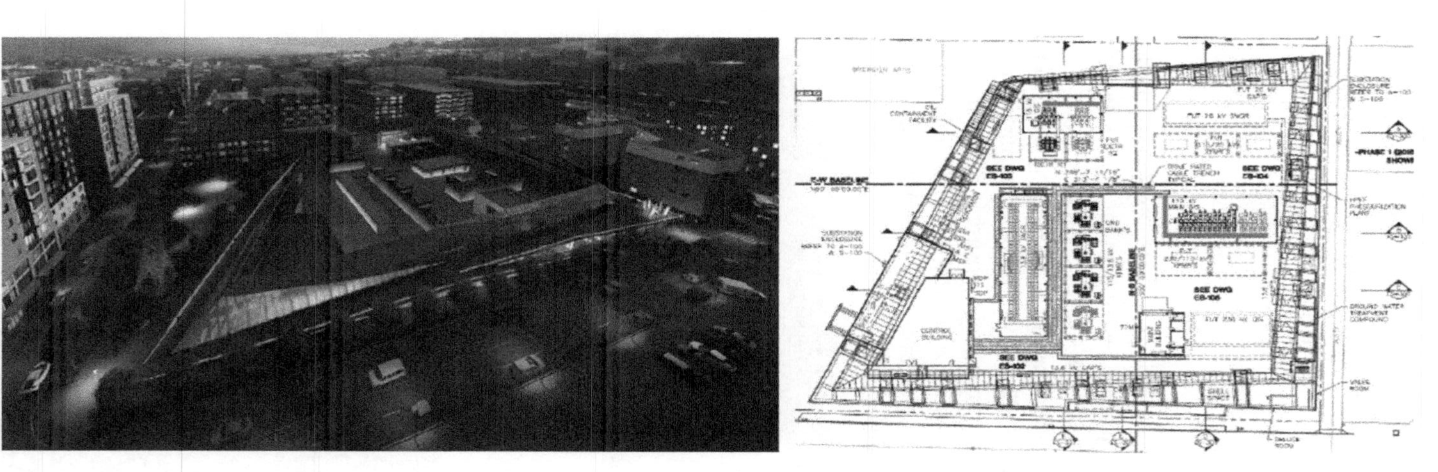

Figure 9.37 GIS substation, resiliency enhanced against both naturally and human-caused physical threats in the city center (Courtesy of IEEE)

- Critical loads that develop in short time periods
- By-passes for transmission switching equipment during repairs or maintenance
- Major storm restoration options
- Temporary switching arrangements to aid in expediting construction of substations and lines
- A "ready backup" for critical infrastructure facilities (i.e., government and public services)

All of these benefits contribute to the overall resiliency of individual substations as well as the resiliency of the transmission system.

9.12.8 Conclusions

1) The use of gas-insulated substation designs is an effective method to protect a substation its associated electrical equipment from severe damage from a naturally occurring catastrophic event or a nefarious human-caused physical intentional attack.
2) The use of gas-insulated substation designs is a cost-effective strategy that results in enhanced substation resiliency and higher reliability due to a reduction in substation area, perimeter, the ability to shield equipment and minimized exposure to external threats.
3) The benefits employing GIS designs in concert with GIS Enabled Construction Methods for use as a hardening tool and a response/recovery tool for substations, contributes to the overall resiliency of the transmission system.

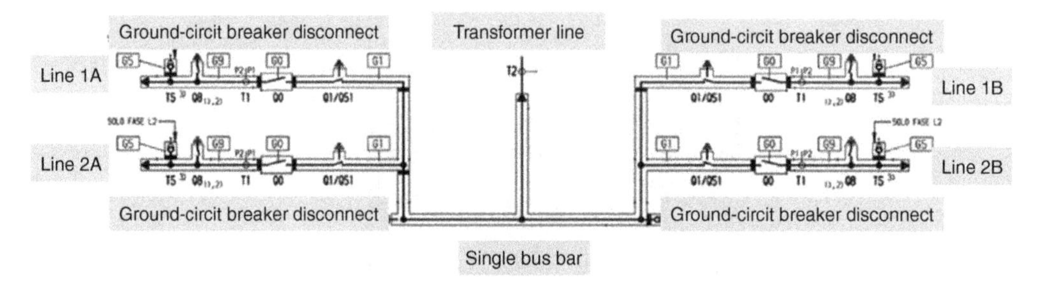

Figure 9.38 Electric circuit of a mobile GIS substation

Figure 9.39 Mobile GIS trailer (source [13])

References Section 9.12

1 Serrano R, Halper E (2014) Sophisticated but low-tech power grid attack baffles authorities. Los Angeles Times. Archived from the original on 2014-05-07. Retrieved 2018-01-09.
2 Sniper Attack On Calif. Power Station Raises Terrorism Fears. NPR. 5 Feb 2014
3 IEEE 1402-2021 - IEEE Approved Draft Guide for Physical Security of Electric Power Substations.
4 Electric Transmission Resiliency, NESC Summit 2015, Bob Bradisch AEP
5 Chris Land sea (1998). How does the damage that hurricanes cause increase as a function of wind speed?. Hurricane Research Division. Archived from the original on 2007-03-09. Retrieved 2007-02-24.
6 Shultz JM, Russell J, Espinel Z (2005) Epidemiology of Tropical Cyclones: The Dynamics of Disaster, Disease, and Development. Oxford Journal. Retrieved 2007-02-24.
7 Rappaport (May 2006). Inland Flooding. National Hurricane Preparedness Week. National Hurricane Center. Retrieved 2006-03-31.
8 Becker G, Boggess J, Mitchell M (2014) Storm & Flood Hardening of Electrical Substations, IEEE Conference Paper, PES T&D Conference and Exposition 2014.
9 National Oceanic and Atmospheric Administration (NOAA) website, www.noaa.gov
10 "Air Insulated Substation Design for Severe Climate Conditions" (April 2015) CIGRE Working Group B3.31.
11 White House Office of Science & Technology (August 2013) Economic Benefits Of Increasing Electric Grid Resilience To Weather Outages.
12 Public internet www.itdzar.blogspot.com
13 Public internet https://www.gegridsolutions.com/hvmv_equipment/gis.htm
14 US National Oceanic & Atmospheric Adminstration (NOAA), https://www.noaa.gov/
15 Backhaus K, Speck J, Hering M, Großmann S, Fritsche R (2014) Nonlinear dielectric behaviour of insulating oil under HVDC stress as a result of ion drift. Paper presented at *Inter'l. Conf. High Voltage Eng. Appl.*, Poznan, 8–11
16 Weedy B (1985) DC conductivity of voltalit epoxy spacers in SF6. IEE Proc. A – Phys. Sc., Measurement and Instrumentation, Management and Education – Reviews 132(7): 450–454

9.13 Vacuum High Voltage Switching

Author: Hermann Koch, Co-Author: George Becker
Reviewer: Dirk Helbig

9.13.1 Introduction

Vacuum switching technology is well known for decades and it is used in millions of load and fault interrupting switching applications. Use of very high vacuum far below the Paschen' Minimum of high voltage dielectric strength makes the vacuum switch a compact and powerful switching device. Vacuum interrupters for high voltage switching to replace SF_6 gas interrupters provide a technology that is simpler in design, requires less maintenance and does not produce toxic by-products of high energy interruption. The economical use of vacuum switching in high voltage applications above 1 kV was restricted to distribution voltage levels typically up to 20 kV for most applications. Above these voltage levels, the mechanical realization was limited, because of the increasing insulation distance of the contact system with increasing voltages. The large distance for the opening gap, increases linearly with the voltage above 20 kV and makes it impossible to interrupt the arc in a short time to avoid damage of the contact system [1–5].

For some applications of higher voltages, vacuum tubes have been connected in series with two interrupter units or even four to cover voltage level of up to 100 kV. The problem with this solution is the synchronization of the interrupters to open at the same time. Small time differences between each interrupter unit typically milliseconds, could destroy the "last vacuum tube" interrupting. Also, such solutions are much more expensive than available SF_6 circuit breaker equipment.

New designs of the vacuum technology were required to handle the high voltage levels of 72.5 kV and above. These developments will be explained in this section.

In addition to the technical advantages of vacuum interrupters for high voltage switching (i.e., simpler design, less maintenance and does not produce toxic by-products of high energy interruption), a second driver for replacement of SF_6 came from the high global warming potential (GWP) of SF_6. World-wide activities to replace SF_6 have been launched by the leading manufacturers of high voltage switchgear, at the request of the users of such equipment in order to reduce the CO_2 equivalent emissions of the power transmission system, increase the public acceptance and comply with the regulations find solutions for reducing the installed capacity of SF_6 on the electric power network. There are different solutions that have been announced by different manufacturers to address these regulatory pressures. One technical solution is based on vacuum switching technology and pressurized air for high voltage dielectric insulation.

In this chapter the advantages of high voltage vacuum switching technology will be presented, a historical overview will be given on how the vacuum technology developed over the time, where we are today and what can be expected for the future.

Modeling and simulation of physical solutions to solve the mechanical problem of insulation distances needed at higher voltage level are explained and principles of plasma flow and interruption processes are presented and simulation results are discussed.

The control of the switching operation and how the dielectric strength is built up after interruption is described. Design features to control the high voltage interruption plasma flow and the measures to reach the requirements for switchgear from the international standards are presented.

The limitation X-radiation emission of high-voltage vacuum interrupters is an important requirement to fulfill the given X-radiation limits and values for safe operation and personal protection. Stray radiation under testing and operating conditions compared with dose rate limits from the relevant standards will be presented and discussed in this chapter.

The major driver behind the technical development of high voltage vacuum switching technology for higher voltage levels greater than 145 kV, including the transmission voltages of 230 kV, 345 kV, 400 kV, and 500 kV is coming from regulatory agencies on users to find solutions for reducing the installed capacity of SF_6 in the electric power network.

Solutions based on the vacuum switching technology will allow new concepts of network operation for the long term and will reduce the use of technical insulating gases which may contribute to the global warming potential.

In this chapter information is given to understand the principle of vacuum switching technology and to give guidance on how it can be implemented into the electric power network.

9.13.2 Advantages

What are the main advantages of high voltage vacuum switching technology?

One of the primary advantages of vacuum switching technology is the high reliability of the equipment and the lack of interruption by-products. In principle, vacuum leaves no material in the interruption chamber to allow any chemical and physical reactions as compared to dielectric fluids such as air, oil, gas, or any gas mixture used to insulate the interrupting unit and quench the arc.

The vacuum required for high voltage, high current interruption is a very high vacuum and this very high vacuum needs to exist for extended periods of time in a hermetically sealed vacuum interruption chamber. The vacuum required is very close to zero pressure.

This high vacuum is a premise to eliminating any influence of decomposition products in the vacuum chamber, and with this to eliminated any aging process of the switching contacts and their capability. In practice, vacuum switching can perform tens of thousands of interruptions of rated currents without any reduction in performance.

The high-performance of vacuum switching technology is proven in millions of applications in the global electric power networks at distribution voltage levels up to 20 kV, for many years.

The switching performance of the vacuum interrupter makes it perfect for frequent switching applications. The excellent interrupting performance is proven for rated nominal current and short-circuit currents throughout lifetime of the vacuum circuit-breaker. Up to 30 short-circuit interruptions are possible without any need for maintenance.

Vacuum technology is perfect for low temperature application were other insulating gases have the limitations such as liquefaction of the insulating fluid. The only limitation for vacuum interrupters based on low temperature operation is the mechanical functionality of the driving mechanism at low temperatures.

Vacuum tube technology uses a design concept which is sealed for the life of the interrupter tube. The tube and the two metallic end plates are connected by a soldering process which provides lifelong gas tightness at the high quality required for the very high vacuum required by the vacuum switch. There is no maintenance required during the operating life of the vacuum tube, and the vacuum breaker tubes rigorously quality controlled during the manufacturing process.

The users of vacuum switching technology enjoy a high level of reliability, minimal maintenance and no requirement for spare parts for the interrupter.

With respect to CO_2 equivalent emissions which are accounted for in the calculation of global warming potential (GWP), vacuum switching technology has a calculated GWP=0.

The environmental benefits of reducing the GWP contribution coming from equipment used in the electric power system is one of the major driving forces for the technical development of vacuum technology for higher voltage levels for use in the distribution and transmission network. The transition of our energy system towards CO_2 neutrality and the related regulations from regulatory agencies in the form of documented directives and laws coming from national, regional, or international organizations, states, or countries, are one of the catalysts for the technical development of vacuum technology.

Certain rules for evaluating different high voltage equipment technologies are being included in the bidding process and will influence the use of design types in the future.

Combining vacuum technology for high current switching capability with technical air for compartments which only require insulation for dielectric integrity, and do not have any switching functions like rated and short-circuit interruption, disconnecting, and grounding/earthing offer a viable option to provide a GWP=0 for all components of the switchgear.

Vacuum switching technology has been used in electric networks for several decades starting in the 1960s, with millions of applications at the distribution voltage level up to 52 kV. During this time the performance has continuously improved with respect to switching capabilities and lower production cost. Recently, vacuum switching technology has entered the high voltage transmission voltage levels at 72.5 and 145 kV, with possible applications at the 245, 400 and 500 kV voltage levels in the future.

9.13.3 History of Vacuum Circuit Breakers

The history of vacuum circuit breaker technology started in the 1960s in parallel with other circuit breaker technologies. At the 47th CIGRE Session 2018 in Paris an overview graphic shows the progress of vacuum circuit breaker development, see Figure 9.40 [6].

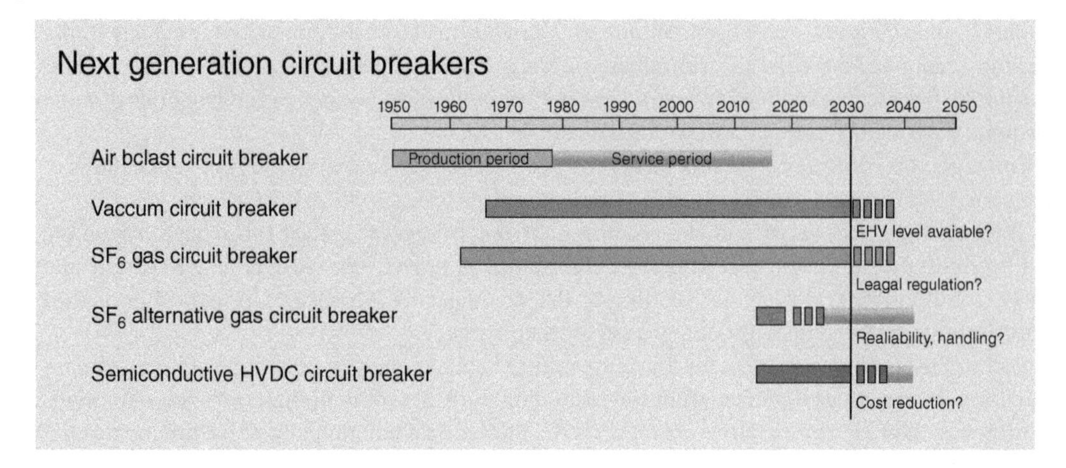

Figure 9.40 Next generation circuit breaker [6]

Not shown are the circuit breaker technologies of bulk oil or minimum oil circuit breakers which have been used since the 1920s and play a less important role in today's electric power network as time progresses.

Starting in 1950s with air blast circuit breakers, where the interruption of the arc was cooled by an air blast of ambient air, this technology is still used today in some installations. The switching capability is limited for rated current and short-circuit current due to the load of the contact system and the limited cooling properties of air. Also, the insulation capability is limited, such that multiple interrupter units have to be connected in series which makes synchronous operation of the different units important to avoid overload of individual breaker interrupting units. This is related to the low insulation capability of air.

In the 1960s the development of SF_6 gas-insulated circuit breakers and vacuum circuit breakers came into focus to provide higher current switching capabilities and operation at higher voltage levels.

The SF_6 gas-insulated circuit breaker technology offered higher current interruption capability at higher voltage levels of 100 kV and above. The SF_6 gas-insulated technology also provided a higher level of reliability and a compact design. The gas tightness level of the of SF_6 gas-insulated technology has developed over time, the leakage rates have decreased from 3–% to 0.5% to today's high-performance 0.1% per year during operation. The compact design of modern gas-insulated switching technologies offers single interrupters up to 500 kV and serves transmission voltage levels of 800 kV up to 1200 kV. Reliability and availability of the SF_6 gas-insulated circuit breakers is very high with statistical mean time between failures (MTFB) of more than 1000 years for 145 kV voltage class equipment. The operating cost has been reduced with first major maintenance activities occurring after 25 years of operation. In most cases, only a visual inspection required for preventive maintenance activities prior to 25 years of service. The expected life of modern gas-insulated switching equipment is 50 years for indoor installations and the practical use of the equipment can be longer for indoor applications with state-of-the art SF6 emission rates below 0.5. . .1%/a. The main downside of SF_6 technology is the high global warming potential of SF_6 which is 23 000 time higher than CO_2. This makes SF_6 equipment a target of regulatory agencies in many countries world-wide. Until 2020 most fluorinated gas regulations allow for exemptions for the use of SF_6 in high voltage equipment, because there are limited technical and economical alternatives available. First regulations plan to phase-out SF6 stepwise to achieve the CO2e emission targets.

The use of vacuum circuit breaker technology started in 1960s and was focused from the beginning on the distribution voltage levels of typical 5–38 kV. It was expected and proven that at these voltage levels, the vacuum technology offered more economical solutions compared to SF_6. Vacuum circuit breakers offer the same high current interruption capability for rated and short-circuit currents as do SF_6 circuit breakers.

The design principle for vacuum interrupters has very few parts (only two copper contacts inside the vacuum tube) provides for a high reliability and less (effectively no) maintenance requirements. The limitation of vacuum circuit breaker technology is associated with the limited dielectric capability based on the size of the interrupter. The classical design of the vacuum tube limited the dielectric insulation because of the increasing distance needed for safe, reignition free interruption. To overcome this limitation new design solutions were required to enter the extra high voltage levels of above 52 kV with only one interrupter unit. These new design solutions in addition to new developments designed to replace the SF_6 interrupter use in high voltage circuit breaker technology were started in the 2010s using alternative gas and using semiconductor HVDC circuit breaker technology. The alternative gas mixtures that being investigated are based on C4-Fluoronitrile and C5-Fluoroketone as the base dielectric gases using CO_2 as the carrier interrupting gas and O_2 as the anti-corrosive. Both C4-Fluoronitrile and C5-Fluoroketone have a GWP that is less than SF_6.

Except for reduced GWP, the relevant physical and electrical performance characteristics of existing and practical SF_6 alternatives, in their mixed form, are inferior in most respects to SF_6 – especially at maximum operating voltages greater than 145 kV. Extension of the basic electrical ratings demonstrated by alternative gases will require extensive research and development. The base gases of C4-Fluoronitrile and C5-Fluoroketone alternative mixtures do not recombine after disassociation because of high temperature arcs. This has to be compensated by design margins and adequate maintenance. There is no measurable improvement in safety related to the handling of SF_6 alternatives as compared to pure SF_6. New gas mixtures generate new arc by-products, and some compatibility testing with existing switchgear materials has led to acceptable solutions. Toxicity of both the base gas and arc by-products must be considered, and evaluation of these compounds for health and safety is paramount. The interrupting capability of switchgear using SF_6 alternatives is largely based upon CO_2 gas mixtures. CO_2 is orders of magnitude more permeable than SF_6, and the sealing technologies of switchgear will be affected.

SF_6 alternative equipment using "synthetic air" and vacuum interrupters operates at a slightly higher pressure (0.5. . .0.7 MPa relative) than SF_6 equipment. The static seals and dynamic seals presently used in switchgear cope with slightly the higher pressures. Vacuum interrupters in equipment using "synthetic air" for maximum operating voltages of 145 kV and below and interrupting ratings of 40 kA are in series production and operation since 2017. SF_6 free GIS with vacuum-interrupters and synthetic air insulation up to 170 kV and 50 kA were type tested in 2020 The design, manufacture, and test of reliable pressure vessels will continue to be an important consideration no matter what gas or gas mixture is employed.

The existing SF_6 density switches can be recalibrated to perform the same functions with the SF_6 alternative mixtures. All of the new gas mixtures operate at higher pressures as compared to SF_6. This presents two challenges: first, the design of pressure vessels that can operate safely and reliably for long periods of time; and second, customer acceptance of switchgear with higher operating pressures. Higher gas pressures can also mean higher leakage rates. New gases, gas mixtures and pressure schemes will require the development of multiple new gas handling systems. The new SF_6 free design has leakage rate of 0.1% and the slightly higher gas-pressure of 10–20% is accepted by operators [7, 8].

The semiconductor HVDC circuit breakers offer very fast interruption of rated or short-circuit currents as the semiconductor control operates in microseconds (μs). Compared to mechanical switching devices like SF_6 or vacuum breakers the semiconductor circuit breaker requires many more devices and needs well-configured software, which is a complex solution. Experiences are available from HVDC technology in principle, but for the circuit breaker function only this technology is still far too expensive.

9.13.3.1 History of Vacuum Switching Capabilities

Vacuum switching technology entered the medium voltage markets in the 1970s. The first serial manufacturing of vacuum interrupters for contactors started in 1971. The new technology proved as very reliable and the statistical failure rate for the mean time to failure (MTTF) was calculated to 68 500 years, based on several hundred-thousand installed vacuum interrupters, see Figure 9.41. More than 40 years of operational experience in the field of medium voltage applications and with more than 5.5 million vacuum interrupters installed (number covers the delivery of only one leading manufacturer), vacuum technology is well proven and accepted.

The advances of product and manufacturing technologies provided continuous improvement of switching capabilities with high current ratings of up to 100 kA in medium voltage up to and including 52 kV. In 2010, the development of 72.5 kV and in 2018, the voltage level of 145 kV are covered by vacuum interrupter using only one interrupter unit. The principle vacuum circuit breaker technology turned in the 1990s from manyfold pumping technology to the process used today, one-shot brazing technology. This is a major step in cost reduction, reliability of the manufacturing process and application for higher voltage levels, see Figure 9.41. The next voltage step of 245 kV with one vacuum interrupter unit has been presented at the CIGRE session 2018 in Paris. The potential for the full range of transmission voltage level classes up to 400/500 kV can be realized with vacuum switching technology. The 400 kV and 500 kV voltage level would use two interrupters rated 245 kV and one synchronized drive operator.

The vacuum circuit breaker technology for current ratings of up to and including 63 kA and voltage levels up to 245 kV are specified in accordance with the international standards IEC 62271-1 [8] and IEC 62271-100 [9] and have been certified by independent test laboratories like KEMA; CESI or PEHLA.

In Figure 9.42, the development steps of vacuum circuit breakers are shown. Based on the 15 kV-rated voltage, vacuum circuit breaker for 3150 A and 50 kA short-circuit interruption capability the up-grading development for higher-rated voltages was started. The KEMA type tested 15 kV version passed all type test requirements in 2001 and also provided a wide basis of installed vacuum circuit breakers in operation.

The next step of development was the 24 kV vacuum circuit breaker for 4000 A rated current and 100 kA short-circuit current type tested in 2015 and certificated by KEMA to fulfill all IEC

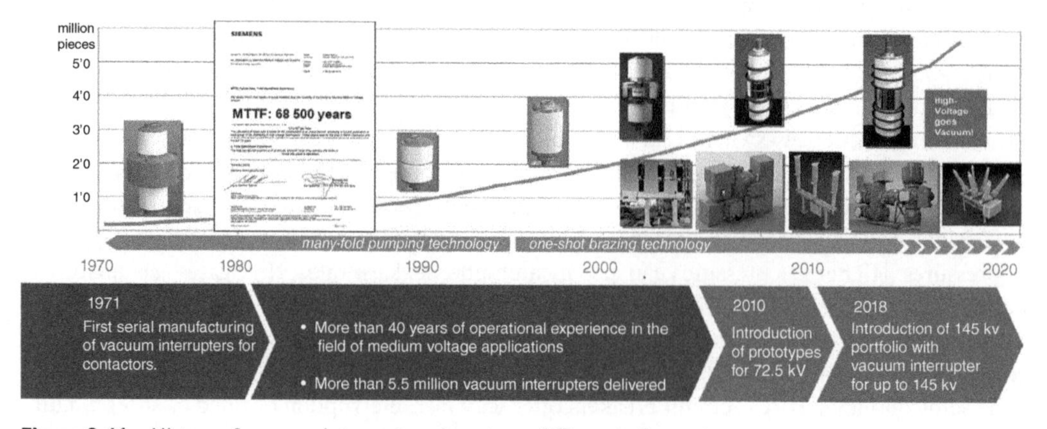

Figure 9.41 History of vacuum interrupters (courtesy of Siemens Energy)

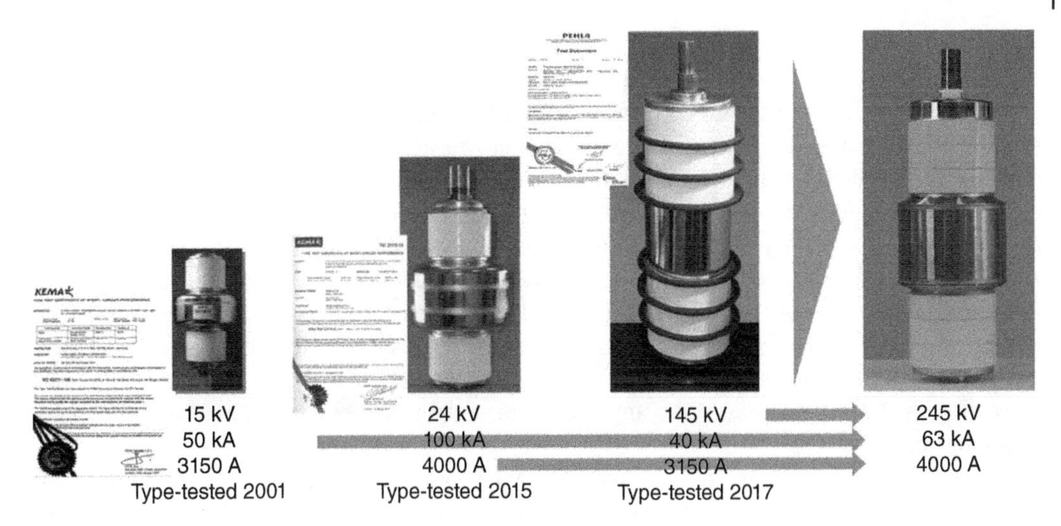

Figure 9.42 Vacuum circuit breaker development steps 15 kV up to 245 kV (courtesy of Siemens Energy)

62271-100 [9] requirements. Using the experiences made with the 24 kV version in operation the next development voltage step for vacuum circuit breakers was the passed type tests following IEC 62271-100 in 2017 with certification of PEHLA. The vacuum circuit breaker with only one interrupter switched 145 kV at a rated current of 3150 A and for short-circuit currents of 40 kA.

All good experiences made at each development step were used to set up the next voltage level design and bring in the proven technology of the 50 kA/3150 A contact system and 145 kV type tested ceramic insulation parts to be combined for the 245 kV rated voltage single interrupter vacuum circuit breaker in 2018, presented at the CIGRE Session in August in Paris.

The development steps from 15 kV as the base vacuum switching technology in 2001 took almost 20 years before the 245 kV voltage level could be presented. A long process for such a basic development and modeling and simulations were the key issues for the final success. Understanding the impact of so many parameters cannot be optimized with a trial-and-error approach in the test field, this would be too expensive and would take too long.

With experience and knowledge of the vacuum interruption process and the behavior of the arc during interruption having been studied in models and simulations, this provided important information for the vacuum chamber design. This will be discussed in the following Section 9.13.4.

9.13.4 Modeling and Simulation of the Vacuum Switching Electric Arc

To understand the theory and practical behavior of electric arc inside the vacuum circuit breaker, modeling, and simulations are required. The electric arc plasma flow, post-arc behavior, ejection of vapor and droplets are topics of the simulation studies. The simulation process employs two-temperature magneto-hydrodynamic (MHD) simulation of using the axial magnetic field plasma (AMF) flow [10].

Vacuum arcs with superimposed axial magnetic field (AMF) are well established for medium-voltage to stabilize the electric arc in its diffuse mode. The extension of the AMF technique to higher voltages levels of 52 kV and above is a challenging task. It requires advanced concepts of contact designs to establish adequate AMF amplitudes for large contact gaps. The influence of AMF on arc evolution has been studied intensively for large-gap contacts in the past. But no systematic studies have been reported so far, that combine arc observation with modeling and simulation of such plasmas.

9.13.4.1 Experimental Setup and Conditions

The experimental setup measures the signal of the contact travel time and the current amplitude of the 50 Hz synthetic test circuit with DC precurrent. A high-speed video camera attached to the vacuum chamber with optical access to the inside provides images of the electric discharge phenomena during the interruption process. The measured data is used for a radial AMF profile at current peak in mid-plane of fully opened contact gap. The locations of Helmholtz current loops are outside the vacuum chamber and are positioned coaxially to the AMF contact system. Adjustable AMF amplitudes are superposed to the AMF profile of the contact system by means of Helmholtz loops carrying currents in phase with the main current. The AMF profiles in the contact gap are implemented in plasma simulation as magnetic background field. The measurements showed constricted and unstable arc appearance for the standard AMF field without the Helmholtz current loops. Diffuse and stable arc appearance for enhanced AMF field have been seen with Helmholtz current loops.

The two-temperature magneto-hydrodynamic (MHD) approach of quasi-neutral arc plasma is used to set up the model. The model covered compressible, subsonic, and laminar two-fluid flow of electrons and multiply charged copper ions. The electron and ion fluids are modeled as ideal gases. Simulation solutions of conservation laws for plasma mass, plasma momentum, ion energy, electron energy is generated. The arc plasma column is coupled to the cathode surface by a self-consistent cathode spot model. Realistic AMF profiles are superposed as used in the experiment setup. Transient 3D simulations are performed at peak of 50 Hz current cycle for full contact gap length.

Ring structures in electric arc column and near the anode at a peak of 40 kA (rms) current cycle and the influence of the AMF on electric arc constriction at the first peak of the used 40 kA cycle current could be simulated in the model and are shown in the experimental test results. This combined experimental and theoretical studies of high-current AMF vacuum electric arcs in large-gap contact system provided the basic information for the design of vacuum circuit breakers above 52 kV.

The results of the experiments were proven by ideography and simulation of electric arc appearances and results showed agreement on a quantitative level. The results of the experiments and simulations permit the specification of electric arc constriction threshold design. The simulations based on the MDH model are capable to identify details of variations of properties inside the electric arc column. The MHD model is suitable to optimize contact designs for high voltage applications.

In Figure 9.43 the parameters of the vacuum circuit breaker for the two-temperature magneto-hydrodynamic (MHD) method is shown. The anode of the vacuum tube is at the top and the cathode at the bottom of the left graphic. The electric current flows from anode to the cathode in the model and is indicated by black lines. The structure of the contacts creates a spiral type of flow of the electric current, indicated by black lines. This will give a rotating force in the gap between the contacts and the plasma of the electric arc starts to rotate. This distributes the electric current over the surface of the contact system, a main reason for the higher ability of dielectric strength of the interruption chamber.

The axial magnetic field plasma (AMF) between the contacts describes the plasma flow in the gap.

The right graphic of Figure 9.43 shows a plot of streamlines of the ion velocity between the anode and the cathode. The plasma flow is in the opposite direction of the current flow starting from the cathode. Because of the contact structure and spiral rotation of the electric arc the ions are emerging from many locations of the surface of the cathode contact (bottom of the graphic) and concentrated into a concentrated plasma canal in the center.

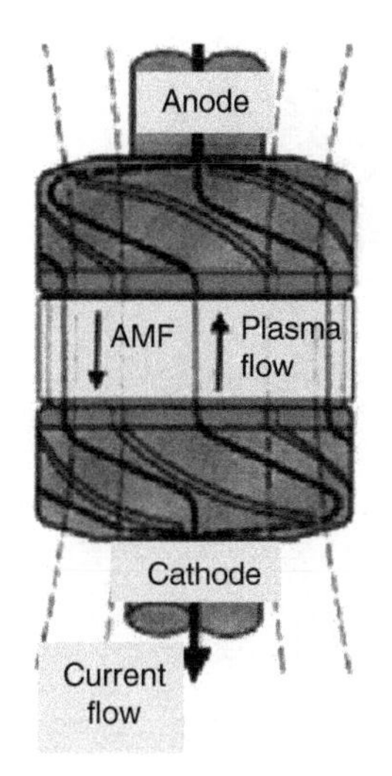

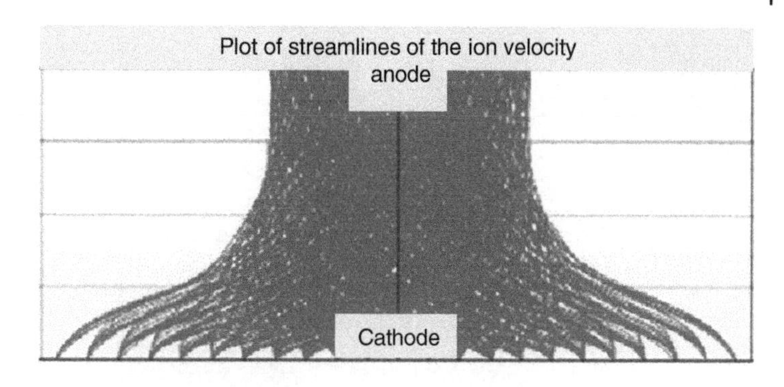

Figure 9.43 Vacuum circuit breaker modeling left: MHD model parameters right: streamlines of the ion velocity (courtesy of Siemens)

This is one result of the simulations based on the MHD model. The design of the structure of the contact system, the current flow through the contacts, the materials used for the contacts and the most important the material used for the surface of the contact have a strong impact of the distribution of the plasma flow during the interruption process. The plasma canal can be more diffuse or more concentrated and the impact on insulation capability and building up a dielectric strength for higher operating voltages is directly influenced by these parameters. The model and the simulation work have the task to vary all the parameters for finding an optimum. Thousands of variations as necessary. The design of the contact system and the choice of the right material for the contacts and their surfaces is the key issue in creating a vacuum interruption chamber for voltage above 52 kV.

9.13.4.2 Modeling Post-Arc Current

For higher-voltage applications above 52 kV, the voltage recovery and breakdown in post-arc plasma after current-zero is of great importance. Post arc period and voltage recovery is shown in Figure 9.44 in the left graphic. The current (I line) zero point is reached after 10 ms. The contact movement starts at 5 ms with the arcing time t_{arc} which ends at 10 ms with the current zero point. At 10 ms the dielectric withstand voltage (U_{diel} line) starts with 5 kV/μs, while the recovery voltage starts also at 10 ms but with only one tens of the speed with 0.5 kV/μs. This shows that the dielectric strength of the gap between the contacts after the arc is coming back faster than the recovery voltage builds up.

The modeling of post-arc current is shown in Figure 9.44 in the right graphic. The boundary sheath at the cathode, where the plasma flow starts is from great impact for the operation recovery voltage and the build-up of the dielectric strength between the anode and the cathode. The boundary sheath has a length (s) which generates a sheath voltage (U_s). Several boundary sheath can be defined in the plasma region.

The voltage withstand ability of the gap is reached after residual ions are collected by the transient recovery voltage (TRV) and neutral metal vapor and droplets are dissipated. The simulation of the dielectric strength after current zero is based on a physical modeling of voltage recovery and thermal failure in post-arc plasma. Modeling "gaseous" breakdown in metal vapor between arbitrary 3D metallic structures and along insulator surfaces are required to explain the interruption process and to find a design solution to fulfill the requirements of higher voltage applications above 52 kV [11].

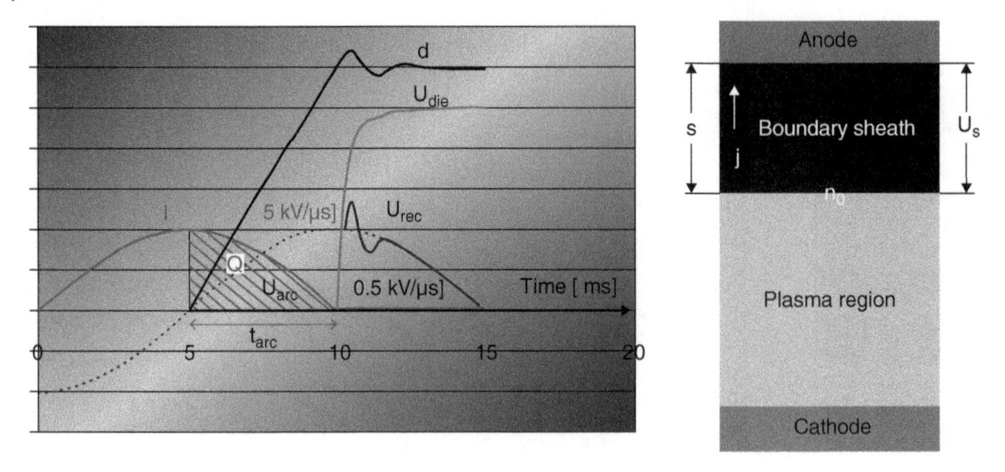

Figure 9.44 Modeling of the post-arc current left: post-arc recovery voltage, right: model for the post-arc current simulation (courtesy of Siemens).

The concept of vacuum switching is following the sequence of arc current interruption and voltage recovery. The vacuum arc ignition starts at contact separation. The first step is the rupture of molten metal builds a bridge across gap between the contacts which results in regions of metal vapor, and metal vapor arc (= "vacuum arc").

The second step is the arcing period and magnetic arc control. The vacuum arc is burning in the contact gap and is controlled by radial magnetic field (RMF) or the axial magnetic field (AMF).

The third step is the post arc period and voltage recovery. The voltage withstand ability of the gap is reached after residual ions are collected by transient recover voltages (TRV). Neutral metal vapor and droplets are dissipated in the vacuum chamber.

DSMC Simulation of Vapor Expansion

The Direct Simulation Monte Carlo (DSCM) is used to simulate the vapor expansion within the vacuum interrupter (VI) unit.

Contact stroke and current amplitude have a strong impact on the amplitude and the spatial profile of the axial magnetic field (AMF) flux density between the contacts. And with this, the switching performance of the VI unit is defined. Nevertheless, the electric arc has to be held in a diffuse mode for all possible combinations of current amplitude and contact gap or arcing time, respectively – without commutation to the chamber wall or the metal vapor shields of the VI unit. In order to assure that the studied contact system controls every possible combination, AC currents of up to 50 kA for different contact gap lengths at current peak [12].

From the simulation data of dielectric tests, we conclude that intelligently designed insulating components and field controlling shields of the vacuum tube are key dielectric characteristics of a vacuum interrupter unit enabling excellent LIV and AC withstand performance at high voltages. We find that properly designed metal shields inside the vacuum tube enable sufficient protection of the inner insulating surfaces against the impact of metal vapor expansion and droplet emission of the vacuum arc [13].

In Figure 9.45 the DSMC simulation of vapor expansion is shown. On the left the interactor history of adsorption levels 0, −1, −2 and −3 for a current of 3000 A. On the right the same adsorption levels of 100 kA current.

The ejection of metal vapor and droplets are modeled for neutral vapor expansion and droplet propagation into vacuum. It shows the metal vapor adsorption and reflection on cool, solid surfaces.

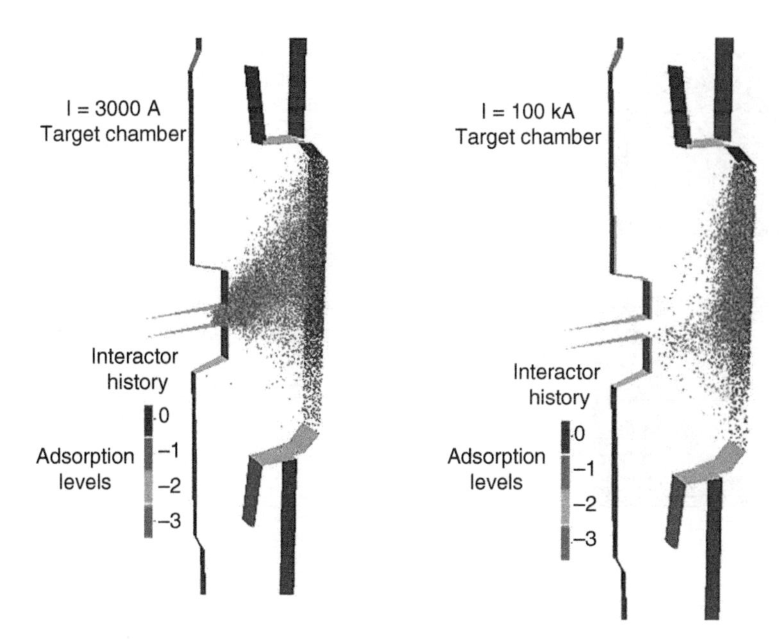

Figure 9.45 DSMC simulation of vapor expansion left: adsorption level of 3000 A current, right: adsorption level of 100 kA current (courtesy of Siemens)

From the data of dielectric tests, it is concluded that intelligently designed insulating components and field controlling shields of the vacuum tube are key dielectric characteristics of a vacuum interrupter unit enabling excellent lightning impulse (LIV) and power frequency withstand voltage performance at high voltages. The properly designed metal shields inside the vacuum tube enable sufficient protection of the inner insulating surfaces against the impact of metal vapor voltages expansion and droplet emission of the vacuum arc [15].

9.13.5 Control of Switching and Dielectric Strength

The control of switching electric arc in high-voltage vacuum interrupters needs an optimization of magnetic field and the travel curve of the contacts during opening of large gaps. The arc plasma in large gap creates an axial magnetic field between the contacts leads to a concentrated plasma canal which is stressing the contact material and will cause damage to the contact surface. An insufficient axial magnetic field (AMF) profile will lead to an unstable and constricted arc plasma as shown in Figure 9.46 for an arc current of 28 kA [14].

The suitable axial magnetic field (AMF) profile for a stable and diffuse arc plasma (50 kA) is shown in Figure 9.47. The arc discharge between cathode and anode is diffuse and with this the maximum temperature on the contact surface is reduce which lead to less stress to the contact material and a better performance of the vacuum interrupter.

The travel curve of the contact system during opening needs to be adjusted to switching operation and to provide the dielectric gap for safe insulation after interrupting the current. The travel curve can be described in three phases: the fast opening, the slow opening and the arbitrary opening over the time. In the fast-opening sequence, the switching arc will see a fast increase of distance between the contacts, which may vary in time, upper and lower curves in the diagram. In the slow-opening sequence, the switching arc is still standing but no more distance between the contacts is developing. In the third sequence of arbitrary-opening speed, the final dielectric gap related to the basic insulation level (BIL) will quench the arc by transient recovery voltage (TRV) effect in

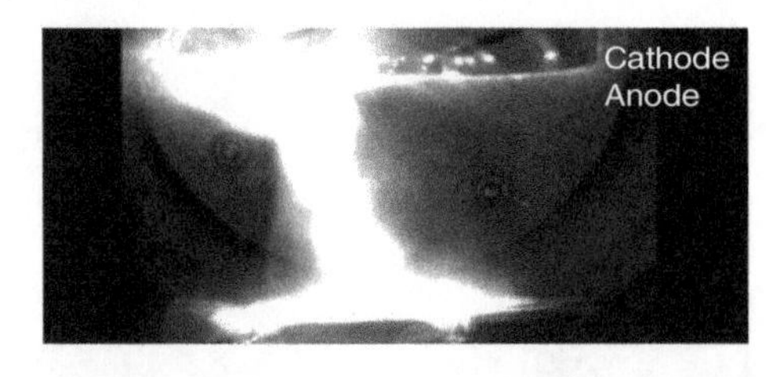

Figure 9.46 Insufficient axial magnetic field profile at 28 kA with an unstable and concentrated arc plasma (courtesy of Siemens Energy)

Figure 9.47 Suitable axial magnetic field profile at 50 kA with a stable and diffuse arc plasma (courtesy of Siemens Energy)

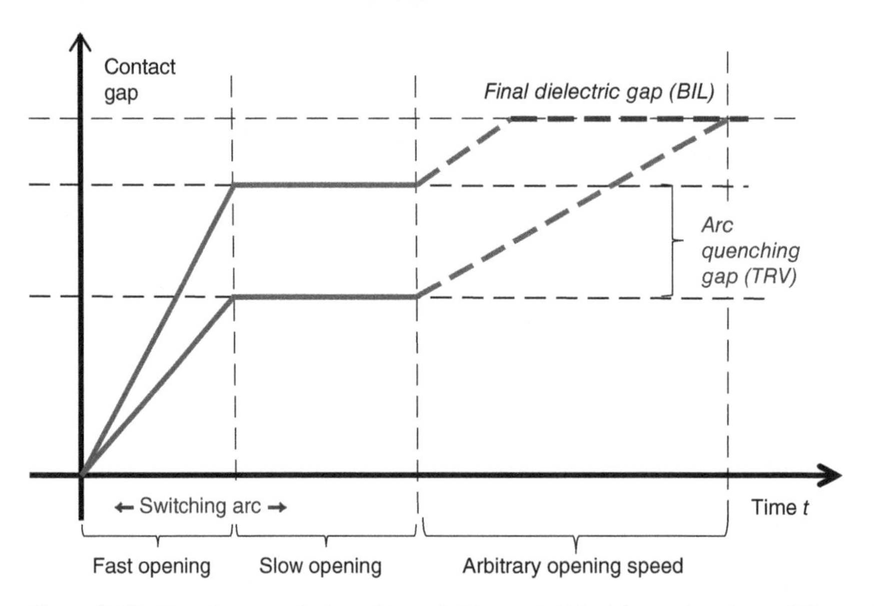

Figure 9.48 Travel curve adjustment to switching and dielectric gap (courtesy of Siemens)

the vacuum tube [15]. This complex functionality is required to interrupt higher voltage levels of above 52 kV with only one vacuum chamber for the safe and reignition free circuit breaker function. The design of the driving mechanism of the vacuum interrupter unit is from great importance of current switching capability of the vacuum tube. This is shown in Figure 9.48.

The dielectric strength of high-voltage vacuum interrupters requires adequate voltage insulation inside and outside the tube, with open contacts. The international standards [16] demand dielectric power withstand voltages and lightning impulse voltage as shown in Table 9.12.

The control of dielectric breakdown channels of the vacuum interrupter tube and the contact system is influenced by the breakdown of the discharge channels, the breakdown between the metal parts, the breakdown within the ceramic tube and the surface breakdown at the ceramic tube. A complex process which needs optimization. The breakdown of the discharge channels of the electric arc plasma inside the vacuum interrupter is one channel. The second channel is outside the interrupter tube in pressurized dry air. Both need to fulfill the electric arc interruption

Table 9.12 Power frequency withstand voltage (AC) and Lightning impulse voltage (LIV)

Demanding dielectric standards	145 kV rated voltage	170 kV rated voltage
Power frequency withstand voltage (AC)	275 kV	325 kV
Lightning impulse voltage (LIV)	650 kV peak	650 kV peak

successfully. The volume breakdown includes the metal parts of the contact electrodes, the chamber wall and the vapor shield inside the vacuum tube. The second volume breakdown is within the ceramic bulk. The surface breakdown along the ceramic insulation of the vacuum tube delivers the fourth parameter for the dielectric breakdown process which need to be managed.

9.13.6 X-radiation Emission

Most of us are travelling by plane! What does this mean in relation to possible exposure with X-radiation for the passengers? In average of all passengers, we will receive 6 μSv/h, with a minimum of 4 μSv/h and a maximum of 7 μSv/h.

Why do we discuss the topic of X-radiation emission in the context with vacuum interrupters? X-rays are produced when high energetic electrons interact with matter. A fraction of the kinetic energy of the electrons is converted into electromagnetic energy (Bremsstrahlung/braking radiation). Vacuum Interrupters in open position with applied high-voltage offer the possibility to emit X-radiation! The vacuum interrupter is a stray radiation source because the possible X-radiation is a by-product of the circuit breaker operation.

The basis of all possible X-radiation is electron emission (field emission) and the theoretical description is given by the Fowler Nordheim equation:

$$j_{FE} = C_1 \cdot (\beta \cdot E)^2 \cdot e^{-\frac{C_2}{(\beta \cdot E)}}$$

The power of the emitted X-radiation is dependent on the field emission current and the applied voltage:

$$P \sim I_{EF} \cdot U^2$$

Consequently, the equivalent dose rate shows an U4 dependency on voltage and a 1/g2 dependency on contact gap:

$$\frac{dH}{dt} = \frac{a_1}{R^2} \cdot \beta^2 \cdot \frac{1}{g^2} \cdot U^4 \cdot e^{\frac{-a_2 \cdot g}{\beta \cdot U}}$$

Values and limits of X-radiation emissions of high-voltage vacuum interrupters are presented in Figure 9.49. Stray radiation under testing and operating conditions are compared with dose rate limits from the relevant standards.

The graphic shows background radiation of 2.1 μSv/h as a dashes line. Measured values of x-radiation are shown as cycle for new vacuum interrupters and as triangle for used vacuum interrupters. The limiting values given in IEC 62271-1 [8] of 5 μSv/h for 145 kV equipment and 150 μSv/h for 275 kV are shown as line in the graphic.

The 145 kV vacuum interrupters are a potential source of stray radiation (X-ray), if voltage is applied to open contacts.

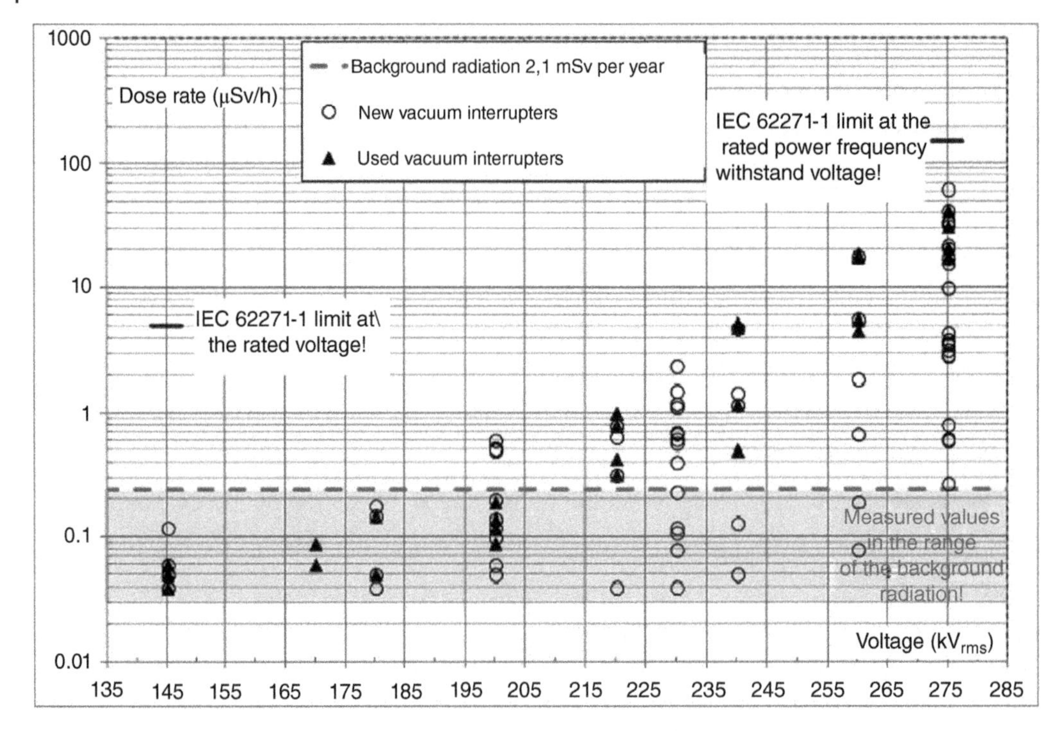

Figure 9.49 X-radiation emission of high-voltage vacuum interrupters (courtesy of Siemens Energy)

The x-radiation measurement of a 145 kV vacuum interrupter is measured with 1 μSv/h at 0.1 m distance which shows conformity certificate for Germany ("Störstrahler") from PTB ("European Council Directive 2013/59/EURATOM, 2013/12/05") [19] limit for AC voltage. The value is below the IEC standard equipment of 5 μSv/h.

The measure values of new and used vacuum interrupters of 245 kV rated equipment also meets the IEC standard requirement of below 5 μSv/h for the 145 kV rated equipment and is far below the limiting values of 150 μSv/h for 275 kV equipment required by IEC 62271-1, [17].

Vacuum interrupters are a source of stray radiation when a sufficient high voltage is applied to open contacts. In this case, the dose rate is in dependent on the applied voltage (U_r) and the energy of the contact stroke ($1/g^2$). Not only the main contact system contributes to the overall X-radiation emission but also the shield grading elements of the vacuum interrupter.

The measurements do not show any differences in X-radiation behavior for new and prestressed vacuum interrupters (the measurement points are within the same scatter band). For 145 kV and 245 kV vacuum interrupters, the measured X-radiation emissions are in the limits given by IEC standards [8].

9.13.7 Vacuum Switching at Rated Voltages up to 550 kV and Beyond

The use of single interrupter units for vacuum circuit breakers up to 245 kV covers the requirements of the lower half of transmission voltage levels. The upper half of the transmission voltages of 300 kV, 400 kV, 500 kV, and 800 kV may be addressed in the future, however it is not foreseen that these high voltage level switching devices will be designed and developed using only one interrupter unit. It is anticipated that a series connection of vacuum interrupter units will be required to achieve switching device designs 300 kV and above. Basic experiments on voltage distribution in dead-tank circuit breaker test set-ups with for 550 kV have been carried out and have produced the following results.

The voltage distribution over both interrupter units has been realized with grading capacitors of a value of C = 825 pF. The measured arcing times of the two interrupters are given in Table 9.13. The vacuum interrupter unit VU1 interrupts after 2.57 ms and VI2 interrupts after 2.27 ms. This results in a voltage distribution after the current zero ($i = 0$) of 50.5% for VI1 and 49.5% for VI2, as shown in Table 9.14.

The investigations showed that with longer arcing times' similar voltage distributions are found. Tests were performed with 145 kV/40 kA vacuum interrupters.

The voltage and current diagram in Figure 9.50 show the time sequences of an interruption. Two single-break vacuum interrupter units designed up to 245 kV and 63 kA are connected in series. They constitute the basic building block to approach rated voltages of 550 kV and in future may be beyond. The arc ignition starts at 9 ms in VI1 and VI2 with only 0.03 ms difference. The current zero point is reached at 12 ms at time (t).

Table 9.13 Arcing times of vacuum interrupter (VI) units

Vacuum Interrupter unit	t_{arc} [ms]
VI1	2.57
VI2	2.27

Table 9.14 Voltage distribution at time t after the current zero ($i = 0$)

Vacuum Interrupter unit	%
VI1	50.5
VI2	49.5

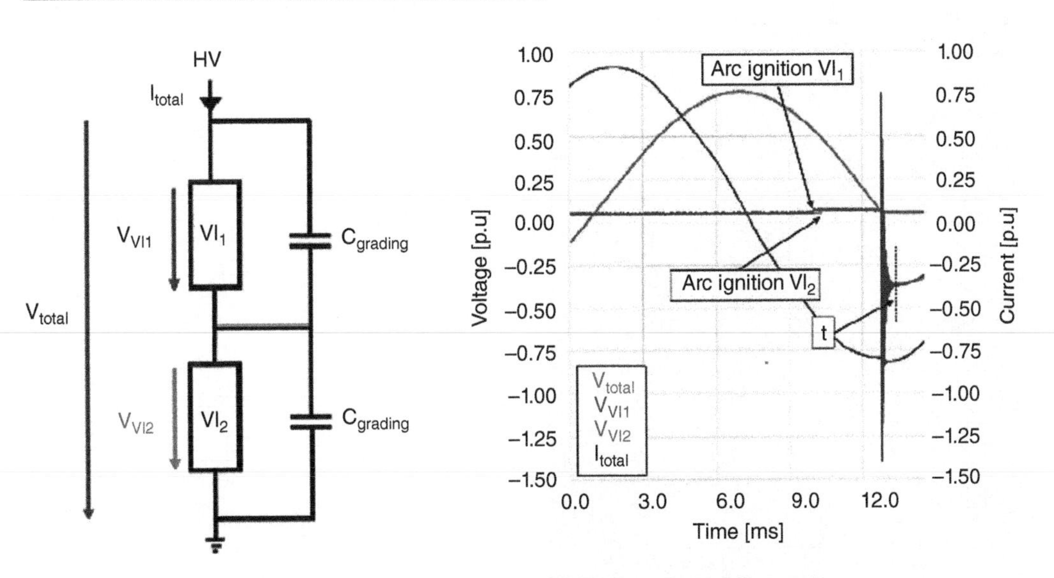

Figure 9.50 Vacuum switching at rated voltages up to 550 kV (courtesy of Siemens)

9.13.8 Market Potential

Mitigation of possible human-caused contributions to climate change is a global goal that is documented in the protocol of the UN Climate Conference in Paris in 2015. Member nations of the Paris Climate Accord agreed to attempt to achieve a limit of anthropogenic global warming temperature increase to under 2° centigrade. Any perceived contribution to increase the global warming potential has been brought into question. SF_6 with its high GWP value of 23 000 times the CO_2 equivalent has been brought into focus even though, the absolute quantity SF_6 emitted is as low as 0.1% of all sources of CO_2 equivalent emissions.

The global warming potential (GWP) neutral design of high voltage switchgear can use technical air for insulation and can meet dielectric requirements with little increase of dimensions for designs up to 145 kV with an interrupting capability of 40 kA.

The next challenge is the higher voltage and higher current interrupter designs for GWP neutral equipment. Vacuum seems to be a viable solution. The challenge for switching technologies at high currents and high voltages has been potentially solved and undergoing testing for 245 kV rated voltage with 63 kA short-circuit interruption current and 170 kV rated voltage and 50 kA short-circuit interruption current, see Figure 9.51.

The vacuum circuit breaker is used in gas-insulated switchgear (GIS) replacing the SF_6 circuit breaker. It combines high-tech performance, easy operation, and sustainability with current breaking and making capability in vacuum. Dielectric insulation in GIS is using "technical grade" air, which is a mixture of N_2/O_2 or N_2/CO_2.

First orders started 2019 for this type of equipment in the USA, Norway, Denmark, Germany, UK and Spain using vacuum circuit breakers for climate-neutral switching of high currents at high voltages have been placed [2–4, 16] (Figure 9.52).

The advantages for GIS using vacuum circuit breakers are the high reliability and high performance of the interrupters for switching operations. The operation of the switchgear is no different from SF_6-insulated switchgear. The vacuum interrupters require minimal maintenance over their

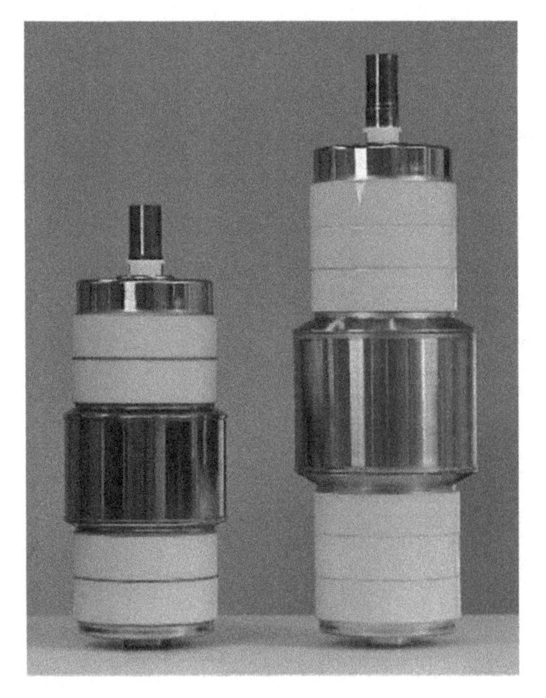

Figure 9.51 Vacuum interrupters Rated Voltage left: 170 kV/right: 245 kV and short-circuit current left: 50 kA/right: 63 kA (courtesy of Siemens Energy)

Figure 9.52 GIS for 145 kV with vacuum circuit breaker and technical air insulation (source [18])

life. The switchgear uses no greenhouse gas and can adhere to national and regional climate change regulations.

Market requirements are the basis of economic development of innovations into market ready products. The research and development costs of SF_6 alternative innovations must be balanced with the appetite of the market for the innovations. Therefore, if the cost of the research, development and production costs of an SF6 alternative technology outweighs the potential revenue from the market appetite for the product, then the product will not become market ready.

9.13.9 Conclusions

The advantages of high voltage vacuum switching technology is explained. Long operational experiences with millions of vacuum circuit breakers in service, world-wide and in many different applications prove the vacuum switching technology to be a very reliable switching technology.

A history overview of the technical development of vacuum circuit breakers since 1970s shows the gradient steps of development with increasing operational voltages, rated currents and short-circuit interruption capability.

There have been several circuit breaker technologies developed since the 1950s starting with air blast circuit breaker technology. In the 1960s, the SF_6 circuit breaker and vacuum interrupter technologies entered the development and went into production. Vacuum interrupters have been applied at the distribution voltage level up to 40 kV. SF_6 interrupters have been applied more predominantly at the transmission voltage levels above 100 kV. In the 2010s, development of alternative gases for circuit breakers and high voltage DC switches using semiconductors entered the circuit breaker market.

Vacuum switching capabilities developed over time to higher voltage levels and higher current ratings. The history of vacuum circuit breakers shows the steps of development starting with 15 kV in 1971 up to 245 kV introduced in 2018 at the CIGRE Session in Paris.

Modeling and simulation of the vacuum switching electric arc process is a key element to understand the functionality and physics behind vacuum switching. To many parameters influence the design of a vacuum circuit breaker which cannot be solved by testing and measurements only. Without the principle understanding of all effects the optimized design cannot be achieved.

Modeling of the post-arc current for the post-arc transient recovery voltage requires a model for the post-arc current simulation. The vacuum circuit breaker modeling using the two-temperature magneto-hydrodynamic (MHD) simulation to simulate the axial magnetic field plasma (AMF) flow inside the vacuum interrupter gives important information for the design of the contact system and shielding plates. The MHD model parameters show the streamlines of the ion velocity during the interruption process and gives information on how to design the interrupter for a diffuse plasma flow.

The Direct Simulation Monte Carlo (DSCM) is used to simulate the vapor expansion within the vacuum interrupter (VI) unit. Vapor expansion is an effect inside the vacuum interrupter where metal vapor is contributing to the destabilization of the dielectric strength of the gap between the contact system. This effect is used to increase the switching voltage and to reach the reignition free vacuum circuit breaker function at voltage above 52 kV using only one interrupter.

The control of switching processes in terms of regulating the right speed of the contact system and the forces of the operating mechanism is a key feature to reach dielectric strength in the switching gap between the contacts. This is reflected in the axial magnetic field. An insufficient axial magnetic field profile will lead to an unstable and concentrated arc plasma generating a constricted arc column and will reduce the switching capability and the dielectric strength within the gap of the contacts. A suitable axial magnetic field profile, for example, at 50 kA will lead to a stable and diffuse arc plasma in the switching gap between the contacts and with this an improved high switching capability and a high dielectric strength.

Time and travel curve adjustments of the switching process of speed and force are important for the strong dielectric gap of the contacts after the electric arc is extinguished. Power frequency withstand voltage (AC) and lightning impulse voltage (LIV) applied to the vacuum circuit breaker are required by international standards and reflect the requirements from electric network.

X-radiation emission of high-voltage vacuum interrupters is related to the voltage level and emissions from open contacts and is limited by values given in IEC 62271-1. The higher voltage level the higher the x-radiation. For 145 kV and 245 kV, the measured values for vacuum circuit breakers are below these limiting values.

Vacuum switching at rated voltages up to 550 kV and beyond is now under investigation using two 245 kV vacuum interrupter units in series. In this case, the arcing must be equally distributed across the two interrupter units to synchronize the arcing times of each. In a dead-tank circuit breaker test set-up with two medium-voltage vacuum interrupters (AMF contacts) in a V–layout, the arcing times of the two vacuum interrupter units were measured with a split of 2.57ms and 2.27 ms; while the voltage distribution for each interrupter unit was measured with a split of 50.5 and 49.5%.

Market requirements are the drivers for each technical development. In this case, one of the market drivers is the perceived contribution of human-caused climate change activities to global warming potential. New designs vacuum interrupters of 170 kV rated voltage and 5 kA short-circuit current or 245 kV rated voltage and 63 kV short-circuit current switching capability are being brought to the market. Market requirements are the basis of economic development of innovations into market ready products. The research and development costs of SF_6 alternative innovations must be balanced with the appetite of the market for the innovations. Therefore, if the cost of the research, development, and production costs of an SF_6 alternative technology outweighs the potential revenue from the market appetite for the product, then the product will not become market ready.

References Setion 9.13

1 CIGRE (2014) Technical Brochure 589: The Impact of the Application of Vacuum Switchgear at Transmission Voltages, Cigre, Paris.

2 Kuschel M, et al. (2020) First 145 kV / 40 kA gas-insulated switchgear with climate-neutral insulating gas and vacuum interrupter as an alternative to SF6, Design, Manufacturing, Qualification and Operational Experience; Paper B3-107, 48th CIGRE Session Paris.

3 Kuschel M., et al. (2020) First 170 kV / 50 kA GIS with Clean Air and Vacuum Interrupter Technology as a Climate-neutral Alternative to SF6, Paper A3-301, 48th CIGRE Session Paris.

4 Nikolic T, et al. (2020) Basic aspects of switching with series-connected vacuum interrupter units in high-voltage metal-enclosed and live tank arrangements, Paper A3-112, 48th CIGRE Session Paris.

5 Shen W et al. (2020) Environment, Health and Safety Aspects of Gas-Insulated Electric Power Equipment Containing Non-SF6 Gases and Gas Mixtures, Paper C3, PS2, 48th CIGRE Session Paris.

6 Ito H (2019) CIGRE Greenbook: Switching Equipment. Future Interrupter Technologies. Study Committee A3: High voltage equipment, Springer International Publishing

7 Becker G (2020) Alternative Insulating Fluids to SF6 Gas Status and Strategy Considerations. IEEE Conference Paper, PES T&D Conference and Exposition 2020.

8 IEC 62271-1 Edition 2.0 (2017) High-voltage switchgear and controlgear, Part 1: Common specifications for alternating current switchgear and controlgear.

9 IEC 62271-100:2008, Edition 2.0 (2008), High-voltage switchgear and controlgear, Part 100: Alternating current circuit-breakers.

10 Wenzel N, Lawall A, Schümann U, Wethekam S (2014) Combined experimental and theoretical study of constriction threshold of large-gap AMF vacuum arcs, XXVI International Symposium on Discharges and Electrical Insulation in Vacuum, September 30th, 2014, Oral Session 4, Paper B3_O_74, Mumbai, India.

11 Wenzel N (2015) Physical Modelling and Numerical Simulation of Vacuum Switch Arcs, DPG-Frühjahrstagung 2015 – Plasmaphysik, Bochum, March 2 – 5, 2015

12 Giere S, Helbig D, Koletzko M, Kosse S, Rettenmaier T, Stiehler C, Wenzel N (2018) Vacuum interrupter unit for CO2-neutral 170kV/50kA switchgear – Vakuumschaltröhre für CO2-neutrale 170kV/50kA Schaltanlagen, VDE Hochspannungstechnik, Berlin.

13 Smeets RPP, Wiggers R, Kuivenhoven S, Bannink H, Chakraborty S, Sandolache G (2012) The Impact of Switching Capacitor Banks with Very High Inrush Current on Switchgear, CIGRE Conference, paper A3-201

14 Wenzel N, Geisler AE, Goebel PT, Nikolic G, Giere S, Kuschel M, Kosse S (2019) XXIII Symposium on Physics of Switching Arc, Nové Město na Moravě, Czech Republic, September 9–13, 2019.

15 Heinz T, Giere S, Wethekam S, Koletzko M, Wenzel N (2018) Control of Vacuum Arcs in HV Vacuum Interrupters by Suitable Stroke Trajectories of Opening AMF Contacts, Proc. 28th ISDEIV, Greifswald, Germany.

16 Helbig D, et al. (2020) First CO2-neutral 145 kV and up to 63 kA Dead Tank Circuit Breakers based on Vacuum Switching and Clean Air Insulation Technology, Paper B3-106, 48th CIGRE Session Paris.

17 Giere S, Heinz T, Lawall A, Stiehler C, Taylor ED, Wenzel N, Wethekam S (2018) X-Radiation Emission of High-Voltage Vacuum Interrupters: Dose Rate Control under Testing and Operating Conditions, Proc. 28th ISDEIV, Greifswald, Germany.

18 Gas-insulated switchgear. Portfolio. Siemens Energy Global. siemens-energy.com.

19 Directive 2013/59 - Basic safety standards for protection against the dangers arising from exposure to ionising radiation, and repealing Directives 89/618/Euratom, 90/641/Euratom, 96/29/Euratom, 97/43/Euratom and 2003/122/Euratom.

9.14 Low Power Instrument Transformer

Author: Hermann Koch
Co-Author: Mark Kuschel
Reviewers: Arnaud Ficheux, Robert Lüscher

9.14.1 Introduction

The Low Power Instrument Transformers (LPIT) offer new performances of gas-insulated switchgear assemblies (GIS) replacing conventional instrument transformers on the electro-magnetic principle. On basis of linear sensors and digital electronic LPIT are integrated into the protection and automation system of the high voltage substation and offer additional features needed in a smart grid world.

The LPIT development started in the 1990s and has reached a development stage for wider use in the power transmission and distribution network. References will show examples of applications.

9.14.1.1 Advantages of Low Power Instrument Transformers

One first advantage of low power instrument transformers over conventional instrument transformers is the reduced size and weight. Weight reduction of conventional CTs and VTs can reach 1.500 kg on one bay of 145 kV GIS and a reduced footprint of a bay up 40%.

LPIT are environmentally friendly do not need additional gas fill and contribute with less CO_2 global warming potential.

LPIT provide an improved measurement performance. The current and voltage measurement is linear, does not have saturation effects, contributes with no magnetic losses and electro-magnetic forces, no ferro-resonance can occur and it performs with a higher-frequency bandwidth of measured current and voltage signals.

LPIT provides simplified engineering and approval processes. One standardized multi-purpose device covering all possible current ratings, requires less adaptations in hardware. No CT rating approval and no multiple CT cores are necessary. This allows higher flexibility with regard to late changes of the GIS configuration.

LPIT are easy and safe to connect with fewer and thinner cables, easier cable works, smaller cable ducts. LPIT are usually connected by an Integrated Electronic Device (IED) direct to the optical fiber communication system of the bay. This reduces the risk of misconnection, no need for disconnection during tests, improved personal safety, because of no hazardous overvoltage at open terminals.

LPIT provide simplified logistics and fast spare part delivery with shorter delivery times, off the shelf spare parts. It is robust and durable because of only passive components inside the partition, sealed for lifetime and maintenance free.

9.14.2 Basics of Current and Voltage Instrument Transformer

Conventional current and voltage transformers use high output signals. The current transformer typically provide 1 A output signal and the voltage transformer produces typically 100 V.

Low power instrument transformers operate at much lower levels. LPIT are current and/or voltage measurement devices which do not provide a significant output power (>=2.5 VA).

9.14.2.1 Current Measurement

Figure 9.53 shows the principles of measurement for conventional and low power instrument to measure the current. Conventional methods use inductive principles with iron core and high-power output. The signal for the measurement is generated by hall effect or with a shunt. Low power instrument transformers use the iron free Rogowski coil, a inductive coupling with iron coil but lower power output or the optical Faraday effect.

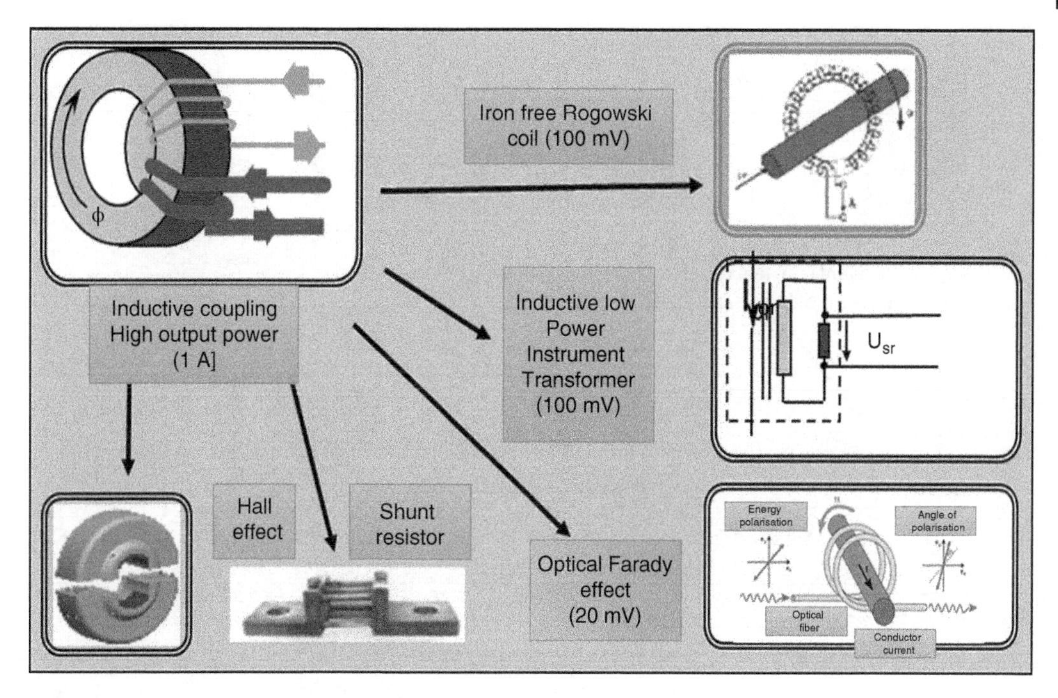

Figure 9.53 Principle measuring method of conventional and low power instrument transformers to measure content (*source* [1–3])

9.14.2.2 Voltage Measurement

Figure 9.54 shows the principles of measurement for conventional and low power instrument to measure the voltage. Conventional methods use inductive or capacitive principles with high-power output. The voltage signal for the measurement is generated by inductive coupling of coils with iron core or a capacitance with high power output. Low power instrument transformers use R-divider, C-divider or RC-divider with low power out or for optical measuring method the Pockels effect [9].

9.14.3 Low Power Current Sensor with Rogowski Coil

The Rogowski Coil is an electrical device for measurement of an alternating current (AC). It is sensitive to magnetic field generated by the flow of AC current in the conductor. It consists of a helical coil of wire with the lead from one end returning through the center of the coil to the other end, so that both terminals are at the same end of the coil [10].

There is no metal (iron) core and therefor no saturation. The secondary output signal is voltage [mV]. Typically, 60 mV for 1 kA of current flow in the conductor.

The output signal is depending on the area of the loop (A), the number of turns, the circumference of the coil and the specific magnetic constant (μ_0), as shown in the formula in Figure 9.55.

The connection box includes inductances and capacitance for signal filters and surge arresters for voltage limitations at the output signal. The secondary cable shall be less than 100 m before connected to the electronic merging unit located inside the bay control cubicle.

The electronic device of the merging unit connects the measure current signal as a digital communication protocol to station or process bus.

The Rogowski coil produces a linear voltage related to the current in the conductor over a wide range. It can measure rated currents from amperes up to short-circuit currents of some thousand

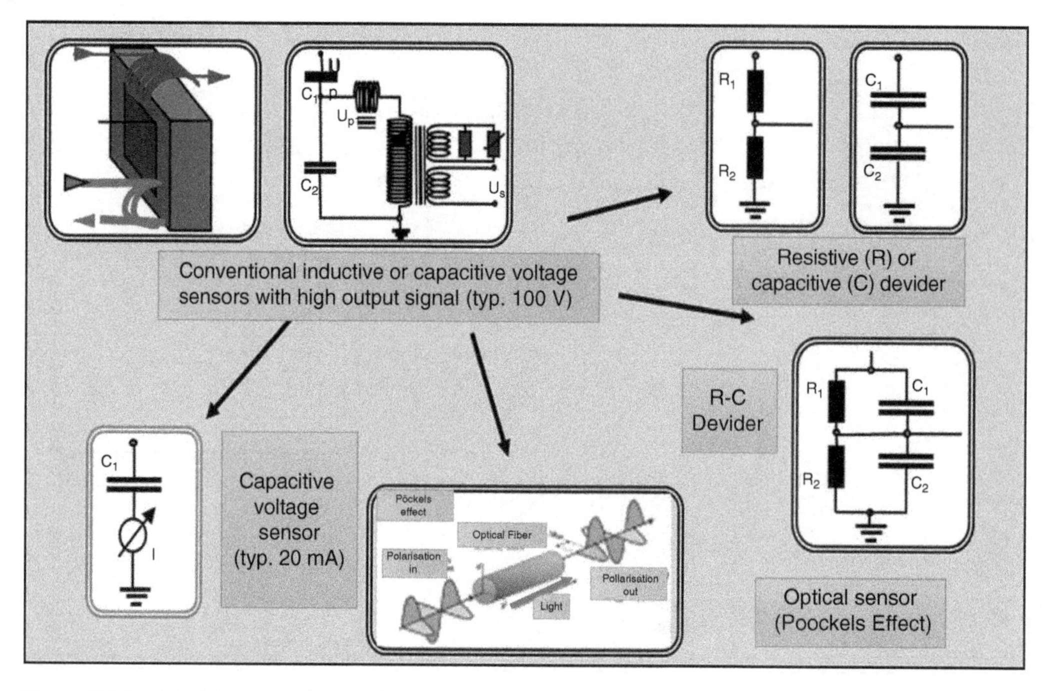

Figure 9.54 Principle measuring method of conventional and low power instrument transformers to measure voltage (*source* [1–3])

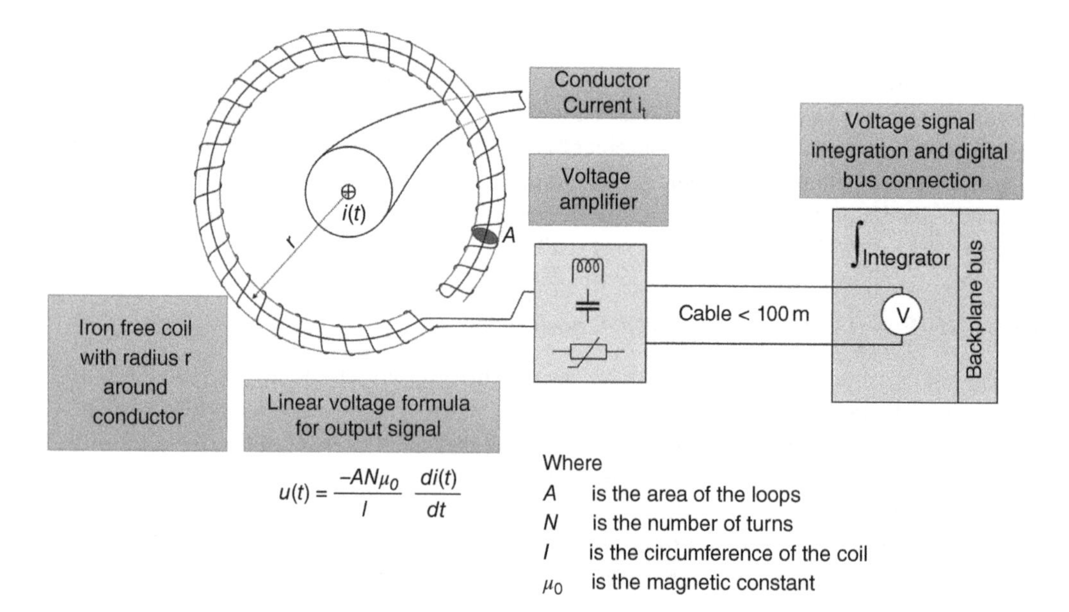

Figure 9.55 Principle measuring method of Rogowski Coil (*source* [1–3])

amperes with the same device. The software in the merging unit will handle the analog signal and transfer in a digital communication.

Also, the Rogowski Coil can measure a wide range of current frequencies from 50 Hz up to kHz with the same device. This helps for better failure analysis or high-frequency harmonics in the network. The evaluation of this high-frequency signals is done by software in the merging unit.

The Rogowski coil offers new features of current measurements for protection and control of the power network. In regenerative energy generation scenario with fast changing power flows, this will be a great help for future network operators.

9.14.4 Low Power Voltage Sensor with Capacitive Dividers

A capacitive voltage sensor measures the primary AC voltage using the principle of capacitive coupling. The capacitor is formed by an electrode-ring and the primary conductor itself. The displacement current through this capacitor is measured.

The secondary output signal is a current of typically 1 mA at rated voltages.

Figure 9.56 shows the measuring principle. The high voltage at the primary conductor is one side of the capacitance, the dielectric material of cast resin insulators is the electrical insulation and the metallic electrode-ring typically aluminum at the out enclosure, but insulated to the enclosure forms the second side of the coupling capacitance.

The connection box includes inductances and capacitance for signal filters, and surge arresters for voltage limitations at the output signal. The secondary cable shall be less than 100 m before connected to the electronic merging unit located inside the bay control cubicle.

The electronic device of the merging unit connects the measure current signal as a digital communication protocol to station or process bus.

The capacitive voltage sensor produces a linear voltage related to the voltage at the conductor over a wide range. It can measure rated voltages of some volts up to transient overvoltage some thousand volts with the same device. The software in the merging unit will handle the analog signal and transfer in a digital communication.

Also, the capacitive voltage sensor can measure a wide range of voltage frequencies from 50 Hz up to MHz with the same device. This helps for better failure analysis or high-frequency harmonics in the network. The evaluation of these high-frequency signals is done by software in the merging unit.

The capacitive voltage sensor offers new features of voltage measurements for protection and control of the power network. In regenerative energy generation scenario with fast changing power flows, this will be a great help for future network operators.

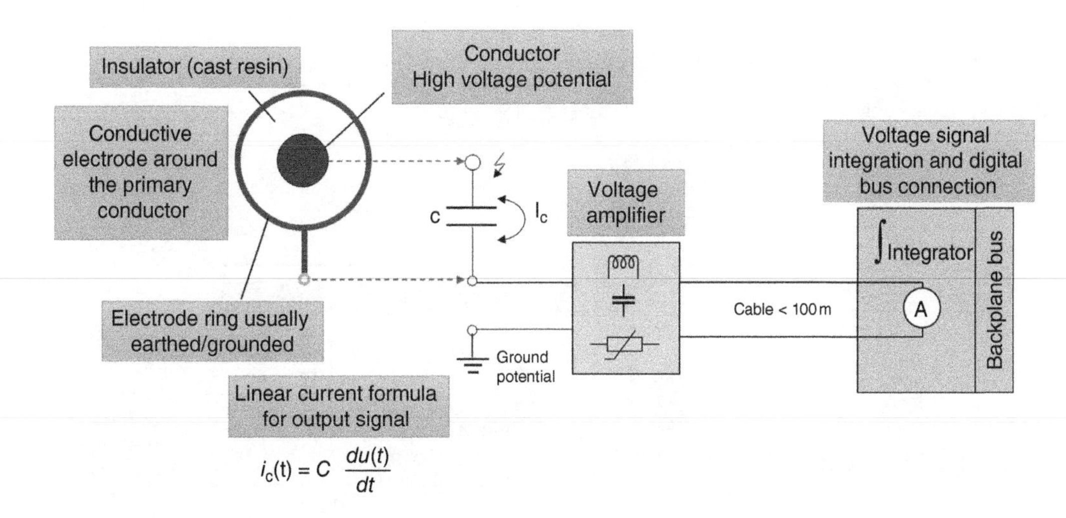

Figure 9.56 Principle measuring method of capacitive divider (*source* [1–3])

9.14.5 Comparison Low Voltage and Conventional Instrument Transformers

Figure 9.57 shows the difference between low voltage and conventional instrument transformers. The upper half of the shows an inductive voltage (VT) and current (CT) transformer using conventional coils with iron core for magnetic amplification. The VT typically is attached, for example, at the cable bay on incoming or outgoing line, while the CT is attached to the circuit breaker module to connect the cable bay. The dimensions given of 3.1 m height and 5.5 m depth of the GIS bay represent an impression of the large size of the VT and CT.

The lower part of Figure 9.57 shows a capacitive voltage sensor for voltage measurements and two Rogowski coils for current measurement located inside a GIS conical insulator. The capacitive voltage sensor uses the capacitance between the electrode ring and the conductor to measure the voltage. The Rogowski coil (an iron free coil) uses the induced current to measure the current in the conductor. The voltage generated by the capacitive voltage sensor is in the range of mV and the induced current in the iron free Rogowski coil is in mA. Both measuring signals need to amplified. Therefore, the size of the capacitive voltage sensor and the Rogowski coil is small and can be installed inside a cast resin partition. This reduces the bay size of the GIS down 3.0 m height and 3.7 m depth.

The combined electronic voltage and current sensor is designed according to IEC 61869 standards part 10 and 11 [5, 6]. Two current sensors and one voltage sensor with two taps in each bay provide redundancy in providing voltage and current signals for the protection and control system. This redundant can be offered because of the small size of the sensors. The current and voltage sensors are multi-purpose devices with electronic circuit for protection and measurement functions. The standardized current sensor uses the same device for every possible rated current which allows more flexible installation everywhere in the GIS.

9.14.5.1 LPIT Integration

Figure 9.58 shows the low power instrument transformer integrated into a GIS bay between the circuit breaker module and the cable connection interface. The current and voltage sensor are part of the cast resin partition and is connected by analog signals to the control and protection device inside the local control cubicle, the so-called merging unit. The merging unit communicates the voltage and current measurements by digital data to the station and process bus [1, 7].

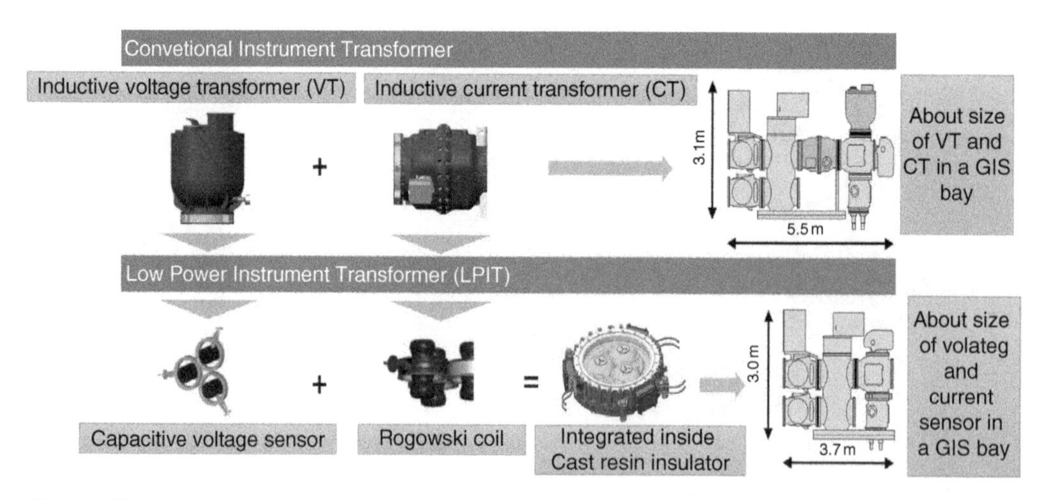

Figure 9.57 Comparison of low voltage and conventional instrument transformers (*source* [1–4])

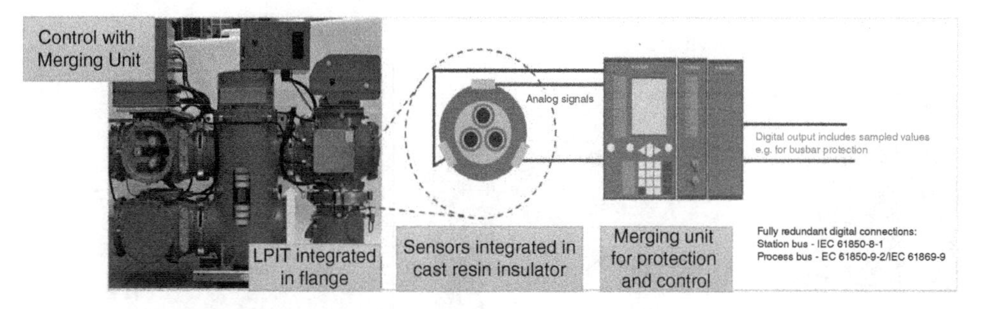

Figure 9.58 LPIT integrated in a GIS bay and connected to digital protection and control (*source* [1–4])

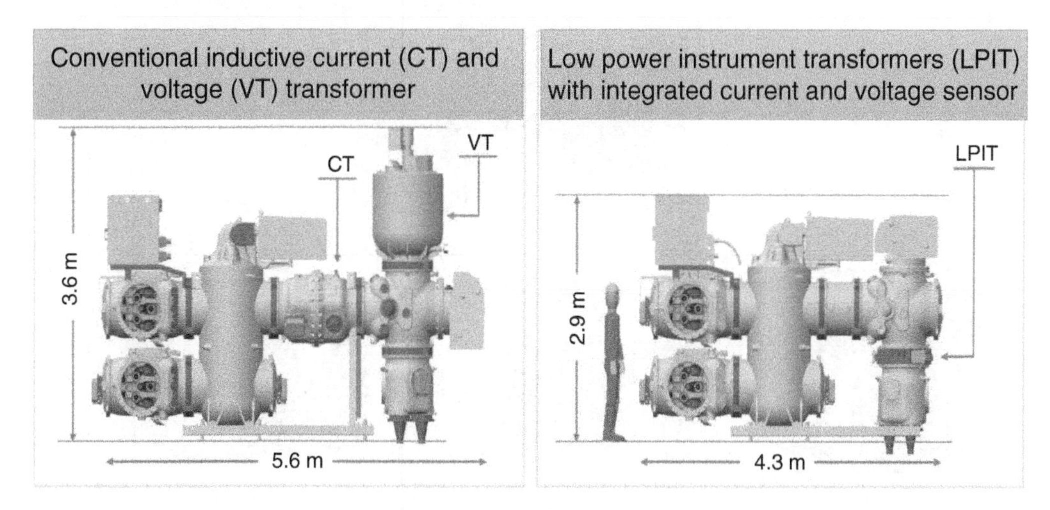

Figure 9.59 Size and weight reduction of GIS bay with low power instrument transformers LPIT . Right grahic for LPIT GIS Bay: 145 kV GIS 1400 kg and 170 kV 1500 kg less weight. (*source* [1–4])

9.14.5.2 Size and Weight Reduction

The small size of the low power instrument transformer reduces the size and the weight of each GIS bay as shown in Figure 9.59. The bay length reduction of the example 170 kV GIS bay with a double bus bar system, a current transformer between the circuit breaker module and the cable interface, with a voltage transformer attached to the cable connection module provides 23% reduced the weight of the bay. This is about 1500 kg less weight for a 170 kV GIS bay.

At 145 kV voltage level, the same configuration shows a reduction in size of 25% and with this a reduced weight if 1400 kg. For the 145 kV technical air-insulated GIS bay of the same configuration, the size reduction is even at 30% which results in a weight reduction of 1500 kg per bay.

Reduction in size and weight is linear with the reduction in cost, because of less material used. Material saving is with aluminum of the enclosures, copper for the coils and iron for the cores.

The standardized voltage and current sensors require less secondary wiring. There is no need for different coils for the various current ratings, for protection coils or measurement coils. The analog signal is only once to be connected to the merging unit within the bay. From the merging unit the voltage and current data is connected to the station bus by digital data communication via an optical fiber link.

Figure 9.60 shows the impact of seize reduction by using low power current and voltage sensors instead of conventional, inductive solutions. Three different GIS designs of 145 kV class equipment of three manufacturers are presented for a typical GIS bay.

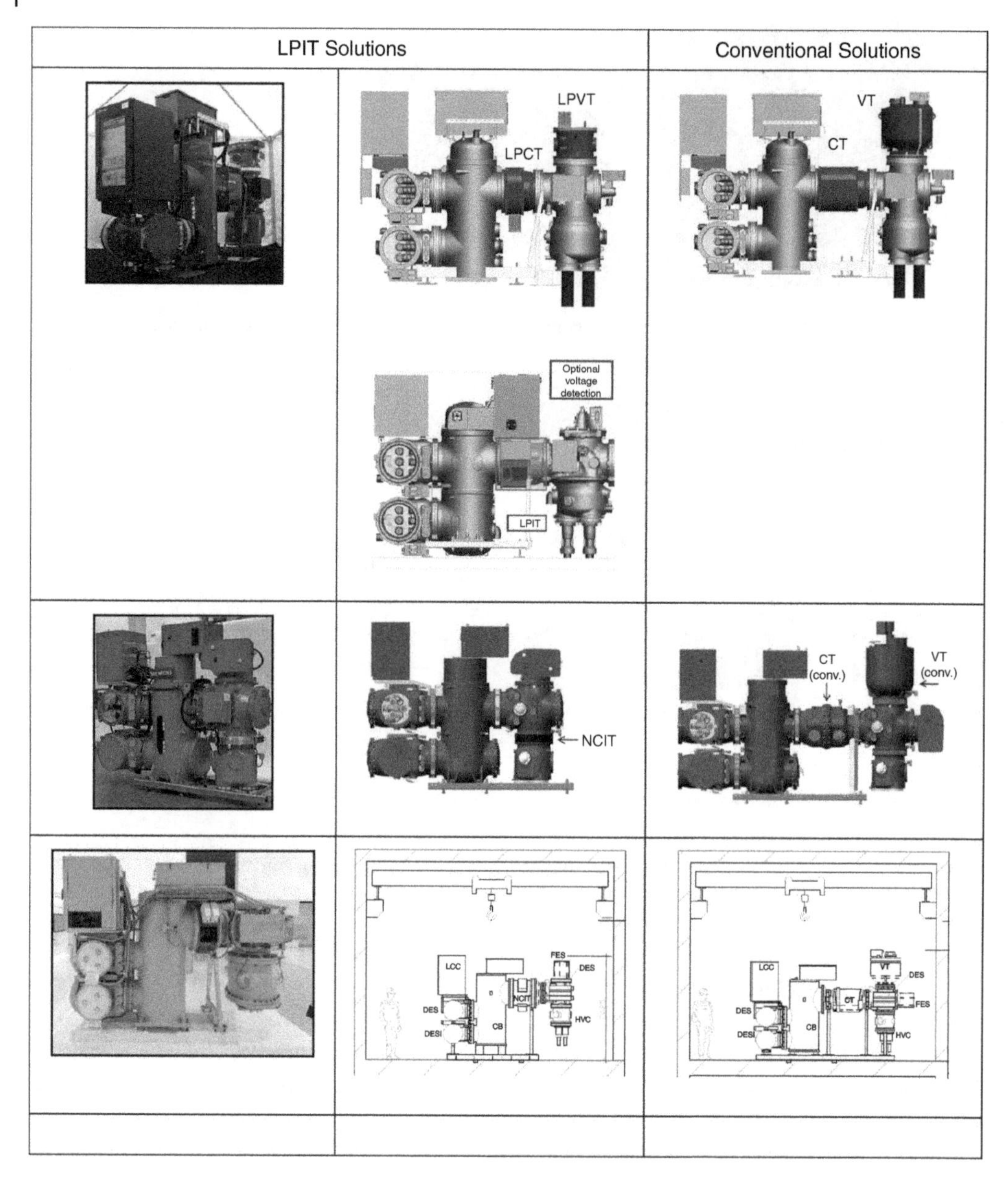

Figure 9.60 Overview of size reduction of three different GIS designs (courtesy of CIGRE [5], Figure 4.4)

The first column shows the physical assembly of the GIS bay, the second column the LPIT version and the third column the conventional solution. In each case, the seize reduction is well visible. The reduction in the depth of the bay is mainly related to the current transformer between the circuit breaker encloser and the cable connection interface. The reduction in height is mainly related to the conventional-type voltage transformer on top of the cable connection interface which is no more needed in case of low power instrument transformers.

This overview of GIS bays using LPITs shows that in cases of additional need of conventional current and voltage transformers because of protection relay requirements or for calibrated meters the seize reduction may be much larger than show here.

Figure 9.61 and 9.62. shows the impact of using low power instrument transformers instead of conventional, inductive instrument transformers on the footprint, GIS equipment weight and SF_6 weight. of a GIS substation. Figure 9.61 shows a substation with 13 GIS bays using conventional, inductive current and voltage transformers. The footprint of the complete installation is 14.2 m wide and 6.1 m deep. The total weight of the installation is 45 t and the amount of SF_6 is 1.2 t.

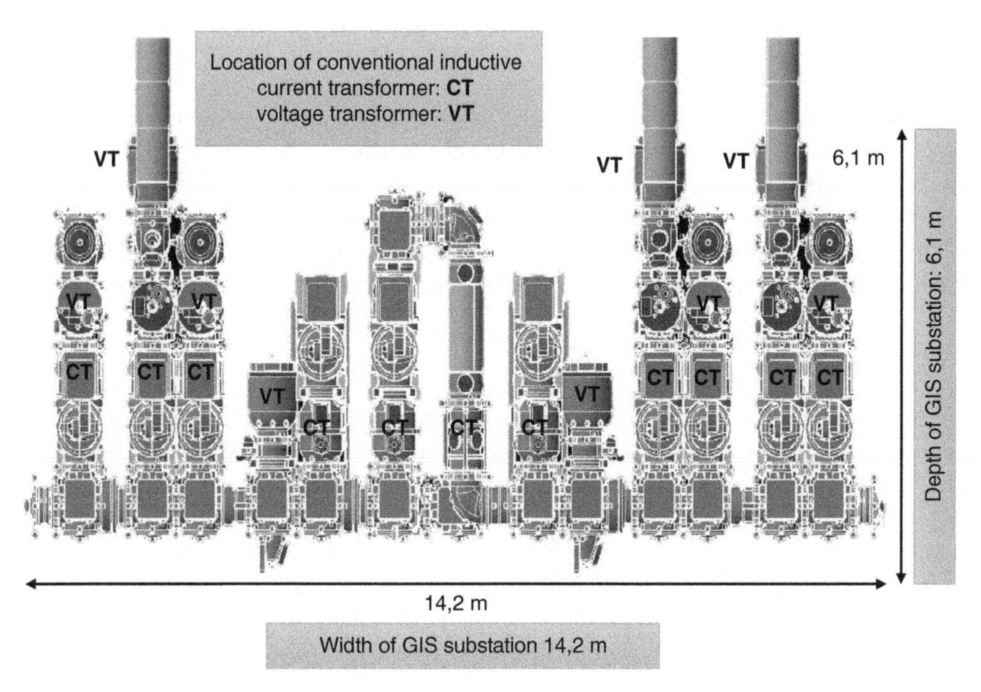

Figure 9.61 Footprint of a 13 bay 145 kV GIS substation with conventional CTs and VTs of 14.2 m to 6.1 m (source [1–4])

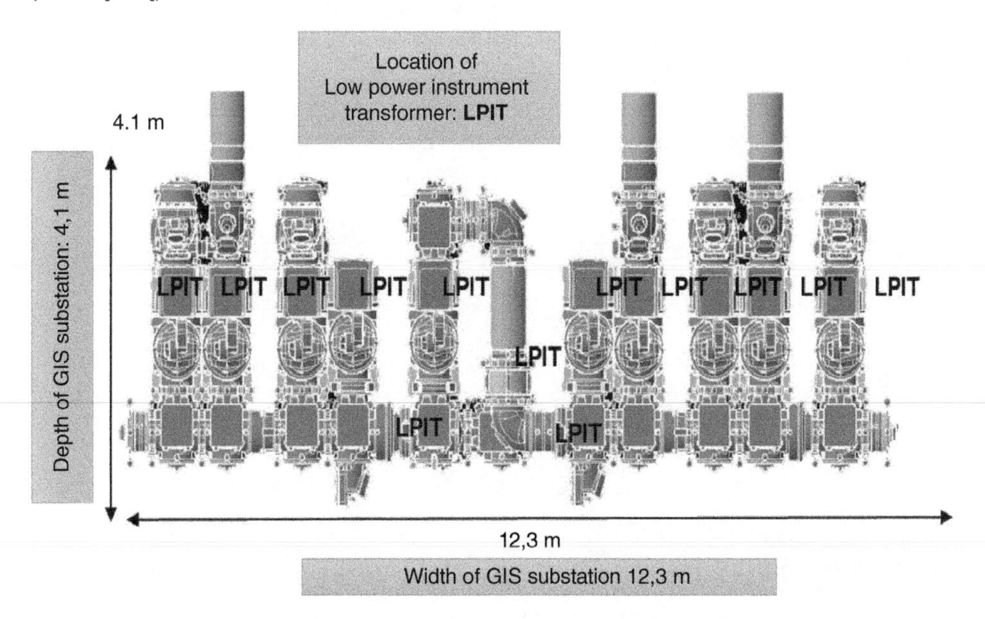

Figure 9.62 Footprint of a 13 bay 145 kV GIS substation with low power instrument transformers of 12.3 m to 4.1 m (*source* [1–4])

Table 9.15 Footprint, weight, and SF_6 amount for conventional and LPIT GIS bays with reductions

	With conventional CT/VT	With LPIT	Reductions (%)
Footprint	14.2×6.1 m	12.3×4.1 m	40
Weight	45 t	34 t	25
SF_6 amount	1.2 t	1.0 t	15

Figure 9.62 shows the footprint of the same 13 GIS bays using low power instrument transformers. The footprint is reduced to 12.3 m width and 4.1 m depth. The total weight of the 13 GIS bay substation is 34 t and the amount of SF_6 is 1.0 t.

The reduction numbers are presented in Table 9.15. The footprint is reduced by 40%, the total weight of the 13 bay GIS substation is reduced by 25% and total weight of SF_6 is reduced by 15%. This reductions in material will reduce also the cost of the equipment and the building installations.

9.14.6 Interfaces

The interface between the low power sensors and the protection and control system of the GIS bay and substation is defined by international standards. For digital interfaces of instrument transformers IEC 61869-9 [6], for additional requirements for low-power passive current transformers IEC 61869-10 [8] and for additional requirements for low power passive voltage transformers IEC 61869-11 [7].

The share of responsibilities in standards is shown in Figure 9.63. The digital interface of the sensor and the merging unit is following IEC 61869-9, left side of the graphic. The digital communication between the sensor merging unit and the protection and control system is defined in IEC 61850-9-2 [11].

The merging unit (MU) is a universal device used for voltage and current sensors. It offers several methods to connect a sensor of the primary power system following the IEC 61869 series. The analog signals are transferred in digital protocols which communicate with the protection and control system of the GIS bay and substation, see Figure 9.64.

In Figure 9.65 an overview of the IEC 61869 series, its scope and relation to the old instrument transformer standards IEC 60044 series is presented. The IEC 61869 series covers general

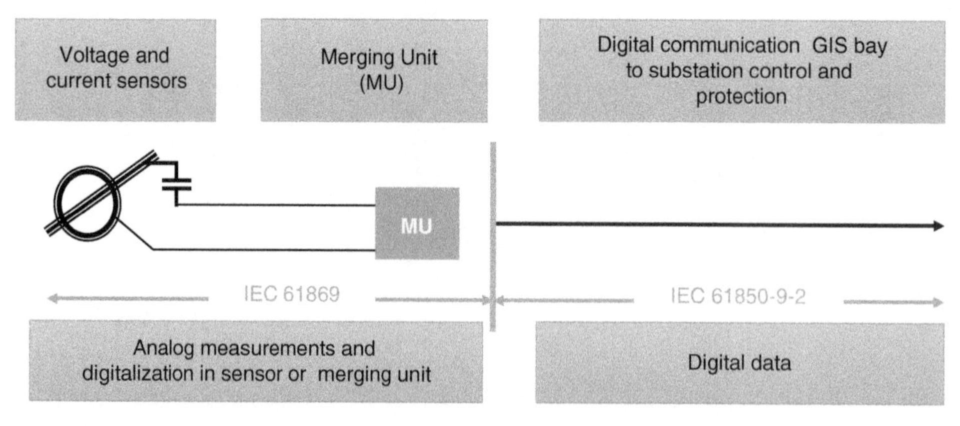

Figure 9.63 Responsible standards: IEC 61869 for sensors and merging unit (left) and IEC 61850-9-2 for digital communication (right)

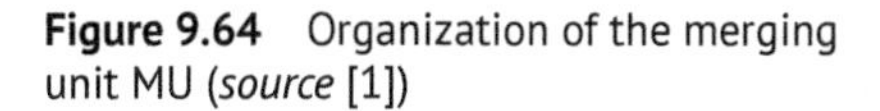
Figure 9.64 Organization of the merging unit MU (*source* [1])

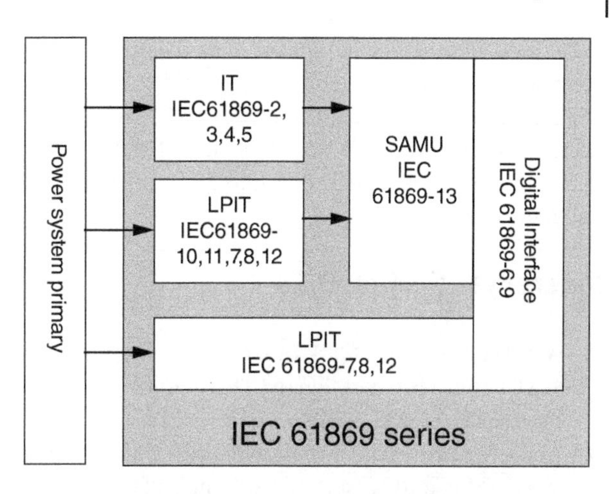

Product family standards		Product standard	Products	Old standard
61869-1 General requirements for instrument transformers		**61869-2**	Additional requirements for current transformers	**60044-1/-6**
		61869-3	Additional requirements for inductive voltage transformers	**60044-2**
		61869-4	Additional requirements for combined transformers	**60044-3**
		61869-5	Additional requirements for capacitive voltage transformers	**60044-5**
	61869-6 Additional general requirement for electronic instrument transformers and low power stand alone sensors	**61869-7**	Additional requirements for electronic voltage transformers	**60044-7**
		61869-8	Additional requirements for electronic current transformers	**60044-8**
		61869-9	Digital interface for instrument transformers	**60044-8**
		61869-10	Additional requirements for low power stand alone current sensors	**60044-8**
		61869-11	Additional requirements for low power stand alone voltage sensors	**60044-7**
		61869-12 (not yet published)	Additional requirements for combined electronic instrument transformer or combined stand alone sensors	**60044-7/-8**
		61869-13	Stand alone merging unit	

Figure 9.65 Overview the IEC 61869 series of standards, its scope and relation to the old IEC standard series IEC 60044

requirements for instrument transformers in part 1 and additional general requirement for electronic instrument transformers and low power standalone sensors in part 6. In parts 2–5, additional requirements for current transformers, inductive voltage transformers, combined transformers and capacitive voltage transformers are defined. Electronic voltage and current transformers are in the scope of parts 7 and 8. In part 9, the digital interface for instrument transformers is standardized. Additional requirements for low power stand-alone current and voltage sensors are in the scope of part 10 and 11. In part 12, additional requirements for combined electronic instrument transformer or combined stand-alone sensors and in part 13 stand-alone merging unit are defined.

9.14.7 Conclusion

The low power instrument transformers (LPIT) are in practical use today and provide new options in voltage and current sensors for the future requirements of the power network. The Rogowski coils and Capacitive voltage sensors are established technologies for GIS.

LPITs may be embedded in cast resin partitions or other insulators bring significant benefits in compactness of GIS and reduce engineering and logistics efforts.

The standardized configurations of LPIT-sensors and Merging Units allow interoperable connections to Integrated Electronic Devices (IED) of different suppliers.

The merging unit integration into bay control and/or protection further simplifies the system and reduces costs.

References Section 9.14

1 Wojciech Olszewski, Mark Kuschel, 'New smart approach for a U/I-measuring system integrated in a GIS cast resin partition (NCIT) – design, manufacturing, qualification and operational experience', VDE ETG 2017

2 Wojciech Olszewski, Mark Kuschel, et al, New smart approach for a U/I-measuring system integrated in a GIS cast resin partition (LPIT) – design, manufacturing, qualification and operational experience, CIGRE B3 Puplic internet Session 2018, Paris

3 www.siemens-energy/gis 2020

4 IEC standard 61869 series part 1 to part 15 Instrument Transformers (2021)

5 TB 814 "LPIT applications in HV Gas Insulated Switchgear," B3 Substations and electrical installations, CIGRE, 21, rue d'Artois 75008 Paris – FRANCE, September 2020.

6 IEC 61869-9:2016, Instrument transformers, Part 9: Digital interface for instrument transformers, IEC TC 38, Geneva, Switzerland.

7 IEC 61869-11:2017, Instrument transformers, Part 11: Additional requirements for low power passive voltage transformers, IEC TC 38, Geneva, Switzerland.

8 IEC 61869-10:2017, Instrument transformers, Part 10: Additional requirements for low-power passive current transformers, IEC TC 38, Geneva, Switzerland.

9 https://en.wikipedia.org/wiki/Pockels_effect

10 https://en.wikipedia.org/wiki/Rogowski_coil

11 IEC 61850-9-2:2011+AMD1:2020 CSV Communication networks and systems for power utility automation - Part 9-2: Specific communication service mapping (SCSM) - Sampled values over ISO/IEC 8802-3.

9.15 Digital Twin of GIS and GIL

Author: Hermann Koch
Co-author: Aron Heck
Reviewer: Dirk Helbig

9.15.1 Introduction

The digitalization of data related to GIS or GIL offers the possibility to create a digital mirror of the real equipment. The digital twin [1–3].

The basis of the digital twin is a data base collecting all the data from the starting point of manufacturing, over the implementation into the network, for the complete operation time, covering all changes, repair, or maintenance made on the equipment until the final dismantling of GIS or GIL including all recycling of materials.

The first technical area looking for such a close watching all steps of manufacturing was the consumer electronic industry. Here, a process failure on integrated circuit manufacturing would have a very wide impact and cost. Used by millions of electronic boards in radios, TV's or similar such a failure would distribute quickly and in wide range of the integrated circuit users. For this reason, the quality system in the semi-conductor industry developed methods to protocol and control materials and process at each step of manufacturing. A quality assurance system has been developed and implemented in the factories.

The next step for digitalization of equipment live was to continue this quality assurance system into the operation time of equipment and until his final dismantling and end of lifetime. The basic information needed for this approach is defined in international standards from IEC [4–6]. Standard data element types are associated with classification schemes for: principles and methods, an EXPRESS dictionary schema or IEC Common Data Dictionary (IEC CDD) quality guideline, to mention some of the many standards covering this field.

How a digital twin may look like and what aspects of the real asset it represents mainly depends on the asset and its value proposition for a certain user. Potential users of a digital twin are spread over the entire lifecycle and value chain of an asset. A general approach to cluster digital twin applications is distinguishing between asset design, asset production and delivery and asset operation.

In Figure 9.66 the principles of digital twin approach are shown. Starting with the digital description of product to form a virtual product which covers all characteristics and functions of the product. For a GIS or GIL, this would be the voltage level, the current rating and other technical data and functional elements like the circuit breaker or disconnector or others.

The second digital description is related to the construction which will deliver a virtual building with structural and functional data. This digital information covers elements like indoor GIS, the building, the grounding system, the steps of construction, and all functionalities like air conditioning, heating, door locking system, fire protection and so on. All related elements and functions are defined by specialist planners in their models, which federate the data base system of the digital twin.

The third step of digitalization is related to the performance of the digital twin. In case of a GIS or GIL, this would be the operational requirements like transmitted power, over-voltages, number of switching operations, insulating gas loss, short-circuit interruption, operation temperatures of conductor or ambient, partial discharge activities, and others. These data will be stored and collected for life time of the digital twin to provide information about the operational history of the equipment.

The digital twin data base is a living system. During all step of production of the equipment, installing the equipment, following the equipment performance and finally for the dismantle and recycle processes the data base of the digital twin will be continuously updated and each change will be validated.

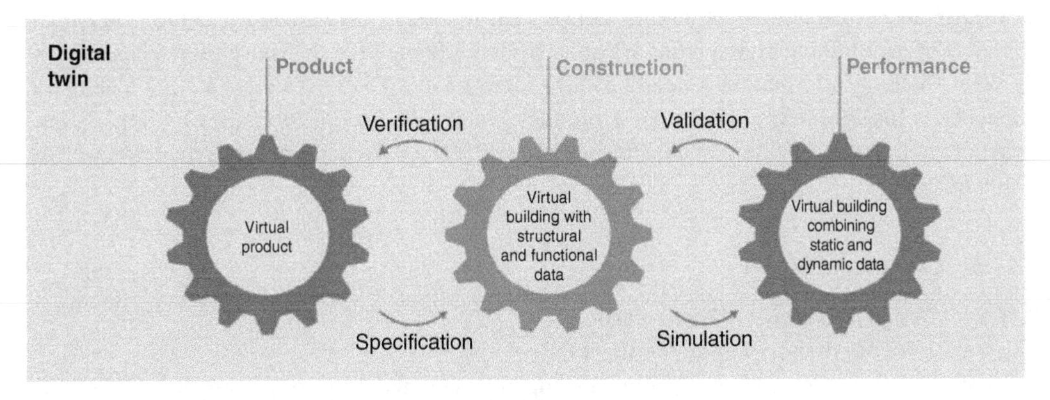

Figure 9.66 Principles of digital twin approach (courtesy of Siemens)

The digital twin of a GIS or GIL reflects the information of the manufacturing process with all material used and factory test made. It documents all parts which come into the factory and how they get assembled to the GIS bay. The transport to the construction site, the erection of the GIS or GIL on site with all its specific aspects will be digitalized in specialist models which reflect the parameters relevant for this step. The commission testing will deliver information about the high voltage testing and all other on-site test. Finally, the whole lifetime data of the GIS or GIL project is stored in the digital data twin.

9.15.2 Basic definitions

9.15.2.1 Asset Design Digital Twin Target: Figure Out Most Suitable Design

The asset design of the digital twin consists of tools and processes that allow a GIS or GIL design team to collect and process customer requirements and onshore constraints in order to build consistent design drafts represented in digital models. Those models represent certain aspects of a design draft like for example geometries, locations, mechanical, or electrical behavior, cost, schedules, or other aspects of a GIS/GIL a customer is interested in.

The asset design and so its twin starts with a demand. Agents of the prospective operator, consultants, and suppliers start figuring out solutions meeting requirements to deliver the demand and a certain value to fit into a current or future operation and asset lifecycle management environment.

9.15.2.2 Asset Production Digital Twin Target: Increase Delivery Performance

The asset production and delivery of the digital twin is derived from the asset design digital twin chosen by the customer and focuses on optimizing the value chain of a project that delivers a GIS/GIL to an operator. Examples for production and delivery twins are a common data environment [7, 8] quantity take-offs from design models for quotation requests, paperless quality assurance documentation, simulations in production and logistics for factory and site processes [9] and more.

9.15.2.3 Asset Operation Digital Twin Target: Increase Asset Performance

The asset operation digital twin serves the asset owner to maximize the asset value until end of lifetime. Asset operation digital twins focus on increasing the performance of the asset itself (e.g., deriving operational asset capabilities from current condition), reducing the overall effort to maintain the asset (e.g., predictive maintenance, digitalization of maintenance activities) and an as-built model to retrieve information for nearby construction projects or in case of an incidence and to append information about asset modifications.

Most organizations already build and utilize such models. These models and the conceptual approach of maintaining digital twins are quite similar compared to the concepts associated with the term Building Information Modelling (BIM) defined in ISO 19650 series [10–12]. The main difference is that the term digital twin is product focused whereas BIM focuses on built assets. Depending on the organization and its role within a project a GIS can be both, a product as well as a built asset.

9.15.3 Data model set up

9.15.3.1 Design Data Model of Environment

Lidar, remodel, put design draft into virtual model.

The first step is to record the environment using Lidar (Light detection and ranging) using some equipment that generates geo-referenced point clouds. This laser scan reflects an object in the project area and give basic information for the obstacles at the site and the availability of space to be used for the GIS or GIL. The Lidar produces a 3D picture to be remodeled into virtual model by given colors to the objects so that they can be identified and used for site evaluations. See Figure 9.67.

The next step is processing this Lidar data to get a geo-referenced digital terrain model (DTM) raster and a digital surface model (DSM) raster, which provides the terrain and surface elevation absent of vegetation for each raster cell.

The file format of these models has to be appropriate for further processing in the described use-cases.

With this digital terrain and surface models the project site for GIS or GIL can be evaluated and the best location for installing the GIS or routing the GIL can be found.

The Lidar process is using drones or a backpack version for the scan and is highly automated to process the surface and terrain data (see Figure 9.68).

The DTM raster have to be converted to a TIN (triangulated irregular network) to use them in the Review System (Autodesk Navisworks). Non-terrain objects like vegetation, buildings etc. can be saved as point clouds.

The inversion of the surface and terrain data from the lidar is needed to generate standard software structures using triangulated irregular network (TIN) model for further use in the data base of the digital twin, see Figure 9.69.

Not all of the objects can be transferred into TIN model which provides strong reduced data size compared to pixel graphics of lidar scan. Such objects are, for example, trees, and they need to be reduced in point density.

Color points are used for geo-referenced orthophoto on basis of the digital terrain or surface model using the DTM mesh.

The data generated for the digital surface and terrain model are generated on different ways by lidar drones or lidar backpack systems. The different data is combined into one data set.

Figure 9.67 Lidar scan of a possible GIS/GIL project site (courtesy of Siemens)

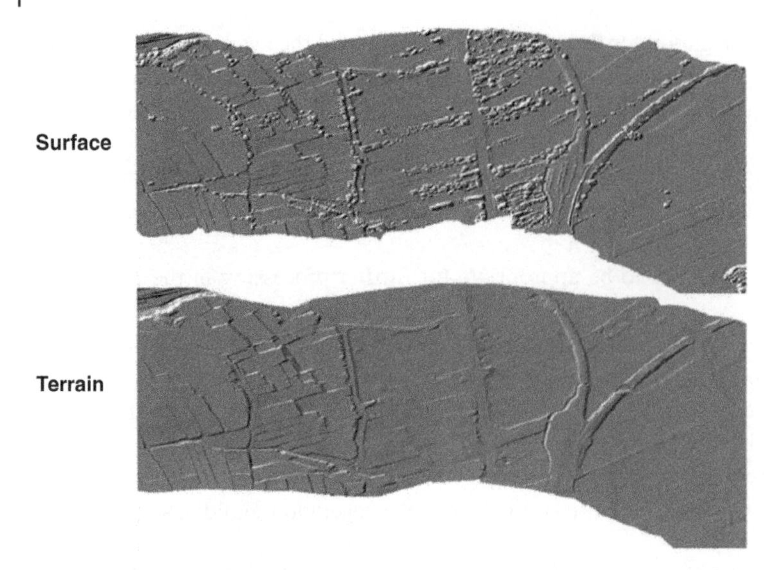

Figure 9.68 Digital surface model (DSM) and digital terrain model (DTM) (courtesy of Siemens)

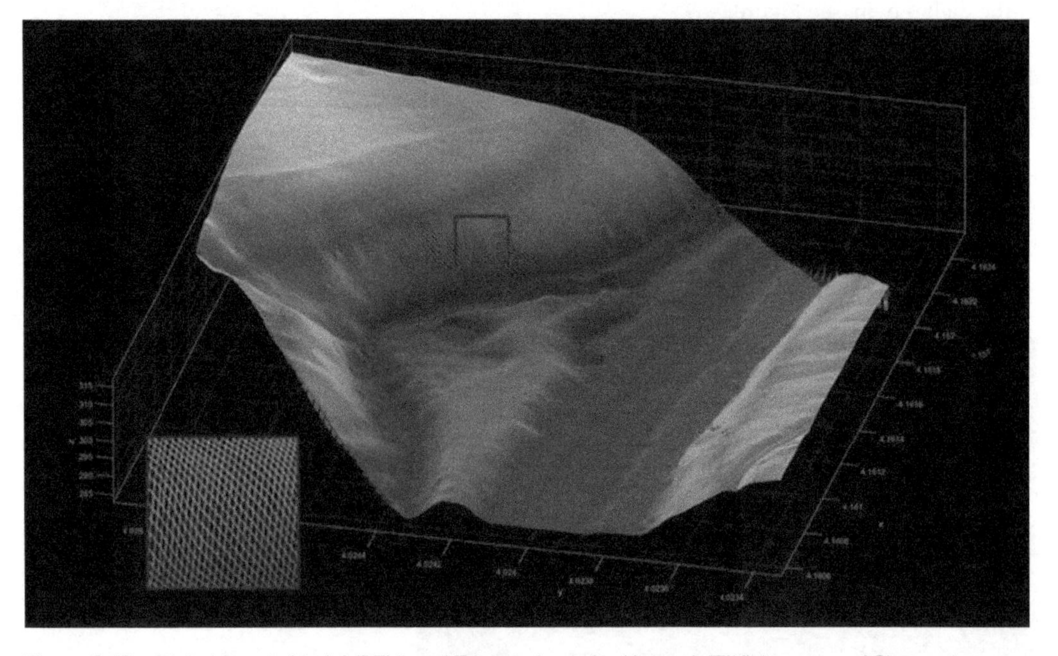

Figure 9.69 Digital Terrain Model (DTM) and Triangle Irregular Network (TIN) (courtesy of Siemens)

Clip point clouds are used with respect to georeferenced buffered polylines or with respect to georeferenced shapefiles.

9.15.3.2 Data model of Production

The data model of the production sequence of GIS or GIL is delivering the product design data for the specific project. The dimension data model of the GIS or GIL are connected to geo-reference data model for the surface (DSM) and terrain (DTM).

The electrical requirements of the GIS or GIL are defined in the project specification including any environmental, mechanical, thermal, EMF or other requirements. The manufacturer of GIS or

GIL will turn these requirements in a physical solution. This physical solution has certain number of bays, in- and outgoing lines for GIS and a certain length and directional changes for GIL. The solution has specific dimensions. In case of the GIS, this are the fixing points for each bay, the three dimensions of each functional unit (circuit breaker, disconnect, or grounding switch, voltage, and current transformer) are integrated into the digital twin model. The position of each flange, each bolt and any connection are covered by these data. In case for GIL, the diameter and length of the segments including any directional change is defined in the data base.

With the definition of the GIS bays and units the data base of the digital twin gets the information of the interior of the functional unit. In case of the circuit breaker unit, this would be the detail dimensions of the enclosure, the fixing points, the interruption unit, the contacts of the interruption unit and all part related to the circuit breaker unit. These are construction details which are drawn from the manufacturing drawings.

For GIL, the dimension of the enclosure pipe, the conductor and insulators are part of the digital twin data base. The position of the insulators inside the GIL enclosure, of the welding or flanges of enclosure or conductor and the location of the gas-tight insulator forming a gas compartment are stored in the data model of the GIL digital twin.

With this information each element used in the project is fixed and can be traced down its production data. For example, an insulator is linked by its number to a GIS or GIL unit. This number links it to certain production unit in the factory of the insulator. The production data and the factory test protocols are also linked to this insulator. With the information available in the digital twin, it is possible to get access to this information at any stage of production. And the information will remain in the digital twin data base for any question in the future.

It is from great importance to up-data this data base with any changes made during production. And changes are always necessary for many reasons. Change of a sub-supplier, change of electrical requirements, change of environmental requirements which will come in most cases of the project developing process. The importance of data in the data base of the digital twin to be up to date is essential. Otherwise, failures could occur and cause large damage.

The production sequence is finished after the installation is completed and the commissioning test have been passed. During the erection of GIS or GIL many data is generated and, in many cases, adapted to local requirements not covered in the project planning so fare. This information needs to be implemented to the data base of the digital twin.

In addition, the data base for the digital twin can be used for the installation planning process by digital simulation. The GIS or GIL model can be used to simulate the installation process by using the surface and terrain models. This can avoid problems during the on-site assembly process and safes additional time and cost.

The advantages for the project execution of using digital twin data base information is to be much more aware of any mismatch in the project. Each step added to the project process will be verified through all data in the data base. An example, a GIS bay may be specified for 3150 A rated current. During the production in the factory a conductor has been added qualified for 2500 A. The digital twin data base will check this and give a warning. This is a simple example and is part of good project management today to avoid such failures. But the data base offers a higher level of project wide quality control.

9.15.3.3 Operation

The digital substation concept using the digital twin is to have all relevant date of the substation available at any time along the life time of the substation, which is usually very long.

For GIS and GIL, this means that all data coming from the production sequence will be integrated into the digital model of the surrounding including the building and for the operation time all operational data are linked to the same digital twin model.

During operation data is generated from operation conditions and actual, real requirement seen by the GIS or GIL. The specification is usually based on calculated and simulated values which have added some safety margins. In real operation, these data may be different. It is from great interest for life time expectations or maintenance requirements to know the real stress of the GIS or GIL.

For example, the rated current. If a 3150 A-rated current GIS or GIL is operated at average 2000 A, the impact for maintenance a life time is different, when operated 3150 A the type tested value and sometime at 3500 A. Same is with rated voltage or number of switching operations. These data are in principle available in the substation. The digital twin data base collects this information and can be used for evaluating the status of the GIS or GIL.

The digital twin of GIS or GIL reflects the real history of requirements. The current flow over the time, the short-circuit fault current interruption by current value and number of occasions, the over-voltage operation situations by voltage values and time line, the temperatures of the conductor inside, the enclosure and the insulating gas, the temporary measure partial discharge measurements of the high voltage insulation system, the gas pressure or in case of leakages the losses of insulating gas and any activity coming from the protection and control, of the GIS or GIL related to network instabilities, failure, lightning strokes to connected overhead lines or overload conditions related to network restructuring operations. All this information will be collected during the time of operation and is available at any time for the condition evaluation of the equipment.

The operation schedule can be planned based on this information for maintenance work or even extension of the substation. This is a great help to reduce down time of the network and to prevent from failures of the GIS and GIL relating to any overload or abnormal operation condition which may harm the equipment.

The whole environment is also connected to the digital twin data base and collaborates with the equipment data of the GIS or GIL digital twins. There are two methods for covering the environmental impact. One conventional method, where the user-oriented runtime of the digital twin for environmental data which isolates the digital twin models and the user equipment model with separate tools.

The other method is the model/project-oriented runtime of the digital twin for environmental data and equipment data to the used shares models and tools.

The digital twin technology is at the first development stage and depending on the current business strategy or market demands it might be reasonable to develop or utilize new systems to use the data base more efficient for the purpose of the substation production, construction, and performance during operation. The future development of effective digital twin processes for GIS and GIL will influence the project and the business organization.

9.15.4 Used Case Digital Twin of GIL

9.15.4.1 Motivation

GIL projects are connecting point A and point B in an electric network. The installation is typical underground, direct buried or in a tunnel, and the project has to find technical solutions for each section of the underground transmission line. Typical underground routings show obstacles in 1–2 km distances. These are road, high ways, rivers, rail roads, gas pipe lines, other electric underground cable, constructions, natural protected areas, swamp, rock, mixed soil and many more parameters to be solved for the connection.

Route planning is an intensive and for the success of the project from great importance. All data collected during this time will be gained on different levels of detail and many data is used during the complete project process. The soil condition and type of soil is important to know for the excavation works for the construction but also the soil type and its thermal conductivity is important for the operation of the GIL. This data store in the digital twin data base will then be used for both tasks at very different stages of the project.

The second important contribution of the digital twin in the GIL project is to deliver data for construction process. The assembly of GIL on site receives the single GIL units of encloser pipe, conductor pipe, insulators, and disconnecting units and assembles them by an automated welding process direct at the trench or tunnel on site. All the information of the GIL units is available in dimensions and time to assemble to allow a digital construction planning before the on-site construction has started. This digital project construction possibility gives an important advantage to the complete project planning and execution process. The GIL project can be simulated in advance and for each sequence of the complete transmission line the obstacles and possible problems can be solved before work has started on site. This avoids high additional construction cost and increases the project execution quality.

The use of digital twin will require a full digital availability of data. Staring with all environmental data of the surface and terrain, the digital information of anything on and under the surface like buildings, trees, lakes, river, and others, the digital availability of all the equipment data in data format following the international standard requirements of digital documentation, the data for the preassembly, assembly, and installation process in digital data for the function, parameter and process steps, the connection to the digital data coming from the operation of the GIL and finally the access to all this data in the digital twin data base will be a great challenge for such digitalized projects.

The digital twin process is not easy to use in existing network structures because many of this data required is not available in digital form, following international data standards and must be gained by using paper copy to be copied in digital data. This is in most cases a burden to high. It is expected that digital twin GIL projects will first be applied for new project.

In the international project business, the requirement for localization of project works is essential for being successful. The GIL offers a large part of the work to be localized and carried out by local companies and workers. The automated assembly process on site needs mechanical and technical skills from working staff which is available in every country.

In the following a digital twin GIL project is explained for its main steps of data model generation and local manufacturing process.

9.15.4.2 Project Information Model

The project information model needs to be developed for each project to reflect the different structures and requirements.

Every country in the world has options to localize parts of GIL in their specific value chain. This value chain includes materials like enclosure pipes, conductors, insulators, contact system or steel structures. But also, services like welding works, civil works, assembly works or testing works. All these project elements follow different resources and need to be implemented into the digital twin data base structure by technical parameters and models. These individual value chains in GIL projects consist of different entities working together on different processes, having different resources available. In order to figure out most suitable processes and partners when planning a value chain for an individual GIL project, digital twins provide information from simulations and data models.

For brownfield or underground GIL projects, the landscape can be recorded to generate a 3D terrain and surface models to use it in the GIL civil model for earthworks optimization. Adding results from over ground and underground surveys using data processing or remodeling generates the GIS 3D project information model. In Figure 9.70, a result from a ground investigation report clashing with the design of an underground GIS is shown.

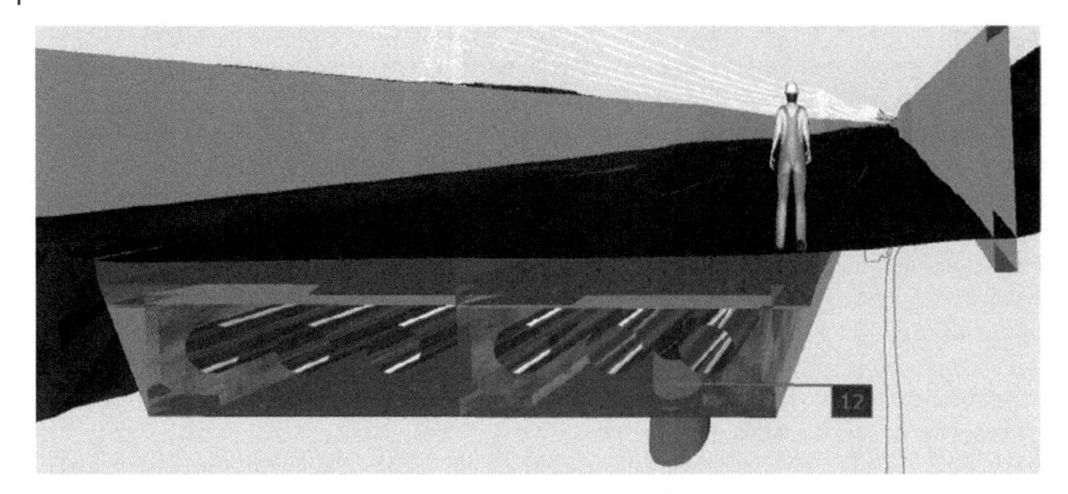

Figure 9.70 Two electric three-phase systems of GIL in a 3D project information model (courtesy of Siemens)

The GIL 3D project information model provides quantity take-offs of modeled materials, parametric generation of section drawings, clash detection and is a great discussion base to align the design with internal and external stakeholders. The total information of the GIL project can be made visible in routing, local installing facilities or details of the construction. The digital twin combines all project information and make them accessible.

Once the GIL is built, measures can be updated to as-built and the final model is handed over to the operator and owner for further use in the GIL lifecycle. The digital twin is the data base for the GIL project information and any change or any given operation parameter can be linked to the digital twin of this specific GIL project.

A GIL Construction Site Simulation [9] investigates different execution variants of a GIL project. It includes a data base for saving relevant project data, a simulation tool kit for modeling individual GIS on-site manufacturing facilities and an analysis tool to evaluate different configurations. It is used in projects in order to determine site setups with associated process and material sequences that meet installation dates and realize an acceptable utilization of workers and resources.

Other digital twin approaches like true to scale animations of manufacturing and installation processes are utilized in projects, if doubts on feasibility arise.

For mechanical and thermal design of GIL routes, finite element method (FE) is used. To meet local requirements GIL is designed for various laying depth. This is possible due to the use of temporary flowable backfill around the enclosure tubes. It provides mechanical strength and a low thermal resistance without dry out effects for a wide temperature range. The mechanical calculation is based on German standards for sewers and district heating. For special cases not covered in those standards, FE simulation is used.

FE-Simulation is also used for the thermal design in regards to the center distance of the phases required for the individual current rating. An example is given in Figure 9.71.

9.15.4.3 Local Manufacturing

Local manufacturing is an important part of the GIL project execution concept. GIL is using long sections of pipes for enclosure and conductors which are delivered directly to the construction where they get assembled with the other part of the GIL like insulators, contact systems and connecting housings for interfaces. The local manufacturing concept brings these elements into the GIL with local manufacturing tools. This local manufacturing process can be modeled and implemented into the digital twin of the GIL project.

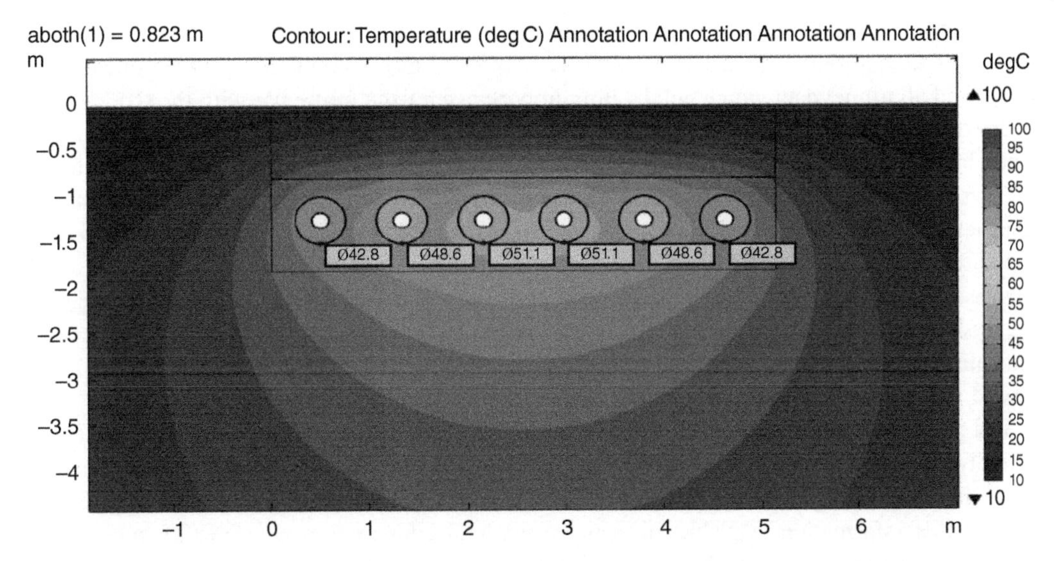

Figure 9.71 Thermal calculation of two three-phase electric systems of GIL directly laid in the soil (courtesy of Siemens)

Some enclosure aluminum pipes consist of casted parts, that require qualified welding, x-ray and pressure testing to become a GIL module. Based on the technical data the factory is providing this sequence of providing the GIL module at local manufacturing the digital twin data base will cover this step too and integrate in the GIL project model. Technology and process data which are available for production, construction, and performance of GIL modules can also be localized in the local manufacturing model.

The GIL conductor aluminum pipes are extruded in the factory of the pipe manufacturer. These aluminum pipes need to fulfill geometric dimension with high tolerances and need to provide a specific surface quality required by high voltage limitations of surface roughness. These quality requirements are part of the quality assurance system installed by the manufacturer of these pipes. The technical data of each pipe for diameter, wall thickness of the pipe, roundness of the pipe, roughness, and integrity of the surface are available with the production of each pipe. This data can be modeled and made available in the digital twin injuring the quality of the complete GIL project in the digital twin.

Conical- and post-type insulators as well as other GIL modules used in GIL projects like elbow modules for directional changes, expansion joints for thermal expansion of the enclosure pipe and connection modules for the interfaces to overhead lines, cables, or GIS are also delivered and assembled on site. The technical data of these element are modeled for the digital data base of the digital twin and can be used for complete GIL project.

Above ground installed GIL projects require steel structures and civil works to build the foundation and steel structures to hold the GIL. These civil and steel works are part of the project planning process and hold about 20–30% of the cost. They have a great impact to the GIL project economic and ecology success. Planning the foundation at the right location with the required strength and accuracy can avoid high additional cost if failures occur and changes are required. The digitalization of this work will help to provide a better planning and avoid problems. The models for the civil works and steel structures are integrated into the surface and terrain model and can be validated for correctness and best solution. The digital twin offers a high planning security and avoids costly changes of civil works on site.

Underground installations in tunnels or directly buried in trenches require even more civil works that are traditionally carried out by local suppliers or special companies. The routing of the tunnel and all tunnel data which will be from importance for the future use with the GIL can be modeled and made available to the digital twin data base. There is information on dimensioning, width, height, the quality of the tunnel wall, the requirement for fixing to be brought to the tunnel wall, water prove against ground water, the electric grounding system of the tunnel and many more detailed technical data will come with the tunnel model. The digital twin data base will verify each future step in the tunnel and GIL building process to avoid failures.

In case of directly buried GIL the information of the type of soil and its conditions including the prepared backfill material around the GIL pipes are part of the digital twin data. The most important data is related to the thermal conductivity of the soil around the GIL pipes which is an indicator and can be the limitation of power transmission capability during operation. The backfill is created by mixing the excavated soil other adhesives or bentonite to reach a required thermal conductivity. This mixture may change along the routing depending on the depth of laying or thermal impact from the surrounding (crossing a street or a river). The model of the digital twin project will have all this information available and can be used for construction and later for operation. Thermal calculations can be made based on this data and measurements of actual temperatures of the GIL routing to be used for operation in the network and to prevent from overheating.

The second important information for direct buried GIL is related to the mechanical stability in the soil. The density of the soil, the pressure of the soil to the aluminum pipes of the GIL enclosure and the surface friction to hold the GIL pipes in the soil fixed are parameters which are given by the calculations of the project planning from engineering studies to fix the requirements. During construction the required data will be controlled and measured. All of this data can be part of the digital twin and made available as digital model in the data base for use during operation. The mechanical data, in relationship with the operational data of real current flow and thermal heat up of the GIL will give information of the correct function of the transmission line in the network.

This planning sequences start with the project planning and continue for the time of construction at the project site and along the routing of the GIL. The information related to the different steps of simulations, calculations, and measurements are generated at different project times and locations. They all will go into the digital twin data base to fill their model with project data. Some data can be used in advanced to plan and simulate the construction phase of GIL project. This can avoid unforeseen obstacles and avoid additional cost at times when the construction team is at site and the machines are working in the project schedule. Any required change made at this stage is most costly. Any change avoided before this stage in the project can safe high additional cost. This is one big advantage of the digital twin process.

The process of GIL laying on site requires a large number of machinery and personal to operate it. Each project requires its own set up of machines and work team to be optimized for on-site installation cost and total project time. There could be one set of machinery and work team in one shift, or two shifts, or two sets of machinery in one or two shifts. This will require longer or shorter construction times in a relationship of 6 month or 9 month or 12 months, which will come from the project requirements to be ready for the network connection or to fulfill the timing of accessibility in terms of weather condition (winter with snow and freezing temperature or heat and storm in the summer with flooding's).

The digital twin model data of the on-site manufacturing process can optimize these preconditions in a best project time schedule and investment cost solution by simulating the different possibilities. For this, the local manufacturing process needs to be modeled for the digital twin.

The GIL on-site manufacturing facility and works cover the delivery of parts on site, the storage and logistics on site, the preassembly works to prepare the GIL units for final laying, the positioning and welding of the GIL segments at the trench or tunnel, the ultra-sonic weld tests, the pulling in the trench or tunnel and the final high voltage testing of the GIL.

The GIL module assembly and the final installation of modules is shown in Figure 9.72. as a model of the digital twin data base. It is designed virtually in advance during the planning process of the GIL project to ensure the productivity of the GIL laying process in order to meet the GIL project construction within the given time frame.

Personnel skills and required equipment functions for different proven construction methods (welded or bolded joints) are available in the databases to choose most suitable ones in terms of localization and overall project efficiency.

Personal skills for the work at the local manufacturing of GIL needs to have basic mechanical knowledge and will get special training to handle the automated pipe positioning, welding, and ultrasonic control equipment. Each step of the local manufacturing is planned with work time plane and available in the data base.

Supervisors need to have special skills and will come from the GIL manufacturer to ensure correct assembly of the high voltage parts and the correct welding process and non-destructive weld testing are sent on site to ensure crucial quality of the GIL. The integrated quality assurance system for each weld, provides a 100 % quality control and documentation of each joint.

In Figure 9.72 a local manufacturing is shown for two lines of GIL to be assembled in parallel. This would be on working location and the team will work in alternative welding a conductor pipe on one line and an enclosure pipe at the other line. At the next location in this work shop the ultrasonic text of the weld is made also in alternatives on the conductor and the enclosure.

9.15.5 Used Case Digital Twin of GIS

9.15.5.1 Merging Real and Virtual Operation

The digital twin of the GIS is merging the real and virtual operation. The future electric power grid and its substations have to face new challenges in fluctuating power generation by renewable energy and with this a strong impact on power flow in the electric network. Changing local generation will require different power flow on transmission and distribution lines, managed and

Figure 9.72 GIL on-site manufacturing facility with two working places to assemble and finally weld the GIL pipes (courtesy of Siemens Energy)

controlled by the switching operations in the substation. To avoid congestions, overload conditions, high currents causing high temperatures in the equipment and the potential danger of over-heating as limiting factors for the well managed operation of the substations and the electric network.

The knowledge of the influencing factors related to the material limits of the GIS is coming from the digital twin data base of the GIS covering mechanical, thermal, and electric parameters to optimize the use of the GIS and to avoid reaching the limits of the GIS. The digital twin of the GIS provides data of the GIS design, testing, production, and operation.

Most data are available today in the substation and with the GIS but not used for the digital twin data base. To combine operational data with the digital twin can be used for virtual operation simulation and factory testing data to manage congestions actively in the network when they occur. This enables grid operators to transmit additional power through higher current for a certain period of time depending on ambient temperature and actual current flow. In Figure 9.73, the interaction between the virtual world of the digital twin and the real world of the substation equipment is shown.

The real data collected in the substation gives information of the actual situation and load condition of the GIS or transformer. This performance data is transmitted to the digital twin stored in the data base of the virtual substation equivalent.

The intelligence of the digital twin is providing simulations, predictions, and makes proposals for the optimized operation of the GIS under the changing conditions of current flow in the network. The realized and optimized changes in the substation condition and operation again produces data of the substation performance back to the digital twin for further simulations and new proposals.

This cycle of information between the real world with its equipment and the virtual world with its software and data base stored information allow to operate the substation more precise, avoid overload conditions and reach at the end higher reliability of the substation and electric network [13–20].

Connectivity Concept

The data base of the digital twin of the GIS collects and provides the following information. First is the GPS-signal to allow the operator of the substation the exact identification of the equipment

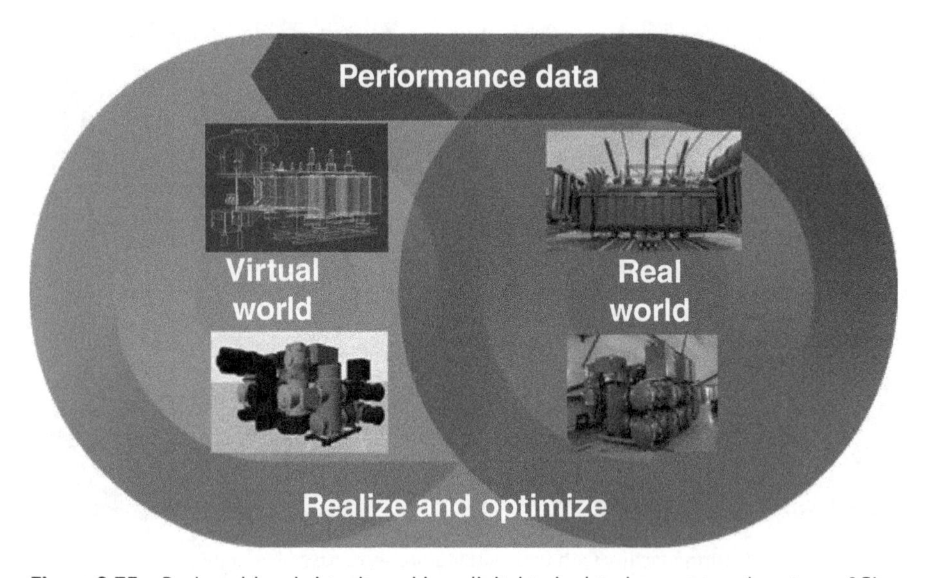

Figure 9.73 Real world and virtual word in a digital twin data base system (courtesy of Siemens Energy)

in his network. Second are the local weather information which have a great impact on the thermal condition of the equipment in combination with the actual current flow which is basis for the simulation calculations for any change in the network operation and power flow.

The relevant product data of the data based on type test values and the measured data are transferred to the data base in the digital twin cloud. Transfer data from the substation equipment to the data cloud of the digital twin need to follow secure data transmission protocols.

Such data is coming from the gas-insulated switchgear (GIS) and circuit breaker (CB) as gas density, contact and enclosure temperature, breaker counter, breaker switching position, and breaker readiness for switching operations.

From the surge arrester the surge counter and leakage current are measured and transferred to the data cloud. The disconnector drive current is measured and gives information of the functionality of the device. From the instrument transformers the gas density or oil-level alarm are protocolled and send to the digital twin data base in the cloud. In case of arc suppression coils, the oil-level alarm, top oil temperature, zero sequence voltage and coil setpoint are measured and transmitted.

Integrated digital sensors and devices are connected to the Internet of Things (IoT) as shown in Figure 9.74. The hardware devices and data sensors are well-proven to fulfill the specific and hard environmental conditions of the high voltage substations following the international standard presented in Table 9.16 For more functional requirements to the digital GIS, see Section 9.18.

In Figure 9.74 the location of the IoT devices within the GIS bay are indicated. The IoT device is connected by data lines usually wireless to the digital twin data cloud platform with encrypted data. The GPS mapping is using an GPS antenna including sensors for weather data. Gas density sensors are connected to gas compartment to monitor the actual gas density data.

Auxiliary and line switching devices are connected to the data cloud to provide monitoring data of the ON or OFF switch position, the mode condition of the spring-operated drive mechanism on the spring position and the counted operation cycles of the switching device.

The temperature sensor inside the local control cabinet monitors the temperature condition of the GIS bay and the ambient temperature sensor outside the cabinet monitors the thermal condition around the GIS bay. Both temperature monitoring data is used in the digital twin as basis for thermal calculations and limitation in load or overload values and times.

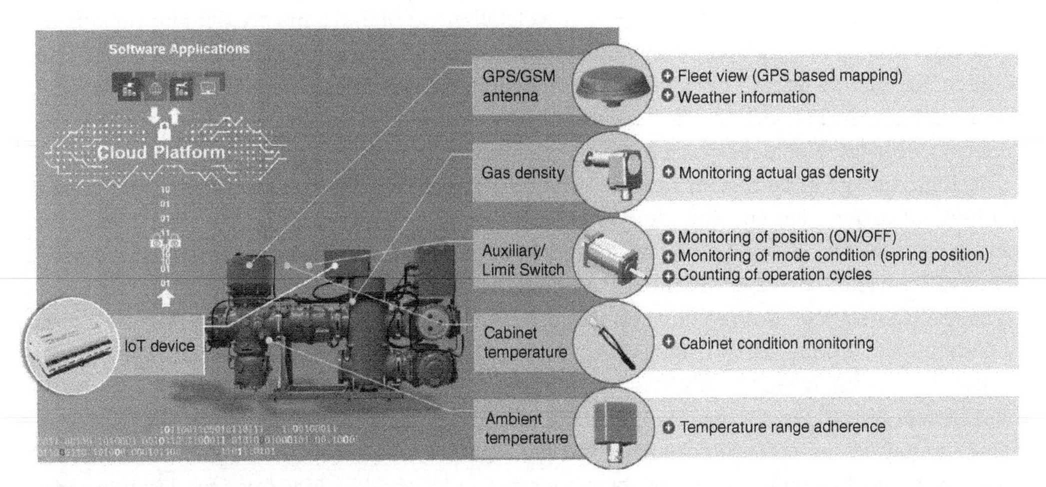

Figure 9.74 Integrated digital sensors and devices are connected to the Internet of Things (IoT) (courtesy of Siemens Energy)

Table 9.16 Overview IoT devices qualification tests

✓ **Mechanical Endurance Test:**
- 10.000 switching cycles on circuit breaker (M2 testing)
- Oscillating testing in three axis Sinusoidal oscillating testing, Shock testing, Switching impulse testing (± 35 g)

✓ **High Voltage, High Power Testing on Circuit Breaker and Gas Insulated Switchgears** (up to 420 kV/63 kA)

✓ **Disconnector Switching Testing**

✓ **EMC Testing according to IEC 61000-6-5:2015 (Power Station and Substation Environment)**
- IEC 61000-4-2: Testing of the immunity to the discharge of static electricity
- IEC 61000-4-3: Testing of the immunity to high-frequency electromagnetic fields
- IEC 61000-4-4: Testing of the immunity to fast transient electrical interference
- IEC 61000-4-16: Testing of immunity to conducted, asymmetric interference in frequency range from 0 Hz to 150 kHz
- IEC 61000-4-18: Testing of the immunity to damped periodic waves
- IEC 61000-4-29: Testing of the immunity to voltage dips, short-term interruptions and voltage supply deviations at direct current supply inputs

✓ **Climate Testing:**
- Temperature change testing between −60 °C and +70 °C, humidity

As shown in Figure 9.74 the IoT devices are integrated into the GIS bay and focus is given to the most relevant signals required to optimize the management and operation of the digital GIS. It is not recommended to follow the way to measure any physical data possible. This would end in an overload of the monitoring, simulation, and prediction software in the data base of the digital twin to optimize the performance for high reliability, availability, and long-life time of the GIS equipment. Information on transmitted power, current and other electric data from operation may be added.

The use of electronic devices within a high voltage GIS requires high mechanical, electrical, EMC/EMF and climatic condition coming with GIS operations, see Table 9.16. Mechanical requirements are related tom the dynamic mechanical forces during switching operations for 10 000 operation cycles of the circuit breaker. Oscillation tests are for all three-axis covering mechanical sinusoidal oscillations, shock tests and switching impulse test of ± 35g acceleration. e.g. of the circuit breaker interruption unit.

The high voltage and high-power switching conditions of the circuit breaker function for the test of the IoT devices switches 420 kV as maximum voltage of the equipment and 63 kA for the short-circuit rating in the network. The IoT devices need to cover these requirements including the transient over voltages related to these switching operations.

The high voltage disconnector operates after the circuit breaker has interrupted the rated or short-circuit current. The current is low and the voltage is rated high voltage. With the slow contact movement, the disconnector produces a high level of very fast transient over voltages (VFTO). These VFTO need to managed by the IoT without getting into failure mode.

The electro-magnetic compatibility (EMC) proves the immunity of electronic device to manage all disturbances around the IoT. This covers immunity to the discharge of static electricity, high-frequency electromagnetic fields, fast transient electric interferences, conducted, asymmetric interferences 0–150 kHz, damped periodic waves and voltage dips, short-term interruptions and voltage supply deviations at direct current supply inputs.

The operational climatic conditions for indoor and outdoor GIS varies from the location. Arctic temperatures or tropic condition need to be fulfilled by GIS equipment. For this reason, the IoT devices are designed and test to cover the temperature range from −60 °C up to +70 °C and 100% humidity.

In Table 9.16 an overview on performed qualification tests of GIS devices are shown, which are used for IoT.

9.15.5.2 Digital Twin Operated & Intelligent

A digital twin operated GIS adds the intelligence of the software for evaluations and simulations of the real operation to increase the performance of the equipment. In Figure 9.75, it is shown how the digital twin operation is merging the real and virtual world for advanced GIS applications. The main functionalities of advanced GIS are active load and overload prediction based on real and virtual sensors intelligence [15]. The following functions are part of IoT evaluation process:

- Active load and overload prediction & management
- Each module of the GIS is stored with its data in the digital twin as 3D model with the geometrical, chemical, physical
- And thermo-dynamic properties. Based on this data 3D calculations of the thermal losses are carried out using the type test data of temperature rise test with temperatures measured at each critical part.
- For short-time overload calculations, thermal simulations are provided for the operator's information on how high and how long an overload condition can be accepted without reaching the thermal limits of the equipment. This gives benefits to the network operation and provide higher power transmission for a given time, if needed.

The digital twin operated GIS provides the following advantages for the network operator:

- Transparency on thermal utilization of the switchgear and its modules based on digital thermal twin and ambient temperature: virtual sensors
- Indication of continuous and temporary overload current capabilities without reducing reliability and lifetime, see Figure 9.75.
- Advice for additional power transmission potential through overload current
- Prediction of overload current capabilities based on ambient temperature prediction deducted from weather forecast
- Cost reduction potential through reduced redispatch

Advanced Intelligence

Huge opportunities are possible by applying Artificial Intelligence (AI) to products and systems. From trending and prediction functionalities to complex decisions based on expert knowledge on product functionalities, behavior, predictions, and prescriptions. A first realized example is:

Gas Density Trending & Prediction

Applying artificial intelligence to optimize gas monitoring

Conventional trending models require long data history (weeks) before performing linear interpolation. Faster and more accurate gas trending is achieved by applying neural networks to compensate sensor data with regard to weather influences. The AI model requires less historical data (days). Benefits for the grid operators are:

- Reduced $SF_6 = CO_2e$ emissions
- Cost savings for unplanned SF_6 leakage repairs
- Less risk contingencies and penalties for SF_6 emissions

In Figure 9.75 it is shown how the digital twin is interacting with the GIS in operation. In a continues cycle, data from the real operation is collected and transferred to the digital twin data base. This data is used to optimize the operation of the GIS in performance, cost, and safety.

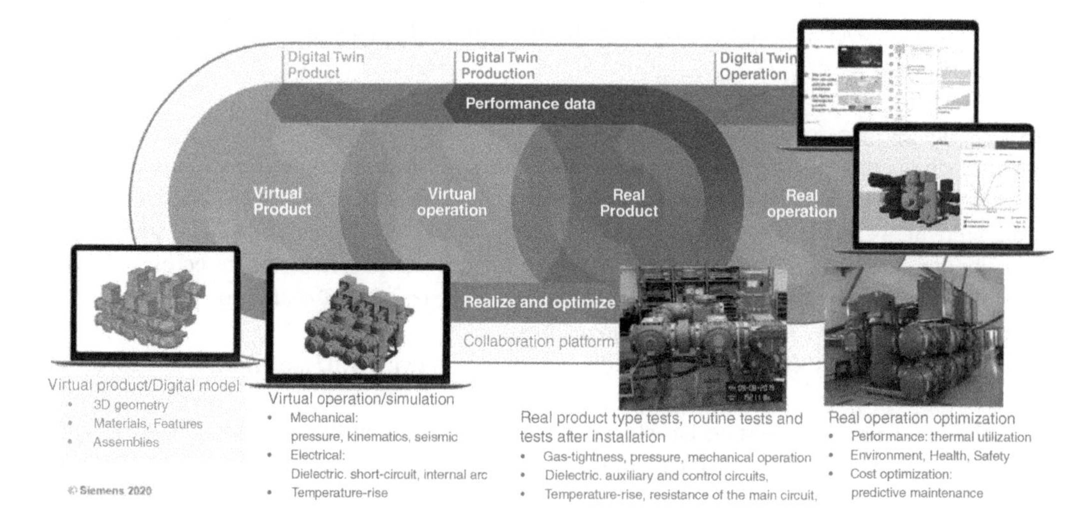

Figure 9.75 Digital Twin Operation merging the real and virtual world – example advanced GIS (photos in courtesy of Siemens)

The real product data generated with the type test of the GIS are giving the real product data gas tightness, pressure, mechanical operations, dielectric performance, auxiliary and control circuits, temperature rise test, and resistance of the main circuit.

At the virtual side, the digital twin virtual operational data and simulations provide calculations on mechanical performances like gas pressure, kinematics of the operation drives and seismic impact to the GIS. For electrical calculations and simulations, the dielectric, short-circuit and internal arc performance are evaluated by the digital twin software to predict operational impact based on actual changes on the real side of the GIS. Temperature-rise simulations are provided continuous taking into account the real-world side of temperature and current flow.

This part is the digital twin production where performance data from the real GIS is taken to calculate and simulate on the virtual side its impact to next future of the GIS and give advice and make recommendations for the network operator.

Behind this digital twin production level, the digital twin product level has a more detailed view into the GIS and its modules. The virtual product is based on digital models of each GIS module covering any detail in a 3D geometry. All materials and its parameters are stored in the digital twin data base for calculation at the virtual operation level to optimize the GIS operation for changing network conditions.

The virtual product level combines all the features of the assemblies of each GIS installation. This combines data from the product design, the prove during type and routine test, the installation and commissioning data. The digital twin is therefore a power full tool to evaluate the GIS status at each phase of its operation. To make predictions on instant changes in the network operation how high an overload condition can be acceptable and how long it can stay without damaging the GIS and reduce its life time. An example of such evaluation and information to operation staff is shown in Figure 9.75.

With fluctuating power generation using regenerative energy source like wind or solar power the consequence in the power transmission and distribution network is that power flow is changing within minutes. From full wind generation to no wind the transition phase is less than 30 minutes. For solar power, this transmission phase is even shorter, within 1 minute the solar power generation can drop from 100% down to 10% because of clouds coming up. While the load is much more constant for a safe and reliable power supply the electric power needs to change to other resources in the network at other locations. With this the power flow in the network can vary from 100% load to 10% within minutes or the other way around. This can cause temporary over load situations at some network branches. For the operator of the network, this fast-transient times for changes of power flow is too fast for manual control. Therefore, the digital twin is offering an operational support giving recommendations and makes proposal for optimized operation of the GIS.

In Figure 9.76 the digital twin operation cockpit is shown for a temporary overload situation. All modules of the GIS assembly are available in the digital twin as digital model. This covers any hot spot of the assembly based on type test protocols and temperature rise test measurement of the GIS design.

The digital twin operation cockpit indicates in this view the overload situation of general utilization of the GIS related to the environmental temperature. The graphic shows the past six hours and make predictions for the next 16 hours. The general utilization is calculated to reach 90.84% at its peak at 1:00 pm and will go down for the next hours below 20% even with a rising environmental temperature rising from 10.5 °C up to 24 °C.

This prediction will help the operator in the moment of temporary overload condition to see that the given temperature limits of the GIS are not reached. This avoids damage and possible life time reductions because of thermal aging.

9.15.5.3 User interface, App and Cyber Security

The digital twin is the synonym for the digital storage of GIS design and feature data in the data base of the cloud-based information system. Digital twin also includes intelligent software for calculations, evaluations, simulations of operation condition of the GIS in the substation of the network [21, 22].

What the operator or user will see and how he can react to the digital twin is defined with the user interface. The communication between digital twin and operator is based on a software application (APP) using computer, laptop, or smart phone as communication device. The data traffic between the operator or user with the digital twin needs to be protected from third parties and must use cyber security systems proven by international authorities and organizations.

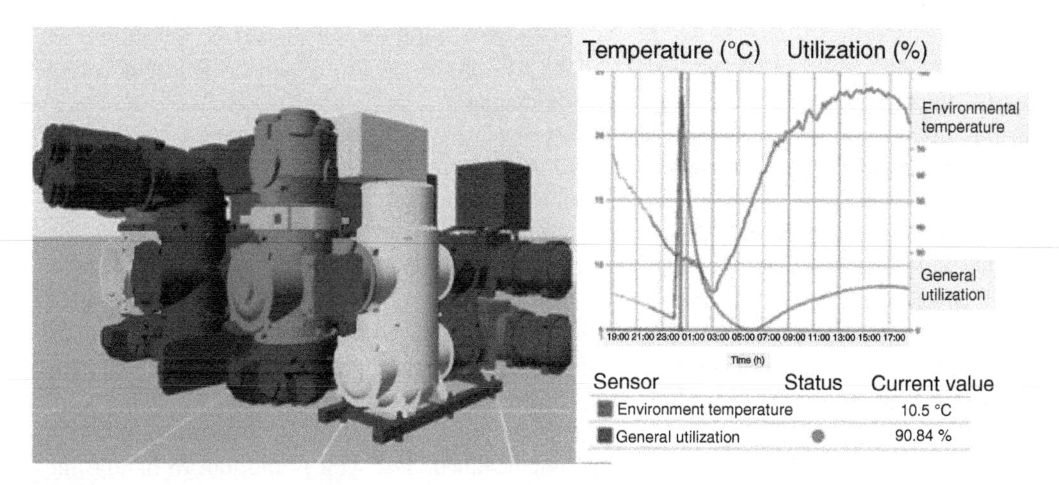

Figure 9.76 Digital Twin Operation Cockpit – example overload situation (Courtesy of Siemens)

User Interface

The main task of the user interface is to select information of importance for the operator at each given time and during each possible operational situation. This will cover the normal operation when all system works proper and no failures are in the system. In this situation, an overview to the operator is sufficient to see no action required.

If changes occur the user interface shall react to the specific change in the network. Change of power flow, failure of a device, missing signals or faults on the line or in the substation the user interface shall focus the information on the event and the consequences out of it. Recommendations are presented to the operator and clear activity schedules are presented to help the operator for right decisions.

A third situation is during maintenance and repair works in the substation or GIS. In this situation, scheduled switching operations are prepared by the digital twin to optimize the functionality of the GIS and the substations and to maintain as much as possible of power transmission availability.

The user interface plays a key role in the efficiency of the use of the digital twin. The final decision is with the operator of the substation or network. The digital twin can only provide recommendations, not all of the recommendations found by the digital twin can be activated automatically. This requires a clear structure of the user interface and a reduction of information down to the necessary.

APP

The digital twin-based equipment like GIS will be connected by software applications (APP) for visualization and analytics. The APP platform is standardized for the user in the complete fleet of his electric power network. Each single device at each substation location is connected to the internet and can be addressed by the APP from remote position. The APP allows to build up contact to each device and to check for operational data at any time. This gives transparency of the situation at each substation with its equipment.

In case of any abnormal situation the APP will inform the operator about the circumstances and will provide proposals for activities. Based on the alert-level serious, minor or can be done later the software will priories the data transmitted to the APP to avoid overflow of non-relevant information for the operator.

The data security is a basic requirement for using APPs through the internet in communicating with electric power system installations. The remote access to the substation devices allows the transmission of near real-time data and gives an overview on the actual situation in the substation and for the equipment.

Figure 9.77 General Overview of the APP for digital equipment installed at substations in Netherlands, Germany and Poland is shown. The APP shows the location of the equipment connected to the digital twin GIS and transformer in the map.

Standardized warnings and status messaged are shown at the bottom of the graphic. In this case, five equipment data indicate the status OK, two gas-density alarm or warnings, zero mechanical life time warnings, three cabinet heater warnings, one ambient temperature warning and zero circuit breaker idle time warning are given.

Such a wide area approach of covering the complete fleet of equipment installed in the substations of the network operator is only possible by using the internet as a link and communication tool. Data from the equipment, at each location, will be received by the digital twin virtual data base installed in a cloud-based system. This big data concentration will allow to get access to any information in the network of any equipment connected.

The software tools of the cloud data base provide intelligent analysis possibilities to optimize the operation of the network and of any equipment installed. The APP is the tool to handle the communication between the data base software and the operator.

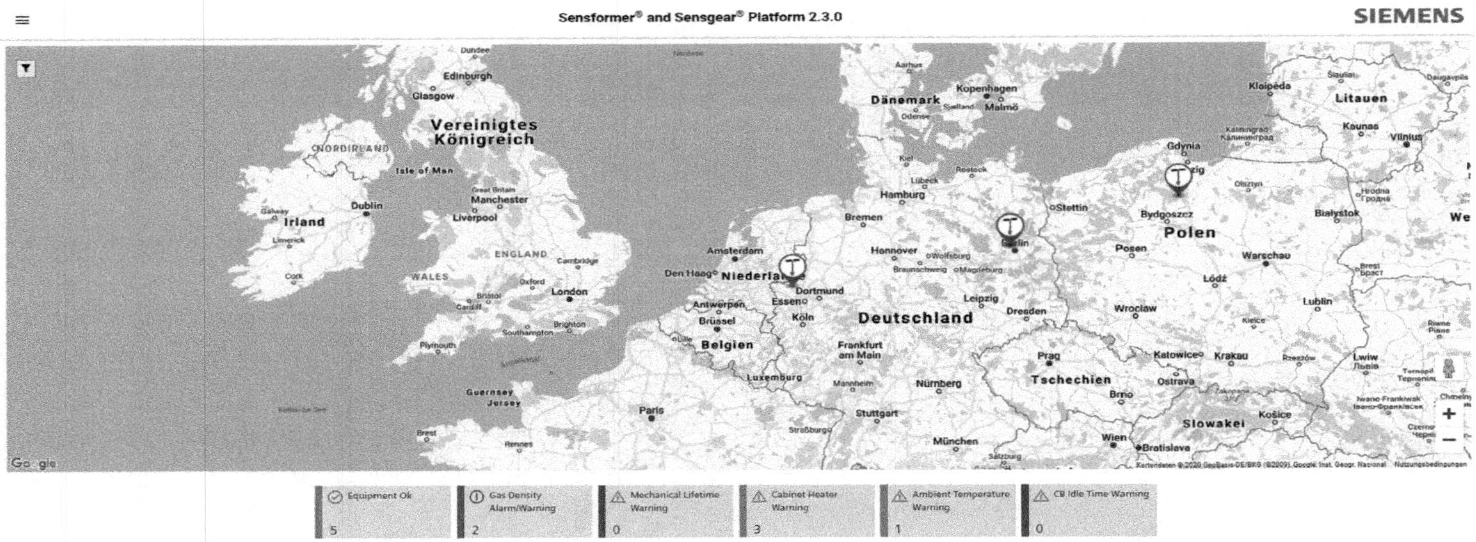
Sensformer® and Sensgear® Platform 2.3.0
SIEMENS
Vereinigtes Königreich
Irland
Dänemark
Deutschland
Polen
Niederla
Belgien
Tschechien
Slowakei
Litauen
Equipment Ok
5
Gas Density Alarm/Warning
2
Mechanical Lifetime Warning
0
Cabinet Heater Warning
3
Ambient Temperature Warning
1
CB Idle Time Warning
0

Figure 9.77 Exemplary user interface of APP (courtesy of Siemens Energy)

At the moment of this writing the digital twin-based equipment installed is still at its proto type status and experiences are now collected. The powerful tools of software in the cloud-based system for optimized operation will be the driving force for implementing this technology in the future.

The next step for more details of information comes with clicking to an asset on the application map. Selected detailed information will be shown. In this example, the circuit breaker of a substation was chosen. The first information is on mechanical lifetime of the circuit breaker, in this case a new one with only 0.02% of mechanical lifetime used.

Second information is the switching position as OPEN and with the indication "ready to close." The idle time of the circuit breaker is given with 8 days. The cabinet heater is indicated as OK and the temperature inside the cabinet is 34.8 °C. The ambient temperature range for the equipment is OK with an actual ambient temperature of 24.15 °C.

The weather forecast for the substation location in Berlin, Germany is for the day and the following three days.

The APP enables for key performance indicators (KPIs) of the equipment by push messages. In case of alarms, the APP does simplify the service messages and the information for asset management (Figure 9.78).

Cyber Security

Power transmission and distribution systems are relevant for the society and the country. Cyber-attacks on power systems can cause big damage and without electrical supply many daily important things do not work anymore. For this reason, the complete data chain from the sensors and actors, through the internet and to the cloud data base needs to secure protected against third party access.

State-of-the-art security and encryption technologies are the base for secure data handling. For data transmission to the cloud storage, an end-to-end encryption is used. Each transmission product has a unique ID, which is also used for encryption. The transmission is via HTTPS with a 256-bit TSL encryption. It complies with state-of-the-art data handling and management guidelines to ensure your data privacy and secure segregation in the cloud. IoT device and cloud services fulfill the highest cyber security standards.

The IoT devices installed in the equipment are preconfigured and preinstalled as part of the product manufacturing process in the factory. The IoT device installed in the transmission equipment is predominantly collecting and transmitting data to a secure storage at the digital twin cloud

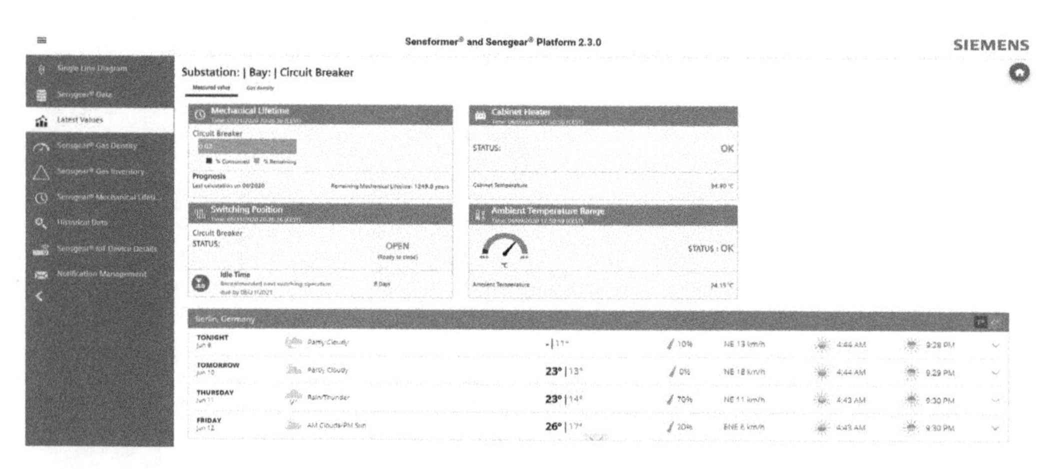

Figure 9.78 General Overview of APP – example digital equipment in Netherlands, Germany and Poland (courtesy of Siemens Energy)

data base and is using analytical software tools. For operational safety and security reasons, there are only monitoring functions related to the digital and no automated remote-control functionality to operate the equipment. Between the analysis and proposal made by the digital twin virtual operation system and the activating of switching in the network the operator has to verify the commands.

Figure 9.79 shows the cyber security organization of a digital twin cloud data base orientated high voltage substation here for example for transformers and switchgear. The IoT devices are installed at the equipment to get the required data encrypted as TSL1.2 for data communication and system login. The data protection is using trusted certificates and 4096-bit RSA keys. The firmware of the IoT is using SHA encryption. The IoT is protected against attacks on the DoS by a firewall. The resource utilization is monitored by a watchdog system. Verification and changes of configurations are protected by private RSA keys. An automated vulnerability monitoring system protects against third-party software at firmware level. The firmware can be up-dated by remote control. This rounds the cyber security at the IoT level installed at the equipment.

At cloud service stored data is protected in private zones using secure database clusters. The data access is controlled by role-based access models with multifactor authentication. The data is stored with a back-up concept to prevent from data loss and secure a disaster data discovery. Security events are automatically evaluated. Malware of the software will be detected and the system will be protected by automated health checks of the software. The low-level network is separated from the data base storage; incident handling and vulnerability of data are defined. Data is always stored and sent encrypted.

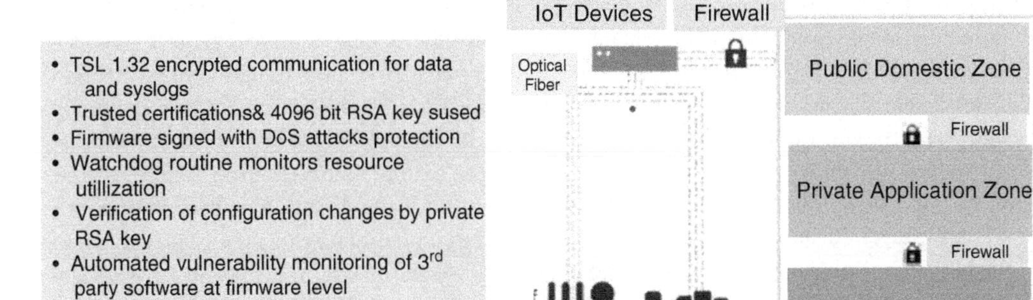

• All data stored in protected private zones
• Secure database cluster
• Role based access model with multifactor authentication
• Backup concept with desaster recovery
• Automeated evaluation of security events
• Malware protection & automated health checks
• Low level network separation
• Defined vulnerability and incident handling
• Data is always stored and sent encrypted

Figure 9.79 Customer data, IoT device and Cloud service safeguarded by latest cyber security standards (courtesy of Siemens)

9.15.6 Conclusions

The digital twin approach is a new concept to use the digitalization of technical project data and create a software model which reflects to function and performance of a technical system. In this case, the application for GIS and GIL is explained.

In principle the of digital twin approach covers the digitalization for production, construction, and performance during operation from the beginning over the complete live cycle of operation until the dismantling after end of using time of the GIS or GIL. The digital twin collects data, holds models and allows access to all data in the data base at any time of the GIS and GIL project.

The asset production and delivery of the digital twin is derived from the asset design of the digital twin chosen by the customer and focuses on optimizing the modeling process. The asset operation of the digital twin serves the asset owner to maximize the asset value until end of lifetime.

The data model set up is design for a unique data model of environment to be transparent and useable for all data base user and performances.

Using lidar scanning methods allows to build up surface and terrain models for GIS or GIL projects covering all requirements and local situations on site. The digital surface model (DSM) and digital terrain model (DTM) are used hold the data available in the data base. The triangle irregular network (TIN) data storage method allows to reduce the data volume of the digital surface (DTS) and terrain model (DTM).

In the data model of production all GIS or GIL project data required to mirror the technical specification and performance are available for implementing into the surface and terrain data. All functions and technical limitation for the production of the GIS and GIL are getting validated of consistency and correctness to prove the overall quality of the project.

During the operation of the GIS or GIL project data is collected and measured to provide a safe operation in the network. This data related to real currents, voltages, transient over-voltages, switching operations and other information related to the network and the functionality of the GIS or GIL are collected over the live time of the equipment and presents the equipment history for any time and moment. This information is used to maintain the equipment and give identifications for end of live time and replacement planning's.

The used case digital twin of GIL shows the motivation for using digital twin concept and presents its advantages during all stages of the GIL project. From planning status, with simulations to find the best technical and economical solutions until the use of the GIL in the power network the different features of digital twin are explained.

The project information model for production of the GIL explains the aspects of how the different elements of the GIL are implemented and used in the digital twin data base. Here it can be used for different purposes of coordinating the part to be delivered in time on site, to set up the quality control on-site and to validate all parameters of correctness and fitting to the GIL technical requirements and specification.

The local manufacturing process of GIL is an important part of project efficiency and temporally coordination of all logistics for a continuous on-site assembly and installation process. The digitalization of these work steps and the adaptation of the complete project schedule and works allows an optimization of the GIL project under the given local conditions and requirements. A simulation and optimization process will be carried out before the work on-site starts to find obstacles and increase local efficiency.

The digital twin concept allows new ways of project execution. Starting from the planning stage the digitized project can use the models of all elements, parts, and function to go through the steps of project execution before the work has started on site. Using the simulation software with the parameters and data models stored in the data base of the digital twin will allow the project management to identify obstacle, short comings, time bottlenecks, leak of materials, leak of personal, safety restrictions, delivery short comes, delivery over load on site, validation of delivered parts and elements according to the project specification and requirements and many other project parameters relevant for construction.

The digital twin concept will provide on-time and Up-to-date information on the performance of the GIS or GIL during the complete time of operation in the network. All related parameters will be measured, stored, and evaluated by software to prevent overload and to give indication for maintenance, repair, and replacement.

To set up a digital twin concept in a GIS or GIL project requires a large commitment and lots of steps need to be solved before the advantages can be used. But once done the digital twin concept can be used for more project using the same base of models and evaluation software. The availability of software and digital storage capacity will drive this new concept. The advantages for manufacturer are related to better technical solutions, optimized for specific local requirements and less failure by using the simulation possibilities before project execution. This makes digital twin interesting for manufacturers.

On the user side the digital twin brings more information into the performance stage during the operation of the GIS or GIL. At any time, a much better picture is available to the operator and user to find the right decisions. Data stored in the digital twin data base can be used for state estimations and planned actions for maintenance, repair, or replacement of GIS or GIL.

The used case digital twin of GIS shows first applications of this advanced technology. Merging real and virtual operation improves the use of GIS under the new requirements of fluctuating renewable energy generation and its ever-changing power flow in the electric power network.

Real world and virtual word in a digital twin data base system are connected to optimize the operation of GIS. Sensors in the GIS connected by internet to the digital twin cloud data base measure continuously the status of the GIS of key values like temperature, gas density, current flow, weather conditions and status of switches including their drive mechanism. In the virtual world of digital twin data base software can analyze, evaluate, and make proposal on the operation of the GIS.

Integrated digital sensors and devices are connected to the Internet of Things (IoT) are designed and proven according to international qualification tests of IEC, ISO, NIST, NERC and others to fulfill the specific high voltage requirements and the data security levels required for a control system of an electric power supply system.

The digital twin operation cockpit is the user interface to the operator. It is using APPs to handle the communication between the cloud data base, the analysis software tools and the operator's computer, laptop, or smart phone screen. All data communication is following cyber security rules.

A general overview APP for a first application with substations involved in Netherlands, Germany and Poland shows the power full features of the digital twin cloud data base system for advance substation control and management. Customer data, IoT devices and cloud service safeguarded by latest cyber security standards.

References Section 9.15

1 Helbig D, Fritsche R, Heinecke M, Singh P (2020) Intelligent IoT-connected transmission equipment in substations. Technical paper B3-305, Cigre Session, Paris.

2 Engel M, Oechsle F, Helbig D, Kuschel M (2020) Hochspannungsanlagen im 21. Jahrhundert. Fachtagung Hochspannungs-Schaltanlagen, TU Darmstadt.

3 Dohnke O, Groth R, Demmer D (2020) Edge Devices, Sensors, Cloud for High Voltage Switchgears. VDE-Fachtagung Hochspannungstechnik, Berlin.

4 IEC 61360-1:2017 Edition 4.0 (2017) Standard data element types with associated classification scheme, Part 1: Definitions – Principles and methods.
5 IEC 61360-2:2012 Edition 3.0 (2012) Standard data element types with associated classification scheme for electric components, Part 2: EXPRESS dictionary schema.
6 IEC 61360-6:2016 Edition 1.0 (2016) Standard data element types with associated classification scheme for electric components, Part 6: IEC Common Data Dictionary (IEC CDD) quality guidelines.
7 DIN SPEC 91391-1:2019-04 Gemeinsame Datenumgebungen (CDE) für BIM-Projekte – Funktionen und offener Datenaustausch zwischen Plattformen unterschiedlicher Hersteller – Teil 1: Module und Funktionen einer Gemeinsamen Datenumgebung; mit digitalem Anhang
8 DIN SPEC 91391-2:2019-04 Gemeinsame Datenumgebungen (CDE) für BIM-Projekte – Funktionen und offener Datenaustausch zwischen Plattformen unterschiedlicher Hersteller – Teil 2: Offener Datenaustausch mit Gemeinsamen Datenumgebungen
9 Heck A, Habenicht I (2015) GIL-Baustellensimulation – Ein entscheidungsunterstützendes Planungswerkzeug für Baustellen zur Errichtung kundenindividueller Anlagen. In: Markus Rabe und Uwe Clausen (Hg.): Simulation in Production and Logistics 2015. Stuttgart: Fraunhofer Verlag, S. 513–522.
10 ISO 19650-1:2018 Organization and digitization of information about buildings and civil engineering works, including building information modelling (bim) – information management using building information modelling – part 1: concepts and principles
11 ISO 19650-2:2018 Organization and digitization of information about buildings and civil engineering works, including building information modelling (bim) – information management using building information modelling – part 2: delivery phase of the assets
12 ISO/FDIS 19650-3 Organization and digitization of information about buildings and civil engineering works, including building information modelling (bim) – information management using building information modelling – part 3: operational phase of the assets
13 Kuschel M, Helbig D, Heinecke M, Singh P (2019) Sensformer and Sensgear – Intelligent IoT-connected HV T&D Products. Proceedings of International Conference on Condition Monitoring, Diagnosis and Maintenance; CMDM 2019; Bucharest, Romania; September 9th – 11th, 2019.
14 Heinecke M, Doring M, Fritsche R, Helbig D, Schulz R, Singh R, Singh P (2020) Transmission Products and Systems for Flexible Grids of the Future – IoT connected, Digital Twin operated, intelligent, VDE-Fachtagung Hochspannungstechnik, Berlin.
15 Cigre Study Committee B3 Substations and electrical installations: Expected impact on substation management from future grids. Cigre Technical Brochure 764, 2019.
16 Cigre Study Committee A2 Power transformers and reactors: Condition assessment of power transformers. Cigre Technical Brochure 761, 2019.
17 Cigre Study Committee A2 Power transformers and reactors: Transformer bushing reliability. Cigre Technical Brochure 755, 2019.
18 Cigre Study Committee A3 Transmission and distribution equipment: Non-intrusive methods for condition assessment for distribution and transmission switchgear. Cigre TB 737, 2018.
19 Cigre Study Committee A3 Transmission and distribution equipment: Ageing high voltage substation equipment and possible mitigation techniques. Cigre Technical Brochure 725, 2018.
20 Sensformer® and Sensgear® info: https://new.siemens.com/global/en/products/energy/high-voltage/transmission-products/sensgeartm.html
21 Kuschel M, Albert A, Ehrlich F, Nesheim N, Pohlink K, Rank T and Skar J (2020) First 145 kV / 40 kA gas-insulated switchgear with climate-neutral insulating gas and vacuum interrupter as an alternative. to SF6 -Design, Manufacturing., Qualification and Operational Experience. CIGRE session, Paris.
22 Wold Economic Forum: Global Innovations from the Energy Sector 2010-2020. WEF White Paper, Geneva 2020.

9.16 Offshore GIS

Author: Hermann Koch
Co-author: Dirk Helbig,
Review: Michael Novev, Bala Kotharu

9.16.1 Introduction

Offshore wind farms are connected to the onshore electric network by AC or DC cables. The decision to use one or the other is measured by the distance of the offshore wind farm from the coast line. AC is more economical for shorter distance in the range ranges of between 20 km and 30 km. Due to the fact that DC cables do not have a capacitive load, DC is chosen for distances longer than 30 km with DC connection length of 100–200 km from the offshore collecting platform at the wind farm to the coast line have been realized, mainly at projects in the North Sea of Europe.

Wind farms typical use wind turbines of typical 3–6 MW and up to 10–14 MW for the largest turbines. The connecting AC voltage for each wind tower is typically 33 kV and for the higher power 66 kV is used. One string, which typical consists of 8–10 wind towers is connected with the offshore platform using sea cables of AC voltages from 132 to 170 kV. The offshore station collecting platform typically collects 5–10 strings of wind turbines.

The power generation capacity of windfarm is typically in the range of 100–500 MW for AC and the for 700–1200 MW for DC connected windfarm. This depends on the power generation capability of each wind engine, the number of wind turbines on one string and the number of wind engine strings connected the offshore collection platform.

Cables rated from 220 to 330 kV are typically used in case of an AC connection to the onshore substation. For the DC connection, a 320 kV DC cable is used.

Figure 9.80 represents three principal ways of offshore power transmission installations. Two types of installations are collecting wind energy from the offshore wind park to the onshore

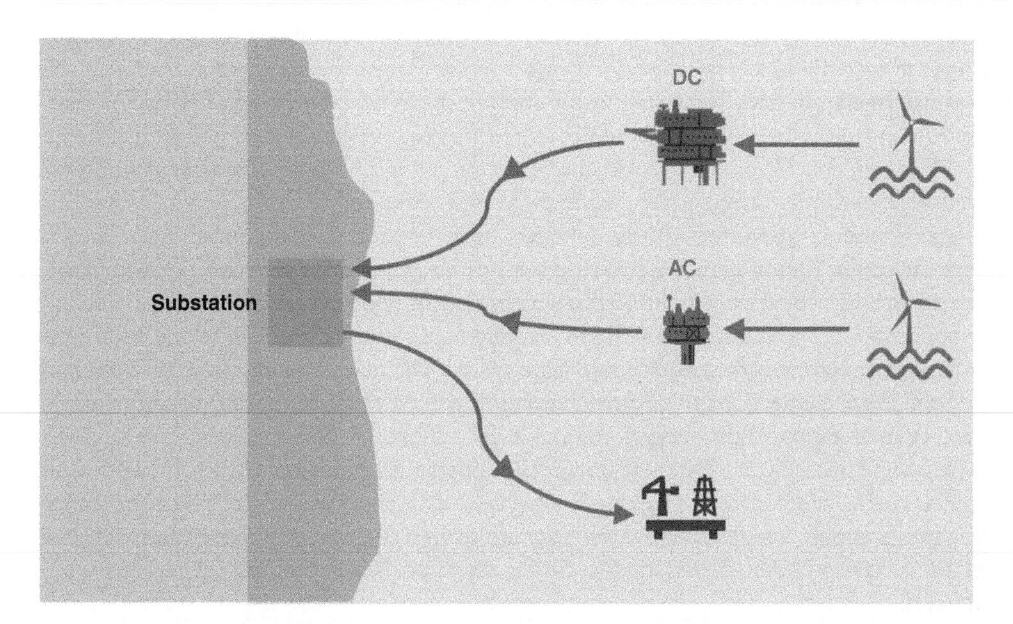

Figure 9.80 Offshore windfarm and platform connections to onshore network, upper section: DC cables middle section: AC cables lower section AC or DC power supply of oil/gas platforms (courtesy of Siemens Energy)

substation connected to the electric power network [1, 2]. One type of installation is shown for the direction of power supply for offshore platforms from the onshore substation

The upper path shows the connection of wind turbines through the DC connecting platform to the onshore substation. Rated voltages of 33 kV or 66 kV are used within the wind park 132–170 kV rated voltages are uses to connect strings of wind turbines by sea cable collectors to the DC platform. The DC platform converts the AC voltage to DC voltage of 320 kV rating for the transmission of wind energy to the onshore substation. This DC solution is used for long distances of 100–200 km from the DC platform to the onshore substation. The DC wind farm connections show high power transmission capabilities of 700–1200 MW

The middle part shows the connection of the wind turbines in the wind park through an AC collection station platform to the onshore substation. The rated voltage inside the wind park is using 33 kV or 66 kV sea cables to connect each string, which typically consists of 10 wind turbines. Each string of the wind park turbines is connected by sea cables of rated voltage 132–170 kV to the AC offshore colleting station platform. Transport cables of rated voltage 220–330 kV are used to connect the AC platform to the onshore substation. This type of AC collecting station platform is used for shorter distances typical below 50 km from the AC platform to the onshore substation. The AC wind farm connections show power transmission capabilities of typical up to 700 MVA.

The rated voltage for onshore the AC substations is 420 kV or 550 kV to transforms the electric wind energy to the rated voltage of transmission network of.

The lower part shows the power supply of an offshore platform from the onshore substation used, for example, for oil and gas platforms. Depending on the distance from the onshore substation to the offshore station platform, an AC or a DC solution could be implemented. The electric power supply from the land replaces the oil or gas power generation, which is typically used.

GIS is used in all of the offshore wind farm examples of principle stations for both AC and DC connection to onshore. Starting with the wind turbine tower a three-phase 33 kV or 66 kV GIS is located inside the tower for the connection of the wind engine to the electric collecting string, which is typically of typical 8–10 wind turbines. The end of each string is equipped with a step-up transformer for the power transmission to both, the AC and DC type of offshore energy collection station platform. The step-up voltages are typically 132–170 kV depending on the sea cable. They will run over tens of kilometers from the wind engine string to the collecting station platform. The step-up transformers are usually located inside the last tower of each string of wind turbines. The connection made by 132–170 kV GIS equipment, which is three-phase insulated and offers the functional requirements for circuit breaker, disconnector and earthing/grounding switches, voltage and current transformers and direct plug-in cable connections.

GIS offers compact solutions for offshore applications in the wind tower or on the offshore electric energy collection station platform where space is limited and ambient condition are harsh. Special design has been developed to fulfill these requirements and have been used for AC and DC offshore applications. The future use of GIS in offshore applications is foreseen world-wide with ever increasing windfarms capacity reaching Gigawatt sizes. GIS plays a major role in this development. AC windfarm connections to onshore substation will be used for shorter distances, while HVDC will be used for longer distances, for example, 50–100 km.

In a wider long-term view of offshore wind farm development an offshore electric network with inter-connections of windfarms will be increasingly used to increase availability and redundancy of electric power supply. The most activities today are seen in the European North Sea, the USA coast of the northern Atlantic and in the China Sea with large resources of electric wind generation [3].

Figure 9.81 shows both options, power to shore and power from shore. The left photo shows an offshore collecting station platform for the energy coming from the wind park to be transported

Oil platform offshore with electric
power supply from landside

Figure 9.81 Left: Power to shore by AC/DC right: Power from shore by AC/DC (*source* [4, 5])

onshore. This platform could have AC or DC installations depending on the distance from offshore to onshore.

The right photo shows an oil or gas platform which is connected to onshore for electric power supply. This connection may be made by AC or DC depending on the required energy and the distance from the platform to onshore.

9.16.2 Types of Collection Platforms

There are different types of collecting platforms used. One type is the monopile type and one is the jacket type, see Figure 9.82. The standards foundation of the monopiles platform is a modular design, with a single circuit offshore AC connection to the onshore substation. The power rating of the offshore platform is in the range of up to 400 MVA with one main transformer. The offshore AC voltage GIS substation is rated for 33 kV or 66 kV and typically offers 9 feeders and a compensation reactor for sea cables of up to 100 MVAr. The topside is 26 × 26 × 15 m in dimension and the weight including equipment is about 1300 t.

Second type is the standard foundation of jacket-type platform is using modular design and a symmetrical monopole AC/DC converter station to connect the offshore platform to the onshore substation. The power rating of the offshore platform is in the range of up to 900 MW using two main transformers. The offshore AC voltage GIS uses 66 kV or 155 kV rated voltage with up to 24 feeders. The export circuit from the platform to the onshore DC/AC converter station is rated for ±320 kV DC voltage. The reactive power supply for the export circuit is integrated to the converter station. The topside is 72 × 32 × 32 m in dimension and the weight including the equipment is about 7500 t.

In addition to the standard 400 MW AC and 900 MW DC offshore collection platform larger platform types are also. Today AC platforms reach up to 800 MW, while DC platforms reach 1200 MW. The rating limitation is mainly given by the availability of the sea cables for AC and DC applications.

9.16.3 General Offshore Requirements

9.16.3.1 Risk Management

Offshore substation design has to follow more stringent reliability, maintenance and certification requirements than those onshore. This is based on the remote location and limited accessibility of the offshore substation. Only trained personnel can be allowed to enter a wind tower or offshore platform. The weather conditions are crucial as access to offshore installations is only possible by boat, ship or helicopter. There can be days and weeks with uncertain weather conditions which limit accessibility to and from the offshore installation.

Figure 9.82 Left: Monopile type for AC collecting platform right: Jacket type for DC collecting platform (courtesy of Siemens Energy)

The risk management needs to analyze, identified, evaluated and have mitigation measures in place in case of a failure in offshore installations Based on this risk analysis a system is established. It is based on the operators of offshore installation and equipment provided data of failure analysis or maintenance requirements. The GIS needs to have sensors to deliver such data. A repair and maintenance plan will then cover all required actions. The goal is to be able to do most of this work remote without the need to go to offshore installation. Since GIS offers high equipment reliability it is good solution to meet these offshore risk requirements. In case of an event on the platform personnel is required to go to the offshore installation, these specialists should be able to handle several types of problems of the primary and secondary GIS installation to reduce the number of people required on the offshore platform. For this reason, telecommunication via monitoring cameras and software is key in establishing a connection between the specialist offshore with the expert onshore.

9.16.3.2 Redundancy

The redundancy of the windfarm is usually defined by the connection to land with only one sea cable system. If this sea cable fails, the energy delivery stops. Because of the high cost of the sea cable and the high reliability of the cables (usually failure of sea cable is due to ship anchoring) the interruption time for cable repair (could be month) is less costly than to provide a second sea cable connection.

For GIS in the wind tower or on the offshore collection, platform requirements differ. Here, measures are needed to improve the reliability and availability of the GIS: corrosion protection, over-voltage protection, over-load protection, redundancy in the electric switching scheme, redundancy for secondary systems and remote supervision capabilities are some relatively low-cost additions which will improve the long-time reliability of the GIS. This is done to avoid a failure or malfunction of the GIS, which will interrupt the power delivery of the windfarm.

9.16.3.3 Electric Circuit Rating

The electric circuit rating of the GIS for offshore windfarm needs to be based on the rated generation power in the windfarm. This differs from substations on land, as the power flow in the offshore windfarm and on the offshore collection platform substation will be at rated power of the wind turbines for long periods of time. The wind in offshore installations is more stable for a longer period of time when available and can last up to a month. This means that the rate current is continuously flowing and will heat the GIS. This differs from onshore substations where a day profile typically has a peak load and down time that is constantly changing [6, 7].

In a consequence, the rated power of the wind farm is creating a continues rated current for a long time of operation. Not to operate the GIS at its rated limits for long time periods and may be reducing life time of the equipment some power reserves shall be covered by the GIS design.

9.16.3.4 Personnel Safety

The personnel safety starts with the transportation of GIS to the offshore wind tower or collection platform. Decisions in regards to the method of transportation need to be analyzed and made, whether it be to transport by boat or ship or even helicopter. The planning of these steps shall follow the rules given by the Douglas Sea Scale [29] to evaluate the sea conditions or the Beaufort Scale [30] for wind speed evaluations.

Emergency evacuation plan by sea or by air are required to be planned before the installation commence. Also, personnel are required to stay on the platform or even inside the wind tower when weather condition quickly change at sea for weeks or even month.

Additionally, an emergency evacuation plan is created in case of an injury.

9.16.3.5 Asset Safety

The GIS asset in the wind tower and on the collection, platform needs protection against fire, explosion, ship collisions and secure access to the GIS operation rooms only for high voltage qualified personnel. During operation the ventilation and/or cooling system needs to be monitored for safe function. The high voltage protection and control system proper operation is essential and needs supervision during operation.

Spare parts management and storage is important to reduce the down times for the GIS if some replacements are necessary. The recommended spare parts to be stored either offshore in the wind tower or collection platform onshore or kept in the manufacturer's factory on demand need to be discussed and decided with the manufacturer.

Asset safety planning requires the evaluation of the cost of preventive measures versus against the cost of down time and repair and is part of the total windfarm power delivery reliability. The cost of maintenance work is and evaluated against the cost of the power interruption for repair.

Accessibility to repair parts for the GIS are a crucial requirement in the environment of space limitations available in the wind tower or at the collection platform.

Remote diagnostics of the condition of the GIS are of great importance for correct evaluation of GIS condition and prevention of unforeseen failures. Planned maintenance and repair are much more efficient than unplanned event with power delivery interruptions.

Using remote diagnostic methods will reduce routine maintenance and technical over-design of the GIS will reduce unplanned maintenance. Transportation equipment and personnel, maintenance specialists with offshore training and required certifications and verifications will assure fast reaction in case when needed. International standard requirement for wind energy generation systems is given in IEC 61400 series [8].

9.16.3.6 Engineering Studies

Safety considerations for offshore GIS windfarms include potential risk to personnel, damage to equipment, failing transportation or costs associated with unforeseen and unplanned environmental impacts.

Many regulations for offshore installations are come from the oil and gas industry which need to be adapted. The main difference between oil and gas installations and windfarms is that in the windfarm there is no handling of explosives, like natural gas or oil dwells. This means that many requirements for oil and gas platform are not needed for windfarm. The engineering studies should focus on the offshore high voltage substation and eliminate requirements that are not relevant to windfarm in order to reduce the cost.

9.16.4 System Offshore Requirements

GIS is a structural part of the offshore windfarm. As such more consideration should be given to the electric scheme and structure of the windfarm, as reliability, availability and maintenance have a lower impact with GIS. The GIS has a very compact design and can be used inside the wind towers with different electric schemes to increase windfarm reliability and availability.

GIS provides a highly reliable option for windfarms, and do not require redundancy. The reason is its very low failure rate (statistical failure rate MTBF of 1000 years), minimizes any installation and maintenance work offshore (GIS uses pretested and assemble modules and switching bays), smart maintenance to prevent rather than repair and maximize the availability of electric power transmission (not to focus on time for availability).

9.16.4.1 Cable Collector System of the Wind Farm

The simplest method to organize a windfarm is the radial cable collector system which connects several strings of wind turbines to one collection station platform. The strings are not interconnected.

The second method is the full power rated ring network which connects the wind tower strings and can carry the full generation load of the wind generators. This allows for two independent circuits from each wind tower to the offshore collection station platform.

The third method is rings network with reduced power transmission capability as it only allows an emergency power transmission through the ring with limited power transmission capability at the end of each string. This will allow for the use of smaller cables with less current carrying capability to connects the string ends.

The combined ring network divides the windfarm in sections. For example, instead of 8 strings with 8 wind tower per string, the wind farm is divided into 16 strings with 4 wind towers each. This will require more sea cable length with less conductor cross section. Each ring has a total of 8 wind towers from two cable strings.

This choice of more redundancy in the windfarm collector system will require more investment in cables and GIS and should be evaluated in relation to the total cost. The additional cost associated with GIS related to the end of the cable strings with an additional bay to connect the cable and close the ring. Additional sea cable is also needed at the end of each string to connect with the other string.

This solution provides for an option to have multiple wind farm interconnect and create an offshore network. The interconnections can be made at the end of the cable strings of wind towers of one windfarm with another by using a cable connection at voltage levels of 33 kV or 66 kV. Another option is by inter-connecting the offshore collection station platform of one windfarm with another by using higher sea cable voltage levels of 132 kV or 170 kV.

9.16.4.2 Overload Operation

Operation of the offshore substation at or close to their maximum rated transmission capability helps to reduce investment cost, but in turn also reduces the live expectancy of the equipment. Transformers and cables are most sensitive to overload conditions, which leads to higher temperatures during operation. GIS and GIL are less sensitive to temporary overload condition due to the time it takes to heat up the equipment, it typically takes days to reach the limiting temperatures.

A situation like that can occur due to prolonged wind conditions. To avoid such conditions, it may be economically beneficial to built-in some overload capability. This will allow for redundancy in case of N-1 state in the windfarm and its connection to the onshore network onshore.

9.16.4.3 Substation Design

Offshore windfarms are planned based on the geographical conditions at the proposed area for the installation. Substation design for the GIS depend on electric power of available wind turbines (4, 6, 10, 12 or 14 MW), length of cable collector strings for gathering generated electrical energy (40–120 MW per cable collector string) and the total electric power collected at the offshore platform (300, 700, 1000 or 1200 MW), from several cable collector strings (4, 6 or 8 strings of wind turbines).

The system code connection requirements at the point of interconnection of the offshore windfarm to the onshore power transmission network add requirements to the windfarm electrical design.

Characteristics of the electrical conditions in case of a failure in the windfarm or the connection onshore are based on the electrical system structure at the connection point onshore. The fault ride through or recovery requirements at the onshore point of interconnection give maximum voltage recovery times, handle the frequency responses or the requirement for reactive and active power control. This might require remote control of the substation operation from the onshore control center and may need additional switching capabilities for reactor on the platform.

The choice of wind turbine is also influencing the substation design. For example, fixed speed wind turbines using Squirrel Cage Induction Generator (SCIG), variable speed wind turbine with partial-scale power converter using a Doubly Fed Induction Generator (DFIG), Variable Speed Wind Turbines with Full-Scale Power Converter or variable speed wind turbines with Full-Scale Power Converter (FCG) using a synchronous generator influences the type of GIS in each wind tower. Different types of transformers, capacitive banks and voltage converters are used.

The offshore windfarm and its substation need to fulfill certain requirements for the dynamic voltage control at the onshore connecting point. The offshore load tap changer of the transformers, the ability of wind turbines to adapt to reactive power and voltage control requirements or reactive power compensation will influence the substitution and GIS design as well. Switching reactors may be installed at the wind tower or at the collection station platform, alternatively static var compensation (SVC/STATCOM) electronic equipment may be used for reactive and active power control. This may require harmonic filters at the offshore platform, the onshore substation or converter station.

Several technologies can be used to fulfill the requirements for the fault recovery in windfarms and their collection stations *r*. One option is to use a so-called crowbar to short circuit the dually fed induction generator (DFIG) in the wind tower at the electric generator. To adjust the terminal voltage, transformer tap changer are used. With electronic equipment a full power flow control is given by Flexible AC power Transmission Systems (FACTS), Mechanical Switched Capacitor (MSC), Mechanically switched capacitive damping networks (MSCDN), Mechanically switched reactors (MSR), Thyristor-based Static VAR Compensators (SVC) and Synchronous Static Compensators (STATCOM). Using electronic power flow control systems involves electric filters for harmonic frequencies in order to reduce the impact of higher frequencies and avoid impact to the electric power control and protection system. The substation design needs to incorporate passive and active filters.

The specification of the fault current making and breaking capability of the GIS in windfarms is based on the fault level for three phase or single phase. The requirements are outlined in international standard IEC 60909-0 [9]. The short-circuit fault current is dependent on the energy from the onshore system connection and from the contribution by the wind turbines in the offshore windfarm. The short-circuit-limiting requirements will also influence the use of two or three winding transformers.

The effect of the capacitive storage of the high voltage power cables in the windfarm cable collecting network and the connecting cables to the onshore substation shall be taken into account by installing loading and de-loading resistances in the substation. Additional switching options of the GIS may be required.

Substation configuration for onshore are mainly chosen by the importance of the substation in the system. The decision for a single bus bar, double bus bar, ring bus, breaker and a half or even three bus bar configurations, will be based on economics and on the substation importance. The cost associated with a 1 GW power plant failure can quickly add up to millions of $ or €. As such, the additional cost associated with the investment outweighs the potential risk.

In the past the impact of the energy loss offshore windfarms was considered less important and a redundancy seemed too expensive. This led to substation design with less or no redundancy. Now after some years of experiences and increased power generation capability of offshore windfarms reaching 1 to 1.2 GW the considerations of redundancy are changing. Higher redundancy becomes more important which is reflected in substation and windfarm design.

This first design considerations are the choice of the voltage levels for the wind tower, the cable collector system in the windfarm and connecting cable to the onshore network. The decision is influenced by the availability of equipment mainly the sea cables. The power transmission losses are lower with the use of higher voltage ratings in the windfarm sea cable collector and for the connecting cables to the onshore substation.

At the medium AC voltage level of the wind turbines, the transformer and switchgear are installed inside tower of the wind turbine. Due to the limited space only compact equipment design can be used. Equipment in the tower is installed at different levels and GIS is used for the switchgear. For electric generators of wind turbines of up to 6 MW of power, the voltage level is 33 kV. For electric power generation of above 6 MW up to 14 MW of one wind turbine, the voltage used is 66 kV, and higher voltage ratings are under discussion. GIS equipment at this voltage level is typical of three-phase insulated design. The current rating in the full power ring configuration of the windfarm may be from 2000 A up to 4000 A rated current.

The medium voltage substation scheme uses single bus-bar design with two power transformers. The bus bar is divided in two sections to increase redundancy. The rationale for choosing a single bus-bar design is the high reliability of the medium voltage GIS equipment.

High AC voltage levels of 132–170 kV are used to connect the cable collector string with the collection station platform and from the collecting station to the onshore substation. In case of a DC connection from the collecting station to the onshore converter station, a 320 kV DC sea cable is used for maximum power transmission of 1200 MW. The electric system connection point onshore is 420 kV or 550 kV transmission lines to bring the electric energy to the load centers. There are technical limitations for the voltage levels used for sea cables. The very limited space in the wind tower and at the collection platform makes the GIS equipment most practical for this application. Beside the compact design of GIS, the high electrical safety with fully encapsulated high voltage conductors and a solid grounded metallic enclosure enables access to the switchgear room during operation. Redundancy is provided by divided single bus bar, double bus bar or ring bus design to allow switching operation in the substation.

The power transformers are connected to the GIS using high voltage cables or gas-insulated bus to avoid the large space required for connections made by air bushings. At the medium voltage levels (33 kV) and lower high voltage levels (66, 123 and 132 kV) plug-in-type solid-insulated XLPE cable connections are used.

9.16.4.4 Earthing/Grounding

The earthing/grounding concept of offshore windfarms is dependent form the specific requirements in the wind tower, the sea cable connections and the offshore collection station. There are four different principal methods used: isolated neutral, solid grounded, resistance grounding and switchable between solid and resistance.

The solid grounding/earthing is used for high voltage applications above 100 kV in most cases to identify phase to ground fault quickly and prevent damage from currents possible transient overvoltages. The high short-circuit current may generate high touch and step potential voltages and effective grounding of any metallic part in the wind tower or at the collection platform is needed.

The isolated or resistance grounded neutral is used in low (LV) and medium (MV) voltage applications which requires higher insulation rating but less short-circuit current rating. By using a Peterson Coil in series to the ground connection the short-circuit current is limited. In high capacitive systems, such as in windfarm cable collector networks, this resistive grounding may form a resonance grounding reducing the fault current to a very small resistive value. The disadvantage of the isolated or resistive grounding is that it is difficult to identify short-circuit situations because of the low current values and to manage a 1.73 overvoltage for the healthy phases in the system.

Trapped charges in long cables lead to high inrush currents when the open ends are closed and the capacitive current discharges the cable. To avoid damage to the switchgear cables are disconnected via the circuit breaker before disconnector and ground switch operations. Additionally, suitable interlocks are designed as part of protection and control system of the substation and GIS. The cable discharge can also be managed by using discharge current carrying voltage transformers or by using the power transformer on the offshore side for cable discharging. This also requires suitable GIS schemes.

9.16.4.5 Insulation Co-ordination

Fast and slow front over voltages needs to be analyzed for the offshore windfarm and its connections in order to maintain a safe function. IEC 60071 [10, 11] is used to calculate the over voltages and chose the required power frequency and impulse withstand voltage levels. In North America, the insulation coordination guiding standards are C62.82.1 [12] and IEEE 1313.2 [13], later superseded by C62.82.2 [14] soon.

The structure of offshore windfarms with several setup transformers and different line impedances defines a complex overvoltage scenario with several limitations given by the equipment. Overvoltage surge arresters are required in most case to limit the over voltage stress to an expectable level.

The continuous operating voltage at the open ends of the cable collector may be high when the power output of the windfarm is high. This needs to be identified during the design of the cables and connected GIS. Surge arrestors may be required when very fast front over voltages occur through switching operations of disconnectors, ground switches and circuit breakers of the GIS and when the voltage peak increases which may generate values not covered by standard equipment.

Due to the fact that the transmission utilized cables and encapsulated GIS fast over voltages from lightning strikes to the transmission lines or substations is unlikely and therefore not further over voltage protection is required.

Slow front over voltage happens from switching operation with charging currents of cables and may interact with electric filters for harmonic frequencies. Switching studies according IEC 60071 [11] or IEEE C62.82.1 [12] are required to provide stresses on the equipment from switching over voltages of the wind farm and to limit damaging over voltages. The study shall cover charging the electric cable system, reactor switching operations, cable collector disconnecting, ground fault situations and phase to phase faults.

Temporary over voltage may last several seconds and in offshore windfarms with delay to transformer energization, ground faults and load rejection. Offshore windfarms are operating with a large number of transformers (high inductance) and long cable lines (high capacitance) which results in a weak electric system and higher probability of transient over voltages and causes a longer duration of resonance conditions. The same is true for ground faults initiating temporary over voltages lasting for seconds. Load rejections on the onshore side of the windfarm connection may also result in a temporary overvoltage through loading the cable network with capacitive energy coming from the wind turbines until they are interrupter by the protection relays. This results in high over voltages to the windfarm equipment.

9.16.4.6 Voltage Fluctuations

Offshore windfarms are electrical weak systems with high capacitances of the cable and high inductance of the transformers. In addition, the power generation is fluctuating due to the availability of wind. Wind turbulences and gusts result in fluctuation voltages at the wind turbine generator with each rotation. This fluctuation of uneven power output is the source of flicker voltage variations. Flicker voltage variations are felt by humans and can cause irritation problems. The flicker voltage values are limited by international standards IEC 61400 [15, 16].

Several measures can be implemented to control variation of wind speed at the output voltage of the generator, such as the use of static var compensation systems; limit switching operations in the windfarm; increase the short-circuit level at the point of common coupling; control energization of transformers and cables.

9.16.4.7 System Studies

It is recommended to perform system studies of the offshore windfarm before it is designed and constructed. The following studies are recommended for the design depending on the system code requirements covering reactive power, harmonic voltage performance, static and dynamic stability performance. For wind power generation and the export of energy from the offshore windfarm to the onshore substation and electric system, the component ratings need to be specified, protection setting to be defined and safety requirements are to be addressed. See Table 9.17.

9.16.5 Equipment Offshore Requirements

The specifications for the offshore windfarm need to be adopted to the local conditions and are based on the engineering and system studies, operators and maintenance requirements, windfarm requirements and equipment supplier parameters.

9.16.5.1 Medium Voltage Switchgear

Gas-insulated type of switchgear with vacuum circuit breakers are used for medium voltage at offshore windfarms both in the wind tower or at the collection stations. The gases used are SF_6 and air for dielectric insulation.

The voltage and current rating of the medium voltage equipment is defined by the structure of the windfarm and the number of power generation connected to the transformers. A typical transformer power value is 40 MVA, which is a standard-type transformer. With this power rating at 33 kV equipment will be 1250 A. The fault current rating in the windfarm is relatively low and rated to 20 kA in most cases. This requirement can be met by standard medium voltage GIS of 40 kV short-circuit rating.

The lightning impulse withstand level (LIWL) of standard IEC MV GIS is 170 kV for 36 kV rated equipment. Transient overvoltage studies often result in higher transient impulse voltages of 185 kV. This will require additional surge arresters to protect the equipment.

Table 9.17 Recommended studies for offshore windfarms

Load flow study	Determine reactive power requirement Current ratings of equipment Calculate voltage limits Calculate active power losses Define tap changer of transformers
Short-circuit study	Maximum short-circuit currents Define settings of overcurrent protection Maximum phase to ground currents Short-circuit contribution of wind turbines to point of common coupling
Harmonics study	Define the resonance behavior Calculate harmonic impedance frequency scan Verify measure for limit harmonic distortion Evaluate wind turbine generator with converter
Insulation Coordination Study	Calculated maximum voltage stress Identify the use and rating of surge arresters
Electromagnetic Transient Study	Calculate limits, wave shape and values of over voltages Simulate different scenarios of the windfarm operation Simulate energization of windfarm
High Voltage Export Network Transient Study	Simulate energization of export cables Simulate inrush currents Simulate resonances Simulate the impact of transient voltages
Voltage Fluctuation Study	Calculate the voltage flicker emission Calculate the voltage drop at energization Calculate specific measure (point-to-wave switching, preinserted resistor switchgear
Dynamic Stability Study	Simulate reactive power output from wind turbines Simulate reactive power from windfarm to the electric network Simulate stability of the windfarm beyond reactive power limits Simulate windfarm on symmetrical and asymmetrical voltage sags
Safety Earthing/grounding Study	Calculation of cross sections of earthing/grounding cables Calculate touch voltages Calculate dissipation of fault currents to ground Determine the earth/ground impedance Calculate ground potential
Neutral Grounding Study	Check rating for earthing/grounding transformers Calculate zero sequence current contribution Calculate power frequency voltage stress during first phase short circuit
Protection Coordination Study	Design protection philosophy Select individual protection devices Define current transformers Determine settings of protection
Electro Magnetic Field Study	Calculate magnetic flux density Calculate electric field strength

Medium voltage electric circuit configuration is single or double bus bar. Single bus bar has no redundancy and double bus bar offers redundancy in case of failure or for maintenance or repair work. The decision must be analyzed by a system study for availability and reliability of power transmission. System power flow and reliability studies are used to find the optimum solution for the windfarm.

Medium voltage GIS has to switch the transformers, the cable collector, reactors or capacitor banks for electric filters.

The operation and maintenance aspects for medium voltage GIS in the offshore windfarm requires very high equipment reliability to ensure that the operation is uninterrupted for life time of the windfarm. Key considerations for offshore applications are the minimum level of required maintenance, remote monitoring and preventive actions in order to limit personnel need to go to the windfarm. The life expectancy is 25–40 years depending on the windfarm developer requirements.

Environmental conditions are harsh because of high saline air and high moisture levels at offshore installations. Closed rooms and controlled air exchange with filter systems are required for medium voltage GIS and its secondary and control equipment.

Vibrations and transportation requirements are another key design consideration for the specification of medium voltage GIS. International standard IEC 62721-1 gives performance values [17]. Vibrations are coming from the wind turbines, the water movement and the waves at sea and give permanent stress to the equipment. Transportation stresses are related to the long ways from the factory to the installation in the wind tower or at the collection platform. Medium voltage GIS is prefabricated and pre tested before delivered to the offshore windfarm.

Circuit breaker used in medium voltage GIS are of SF_6 or vacuum interruption type. Both types are design according the requirements of IEC 61271-100 [18]. In North America, see IEEE C37.06 [19].

9.16.5.2 Transformers and Reactors

The main transformer is the largest equipment used at an offshore windfarm connecting station and is used to step up the electric energy to transmission level for the connection by sea cables to the onshore substation. Reactors are used for power factor compensation. Due to the limited space and the high-power transmission capability, oil-insulated types of transformers and reactors are used. The large amount of oil on the platform carries a high fire risk and special hazard protecting measured are required. The offshore installation of transformers and reactors differs greatly from onshore installations. Beside the fire hazard, the oil poses an environmental risk and can cause sea pollution should a leak occur.

Natural and synthetic ester oil can be used as an alternative in order to reduce those risks. They pose a lower fire hazard and have biodegradation rates of 90% degradation within 20 days.

Transformers and reactors for offshore platforms need to be light weight, small in size, with low maintenance requirements and high in reliability.

The operating voltages of the main transformer are related to the availability of sea cables and the required power transmission capability. On the low voltage side where the collecting cable array connects the strings of wind generators to the collecting station, 33 kV or 66 kV are used and for longer distances 110 kV, 132 kV or 150 kV are used. On the high voltage side of the main transformer 170 kV, 220 kV or 330 kV are used. To connect to the onshore network another transformation to 420 kV is required. In case of DC cable connection, the rated voltage is ± 320 kV DC.

The transformers and reactors may be installed outdoor on top of the platform. In this case, the transformer is exposed to the environmental conditions and needs an effective corrosion protection to with stand the saline and wet environment. In the cases when, the transformers and reactors are located indoor on the platform, e the cooling and ventilation of the transformer needs to be managed by radiators which are installed outside on the platform and which require protection from the weather conditions. Any plastic materials used need to be able to withstand the harsher weather conditions, as ultra violet ray at sea is stronger than on land.

Further, the typical vibrations from the power frequency which are 50 or 60 Hz are 100 Hz or 120 Hz in offshore windfarms, which cause additional mechanical stress to the platform. Anti-vibration insulators are used in the transformer to avoid malfunctions of other equipment installed close to it. Specially the medium and high voltage GIS shall be protected from long time transformer vibrations.

The connection of the transformer to the GIS will require protection against transient overvoltage coming from the switching operations of the GIS. An insulated coordination study is performed in order to determine the type of surge arresters to be used to limit the over voltage values. In case of a direct connection between transformer and the GIS using gas-insulated bus ducts or solid-insulated cables, the overvoltage protection is mandatory to protect the transformer from transient overvoltage and damage of the winding.

9.16.5.3 High Voltage Switchgear

High voltage switchgear used at offshore windfarms in the wind tower or at the collection station is from the gas-insulated type with vacuum or SF_6 circuit breakers. Today, vacuum circuit breakers are used at the lower voltage levels of 66 kV up to 145 kV, and in future it is expected that this technology will be used at higher voltage levels of up to 245 kV. SF_6 or technical air is used for dielectric insulation.

The voltage rating of the high voltage part of the offshore windfarm that connects to the onshore substation is determined by the available sea cables, which can range up to 330 kV. The current rating of the high voltage switchgear depends on the power to transmit between 2000 A and 4000 A.

The fault current rating of the windfarm is relatively low at less than 25 kA. But the direct connection of the windfarm to the onshore system could require higher short-circuit rating of up to 63 kA.

The simplest high voltage switchgear configuration is only a disconnector and earthing/grounding switch for the sea cable. In this case, the circuit breaker would be installed at the onshore substation to trip in case of failure of the sea cable.

The most common configuration is the switchgear with a circuit breaker, disconnector and earthing/grounding switch at the windfarm collection platform, which can trip, disconnect and earth/ground the sea cables.

The collection station could have one or two transformers to feed into on sea cable. In case of one transformer, the GIS would have one bay to connect the transformer. In case of two transformers, the GIS would need two bays for the transformer connection and a single bus bar to connect both transformers. There are numerous different design possibilities including those switching the transformers from the medium voltage side.

High voltage GIS is high reliable and is maintenance free for long periods of time. For onshore installations, the leading manufacturers offer 25 years of operation before the first maintenance is required. This has been proven by experience with onshore GIS. Offshore applications differ in that they combine long cables (high capacity) and transformers (high inductance), and the GIS is exposed to high electro-magnetic stress which may require more frequent maintenance.

Maintenance of any equipment installed offshore is expensive and should be part of the project planning process. High voltage GIS is very compact and can be located within small spaces offshore, which saves on the costs associated with the steel structure of the platform. However, the smaller size makes maintenance and repair more complex. Therefore, the GIS exchange module plan should be developed before construction commences in order to make repair easier.

SF6 management of GIS involves monitoring the gas density of each GIS gas compartment and provides indication of any gas leak at an early stage to prevent a lockout. Gas handling work preparation includes having the required tools and equipment, including some spare SF_6 readily available at the collection station. Any gas works on the GIS need to follow the standard requirements of IEC 62271-4 to avoid SF_6 release to the atmosphere [20]. The volume of SF_6 gas leakage and the quantity of gas refilled need to be documented.

Air-insulated GIS use a gas mixture of N_2/O_2 or N_2/CO_2 which are natural gases and a leak to the atmosphere does not contribute to global warming. This is why it is possible to use vacuum switching technology and these types of GIS is preferred for offshore applications.

In order to make GIS more available and reliable its condition is monitoring in offshore windfarms Online monitoring is implemented in order to avoid un-necessary maintenance works and unexpected failures. It is important to keep the sensors at the offshore GIS as simple as possible and to indicated only those functions which are needed to avoid false alarms.

Since most spare parts are standard, it is not required that they be stored offshore, where environmental conditions are a factor. A service agreement can be made with the manufacturer where spare parts can be delivered directly from the GIS manufacturer factory without delay.

High voltage GIS for offshore installations is exposed to harsh environmental conditions with wide temperature ranges, high solar radiation, high saline pollution, ice coating, wind and vibrations of the platform. By installing the GIS indoor in an air-controlled room on the platform or inside the wind tower, most of those issues are addressed. This is the most common solution for offshore GIS installations and is addressed in the international standard for onshore installation IEC 62271-203 can be used [21].

Additionally, vibrations, transportation and seismic forces to the GIS need to be monitored and kept within limits. The values of these limits are provided in IEC 62271-1 [17]. The vibration at the platforms depends on the type of platform installation. Monopile structure generate different vibration frequencies than jacket foundations. The upper part of the bay of the GIS is affected by the highest forces, due to the vibrations from the wave motion and the design needs to address those

The seismic requirements for the mechanical stress on the GIS design are addressed in standard IEC 62271-207 [22], for North America IEEE 693 [23].

9.16.6 Example Applications of Offshore GIS

9.16.6.1 AC GIS for 12 MW Wind Turbine

The offshore windfarm projects in US "SKIPJACK" and "OCEAN WIND" will introduce 12 MW wind turbines. The projects reflect the increasing importance of wind energy generated offshore for the electric power supply. By using 12 MW wind turbine class the electric power generation is will reach a new level and will reduced cost per generated kWh, see Figure 9.83.

The GIS used for the 12 MW wind turbine is "plug & play" type with modules assembled and tested onshore prior to being shipped offshore along with the wind tower. This reduces the installation cost of the GIS.

GIS modules are tailored to the requirements of the network operators, and into consideration the multiple elements, breaker, disconnector, earthing/grounding switches, fast acting earth/ground switches, current (CT) and voltage (VT) transformers and service voltage transformers (SVT) to provide station service to the wind turbine in case of low wind.

The assembly and testing of the GIS module are to be done onshore in a horizontal position. After the assembly and tests are completed the GIS module will be inserted into the wind turbine tower in a vertical position. The wind turbine tower including the GIS will then be transported to the wind turbine foundation offshore and installed at their final position, see Figure 9.84.

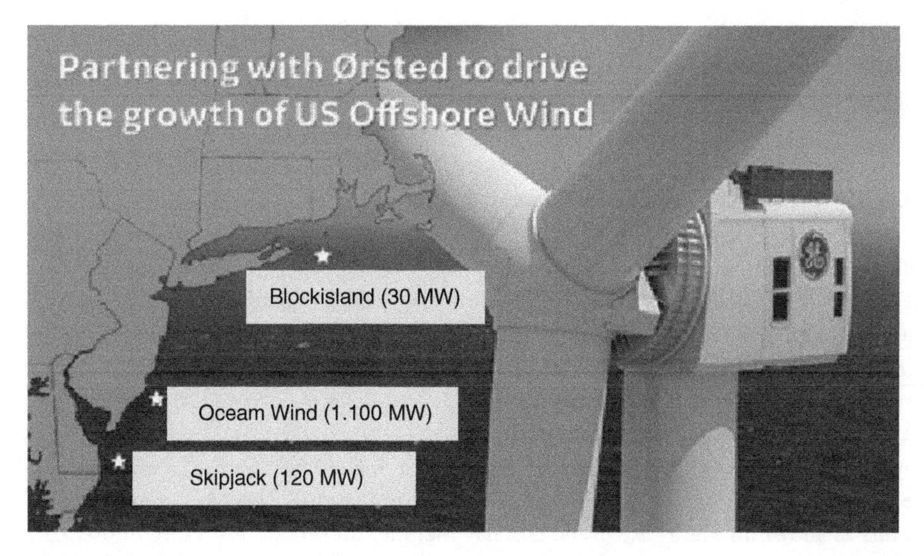

Figure 9.83 Offshore windfarms planned at the east coast of the USA using 12 MW wind turbines (*source* [24, 25])

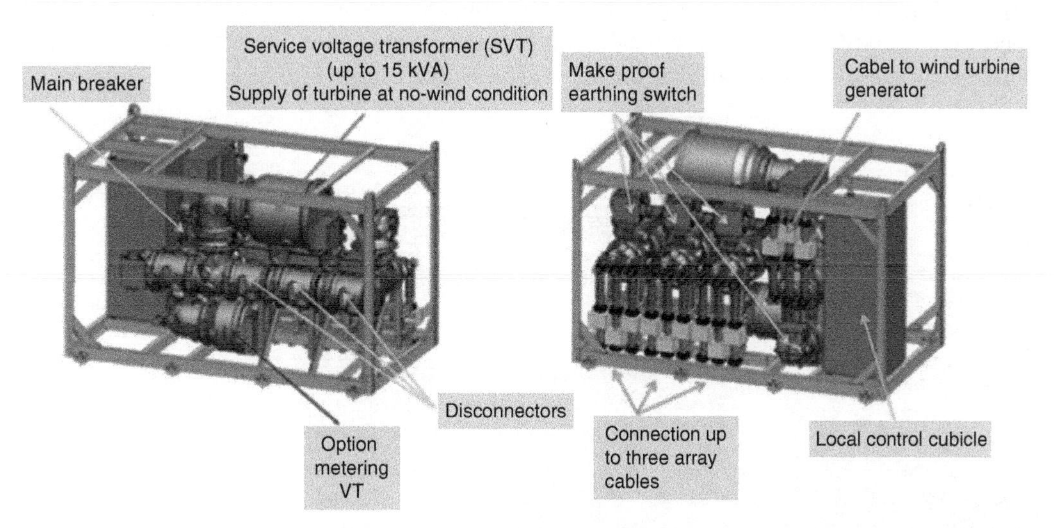

Figure 9.84 GIS modules for the offshore wind turbine of 12 MW left: front side with main breaker, service voltage transformer (SVT), disconnectors and optional metering voltage transformer (VT) right: back side with fast acting earthing/grounding switches, cable connection to wind turbine generator, cable connection for three cable collectors and the local control cubicle (*source* [24, 25])

GIS modules for the offshore wind turbine for 12 MW is mounted in a steel frame in the factory. It includes the main circuit breaker, service voltage transformer (SVT), disconnectors and optional metering voltage transformer (VT), fast acting earthing/grounding switches, cable connection to wind turbine generator, cable connection for three cable collectors and the local control cubicle.

The tested GIS module will then be transported to the wind turbine tower in a horizontal position, as shown in Figure 9.84. The GIS module will then be inserted into the tower base and the tower will be lifted into its final vertical position. The GIS module will turn at 180°. This works is done in the harbor where the wind tower is installed before shipped offshore to final location.

In Figure 9.85 the GIS module mounted in a steel frame ready for installation in the wind tower is shown.

9.16.6.2 AC GIS Using Clean-Air Insulation and Vacuum Interrupter

Nissum Bredning Wind Farm

The technical-air which is called clean-air in this application for insulation and the vacuum circuit breaker for interruption forms AC GIS technology was first installed offshore in 2018. The "Nissum Bredning" wind farm went online on 17.02.2018. The energy generated is delivered to the mainland of Denmark, see Figure 9.86. The Nissum Bredning wind farm is close to the coast line of Denmark and the connection is made by AC cables.

The GIS bays for the cable connection from the wind farm to the main land of Denmark are located in the basement of the wind tower with jacket-type foundations. The GIS room provides indoor conditions as shown in Figure 9.87. The GIS bay includes all functions for circuit breaker with vacuum switch, disconnector and earth/grounding switches, voltage and current transformer and the control and protection devices inside the integrated control cubicles.

The cable-collector connection enters the GIS at the bottom with three single-phase solid-insulated high voltage cables, see right photo of Figure 9.87. The cable connecting the machine transformer is located on top of the wind tower in the machine house. See left photo with three single-phase insulated cables entering the upper cable box of the GIS bay.

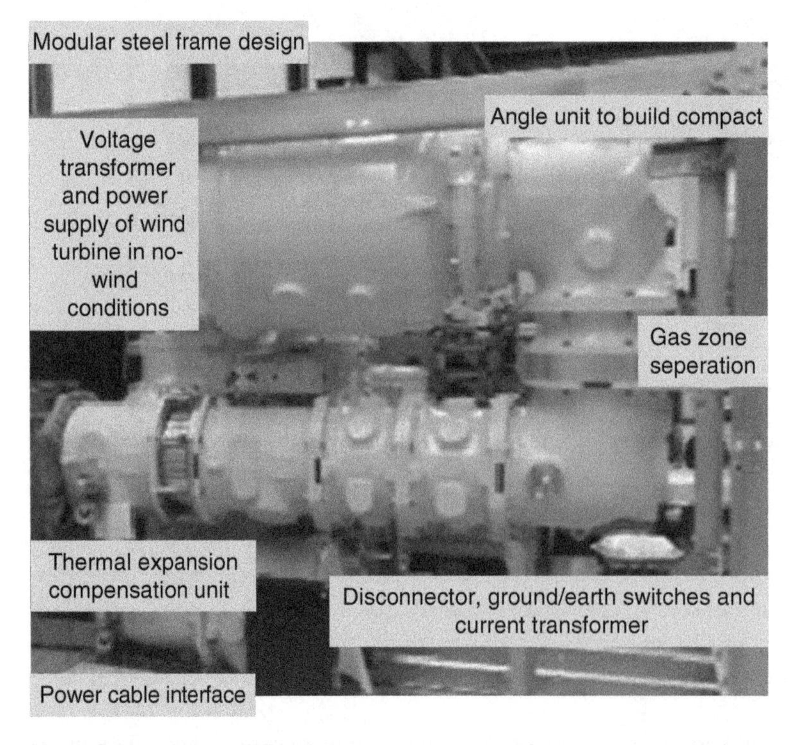

Figure 9.85 Modular GIS build in a steel frame for 12 MW wind turbines (*source* [24, 25])

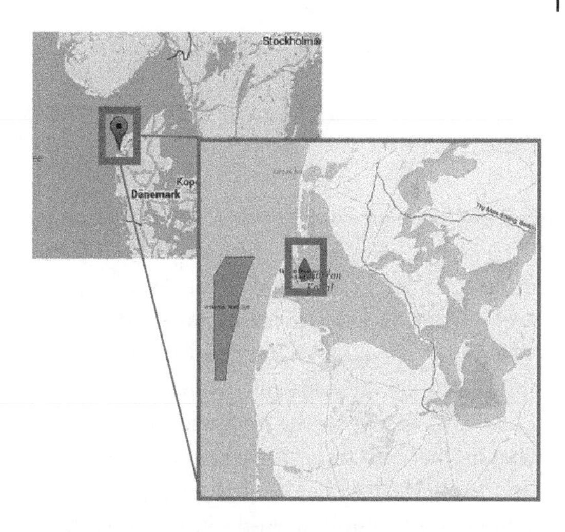

Figure 9.86 Nissum Bredning wind farm close to the coast line of Denmark (courtesy of Siemens Energy)

Figure 9.87 Indoor conditions for the GIS in the basement of the wind tower (left photo) and plug-in cable-collector connections at the bottom and transformer cable on top (right photo) (courtesy of Siemens Energy)

This represents a typical offshore application of the connection of the wind turbine and generator to the wind farm inter-collector cables. Typical voltage ratings used are 33 and 66 kV depending on the power generated by one wind tower and the number of wind towers in one string of the cable collector.

Until 2020 more than one hundred GIS bays of clean-air and vacuum interrupter type have been installed world-wide with project in Denmark, United Kingdom, United States of America,

Netherlands, Taiwan, China and France. Until 2020 more than 1000 units of GIS bays have been ordered.

In Figure 9.88 an impression is given of Nissum Bredning wind farm close to the coast line of Denmark.

GIS Bay of Clean Air Insulation and Vacuum Interrupter

This type of GIS utilizes clean-air as an insulating gas in order to avoid contribution to global warming. With no additional Global Warming Potential (GWP) the technical air mixture of 80/20% uses N_2/O_2 or N_2/CO_2. A vacuum interrupter technology for the circuit breaker is used for voltages up to 72.5 kV in order to interrupt the rated currents and short-circuit currents. The operational and environmental values of this technology are a preferred option for offshore wind parks. This technology is hermetically tight and thus reduces the degradation of the product. Further, the high switching performance of the vacuum interrupter requires less maintenance, thus reducing any costs associates with it. It can withstand to 30 maximum short-circuit current interruption before maintenance is required. Additionally, no liquefaction of the insulating gas will occur under the severe environmental conditions offshore.

The clean-air mixture does not contribute to global warming potential (GWP=0) and has no ozone depletion potential (ODP=0). The gas mixture is non-toxic, non-flammable with low fire protection costs, has a high molecular stability with no gas replacement needed, low liquefaction temperature with no need for heating in operation, not related to the F-gas regulations and free of C-gases, no documentation cost, no CO_2 compensation cost for future tax risks and no gas recycling cost at the end of the life of the GIS.

The vacuum interrupter technology has been tested as a reliable option and has been in service for more than 40 years with more than 4 million applications. It is perfect for frequent switching

Figure 9.88 Impression of Nissum Bredning wind farm close to the coast line of Denmark (courtesy of Siemens Energy)

and at low temperature applications. The type test equipment offers low maintenance cost and is sealed for life.

Figure 9.89 shows a bay of GIS for 72.5 kV rated voltage and 25 kA short-circuit current interruption capability using clean air for electrical insulation and a vacuum circuit breaker. The GIS enclosure is filled with clean air which is also called technical air and has a gas mixture of 80 N_2 and 20% O_2, which is the same ratio as that in the atmosphere. The GIS is three-phase in one enclosure type and offer any function need for a substation bay including circuit breaker, disconnector and earthing/grounding switch, voltage and current transformer, pluggable power cable connection and an integrated local control cabinet. The enclosure is represented open in the graphic to show the location of vacuum interruption unit in the center of the bay.

Technical Data

The technical date of the clean air GIS with vacuum interrupter is given in Table 9.18. The three-phase encapsulation is for 72.5 kV rated voltage at 50 or 60 Hz. The clean air insulation is type tested in accordance with IEC 62271-1 and IEC 62271-203 [17, 21] and fulfills the requirements of 140 kV Rated short-duration power-frequency withstand voltage (1 minute) and 325 kV Rated lightning impulse withstand voltage (1.2/50 μs), The rated continuous current is 1250 A with a rated short-circuit breaking current of 25 kA, a rated peak withstand current of 68 kA and a rated peak withstand current of 25 kA.

The type-tests and long-time measurements for gas leakage rate per year are below 0.1%. The driving mechanism for the circuit-breaker is the use of the stored-energy spring operation. The energy for the rated operating sequence is stored in the spring and can switch O-0.3 s-CO-3 minute-CO.

The interrupter unit is of vacuum tube type while the high voltage insulation and switching of disconnector and earthing/grounding switch uses clean air. This makes the weight of SF_6 or other fluorinated greenhouse gases to 0 kg and with this the CO_2 equivalent is also 0. The rated filling pressure is 0.56 MPa abs for clean air gas mixture.

The bay width for a common pole drive of all three vacuum interruption units is 1200 mm, the bay height is 2435 mm and the bay depth is 2940 mm. The ambient temperature range for operation is −25 °C up to +45 °C and the application is for indoor use.

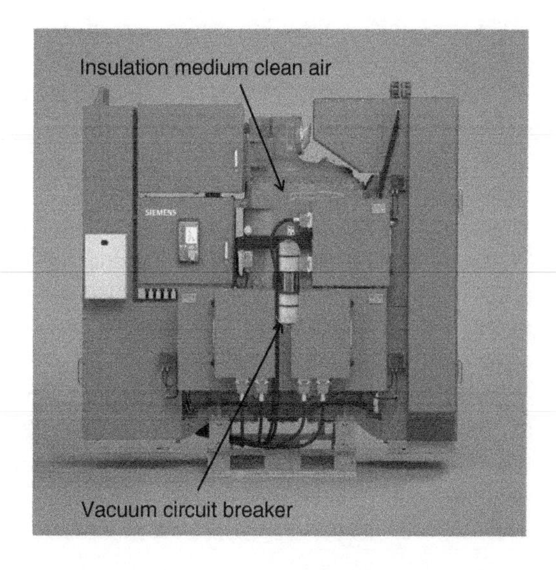

Figure 9.89 GIS bay of clean air insulation and vacuum interrupter for 72.5 kV rated voltage and 25 kA short-circuit interruption capability (courtesy of Siemens Energy)

Table 9.18 Technical Date of the clean-air GIS with vacuum interrupter

Rated voltage	up to	72.5 kV
Rated frequency		50/60 Hz
Rated short-duration power-frequency withstand voltage (1 min)	up to	140 kV
Rated lightning impulse withstand voltage (1.2/50 µs)	up to	325 kV
Rated continuous current	up to	1250 A
Rated short-circuit breaking current	up to	25 kA
Rated peak withstands current	up to	68 kA
Rated short-time withstand current (up to 1 s)	up to	25 kA
Leakage rate per year and gas compartment (type-tested)		< 0.1%
Driving mechanism of circuit-breaker		stored-energy spring
Rated operating sequence		O-0.3 s-CO-3 min
Interrupter technology		Vacuum
Insulation medium		Clean air
Weight of SF_6 or other fluorinated greenhouse gases		0 kg
GWP Global Warming Potential		0
CO_2 equivalent		0 kg
Rated filling pressure		0.56 MPa abs
Bay width common pole drive		1050 mm
Bay height, depth		2330 × 2500 mm
Bay weight		1.6 t
Ambient temperature range		−30 °C up to +45 °C
Installation		indoor
First major inspection		> 25 years
Expected lifetime		> 50 years
Standards		IEC/IEEE

This technology has a life expectancy of over 50 years, with the first major inspection being expected in more than 25 years from service date

All tests followed IEC [17, 21] and IEEE [26] standards for the type tests, routine, test, factory test and on-site tests.

Typical Configurations

The typical configurations for clean air GIS used in offshore wind farms are shown in Figure 9.90. The left side of the graphic shows the application when a collector sea cable is looped into the wind tower to connect the high voltage transformer with the cable collector. There are two cables from the wind farm collector one incoming and one outgoing. There are typically 8–10 wind towers connected in one string of a sea cable collector.

The incoming collector cable is connected to the GIS with the pluggable cable termination -XB3 at the disconnecting and earth/ground switch -QZ93. Voltage detecting device can be connected to the cable termination unit to check for voltage clearance in case of disconnected and earthed/grounded cable. The incoming collector cable is connected to the outgoing array cable by disconnect and earth switch -QZ92. A voltage detection device can be connected at the outgoing cable

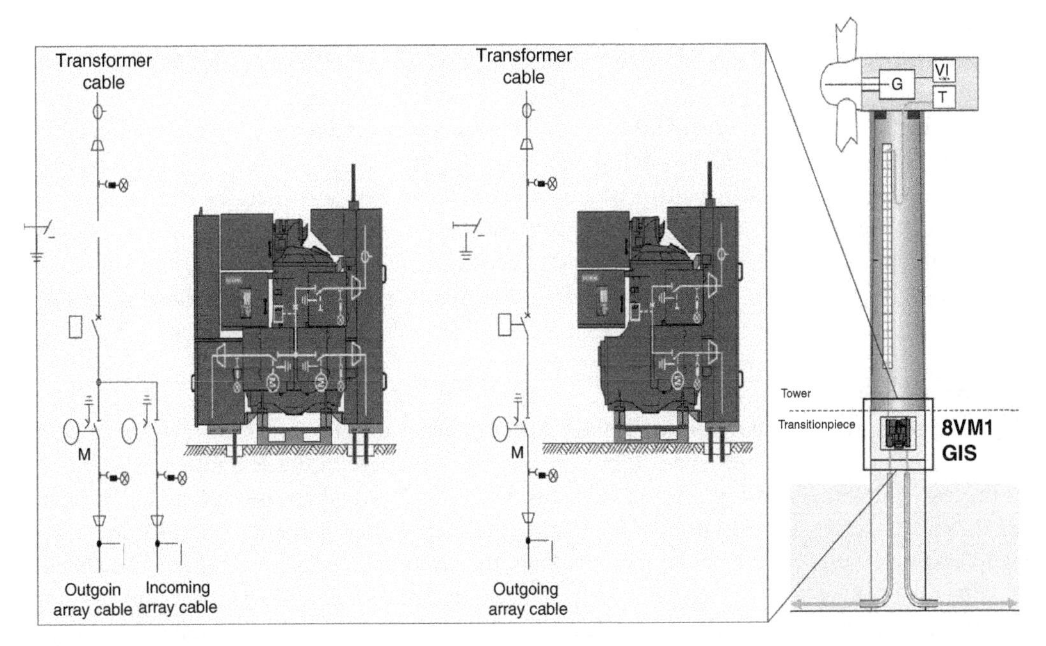

Figure 9.90 Typical circuits of GIS in offshore wind farms left: incoming and outgoing cables, right: only outgoing cables (courtesy of Siemens Energy)

side to the check the clearance of high voltage in case of disconnecting and earthing/grounding the outgoing cables.

The circuit breaker -Q1 is connected by the transformer side disconnecting and earthing/grounding switch -QZ91 to the transformer cable termination -XB1 and the current transformer -BC1 is connected to the transformer at the machine house on top of the wind tower by the transformer cable.

In case of a failure in the wind tower, the transformer can be disconnected while the collector cable continues power transmission in the string of cable collector of the wind farm.

On the right side of the graphic the circuit shows the direct connection of the wind tower to the collection platform. In this case, only one outgoing cable is needed to connect the wind turbine to the high voltage transformer.

The principal wind tower graphic on the right represents the location of the GIS and current, which is the so-called transition piece of the wind tower, as part of the foundation the high voltage transformer and the voltage current control of the electric generator are located on top of the wind tower in the machine house and the propeller is directly connected to the generator.

Typical wind farms of 500–1200 MVA are divided in sections of cable collectors. Each cable collector has 8–10 wind turbines of typical 6–10 MVA power. This requires about 50–100 wind towers to be connected. Each tower will require a high voltage GIS in order to be able to interrupt a faulty electric generator or transformer in one of the wind towers. This provides the possibility to continue operation in the wind farm and continuously collect the energy generated.

There are scheduled maintenance cycles to bringing maintenance personnel to the wind tower by boat or by helicopter. This requires large efforts and specialized personnel training for the environment several hundred kilometers offshore. Safety and protection measure are costly and precautions must be taken in order to ensure personnel is trained to withstand prolonged exposure to the weather conditions, as boats and helicopters are sometimes not available for days or weeks.

This is why the electric equipment installed in the wind tower needs a very high level of reliability and a minimum maintenance requirement.

The cost of equipment with built in redundancy is compared and evaluated to the cost of potential energy production loss. GIS offers high reliability and can be configured with redundant devices and remote monitoring equipment to avoid costly travelling to the offshore wind farm. To balance these requirements simulations and studies in the project planning phase shall carried out.

Transportation of GIS

A big task in offshore wind farm project is the transportation of the GIS from the factory to its final destination, which can be sometimes hundred kilometers offshore in the sea. Figure 9.91 shows this process. The transportation process starts in the GIS factory and requires planning for each step ahead. Nowadays, the GIS bays of leading manufacturers at voltage level 72.5 kV are completely assembled and tested in the factory. After the factory acceptance tests are completed, including high voltage tests the gas pressure is reduced to a value below 0.5 bar over pressure, for transportation in order to keep the high voltage interior of the GIS in controlled and stable condition. The GIS bay is sealed and packed for transportation to the harbor.

The GIS is put into a so-called transition piece at the dockyard in order to be transported to the wind tower. The transition piece is located in the basement of the wind tower and will be positioned on top of the wind tower foundation. The GIS will be filled with N_2/O_2 gas mixture to its nominal filling pressure of up-to 4.6 bar abs. The transition piece including the GIS are shipped to the wind tower at its final destination.

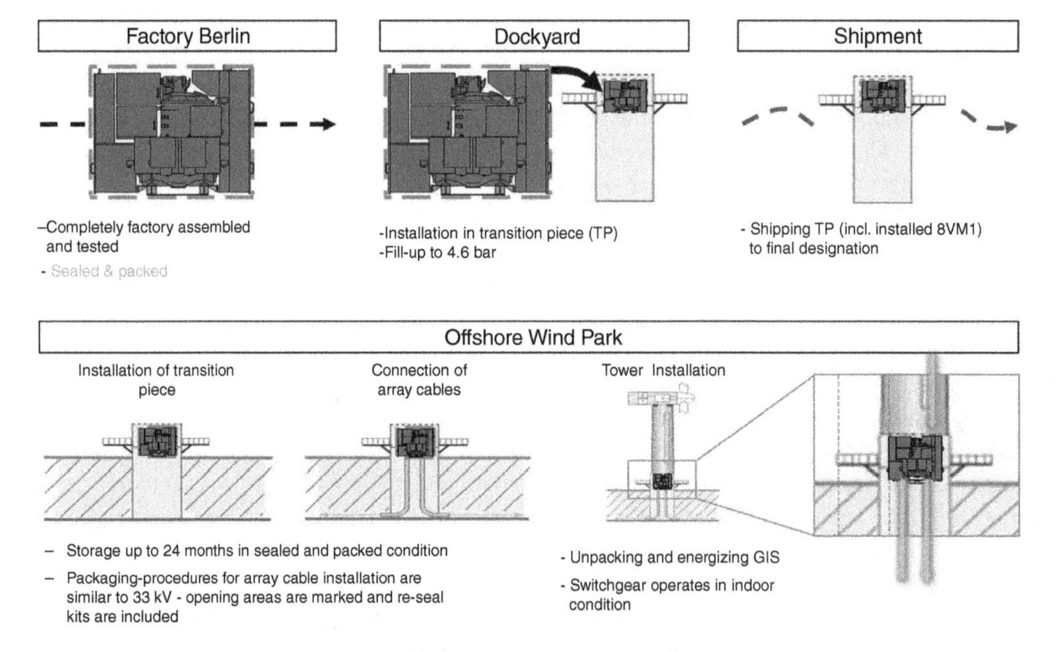

Figure 9.91 Transportation process for GIS to be installed in offshore wind farms (courtesy of Siemens Energy)

At the offshore wind park, the transition piece is mounted on top of the wind tower foundation. The sea cable collector is then connected to the GIS by cable joints inside the wind tower for the incoming and outgoing cables.

The GIS is packed and sealed with seaworthy packing material which allows up to 24 months storage under harsh environmental conditions Special packing procedures are also followed for the connection to the sea cables, where opening areas are clearly marked and can be resealed once connection is complete.

Finally, the GIS will be unpacked and the high voltage cables to the transformer, which is located at the machine house on top of the wind tower, will be finalized. The secondary control and protection devices will then be connected and functionally tested during the commissioning process prior to energization of the GIS.

Figure 9.92 shows the installation of the GIS within the transition unit. The picture on the left shows the moment when the transition unit is placed on top of the wind tower foundation. In this case, it is a three-leg jacket-type foundation. The weight of this unit is several hundred tones and weather conditions at sea should be calm when installing. These two requirements may change fast during the installation process and impose delays of days or even weeks. Ship cranes of these sizes are hard to find and thus delays in the installation time can occur. This is why it is so important to ensure that GIS equipment is packaged properly, in order to withstand prolonged exposure to the elements due to delays in installation. Delays can sometimes last months and equipment might need to be stored during the entire winter.

The right picture shows the wind tower after the installation is completed. The sea cable collector is connected to the GIS. The wind turbine generates electric energy through the generator which is connected to the power transformer in the machine house, which is connected to the GIS in the basement of the wind tower. The operational condition for indoor GIS is in accordance with IEC [17, 21] and IEEE [26]. To meet the indoor requirements under offshore conditions the GIS room needs to be hermetically close and ventilated through air filters to prevent exposure of the equipment to moisture and salt. The life expectancy of the GIS under these conditions is over 50 years.

Packaging System for GIS

The packaging system for GIS used for offshore wind farm conditions will also follow specific requirements. The packaging must be suitable for land transport by truck, sea transport via

Figure 9.92 Installation of the transition piece including the GIS at the three-leg jacket foundation of the wind tower (courtesy of Siemens Energy)

container and sea transport in an offshore wind turbine transition piece (TP). During this process the packaging must be suitable for handling via forklifts and/or cranes. The packaging system must withstand mechanical and environmental impacts related to transportation.

Special packaging requirements for GIS arise from the necessity to install the transition piece (TP) at the offshore site for up to 18 months without having the environmental protection of a fully commissioned wind tower. The installation process can take this long from receiving the GIS to wind tower installation.

Another important consideration is the installation of the collector HV cables to the GIS with simple opening and closing/sealing of the packaging. This avoids long exposure to harsh sea environment until the sea collector cables are connected to the GIS.

The installation process of the wind farm requires the option to the GIS being operated without the completion of the wind tower in which it is installed. This will allow for partially energized GIS to operate the collector cables and facilitating the operation of those wind towers which are completed and are generating energy.

To be prepared for any project delay storage of GIS under both indoor and outdoor conditions is required. The ambient temperature range is min. −25 °C/max. +55 °C and the humidity: ≤100%.

All equipment requires labeling and documentation containing information about the GIS for the transport from factory to wind farm.

The packaging system can be divided into a basic configuration, a sea transport configuration and the collector high voltage cable installation.

The basic packaging configuration is shown on Figure 9.93. The upper left picture shows the GIS packed for land transportation. This packaging gives a minimum protection with tarpaulin cover. The cover will be kept for additional packing when transported over sea to the wind tower later. It is suitable for land transport and for the time when the GIS will be installed in the wind tower and will provide environmental impact protection until the wind turbine generator is fully commissioned.

The two-bottom picture in Figure 9.93 show a fully protected GIS package which utilizes special sealed air-tight and moisture repellent aluminum foil packaging, left picture. The basic packaging with the tarpaulin cover is then put on top the sealed special air-tight and moister repellent aluminum-foil package. This double coverage is suitable for land transport as well as for the GIS installation in transportation piece (TP) at the offshore site in the wind tower. The protection lasts up to 18 months to avoid environmental impact until final commissioning of the wind tower.

The packaging system of GIS for sea transport in a container is shown on Figure 9.94. Containers are used for sea transport in order to provide addition protection for handling in the harbors and on the ships. Additionally, a wooden box with the dimensions of the container is used to provide extra protection for the GIS, see upper picture on Figure 9.94. The full protection packaging configuration is using the tarpaulin cover (left picture), the sealed special air-tight and moisture repellent aluminum foil packaging (middle picture) and the wooden box (right picture).

The packaging system of the collector high voltage cable installation has to fulfill specific requirements related to the environmental condition of the cable connection terminations at the GIS, which are located inside the base of the wind tower. The monopile tower types have an open

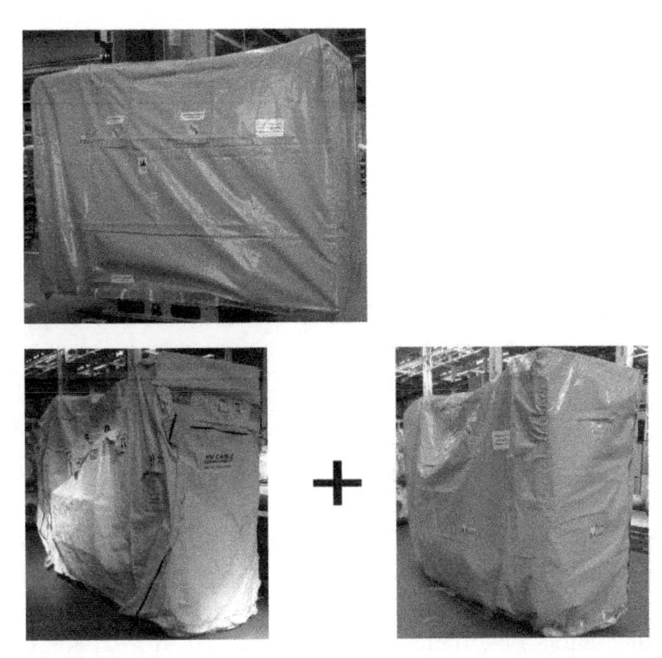

Figure 9.93 Packaging of GIS basic configuration, top: minimum protection using tarpaulin cover bottom: sealed special air-tight and moister repellent aluminum foil (left) and basic packaging on top (right) (courtesy of Siemens Energy)

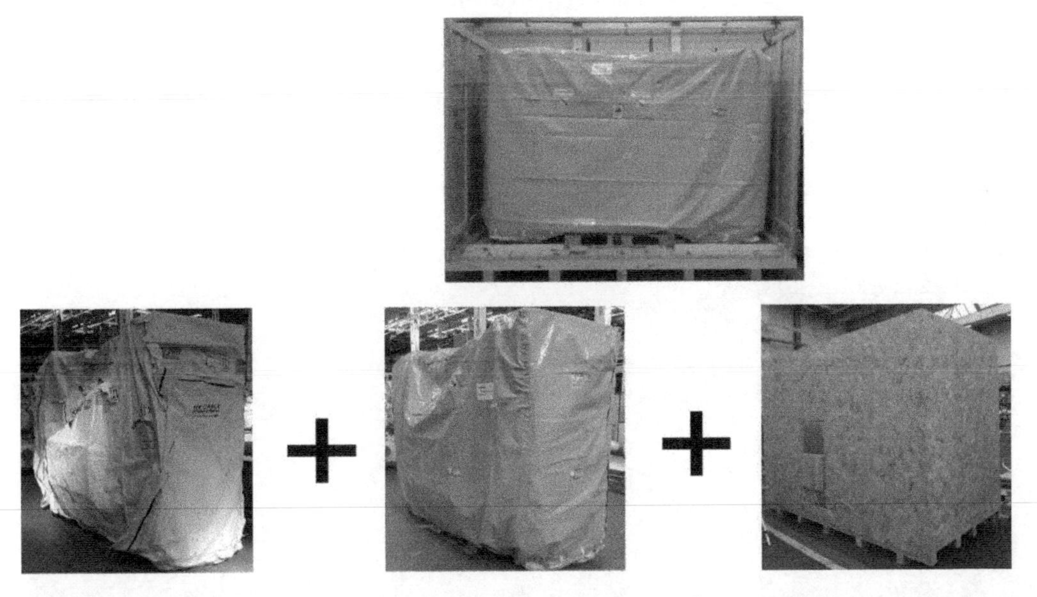

Figure 9.94 Packaging system for sea transport in a container, upper photo: container size wooden box, below left: sealed special air-tight and moisture repellent aluminum foil, below middle: tarpaulin cover, below right: wooden box (courtesy of Siemens Energy)

space and are thus exposed to the sea and its environmental conditions, as the sea cables enter the GIS room from below. For jacket-type foundations, the sea cables are connected to the legs of the jacket and enter the GIS room right from below at sea environmental conditions.

For this reason, the GIS room and the cable terminations need to be protected against the impact of moisture, sea water and salt and the high voltage parts need to be kept away from them. The packaging is designed in a way in which the GIS opens only as much as needed to connect the cables.

The sealed special air-tight and moisture repellent aluminum-foil packaging offers zippers at the needed areas for simple access to the installation areas of the switch gear as shown on Figure 9.95.

The cutting areas required for access are reduced to a minimum and are appropriately marked with red dotted lines on the aluminum-foil. In Figure 9.96, the picture on the left shows the three-phase plug-able cable joint on the top of the cable box. At the bottom of the picture the inlet and fixing point for the sea cable are shown. The picture at the middle of Figure 9.96 shows the closed aluminum foil for moisture protection.

The cable box at the GIS is open to the environmental conditions at the offshore wind tower. There is the open sea below the GIS where the cable collectors are coming in with three phases of one circuit and going out with three phases of another circuit. These are six single-phase insulated cable to be connected under harsh environmental conditions and in a complex environment.

The installation of the collector HV cables are performed by qualified personnel only and strictly follow the dedicated manual. It is necessary to train the staff in the factory or at suitable locations on land before starting work offshore in the wind tower.

Each step of the collector cable installation shall be described in a manual to verify the installation process. Beside the technical training it is essential to train the personnel in the environmental conditions in the wind tower, weather conditions and working at heights above sea level. There are 20–30 m height below the GIS cable box and it is required to install the plug-in type cable connection and fix the cable securely from the sea bottom up to the GIS in the wind tower base.

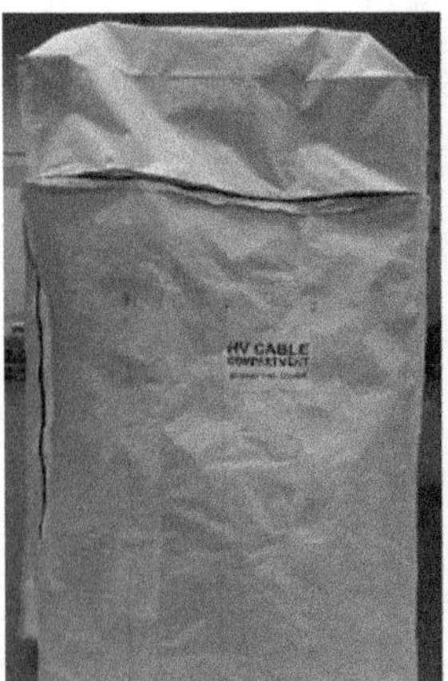

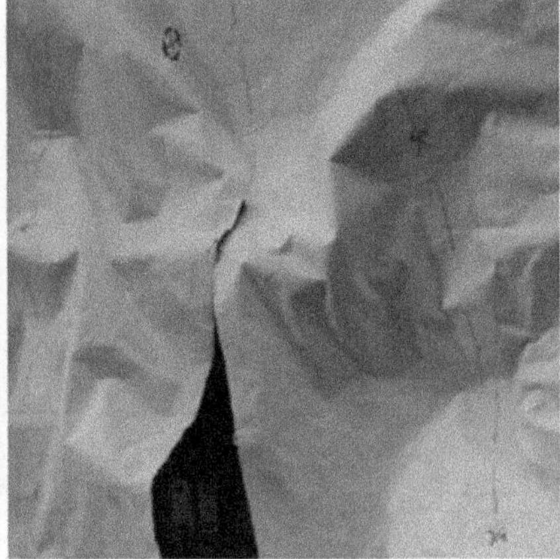

Figure 9.95 Packaging system for the collector cable installation, left: sealed special air-tight and moisture repellent aluminum-foil packaging, middle: zippers to get access to GIS cable termination, right: open cable box (courtesy of Siemens Energy)

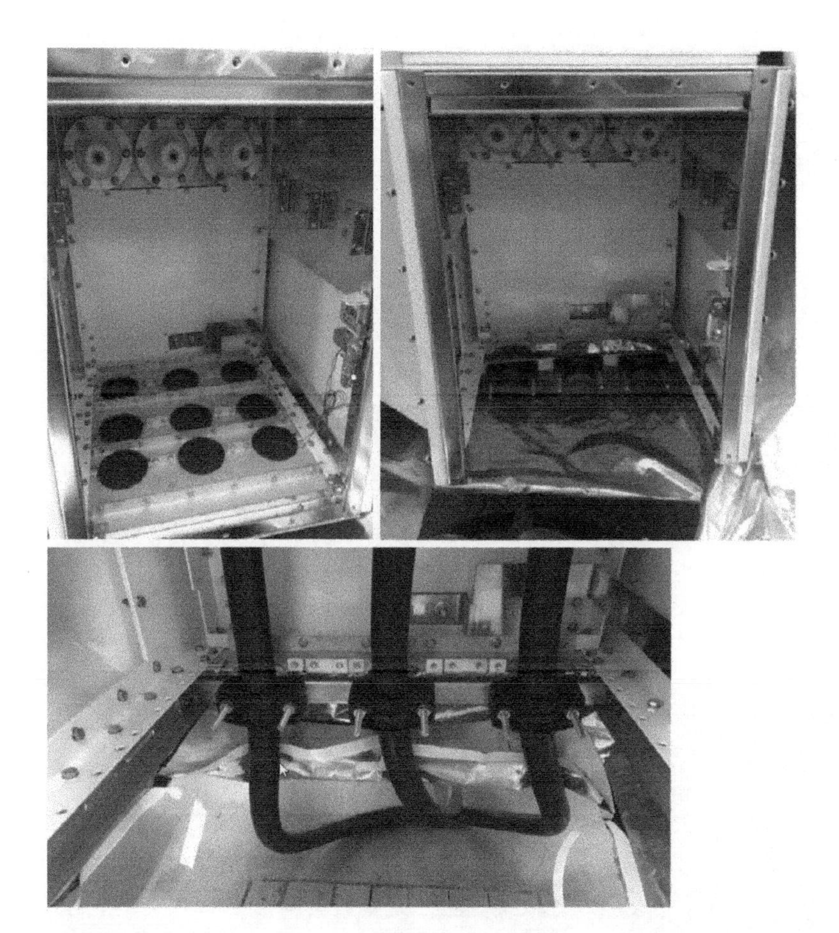

Figure 9.96 Packaging system for the collector cable installation, left: three-phase pluggable cable joint on the top of the cable box and at the bottom of the photo the inlet and fixing point for the sea cable, middle: closed aluminum foil for moisture protection, right: three-phases of single-phase insulated cables connected to the GIS (courtesy of Siemens Energy)

The right picture on Figure 9.96 shows three-phases of single-phase insulated cables connected to the GIS in the base of wind tower. The cables enter the cable box through the aluminum foil, are fixed by clamps to the cable box and (outside the picture) are plugged into the plug-in cable termination.

Each HV cable compartment of the collector includes a repair and reseal kit, as shown on Figure 9.97. This repair and reseal set is provided with the delivery of the GIS and is located inside the cable termination box in case of an issue during the installation process of the collector cables. This kit is necessary, as the delivery of spares takes time due to the location of the windfarm and will cause delays.

9.16.6.3 DC GIS for HV DC Offshore Converter Platform

DC offshore converter station platforms are used in cases with long distances between the windfarm to the onshore substation connected to the electric power grid. Another reason for DC connection of the export cable is the high-power transmission capability of the DC cable. DC connections are typically used in windfarms of 700–1200 MW.

Figure 9.98 represents the typical connection of the windfarm to the substation onshore and the power grid.

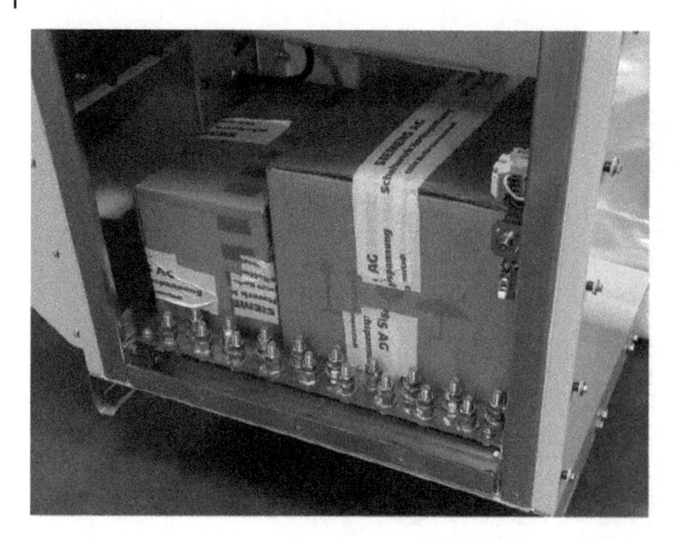

Figure 9.97 Dedicated repair and reseal kit for the cable-collector installation (courtesy of Siemens Energy)

Figure 9.98 Overview of types for offshore windfarm connections, lower part: AC voltage connection to onshore, upper part: AC collection of power in the windfarm and DC connection onshore (courtesy of Siemens Energy)

The lower part of the graphic shows the AC voltage connection of the offshore windfarm. The windfarm is connected by the inter-windfarm collector sea cables (left side) to the collection station platform (yellow) using cables of 33 and 66 kV within the windfarm to connect the wind towers and 132–170 kV sea cables for the connection of sea cable strings of the inter-windfarm collector to the collection station platform. The onshore connection from the AC wind power collecting station platform uses 220–330 kV AC sea cables. The distance from the AC wind power collecting platform to the coastline is typical no more than 30 km.

The upper part of the graphic on Figure 9.98 shows the DC windfarm connection to the onshore converter station and power transmission network connection. The graphic represents the typical connection of the AC power collection station platform that connects several windfarm sections, to a DC platform. This type of connection requires an AC/DC converter station and a DC export cable from the connection onshore.

The inter-windfarm collector uses 33 kV or 66 kV cables to connection of the wind towers. Several wind towers are connected in strings of 8–10 wind towers. The single strings of connected wind tower are than connected by 132 and 170 kV sea cables to the AC collecting station platforms (small yellow platforms). Two of these AC collecting station platforms are show in this graphic and they are connected to an AC/DC converter station platform (large yellow platform). There could be more AC connection platforms involved to collect up to 1200 MW of wind power.

The AC/DC converter station platform is supplied on the AC side by 132–170 kV voltage cables which are transformed to ±320 kV DC for the export cable to the onshore station. The size of AC/DC converter platforms are much larger than the AC type because of the space needed for the AC switchyard and the equipment of AC/DC converter. The distance between the offshore AC/DC converter station and the onshore converter station is of several hundred kilometers. Most such applications are installed in the European North Sea.

An overview of an AC/DC offshore converter stations is shown on Figure 9.99, using air-insulated substation (AIS) equipment. The left half of the platform uses the plus and minus polarity of the AC/DC converter. In the right half of the platform the DC switchyard is shown using air-insulated substation (AIS) equipment. High voltage clearances require large spaces on the platform. By using GIS, the required space can be reduced by 10% for the AC switchyard at the platform. The space reduction between GIS versus AIS is approximately 90% for the switchyard function. The compact design of GIS allows for an installation at the side of the converter room.

AIS option led to higher maintenance and service costs compared to the much smaller GIS installation, as environmental factors at offshore platforms are a major concern and GIS allows for an environmentally controlled room with moisture filters and air conditioning.

An overview of AC/DC offshore converter stations is shown on Figure 9.100, using gas-insulated substation (GIS) equipment for AC and DC switching.

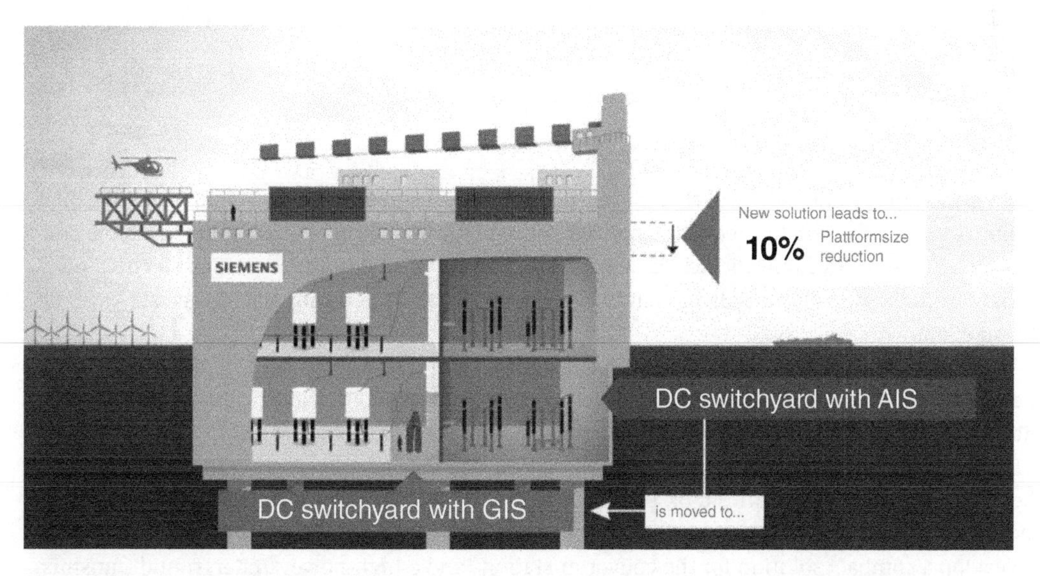

Figure 9.99 Overview of AC/DC offshore converter platform using air-insulated substation (AIS) equipment (courtesy of Siemens Energy)

Figure 9.100 Overview of AC/DC offshore converter platform using gas-insulated substation (GIS) equipment left: inter-windfarm sea cable collector, AC GIS and DC transformer, center: two converters ±320 kV at two levels, right: DC switchyard and DC export cable ±320 kV (courtesy of Siemens)

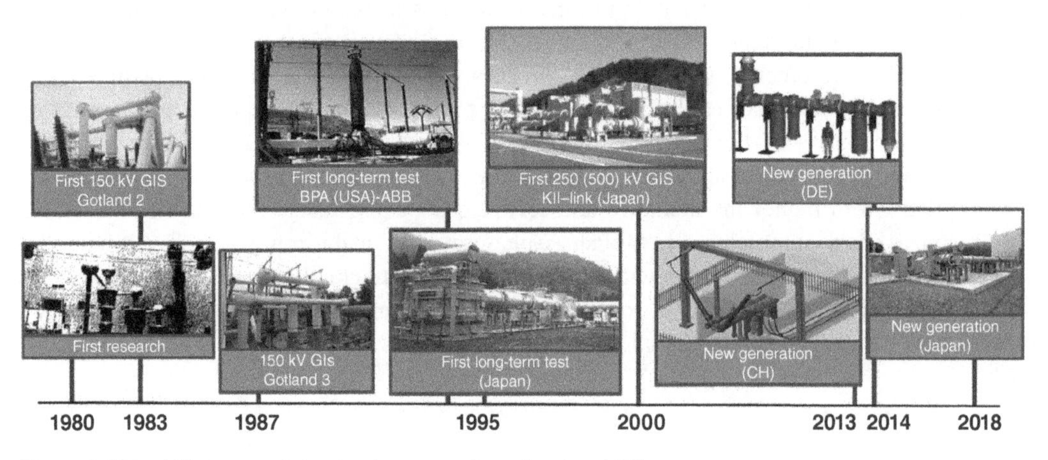

Figure 9.101 Milestones of the development of gas-insulated DC systems

The wind energy enters the AC/DC platform from the left side by the inter-windfarm sea cables connected to the AC GIS located at the bottom of the platform directly on top of the sea cable connections. The AC GIS connects the DC step up transformers at the upper level. The electric power coming from the wind farm is divided in two parallel VSC converters and reactors at two levels on the platform.

The DC switchyard for each converter level is located at the right end of the platform to connect the wind farm energy to the DC export cable using ±320 kV. The DC export cable is connected to the DC/AC converter station onshore for the connection to the AC power grid.

Direct current gas-insulated technology (DC-GIS) has a long history starting in the 1980s when the with first research activities took place, see Figure 9.101. Different research laboratories have been involved in the DC gas-insulated technology research programs. The goal of this research was to develop a compact solution for the converter stations and a high-power underground transmission line (DC GIL). In 1987, the first DC GIS of 150 kV went into service at the Gotland 3 Project.

Long duration field tests in Japan and the USA starting in 1995 provided the first experiences for the DC GIS installation; the first being at Kii Chanel in Japan. In this case, the available space at the landing point of the DC sea cable required a very compact solution for the ±250 kV DC link to the converter station several kilometers in land. The design of the DC GIS was for ±500 kV applications and is operated since the year 2000 successfully with ±250 kV.

On the basis of this experiences the new generation of DC GIS have been designed and were introduced and passed the tests in 2013 in Switzerland, 2014 in Germany and 2018 in Japan. The new generation of DC GIS has been successfully tested is using the test requirements given by IEC 62271 series and CIGRE recommendations of JWG D1/B3.57.

In Table 9.19 the technical data of the DC GIS is given for 550 kV voltage rating. The focus of the applications is on nominal voltage (U_{ndc}) of ±500 kV with rated continuous DC voltage as maximum continuous operating voltage (U_{rdc}) of ±550 kV. This covers the operational conditions of the DC transmission.

The type test requirements are related to the abnormal operation conditions including lightning strikes on connected overhead lines and switching operations. The rated lightning impulse withstand voltage (U_{LI}) uses the impulse shape of 1.2/50 μs. To reflect the possibility of a lightning strikes on the plus and minus pole of the DC line of ±550 kV DC a superimposed lightning impulse voltage of ±1550 kV is applied to the high voltage insulation from pole to earth/ground and across the open isolating gap of the disconnector.

The rated switching impulse withstand voltage (U_{SI}) uses the impulse shape of 250/2500 μs. To reflect the possibility of a switching operations to the plus and minus pole of the DC line of ±550 kV DC a superimposed switching impulse voltage of ±1550 kV is applied to the high voltage insulation from pole to earth/ground and across the open isolating gap of the disconnector.

The DC rated nominal current (I_{rdc}) is 5000 A for continuous operation at maximum ambient temperature 50 °C. The type test requirements of IEC for the maximum current provides maximum temperature for contact material and insulation material.

The IEC standard requirement for the rated short-time withstand current (AC) is defined to 50 kA for 1 s. This proves the maximum thermal impact of the current and mechanical forces to the conductors.

The ambient temperature range for the DC GIS is related to the offshore application with −30 °C at the lowest and +50 °C at the highest and with this the requirement is higher than the standard ambient temperature requirement according to IEC standards which −25 °C and +40 °C.

Table 9.19 Technical data of the DC GIS for 550 kV voltage rating

U_{ndc} nominal DC voltage	±500 kV
U_{rdc} rated continuous DC voltage (max. continuous operating voltage)	±550 kV
U_{LI} rated lightning impulse withstand voltage (1.2/50 μs)	±1550 kV LI
• superimposed with DC	+/-550 kV DC & +/-1550 kV LI
• across the open isolating distance	+/-550 kV DC & +/-1550 kV LI
U_{SI} rated switching impulse withstand voltage (250/2500 μs)	±1175 kV SI
• superimposed with DC	+/-550 kV DC & +/-1550 kV LI
• across the open isolating distance	±550 kV DC & ±1175 kV SI
I_{rdc} rated nominal current (DC)	5000 A
Rated short-time withstand current (AC)	50 kA (1 s)
Ambient temperature range	(−30. . .+50) °C

The IEC standard required type test have all been performed for the ±500 kV DC GIS according to IEC 62271-1, -102, -103 and -203 [47–50] and in addition to the requirements of CIGRE JWG D1/B3.57, which is currently used for the new IEC standards for DC GIS of IEC 62271-5 Common specifications for DC switchgear and IEC 62271-318 DC gas-insulated switchgear assemblies [51, 52].

The following type test have been applied: Dielectric tests, temperature rise tests, mechanical endurance test with low and high temperature tests, short-time withstand current and peak withstand current tests, insulator tests, internal arc tests and insulation system test (with $\Delta\vartheta$) with DC superimposed to lightning impulse (LI) and switching impulse (SI) tests.

The DC GIS does not include the circuit breaker function. The functions of disconnector and earth/ground switch, voltage and current measurement, overvoltage surge arrester, cable termination, bushing connection and DC GIL connection are shown on Figure 9.102.

The first installation of DC GIS at an offshore converter station is now in the project execution stage and will be install at the offshore platform of DolWin6 in the European North Sea to provide grid access for the 900 MW of renewable wind energy. The DolWin6 project is one of six new 900–1200 MW wind farms added to existing four, as shown in Figure 9.103. This covers only the Siemens projects.

The DolWin6 offshore windfarm project will enter a new field of gas-insulated switchgear assemblies (DC GIS). The growing rates of offshore wind farm installation are indicating a world interest in this type of renewable electric energy installations. The cost one of each wind generated kW/h is decreasing constantly and has reached the levels equal and below of thermal power plants and below.

The innovative solutions for offshore applications have increasing reliability of the offshore wind energy generation. The high availability of wind with up to 8000 hours per years provides high total power generation results for each year.

The two last pictures of this chapter (Figures 9.104 and 9.105) give an impression of the technology used for offshore wind farms. All aspects of offshore applications differ from those onshore

Figure 9.102 Type test set up for DC GIS of following functional units from left to right: bushing or DC GIL connecting flange, angle unit, thermal compensator, surge arrestor, RC divider, disconnecting and earthing/grounding switches, surge arrestor and cable connection module (courtesy of Siemens Energy)

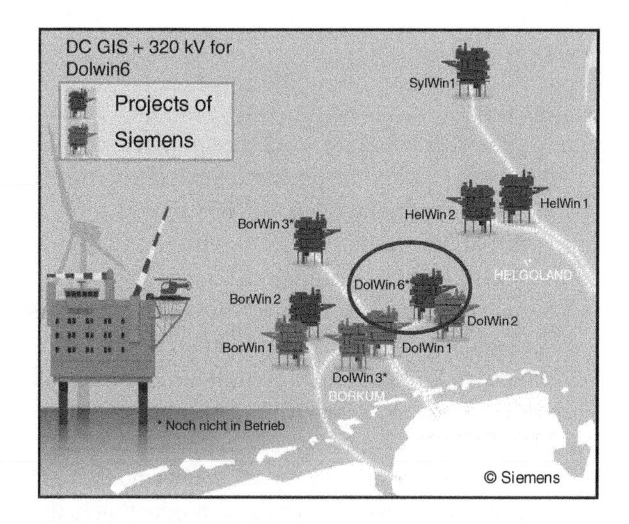

Figure 9.103 Overview of the offshore windfarms with DC cable connection to the onshore grid installation (courtesy of Siemens Energy)

Figure 9.104 Wind towers next to the AC/DC converter platform in the windfarm (courtesy of Siemens Energy)

Figure 9.105 AC/DC converter platform for 700–1200 MW windfarms with long distance DC sea cables for onshore connection (courtesy of Siemens Energy)

applications. The complete process of planning, engineering, studying, installation and operation require new skills to ensure successful project execution and a successful operation.

In the planning stage the environmental condition is a dominant factor. Climatic conditions can be very harsh, especially in cases when the platform is installed some hundred kilometers away from the coastline. Sudden weather change has a big impact on the project execution. Strong wind, high waves, sometime coming from a storm far away, may stop the work for weeks.

Safety of personnel dealing with the construction work has an important impact to all work processes and is a priority. The availability of equipment is limited, mainly the heavy weight off-shore cranes to lift hundred or even thousands ton platforms to the foundations. Such equipment must be rented for a longer period, sometimes years in advance and if the time slot is missed it may take a long time before new rental becomes available.

The ecological condition concerning the wild life in the water or in the air have a strong impact on the approval process. The whale's reproduction in the North Sea, for example, is related to trave-ling of the fish to certain areas in North Sea, which restricts any construction works for months. This could cause delay in the project process. The same applies for bird flight routes which might be disturbed from Wind tower construction work at a certain time. Shifting the construction by some weeks or months could result in getting into bad weather from October to April and work may stop for an entire year.

All of these factors and past experiences have become the knowledge base for the windfarm planning, engineering, execution works and operation. This has improved the offshore windfarm processes and allows offshore windfarm to provide reliable wind energy.

Figure 9.104 shows some wind towers close to the wind farm collecting station platform. It shows that offshore large areas can be used for windfarm installations.

The wind towers with the wind turbine, the electric generator and the sea cable connection to the collection station platform is an impressive technology used offshore. The gigantic size of the offshore AC/DC converter platform is another impressive construction at sea.

These platforms are built in harbors around the world and float at harbors close to the location of installation at sea.

At the harbor the electrical equipment of transformers, AC and DC GIS and the high voltage converter station will be installed on the platform. For a period of 1–2 years, this construction site is a busy place. Testing and checking the functionality of the installation will reduce the possible failures prior to installation offshore and thus reduce the cost.

Finally, the platform is transported by floating it from the harbor to the location of installation which may be one or two hundred kilometers from the coast line. The platform will be fixed to the sea bed and connected to the power cables.

Figure 9.105 gives an impression of an AC/DC converter platform for 700–1200 MW windfarms with long distance DC sea cables for onshore connection

14MW Wind Turbine

The innovation in wind turbine technology continues to improve. In 2020, Siemens Gamesa launched 14 MW offshore Direct Drive turbines with 222-meter rotor diameter. The offshore wind turbine was released with nameplate capacity of 14 MW but can reach 15 MW by power boost. The blades are 108 meters long.

This increases the annual energy production by 25% compared the predecessor wind turbine of 10 MW. Within the lifetime of the wind turbine approximately 1.4 million tons of CO_2 emissions per unit will be saved compared to coal-fired electric power generation.

The light 500-ton nacelle weight enables an optimized substructure of the wind tower at lower cost. The prototype will be ready in 2021 and commercially is the 14 MW turbine will be available in 2024.

With an unprecedented 14-megawatt (MW) capacity – reaching up to 15 MW using the power boost function, a 222-meter diameter rotor, 108-meter-long blades, and an astounding 39 000 m^2 swept area, see Figure 9.106.

The 14 MW capacity allows one turbine to provide enough energy to power approximately 18 000 average European households every year. Approximately 30 offshore wind turbines could furthermore cover the annual electricity consumption of Bilbao, Spain.

The 222-meter diameter rotor uses the new blades, which are almost as long as three Space Shuttles placed end-to-end, each blade 108-meter long. The integrated blade design is cast in one piece.

Furthermore, the new offshore giant features a low nacelle weight at 500 metric tons. This light weight enables to safely utilize an optimized tower and foundation substructure design compared to a heavier nacelle. Benefits thus arise in the form of lower costs per turbine by minimizing sourced materials and reducing transportation price.

Figure 9.107 represents wind turbine with rotor diameter (222 m) and swept area of wind of the 14 MW.

Table 9.20 give the technical data of the 14 MW wind turbine.

The direct drive technology reduces the number of different rating and wear-prone components in offshore turbines, making them simpler to maintain. Efficiency is improved by using a permanent magnet generator, which needs no excitation. These and other design simplifications keep the weight low and dimensions small to lower the transport and installation costs per MW.

The wind turbine uses a high wind ride through system (HWRT) which stabilizes the energy output. When the wind speed is higher than 25 meters per second, wind turbines typically shut down for self-protection. Those equipped with the HWRT system will slowly ramp down power output, enabling smoother production decrease and improve electrical grid reliability.

The integrated blade technology enables each fiberglass-reinforced epoxy blade to be cast in one piece. This process eliminates weaker areas at glued joints and produces blades of optimum quality, strength, and reliability.

Turbine sensors continuously transmit data to the Diagnostic Centre, enabling the early detection of anomalies and preventing potential failures. By analyzing vibration patterns, the service plan is optimized and anticipate repairs scheduled before serious damage occurs.

Figure 9.106 14 MW offshore direct drive wind turbine with 222-meter rotor diameter (courtesy of Siemens Energy)

Figure 9.107 14 MW wind turbine rotor diameter (222 m) and swept area of wind (39 000 m^2) (courtesy of Siemens Energy)

Table 9.20 Technical specifications of the 14 MW wind turbine

IEC class	I, S
Nominal power	14 MW
Booster power	15 MW
Rotor diameter	222 m
Blade length	108 m
Swept area	39 000 m^2
Hub height	Site-specific
Power regulation	Pitch-regulated, variable speed

Applications

First applications using the 14 MW wind turbines are contracted in UK and USA.

The offshore windfarm in the Virginia, USA coast will generate 2640 MW and will be the biggest one in the USA. The exact quantity of turbines to be deployed is subject to final project site conditions. All turbine installations are expected to be completed by 2026.

In UK the Sofia offshore wind farm project is located at Dogger Bank in the North Sea. It is one of four projects planned at Dogger Bank, about 100–200 km offshore from the coast line. Sofia wind farm will generate 1400 MW of power and will be in operation in 2026. 1.2 million households can be supplied with electric energy from this new wind farm.

There are hundreds of wind towers in each of these projects and each wind tower needs a high voltage GIS of 66 kV or higher rated voltage. There is a big market for AC and DC GIS developing at offshore wind farms.

References Section 9.16

1 Kosse M, Klein C, Frey K, Küttinger C, Haupt T, Foehr M(2020) Investigation of the gas-solid insulating system in DC GIS exposed to AC-DC mixed voltage for offshore bipole applications at ±525 kV DC, VDE High Voltage Conference, 2020-11, Berlin, Germany
2 Juhre K, Geske M, Kosse M, Plath R (2020) Feasibility Study on the Applicability of Clean Air in Gas-insulated DC Systems, VDE High Voltage Conference, 2020-11, Berlin, Germany
3 CIGRE Technical Brochure TB 430 Guidelines for the Design and Construction of AC Offshore Substations for Wind Power Plants Working Group B3.26 December 2011.
4 Public internet DolWin2 TenneT
5 Public internet BorWin2 TenneT
6 IEEE C37.100.1, Standard of Common Requirements for High-Voltage Power Switchgear Rated Above 1,000V.
7 IEEE C37.122.3, Guide for Sulphur Hexafluoride (SF6) Gas Handling for High-Voltage (over 1,000Vac) Equipment.
8 IEC 61400-series Wind Energy Generation Systems
9 IEC 60909-0:2016 Second Edition Short-circuit currents in three-phase a.c. systems –Part 0: Calculation of currents
10 IEC 60071-1:2019 Edition 9.0, 2019-08 Insulation co-ordination – Part 1: Definitions, principles and rules
11 IEC 60071-2: 2018 Edition 4.0 2018-03 Insulation co-ordination – Part 2: Application guidelines
12 IEEE C62.82.1, Insulation Coordination – Definitions, Principles, and Rules.
13 IEEE 1313.2, Guide for the Application of Insulation Coordination
14 IEEE C62.82.2, Guide for the Application of Insulation Coordination
15 IEC 61400-4-15:2010 Edition 2.0 2010-08 Electromagnetic compatibility (EMC) – Part 4-15: Testing and measurement techniques – Flickermeter – Functional and design specifications
16 IEC 61400-21-1:2019 Edition 1.0 2019-05 Wind energy generation systems – Part 21-1: Measurement and assessment of electrical characteristics – Wind turbines
17 IEC 61271-1:2017 Edition 2.0 2017-07 High-voltage switchgear and controlgear – Part 1: Common specifications for alternating current switchgear and controlgear
18 IEC 62271-100:2008 Edition 2.0 2008-04 High-voltage switchgear and controlgear – Part 100: Alternating current circuit-breakers
19 IEEE C37.06, AC High-Voltage Circuit Breakers Rated on a Symmetrical Current Basis – Preferred Ratings and Related Required Capabilities for Voltages above 1,000V.
20 IEC 62271-4:2013 Edition 1.0 2013-08 High-voltage switchgear and controlgear – Part 4: Handling procedures for Sulphur hexafluoride (SF6) and its mixtures
21 IEC 62271-203:2013 Edition 2.0 2013-09 High-voltage switchgear and controlgear – Part 203: Gas-insulated metal-enclosed switchgear for rated voltages above 52 kV
22 IEC 62271-207:2012 Edition 2.0 2012-04 High-voltage switchgear and controlgear – Part 207: Seismic qualification for gas-insulated switchgear assemblies for rated voltages above 52 kV
23 IEEE 693, Recommended Practice for Seismic Design of Substations.
24 12 MW wind turbine (2020). www.ge.com/renewableenergy.
25 R. Lüscher (2019). Integration von GIS in Offshore Anwendungen, GIS Anwenderforum, TU Darmstadt.
26 IEEE C37.122-2010, IEEE Standard for High Voltage Gas-Insulated Substations Rated Above 52 kV, 2011-01-21.
27 IEC TS 62271-318 ED1 High-voltage switchgear and control gear- Part 318- DC gas-insulated switchgear assemblies, IEC SC 17C, WG 42, to be published in 2022.
28 IEC 62271-5 Ed1, High-voltage switchgear and controlgear, Part 5: Common specifications for DC switchgear.
29 Douglas <Sea Scale, https://www.encyclopedia.com/science/encyclopedias-almanacs-transcripts-and-maps/douglas-sea-scale.
30 Beaufort wind force scale, https://www.rmets.org/resource/beaufort-scale.

9.17 HVDC GIS

Author: Maria Kosse
Co-Author: Uwe Riechert
Review: Hermann Koch

9.17.1 General

High-voltage direct current (HVDC) technology for long-distance transmission of very high power has also made its way into grid access of offshore windfarms and interconnectors between different countries. Gas-insulated HVDC systems for converter and transition stations offer further advantages for integrating this technology, particularly compared to air-insulated systems [1].

High DC voltage ratings require large area converter stations, especially for air-insulated equipment in HVDC yards used for switching and reconfiguration, such as disconnectors, earthing switches, current and voltage measuring systems, surge arrestors, gas-to-air (Overhead line – OHL) bushings, and cable connections. HVDC GIS technology enables significantly decreasing the footprint of onshore converter stations, but its impact on cost savings of offshore installations is even more substantial. HVDC GIS technology enables significantly decreasing the footprint of onshore converter stations, but its impact on cost savings of offshore installations is even more substantial. Using insulating gas under pressure, space requirements of HVDC switchyards can be reduced by (70 . . . 95) % – an obvious advantage given the extreme limits imposed by offshore platforms. Since both overhead lines and underground transmission by HVDC power cables (or HVDC gas-insulated lines) are expected in the upcoming HVDC links and overlay networks, HVDC GIS may also be applied at transition stations between different transmission media to minimize the footprint of these stations, while integrating all relevant functionalities in gas-insulated technology. Besides their great space-saving advantages, HVDC gas-insulated switchgear assemblies provide high reliability and availability – being totally enclosed, the high voltage live parts are not affected by ambient conditions such as dust, salty air/salt-spray, ice, or rain. Up to now, no standards for testing of gas-insulated HVDC systems are available, although prestandardization work is in progress within CIGRE. Some tests can be performed as required in AC GIS standards. Special aspects of DC voltage stress, like the electric field distribution of insulators influenced by the accumulation of electrical charge carriers and the operation-related inhomogeneous temperature distribution, must be considered by additional electric and thermoelectric tests. For HVDC GIS, the experience of long-term performance is limited today. Although ageing is expected to be of lower importance, tests are recommended. This chapter summarizes the physical and technical background to design and develop compact HVDC switchgear assemblies using gas-insulated technology. It explains the developed modules of the substation and gives an overview of the performed tests. Furthermore, it provides an insight in the on-going standardization activities and describes applications in converter and transition stations, showing its space-saving characteristics.

9.17.2 Introduction

With the growing demand of transporting higher power ratings over very long distances, the HVDC technology is technically superior over conventional HVAC technology. One worldwide driver of HVDC technology is the integration of renewable electric energy resources, resulting in a change of the existing electric power transmission system. Based on an increasing demand for space-saving and reliable HVDC solutions, compact gas-insulated systems for HVDC applications are under development worldwide. While gas-insulated systems for AC applications are common standard since decades [2], the experience in the development and testing of gas-insulated HVDC systems is limited today. Insulation requirements and testing procedures for gas-insulated AC systems are fixed in standards as in IEC 62271 series since several tens of years. For HVDC GIS products, specific standards are not available today. Since 2014, recommendations have been developed by the CIGRE JWG D1/B3.57, "Dielectric testing of gas-insulated HVDC systems." In principle, the electrical and mechanical requirements must be fulfilled according to IEC 62271-203 [47], which are independent of the operating voltage (AC or DC). Special aspects of DC voltage must be considered by means of additional electric and thermoelectric tests. [3].

While in gas-insulated HVAC systems the electric field distribution remains unchanged after energization, in HVDC systems it changes from electrostatic field to electric flow field. Intensive research on this field during the last years shows a growing understanding of the basic characteristics of HVDC insulation, including charge carrier-based processes in the gaseous and solid insulating media and at the gas-solid interface [4–10].

These phenomena must be considered during design and development of gas-insulated HVDC systems to ensure a high reliability, as well as for the determination of appropriate testing strategies.

9.17.3 History of Gas-Insulated HVDC Systems

The potential of gas-insulated HVDC systems was recognized and studied in the 1960s following the first installation in 1983 [11, 12]. Nevertheless, the commercial application of HVDC GIS was limited to only few applications. The further use of gas-insulated systems was hampered by a tendency for the insulating materials to fail during polarity reversal tests. This was generally attributed to the presence of space charges trapped within the insulation [13]. Today, the increasing demand for HVDC connections for both submarine and land applications was the reason to develop new HVDC gas-insulated systems (Figure 9.108) [14].

The first commercial HVDC-GIS was installed in the year 2000 in Japan. The ± 500 kV HVDC-GIS Anan Converter station of Shikoku Electric Power consists of disconnectors and one bus bar. Since commissioning, the operating voltage is only ±250 kV. A DC bus bar with superimposed DC

Year	Project	Type
1980	First Research (Sweden/Switzerland)	Basic HVDC GIS studies
1983	Worldwide first HVDC busbar installation First generation busbar design (Sweden)	Gotland 2 link 150 kV GIS
1987	First generation design (Sweden)	Gotland 3 link 150 kV GIS
1995	BPA Laboratory (USA)	First long-term test (500/600 kV)
2000	Worldwide first HVDC GIS installation First generation design (Japan)	Kii-Channel HVDC link 250 (500) kV GIS
2013	New generation design (Switzerland)	Type tests for 320 / 350 kV GIS
2014	New generation design (Germany)	Type tests for 320 kV GIS
2018	First Prototype Installation Test (NL/CH)	First standardized long-term test 350 kV GIS
2018	New generation design (Japan)	First project Hokkaido-Honshu HVDC Link 250 kV
2020	New generation design (Japan)	First project Hida-Shinano HVDC Link 200 kV
2021	New generation design (China)	Type tests for 200 kV
2023 (expected)	1st offshore installation (Germany)	DolWin6 offshore converter platform with 320 kV DC GIS
2024 (expected)	2nd offshore installation (USA)	Offshore converter platform with 320 kV DC GIS
2025 (expected)	3rd offshore installation (Germany)	BorWin5 offshore converter platform with 320 kV DC GIS

Figure 9.108 HVDC GIS projects

voltage of ±150 kV is in operation since 1983/1987 in Gotland (Sweden). In 1986, ABB and BPA have performed together a development of ±500 kV HVDC-GIS. From 1990 until 1995, long-term tests at BPA's test center were carried out. The project involved energizing a test pole containing the elements of an SF_6-insulated station for duration of approximately 2 years. [13] The elements of the test pole consisted of GIS spacers, SF_6 air bushings, air-insulated arrester, SF_6-insulated arrester, and SF_6-oil bushing. The long-term tests were successfully completed in 1996 [14].

Today, the increasing demand for HVDC connections for both offshore and onshore applications connected with cost reduction efforts, the goal to be more environmentally friendly is the reason to develop new HVDC gas-insulated systems. The high level of quality of the GIS technology provides security of supply and high availability of electricity. In 2019, a new installation in Japan is going into operation with a rated voltage of ±250 kV (Hokuto converter station). The first offshore installation with ±320 kV rated voltage is expected to be in operation in 2023 (Grid Access DolWin6 [15]).

9.17.4 Special Aspects of HVDC Insulation

Dimensioning of HVDC GIS requires the knowledge of electric field distributions occurring in the case of DC conditions. Starting from a capacitive field distribution at the moment of energization with DC voltage, the electric field distribution continuously evolves over time. This field transition depends on the surface and bulk currents in the solid insulation, defined by the strongly temperature-dependent volume conductivity as well as ionic currents in the gas. It results in surface charging, particularly of the solid-gas interface of insulator surfaces, and space charge accumulation in the solid insulation, until the steady state under DC voltage stress is reached. Therefore, for gas-insulated HVDC equipment, purpose-specific design of both components and test procedures are of high importance to adequately verify the electrical insulation behavior and long-term reliability [3, 6, 9].

Conventional models have described the behavior of the insulating system with RC elements and with using distinct values of the conductivity. Taking practical field strength into account, investigations in [3, 6, 9, 16] have shown that besides natural ionization, additional charge carriers from voltage dependent sources are present. Therefore, conduction has to be understood as the movement of different charge carriers within the insulating material [9, 16]. The determining complex physical processes are influenced by several parameters, such as non-linear material properties dependent on temperature, humidity and electrical field strength. Besides, the duration of the field transition is depending on the conductivity. Results confirm that a higher conductivity is reducing the time for the field transition [9, 17–20].

Relevant electrical effects in solid and gas insulation and at their interfaces are summarized in Figure 9.109. Effects (and *quantities*) can be categorized as [21]:

1) Electrical effects in solid insulation: electronic and/or ionic conduction (*effective conductivity* σ), polarization (*permittivity* ε), space charge accumulation (*space charge density* ρ), in gas insulation ions generation by:
   a) natural radiation,
   b) field electron emission from electrodes,
   c) direct ionization by high local field,
   d) charged metallic particles,
   e) ion drift
   f) on interfaces/ surfaces:
   g) Surface charge accumulation (*surface charge density* ρ_s),
   h) surface conduction (*surface conductivity* σ_s),

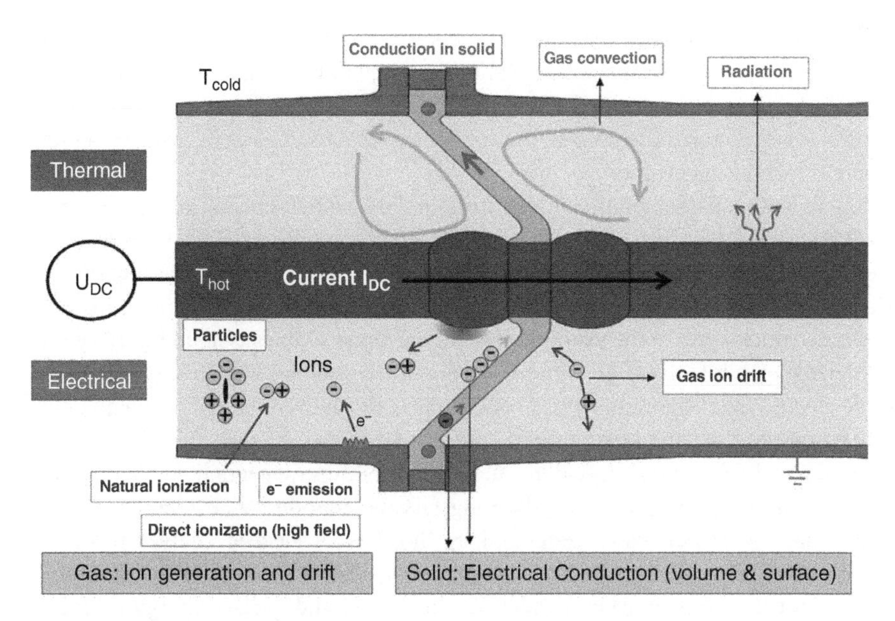

Figure 9.109 Important effects for the insulation design of a HVDC GIS (courtesy of Hitachi Power Grid)

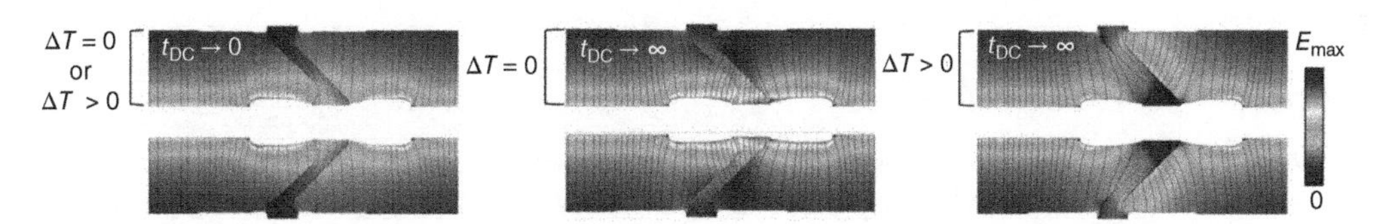

Figure 9.110 Electric field distribution of a conical insulator shortly after energizing with DC (left), after reaching DC steady state without (middle) and with (right) a temperature gradient between conductor and enclosure (assuming a rotationally symmetrical temperature distribution) (*Source* [18])

   i) charge transfer into solid,
   j) charge recombination with gas ions,
   k) injection/ emission at electrode/ solid insulation interface.
2) Thermal effects heat generated by ohmic losses of the nominal current I_{DC} in the conductor is transferred to the ambient by
   l) gas convection,
   m) conduction through the solid insulation and
   n) radiation leading to a thermal gradient (T_{hot} -T_{cold}) across insulation.

Since the operating currents cause an inhomogeneous temperature distribution between conductor and enclosure, the solid insulator experiences a high temperature gradient. The conductivity of the material is varying by orders of magnitude in the relevant temperature range [22]. Hence, the conductivity is increasing in the vicinity of the heated conductor, usually resulting in an enhanced electric field strength in the colder regions of the insulation (Figure 9.110) [9, 18, 23].

Under AC condition (Figure 9.110, left) the electric field distribution is determined by the permittivities. Any transient behavior after energizing with a constant voltage and any dependency on the temperature is neglectable. The highest field strength is occurring at the inner conductor in the insulating gas. This condition, also occurring shortly after energizing with DC, and has to be tested with a typical dielectric voltage withstand test with AC and DC.

Under DC condition, the electric field distribution is determined by the charge carrier-based conduction processes in the gas and in the solid. Without a temperature gradient (Figure 9.110, middle) in cold condition, the maximum field strength occurs close to the conductor within the solid insulator. Hence, it is recommended to perform long-term tests to investigate possible aging processes, since there is lacking experience.

With a current flowing in the conductor, the temperature in the system increases and a temperature gradient arises (Figure 9.110, right). Since the conductivity of the solid material is increasing with the higher temperature, the electric field is shifted to colder regions. Hence, the maximum field strength is displaced away from the center of the insulator toward the outer part of the insulator, close to the grounded enclosure. In the example shown, the stress to the gas-solid interface on the convex side is increasing, while the field strength close to the conductor is reduced. This thermal transition of the electric field represents typical operating conditions and must be adequately represented and therefore must be emulated in thermoelectric tests.

An important criterion is the insulator's resistivity and its temperature-dependent behavior. The higher the resistivity of the insulating material, the longer is the charging time. The lower the resistivity, the higher are the power losses generated by the flow of electric charges through the insulators. Both limits must be taken into account when dimensioning HVDC systems. Thermal overloading of the insulators must be avoided. On the other hand, surface charging effects must be minimized. This can be achieved through increased conductivity or optimized surfaces shapes to avoid surface charges by minimized electrical fields [24, 25].

9.17.5 Modular Design of HVDC GIS

The development of HVDC GIS is based on the technology used for HVAC GIS with, in principle, the same components and a newly designed HVDC insulator to meet the requirements of the electric field distribution under DC voltage stress. The elementary functions of gas-insulated switchgear assemblies without a circuit-breaker under DC operation are:

- Set up a safe isolating distance with earthing on both sides
- Measurement of current and voltage
- Protection against overvoltage
- Connection to other transmission media (HVDC cable, HVDC overhead line, HVDC GIL)

The modular structure of HVAC GIS has been proven in many applications worldwide with high reliability and has been optimized over the last 50 years since the first HVAC GIS has been installed in Germany in 1968 [2].

The gas-insulated system is independent from external ambient conditions, like humidity, dust, ice, wind, corrosion, etc., because of its gas and watertight encapsulation. The grounded enclosure can be located inside a building. Underground installation is possible as well.

Respecting all relevant design requirements including geometry and material characteristics, development has led to the modules shown in Figure 9.111, that can flexibly be arranged according to the project's needs, for example, in a bay layout for a HVDC gas-insulated system (Figure 9.112).

9.17.5.1 Disconnector and Earthing Switches

At the core of the switchgear assemblies is the disconnector. Together with the earthing switches on either side of the isolating gap, the disconnector ensures the safe insulation and earthing of de-energized circuits. The fast-acting earthing switch also enables the safe discharge of potential resulting from residual DC charges or cables.

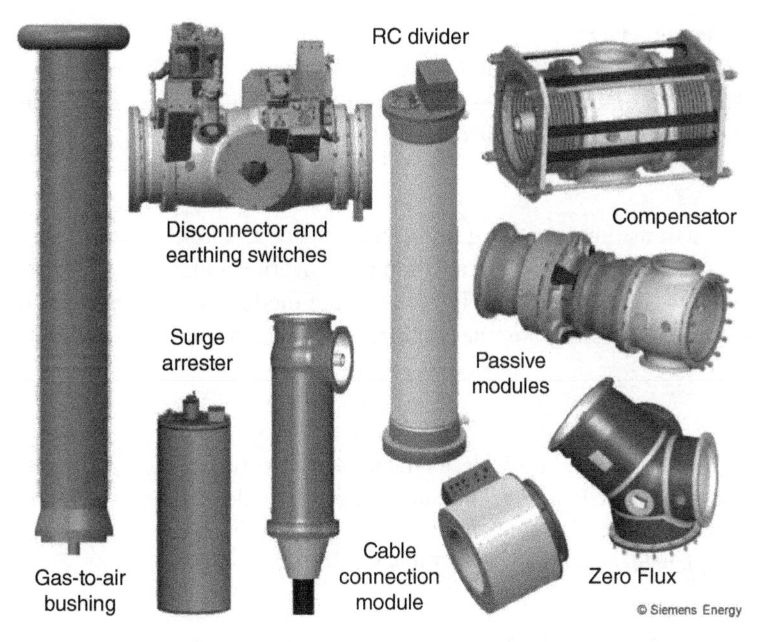

Figure 9.111 Modules of HVDC gas-insulated switchgear assemblies (*source* [30])

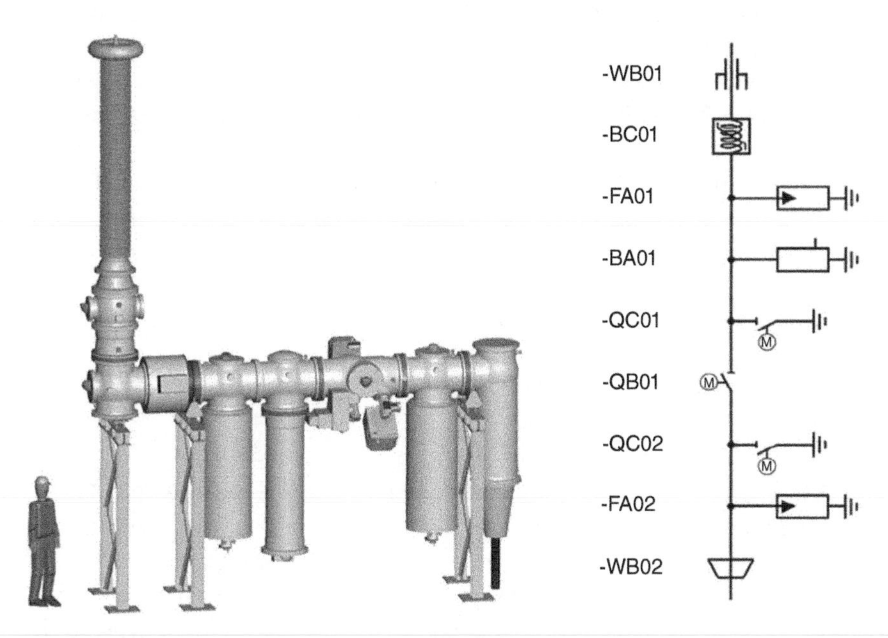

Figure 9.112 Principal structure of a compact HVDC substation – converter pole feeder as HVDC substation in/out bay (*source* [30])

9.17.5.2 Surge Arrester

Encapsulated surge arresters ensure the protection from overvoltage. Their active parts consist of metal oxide varistors with a strongly nonlinear current-voltage characteristic. Depending on the overvoltage surge requirements, more than one surge arrester may be used.

9.17.5.3 Voltage and Current Measurement

Gas-insulated RC voltage dividers map high-voltage linearly over a frequency range from DC up to 30 kHz. They are designed also for an optimal transient behavior [26]. Current detection relies on the zero-flux measurement principle [27] for rated currents up to 5000 A.

9.17.5.4 Interface Modules

The interface modules enable the transition from the gas-insulated switchgear assemblies to other equipment. Gas-to-air bushings are available for the transition to an overhead line. The cable terminations follow IEC 62271-209 [45]. The conductor link within the module can be removed to separate cable and GIS for on-site cable testing. Another interface module is available for the transition to a gas-insulated line.

9.17.5.5 Passive Modules

Several passive modules (e.g., T-shaped) are available for flexible configuration. Compensation modules provide options to deal with heat dilatation and make access to single modules within the arrangement for service and repairs.

Technical data for two voltage levels is summarized in Table 9.21.

9.17.6 Testing of Gas-Insulated HVDC Systems

The testing requirements for gas-insulated HVDC systems are currently not standardized. Recommendations are under development in CIGRE JWG D1/B3.57, TB 842 [46]. According to these, the electrical and mechanical requirements of IEC 62271 series, that are independent of the type of operating voltage, should be fulfilled also [28, 29] [45]. Beside DC withstand voltage tests, composite voltage tests with lightning impulse and switching impulse voltage, superimposed to

Table 9.21 Exemplary technical data of HVDC GIS of the latest generation TB 842 [46].

Technical data		
U_{ndc} nominal DC voltage	±320 kV	±500 kV
U_{rdc} rated continuous DC voltage (max. continuous operating voltage)	±336 kV ±350 kV ±352 kV	±550 kV
U_{LI} rated lightning impulse withstand voltage (1.2/50 μs)	1050 kV LI 1175 kV LI	1425 kV LI 1550 kV LI
• superimposed with DC	$\pm U_{rdc}$ & ±1175 kV LI	$\pm U_{rdc}$ & ±1550 kV LI
• across the open isolating distance	$\pm U_{rdc}$ & ∓1175 kV LI	$\pm U_{rdc}$ & ∓1550 kV LI
U_{SI} rated switching impulse withstand voltage (250/2500 μs)	850 kV SI 950 kV SI	1050 kV SI 1175 kV SI
• superimposed with DC	$\pm U_{rdc}$ & ±950 kV SI	$\pm U_{rdc}$ DC & ±1175 kV SI
• across the open isolating distance	$\pm U_{rdc}$ & ∓950 kV SI	$\pm U_{rdc}$ DC & ∓1175 kV SI
I_{rdc} rated nominal current	2000 A DC 4000 A DC	2000 A DC 4000 A DC 5000 A DC
I_k rated short-time withstand current	50 kA AC (1 s) 63 kA AC (1 s)	50 kA AC (1 s) 63 kA AC (1 s)
Ambient temperature range	(−30 ... +50) °C	(−30 ... +50) °C

DC voltage, have to be conducted after two hours of DC prestress. DC prestress longer than two hours is not required, as the main objective of the tests is to verify the gas insulation only. The isolating distance of the disconnectors has to be stressed with combined voltage tests, consisting of DC voltage at one terminal and lightning or switching impulse voltage at the other. Additional electric and thermo-electric tests must be performed to consider the special aspect of DC voltage in terms of electric field distribution of insulators, influenced by the accumulation of electrical charge carriers and the operation-related inhomogeneous temperature distribution.

An overview about all type, routine, on-site and special tests acc. to the CIGRE JWG D1/B3.57 TB 842 [46] is given in Table 9.22. The order or sequence of tests is not specified and could be adopted to the needs. Selected tests will be described in the following sections using examples.

9.17.6.1 Type and Special Tests

Dielectric-Type Tests

For the dielectric test, all relevant HVDC GIS components were assembled [30]. Beside DC withstand voltage tests, composite voltage tests with LI and SI voltage, superimposed to DC voltage, were conducted after two hours of DC prestress. To ensure adequate testing, an optimized test-circuit, including a sphere gap for the superposition, was applied [31]. DC prestress longer than two hours was not required, as the main objective of the tests is the gaseous insulation only. The isolating distance of the two types of disconnector switches was stressed with combined voltage tests, consisting of DC voltage at one terminal and lightning or switching impulse voltage at the other. An example of the performed dielectric-type tests is given in Table 9.23 [32].

Insulation System Test

For HVAC GIS, this type of test is not required, as there is no relevant dependence of the electric field distribution on the temperature distribution inside the GIS. On the contrary, electro-thermal tests are of high importance for HVDC GIS, due to the temperature-dependent field transition.

Table 9.22 Type, routine, and special tests

	Clause			
Test	**Type**	**Routine**	**On-site**	**Special**
Non dielectric tests for insulators	×	–	–	–
DC withstand voltage tests	×	–	–	–
Power-frequency voltage tests	–	×	×	–
Polarity reversal tests	Not mandatory	–	–	–
Superimposed voltage tests	×	–	–	–
Switching impulse voltage tests	Not mandatory	–	–	–
Lightning impulse voltage tests	Not mandatory	–	–	–
Partial discharge tests	×	×	×	–
Voltage test across open switching device	×	–	–	–
Load conditions tests	Not mandatory	–	–	–
DC Insulation system tests	×	–	–	–
Prototype installation test	–	–	–	Not mandatory

Table 9.23 Dielectric tests for HVDC GIS

Test	Details
AC withstand voltage test	with PD measurement
DC withstand voltage test	
LI withstand voltage test	15 impulses
SI withstand voltage test	15 impulses
Composite voltage test[a] DC + LI	with 2 h DC prestress (15 impulses)
Composite voltage test[a] DC + SI	with 2 h DC prestress (15 impulses)
Combined voltage test, DC + LI	across the isolating distance (15 impulses)
Combined voltage test, DC + SI	across the isolating distance (15 impulses)

[a] All polarity combinations

The transition of a gas-solid insulation system from a capacitive field distribution at the moment of energization with DC voltage to a resistive field distribution, results in interface charging, particularly of the solid-gas surface and space charge accumulation in the solid insulation until the resistive steady state is reached. The design of solid insulating components as well the right testing procedure to verify the electrical insulation behavior are therefore of high importance.

The time duration d_{DC} of the long duration continuous DC voltage test depends on the transition time from capacitive to resistive field conditions and has to be determined before starting the tests. If a degree of charging of 90% has been passed at each location of the insulator, the transition time is reached. As the DC-stress is only reached asymptotically, those 90% seem to be a reasonable compromise for testing purposes. The transition itself depends on the local temperature distribution and on the lowest temperature. The transition to a DC field distribution could lead to a long test duration of hours to months.

The duration of the test can be approximated with the following methods [33]:

- Worst case approximation of dielectric time constant $\tau_m = \varepsilon_0\varepsilon_r/\sigma$ (taking the longest value over the whole insulation system, that is, typically the location with lowest conductivity). For the estimation of the dielectric time constant, it is necessary to measure permittivity and electrical conductivity of the insulating material (bulk).
- DC electrical field simulation. A simulation verified by experiments (at least by means of model-arrangements) has to be done [34]. Material and gas characterizations, which are relevant for the numerical model, have to be provided [35]. The scalability of the model has to be demonstrated. Of importance are the electric field strength, temperatures and temperature gradients. These parameters have to represent realistic service conditions.
- Measurement of the DC potential field on the actual insulator surface in energized state on different locations along the insulator radius. The temperature gradient across the insulators shall represent the worst case under service conditions with tolerances less than 20%.

The accurate definition of the transition time and the verification of the transition behavior is essential to derive a reliable testing procedure of HVDC gas-insulated systems. Also, the surface conductivity of the insulating material shall be measured. If the charge transport in the solid insulation is dominated by the bulk conductivity compared to the charge transport caused by the surface conductivity, only the bulk material has to be taken into consideration for calculation of the dielectric time constant (in case of dominant bulk conductivity).

Especially in the case of the third method, the time to reach the DC steady state d_{DC} is only indirectly determined. In those cases, d_{DC} is considered to be fulfilled, if the rate of change of the measured potential field is lower than 10% of the initial rate (initial time delays shall be ignored). The accurate definition of the transition time and the verification of the transition behavior is essential to derive a reliable testing procedure for HVDC gas-insulated systems.

The insulation system test has similarity in the test procedure to the prequalification test for polymeric-insulated HVDC cables [36]. However, the prequalification test for HVDC extruded cable systems is intended to indicate the electrical long-term performance test on the voltage-time characteristic of the insulating material and to cover thermo-mechanical aspects. For HVDC gas-insulated systems, a verification of the electrical lifetime is of insignificant importance, because the electrical lifetime of the solid insulating material used in GIS/GIL is equal or even better under DC voltage stress compared to AC at typical service stress [37, 38]. Experiences with AC gas-insulated systems show that all parts made of solid insulating material do not reveal any critical ageing, assuming a manufacturing in sound condition according to the quality requirements. Additional mechanical stresses caused by load cycles are of minor interest for GIS/GIL applications and are covered by thermal cycles performance tests of each insulator design according to IEC 62271-203 [47]. Nevertheless, a thermo-mechanical pretest or a combination with the insulation system test under high load conditions could be possible.

The duration of the long duration continuous DC voltage test depends on the transition time from capacitive to resistive field conditions and has to be calculated before starting the tests. The transition itself depends on the local temperature distribution and on the lowest temperature. The temperature difference across the insulators is of high importance and must be defined and verified for testing. ΔT_{max} is the maximum temperature difference over the gas insulation in thermal steady state at which the gas-insulated HVDC system is designed to operate at rated current. This value is also to be calculated or measured during pretests and stated by the supplier, who should also provide evidence of the correlation between this design value and data measured during testing. $T_{cond,max}$ is defined as the maximum temperature at which the conductor is designed to operate at rated current. This value is to be stated by the supplier. For superimposed voltage tests after long DC prestress to reach the defined DC steady state and worst-case temperature conditions with maximum temperature gradient across the insulators, high load (HL) conditions must be used. High load consists of a continuous heating period at rated current up to the thermal steady state (duration $d\vartheta$). For the insulation system test, a higher equivalent DC or AC current compared to the rated DC current is allowed, because the maximum conductor temperature and maximum temperature drop across the insulation has to be safely achieved.

The duration $d\vartheta$ is defined as the time sufficient for the temperature rise to reach a stable value and should be evaluated by pretests on the same test arrangement as used for dielectric HL tests as preferred procedure (thermal calibration). The test shall be made over a period of time, sufficient for the temperature rise to reach a stable value. This condition is deemed to be obtained when the increase of temperature rise does not exceed 1 K in one hour. This criterion will normally be met after a test duration of five times the thermal time constant of the tested device. Duration $d\vartheta$ for gas-insulated components is typically in the range of some hours. The time for the whole test may be shortened by preheating the circuit with a higher value of current, provided that sufficient test data is recorded to enable calculation of the thermal time constant [35].

As described the time duration d_{DC} is defined as time required to reach at least 90% of the DC steady state electric field at the major locations (i.e., locations with at least 20% of E_{max}) on the insulator surface. One example for a DC transition simulation result is shown in Figure 9.113.

The transition depends on the local temperature and the process time varies with the temperature at the specific location. In the case shown, a transition duration of some months has to be

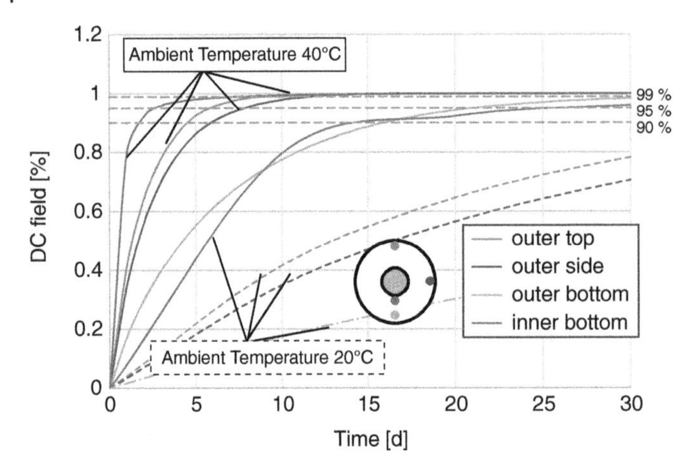

Figure 9.113 Transition time for the insulator, depending on the location and ambient temperature; DC field strength normalized [35]

chosen, following the worst-case consideration in the coldest region on the outer bottom surface of the enclosure. The time duration could be reduced by increasing the temperature due to higher currents or an increase of the ambient temperature. The ambient temperature could be realized by an additional housing around the test device along with forced ventilation to achieve a homogeneous temperature distribution throughout the interior of the housing. In this case, a time duration d_{DC} of 30 days has to be chosen to reach at least 90 % of the DC steady state electric field at each location of the insulator surface.

As continuous DC voltage the rated DC voltage U_{rdc} has to be used. As composite voltage test of DC voltage and impulse voltage, the rated lightning and switching impulse voltages shall be superimposed to the rated DC voltage.

Regarding the number of impulses, the test procedure A of IEC 60060-1 is the preferred test procedure [39], because a flashover in the self-restoring insulation (gas) could be of influence on the electric field distribution. The gas-insulated system has passed the impulse tests if the following conditions are fulfilled:

- Each series has at least 3 impulses.
- Three impulses of the specified shape and polarity at the specified withstand voltage level are applied to the test object. The requirements of the test are satisfied if no indication of failure is obtained.

The duration of lightning and switching impulse voltage stress is assumed to be too short to affect the electrical field distribution under DC stress significantly. Therefore, the time between two successive impulses shall be short, but not less than 1 minute.

The insulation system test has to be performed under high load conditions. After a heating period, the maximum conductor temperature and maximum temperature drop across the insulation has to be achieved and maintained for the complete test duration. It is of advantage to measure partial discharges (UHF PD monitoring), temperature (ambient and test device), test current and test voltage during the entire test and record the data. The measured temperatures have to be compared to data obtained from calibration measurements from previous continuous current tests. As insulation system test, a sufficient number of insulators assembled in realistic arrangements must be tested. A dielectric routine test or preconditioning could be considered before starting the insulation system test. The normal sequence of tests is described in Table 9.24. An example of a test set-up is given in Figure 9.114.

Table 9.24 Sequence of insulation system test acc. CIGRE JWG D1/B3.57 TB 842 [46].

Test	Conditions	Load
Thermal Pretest	Heating @ defined temperature ±5 K	HL
Dielectric Pretest	PD test with AC or DC voltage	ZL
Long-duration continuous DC voltage test	Rated DC voltage U_{rDC} (one polarity)	HL
Superimposed lightning impulse voltage test (bipolar and unipolar)	Rated LIWV values, superimposed to the rated DC voltage U_{rDC}	HL
Superimposed switching impulse voltage test (bipolar and unipolar)	Rated SIWV values, superimposed to the rated DC voltage U_{rDC}	HL
Long-duration continuous DC voltage test	Rated DC voltage U_{rDC} (other polarity)	HL
Superimposed lightning impulse voltage test (bipolar and unipolar)	Rated LIWV values, superimposed to the rated DC voltage U_{rDC}	HL
Superimposed switching impulse voltage test (bipolar and unipolar)	Rated SIWV values, superimposed to the rated DC voltage U_{rDC}	HL

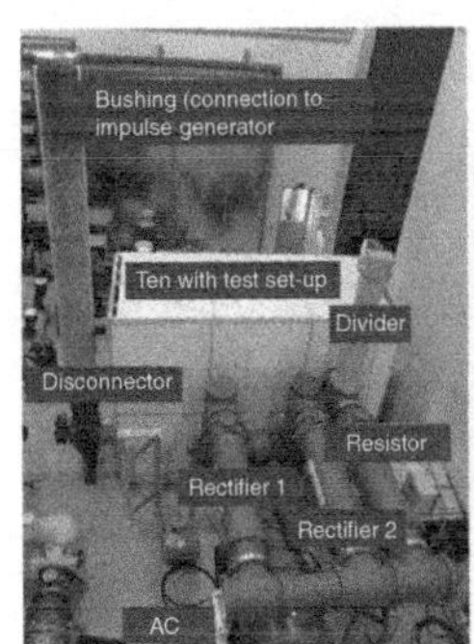

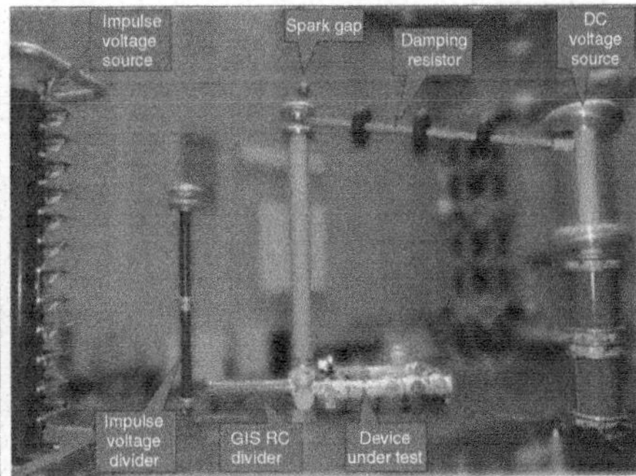

Figure 9.114 Examples of test arrangements for insulation system tests, completely gas insulated (left) and air insulated (right) [30] (courtesy of Hitachi Power Grid)

Prototype Installation Test

Based on service experience, gas-insulated HVAC systems feature a high degree of reliability and an excellent long-term performance. In comparison, HVDC GIS is a new technology with limited operational experience [31]. The user who intends to apply gas-insulated HVDC systems does expect the same reliability and long-term performance as in HVAC GIS.

From a high-voltage engineering perspective the main differences between HVAC and HVDC GIS are the long periods required to reach steady state DC electric fields and charge accumulation phenomena. Thus, HVDC GIS requires adapted design and testing. Until today, no tests have been standardized for HVDC GIS systems. However, the work of CIGRE JWG D1/B3.57 TB 842 [46] is proposing a "prototype installation test" as a demonstration test.

The main intention of the prototype installation test for gas-insulated systems is to confirm the reliability of the system under real service conditions. Real service conditions refer to:

- the components included in the test object
- installation and commissioning procedures
- as well as the dielectric, thermal and mechanical stresses applied in the test itself.
- Testing with representative stresses for real life operation
- Inclusion of monitoring and diagnostics during the test

Thus, the prototype installation test is not an additional type test, but rather a one-time, non-mandatory test performed after successfully completing the type tests to verify the effectiveness of the HVDC GIS specific type and routine test procedures [40, 41].

An example is shown in Figure 9.115. The HVDC GIS test pole consists of a bus-duct ring that is connected to high-voltage via SF_6 to air bushings. To achieve a thermal regime typical for high load condition, an AC heating current can be induced in the ring using conventional current transformers. The dielectrically decisive quantity – temperature difference inner conductor to enclosure (at the insulator) – will be generated with at least the same magnitude compared to a DC current load [42] (see Figure 9.116).

9.17.6.2 Routine and On-Site Tests

Routine Testing

The intense of routine tests is to demonstrate the integrity of the manufactured assembly units. They should reveal faults in material and assembly. Based on IEC 62271-1 [4], routine testing comprises:

a) dielectric test on the main circuit
b) tests on auxiliary and control circuits
c) measurement of the resistance of main circuit
d) tightness test
e) design and visual checks

Since the tests b) to e) are not depending on the type of voltage, they can be performed without any constraints.

The experience of using DC voltage for dielectric routine testing of gas-insulated systems is limited. Therefore, the recommendation of CIGRE JWG D1/B3.57 TB 842 [46] foresees to carry out the

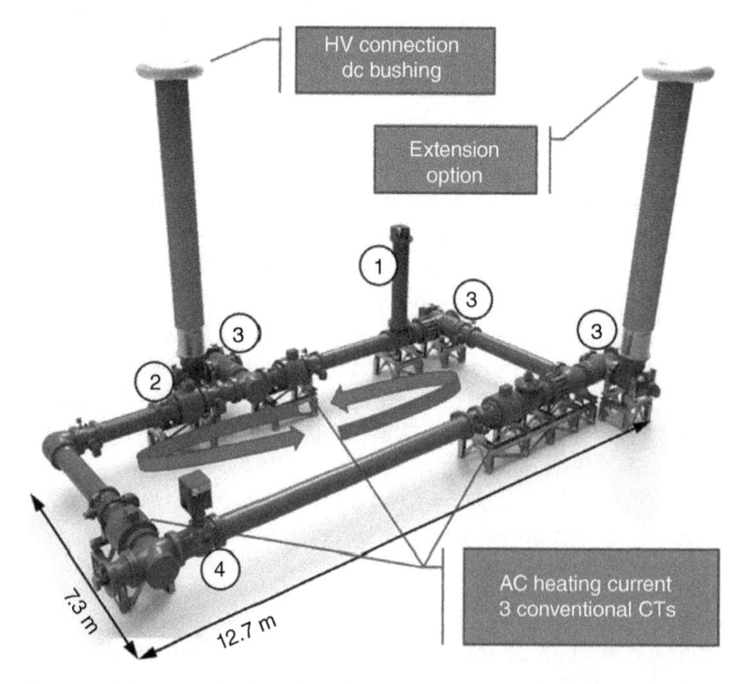

Figure 9.115 Example of an HVDC GIS prototype installation: ① VT: RC-divider; ② CT: Zero-Flux Sensor; ③ Combined disconnector and earthing switches; ④ Fast-acting earthing switch [42] (courtesy of Hitachi Power Grid)

Figure 9.116 Test set-up of a prototype installation test [32, 43] (courtesy of Hitachi Power Grid)

dielectric tests of the main circuit with AC voltage, according to IEC 62271-203 [47]. Accompanying partial discharge (PD) measurements should detect possible material and manufacturing defects on all transport units. Especially, the extensive experience to judge possible partial discharges under AC voltage stress by patterns is very valuable.

The JWG recommends for AC prestress (1 minute) before PD measurement

$$\widehat{U}_{\text{prestress AC}} = 1.5 \cdot U_{\text{rdc}} \tag{9.1}$$

and for the PD measurement (>1 minute)

$$\widehat{U}_{\text{pd-test AC}} = 1.2 \cdot U_{\text{rdc}} \tag{9.2}$$

U_{rdc} is the rated DC voltage, that is, the maximum continuous operating voltage.

On-Site Testing

The intense of dielectric testing on-site is to check the dielectric integrity of the equipment after transport and erection, in order to eliminate fortuitous causes, which might give rise to an internal fault in the future. Best experience with reliable techniques to detect and to identify these causes has been made with AC voltage testing, accompanied by PD measurement. Since the experience of using PD measurement at DC voltage is very limited and reliable measurement techniques are lacking, on-site tests should be completely carried out at AC voltage, with regard to the recommendations of CIGRE JWG D1/B3.57 TB 842 [46].

The tests are performed at nominal gas filling pressure. The test voltage will usually be supplied by a resonant test system. The test is accompanied by PD measurement, using UHF PD sensors.

As HVDC GIS are assembled under clean factory conditions, the gas compartments are generally free from free-moving particles, as confirmed in the high-voltage routine tests with accompanying PD measurement. Nevertheless, the integrity and cleanliness after transport, full assembly on-site and gas works should be confirmed by high-voltage tests and PD measurements after erection on-site. For HVDC GIS, CIGRE JWG D1/B3.57 TB 842 [46] recommends the application of the same procedures as for HVAC GIS, as given in IEC 62271-203 [47]. The AC test voltage values for PD measurement and high-voltage test should be calculated as for the HVDC GIS routine tests, as given in the previous section. The voltage application is typically performed in individual steps, including a potential conditioning phase and the main high-voltage test, followed by partial discharge test.

9.17.7 Standardization

There are no standards available up to now, for testing gas-insulated HVDC systems such as HVDC GIS and HVDC GIL. Since 2014, recommendations have been developed by CIGRE JWG D1/B3.57 TB 842 [46] "Dielectric testing of gas-insulated HVDC systems".

In 2017, international standardization has been started in IEC Technical Committee 17. Therefore, a task force has investigated the necessity of standardization for HVDC GIS and HVDC GIL. Firstly, the business and market needs were investigated. An increasing market for HVDC GIS and HVDC GIL over the next years was identified, due to the developments of offshore installations in the North Sea in Europe, of HVDC converter stations onshore under space limitations in Japan etc. In 2019, a new HVDC GIS installation is planned in Japan, to go into operation as part of the Hokuto converter station with a rated voltage of ± 250 kV. The first offshore application (DolWin6 grid access) with a HVDC GIS of ± 320 kV is already announced [15]. Due to these upcoming installations, the task force concluded in 2018, that starting standardization now would be just in time.

Furthermore, the task force compared existing AC IEC standards for GIS and GIL with the requirements resulting from HVDC applications. They found major differences of the dielectric requirements and testing between AC and DC stress. Some sections of IEC 62271-203 (AC GIS) [47] and IEC 62271-204 (AC GIL) [53] can be used after adaptations of the text, other sections related to dielectric testing need to be completely recreated with respect to the results of CIGRE JWG D1/B3.57 TB 842 [46]. Hence, new standards are necessary. The decision in 2018 of IEC SC17C for HVDC high-voltage switchgear assemblies and transmission lines is to have several new standards, amongst them these two:

- IEC 62271-5:
  Technical Specification on general requirements for DC switchgear devices & DC switchgear assemblies [52]
- IEC TS 62271-318:
  DC Gas-Insulated Switchgear Assemblies (DC GIS) [51].

9.17.8 Application of HVDC GIS

Whenever requirements exist that dictate space-saving designs, an independence from external conditions, the prevention of transient interference caused by lightning strikes and an aesthetic station planning with high visual amenity, a gas-insulated design of HVDC systems in converter and transition stations can have advantages over air-insulated systems. At higher voltages, the benefits of HVDC GIS to air-insulated switchgear increase. The ability to extend HVDC into cities or populated areas will favor compactness and thus HVDC GIS.

9.17.8.1 Converter Station

Interest in developing gas-insulated systems for HVDC applications in the last ten years basically arose as a solution to help provide grid access of remote offshore wind farms located far away from the coast. A gas-insulated switchgear design can reduce the volume of the HVDC substation by up to 95%, resulting in a 10% smaller converter platform and decreased weight, Capex and CO_2 footprint. Furthermore, high reliability and low maintenance requirements, typical features of gas-insulated systems, are essential due to poor accessibility and inherently hostile weather offshore.

With respect to the air clearances of air-insulated equipment, the HVDC substation with disconnectors and earthing switches, measuring devices, surge arrestors and cable connection is quite space consuming (see Figure 9.117 left). Using HVDC GIS instead with the same functional properties, it can be placed close to the walls, for example, at the wall between converter and reactor room (Figure 9.117 left). Depending on the overall platform design, this can save, in this case, up to 10% of the platform length and hence 10% of the whole platform size. Although, space constraints are less important for onshore than for offshore converter stations, HVDC GIS remain an attractive option for specific cases (Figure 9.117 right) with strict space constraints, high risk of lightning strikes, or installations at high altitude. However, depending on the HVDC transmission system topology used, reconfigurations between two systems can help to make redundancies available in the event of single-pole converter failure; the necessary interconnections require more space in the converter station, which can again be substantially reduced by using gas-insulated HVDC components.

9.17.8.2 Transition Station

Power transmission media (overhead-line (OHL), different cables, gas-insulated line) vary depending on different requirements, locations, or surroundings. These transition stations (Figure 9.118). enable separation and the safe earthing of each line section. They can include measuring devices and surge arrestors. Especially in the case of long HVDC cable lines, transition stations are required between long cable runs in order to enable [32].

- cable fault location online and offline with access to the cable core,
- commissioning tests using AC voltage and
- separation of different cable suppliers.

In parallel to the transition stations between different transmission media, especially very long cable lines require cable-cable transition stations. These stations integrate measuring systems for line-fault location by identifying transient voltages which are related to high-voltage cable failures. In this case, elements for the application of online and offline location systems are foreseen.

In Germany, the government has decided to privilege underground installations for the north-south HVDC transmission lines of about 700 km length each, to connect renewable energy sources of wind power in the north with load centers in the south. These long underground HVDC transmission lines need a separation in length to, if necessary, enable sections for repair and high-voltage testing. HVDC GIS offer a space-saving installation, while integrating all necessary functionalities, independent of the different transmission media that will be connected. Due to the modular design, even complex arrangements and layouts can be implemented with very little interface engineering. HVDC GIS can also be installed underground to keep the visual impact of a transition station to an absolute minimum and reliably prevents unauthorized access.

At the same time, these transition stations represent a clear separation between the cable sections, which can be relevant if the cables are provided by different cable manufacturers. For maintenance and repair purposes, disconnection and earthing of the sections have to be ensured.

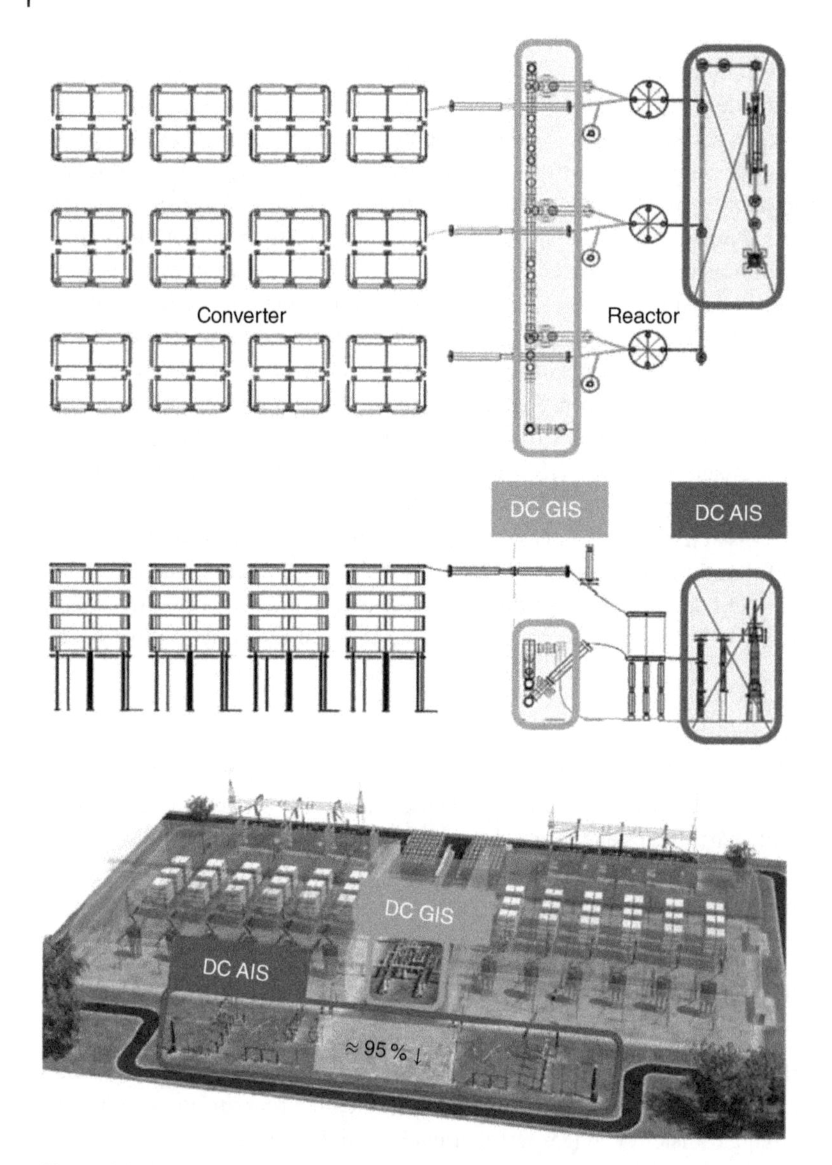

Figure 9.117 Top and front view of converter and reactor room in offshore platform using HVDC AIS or HVDC GIS (left) (*Source* [44]). Space-reduction of onshore converter station due to HVDC GIS (right) (*Source* [30]).

The layout of the cable-cable transition includes cable-sealing ends, a voltage divider, a decoupling option for transients (e.g., a capacitor) as well as disconnector and earthing switches (Figure 9.119).

In order to prevent the incidence of transient interference caused by lightning strikes, the air-insulated systems might be enclosed with a building. As an example: For the connection of the HVDC cables of two systems of Un = ±320 kV, a hall for the primary equipment of about 50 × 50 m and about 20 m in height is required, considering relevant air clearances (Figure 9.120). The footprint is 2500 m^2.

For a much more compact (and aesthetic) station design, all of the necessary functions already mentioned are also available in gas-insulated modules for HVDC GIS installations. The necessary components of the same functionality are available up to a voltage of ±550 kV.

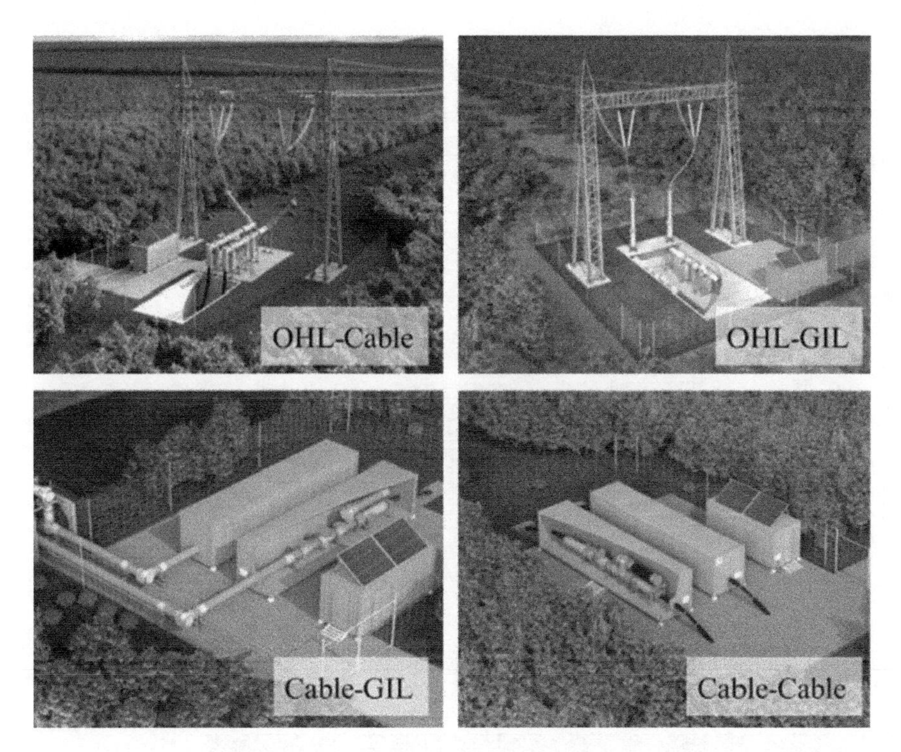

Figure 9.118 Example of transition station layout with HVDC GIS for connecting different transmission media (*Source* [44])

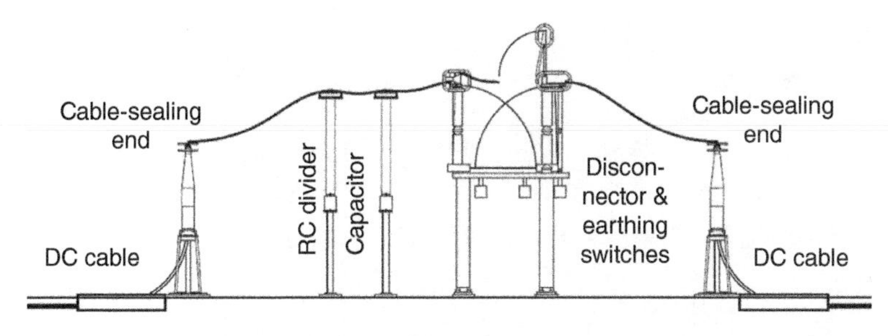

Figure 9.119 **HV**DC cable to cable transition in air-insulated technology (HVDC AIS) (*Source* [30])

The gas-insulated RC divider with a bandwidth from DC to approximately 1 MHz, including high amplitude accuracy, can be used for decoupling transient signals, for example, for online cable-fault location or even for partial discharge measurement. After gas works, measurement systems can be connected to the high-voltage conductor via T-shaped passive modules for offline fault location.

Furthermore, the effort for civil works, erection and commissioning can be reduced by prefabricated containers and delivering them to the construction site already pretested (Figure 9.121). This increases the installation speed on-site and reduces the risk of failures. Underground installation is also possible. There is no danger of lightning strike to the high-voltage conductor. Each container has a standard length of 13 m (40″) and a height of 3 m, containing HVDC GIS installations for voltages up to ±550 kV.

In the example for two systems of ±320 kV (Figure 9.121): The containerized solution with the HVDC GIS requires for the primary equipment 13 × 18 m = 234 qm and thus reduces the footprint from 2500 qm in air-insulated technology (compare Figure 9.120) by approximately 90%. Especially

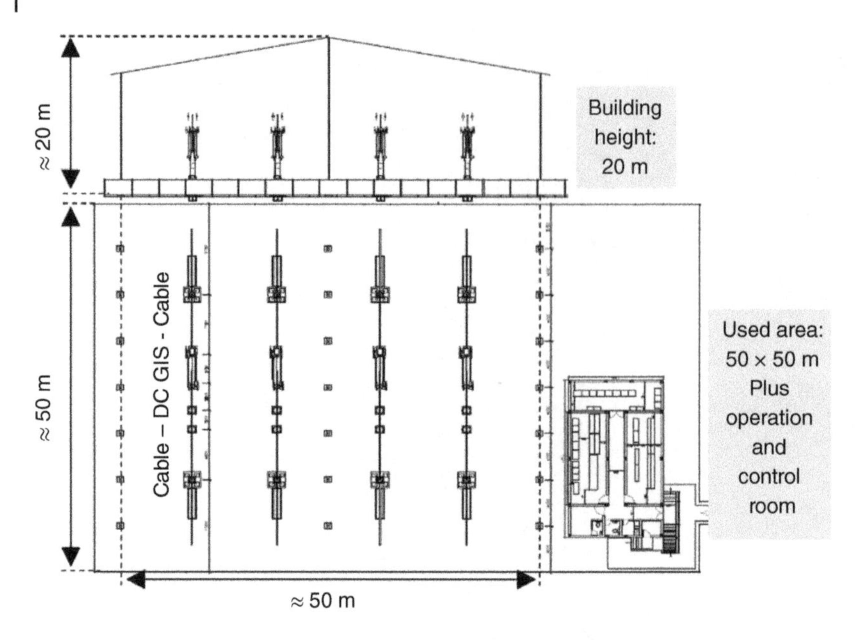

Figure 9.120 Example of a cable-cable transition station for two systems of Un = ±320 kV in air-insulated technology (*Source* [30])

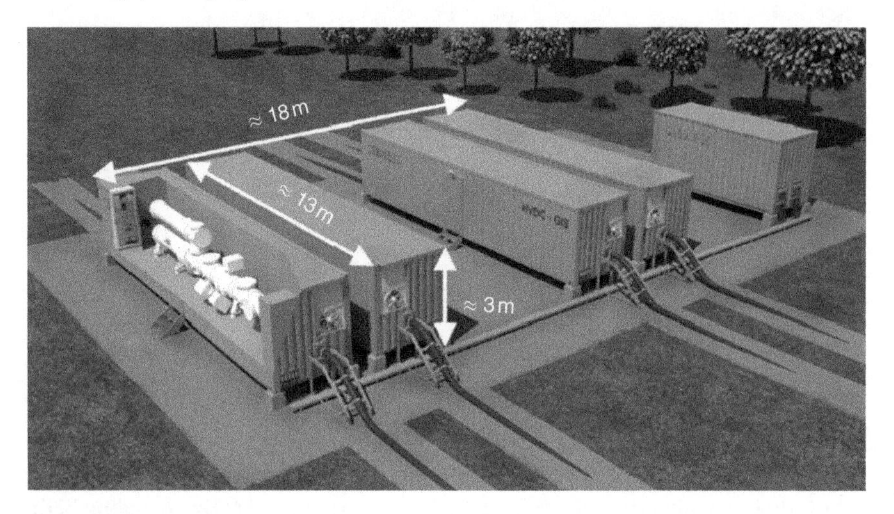

Figure 9.121 Example of a cable-cable transition station for two systems of up to ±550 kV with HVDC GIS (*Source* [30])

for cable-cable transition, the effort for civil works, erection, and commissioning can be reduced by using prefabricated containers and delivering them to the construction site already pretested. This increases the installation speed on-site and reduces the risk of failures. Underground installation is also possible.

9.17.9 Conclusion

High-voltage direct current (HVDC) technology for long-distance transmission of very high power has also made its way into grid access of offshore windfarms and interconnectors between different countries. Gas-insulated HVDC systems for converter and transition stations offer further advantages for integrating this technology, particularly compared to air-insulated systems.

High DC voltage ratings require large area converter stations, especially for air-insulated equipment in HVDC yards used for switching and reconfiguration, such as disconnectors, earthing switches, current and voltage measuring systems, surge arrestors, gas-to-air (Overhead line – OHL) bushings, and cable connections. HVDC GIS technology enables significantly decreasing the footprint of onshore converter stations, but its impact on cost savings of offshore installations is even more substantial. Using insulating gas under pressure, space requirements of HVDC switchyards can be reduced by (70 . . . 95)% – an obvious advantage given the extreme limits imposed by offshore platforms. Since both overhead lines and underground transmission by HVDC power cables (or HVDC gas-insulated lines) are expected in the upcoming HVDC links and overlay networks, HVDC GIS may also be applied at transition stations between different transmission media to minimize the footprint of these stations, while integrating all relevant functionalities in gas-insulated technology. Besides their great space-saving advantages, HVDC gas-insulated switchgear assemblies provide high reliability and availability – being totally enclosed, the high voltage live parts are not affected by ambient conditions such as dust, salty air/salt-spray, ice, or rain.

HVDC GIS can also be delivered prefabricated and pretested in modular containers with low effort for civil works and commissioning. This also helps keeping the visual impact of a transition station to a minimum and reliably prevents unauthorized access.

The technology of high-voltage direct current gas-insulated switchgear assemblies (HVDC GIS) is ready for voltage ratings of up to ±550 kV. Testing followed the recommendations of CIGRE JWG D1/B3.57 TB 842 [46] and relevant parts of IEC 62271 series [54].

References Section 9.17

1 Riechert, U.; Straumann, U.; Gremaud, R.: Compact Gas-insulated Systems for High Voltage Direct Current Transmission: Basic Design. IEEE PES Transmission & Distribution Conference & Exposition, Dallas, 2016.

2 Koch H (eds) (2014) Gas-Insulated Substations. Wiley

3 Neumann C, Juhre K, Riechert U, Schichler U (2019) Basic phenomena in gas-insulated HVDC systems and adequate dielectric testing. Paper presented at CIGRE-IEC Conference on EHV and UHV (AC & DC), Hakodate, 23-26 Apr 2019

4 Winter A, Kindersberger J (2014) Transient field distribution in gas-solid insulation systems under DC voltages. *IEEE Dielectr. Electr. Insul.* 21(1): 116–128

5 Gremaud R, et al. (2016) Solid-gas insulation in HVDC gas- insulated system: Measurement, modeling and experimental validation for reliable operation. Paper presented at 46th CIGRE Session, Paris, 21-26 Aug 2016

6 Zavattoni L (2015) Conduction phenomena through gas and insulating solids in HVDC gas insulated substations, and consequences on electric field distribution. Dissertation, Université de Grenoble

7 Wendel T, Kindersberger J, Hering M, Juhre K (2018) Space charge measurement in epoxy according to the pulsed electro acoustic method under consideration of attenuation and dispersion. Paper presented at VDE-Fachtagung Hochspannungstechnik, Berlin, 14-16 Nov 2018

8 Zhao S, Kindersberger J, Hering M, Juhre K (2018) Measurement of surface potential at the gas-solid interface for validating electric field simulations in gas-insulated DC systems. Paper presented at 47th CIGRE Session, Paris, 26-31 Aug 2018

9 Hering M, Speck J, Großmann S, Riechert U (2017) Field transition in gas-insulated HVDC systems. *IEEE Dielectr. Electr. Insul.* 24(3): 1608–1616

10 Schueller M, Straumann U, Franck C (2014) Role of ion sources for spacer charging in SF_6 gas insulated HVDC systems. *IEEE Dielectr. Electr. Insul.* 21(1): 352–359

11 Riechert, U.; Hama, H.; Endo F.; Juhre, K.; Kindersberger, J; Meijer, S.; Neumann, C.; Okabe, S.; Schichler, U., On behalf of CIGRE Task Force D1.03.11, Gas Insulated Systems for HVDC, ETG Fachtagung: Isoliersysteme bei Gleich- und Mischfeldbeanspruchung 2010, 27. - 28. September 2010, Köln, Germany, 2010, S. 101-107

12 R. Alvinsson, E. Borg, A. Hjortsberg, T. Höglund, and S. Hörnfeldt, GIS for HVDC Converter Stations, CIGRE 1986, Report, 14.02

13 Mendik, M., Lowder, S.M., Elliott, F. (1999) Long term performance verification of high voltage DC GIS, Transmission and Distribution Conference, IEEE, vol.2

14 Riechert, U. (2020) Compact High Voltage Direct Current Gas-insulated Systems 2020 IEEE International Conference on High Voltage Engineering and Application (ICHVE), Beijing, 2020, pp. 1-4, doi: 10.1109/ICHVE49031.2020.9279851.

15 Press release, updated on 12th July 2018. Link: https://www.siemens.com/press/PR2017070370EMEN

16 Backhaus K, Speck J, Hering M, Großmann S, Fritsche R (2014) Nonlinear dielectric behaviour of insulating oil under HVDC stress as a result of ion drift. Paper presented at Inter'l. Conf. High Voltage Eng. Appl., Poznan, 8-11 Sep 2014

17 Gremaud R, et al. (2014) Solid insulation in DC gas-insulated systems. Paper presented at 45th CIGRE Session, Paris, 24-29 Aug 2014

18 Juhre K, Lutz B, Imamovic D (2015) Testing and long term performance of gas-insulated DC compact switchgear. Paper presented at CIGRE Colloquium, Rio de J'o., 13-18 Sep 2015

19 Straumann U et al. (2012) Theoretical investigation of HVDC disc spacer charging in SF6 gas insulated systems. Dielectrics and Electrical Insulation, IEEE Transactions on, vol. 19, pp. 2196–2205, 2012

20 Riechert U, Gremaud R, Thorson S, Callavik M (2017) Application options and electrical field studies as basis for adequate testing of gas-insulated systems for HVDC, CIGRE Winnipeg 2017 Colloquium Study Committees A3, B4 & D1, Winnipeg, Canada, September 30 – October 6, 2017, paper 139

21 Gremaud R, et al. (2014) Solid Insulation in DC Gas-Insulated Systems, in CIGRE Session 45, Paris, France, 2014, p. D1_103_2014.

22 Weedy B (1985) DC conductivity of voltalit epoxy spacers in SF6. IEE Proc. A – Phys. Sc., Measurement and Instrumentation, Management and Education – Reviews 132(7): 450-454

23 Riechert U, Salzer M, Callavik M, Bergelin P (2016) Compact high voltage direct current (HVDC) transmission systems. Paper presented at Stuttgarter Hochspannungssymposium, Stuttgart, 1-2 Mar 2016

24 CIGRE WG D1.03 (2012) Gas Insulated Systems for HVDC: DC Stress at DC and AC Systems. TB No. 506

25 Kosse M, Juhre K (2019) Experience in testing of high-voltage DC gas-insulated systems and potential applications for rated voltages up to ±550 kV. Paper presented at Highvolt-Kolloquium, Dresden, 9-10 May 2019

26 Sperling E, Riechert U (2015) HVDC GIS RC-dividers in new GIS substations with increased dielectric requirements. Paper presented at CIGRE Colloquium, Nagoya, 28 Sep–2 Oct 2015

27 Appelo H, Groenenboom M, Lisser J (1977) The zero-flux DC current transformer – a high precision bipolar wide-band measuring device. *IEEE Trans. Nuclear Science* 24(3): 1810–1811

28 Riechert U, et al. (2018) Experiences in dielectric testing of gas-insulated HVDC systems. Paper presented at 47th CIGRE Session, Paris, 26-31 Aug 2018

29 Juhre K, Reuter M (2019) Composite voltage testing of gas-insulated HVDC systems – basic test circuits and testing experience. Paper presented at 21st Inter'l. Sympos. High Voltage Eng., Budapest, 26-30 Aug 2019

30 Hering M, Koch H, Juhre K (2019) Direct current high-voltage gas-insulated switchgear up to ±550 kV. Paper presented at CIGRE-IEC Conference on EHV and UHV (AC & DC), Hakodate, 23-26 Apr 2019

31 CIGRE Working Group D1.03 (TF11), Endo F, Giboulet A, Girodet A, Hama H, Hanai M, Juhre K, Kindersberger J, Koltunowicz W, Kranz HG, Meijer S, Neumann C, Okabe S, Riechert U, Schichler U (2012) Gas Insulated Systems for HVDC: DC Stress at DC and AC Systems, CIGRE Brochure No. 506, August 2012

32 Riechert U, Kosse M (2020) HVDC gas-insulated systems for compact substation design, 48th CIGRE Session 23 - 28 August 2020, Paris / France, B3-120

33 Straumann U, Riechert U, Gremaud R, Schueller M, Franck CM (2014) HVDC Insulator Charging in SF6 Insulated Systems, DPG Frühjahrstagung, Deutsche Physikalische Gesellschaft e.V., P 21: Plasma Technology II, P21-1, Berlin/Germany, 17-21 Mar 2014

34 Hering M, Speck J, Großmann S, Riechert U (2015) Feldumbildung in gasisolierten Systemen bei Gleichspannung, 6. RCC-Fachtagung, Werkstoffe - Forschung und Entwicklung neuer Technologien zur Anwendung in der elektrischen Energietechnik, 20.-21. Mai 2015, Tagungs- und Veranstaltungszentrum im Johannesstift, Berlin, Germany, RCC-Tagungsbericht 2015, pp. 239-244

35 Riechert U, Gremaud R., Callavik M. (2017) "Application options and electrical field studies as basis for adequate testing of gas-insulated systems for HVDC, CIGRE A3, B4 & D1 International Colloquium", Winnipeg, MB Canada · September 30 – October 6 2017

36 CIGRE Working Group B1.32 (2012) "Recommendations for Testing DC Extruded Cable Systems for Power Transmission at a Rated Voltage up to 500 kV", CIGRE Technical brochure TB 496, 2012.

37 Juhre K, Hering M (2019) Testing and long-term performance of gas-insulated systems for DC application. CIGRE-IEC 2019 Conference on EHV and UHV (AC & DC), April 23-26, 2019, Hakodate, Hokkaido, Japan.

38 Riechert U, Hama H, Endo F, Juhre K, Kindersberger J, Meijer S, Neumann C, Okabe S, Schichler U (2010) On behalf of CIGRE Task Force D1.03.11, Gas Insulated Systems for HVDC, Gasisolierte Systeme für HGÜ, ETG Fachtagung: Isoliersysteme bei Gleich- und Mischfeldbeanspruchung 2010, 27. - 28. September 2010, Köln, Germany, ETG-Fachbericht 125, 2010, S. 101-107, VDE VERLAG GMBH, Berlin und Offenbach, ISBN 978-3-8007-3278-4

39 IEC 60060-1: ed3.0 (2010) High voltage test techniques: Part 1: General specifications and test requirements, Publication date 2010-09-29

40 Riechert U, Gatzsche M, Hassanpoor A, Plet C, Belda N (2019) Compact switchgear for meshed offshore HVDC networks – between vision and reality. Paper presented at Highvolt-Kolloquium, Dresden, 9-10 May 2019

41 Riechert U and Gatzsche M (2019) Dielectric Long-term Behavior and Testing of Gas-insulated HVDC Systems CIGRE Colloquium SC A2 / B2 / D1, 18th-23rd Nov. 2019, New Delhi, India

42 Gatzsche M, Riechert U, He H, Audichya Y, Mor AR, Heredia LC, Muñoz F (2020) Prototype Installation Test of HVDC GIS for Meshed Offshore Grids, 48th CIGRE Session 23 - 28 August 2020, Paris / France, B3-121

43 Public website promotion Offshore D15.6 White- and position papers on pre-standardization of DC GIS testing

44 M. Kosse, D. Li, K. Juhre, M. Kuschel (2019) Overview of development, design, testing and application of compact gas-insulated DC systems up to ±550 kV. *Global Energy Interconnection*, 2(6): 567–577

45 IEC 62271-209:2019 High-voltage switchgear and controlgear - Part 209: Cable connections for gas-insulated metal-enclosed switchgear for rated voltages above 52 kV - Fluid-filled and extruded insulation cables - Fluid-filled and dry-type cable-terminations.

46 CIGRE Technical Brochure TB 842 Dielectric testing of gas-insulated HVDC systems, JWG D1/B3.57, September 2031, Paris.

47 IEC 62271-203:2011 High-voltage switchgear and controlgear - Part 203: Gas-insulated metal-enclosed switchgear for rated voltages above 52 kV.

48 IEC 62271-1:2017+AMD1:2021 CSV High-voltage switchgear and controlgear - Part 1: Common specifications for alternating current switchgear and controlgear.

49 IEC 62271-102:2018 High-voltage switchgear and controlgear - Part 102: Alternating current disconnectors and earthing switches.

50 IEC 62271-103:2021 High-voltage switchgear and controlgear - Part 103: Alternating current switches for rated voltages above 1 kV up to and including 52 kV.

51 IEC 62271-318 to be published in 2023 Common specification for high voltage DC gas-insulated switchgear.

52 IEC 62271.5 to be published in 2022 High voltage DC gas-insulated switchgearassemblies.

53 IEC 62271-204:2011 High-voltage switchgear and controlgear - Part 204: Rigid gas-insulated transmission lines for rated voltage above 52 kV.

54 IEC 62271:2021 SER High-voltage switchgear and controlgear - ALL PARTS.

9.18 Digital Substation

Author: Hermann Koch
Co-author: Dirk Helbig
Review: George Becker

9.18.1 Introduction

The digitalization has reached gas-insulated switchgear as part of the digital high voltage substations.

The electric power transmission system is highly automated because of the need for fast reaction in case of failures. Such high voltage and high current failures at the 400 kV or 500 kV voltage rating at short-circuit currents of 50 kA or 63 kA or even more would cause large destruction in the substations and would be dangerous for personal at operation. For this reason, the protection of high voltage substations are using protective relays first electro-mechanic, later in the 1980s electronic devices and today the Internet of Things (IoT) is entering the transmission substations.

In this chapter explanations are given for technology offering such new functions of substations and it will show which future advantages can be reached with a digital GIS. Sensformer® [7, 8] and Sensgear® [9] are a new class of transmission products merging Cloud- and Sensor- technology and Power Transformers, called Sensformer ®, and Cloud- and Sensor technology with switchgear, called Sensgear®. Target is to improve performance, flexibility, reliability and availability of connected high voltage transmission and distribution equipment in the substation.

Figure 9.122 give an overview on the functionality of the digital Sensgear® GIS as part of the digital substation. The example of the equipment shows from left to right an incoming overhead line,

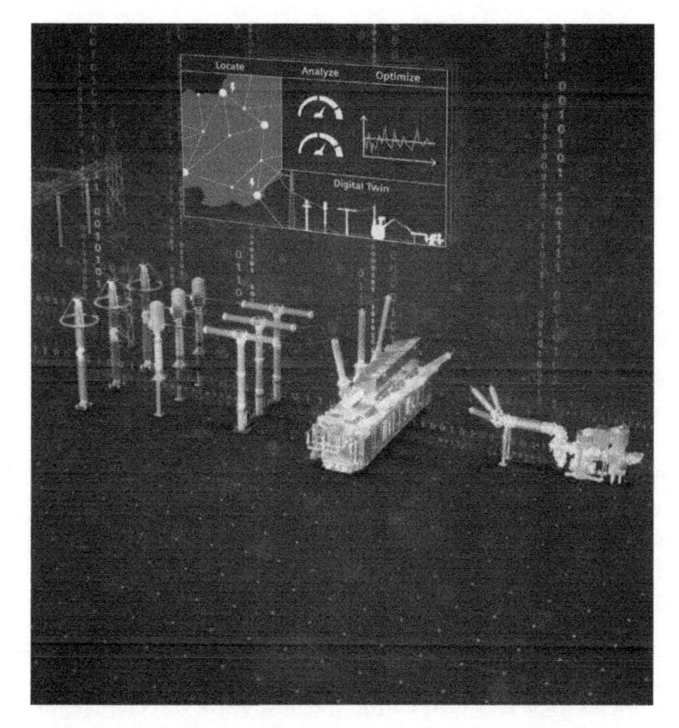

Figure 9.122 Overview on the functionality of the digital Sensgear® GIS as part of the digital substation (courtesy of Siemens Energy)

over-voltage surge arresters, voltage and current transformer, high voltage switchgear of circuit breaker, disconnect and earth switch, the power transformer from transmission to distribution voltage levels and high voltage GIS.

The digital substation, and with this the digital GIS, covers information related to the location of the substation in the network and the functionality and requirement connect with the location. With the location in the electric network the signification of the substation is defined and under which condition a power transmission interruption is acceptable and when not. It defines the redundancy of equipment in the substations and measures for maintenance and if needed repair processes. This information is related to use of the digital GIS in the network at a specific location and will be different at each location.

The second aspect of digital substations is the possibility to analyse the condition of the substation and its equipment. This reflects many parameters from level of current flow, times of over-voltages, number of switching operations for rated or short-circuit currents, gas density of the GIS, partial discharge (PD) activities and patterns, ambient temperature, over-load conditions in value and time, equipment aging effects and maintenance scheduling only to mention some relevant parameters of the digital GIS.

Based on information collected for the analysis of the digital substation the operation condition will be optimised for best use and to avoid failures and unplanned interruptions. The results of the optimisation could be the basis for changes in the network operation. Load flow changes, extension of substation equipment with new lines or replacement strategies for higher performance of the equipment.

The basis of digital substation is the digital twin for all equipment used, see also Section 9.15. The digital twin combines all information of the equipment from manufacturing in the factory, type tests, routine test, on site tests, operation data of voltage and current, environmental data of temperature, ice, wind, water and any value from importance to evaluate the condition and history of the digital substation. The digital twin is a data base for the substation equipment reflecting the complete life time.

Digitalization of substations offers huge benefits to better manage current trends and challenges of the substation and its equipment. Trends, challenges and opportunities of digital substations are shown in Figure 9.123.

9.18.1.1 Trends

Trends are coming from the growing share of renewable energy generation and transmission through the electric network. Fluctuating energy generation by using wind and solar power generation will change the use of substations compared to the classical power generation of large thermal power plant by continuous adaptations to the power flow in the network.

The electric network is growing and turns to be much more complex. Changing power flows in a matrix structure of power generation and loads at each voltage level will require complex management tools and structures. The question of reliability is different to a classical top-down structure from thermal power plants to the load at the consumer's side. Reliability needs to be defined for each section of the network and with this for each substation with its specific equipment.

Decentralized grid structures like micro-grid will operate independent in the electric network with the requirement of power supply in case of internal failures or problems. Prosumers enter the scene with sometime generating electric power by wind or solar and sometime consume electric power.

The fast-growing electrification of vehicles to be expected within the next years will change the use of the electric power network. Power flow control from the transmission level, to distribution and consumer voltage is needed to manage these new requirements.

9.18.1.2 Challenges

The challenges for the design and construction of substations in the future are to meet the required performances. Operation voltage may vary more than it is usual today because of more fluctuation energy in the network. Short-circuit levels may change because of large renewable energy generation at remote locations. Power frequency will change more when over production situations and low production situations of electric energy will change. These requirements will ask for technical values of the substation equipment to provide the performance on operation.

Health, safety and environmental requirements are better to manage by using the digital substation and the sensors available. Detecting leakages of oil or insulating gases, non-functioning

Figure 9.123 Trends, challenges and opportunities of digital substations (courtesy of Siemens Energy)

secondary equipment like ventilators or pumps or over temperatures of equipment in the substation or the connected lines will help to avoid problems with health, safety and environmental aspects. There are much more sensor activities which can be monitored by digital substation to improve substation management and avoid not foreseen problems and power flow interruptions.

Cost reduction is one monetary advantage of digital substations. The cost reduction can be reached by digital simulation of the substation and defining the right technical parameters so that the substation can fulfil its task in the future. The cost reduction is also effective when the substation is in operation and based on the digital analysis and right decisions non-planned interruptions can be avoid. These interruptions are always very costly.

The socio-demographic changes in the society are turning a work force not-used to digital devices with little experiences into a work force with digital experiences of their daily live. Smart Phones, digital communication, banking, dating, the whole living is digital, this will change the operation of electric power substations too. Safety and reliability requirements will for sure be needed to be fulfilled.

9.18.1.3 Opportunities

The digital substation will bring new opportunities to the operator, used and manufacturer. These opportunities are in the data cloud, providing big data on information available on a mouse click, for user who can have access. From a technical point of view allows the cloud data management to handle sensor data (voltage, current) in real-time calculations in the data cloud and give back actor signals to react onto a change in the electric power system. This will improve the operation and management of substations.

Sensors are connected to IoT channels to communicate with the data cloud produce automated reactions or inform the operator on required activities with proposals available. The communication based on the 5G mobile digital communication will allow real-time activities in the operation and management of substations. With reaction time from sensing, data calculations to acting in 1 ms.

Most of the smart data to be required for operating and managing a substation is stored in a data base with software available for any required data treatment. This could be power flow calculations, restructuring of the electric circuit of the substation, asking for maintenance works or even repair to avoid non-planned interruptions.

9.18.2 Functional Concept

The functional concept covers a wide field of sensors of the digital substation including the digital GIS. The features of digital substation are receiving general type of information of the substation and in addition product specific data.

The first global information is coming from the global positioning system (GPS) to identify the exact location of each equipment installed at the substation. The global substation data in a future application will be generated with the digital twin generated during the planning and building of the substation with all equipment digital identified at its location.

The second global information is coming from the ambient weather and climatic conditions. Local weather will impact the operation of the substation concerning weather impact like hot temperature, strong wind, heavy rain, flooding, thunderstorms, hurricanes and any other impact related to weather which might influence the functionality of the substation. The precise local information about weather conditions has a big influence of drawing the right decisions during changes in the electric power operation of the network. Today this knowledge is very general and real conditions at the substation local may be much different.

The experiences of network operators are that in critical situations the time to react in the right way is short. After a heavy loaded line will trip in a mashed system there is about an hour time to react with changing the network power flow to avoid overload conditions at some bottlenecks in the network. In a meshed 400 kV transmission network of about 30 substations operated by one operator, it can be required to check and simulate the planned circuit changes with the simulation of impact to may be 300 transmission substations connect to the network. This can only be done in automated processes which will propose switching operations to the operator.

For the simulations and actions, precise local data are necessary to find the optimal solution in the network. Local weather, temperature and current loading are important data for these simulations and should be provided by the digital substation continuously.

The digital substation is collecting additional product specific sensor data to be available for the management and operation of the substation. This covers data from the transformer, GIS, circuit breaker, disconnector and earthing switch, arresters, instrument transformers and arc suppressing coils.

9.18.2.1 Data Collection

The collection of data needs to be focused on the use of the data. It is not constructive to collect as many as possible data from any possible sensor. This would end in too much data and to less value of using these data for the substation operation and maintenance.

The data collected must follow the requirement of generating a digital picture of the substation to be able to simulate, predict and propose action for any situation of the network operation including occurring failures and power delivery interruptions. The following selection on data give an idea on which data can be used in a digital substation.

9.18.2.2 Transformer

The transformer is turning in a digital substation in two functional parts: to transform and sense. So, could be called a Sensformer with the following sensors for data collection as base version: Oil-Level Alarm, Top Oil Temperature, Low-Voltage Winding Current

The oil-level alarm is giving information tom the operator for maintenance or repair requirements in case the oil level is falling low. This will avoid non planned tripping of the transformer. The top oil temperature is an important information about the load situation of the transformer and in case of failure in the network how long and how high an overload current condition can be accepted. This is a big help in cases of emergency and for fast reactions of network circuit switching operations. The low voltage winding current is another measured data for the load condition of the transformer and can be used to calculate life time expectations, maintenance cycles or overload conditions

9.18.2.3 GIS and Circuit Breaker

The switching operations in a substation is carried out by GIS or AIS equipment. Beside the switching function the sensor turns the GIS to Sensgear GIS and the AIS to Sensgear Circuit Breaker with the following sensors: Gas Density, CB Counter, Position, Readiness, Temperature at the Local Control Cabinet (LCC)

The gas density measurement is the basic monitoring for dielectric functionality of the GIS and the circuit breaker. This data is available today usually as an electric contact giving information for gas loss and interlocking the equipment. A gas density sensor which measures the temperature and the pressure and calculates the gas density will allow to given tendency information for the operator. The circuit breaker counter gives information for maintenance requirements for the equipment depending on the location in the network with very different switching requirements. The switch positions of the disconnecting and earth switches are relevant safety information and help to avoid wrong operations when know in the substation management and operation system. The temperature at the local control cabinet (LCC) give information at the local temperature condition and helps to find the right solution in case of overload and redispatching current flow in the network.

9.18.2.4 Surge Arrester

In the functionality of the substation the surge arresters maintain an important contribution for failures related to transient over-voltages coming from lightning stroke into the overhead lines connected or by switching operations. Transformers, GIS and high voltage cables are usually protected by surge arresters. When using sensors with the function of the arrester it could turn into Sensgear Arrester with following sensors: Surge Counter, Leakage Current

The correct function of the arrester in case when over-voltages apply to the substation can be monitored by counting the number of activations of the surge arrester and the leakage current. The number of activations of surge arresters to transient over-voltages gives an indication on the aging effects and will allow to replace surge arrester before they may fail to operate correctly. The leakage current gives the indication for the correct protection voltage of the surge arrester and will help to optimize the maintenance cycles or replacement before it fails to protect the equipment as calculated.

9.18.2.5 Instrument Transformers

The instrument transformers for voltage and current measurement are key element for the operation and protection of the substation. The failure of voltage and current measurement will lead to wrong switching operations in substations and finally to tripping of line in the network. There are two types of instrument transformers one is gas insulated the other is oil insulated. When sensors are connected to the instrument transformer it will turn into a Sensgear Instrument Transformers with the following sensors: Gas Density (gas insulated), Oil-Level Alarm (oil insulated).

The gas density of gas-insulated instrument transformer and the oil-level alarm of an oil-insulated transformer are key data for the dielectric integrity of the equipment. Today these values are monitored by electric contacts for gas or oil leakage. The sensor will give information on temperature and gas-pressure for the gas-insulated instrument transformer and temperature and oil level for oil-insulated transformer. In both cases, the digital sensor allows to get trend analysis and use this data for the operation of the substation and set up a maintenance or replacement schedule.

9.18.2.6 Arc Suppression Coils

Arc suppression coils are limiting the short-circuit fault current in the substation and network. The coils need to function in case of a failure reliable. When using sensor, it would turn into Sensgear Arc Suppression Coils with the following function: Oil-Level Alarm, Top Oil Temperature, Zero Sequence Voltage, Coil Setpoint

The arc suppression coil is part of the operation method of the network with the goal to continue network operation in case of a line to earth failure by limiting the short-circuit current. To prevent the correct functionality the oil-level alarm and temperature needs to be measured and monitored. The digital sensor can provide a continuous monitoring of the values and allows trend analysis for the coil. The zero-sequence voltage and coil setpoint are important data for the failure location of the line to earth fault.

In Figure 9.124 the functional concept and features of digital sensors in substations is shown. At the bottom of the graphic the substation equipment is shown surge arresters, instrument transformers high voltage circuit breakers, power transformers and high voltage GIS. Each of the devices is equipped with digital sensors of the above explained function. The sensors are connected via a connectivity device to the internet with their own identification as internet of things (IoT).

The sensors handle the data measurement and data treatment for the communication with the cloud platform.

The secure cloud platform provides all the software tools, for data handling, calculations of the substation status, simulating the actual and predicted operation, providing trend analysis for maintenance and repair schedules and to react in case of network changes and up-coming network failures with proposals of changes in the substation circuit for avoiding power supply interruptions.

The data cloud functionality provides real time information on the digital substation with cost prediction for the operation, maintenance and repair. The data in the cloud is the basis for optimization of the substation design and operation.

9.18.3 User-Interface/APP

The user interface for the digital substation is using secure data connections by server connection of the operator of the substation and by secure mobile connection of smart phone app. The sensitive data collected in the data cloud requires high data safety levels to avoid miss-use or access of non-licensed persons. This is one of the most important qualification of the digital substation: data safety.

In general, commercial data clouds offer such data safety for the daily use. The importance of the electric power network and the possible high impact when data is hacked require cyber security of the highest class.

Figure 9.124 Functional concept of a digital substation (courtesy of Siemens Energy)

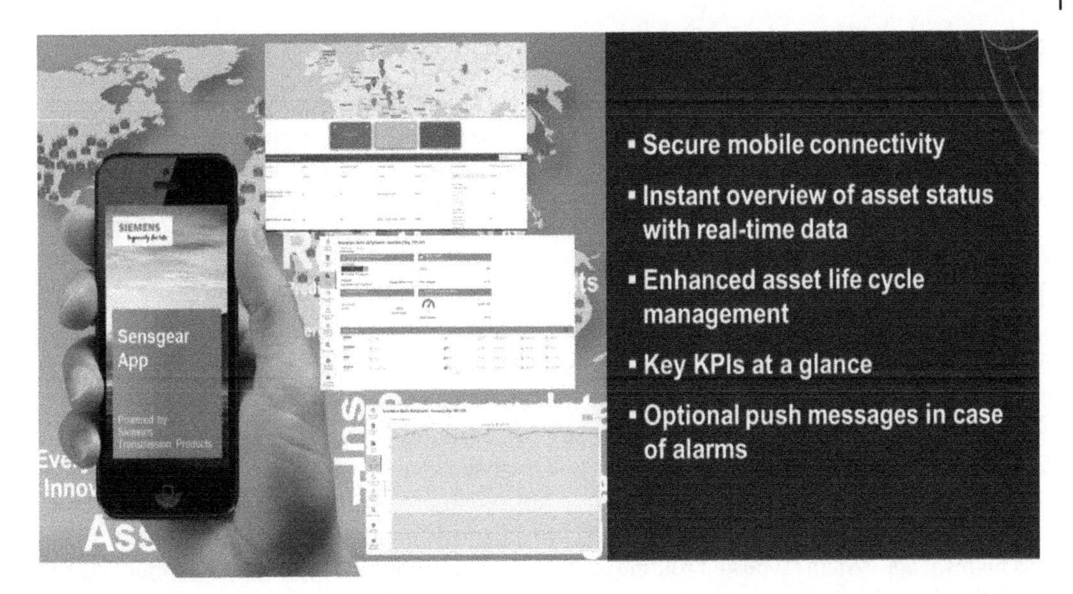

Figure 9.125 User-Interface for the digital substations using APPs (courtesy of Siemens Energy)

In Figure 9.125 the aspects of user-interface for the digital substations using APPs are shown. The digital substation data is available at the operators control centre and by secure APP at any mobile device prepared to handle the secure connectivity of data.

The data cloud will provide an instant overview of the status of all assets of the digital substation with real-time data. The software installed in the data cloud calculated and presents the data in a form the manager and operator of the digital substation can use getting an overview on the condition and in case of action required to have proposal on how to react. This will help to avoid power transmission interruptions and to plan activities as maintenance and repair.

With all the sensor data of the digital substation in the back ground the substation is operated at its optimal conditions and over-load can be avoided or actively managed. This will not only enhance the asset life cycle and improve the operational and maintenance costs of the substation, but also increase the power and revenue stream of the grid operator. Life cycle management is more focused on the real, actual substation conditions.

The sensors will deliver data for the calculation of the Key-Performance-Indicators (KPI) and provides information at a glance. This is an important tool and help for investment decisions like: When do we need to replace equipment? When do we need to up-grade equipment? When to we need to maintain of repair equipment?

The high investment cost and the large impact of failures occurring in the substation and interrupting electric power supply to the customers makes it important to come to the right decision at the right time. The digital substation data will help to reach this goal.

The setting of the alarm or warning messages of changes detected by the sensors in real-time situations, based on real-time data of the substation provides an early information to be prepared for any action. In case of unforeseen changes in the substation, this real-time information is from great importance and the basis for correct decisions for next actions. The analysis of blackout events of the past showed that specially the availability of correct, real-time information was missing. Digital substations can provide.

9.18.4 Exemplary Tests Impressions

The environment of sensors in high voltage substations requires very specific features to be reliable and well-functioning for the long-life time of a substation. There are three main technical requirements to be met by sensors and the connectivity devices in high voltage substations: Electro-magnetic compatibility (EMC), switching operations of GIS disconnectors and mechanical endurance testing.

9.18.4.1 Electro-magnetic compatibility (EMC)

The high voltage environment generates very fast transient over-voltages which sensors and connectivity devices need to withstand and provide correct functionality. The very fast transient over-voltages produce voltages of a frequency spectrum up to some hundreds of megahertz. These disturbances enter the electronic sensors by conductive voltage of the signal wires, over-voltage entering the electronic sensors by potential rise of the earthing system or by radiation through the air. Any of the disturbances need to shielded and earthed/grounded at the electronic sensor.

For the design of the electronic, a disturbance-level concept is applied to find the right shielding and earthing measure to protect against impact to the sensor coming from the very fast transients in a high voltage system. Type and routine tests to prove the functionality of the electronic sensors are defined in IEC standard [1–6]. In Figure 9.126, type tests of electronic sensors for the digital substation are shown. The two photos at the left show the EMC test set-up of the electronic sensor under high voltage conditions.

The photo at the middle of Figure 9.126 shows the test set up of a GIS unit under high voltage condition and with sensors attached. The electronic sensors attached to the GIS switching devices, here disconnector and earth switches, are at close distance. The test proves that no mal function or measurement of the digital sensor will occur when the switches are operating. Under high voltage the disconnector and earth switches produce high voltage impulse of two to three times of the rated voltage and very fast transients. The shielding and earthing design of the electronic sensor needs to withstand these transient over-voltages without failure.

The photo at the right of Figure 9.126 shows the test set up a high voltage circuit breaker for an endurance test process. This endurance test proves the long-time stability of the sensors and connectivity device under its operational conditions. Circuit breaker operation generated high dynamic

Figure 9.126 Type tests of sensors and connectivity devices for the digital substation, left: EMC test of the sensor under high voltage conditions, middle: high voltage switching operation test of GIS unit, right: mechanical duration test of high voltage circuit breaker (courtesy of Siemens Energy)

forces also to the sensor. These forces need to be handled by the electronic sensor for life time of 10000 switching operations. In addition, the ambient climatic conditions need of −25 °C or −40 °C low temperatures or +40 °C or +50 °C temperature to be met by the electronic sensor.

These basic tests are fundamental for the operation of electronic sensors in a digital substation. The sensors are the basis of the digital substation and need to show very high reliability. It is important to design the digital substation with sensors fitting to the specific requirements and only to install those type of sensors which will offer an additional value for the reliability and safety of the digital substation.

9.18.5 Operational Values of Digitalized Products

The operational values of digital substations for the operation and maintenance can be split in four sections: performance increase, health, safety and environment, cost reduction and risk avoidance. See Figure 9.127.

9.18.5.1 Performance Increase

Using digital sensors in the substation the actual load status can be seen in real time with indicating the data in the control centre and be able to use these data for load flow calculations in case of a change in the network structure. Planned or emergency switching's in the network are changing the power flow and the load condition in each branch of the grid.

This can lead to acceptable temporary overload conditions in one branch based on the actual status of power flow before the overload condition and thermal constant for the overload time. There might by several hours of overload operation possible, time in which dispatchment of the load flow can be adapted.

Based on real-time performance data of the substation additional power transmission and revenue can be reached without going above limits of temperature and thermal aging effects. This additional power transmission takes into account the real climatic condition of ambient temperatures and calculates on time the safety margins for maximum current flow.

The real time sensor data reduces the unplanned outage time of the substation because of much better known the limits of operation and to avoid thermal overload conditions. Measure can be planned in advance to reinforce lines or replace equipment before it fails.

The whole asset life cycle cost evaluation can be optimized by sensors in a digital substation with calculating the history of power flow and make prediction for the next future operation conditions. See Figure 9.127 Operational value of digitalized products.

9.18.5.2 Health, Safety and Environment

The health, safety and environment performance of the digital substation provides advantages for increased safety through manhours on the equipment. With more remote available data collected by the sensors, less personal needs to approach the substation in operation and with this the safety risk for personal are less.

SF_6 real-time monitoring equipment will detect gas leakages early and measure to repair can be planned and executed. This will reduce CO_2 equivalent emissions to the atmosphere and prevent the operator from large refilling of SF_6 to the equipment which needs to reported to environmental authorities.

The real-time measurement of the oil levels in the transformers and coils using digital sensors allow to react fast in case of any oil leakage. The reduced oil loss prevents from oil spills to the ground and costly soil exchange works.

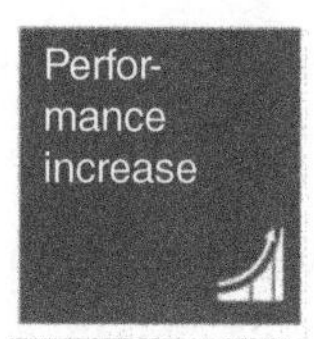

- Load status and prediction of network and products
- Temporary overload management
- Additional power transmission and revenue
- Reduction of unplanned outages
- Asset lifecycle optimization

- Increased safety through reduced manhours on equipment
- SF_6 leakage and $CO2_e$ emission reduction
- Oil leakage reduction

- Reduction of unplanned outages
- Reduced costs for maintenance from time-based to predictive
- Reduced manhours on equipment
- No costs for SF_6 controlling
- Reduced costs for SF_6 reporting

- Less risks and costs for unplanned outages
- Less risks and costs for SF_6 leakages and CO_2e emission penalties
- Less risks and costs for oil leakages

Figure 9.127 Operational values of digitalized products: Performance increase, health, safety and environment, cost reduction and risk avoidance (courtesy of Siemens Energy)

9.18.5.3 Cost Reduction

The main cost reduction effect of digital substations is the reduction of unplanned outages. These unplanned outages generate high costs in non-delivered electric energy in the transmission network because of the high-power rating. As an example, the interruption of a nuclear power plant from the network for just one day would have a cost effect on not delivered energy of close to million Euro, as calculated and required by German TSO's for reimbursement of the shutdown of nuclear power plants.

The second big cost reduction effect is coming from more precise planned maintenance works at the substation. The real-time measurement of the digital sensors for the complete operation time of the substation equipment will deliver data on how much the equipment has been stressed and when maintenance work need to be done. The pure time-based maintenance with certain works defined after 5, 10 or 25 years of operation will be changed in a predictive maintenance based on real data from the substation operation.

With less maintenance need to be done at the substation to control the equipment condition less personal is required. This reduces the manhours on the equipment. Most data of the equipment condition are available at the control enter through the data cloud add of digital twin, see section 9.15. Based on this data automatic condition checks can be calculated and operational personal can be informed about critical condition for some of the substation equipment.

The use of SF_6 type of equipment in substations requires the need of control for gas-leakages of the high voltage gas compartments during the complete time of operation. This data will be collected by the digital gas density sensor and monitor the gas compartments. On-line data from the gas compartments on real gas pressure give information about gas losses and can be used for the SF_6 reporting to the authorities.

9.18.5.4 Risk Avoidance

Digital substations provide more real-time data and with the cloud intelligence more calculations of the status and condition of the substation. This will find much earlier mal functions in the substation reported early to the operational personal. Like in an airplane, which will send information to the airport of arrival the one device is not functioning correctly. At arrival at the airport the device will be available and exchanged. Same to the high-speed train system in Spain where the same systematic is used and all operational data is delivered to a cloud-based management system. The availability of the 25 trains reached 95%, which is compared to the 70% of conventional train systems a much higher value of availability of the trains. Less unplanned outages and less risks in the substation with digital sensors will reduce the operational cost.

The control of SF_6 losses to atmosphere is from importance today and will be even more important in future because of possible penalties for CO_2 equivalent leakages. This implements risks for the operator in unplanned cost.

9.18.6 More Functionality/Advanced Features

The advantages of digital substation equipment are shown in Figure 9.128 with the example of a digital transformer, called sensformer [7]. The advantages can be transferred to any other digital equipment in the substation like digital GIS, circuit breaker, disconnector and earthing switches arrester, overhead lines, line connectors or cables.

The digitalization of the transformer offers maximum flexibility in operating the equipment under daily changing operational conditions. In addition to the basic functionality of transformer protection and control equipment the digital performance offers advanced technology functions. Tailor-made solutions can be implemented with numerous scaling options as there are: active overload prediction, lifetime prediction, advances sensors and virtual sensors. This data is transferred to the digital twin of the transformer and creates a loadable application for the transformer.

The active overload prediction is using digital data in real-time to calculated the thermal reserves of the transformer at any time of operation. This takes into account any impact including actual load, ambient temperature, weather forecast and load predictions to give information to the operator on how much overload can be accepted for how long.

The lifetime prediction is calculating with the digital data available with the digital twin the impact of the real-time operation on the total lifetime of the transformer. The total lifetime of equipment is depending on its load condition's. Over-voltages, over-currents, over-load and over-temperatures are influencing the lifetime of any equipment. The transformer is the most sensitive. The hight and longer the over-stress situation applies to the equipment, the more lifetime of the equipment will be lost. In this aspect the digital twin, with its background of parameter calculations and given predictions and forecasts, the operational value of the digital substation has a great cost impact. Shortening lifetime of equipment can be very costly because of early replacement.

Advanced sensors and virtual sensors build on software using different sources of input data are specific features of the digital equipment in the substation and can provide optimized operational

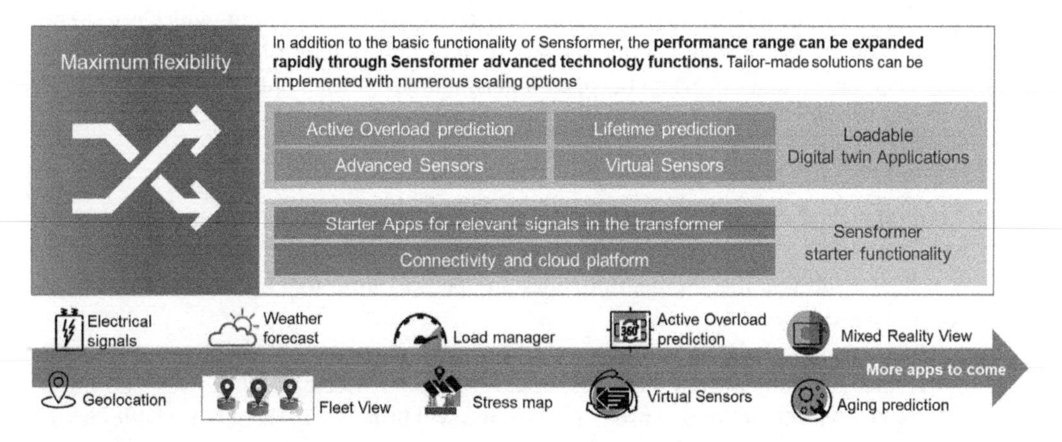

Figure 9.128 Maximum flexibility using basic and advanced functionality and apps in digital substation equipment (courtesy of Siemens Energy)

information for the equipment and substation.

The sensor and virtual sensor signals of the digital equipment are connecting the data through the cloud platform. Applications use the data to generate information for operation, maintenance and asset management. These applications are new to digital substation operation and starts with first sensors of electrical signals, geolocation, weather forecast, fleet view, load management, stress map, virtual sensors, active overload prediction, mixed reality view and aging predictions. More APPs will come.

The electrical signals are available in any substation coming from voltage and current instrument transformers or for transient over-voltages by over-voltage protecting devices like surge arresters. The geolocation data is linked to the location of the substation and can be combined with weather actual or forecast data for calculations of real-time ambient temperatures, lightning risk at thunderstorms, ice on the lines, flooding through heavy rain, very low temperatures or high levels of snow fall. Such information is used to build virtual sensors and to predict the next future of the substation operation.

The operation of substations is to see in connection with the network and a fleet of substation equipment. This fleet view is important to optimize the network operation for the right decisions for load flow and network operation providing high reliability of power supply. The fleet view in operation prevents from switching operations at on location causing over-load conditions at any other location of the network. For this, the digital data in the cloud is used to prepare real-time network calculations for optimized power flow.

Optimized power flow and high reliability of the network is the prior task of the load manager. To handle this with better data and predictions the load management system will use the digital data from the cloud to optimize the operation.

The availability of the digital data in the cloud can be used to set up a stress map for certain network operational conditions, for example, a failure in substation and the impact to others. These calculations are in real-time and the stress map APP will be used to predict overload in the network. The active overload prediction APP is using the stress map data to make recommendation for avoiding over-load condition in the digital substation or network.

As over-load situation of the equipment are directly influencing the expected life time of the equipment the digital twin data of the cloud is used to calculate aging impact to equipment related to over-load conditions. These aging predictions can inform the operator how much impact a given over-load condition may have to the lifetime of the substation equipment.

A mixed reality view of the digital substation condition can be prepared in calculations based on the digital twin cloud data for the real-time situation and for expected future events which can see be forecast data. This allows the operator to find the best solutions to avoid over-load and over-stress of the substation equipment and can extent the operational lifetime of the equipment and substation.

9.18.7 Conclusion

Digitally connected products enable operators to optimize performance, quality and speed of operational decisions as well as to become more flexible, act faster and more efficiently.

Moreover, digital substation equipment provides an open platform that enables co-creation of operator-centric use cases and applications.

The digital substation with digital GIS, power transformer, voltage and current instrument transformers, disconnecting and earthing switches, circuit breakers, arresters, cable, overhead lines, connectors and any type of equipment used in substation will bring a huge benefit for better management and optimized use of the substation.

New sensors will be added to the existing measuring and sensor devices available to generate signals on voltage, current, temperature, gas-pressure, gas-density, vibrations, over-voltages and over-currents. This information will be supplemented with information from actual and local weather conditions and network load to calculated future developments of the substation condition and make proposal on how act as operator. Chancing the substation circuit by switching operations or changing the load flow in the network are some consequences for the operator. The digital data of the digital equipment stored in the digital twin is the basis for such predictions and calculations.

In today's network with ever changing power flow conditions related to the fluctuating renewable energy generation the operator does not have much time for reacting on changes in the network. The control of the network and with this the control of the substations is more complex, faster to react and more sensitive to wrong operations in the network because of no availability of the n-1 criteria.

The digital twin, see also Section 9.15, is using the digital data of the substation equipment to provide new functions for the substation operation.

References Section 9.18

1 IEC 61000:2020 SERIES, Electromagnetic compatibility (EMC) – ALL PARTS.
2 IEC 62271-1:2017 High-voltage switchgear and controlgear – Part 1: Common specifications for alternating current switchgear and controlgear.
3 IEC 62271-100:2008+AMD1:2012+AMD2:2017 CSV, High-voltage switchgear and controlgear – Part 100: Alternating-current circuit-breakers.
4 IEC 62271-102:2018, High-voltage switchgear and controlgear – Part 102: Alternating current disconnectors and earthing switches.
5 IEC 62271-203:2011, High-voltage switchgear and controlgear – Part 203: Gas-insulated metal-enclosed switchgear for rated voltages above 52 kV
6 Kuschel M, Helbig D, Hammer T, Heinecke M, Harminder MS (2019) Sensformer® and Sensgear® – Intelligent IoT-connected High Voltage T&D Products, International Conference on Condition Monitoring, Diagnosis and Maintenance CMDM 2019 (5th edition), September 9th -11th 2019, Radisson Blu Hotel Bucharest, Romania
7 Sensformer ® and cyber secuirity, Sie,ens Energy, 2021, https://www.siemens-energy.com/global/en/offerings/references/sensformer-cyber-security.html.
8 Sensproducts® - from products to system intelligence. Siemens Enetgy, 2021, https://www.siemens-energy.com/global/en/offerings/power-transmission/innovation/sensproducts.html
9 Sensgear®: Born connected 2.0, Added transparency,enhanced productivity,advanced intelligence, https://assets.siemens-energy.com/siemens/assets/api/uuid:982fb983-ba81-4450-9cff-41643de7ff2c/sensgear-4pager-en.pdf

References Sections 9.1 to 9.10

1 Naidu MS (2008) Gas insulated substations, *Sections*, 8.3, 8.6.
2 Guidelines for Life-Cycle Assessment: A 'Code of Practice', Society of Environmental Toxicology and Chemistry (SETAC), Brussels.
3 PAS 2050-2011, updated (2019). Specification for the Assessment of the Life Cycle Greenhouse Gas Emissions of Goods and Services, Metal Packageing Maufacturers Assosiation.

4 Kyoto Protocol. http://unfccc.int/kyoto_protocol/items/2830.php/.
5 CIGRE Brochure No. 430. SF_6 Tightness Guide, CIGRE B3 AA2 WG 18, (2010).
6 C37.122.3- IEEE. Guide for Sulphur Hexaflouride (SF_6) Gas Handling for High-Voltage (over 1000V ac) Equipment.
7 CIGRE Brochure No. 276, Guide for preparation of customized "Practical SF_6 handling instructions," August 2005 edition.
8 Pittroff M, Krahling H, Preisegger E (2001). Product Stewardship for SF_6.
9 Harrison-Edward (Ned), G.P., Maclean, J., Karamanlis, S., and Ochoa, L.F. (2010) Life cycle assessment of the transmission network in Great Britain. *Energy Policy,* voluum 38, pages 3622–3631.
10 Jorge RS, Hawkins TR, Hertwich EG (2012) Life cycle assessment of electricity transmission and distribution – Part 2: Transformers and substation equipment. *International Journal of Life Cycle Assessment*, 17, 184–191.
11 UI Standard (2011) TSS 3.4. *Life Cycle Cost Analysis for Substation Equipment.*
12 Boggs SA, Fujimoto N, Collod M, Thuries E (1984) The modeling of Statistical Operating Parameters and the Computation of Operation-Induced Surge Waveforms for GIS Disconnectors. *CIGRE*, paper 13–15.
13 Boggs SA, Chu FY, Fujimoto N, et al. (1982) Disconnect switch induced transients and trapped charge in gas-insulated substations *IEEE Transactions on Power Apparatus and Systems*, PAS-101 (10).
14 Fujimoto N, Croall SJ, Foty SM (1988) Techniques for the protection of gas-insulated substation to cable interfaces. *IEEE Transactions on Power Delivery*, 3 (4), 1650.
15 Fujimoto N, Boggs SA (1988) Characteristics of GIS disconnector-induced short risetime transients' incident on externally connected power system components. *IEEE Transactions on Power Delivery*, 3 (3), 961.
16 Fujimoto N, Dick EP, Boggs SA, Ford GL (1982) Transient ground potential rise in gas-insulated substations – experimental studies. *IEEE Transactions on Power Apparatus and Systems*, PAS-101 (10).
17 CIGRE Working Group 33/13-09 (1988) Very Fast Transient Phenomena Associated with Gas-Insulated Substations. Session of CIGRE, paper 33-13.
18 Ford GL, Geddes LA (1982) Transient ground potential rise in gas-insulated substations – assessment of shock hazard. *IEEE Transactions on Power Apparatus and Systems*, PAS-101 (10).
19 Naidu MS (2008) Gas Insulated Substations, I.K. International Pvt Ltd.
20 Chapman C, Ward S (2003) Project Risk Management, Processes Techniques and Insights, 2nd edn, John Wiley & Sons, org.
21 Seidl D (2009) Assessing Risk, Purdue University.
22 Chu FY (1986) SF_6 decomposition in gas-insulated equipment. *IEEE Transactions on Electrical Insulation*, EI 21 (5).
23 Byproducts of Sulfur Hexafluoride (SF_6) Use in the Electric Power Industry Prepared for the U.S. Environmental Protection Agency by ICF Consulting January 2002. http://www.epa.gov/electricpower-sf6/documents/ sf6_byproducts.pdf
24 ISO 14040:2006 review (2016), Environmental management — Life cycle assessment — Principles and framework.
25 ISO 14044:2006, reviewd (2016), Environmental management — Life cycle assessment — Requirements and guidelines.
26 TÜV Nord, Test and Certification, Germany, https://www.tuev-nord.de/de.

27 IEEE C37.122.3-2011 - IEEE Guide for Sulphur Hexafluoride (SF_6) Gas Handling for High-Voltage (over 1000 Vac) Equipment.
28 IEC 62271-203:2011 High-voltage switchgear and controlgear - Part 203: Gas-insulated metal-enclosed switchgear for rated voltages above 52 kV.
29 ISO/IEC 62430:2019, Environmentally conscious design (ECD) — Principles, requirements and guidance.
30 CAPIEL Coordinating Committee for the Associations of Manufacturers of Switchgear and Controlgear equipment for industrial, commercial and similar use in the European Union, https://www.capiel.eu/.
31 IEEE C37.122-2021 - IEEE Standard for High Voltage Gas-Insulated Substations Rated Above 52 kV.
32 CIGRE TB 234 SF6 Recycling Guide, revision 2003, Task Force B3.02.01, August 2003.
33 IEEE C37.122.4-2016 - IEEE Guide for Application and User Guide for Gas-Insulated Transmission Lines, Rated 72.5 kV and Above.
34 IEC 62271-204:2021 High-voltage switchgear and controlgear - Part 204: Rigid gas-insulated transmission lines for rated voltage above 52 kV.
35 IEC 62271-4:2013 High-voltage switchgear and controlgear - Part 4: Handling procedures for sulphur hexafluoride (SF6) and its mixtures.
36 CIGRE TB 509 Final report of the 2004-2007 international enquiry on reliability of high voltage equipment, Part 1 Summery and General Matters, Working Group A3.06, October 2012.
37 CIGRE TB 510 Final report of the 2004-2007 international enquiry on reliability of high voltage equipment, Part 2 Reliability of high voltage SF6 circuit breakers, Working Group A3.06, October 2012.
38 CIGRE TB 511 Final report of the 2004-2007 international enquiry on reliability of high voltage equipment, Part 3 Disconnectors and Earthing Switches, Working Group A3.06, October 2012.
39 CIGRE TB 512 Final report of the 2004-2007 international enquiry on reliability of high voltage equipment, Part 4 Instrument transformers, Working Group A3.06, October 2012.
40 CIGRE TB 513 Final report of the 2004-2007 international enquiry on reliability of high voltage equipment, Part 5 Gas insulated switchgear, Working Group A3.06, October 2012.
41 CIGRE TB 514 Final report of the 2004-2007 international enquiry on reliability of high voltage equipment, Part 6 Gas insulated switchgear (Practices), Working Group A3.06, October 2012.
42 IEC 60071:2020 Series, Insulation co-ordination - ALL PARTS.
43 IEEE 1313.2-1999, IEEE Guide for the Application of Insulation Coordination.
44 IEEE 80-2013 - IEEE Guide for Safety in AC Substation Grounding.
45 IEC 61936-1:2021 CMV Power installations exceeding 1 kV AC and 1,5 kV DC - Part 1: AC.
46 IEEE C37.122.1-1993 - IEEE Guide for Gas-Insulated Substations.
47 IEEE C37.122.6-2013 - IEEE Recommended Practice for the Interface of New Gas-Insulated Equipment in Existing Gas-Insulated Substations Rated above 52 kV.
48 IEEE C37.122-1983 - IEEE Standard for Gas-Insulated Substations.

10

Conclusion

Author: Hermann Koch
Reviewer: Dave Solhtalab

Gas-insulated switchgear or substations (GIS) have been successfully in use for more than 50 years with an increasing use world-wide. High reliability and higher efficiency have been reached over the last four decades. Several tens of thousands bays of GIS are in service under a wide range of conditions from indoor to outdoor, high ambient temperature in tropic regions to low temperature in arctic regions, severe environmental conditions in industrial surroundings, high safety concerning grounded enclosure of all high voltage parts for human protection against electric shock, limited space requirements in dense populated areas and cities, high concentration of power with high voltages up to 1100 kV in China, and an 8000 A bus current in Canada are only some aspects that in the end have led to GIS being chosen as the substation switchgear solution. The share of GIS of all project's world-wide is increasing in comparison to air-insulated substations and mixed technology substations reflect this situation in numbers.

The second edition updates the book with information on GIS about the technical development since 2010. New chapter on replacement of SF_6, impact of moister in the insulating gas, determining when to use GIS, on-site test guidance, addressing GIS failures, mobile GIS, condition assessment of GIS, substation resilience using GIS, vacuum switchgear for high voltage ratings, low-power instrument transformers, digital twin of GIS and GIL, underground substations using GIS, offshore GIS applications, DC GIS, digital sensors, and monitoring in GIS have been added.

The extended volume of the GIS book reflects the ongoing technical development in the field of gas insulated technologies. Driven by environmental requirements of reducing the CO_2 footprint of the equipment and its operation have forced to develop new insulating gases which are reducing the global warming potential from 23 000 CO_2 equivalent to numbers between 0 and about 400 with alternative insulating gases.

The changing electric power generation from domination in fusil and nuclear power generation and thermal conversation into electricity into a regenerative generated energy world of dominated solar and wind power generation is changing the transmission and distribution network fundamental in its way to be operated. Fast and often changing power flows according to changes in solar and wind condition and long distances between the power generation areas and the centers of load consumption require fast reaction of the electric network and have to manage higher-rated currents and short-circuit currents. This is driving the GIS design and its ability for fast digital control.

The increasing importance of electric power supply for the society is asking for more resilience of the electric system, while the share of electric power is increasing in relation to fossile resource like oil, coal and gas, Electricity is increasing important for the functionality of the whole society.

Gas Insulated Substations, Second Edition. Edited by Hermann J. Koch.

Medical system, traffic and logistics, grocery and on-line shopping, lighting of cities and pumping drinking water to the houses, there a many dependency on electric power supply. To be more resistive against any impact on the electric power and transmission system in case of flooding, storm, earthquake, terroristic attacks on substations and its equipment, cyber-attacks to switch of the network and many other threats which may occur are in focus. GIS is a good answer to many of these questions for more resilience.

Digitalization of network operation, substation design, and operation and the digital functionality of each device used in the electric power system is a key technology transfer for coming years. Digital twin, integrated electronic devices (IED) with digital input and output connection, the complete network data available in the digital cloud and the intelligent software using artificial intelligence (AI) to optimize network operation. Digital GIS is a future requirement for digital network operation.

This book is written for the practical engineer in charge of planning and building substations. Based on many years of experience, the engineers who have contributed to this book – from users, consultants, and manufacturers – have been brought together to help new engineers in this field avoid failure and to be able to get an optimum solution for their local substation and its specific requirements. The focus is not on theoretical design criteria and detailed technical-specific solutions. It is looking more for the general, typically used GIS solution, with many examples collected in this book.

The changes in the electric powper network of the coming years which are required to fulfil the requirements of sustainable, CO_2 free energy generation with wind and solar in majority will need digital management and control of the equipment in real time modus. Gas-insulated switchgear assemblies (GIS) offers features to reach these goals.

The authors are connected through the IEEE PES Substation Committee GIS Subcommittee and there in Working Group K10 as an open access group to share knowledge and information of this successful technology GIS. Each reader is invited to participate in this work and can contribute with his feedback.

Please contact the editor via email (hermann.koch@drkochconsulting.com).

Index

Gas Insulated Substations, Second Edition. Edited by Hermann J. Koch.

d

h

i

p

q

r